17.95

OPTICS

EUGENE HECHT/ALFRED ZAJAC Adelphi University

OPTICS

 ADDISON-WESLEY PUBLISHING COMPANY

READING, MASSACHUSETTS

MENLO PARK, CALIFORNIA · LONDON · AMSTERDAM · DON MILLS, ONTARIO · SYDNEY

This book is in the
ADDISON-WESLEY SERIES IN PHYSICS

Preface

In recent times, the study of optics has moved into the forefront of scientific and technological thought with a whirlwind of activity, a remarkable series of accomplishments, and an almost dazzling promise of things to come. The old and venerable science, built on the magnificent structure of electromagnetic theory, has never lost its general appeal and applicability. Even so, we are in the midst of an exciting theoretical and technical metamorphosis. Optics is moving in new directions, exemplified by such diversity in form and concept as the photon, spatial filtering, fiber optics, thin films, and, of course, the laser with its myriad theoretical implications and practical potentialities.

The classic treatises of Drude, Sommerfeld, Wood, Rossi, Sears, Ditchburn, Born and Wolf, Jenkins and White, Strong, Towne, and many others, are of lasting value and continued interest. Yet there is a compelling need for a new undergraduate text which will speak in the contemporary jargon of picoseconds, megahertz and nanometers, of Q-switching, coherence length, frequency stability and bandwidth; a text which will embrace the pedagogically valuable classical methods along with the major new developments, techniques, and emphasis.

We have begun our treatment with a brief outline of the historical development of the subject. The modern theory of the nature of light unfolds as the culmination of over two thousand years of activity. Yet it should be appreciated within the perspective of this established pattern of change: We cannot quite see the next rung on the ladder, but we are surely not at the top.

For most optical phenomena, the distinctly quantum mechanical characteristics of light are obscured and its wave nature is the most prevalent manifestation. Accordingly, Chapter 2 deals with the mathematical description of wave motion. The wave equation is developed from very simple considerations which require no knowledge of differential equations. Chapter 3 evolves the electromagnetic theory of light from its most elementary beginnings. At that point, the foundation is set, and the rest of the structure of classical optics (including geometrical optics) is formulated predominantly in terms of wave interactions.

One of the themes which we have woven into the fabric of this book is that optics is physics, and is fundamental to physics. The interrelationships between atomic processes and the associated optical phenomena are explored wherever possible. Rather than isolating optics, we have tried to underscore the remarkable continuity that exists amongst the various fields of physics.

We include numerous descriptions of simple experiments which can be performed away from the laboratory. In many cases, the resulting optical effects are illustrated photographically in order to emphasize the point that elaborate or expensive equipment is not always needed. There is much to be seen with a few microscope slides, and we would encourage the "seeing."

The book is meant to serve as the text for what is often the one and only optics course offered to undergraduates. Its appeal should, therefore, be rather broadly based. To that end, much of the book has been prepared so that it can be used after only a thorough introductory course in both general physics and calculus. More difficult topics are placed at the ends of the chapters in which they appear. Thus, the chapter on diffraction begins with Fraunhofer diffraction via the simple Huygens-Fresnel theory, proceeds to the more involved Fresnel diffraction, and concludes with a discussion of the Kirchhoff treatment and boundary diffraction waves. The advanced student will be adequately stimulated and challenged by some of the more sophisticated techniques, such as the Fourier transform approach to diffraction and image theory, matrix methods in the discussion of polarization, paraxial ray tracing and multilayer films, just to mention a few.

The book provides a fairly extensive selection of material germane to modern optics, from which an instructor can formulate a course reflecting his own emphasis and the needs of his students. For example, an elementary course need not specifically include Chapters 11, 12, and 13, that is, *Fourier Optics*, *Coherence*, and *Quantum Optics*. Even so, pertinent aspects of this material are treated throughout the preceding portion of the text. Moreover, certain sections are adequately self-explanatory and can be offered as reading assignments.

We have given a good deal of attention to being consistently clear, avoiding the temptation to be overly succinct with difficult or subtle points.   Ideas which might profit from further elaboration are either footnoted or included in the problem section, along with guiding comments where necessary.   The *complete* solutions to roughly two-thirds of all problems appear in the back of the book.   (A problem number followed by an *asterisk* indicates the absence of such a solution.)

The student is encouraged to refer to the literature and many "readable" articles are cited, including those which are chosen for their elaborate bibliographies.   Books which are of interest are referred to via author and title—publishers, publication dates, and so on, are given in a complete reference listing at the end of the text.

The authors are fortunate to have profited from the assistance rendered by their friends and colleagues, Professor H. Ahner, Professor D. Albert, and Professor M. Garrell. We also thank Dr. A. Delisa, Dr. J. De Velis, Dr. S. Jacobs, and Dr. M. Scully for helpful discussions and commentary.   We are particularly indebted to Dr. Howard A. Robinson, the translation editor of the *Soviet Journal of Optical Technology*, who carefully read the entire text and made many valuable and insightful suggestions. Our thanks for assistance in the preparation of the manuscript are extended to H. Merkl Villez, M. La Rosa, R. Auerbach, S. Auerbach, and especially to Carolyn Eisen Hecht, whose cooperation and patience sustained the effort.

Finally, we thank our many students who used the early typescript, tried out the experiments, did the problems, took some of the photographs, and served as the medium in which the book grew.

New York   E.H.
September 1973   A.Z.

Contents

A Brief History   1

1.1 PROLEGOMENON

In chapters to come we will evolve a formal treatment of much of the science of optics with particular emphasis on aspects of contemporary interest. The subject embraces a vast body of knowledge accumulated over roughly three thousand years of the human scene. Before embarking on a study of the modern view of things optical, let's briefly trace the road that led us there, if for no other reason than to put it all in perspective.

The complete story has myriad subplots and characters, heroes, quasi-heroes and an occasional villain or two. Yet from our vantage in time, we can sift out of the tangle of millennia perhaps four main themes—the optics of reflection and refraction, and the wave and quantum theories of light.

1.2 IN THE BEGINNING

The origins of optical technology date back to remote antiquity. *Exodus* 38:8 (ca. 1200 B.C.) recounts how Bezaleel, while preparing the ark and tabernacle, recast "the looking-glasses of the women" into a brass laver (a ceremonial basin). Early mirrors were made of polished copper, bronze, and later on of speculum, a copper alloy rich in tin. Specimens have survived from ancient Egypt—a mirror in perfect condition was unearthed along with some tools from the workers' quarters near by the pyramid of Sesostris II (ca. 1900 B.C.) in the Nile valley. The Greek philosophers, Pythagoras, Democritus, Empedocles, Plato, Aristotle and others evolved several theories of the nature of light (that of the last named being quite similar to the ether theory of the nineteenth century). The rectilinear propagation of light was known, as was the law of reflection enunciated by Euclid (300 B.C.) in his book *Catoptrics*. Hero of Alexandria attempted to explain both these phenomena by asserting

that light traverses the shortest allowed path between two points. The burning-glass (a positive lens) was alluded to by Aristophanes in his comic play *The Clouds* (424 B.C.). The apparent bending of objects partly immersed in water is mentioned in Plato's *Republic*. Refraction was studied by Cleomedes (50 A.D.) and later by Claudius Ptolemy (130 A.D.) of Alexandria who tabulated fairly precise measurements of the angles of incidence and refraction for several media. It is clear from the accounts of the historian Pliny (23–79 A.D.) that the Romans also possessed burning-glasses. Several glass and crystal spheres, which were probably used to start fires, have been found amongst Roman ruins, and a planar convex lens was recovered in Pompeii. The Roman philosopher Seneca (3 B.C.–65 A.D.) pointed out that a glass globe filled with water could be used for magnifying purposes. And it is certainly possible that some Roman artisans may have used magnifying glasses to facilitate very fine detailed work.

After the fall of the Western Roman Empire (475 A.D.), which roughly marks the start of the Dark Ages, little or no scientific progress was made in Europe for a great while. The dominance of the Greco-Roman-Christian culture in the lands embracing the Mediterranean soon gave way by conquest to the rule of Allah. Alexandria fell to the Moslems in 642 A.D. and by the end of the seventh century, the lands of Islam extended from Persia across the southern coast of the Mediterranean to Spain. The center of scholarship shifted to the Arab world where the scientific and philosophic treasures of the past were translated and preserved. Rather than lying intact, but dormant, optics was even extended at the hands of one Alhazen (ca. 1000 A.D.). He elaborated on the law of reflection, putting the angles of incidence and reflection in the same plane normal to the interface; he studied spherical and parabolic mirrors and gave a detailed description of the human eye.

By the latter part of the thirteenth century, Europe was only beginning to rouse from its intellectual stupor. Alhazen's work was translated into Latin and it had a great effect on the writings of Robert Grosseteste (1175–1253), Bishop of Lincoln, and on the Polish mathematician Vitello (or Witelo) both of whom were influential in rekindling the study of optics. Their works were known to the Franciscan Roger Bacon (1215–94) who is considered by many to be the first scientist in the modern sense. He seems to have initiated the idea of using lenses for correcting vision and even hinted at the possibility of combining lenses to form a telescope. Bacon also had some understanding of the way in which rays traverse a lens. After his death optics again languished.

Even so, by the mid-thirteen hundreds, European paintings were depicting monks wearing eyeglasses. And alchemists had come up with a liquid amalgam of tin and mercury that was rubbed onto the back of glass plates to make mirrors. Leonardo da Vinci (1452–1519) described the *camera obscura* later popularized by the work of Giovanni Battista Della Porta (1535–1615). Porta discussed multiple mirrors and combinations of positive and negative lenses in his *Magia naturalis* (1589).

This, for the most part, modest array of events constitutes what might be called the first period of optics. It was undoubtedly a beginning—but on the whole a dull one. It was more a time for learning how to play the game than actually scoring points. The whirlwind of accomplishment and excitement was to come later, in the seventeenth century.

1.3 FROM THE SEVENTEENTH CENTURY

It is not clear who actually invented the refracting telescope, but records in the archives at the Hague show that on October 2, 1608 Hans Lippershey (1587–1619), a Dutch spectacle maker, applied for a patent on the device. Galileo Galilei (1564–1642) in Padua, heard about the invention and within several months had built his own instrument, grinding the lenses by hand. The compound microscope was invented at just about the same time, probably by the Dutchman Zacharias Janssen (1588–1632). The microscope's concave eyepiece was replaced with a convex lens by Francisco Fontana (1580–1656) of Naples and a similar change in the telescope was introduced by Johannes Kepler (1571–1630). Turning his telescope skyward, Galileo, on January 7, 1610, discovered the moons of Jupiter. Within the same year he saw Saturn's rings and further concluded that the sun rotated, this after observing moving spots on its surface. But many doubted their eyes and others would not look. In a letter to Kepler, Galileo wrote

> Why are you not here? What shouts of laughter we should have at this glorious folly! And to hear the professor of philosophy at Pisa labouring before the Grand Duke with logical arguments, as if with magical incantations to charm the new planets out of the sky.

In 1611, Kepler published his *Dioptrice*. He had discovered total internal reflection and arrived at the small angle approximation to the law of refraction, in which case the incident and transmission angles are proportional. He goes on to evolve a treatment of first-order optics for thin-lens systems and describes the detailed operation of both the Keplerian

Fig. 1.1 Galileo Galilei (1564–1642)

(positive eyepiece) and Galilean (negative eyepiece) telescopes. Willebrord Snell (1591–1626), professor at Leyden, empirically discovered the long-hidden *law of refraction* in 1621—this was one of the great moments in optics. By learning precisely how rays of light are redirected on traversing a boundary between two media, Snell in one swoop swung open the door to modern applied optics. René Descartes (1596–1650) was the first to publish the now familiar formulation of the law of refraction in terms of sines, and he should perhaps be given equal credit for its discovery. Descartes deduced the law using a model in which light was viewed as a pressure transmitted by an elastic medium; as he put it in his *La Dioptrique* (1637)

> ... recall the nature that I have attributed to light, when I said that it is nothing other than a certain motion or an action conceived in a very subtle matter, which fills the pores of all other bodies ...

the universe was a plenum. Pierre de Fermat (1601–65), taking exception to Descartes' assumptions, rederived the law of reflection from his own *principle of least time* (1657).

Fig. 1.2 René Descartes (1596–1650)

Departing from Hero's shortest-path statement, Fermat maintained that light propagates from one point to another along a route taking the least time, even if it has to vary from the shortest actual path to do it.

The phenomenon of diffraction, i.e. the deviation from rectilinear propagation which occurs when light advances beyond an obstruction, was first noted by Professor Francesco Maria Grimaldi (1618–63) at the Jesuit College in Bologna. He had observed bands of light within the shadow of a rod illuminated by a small source. Robert Hooke (1635–1703), curator of experiments for the Royal Society, London, later also observed diffraction effects. He was the first to study the colored interference patterns generated by thin films (*Micrographia* 1665) and correctly concluded that they were due to an interaction between the light reflected from the front and back surfaces. He proposed the idea that light was a rapid vibratory motion of the medium propagating at a very great speed. Moreover "every pulse or vibration of the luminous body will generate a sphere"—this was the beginning of the wave theory. In the year Galileo died Isaac Newton (1642–1727) was born. The thrust of Newton's scientific effort is clear from his own description of his work in optics as *experimental philosophy*. It was his intent to build on direct observation and avoid speculative hypotheses. Thus he remained ambivalent for a long while about the actual nature of light. Was it corpuscular—a stream of particles, as some maintained? Or was light a wave in an all-pervading medium, the ether? At the

Fig. 1.3 Pierre de Fermat (1608–1665)

Fig. 1.4 Sir saac Newton (1642–1727)

age of twenty-three, he began his now famous experiments on dispersion.

> I procured me a triangular glass prism to try therewith the celebrated phenomena of colours.

Newton concluded that white light was composed of a mixture of a whole range of independent colors. He maintained that the corpuscles of light associated with the various colors excited the ether into characteristic vibrations. Furthermore, the sensation of red corresponded to the longest vibration of the ether and violet to the shortest. Even though his work shows a curious propensity for simultaneously embracing both the wave and emission (corpuscular) theories, he does become more committed to the latter as he grows older. Perhaps his main reason for rejecting the wave theory as it stood then was the blatant problem of explaining rectilinear propagation in terms of waves which spread out in all directions.

After some all too limited experiments, Newton gave up trying to remove chromatic aberration from refracting telescope lenses: erroneously concluding that it could not be done, he turned to the design of reflectors. Sir Isaac's first reflecting telescope, completed in 1668, was only six inches long and one inch in diameter but it magnified some 30 times.

At about the same time that Newton was emphasizing the emission theory in England, Christian Huygens (1629–95), on the continent, was greatly extending the wave theory. Unlike Descartes, Hooke and Newton, Huygens correctly concluded that light effectively slowed down on entering more dense media. He was able to derive the laws of reflection and refraction and even explained the double refraction of calcite, using his wave theory. And it was while working with calcite that he discovered the phenomenon of *polarization*.

> As there are two different refractions, I conceived also that there are two different emanations of the waves of light . . .

Thus light was either a stream of particles or a rapid undulation of ethereal matter. In any case, it was generally agreed that its speed of propagation was exceedingly large. The fact that it was indeed finite was determined in 1676 by the Dane Olaf Römer (1644–1710). Jupiter's nearest moon has an orbit about that planet which is nearly in the plane of Jupiter's own orbit around the sun. Römer found that the interval, T, between successive eclipses of the satellite as it passed into the shadow of Jupiter increased when the Earth–Jupiter distance was increasing, and vice versa. He

Fig. 1.5 Christian Huygens (1629–1695)

correctly deduced that if the planets' separation increases by a distance d during one revolution of the Jovian moon, T will exceed the actual period T_0 by an amount d/c where c is the speed of light, i.e. $T - T_0 = d/c$. The corresponding value of c was found to be about 48,000 leagues per second or roughly 214,000 km/s.

The great weight of Newton's opinion hung as a s̶ over the wave theory during the eighteenth centu̶ stifling its advocates. There were too many content . dogma and too few nonconformist enough to follow their own experimental philosophy, as surely Newton would have had them do. Despite this, the prominent mathematician Leonhard Euler (1707–83) was a devotee of the wave

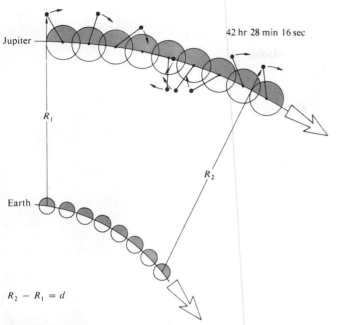

42 hr 28 min 16 sec

R_1

R_2

Earth

$R_2 - R_1 = d$

Fig. 1.6 Römer's measurement of c.

theory, even if an unheeded one. Euler proposed that the undesirable color effects seen in a lens were absent in the eye (which is an erroneous assumption) because the different media present negated dispersion. He suggested that achromatic lenses might be constructed in a similar way. Enthused by this work, Samuel Klingenstjerna (1698–1765), professor at Upsala, reperformed Newton's experiments on achromatism and determined them to be in error. Klingenstjerna was in communication with a London optician, John Dollond (1706–61), who was observing similar results. Dollond finally, in 1758, combined two elements, one of crown and the other of flint glass, to form a single achromatic lens. This was an accomplishment of very great practical importance. Incidentally, Dollond's invention was actually preceded by the unpublished work of the amateur scientist Chester Moor Hall (1703–71) of Moor Hall in Essex.

1.4 THE NINETEENTH CENTURY

The wave theory of light was reborn at the hands of Dr. Thomas Young (1773–1829), one of the truly great minds of the century. On November 12, 1801, July 1, 1802 and November 24, 1803 he read papers before the Royal Society extolling the wave theory and adding to it a new fundamental concept, the so-called *principle of interference.*

> When two undulations, from different origins, coincide either perfectly or very nearly in direction, their joint effect is a combination of the motions belonging to each.

He was able to explain the colored fringes of thin films and determined wavelengths of various colors using Newton's data. Even though Young, time and again, maintained that his conceptions had their very origins in the researches of Newton, he was severely attacked. In a series of articles, probably written by Lord Brougham, in the *Edinburgh Review* Young's papers were said to be "destitute of every species of merit"— and that's going pretty far. Under the pall of Newton's presumed infallibility the pedants of England were not prepared for the wisdom of Young, who in turn became disheartened.

Augustin Jean Fresnel (1788–1827) born in Broglie, Normandy, began his brilliant revival of the wave theory in France unaware of the efforts of Young some thirteen years earlier. Fresnel synthesized the concepts of Huygens' wave description and the interference principle. The mode of propagation of a primary wave was viewed as a succession of stimulated spherical secondary wavelets which overlapped and interfered to reform the advancing primary wave as it would appear an instant later. In Fresnel's words

> The vibrations of a luminous wave in any one of its points may be considered as the sum of the elementary movements conveyed to it at the same moment, from the separate action of all the portions of the unobstructed wave considered in any one of its anterior positions.

These waves were presumed to be longitudinal in analogy with sound waves in air. Dominique François Jean Arago (1786–1853) was an early convert to Fresnel's wave theory and they became fast friends and sometime collaborators. Under criticism from such renowned men and proponents of the emission hypothesis as Pierre Simon de Laplace (1749–1827) and Jean-Baptiste Biot (1774–1862), Fresnel's theory took on a mathematical emphasis. He was able to calculate the diffraction patterns arising from various obstacles and apertures and satisfactorily accounted for rectilinear propagation in homogeneous isotropic media, thus dispelling Newton's main objection to the undulatory theory. When finally apprised of Young's priority to the interference principle, somewhat disappointedly Fresnel nonetheless wrote to Young telling him that he was consoled by finding himself in such good company—the two great men became allies.

Huygens was aware of the phenomenon of polarization arising in calcite crystals, as was Newton. Indeed, the latter in his *Opticks* stated that

Every Ray of Light has therefore two opposite Sides . . .

He further developed this concept of lateral asymmetry even though avoiding any interpretation in terms of the hypothetical nature of light. Yet it was not until 1808 that Étienne Louis Malus (1775–1812) discovered that this two-sidedness of light became apparent upon reflection as well; it was not inherent to crystalline media. Fresnel and Arago then conducted a series of experiments to determine the effect of polarization on interference but the results were utterly inexplicable within the framework of their longitudinal wave picture—this was a dark hour indeed. For several years Young, Arago and Fresnel wrestled with the problem until finally Young suggested that the ethereal vibration might be *transverse* as is a wave on a string. The two-sidedness of light was then simply a manifestation of the two orthogonal vibrations of the ether, transverse to the ray direction. Fresnel went on to evolve a mechanistic description of ether oscillations which led to his now famous formulas for the amplitude of reflected and transmitted light. By 1825 the emission (or corpuscular) theory had only a few tenacious advocates.

The first terrestrial determination of the speed of light was performed by Armand Hippolyte Louis Fizeau (1819–96) in 1849. His apparatus, consisting of a rotating toothed wheel and a distant mirror (8633 m), was set up in the suburbs of Paris from Suresnes to Montmartre. A pulse of light leaving an opening in the wheel struck the mirror and returned. By adjusting the known rotational speed of the wheel, the returning pulse could be made either to pass through an opening and be seen or to be obstructed by a tooth. Fizeau arrived at a value of the speed of light equal to 315,300 km/s. His colleague Jean Bernard Léon Foucault (1819–68) was also involved in research on the speed of light. In 1834 Charles Wheatstone (1802–75) had designed a rotating-mirror arrangement in order to measure the duration of an electric spark. Using this scheme, Arago proposed to measure the speed of light in dense media but was never able to carry out the experiment. Foucault took up the work, which was later to provide material for his doctoral thesis. On May 6, 1850 he reported to the Academy of Sciences that the speed of light in water was *less* than that in air. This result was, of course in direct conflict with Newton's formulation of the emission theory and it was a hard blow to its very few remaining devotees.

Fig. 1.7 James Clerk Maxwell (1831–1879)

While all of this was happening in optics, quite independently the study of electricity and magnetism was also bearing fruit. In 1845 the master experimentalist Michael Faraday (1791–1867) established an interrelationship between electromagnetism and light when he found that the polarization direction of a beam could be altered by a strong magnetic field applied to the medium. James Clerk Maxwell (1831–79) brilliantly summarized and even extended all the then known empirical knowledge on the subject in a single set of mathematical equations. Beginning with this remarkably succinct and beautifully symmetric synthesis, he was able to show, purely theoretically, that the electromagnetic field could propagate as a transverse wave in the luminiferous ether. Solving for the speed of the wave, he arrived at an expression in terms of electric and magnetic properties of the medium ($c = 1/\sqrt{\epsilon_0 \mu_0}$). Upon substituting known empirically determined values for these quantities, he obtained a numerical result equal to the measured speed of light! The conclusion was inescapable—*light was "an electromagnetic disturbance in the form of waves" pro-pagated through the ether.* Maxwell died at the age of forty-eight, eight years too soon to see the experimental confirmation of his insights and far too soon for physics. Heinrich Rudolf Hertz (1857–94) verified the existence of long-wavelength electromagnetic waves by generating and detecting them in an extensive series of experiments published in 1888.

The acceptance of the wave theory of light seemed to necessitate an equal acceptance of the existence of an all-pervading substratum, the luminiferous ether. If there were waves, it seemed obvious that there must be a supporting medium. Quite naturally, a great deal of scientific effort went into determining the physical nature of the ether, yet it would have to possess some rather strange properties. It had to be so tenuous as to allow an apparently unimpeded motion of celestial bodies. At the same time it could support the exceedingly high frequency ($\sim 10^{15}$ Hz) oscillations of light traveling at 186,000 miles/s. That implied remarkably strong restoring forces within the ethereal substance. The speed at which a wave advances through a medium is dependent upon the characteristics of the disturbed substratum and not upon any motion of the source. This is in contrast to the behavior of a stream of particles whose speed with respect to the source is the essential parameter. Certain aspects of the nature of ether intrude when studying the optics of moving objects and it was this area of research, evolving quite quietly on its own, which ultimately led to the next great turning point. In 1725 James Bradley (1693–1762), then Savilian Professor of Astronomy at Oxford, attempted to measure the distance to a star by observing its orientation at two different times of the year. The position of the earth changed as it orbited around the sun and thereby provided a large baseline for triangulation on the star. To his surprise, he found that the "fixed" stars displayed an apparent systematic movement related to the direction of motion of the earth in orbit and not dependent, as had been anticipated, on the earth's position in space. This so-called *stellar aberration* is analogous to the well-known falling-raindrop situation. A raindrop, although traveling vertically with respect to an observer at rest on the earth will appear to change its incident angle when the observer is in motion. Thus a corpuscular model of light could explain stellar aberration rather handily. Alternatively, the wave theory also offers a satisfactory explanation provided that it is assumed that *the ether remains totally undisturbed as the earth plows through it.* Incidentally, Bradley, convinced of the correctness of his analysis, used the observed

aberration data to arrive at an improved value of c, thus confirming Römer's theory of the finite speed of light.

In response to speculation as to whether the earth's motion through the ether might result in an observable difference between light from terrestrial and extraterrestrial sources, Arago set out to examine the problem experimentally. He found that there were no observable differences. Light behaved just as if the earth were at rest with respect to the ether. To explain these results, Fresnel suggested in effect that light was partially dragged along as it traversed a transparent medium in motion. An experiment by Fizeau in which light beams passed down moving columns of water and an experiment in 1871 by Sir George Biddell Airy (1801–92) using a water-filled telescope to examine stellar aberration both seemed to confirm Fresnel's drag hypothesis. Assuming an ether at *absolute rest*, Hendrik Antoon Lorentz (1853–1928) derived a theory which encompassed Fresnel's ideas.

In 1879 in a letter to D. P. Todd of the U.S. Nautical Almanac Office, Maxwell suggested a scheme for measuring the speed at which the solar system moved with respect to the luminiferous ether. The American, Albert Abraham Michelson (1852–1931), then a naval instructor, took up the idea. Michelson, at the tender age of 26, had already established a favorable reputation by performing an extremely precise determination of the speed of light. A few years later, he began an experiment to measure the effect of the earth's motion through the ether. Since the speed of light in ether is constant and the earth, in turn, presumably moves relative to the ether (orbital speed of 67,000 miles/h), the speed of light measured with respect to the earth should be affected by the planet's motion. Michelson's work was begun in Berlin but because of traffic vibrations, it was moved to Potsdam and in 1881, he published his findings. There was no detectable motion of the earth with respect to the ether—the ether was not stationary. But the decisiveness of this surprising result was blunted somewhat when Lorentz pointed out an oversight in the calculation. Several years later Michelson, then professor of physics at Case School of Applied Science in Cleveland, Ohio, joined with Edward Williams Morley (1838–1923), a well-known professor of chemistry at Western Reserve, to redo the experiment with considerably greater precision. Amazingly enough, their results, published in 1887, once again were negative:

It appears from all that precedes reasonably certain that if there be any relative motion between the earth and the luminiferous aether, it must be small: quite small enough entirely to refute Fresnel's explanation of aberration.

Fig. 1.8 Albert Einstein (1879–1955) [Photo by Fred Stein.]

Thus, while an explanation of stellar aberration within the context of the wave theory required the existence of a relative motion between earth and ether, the Michelson–Morley experiment refuted that possibility. Moreover, the findings of Fizeau and Airy necessitated the inclusion of a partial drag of light due to motion of the medium.

1.5 TWENTIETH-CENTURY OPTICS

Jules Henri Poincaré (1854–1912) was perhaps the first to grasp the significance of the experimental inability to observe any effects of motion relative to the ether. In 1899 he began to make his views known and in 1900 he said

> Our aether, does it really exist? I do not believe that more precise observations could ever reveal anything more than *relative* displacements.

In 1905 Albert Einstein (1879–1955) introduced his *special theory of relativity* in which he too, quite independently, rejected the ether hypothesis.

> The introduction of a "luminiferous ether" will prove to be superfluous inasmuch as the view here to be developed will not require an "absolutely stationary space."

He further postulated that

> light is always propagated in empty space with a definite velocity *c* which is independent of the state of motion of the emitting body.

The experiments of Fizeau, Airy and Michelson–Morley were then explained quite naturally within the framework of Einstein's relativistic kinematics.* Deprived of the ether, physicists simply had to get used to the idea that electromagnetic waves could propagate through free space—there was no alternative. Light was now envisaged as a self-sustaining wave with the conceptual emphasis passing from ether to field. The electromagnetic wave became an entity in itself.

On October 19, 1900, Max Karl Ernst Ludwig Planck (1858–1947) read a paper before the German Physical Society in which he introduced the beginnings of what was to become yet another great revolution in scientific thought—*quantum mechanics*, a theory embracing submicroscopic phenomena. In 1905, building on these ideas, Einstein proposed a new form of corpuscular theory in which he asserted that light consisted of globs or "particles" of energy. Each such quantum of radiant energy or photon,† as it came to be called, had an energy proportional to its frequency v, i.e. $\mathscr{E} = hv$ where h is known as Planck's constant. By the end of the nineteen-twenties, through the efforts of such men as Bohr, Born, Heisenberg, Schrödinger, De Broglie, Pauli, Dirac and several others, quantum mechanics had become a well-verified structure. It gradually became evident that the concepts of particle and wave, which in the macroscopic world seem so obviously mutually exclusive, must be merged in the submicroscopic domain. The mental image of an atomic particle (e.g. electrons, neutrons, etc.) as a minute localized lump of matter would no longer suffice. Indeed, it was found that these "particles" could generate interference and diffraction patterns in precisely the same way as would light. Thus photons,

---

* See, for example, *Special Relativity* by French, Chapter 5.

† The word *photon* was coined by G. N. Lewis, *Nature*, December 18, 1926.

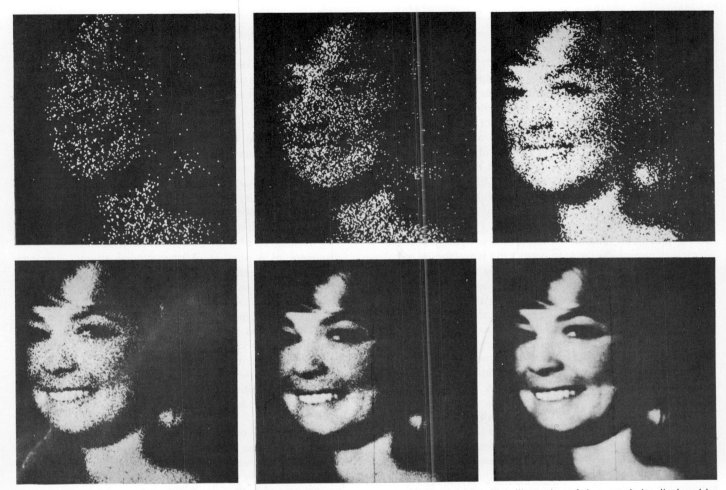

Fig. 1.9 These photos, which were made using electronic amplification techniques, are a compelling illustration of the granularity displayed by light in its interaction with matter. Under exceedingly faint illumination the pattern (each spot corresponding to one photon) seems almost random, but as the light level increases the quantal character of the process gradually becomes obscured. (See *Advances in Biological and Medical Physics* V, 1957, 211–242.) Courtesy Radio Corporation of America.

protons, electrons, neutrons, etc., the whole lot, have both particle and wave manifestations. Relativity liberated light from the ether and showed the kinship between mass and energy (via $\mathscr{E} = mc^2$). What seemed to be two almost antithetic quantities now became interchangeable. Quantum mechanics went on to establish that a particle* of momentum p, had an associated wavelength λ such that $p = h/\lambda$

---

* Perhaps it might help if we just called them all *wavicles*. By the way, how do you envision in your mind's eye the meeting of an electron and a positron and their subsequent annihilation with the creation of two photons?

(whether it had rest mass or not). The neutrino, a neutral particle having zero rest mass, was postulated for theoretical reasons in 1930 by Wolfgang Pauli (1900–58) and verified experimentally later on in the fifties. The easy images of submicroscopic specks of matter became untenable and the wave—particle dichotomy dissolved into a duality.

Quantum mechanics also treats the manner in which light is absorbed and emitted by atoms. Suppose we cause a gas to glow by heating it or passing an electrical discharge through it. The light emitted is characteristic of the very structure of the atoms constituting the gas. Spectroscopy, which is the branch of optics dealing with spectrum analysis, developed from the researches of Newton. William Hyde Wollaston (1766–1828) made the earliest observations of the dark lines in the solar spectrum (1802). Because of the slit-shaped aperture generally used in spectroscopes, the output consisted of narrow colored bands of light, the so-called *spectral lines*. Joseph Fraunhofer (1787–1826) independently greatly extended the subject. After accidentally discovering the double line of sodium, he went on to study sunlight and made the first wavelength determinations using diffraction gratings. Gustav Robert Kirchhoff (1824–87) and Robert Wilhelm Bunsen (1811–99) working conjointly at Heidelberg, established that each kind of atom had its own signature in a characteristic array of spectral lines. And in 1913 Niels Henrik David Bohr (1885–1962) set forth a precursory quantum theory of the hydrogen atom which was nonetheless able to predict the wavelengths of its emission spectrum. The light emitted by an atom is now understood to arise from its outermost electrons. An atom which somehow absorbs energy (e.g. via collisions) changes from its usual configuration, known as the ground state, to what's called an excited state. After some finite time, it relaxes back to the ground state, the electrons return to their original configuration with respect to the nucleus, giving up the excess energy often in the form of light. The process is the domain of modern quantum theory which describes the minutest details with incredible precision and beauty.

The flourishing of applied optics in what has transpired of the second half of the twentieth century represents a renaissance in itself. In the nineteen fifties several workers began to inculcate optics with the mathematical techniques and insights of communications theory. Just as the idea of momentum provides another dimension in which to visualize aspects of mechanics, the concept of spatial frequency offers a rich new way of appreciating a broad range of optical phenomena. Bound together by the mathematical formalism of Fourier analysis, the outgrowths of this contemporary emphasis have been far-reaching. Of particular interest is the theory of image formation and evaluation, the transfer functions and the idea of spatial filtering.

The advent of the high-speed digital computer brought with it a vast improvement in the design of complex optical systems. Aspherical lens elements took on renewed practical significance and the diffraction-limited system with an appreciable field of view became a reality. The technique of ion bombardment polishing, where one atom at a time is chipped away, was introduced to meet the need for extreme precision in the preparation of optical elements. The use of single and multilayer thin-film coatings (reflecting, anti-

reflecting, etc.) became commonplace. Fiber optics evolved into a practical tool and thin-film light guides were being studied. A great deal of attention was paid to the infrared end of the spectrum (surveillance systems, missile guidance, etc.) and this in turn stimulated the development of IR materials. Plastics began to find serious applications in optics (lens elements, replica gratings, fibers, aspherics, etc.). A new class of partially vitrified glass-ceramics with exceedingly low thermal expansion was developed. A resurgence in the construction of astronomical observatories (both terrestrial and extraterrestrial) running across the whole spectrum was well under way by the end of the sixties.

The first laser was built in 1960 and within a decade laser beams spanned the range from infrared to ultraviolet. The availability of high-power coherent sources led to the discovery of a number of new optical effects (harmonic generation, frequency mixing, etc.) and thence to a panorama of marvelous new devices. The technology needed to produce a practicable optical communications system was fast evolving. The sophisticated use of crystals in devices such as second-harmonic generators, electro-optic and acousto-optic modulators and the like spurred a great deal of contemporary research in crystal optics. The wavefront reconstruction technique known as holography, which produces magnificent three-dimensional images, found numerous other applications (nondestructive testing, data storage, etc.).

The military orientation of much of the developmental work of the sixties began to give way in the seventies to the urgency of improving the quality of life. Optical systems are finding increasing uses in health technology, environmental protection and earth resources monitoring.

The vitality of optics in the seventies is in marked contrast to the comparatively dreary state of optical technology even three or four decades ago. The melding of optics and electronics into what is being called *electro-optics* is indicative of the new emphasis.

Profound insights are slow in coming. What few we have took over three thousand years to glean even though the pace is ever quickening. It is marvelous indeed to watch the answer subtly change while the question immutably remains—*what is light?**

---

* For more reading on the history of optics, see F. Cajori, *A History of Physics* and V. Ronchi, *The Nature of Light*. Excerpts from a number of original papers can conveniently be found in W. F. Magie, *A Source Book in Physics*, and in M. H. Shamos, *Great Experiments in Physics*.

The Mathematics of Wave Motion   2

There are a great many, seemingly unrelated, physical processes which can be described in terms of the mathematics of wave motion. In this respect there are fundamental similarities between a pulse traveling along a stretched string (Fig. 2.1), a surface tension ripple in a cup of tea, and the light reaching us from some remote point in the universe. This chapter will develop some of the mathematical techniques needed to treat wave phenomena in general. We will begin with some fairly simple ideas concerning the propagation of disturbances and from these arrive at the three-dimensional differential wave equation. Throughout the study of optics one utilizes plane, spherical and cylindrical waves. Accordingly we'll develop their mathematical representations, showing them to be solutions of the differential wave equation. This chapter will be a completely classical treatment; even so it can be shown, although we will not do so, that our results do indeed obey the requirements of special relativity.

2.1 ONE-DIMENSIONAL WAVES

Envision some disturbance ψ traveling in the positive x-direction with a constant speed v. The specific nature of the disturbance is at the moment unimportant. It might be the vertical displacement of the string in Fig. 2.1, or the magnitude of an electric or magnetic field associated with an electromagnetic wave, or even the quantum-mechanical probability amplitude of a matter wave.

Since the disturbance is moving, it must be a function of both position and time and can therefore be written as

$$\psi = f(x, t). \tag{2.1}$$

The shape of the disturbance at any instant, say $t = 0$, can be found by holding time constant at that value. In this case

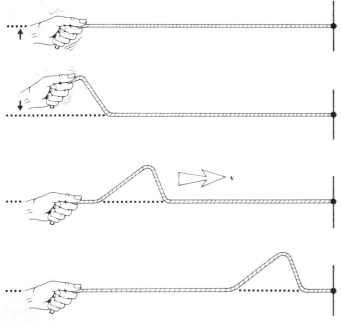

Fig. 2.1 A wave on a string.

$$\psi(x, t)|_{t=0} = f(x, 0) = f(x) \qquad (2.2)$$

represents the shape or *profile* of the wave at that time. The process is analogous to taking a "photograph" of the pulse as it travels by. For the moment we will limit ourselves to a wave which does not change its shape as it progresses through space. Figure 2.2 is a "double exposure" of such a

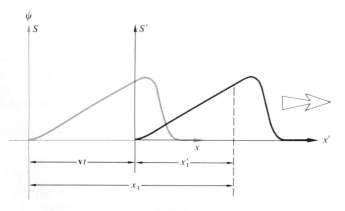

Fig. 2.2 Moving reference frame.

disturbance taken at the beginning and end of a time interval t. The pulse has moved along the x-axis a distance vt, but in all other respects it remains unaltered. We now introduce a coordinate system S' which travels along with the pulse at the speed v. In this system ψ is no longer a function of time and as we move along with S' we see a stationary constant profile with the same functional form as Eq. (2.2). Here, the coordinate is x' rather than x so that

$$\psi = f(x'). \qquad (2.3)$$

The disturbance looks the same at any value of t in S' as it did at $t = 0$ in S when S and S' had a common origin. It follows from Fig. 2.2 that

$$x' = x - vt, \qquad (2.4)$$

so that ψ can be written in terms of the variables associated with the stationary S system as

$$\psi(x, t) = f(x - vt). \qquad (2.5)$$

This then represents the most general form of the one-dimensional *wave function*. To be more specific, *we have only to choose a shape (2.2) and then substitute $(x - vt)$ for x in $f(x)$. The resulting expression describes a moving wave having the desired profile.* If we check the form of Eq. (2.5) by examining ψ after an increase in time of Δt and a corresponding increase of $v\,\Delta t$ in x, we find

$$f[(x + v\,\Delta t) - v(t + \Delta t)] = f(x - vt)$$

and the profile is unaltered.

Similarly, if the wave were traveling in the negative x-direction, i.e. to the left, Eq. (2.5) would become

$$\psi = f(x + vt), \quad \text{with} \quad v > 0. \qquad (2.6)$$

We may conclude therefore that, regardless of the shape of the disturbance, the variables x and t must appear in the function as a unit, i.e. as a single variable in the form $(x \mp vt)$. Equation (2.5) is often expressed equivalently as some function of $(t - x/v)$ since

$$f(x - vt) = F\left(-\frac{x - vt}{v}\right) = F(t - x/v). \qquad (2.7)$$

We wish to use the information derived so far to develop the general form of the one-dimensional differential wave equation. To that end, take the partial derivative of $\psi(x, t)$ with respect to x holding t constant. Using $x' = x \mp vt$ we have

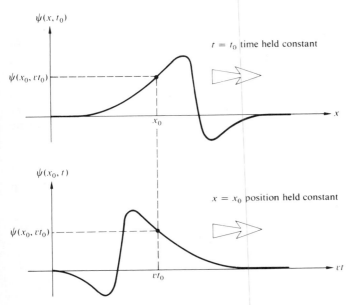

$\psi(x, t_0)$

$t = t_0$ time held constant

$\psi(x_0, vt_0)$

x_0

x

$\psi(x_0, t)$

$x = x_0$ position held constant

$\psi(x_0, vt_0)$

vt_0

vt

Fig. 2.3 Variation of ψ with x and t.

$$\frac{\partial \psi}{\partial x} = \frac{\partial f}{\partial x'}\frac{\partial x'}{\partial x} = \frac{\partial f}{\partial x'} \quad \text{since} \quad \frac{\partial x'}{\partial x} = 1. \quad (2.8)$$

If we hold x constant the partial derivative with respect to time is

$$\frac{\partial \psi}{\partial t} = \frac{\partial f}{\partial x'}\frac{\partial x'}{\partial t} = \mp v\frac{\partial f}{\partial x'}. \quad (2.9)$$

Combining Eqs. (2.8) and (2.9) yields

$$\frac{\partial \psi}{\partial t} = \mp v\frac{\partial \psi}{\partial x}. \quad (2.10)$$

This says that the rate of change of ψ with t and with x are equal, to within a multiplicative constant, as shown in Fig. 2.3. Knowing beforehand that we'll need two constants to specify a wave we can anticipate a second-order wave equation. Taking the second partial derivatives of Eqs. (2.8) and (2.9) yields

$$\frac{\partial^2 \psi}{\partial x^2} = \frac{\partial^2 f}{\partial x'^2}$$

and

$$\frac{\partial^2 \psi}{\partial t^2} = \frac{\partial}{\partial t}\left(\mp v\frac{\partial f}{\partial x'}\right) = \mp v\frac{\partial}{\partial x'}\left(\frac{\partial f}{\partial t}\right).$$

Since

$$\frac{\partial \psi}{\partial t} = \frac{\partial f}{\partial t}$$

it follows using Eq. (2.9) that

$$\frac{\partial^2 \psi}{\partial t^2} = v^2\frac{\partial^2 f}{\partial x'^2}.$$

Combining these equations we obtain

$$\frac{\partial^2 \psi}{\partial x^2} = \frac{1}{v^2}\frac{\partial^2 \psi}{\partial t^2}, \quad (2.11)$$

which is the one-dimensional *differential wave equation*. It is apparent from the form of Eq. (2.11) that if two different wave functions ψ_1 and ψ_2 are each separate solutions then $(\psi_1 + \psi_2)$ is also a solution.* Accordingly, the wave equation is most generally satisfied by a wave function having the form

$$\psi = C_1 f(x - vt) + C_2 g(x + vt) \quad (2.12)$$

where C_1 and C_2 are constants and the functions are twice differentiable. This is clearly a sum of two waves traveling in opposite directions along the x-axis with the same velocity but not necessarily the same profile. The superposition principle is inherent in this equation and we will come back to it in Chapter 7.

2.2 HARMONIC WAVES

Until now we have not given the wave function $\psi(x, t)$ an explicit functional dependence, i.e. we have not specified its shape. Let's now examine the simplest wave form where the profile is a sine or cosine curve. These are variously known as sinusoidal waves, simple harmonic waves, or more succinctly as harmonic waves. We shall see in

---

* Since both ψ_1 and ψ_2 are solutions

$$\frac{\partial^2 \psi_1}{\partial x^2} = \frac{1}{v^2}\frac{\partial^2 \psi_1}{\partial t^2} \quad \text{and} \quad \frac{\partial^2 \psi_2}{\partial x^2} = \frac{1}{v^2}\frac{\partial^2 \psi_2}{\partial t^2}.$$

Adding these, we get

$$\frac{\partial^2 \psi_1}{\partial x^2} + \frac{\partial^2 \psi_2}{\partial x^2} = \frac{\partial^2}{\partial x^2}(\psi_1 + \psi_2) = \frac{1}{v^2}\left[\frac{\partial^2 \psi_1}{\partial t^2} + \frac{\partial^2 \psi_2}{\partial t^2}\right] = \frac{1}{v^2}\frac{\partial^2}{\partial t^2}(\psi_1 + \psi_2),$$

so that $(\psi_1 + \psi_2)$ is also a solution of Eq. (2.11).

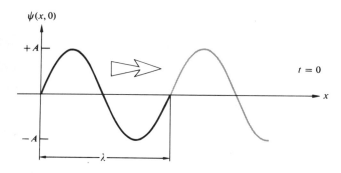

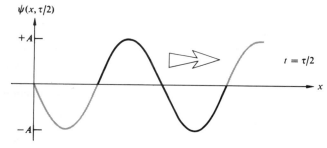

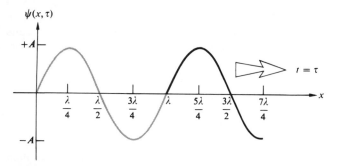

Fig. 2.4 A progressive wave at three different times.

Chapter 7 that any wave shape can be synthesized by a superposition of harmonic waves and they therefore take on a special significance.

Choose as the profile the simple function

$$\psi(x, t)|_{t=0} = \psi(x) = A \sin kx = f(x), \qquad (2.13)$$

where k is a positive constant known as the *propagation number* and kx is in radians. The sine varies from $+1$ to -1 so that the maximum value of $\psi(x)$ is A. This maximum disturbance is known as the *amplitude* of the wave (Fig. 2.4). In order to transform Eq. (2.13) into a *progressive wave* traveling at speed v in the positive x-direction, we need

merely replace x by $(x - vt)$, in which case

$$\psi(x, t) = A \sin k(x - vt) = f(x - vt). \qquad (2.14)$$

This is clearly (Problem 2.4) a solution of the differential wave equation (2.11). Holding either x or t fixed results in a sinusoidal disturbance and so the wave is periodic in both space and time. The *spatial period* is known as the *wavelength* and is denoted by λ. An increase or decrease in x by the amount λ should leave ψ unaltered, i.e.

$$\psi(x, t) = \psi(x \pm \lambda, t). \qquad (2.15)$$

In the case of a harmonic wave, this is equivalent to altering the argument of the sine function by $\pm 2\pi$. Therefore

$$\sin k(x - vt) = \sin k[(x \pm \lambda) - vt] = \sin [k(x - vt) \pm 2\pi]$$

and so

$$|k\lambda| = 2\pi$$

or, since both k and λ are positive numbers

$$k = 2\pi/\lambda. \qquad (2.16)$$

In a completely analogous fashion, we can examine the *temporal period*, τ. This is the amount of time it takes for one complete wave to pass a stationary observer. In this case, it is the repetitive behavior of the wave in time which is of interest, so that

$$\psi(x, t) = \psi(x, t \pm \tau) \qquad (2.17)$$

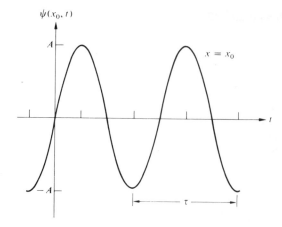

Fig. 2.5 A harmonic wave.

and

$$\sin k(x - vt) = \sin k[x - v(t \pm \tau)] = \sin [k(x - vt) \pm 2\pi].$$

Therefore

$$|kv\tau| = 2\pi.$$

But these are all positive quantities and so

$$kv\tau = 2\pi \qquad (2.18)$$

or

$$\frac{2\pi}{\lambda} v\tau = 2\pi$$

from which it follows that

$$\tau = \frac{\lambda}{v}. \qquad (2.19)$$

The period is the number of units of time per wave (Fig. 2.5), the inverse of which is the *frequency* v or the number of waves per unit of time. Thus

$$v \equiv \frac{1}{\tau} \qquad \text{(cycles/s or Hertz)}$$

and Eq. (2.19) becomes

$$v = v\lambda \qquad \text{(m/s)} \qquad (2.20)$$

There are two other quantities which are often used in the literature of wave motion and these are the *angular frequency*

$$\omega \equiv \frac{2\pi}{\tau} \qquad \text{(radians/s)} \qquad (2.21)$$

and the *wave number*

$$\varkappa \equiv \frac{1}{\lambda} \qquad \text{(m}^{-1}\text{)}. \qquad (2.22)$$

The wavelength, period, frequency, angular frequency, wave number and propagation number all describe aspects of the repetitive nature of a wave in space and time. These concepts are equally well applied to waves which are not harmonic as long as each wave profile is made up of a regularly repeating pattern (Fig. 2.6). We have thus far defined a number of quantities which characterize various aspects of wave motion. There exist, accordingly, a number of equivalent formulations of the progressive harmonic wave. Some of the most common of these are

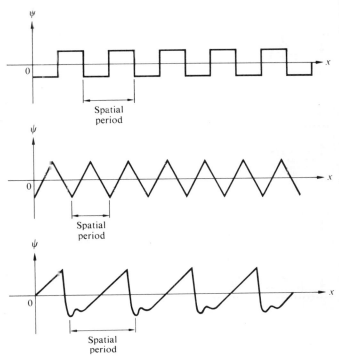

Fig. 2.6 Anharmonic periodic waves.

$$\psi = A \sin k(x \mp vt) \qquad [2.14]$$

$$\psi = A \sin 2\pi \left(\frac{x}{\lambda} \mp \frac{t}{\tau}\right) \qquad (2.23)$$

$$\psi = A \sin 2\pi(\varkappa x \mp vt) \qquad (2.24)$$

$$\psi = A \sin (kx \mp \omega t) \qquad (2.25)$$

$$\psi = A \sin 2\pi v\left(\frac{x}{v} \mp t\right) \qquad (2.26)$$

It should be noted that these waves are all of infinite extent, i.e. for any fixed value of t, x varies from $-\infty$ to $+\infty$. Each wave has a single constant frequency and is therefore said to be *monochromatic*.

2.3 PHASE AND PHASE VELOCITY

Examine any one of the harmonic wave functions, such as

$$\psi(x, t) = A \sin (kx - \omega t).$$

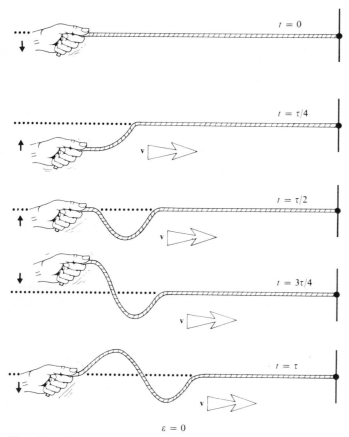

Fig. 2.7 With $\varepsilon = 0$ note that at $x = 0$ and $t = \tau/4 = \pi/2\omega$, $y = A \sin(-\pi/2) = -A$.

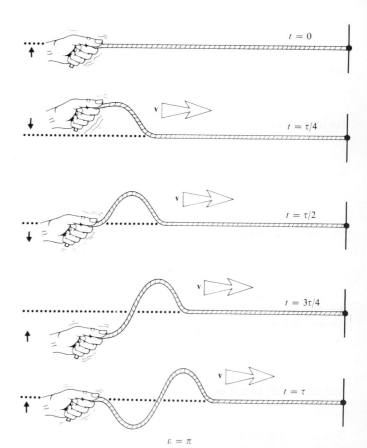

Fig. 2.8 With $\varepsilon = \pi$ note that at $x = 0$ and $t = \tau/4$, $y = A \sin(\pi/2) = A$.

The entire argument of the sine function is known as the *phase* φ of the wave, so that

$$\varphi = (kx - \omega t). \tag{2.27}$$

At $t = x = 0$

$$\psi(x, t)|_{\substack{x=0 \\ t=0}} = \psi(0, 0) = 0,$$

which is certainly a special case. More generally, we can write

$$\psi(x, t) = A \sin(kx - \omega t + \varepsilon), \tag{2.28}$$

where ε is the *initial phase* or *epoch angle*. A physical feel for the meaning of ε can be gotten by imagining that we wish to produce a progressive harmonic wave on a stretched string as in Fig. 2.7. In order to generate harmonic waves,

the hand holding the string would have to move such that its vertical displacement y was proportional to the negative of its acceleration, i.e., in simple harmonic motion (see Problem 2.5). But at $t = 0$ and $x = 0$ the hand certainly need not be on the x-axis about to move downward, as in Fig. 2.7. It could, of course, begin its motion on an upward swing, in which case $\varepsilon = \pi$ as indicated in Fig. 2.8. In this latter case

$$\psi(x, t) = y(x, t) = A \sin(kx - \omega t + \pi)$$

which is equivalent to

$$\psi(x, t) = A \sin(\omega t - kx)$$

or

$$\psi(x, t) = A \cos\left(\omega t - kx - \frac{\pi}{2}\right).$$

The epoch angle is then just the constant contribution to the phase arising at the generator and is independent of how far in space, or how long in time, the wave has traveled.

The phase of a disturbance such as $\psi(x, t)$ given by Eq. (2.28) is

$$\varphi(x, t) = (kx - \omega t + \varepsilon), \qquad (2.29)$$

and is obviously a function of x and t. In fact, the partial derivative of φ with respect to t holding x constant is the *rate of change of phase with time*

$$\left| \left(\frac{\partial \varphi}{\partial t} \right)_x \right| = \omega. \qquad (2.30)$$

Similarly, the *rate of change of phase with distance* holding t constant is

$$\left| \left(\frac{\partial \varphi}{\partial x} \right)_t \right| = k. \qquad (2.31)$$

These two expressions should bring to mind an equation from the theory of partial derivatives, one used quite frequently in thermodynamics, namely,

$$\left(\frac{\partial x}{\partial t} \right)_\varphi = \frac{-(\partial \varphi / \partial t)_x}{(\partial \varphi / \partial x)_t}. \qquad (2.32)$$

The term on the left represents the velocity of propagation of the condition of constant phase. Return for a moment to Fig. 2.8 and choose any point on the profile, e.g. the crest of the wave. As the wave moves through space, the displacement y of the point remains constant. Since the only variable in the harmonic wave function is the phase, it too must be constant. That is, the phase is fixed at such a value as to yield the constant y corresponding to the chosen point. The point moves along with the profile at the speed v and so too does the condition of constant phase.

Taking the appropriate partial derivatives of φ as given e.g. by Eq. (2.29) and substituting them into Eq. (2.32) we get

$$\left(\frac{\partial x}{\partial t} \right)_\varphi = \pm \frac{\omega}{k} = \pm v. \qquad (2.33)$$

This is the *speed* at which the profile moves and is known commonly as the *wave velocity* or, more specifically, as the *phase velocity*. The phase velocity carries a positive sign when the wave moves in the direction of increasing x and a negative one in the direction of decreasing x. This is consistent with our development of v as the magnitude of the wave velocity.

Consider the idea of the propagation of constant phase and how it relates to any one of the harmonic wave equations, say

$$\psi = A \sin k(x \mp vt)$$

with

$$\varphi = k(x - vt) = \text{constant};$$

as t increases, x must increase. Even if $x < 0$ so that $\varphi < 0$, x must increase, i.e. become less negative. Here, then, the condition of constant phase moves in the increasing x-direction. For

$$\varphi = k(x + vt) = \text{constant}$$

as t increases x can be positive and decreasing or negative and becoming more negative. In either case, the constant-phase condition moves in the decreasing x-direction.

2.4 THE COMPLEX REPRESENTATION

As we develop the analysis of wave phenomena, it will become clear that the sine and cosine functions which describe harmonic waves are somewhat awkward for our purposes. As the expressions being formulated become more involved, the trigonometric manipulations required to cope with them become even more unattractive. The complex-number representation of waves offers an alternative description which is mathematically simpler to work with. In fact, the complex exponential form of the wave equation is used extensively in both classical and quantum mechanics as well as in optics.

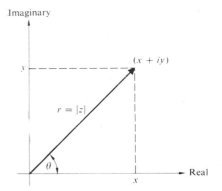

Fig. 2.9 Argand diagram.

The complex number z has the form

$$z = x + iy \qquad (2.34)$$

where $i = \sqrt{-1}$. The real and imaginary parts of z are respectively x and y where both x and y are themselves real numbers. This is illustrated graphically in the Argand diagram of Fig. 2.9. In terms of polar coordinates (r, θ) we have

$$x = r \cos \theta, \qquad y = r \sin \theta$$

and

$$z = x + iy = r(\cos \theta + i \sin \theta)$$

The *Euler* formula*

$$e^{i\theta} = \cos \theta + i \sin \theta$$

allows us to write

$$z = re^{i\theta} = r \cos \theta + ir \sin \theta,$$

where r is the *magnitude* of z, and θ is the *phase angle* of z, in radians. The magnitude is often denoted by $|z|$ and referred to as the *modulus* or *absolute value* of the complex number. The *complex conjugate,* indicated by an asterisk, is found by replacing i wherever it appears, with $-i$, so that

$$z^* = (x + iy)^* = (x - iy)$$

$$z^* = r(\cos \theta - i \sin \theta)$$

and

$$z^* = re^{-i\theta}.$$

The operations of addition and subtraction are quite straightforward:

$$z_1 \pm z_2 = (x_1 + iy_1) \pm (x_2 + iy_2)$$

and therefore

$$z_1 \pm z_2 = (x_1 \pm x_2) + i(y_1 \pm y_2).$$

Notice that this process is very much like the component addition of vectors.

Multiplication and division are most simply expressed in polar form

$$z_1 z_2 = r_1 r_2 e^{i(\theta_1 + \theta_2)}$$

---

* If you have any doubts about this identity, take the differential of $z = \cos \theta + i \sin \theta$ where $r = 1$. This yields $dz = iz \, d\theta$, and integration gives $z = \exp(i\theta)$.

and

$$\frac{z_1}{z_2} = \frac{r_1}{r_2} e^{i(\theta_1 - \theta_2)}.$$

A number of useful facts, which will be of value in future calculations, are well worth mentioning at this point. It follows readily from the ordinary trigonometric addition formulas that

$$e^{z_1 + z_2} = e^{z_1} e^{z_2},$$

whence, if $z_1 = x$ and $z_2 = iy$,

$$e^z = e^{x + iy} = e^x e^{iy}.$$

The modulus of a complex quantity is given by

$$|z| \equiv (zz^*)^{\frac{1}{2}},$$

so that

$$|e^z| = e^x.$$

Inasmuch as $\cos 2\pi = 1$ and $\sin 2\pi = 0$,

$$e^{i2\pi} = 1;$$

similarly

$$e^{i\pi} = e^{-i\pi} = -1 \qquad \text{and} \qquad e^{\pm i\pi/2} = \pm i.$$

The function e^z is periodic, i.e., it repeats itself every $i2\pi$

$$e^{z + i2\pi} = e^z e^{i2\pi} = e^z.$$

Any complex number can be represented as the sum of a real part Re (z) and an imaginary part Im (z)

$$z = \text{Re }(z) + i \, \text{Im }(z)$$

such that

$$\text{Re }(z) = \tfrac{1}{2}(z + z^*) \qquad \text{and} \qquad \text{Im }(z) = \frac{1}{2i}(z - z^*).$$

From the polar form where

$$\text{Re }(z) = r \cos \theta \qquad \text{and} \qquad \text{Im }(z) = r \sin \theta,$$

it is clear that either part could be chosen to describe a harmonic wave. It is customary, however, to choose the real part in which case a harmonic wave is written as

$$\psi(x, t) = \text{Re }[Ae^{i(\omega t - kx + \varepsilon)}], \qquad (2.35)$$

which is, of course, equivalent to

$$\psi(x, t) = A \cos(\omega t - kx + \varepsilon).$$

Henceforth, wherever it's convenient, we shall write the wave function as

$$\psi(x, t) = Ae^{i(\omega t - kx + \varepsilon)} = Ae^{i\varphi}, \qquad (2.36)$$

and utilize this complex form in the required computations. This is done in order to take advantage of the ease of manipulation of the complex exponentials. Only after arriving at a final result, and then only if we want to represent the actual wave, need we take the real part. It has, accordingly, become quite common to write $\psi(x, t)$, as in Eq. (2.36), where it is understood that the actual wave is the real part.

2.5 PLANE WAVES

The plane wave is perhaps the simplest example of a three-dimensional wave. It exists at a given time, when all the surfaces upon which a disturbance has constant phase form a set of planes, each generally perpendicular to the propagation direction. There are quite practical reasons for studying this sort of disturbance, one of which is that by using optical devices we can readily produce light resembling plane waves.

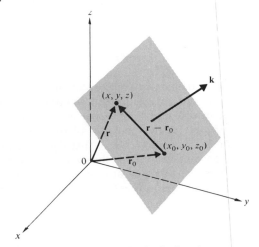

Fig. 2.10   A plane wave moving in the **k**-direction.

The mathematical expression for a plane, which is perpendicular to a given vector **k** and which passes through some point (x_0, y_0, z_0) is rather easy to derive (Fig. 2.10). The position vector, in terms of its components in Cartesian coordinates, is

$$\mathbf{r} \equiv [x, y, z].$$

It begins at some arbitrary origin O and ends at the point (x, y, z) which can, for the moment, be anywhere in space. By setting

$$(\mathbf{r} - \mathbf{r}_0) \cdot \mathbf{k} = 0 \qquad (2.37)$$

we force the vector $(\mathbf{r} - \mathbf{r}_0)$ to sweep out a plane perpendicular to **k** as its endpoint (x, y, z) takes on all allowed values. With

$$\mathbf{k} \equiv [k_x, k_y, k_z] \qquad (2.38)$$

Eq. (2.37) can be expressed in the form

$$k_x(x - x_0) + k_y(y - y_0) + k_z(z - z_0) = 0 \quad (2.39)$$

or as

$$k_x x + k_y y + k_z z = a \qquad (2.40)$$

where

$$a = k_x x_0 + k_y y_0 + k_z z_0 = \text{constant.} \qquad (2.41)$$

The most concise form of the equation of a plane perpendicular to **k** is then just

$$\mathbf{k} \cdot \mathbf{r} = \text{constant} = a. \qquad (2.42)$$

The plane is the locus of all points whose projection onto the **k**-direction (r_k in Fig. 2.12) is a constant.

We can now construct a set of planes over which $\psi(\mathbf{r})$ varies in space sinusoidally, namely

$$\psi(\mathbf{r}) = A \sin(\mathbf{k} \cdot \mathbf{r}) \qquad (2.43)$$

$$\psi(\mathbf{r}) = A \cos(\mathbf{k} \cdot \mathbf{r}) \qquad (2.44)$$

or

$$\psi(\mathbf{r}) = Ae^{i\mathbf{k} \cdot \mathbf{r}}. \qquad (2.45)$$

For each of these expressions $\psi(\mathbf{r})$ is constant over every plane defined by $\mathbf{k} \cdot \mathbf{r} = $ constant. Since we are dealing with harmonic functions, they should repeat themselves in space after a displacement of λ in the direction of **k**. Figure 2.11 is a rather humble representation of this kind of expression. We have drawn only a few of the infinite number of planes, each having a different $\psi(\mathbf{r})$. The planes should also have been drawn with an infinite spatial extent, since no limits were put on **r**. The disturbance clearly occupies all of space.

The spatially repetitive nature of these harmonic functions can be expressed by

$$\psi(\mathbf{r}) = \psi\left(\mathbf{r} + \frac{\lambda\mathbf{k}}{k}\right) \qquad (2.46)$$

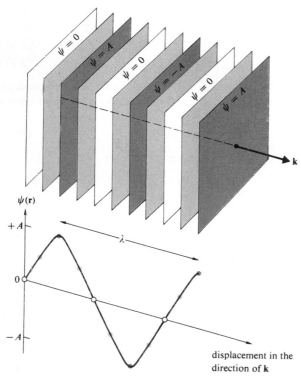

Fig. 2.11 Wavefronts for a harmonic plane wave.

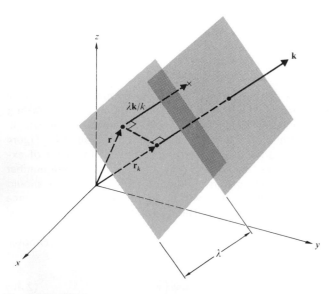

Fig. 2.12 Plane waves.

where k is the magnitude of $\mathbf{k}$ and $\mathbf{k}/k$ is a unit vector parallel to it (Fig. 2.12). In the exponential form, this is equivalent to

$$Ae^{i\mathbf{k}\cdot\mathbf{r}} = Ae^{i\mathbf{k}\cdot(\mathbf{r} + \lambda\mathbf{k}/k)} = Ae^{i\mathbf{k}\cdot\mathbf{r}}e^{i\lambda k}.$$

For this to be true, we must have

$$e^{i\lambda k} = 1 = e^{i2\pi};$$

therefore

$$\lambda k = 2\pi$$

and

$$k = \frac{2\pi}{\lambda}.$$

The vector $\mathbf{k}$, whose magnitude is the *propagation number k* (already introduced), is called the *propagation vector*.

At any fixed point in space where $\mathbf{r}$ is constant, the phase is constant and so too, is $\psi(\mathbf{r})$, in short the planes are motionless. To get things moving, $\psi(\mathbf{r})$ must be made to vary in time, something we can accomplish by introducing the time dependence in an analogous fashion to that of the one-dimensional wave. Here then

$$\psi(\mathbf{r}, t) = Ae^{i(\mathbf{k}\cdot\mathbf{r} \mp \omega t)} \qquad (2.47)$$

with A, ω, and k constant. As this disturbance travels along in the $\mathbf{k}$ direction we can assign a phase corresponding to it at each point in space and time. At any given time, the surfaces joining all points of equal phase are known as *wavefronts* or *wave surfaces*. Note that the wave function will have a constant value over the wavefront only if the amplitude A has a fixed value at every point on the wavefront. In general, A is a function of $\mathbf{r}$ and may not be constant over all space or even over a wavefront. In the latter case, the wave is said to be *inhomogeneous*; but we will not be concerned with this sort of disturbance until later when we consider laser beams and total internal reflection.

The phase velocity of a plane wave given by Eq. (2.47) is equivalent to the propagation velocity of the wavefront. In Fig. 2.12, the scalar component of $\mathbf{r}$ in the direction of $\mathbf{k}$ is r_k. The disturbance on a wavefront is constant, so that after a time dt, if the front moves along $\mathbf{k}$ a distance dr_k, we must have

$$\psi(\mathbf{r}, t) = \psi(r_k + dr_k, t + dt) = \psi(r_k, t). \qquad (2.48)$$

In exponential form, this is

$$Ae^{i(\mathbf{k}\cdot\mathbf{r}\mp\omega t)} = Ae^{i(kr_k + k\,dr_k \mp \omega t \mp \omega\,dt)} = Ae^{i(kr_k \mp \omega t)};$$

therefore

$$k\,dr_k = \pm\omega\,dt$$

and the magnitude of the wave velocity, dr_k/dt, is

$$\frac{dr_k}{dt} = \pm\frac{\omega}{k} = \pm v. \qquad (2.49)$$

We could have anticipated this result by rotating the coordinate system in Fig. 2.12 so that $\mathbf{k}$ was parallel to the x-axis. For that orientation

$$\psi(\mathbf{r}, t) = Ae^{i(kx \mp \omega t)}$$

since $\mathbf{k} \cdot \mathbf{r} = kr_k = kx$. The wave has thereby been effectively reduced to the one-dimensional disturbance already discussed in Section 2.3.

The plane harmonic wave is often written in Cartesian coordinates as

$$\psi(x, y, z, t) = Ae^{i(k_x x + k_y y + k_z z \mp \omega t)} \qquad (2.50)$$

or

$$\psi(x, y, z, t) = Ae^{i[k(\alpha x + \beta y + \gamma z) \mp \omega t]} \qquad (2.51)$$

where α, β and γ are the direction cosines of $\mathbf{k}$ (see Problem 2.8). In terms of its components, the magnitude of the propagation vector is given by

$$|\mathbf{k}| = k = (k_x^2 + k_y^2 + k_z^2)^{\frac{1}{2}} \qquad (2.52)$$

and of course

$$\alpha^2 + \beta^2 + \gamma^2 = 1. \qquad (2.53)$$

We have examined plane waves with a particular emphasis on harmonic functions. The special significance of these waves is twofold: first, physically, sinusoidal waves can be generated relatively simply by using some form of harmonic oscillator; second, any three-dimensional wave can be expressed as a combination of plane waves. each having a distinct amplitude and propagation direction.

We can certainly imagine a series of plane waves like those of Fig. 2.11 where the disturbance varies in some fashion other than harmonically. It will be seen in the next section that harmonic plane waves are, indeed, a special case of a more general plane-wave solution.

2.6 THE THREE-DIMENSIONAL DIFFERENTIAL WAVE EQUATION

Of all the three-dimensional waves, only the plane wave (harmonic or not) moves through space with an unchanging profile. Clearly, then, the idea of a wave being the propagation of a disturbance whose profile is unaltered is somewhat lacking. This difficulty can be overcome by defining a wave as any solution of the differential wave equation. Obviously, what we need now is a three-dimensional wave equation. This should be rather easy to obtain, since we can guess at its form by generalizing the one-dimensional expression (2.11). In Cartesian coordinates, the position variables x, y, and z must certainly appear symmetrically* in the three-dimensional equation, a fact to be kept in mind. The wave function $\psi(x, y, z, t)$ given by Eq. (2.51) is a particular solution of the differential equation we are looking for. In analogy with the derivation of Eq. (2.11) we compute the following partial derivatives from Eq. (2.51)

$$\frac{\partial^2\psi}{\partial x^2} = -\alpha^2 k^2 \psi \qquad (2.54)$$

$$\frac{\partial^2\psi}{\partial y^2} = -\beta^2 k^2 \psi \qquad (2.55)$$

$$\frac{\partial^2\psi}{\partial z^2} = -\gamma^2 k^2 \psi \qquad (2.56)$$

and

$$\frac{\partial^2\psi}{\partial t^2} = -\omega^2 \psi. \qquad (2.57)$$

Adding the three spatial derivatives and utilizing the fact that $\alpha^2 + \beta^2 + \gamma^2 = 1$, we obtain

$$\frac{\partial^2\psi}{\partial x^2} + \frac{\partial^2\psi}{\partial y^2} + \frac{\partial^2\psi}{\partial z^2} = -k^2 \psi. \qquad (2.58)$$

Combining this with the time derivative (2.57) and remembering that $v = \omega/k$, we arrive at

$$\frac{\partial^2\psi}{\partial x^2} + \frac{\partial^2\psi}{\partial y^2} + \frac{\partial^2\psi}{\partial z^2} = \frac{1}{v^2}\frac{\partial^2\psi}{\partial t^2}, \qquad (2.59)$$

the *three-dimensional differential wave equation*. Note that

---

* There is no distinguishing characteristic for any one of the axes in Cartesian coordinates. We should, therefore, be able to change the names of say, x to z, y to x, and z to y (keeping the system right-handed) without altering the differential wave equation.

x, y, and z do appear symmetrically and the form is precisely what one might expect from the generalization of Eq. (2.11).

Equation (2.59) is usually written in a more concise form by introducing the *Laplacian* operator

$$\nabla^2 \equiv \frac{\partial^2}{\partial x^2} + \frac{\partial^2}{\partial y^2} + \frac{\partial^2}{\partial z^2}, \qquad (2.60)$$

whereupon it becomes simply

$$\nabla^2 \psi = \frac{1}{v^2} \frac{\partial^2 \psi}{\partial t^2}. \qquad (2.61)$$

Now that we have this most important equation, let's briefly return to the plane wave and see how it fits into the scheme of things. A function of the form

$$\psi(x, y, z, t) = A e^{ik(\alpha x + \beta y + \gamma z \mp vt)} \qquad (2.62)$$

is equivalent to Eq. (2.51) and, as such, is a solution of Eq. (2.61). It can also be shown (Problem 2.11) that

$$\psi(x, y, z, t) = f(\alpha x + \beta y + \gamma z - vt) \qquad (2.63)$$

and

$$\psi(x, y, z, t) = g(\alpha x + \beta y + \gamma z + vt) \qquad (2.64)$$

are both plane-wave solutions of the differential wave equation. The functions f and g which are twice differentiable are otherwise arbitrary and certainly need not be harmonic. A linear combination of these solutions is also a solution and we can write this in a slightly different manner as

$$\psi(\mathbf{r}, t) = C_1 f(\mathbf{r} \cdot \mathbf{k}/k - vt) + C_2 g(\mathbf{r} \cdot \mathbf{k}/k + vt) \quad (2.65)$$

where C_1 and C_2 are constants.

Cartesian coordinates are particularly suited for describing plane waves. However, as various physical situations arise, we can often take better advantage of existing symmetries by making use of some other coordinate representations.

2.7 SPHERICAL WAVES

Toss a stone into a tank of water. The surface ripples that emanate from the point of impact spread out in two-dimensional circular waves. Extending this imagery to three dimensions, envision a small pulsating sphere surrounded by a fluid. As the source expands and contracts, it generates pressure variations which propagate outward as spherical waves.

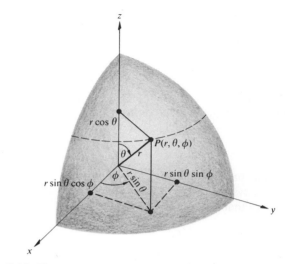

Fig. 2.13 The geometry of spherical coordinates.

Consider now an idealized point source of light. The radiation emanating from it streams out radially, uniformly in all directions. The source is said to be *isotropic* and the resulting wavefronts are again concentric spheres which increase in diameter as they expand out into the surrounding space. The obvious symmetry of the wavefronts suggests that it might be more convenient to describe them mathematically, in terms of spherical polar coordinates (Fig. 2.13). In this representation the Laplacian operator is

$$\nabla^2 \equiv \frac{1}{r^2} \frac{\partial}{\partial r}\left(r^2 \frac{\partial}{\partial r}\right) + \frac{1}{r^2 \sin\theta} \frac{\partial}{\partial \theta}\left(\sin\theta \frac{\partial}{\partial \theta}\right) + \frac{1}{r^2 \sin^2\theta} \frac{\partial^2}{\partial \phi^2} \quad (2.66)$$

where r, θ, ϕ are defined by

$$x = r\sin\theta\cos\phi, \qquad y = r\sin\theta\sin\phi, \qquad z = r\cos\theta.$$

Remember that we are looking for a description of spherical waves, waves which are spherically symmetric, i.e. ones which are characterized by the fact that they do not depend on θ and ϕ so that

$$\psi(\mathbf{r}) = \psi(r, \theta, \phi) = \psi(r). \qquad (2.67)$$

The Laplacian of $\psi(r)$ is then simply

$$\nabla^2 \psi(r) = \frac{1}{r^2} \frac{\partial}{\partial r}\left(r^2 \frac{\partial \psi}{\partial r}\right). \qquad (2.68)$$

This result can be obtained without being familiar with Eq. (2.66). Start with the Cartesian form of the Laplacian (2.60), operate on the spherically symmetric wave function $\psi(r)$

and convert each term to polar coordinates. Examining only the x-dependence, we have

$$\frac{\partial \psi}{\partial x} = \frac{\partial \psi}{\partial r} \frac{\partial r}{\partial x}$$

and

$$\frac{\partial^2 \psi}{\partial x^2} = \frac{\partial^2 \psi}{\partial r^2}\left(\frac{\partial r}{\partial x}\right)^2 + \frac{\partial \psi}{\partial r}\frac{\partial^2 r}{\partial x^2},$$

since

$$\psi(\mathbf{r}) = \psi(r).$$

Using

$$x^2 + y^2 + z^2 = r^2$$

we have

$$\frac{\partial r}{\partial x} = \frac{x}{r}, \qquad \frac{\partial^2 r}{\partial x^2} = \frac{1}{r}\frac{\partial}{\partial x}(x) + x\frac{\partial}{\partial x}\left(\frac{1}{r}\right) = \frac{1}{r}\left(1 - \frac{x^2}{r^2}\right)$$

and

$$\frac{\partial^2 \psi}{\partial x^2} = \frac{x^2}{r^2}\frac{\partial^2 \psi}{\partial r^2} + \frac{1}{r}\left(1 - \frac{x^2}{r^2}\right)\frac{\partial \psi}{\partial r}.$$

Now having $\partial^2\psi/\partial x^2$, we form $\partial^2\psi/\partial y^2$ and $\partial^2\psi/\partial z^2$, and on adding get

$$\nabla^2 \psi(r) = \frac{\partial^2 \psi}{\partial r^2} + \frac{2}{r}\frac{\partial \psi}{\partial r},$$

which is equivalent to Eq. (2.68). This result can be expressed in a slightly different form:

$$\nabla^2 \psi = \frac{1}{r}\frac{\partial^2}{\partial r^2}(r\psi). \qquad (2.69)$$

The differential wave equation (2.61) can then be written as

$$\frac{1}{r}\frac{\partial^2}{\partial r^2}(r\psi) = \frac{1}{v^2}\frac{\partial^2 \psi}{\partial t^2}. \qquad (2.70)$$

Multiplying both sides by r, we obtain

$$\frac{\partial^2}{\partial r^2}(r\psi) = \frac{1}{v^2}\frac{\partial^2}{\partial t^2}(r\psi). \qquad (2.71)$$

Notice that this expression is now just the one-dimensional differential wave equation (2.11), where the space variable is r and the wave function is the product $(r\psi)$. The solution of Eq. (2.71) is then simply

$$r\psi(r, t) = f(r - vt)$$

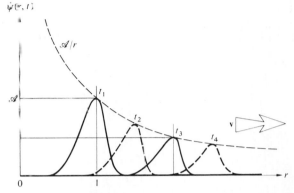

Fig. 2.14 A "quadruple exposure" of a spherical pulse.

or

$$\psi(r, t) = \frac{f(r - vt)}{r}. \qquad (2.72)$$

This represents a spherical wave progressing radially outward from the origin, at a constant speed v, and having an arbitrary functional form f. Another solution is given by

$$\psi(r, t) = \frac{g(r + vt)}{r},$$

and in this case the wave is converging toward the origin.[*] The fact that this expression blows up at $r = 0$ is of little practical concern.

A special case of the general solution

$$\psi(r, t) = C_1\frac{f(r - vt)}{r} + C_2\frac{g(r + vt)}{r}, \qquad (2.73)$$

is the *harmonic spherical wave*

$$\psi(r, t) = \left(\frac{\mathscr{A}}{r}\right)\cos k(r \mp vt) \qquad (2.74)$$

or

$$\psi(r, t) = \left(\frac{\mathscr{A}}{r}\right)e^{ik(r \mp vt)} \qquad (2.75)$$

wherein the constant $\mathscr{A}$ is called the *source strength*. At any fixed value of time, this represents a cluster of concentric

---

[*] Other more complicated solutions exist when the wave is not spherically symmetric. See C. A. Coulson, *Waves* (Chapter 1).

pulse at four different times. The pulse has the same extent in space at any point along any radius r, i.e. the width of the pulse along the r-axis is a constant. Figure 2.15 is an attempt to relate the diagrammatic representation of $\psi(r, t)$ in the previous figure, to its actual form as a spherical wave. It depicts half of the spherical pulse at two different times, as the wave expands outward. Remember that these results would obtain regardless of the direction of r, because of the spherical symmetry. We could also have drawn a harmonic wave, rather than a pulse, in Figs. 2.14 and 2.15. In this case, the sinusoidal disturbance would have been bounded by the curves

$$\psi = \mathscr{A}/r \qquad \text{and} \qquad \psi = -\mathscr{A}/r.$$

The outgoing spherical wave emanating from a point source, and the incoming wave converging to a point, are certainly idealizations. In actuality, light only approximates spherical waves, as it also only approximates plane waves.

As a spherical wavefront propagates out, its radius increases. Far enough away from the source, a small area of the wavefront will closely resemble a portion of a plane wave (Fig. 2.16).

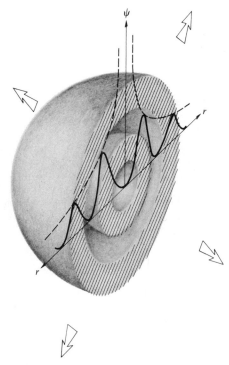

Fig. 2.15 Spherical wavefronts.

spheres filling all space. Each wavefront, or surface of constant phase, is given by

$$kr = \text{constant}.$$

Notice that the amplitude of any spherical wave is a function of r, where the term r^{-1} serves as an attenuation factor. Unlike the plane wave, a spherical wave decreases in amplitude, thereby changing its profile, as it expands and moves out from the origin.* Figure 2.14 illustrates this graphically by showing a "multiple exposure" of a spherical

---

* The attenuation factor is a direct consequence of energy conservation. Chapter 3 contains a discussion of how these ideas apply specifically to electromagnetic radiation.

2.8 CYLINDRICAL WAVES

We will now briefly examine another idealized waveform, the infinite circular cylinder. Unfortunately, a precise mathematical treatment is far too involved to do here. We shall, however, outline the procedure, so that the resulting wave function will evoke no mysticism. The Laplacian of ψ in cylindrical coordinates (Fig. 2.17) is

$$\nabla^2 \psi = \frac{1}{r}\frac{\partial}{\partial r}\left(r\frac{\partial \psi}{\partial r}\right) + \frac{1}{r^2}\frac{\partial^2 \psi}{\partial \theta^2} + \frac{\partial^2 \psi}{\partial z^2}, \qquad (2.76)$$

where

$$x = r\cos\theta, \qquad y = r\sin\theta, \qquad \text{and} \qquad z = z.$$

The simple case of cylindrical symmetry requires that

$$\psi(\mathbf{r}) = \psi(r, \theta, z) = \psi(r).$$

The θ-independence means that a plane perpendicular to the z-axis will intersect the wavefront in a circle, which may

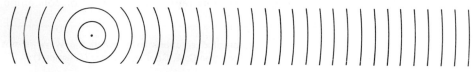

Fig. 2.16 The flattening of spherical waves with distance.

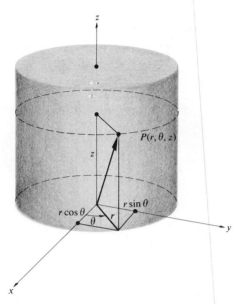

Fig. 2.17 The geometry of cylindrical coordinates.

We are looking for an expression for $\psi(r)$, a solution of this equation. After a bit of manipulation, in which the time dependence is separated out, Eq. (2.77) becomes something called Bessel's equation. The solutions of Bessel's equation for large values of r asymptotically approach simple trigonometric forms. Finally then, when r is sufficiently large, we can write

$$\psi(r) \approx \frac{\mathscr{A}}{\sqrt{r}} e^{ik(r \mp vt)}$$

$$\psi(r) \approx \frac{\mathscr{A}}{\sqrt{r}} \cos k(r \mp vt). \qquad (2.78)$$

This represents a set of coaxial circular cylinders filling all space and traveling toward or away from an infinite line source. No solutions in terms of arbitrary functions can now be found as there were for both spherical (2.73) and plane waves (2.65).

A plane wave impinging on the back of a flat opaque screen containing a long thin slit will result in the emission, from that slit, of a disturbance resembling a cylindrical wave, see Fig. 2.18. Extensive use has been made of this technique to generate cylindrical light waves. Remember that the actual wave, however generated, only resembles the idealized mathematical representation.

vary in r, at different values of z. In addition, the z-independence further restricts the wavefront to a right circular cylinder centered on the z-axis and having infinite length. The differential wave equation is accordingly

$$\frac{1}{r} \frac{\partial}{\partial r} \left(r \frac{\partial \psi}{\partial r} \right) = \frac{1}{v^2} \frac{\partial^2 \psi}{\partial t^2}. \qquad (2.77)$$

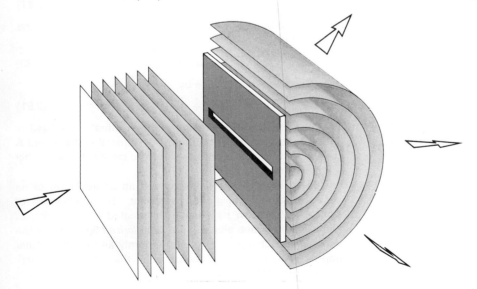

Fig. 2.18 Cylindrical waves emerging from a long narrow slit.

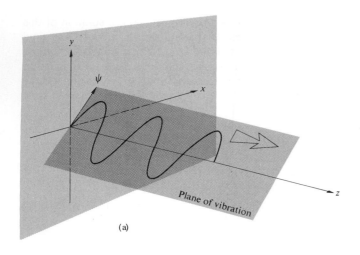

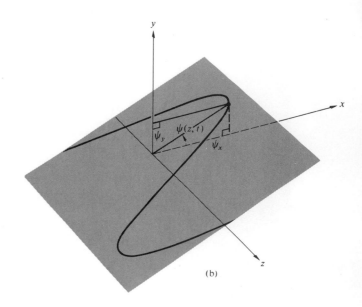

Fig. 2.19 Linearly polarized waves.

2.9 SCALAR AND VECTOR WAVES

There are two general classifications of waves, longitudinal and transverse. The distinction between the two arises from a difference between the direction along which the disturbance occurs and the direction, $\mathbf{k}/k$, in which the disturbance propagates. This is rather easy to visualize when dealing with an elastically deformable material medium. A longitudinal wave occurs when the particles of the medium are displaced from their equilibrium positions, in a direction parallel to $\mathbf{k}/k$. A transverse wave arises when the disturbance, in this case the displacement of the medium, is perpendicular to the propagation direction. Figure 2.19(a) depicts a transverse wave (as on a stretched string) traveling in the z-direction. In this instance, the wave motion is confined to a spatially fixed plane called the *plane of vibration* and the wave is accordingly said to be *linearly* or *plane polarized*. In order to determine the wave completely, we must now specify the orientation of the plane of vibration, as well as the direction of propagation. This is equivalent to resolving the disturbance into components along two mutually perpendicular axes, both normal to z [see Fig. 2.19(b)]. The angle at which the plane of vibration is inclined is a constant, so that at any time ψ_x and ψ_y differ from ψ by a multiplicative constant and are both therefore solutions of the differential wave equation. A most significant fact has evolved: the wave function of a transverse wave behaves somewhat like a vector quantity. With the wave moving along the z-axis, we can write

$$\boldsymbol{\psi}(z, t) = \psi_x(z, t)\hat{\mathbf{i}} + \psi_y(z, t)\hat{\mathbf{j}}, \tag{2.79}$$

where, of course, $\hat{\mathbf{i}}$, $\hat{\mathbf{j}}$ and $\hat{\mathbf{k}}$ are the unit base vectors in Cartesian coordinates.

A scalar harmonic plane wave is given by the expression

$$\psi(\mathbf{r}, t) = Ae^{i(\mathbf{k}\cdot\mathbf{r} \mp \omega t)}. \tag{2.47}$$

A *linearly polarized* harmonic plane wave is given by the *wave vector*

$$\boldsymbol{\psi}(\mathbf{r}, t) = \mathbf{A}e^{i(\mathbf{k}\cdot\mathbf{r} \mp \omega t)}, \tag{2.80}$$

or in Cartesian coordinates by

$$\boldsymbol{\psi}(x, y, z, t) = (A_x\hat{\mathbf{i}} + A_y\hat{\mathbf{j}} + A_z\hat{\mathbf{k}})e^{i(k_x x + k_y y + k_z z \mp \omega t)}. \tag{2.81}$$

For this latter case where the plane of vibration is fixed in space, so too is the orientation of **A**. Remember that ψ and **A** differ only by a scalar and, as such, are parallel to each other and perpendicular to $\mathbf{k}/k$.

Light is a transverse wave and an appreciation of its vectorial nature is of great importance. The phenomena of optical *polarization* can readily be treated in terms of this sort of vector wave picture. For *unpolarized* light, where the wave vector changes direction randomly and rapidly, scalar approximations become useful, as in the theories of interference and diffraction.

PROBLEMS

2.1 How many "yellow" light waves ($\lambda = 580$ nm) will fit into a distance in space equal to the thickness of a piece of paper (0.003 in)? How far will the same number of microwaves ($\nu = 10^{10}$ Hz, i.e. 10 GHz, and $v = 3 \times 10^8$ m/s) extend?

2.2* The speed of light in vacuum is 3×10^8 m/s. Find the wavelength of red light having a frequency of 5×10^{14} Hz. Compare this to the wavelength of a 60 Hz electromagnetic wave.

2.3 Using the wave functions

$$\psi_1 = 4 \sin 2\pi(0.2x - 3t)$$

and

$$\psi_2 = \frac{\sin (7x + 3.5t)}{2.5}$$

determine in each case (a) the frequency, (b) wavelength, (c) period, (d) amplitude, (e) phase velocity and (f) direction of motion. Time is in seconds and x is in meters.

2.4* Show that

$$\psi(x, t) = A \sin k(x - vt) \qquad [2.14]$$

is a solution of the differential wave equation.

2.5 Show that if the displacement of the string in Fig. 2.7 is given by

$$y(x, t) = A \sin [kx - \omega t + \varepsilon],$$

then the hand generating the wave must be moving vertically in simple harmonic motion.

2.6 Consider the pulse described in terms of its displacement at $t = 0$ by

$$y(x, t)\Big|_{t=0} = \frac{C}{2 + x^2}$$

where C is a constant. Draw the wave profile. Write an expression for the wave, having a speed v in the negative x-direction, as a function of time t. If $v = 1$ m/s sketch the profile at $t = 2$ s.

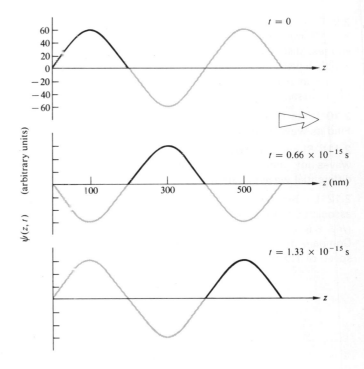

Fig. 2.20 A harmonic wave.

2.7* Determine which of the following describe traveling waves.

$$\psi(y, t) = e^{-(a^2y^2 + b^2t^2 - 2abty)}$$

$$\psi(z, t) = A \sin (az^2 - bt^2)$$

$$\psi(x, t) = A \sin 2\pi\left(\frac{x}{a} + \frac{t}{b}\right)^2$$

$$\psi(x, t) = A \cos^2 2\pi(t - x)$$

Where appropriate draw the profile and find the speed and direction of motion.

2.8 Beginning with Eq. (2.50) verify that

$$\psi(x, y, z, t) = A\, e^{i[k(\alpha x + \beta y + \gamma z) \mp \omega t]}$$

and that

$$\alpha^2 + \beta^2 + \gamma^2 = 1.$$

Draw a sketch showing all of the pertinent quantities.

2.9 Consider a light wave having a phase velocity of 3×10^8 m/s and a frequency of 6×10^{14} Hz. What is the shortest distance along the wave between any two points which have a phase difference of 30°? What phase shift occurs at a given point in 10^{-6} s, and how many waves have passed by in that time?

2.10 Write an expression for the wave shown in Fig. 2.20. Find its wavelength, velocity, frequency and period.

2.11* Show that Eqs. (2.63) and (2.64) which are plane waves of arbitrary form, satisfy the three-dimensional differential wave equation.

2.12 De Broglie's hypothesis states that every particle has associated with it a wavelength given by Planck's constant ($h = 6.6 \times 10^{-34}$ J s) divided by the particle's momentum. Compare the wavelength of a 6.0 kg stone moving with a speed of 1.0 m/s to that of light.

Electromagnetic Theory, Photons, and Light

The work of J. C. Maxwell and subsequent developments since the late eighteen hundreds have made it evident that light is most certainly electromagnetic in nature. Classical electrodynamics, as we shall see, unalterably leads to the picture of a continuous transfer of energy by way of electromagnetic waves. In contrast, the more modern view of quantum electrodynamics describes electromagnetic interactions, and the transport of energy, in terms of massless elementary "particles" known as *photons* which are localized quanta (or globs) of energy. The quantum nature of radiant energy is not always readily apparent, nor indeed is it always of practical concern in optics. There is a whole range of situations in which the detecting equipment is such that it is impossible, and desirably so, to distinguish individual quanta. More often than not, the stream of incident light carries a relatively large amount of energy and the granularity is obscured, in any event.

If the wavelength of light is small in comparison to the size of the apparatus, one may use, as a first approximation, the techniques of *geometrical optics*. A somewhat more precise treatment, which is applicable as well when the dimensions of the apparatus are small, is that of *physical optics*. In physical optics the dominant property of light is its wave nature. It is even possible to develop most of the treatment without ever specifying the kind of wave one is dealing with. Certainly, as far as the classical study of physical optics is concerned, it will suffice admirably to treat light as an electromagnetic wave.

We can think of light as another manifestation of matter. Indeed, one of the basic tenets of quantum mechanics is that both light and material objects each display similar wave—particle properties. As Erwin C. Schrödinger (1887–1961), one of the founders of quantum theory, put it:

> In the new setting of ideas the distinction [between particles and waves] has vanished, because it was discovered that all particles have also wave properties, and *vice versa*. Neither of the two concepts must be discarded, they must be amalgamated. Which aspect obtrudes itself depends not on the physical object, but on the experimental device set up to examine it.[*]

The quantum-mechanical treatment associates a wave equation with a particle, be it a photon, electron, proton, etc. In the case of material particles, the wave aspects are introduced by way of the field equation known as Schrödinger's equation. For photons we have a representation of the wave nature in the form of the classical electromagnetic field equations of Maxwell. With these as a starting point one can construct a quantum-mechanical theory of photons and their interaction with charges. The dual nature of light is evidenced by the fact that it propagates through space in a wave-like fashion and yet *can* display particle-like behavior during emission and absorption processes. Electromagnetic radiant energy is created and destroyed in quanta or photons and not continuously as a classical wave. Nonetheless its motion through a lens, a hole, or a set of slits is governed by wave characteristics. If we're unfamiliar with this kind of behavior in the macroscopic world it's because the wavelength of an object varies inversely with its momentum (Chapter 13) and even a grain of sand (which is barely moving) has a wavelength so small as to be indiscernible in any conceivable experiment.

The photon has several properties which distinguish it from all other subatomic particles. These properties are of considerable interest to us because they are responsible for the fact that quite often the quantum aspects of light are thoroughly obscured. In particular, there are no restrictions on the number of photons which can exist in a region with the same linear and angular momentum. Restrictions of this sort (the Pauli exclusion principle) do exist for most other particles (with the exception for example of the still hypothetical quantum of gravity, i.e. the graviton, He_4 and π mesons). The photon has zero rest mass and therefore exceedingly large numbers of low-energy photons can be envisioned as present in a beam of light. Within that model dense streams of photons (many of which may have essentially the same momentum) act on the average to produce well-defined classical fields. We can draw a rough analogy with the flow of commuters through a train station during a rush hour. Each one presumably behaves individually as a quantum of humanity but all have the same intent and follow fairly similar trajectories. To a distant

---

[*] Erwin C. Schrödinger, *Science Theory and Man.*

myopic observer there is a seemingly smooth and continuous flow. The behavior of the stream en masse is predictable from day to day and so the precise motion of each commuter is unimportant, at least to the observer. The energy transported by a large number of photons is, *on the average*, equivalent to the energy transferred by a classical electromagnetic wave. It is for these reasons that the field representation of electromagnetic phenomena has been, and will continue to be, so useful. It should be noted however that when we speak of overlapping electromagnetic waves, it is essentially a euphemism for the interference of probability amplitudes, but more about that will have to wait for Chapter 13.

Quite pragmatically, then, we can consider light to be a classical electromagnetic wave, keeping in mind the fact that there are situations (on the periphery of our present concern) for which this description is woefully inadequate.

3.1 BASIC LAWS OF ELECTROMAGNETIC THEORY

Our intent in this section is to review and develop, if only briefly, some of the ideas needed to appreciate the concept of electromagnetic waves.

We know from experiments that charges, even though separated in vacuum, experience a mutual interaction. Recall the familiar electrostatics demonstration in which a pith ball somehow senses the presence of a charged rod without actually touching it. As a possible explanation we might speculate that each charge emits (and absorbs) a stream of undetected particles (*virtual photons*). The exchange of these particles amongst the charges may be regarded as the mode of interaction. Alternatively, we can take the classical approach and imagine instead that every charge is surrounded by something called an electric field. We then need only suppose that each charge interacts directly with the electric field in which it is immersed. Thus if a charge q experiences a force $\mathbf{F}_E$ the electric field $\mathbf{E}$ at the position of the charge is defined by $\mathbf{F}_E = q\mathbf{E}$. In addition, we observe that a moving charge may experience another force $\mathbf{F}_M$ which is proportional to its velocity $\mathbf{v}$. We are thus led to define yet another field, namely the *magnetic induction* $\mathbf{B}$, such that $\mathbf{F}_M = q\mathbf{v} \times \mathbf{B}$. If both forces $\mathbf{F}_E$ and $\mathbf{F}_M$ occur concurrently the charge is said to be moving through a region pervaded by both electric and magnetic fields, whereupon $\mathbf{F} = q\mathbf{E} + q\mathbf{v} \times \mathbf{B}$.

There are several other observations which may be interpreted in terms of these fields and in so doing we can get a better idea of the physical properties which must be

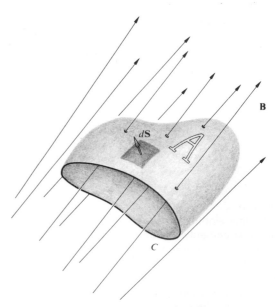

Fig. 3.1 $\mathbf{B}$-field through an open area A.

attributed to $\mathbf{E}$ and $\mathbf{B}$. As we shall see, *electric fields* are generated by both electric charges and by *time-varying magnetic fields*. Similarly, *magnetic fields* are generated by electric currents and by *time-varying electric fields*. This interdependence of $\mathbf{E}$ and $\mathbf{B}$ is a keypoint in the description of light and its elaboration is the motivation for much of what is to follow.

3.1.1 Faraday's Induction Law

Michael Faraday made a number of major contributions to electromagnetic theory, one of the most significant of which was his discovery that a time-varying magnetic flux, passing through a closed conducting loop, results in the generation of a current around that loop. The flux of magnetic induction (or *magnetic flux density*) $\mathbf{B}$ through any open area A, bounded by the conducting loop (Fig. 3.1) is given by

$$\Phi_B = \iint_A \mathbf{B} \cdot d\mathbf{S}. \tag{3.1}$$

The induced *electromotive force* or *emf* developed around the loop is then

$$\text{emf} = -\frac{d\Phi_B}{dt}. \tag{3.2}$$

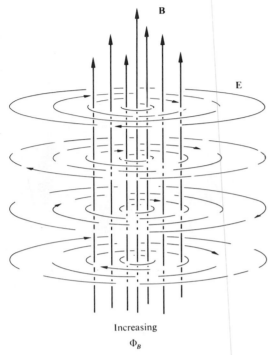

Fig. 3.2 A time-varying **B**-field. Surrounding each point where Φ_B is changing the **E**-field forms closed loops.

We should not, however, get too involved with the image of wires and current and emf. Our present concern is with the electric and magnetic fields themselves. Indeed, the emf exists only as a result of the presence of an electric field given by

$$\text{emf} = \oint_C \mathbf{E} \cdot d\mathbf{l},$$ (3.3)

taken around the closed curve C, corresponding to the loop. Equating Eqs. (3.2) and (3.3), and making use of Eq. (3.1) we get

$$\oint_C \mathbf{E} \cdot d\mathbf{l} = -\frac{d}{dt} \iint_A \mathbf{B} \cdot d\mathbf{S}.$$ (3.4)

We began this discussion by examining a conducting loop and have arrived at Eq. (3.4); this expression, except for the path C, contains no reference to the physical loop. In fact, the path can be chosen quite arbitrarily and need not be within, or anywhere near, a conductor. The electric field in Eq. (3.4) does not arise directly from the presence of electric

charges but rather from the time-varying magnetic field. With no charges to act as sources or sinks, the field lines close on themselves, forming loops (Fig. 3.2). For the case in which the path is fixed in space and unchanging in shape, the *induction law* (3.4) can be rewritten as

$$\oint_C \mathbf{E} \cdot d\mathbf{l} = -\iint_A \frac{\partial \mathbf{B}}{\partial t} \cdot d\mathbf{S}.$$ (3.5)

This, in itself, is a rather fascinating expression since it indicates that a time-varying magnetic field will have an electric field associated with it.

3.1.2 Gauss's Law—Electric

Another of the fundamental laws of electromagnetism is named after the German mathematician Karl Friedrich Gauss (1777–1855). It relates the flux of electric field intensity through a closed surface A

$$\Phi_E = \oiint_A \mathbf{E} \cdot d\mathbf{S}$$ (3.6)

to the total enclosed charge. The circled double integral is meant to serve as a reminder that the surface is closed. The vector $d\mathbf{S}$ is in the direction of an *outward normal* as shown in Fig. 3.3. If the volume enclosed by A is V, and if within it there is a continuous charge distribution of density ρ, then Gauss's law is

$$\oiint_A \mathbf{E} \cdot d\mathbf{S} = \frac{1}{\epsilon} \iiint_V \rho \, dV.$$ (3.7)

The integral on the left is the difference between the amount of flux flowing into, and out of, any closed surface A. If there is a difference, it will be due to the presence of sources or sinks of the electric field within A. Clearly then, the integral must be proportional to the total enclosed charge. The constant ϵ is known as the *electric permittivity* of the medium. For the special case of a vacuum, the *permittivity of free space* is given by $\epsilon_0 = 8.8542 \times 10^{-12} C^2 \, N^{-1} \, m^{-2}$. The permittivity of a material can be expressed in terms of ϵ_0 as

$$\epsilon = K_e \epsilon_0,$$ (3.8)

where K_e, the *dielectric constant* (or *relative permittivity*), is a dimensionless quantity, and is the same for all systems of units. Our interest in K_e anticipates the fact that the permittivity is related to the speed of light in dielectric materials, like glass, air, quartz, etc.

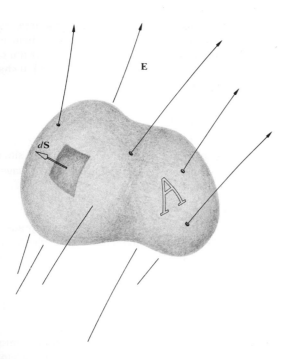

Fig. 3.3 E-field through a closed area A.

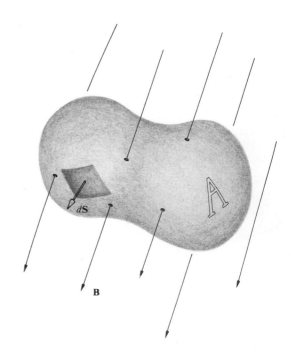

Fig. 3.4 B-field through a closed area A.

3.1.3 Gauss's Law—Magnetic

There is no known magnetic counterpart to the electric charge, i.e. no isolated magnetic poles have ever been found, although they've been looked for extensively, even in lunar soil samples. Unlike the electric field, the magnetic induction **B** does not diverge from or converge toward some kind of magnetic charge (a monopole source or sink). Magnetic induction fields can be described in terms of current distributions. Indeed we might envision an elementary magnet as a small current loop where the lines of **B** are themselves continuous and closed. Any closed surface in a region of magnetic field would accordingly have an equal number of lines of **B** entering and emerging from it (Fig. 3.4). This situation arises from the absence of any monopoles within the enclosed volume. The flux of magnetic induction Φ_B through such a surface is zero, and we have the magnetic equivalent of Gauss's law:

$$\Phi_B = \oiint_A \mathbf{B} \cdot d\mathbf{S} = 0. \tag{3.9}$$

3.1.4 Ampere's Circuital Law

Another equation which will be of great interest to us is due to André Marie Ampère (1775–1836). Known as the *circuital law*, it relates a line integral of **B** tangent to a closed curve C, with the total current i passing within the confines of C:

$$\oint_C \mathbf{B} \cdot d\mathbf{l} = \mu \iint_A \mathbf{J} \cdot d\mathbf{S} = \mu i. \tag{3.10}$$

The open surface A is bounded by C and J is the current per unit area (Fig. 3.5). The quantity μ is called the *permeability* of the particular medium. For a vacuum $\mu = \mu_0$ (the *permeability of free space*), which is defined as $4\pi \times 10^{-7}$ N s² C⁻².

As in Eq. (3.8)

$$\mu = K_m \mu_0 \tag{3.11}$$

with K_m being the dimensionless *relative permeability*.

Equation (3.10), although often adequate, is not the whole truth. Moving charges are not the only source of a magnetic field. This is evidenced by the fact that while

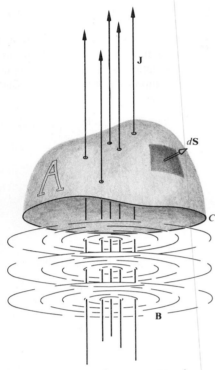

Fig. 3.5 Current density through an open area *A*.

charging or discharging a capacitor, one can measure a **B** field in the region between its plates (Fig. 3.6). This field is indistinguishable from that surrounding the leads even though no current actually traverses the capacitor. Notice, however, that if *A* is the area of each plate and *Q* the charge on it

$$E = \frac{Q}{\epsilon A}.$$

As the charge varies, the electric field changes and

$$\epsilon \frac{\partial E}{\partial t} = \frac{i}{A}$$

is effectively a current density. James C. Maxwell hypothesized the existence of just such a mechanism, which he called the *displacement current density*,[*] defined by

$$\mathbf{J}_D \equiv \epsilon \frac{\partial \mathbf{E}}{\partial t}. \qquad (3.12)$$

The restatement of Ampere's law as

$$\oint_C \mathbf{B} \cdot d\mathbf{l} = \mu \iint_A \left(\mathbf{J} + \epsilon \frac{\partial \mathbf{E}}{\partial t} \right) \cdot d\mathbf{S} \qquad (3.13)$$

was one of Maxwell's greatest contributions. It points out that even when $\mathbf{J} = 0$, a time-varying **E**-field will be accompanied by a **B**-field (Fig. 3.7).

---

[*] Maxwell's own words and ideas concerning this are examined in an article by A. M. Bork, *Am. J. Phys.* **31**, 854 (1963).

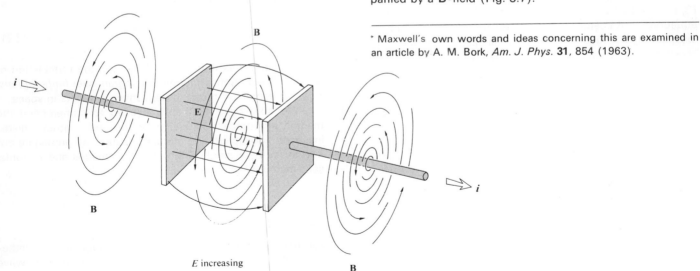

Fig. 3.6 **B**-field concomitant with a time varying **E**-field in the gap of a capacitor.

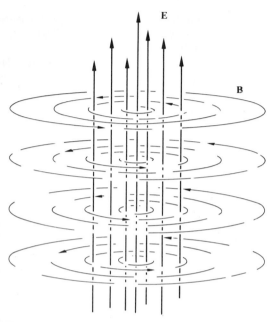

Fig. 3.7 A time-varying **E**-field. Surrounding each point where Φ_E is changing the **B**-field forms closed loops.

3.1.5 Maxwell's Equations

The set of integral expressions given by Eqs. (3.5), (3.7), (3.9) and (3.13) have come to be known as Maxwell's equations. Remember that these are generalizations of experimental results. The very simplest statement of Maxwell's equations governs the behavior of the electric and magnetic fields in free space where $\epsilon = \epsilon_0$, $\mu = \mu_0$ and both ρ and **J** are zero. In that instance

$$\oint_C \mathbf{E} \cdot d\mathbf{l} = -\iint_A \frac{\partial \mathbf{B}}{\partial t} \cdot d\mathbf{S}, \qquad (3.14)$$

$$\oint_C \mathbf{B} \cdot d\mathbf{l} = \mu_0 \epsilon_0 \iint_A \frac{\partial \mathbf{E}}{\partial t} \cdot d\mathbf{S}, \qquad (3.15)$$

$$\oiint_A \mathbf{B} \cdot d\mathbf{S} = 0, \qquad (3.16)$$

$$\oiint_A \mathbf{E} \cdot d\mathbf{S} = 0. \qquad (3.17)$$

Observe that except for a multiplicative scalar, the electric and magnetic fields appear in the equations with a remarkable symmetry. However **E** affects **B**, **B** will in turn

affect **E**. The mathematical symmetry implies a good deal of physical symmetry.

Maxwell's equations can be written in a differential form which will be somewhat more useful for our purposes. The appropriate calculation is carried out in Appendix 1 and the consequent equations for *free space*, in Cartesian co-ordinates, are as follows:

$$\frac{\partial \mathbf{E}_z}{\partial y} - \frac{\partial \mathbf{E}_y}{\partial z} = -\frac{\partial \mathbf{B}_x}{\partial t}, \quad \text{(i)}$$

$$\frac{\partial \mathbf{E}_x}{\partial z} - \frac{\partial \mathbf{E}_z}{\partial x} = -\frac{\partial \mathbf{B}_y}{\partial t}, \quad \text{(ii)} \qquad (3.18)$$

$$\frac{\partial \mathbf{E}_y}{\partial x} - \frac{\partial \mathbf{E}_x}{\partial y} = -\frac{\partial \mathbf{B}_z}{\partial t}, \quad \text{(iii)}$$

$$\frac{\partial \mathbf{B}_z}{\partial y} - \frac{\partial \mathbf{B}_y}{\partial z} = \mu_0 \epsilon_0 \frac{\partial \mathbf{E}_x}{\partial t}, \quad \text{(i)}$$

$$\frac{\partial \mathbf{B}_x}{\partial z} - \frac{\partial \mathbf{B}_z}{\partial x} = \mu_0 \epsilon_0 \frac{\partial \mathbf{E}_y}{\partial t}, \quad \text{(ii)} \qquad (3.19)$$

$$\frac{\partial \mathbf{B}_y}{\partial x} - \frac{\partial \mathbf{B}_x}{\partial y} = \mu_0 \epsilon_0 \frac{\partial \mathbf{E}_z}{\partial t}, \quad \text{(iii)}$$

$$\frac{\partial \mathbf{B}_x}{\partial x} + \frac{\partial \mathbf{B}_y}{\partial y} + \frac{\partial \mathbf{B}_z}{\partial z} = 0, \qquad (3.20)$$

$$\frac{\partial \mathbf{E}_x}{\partial x} + \frac{\partial \mathbf{E}_y}{\partial y} + \frac{\partial \mathbf{E}_z}{\partial z} = 0. \qquad (3.21)$$

The transition has thus been made from the formulation of Maxwell's equations in terms of integrals over *finite regions*, to a restatement in terms of derivatives at *points* in space.

We now have all that is needed to comprehend the magnificent process whereby electric and magnetic fields, inseparably coupled, and mutually sustaining, propagate out into space as a single entity, free of charges and currents, sans matter, sans ether.

3.2 ELECTROMAGNETIC WAVES

We have relegated to Appendix 1 a complete and mathematically elegant derivation of the electromagnetic wave equation. We will spend some time here at the equally important task of developing a more intuitive appreciation of the physical processes involved. Three observations, from which we might build a qualitative picture, are readily available to us and these are the general perpendicularity of the fields, the symmetry of Maxwell's equations, and the interdependence of **E** and **B** in those equations.

In studying electricity and magnetism one soon becomes aware of the fact that there are a number of relationships which are described by vector cross products, or if you like, right-hand rules. In other words, an occurrence of one sort produces a related perpendicularly directed response. Of immediate interest is the fact that a time-varying **E**-field generates a **B**-field which is everywhere perpendicular to the direction in which **E** changes (Fig. 3.7). In the same way, a time-varying **B**-field generates an **E**-field which is everywhere perpendicular to the direction in which **B** changes (Fig. 3.2). We might, accordingly, anticipate the general transverse nature of the **E**- and **B**-fields in an electromagnetic disturbance.

Now, consider a charge which is somehow caused to accelerate from rest. When the charge is motionless, it has associated with it a radial, uniform electric field extending out to infinity. At the instant the charge begins to move, the **E**-field is altered in the vicinity of the charge and this alteration propagates out into space at some finite speed. The time-varying electric field induces a magnetic field by way of Eq. (3.15) or (3.19). But the charge is accelerating, $\partial \mathbf{E}/\partial t$ is itself not constant and so the induced **B**-field is time dependent. The time-varying **B**-field generates an **E**-field, (3.14) or (3.18), and the process continues with **E** and **B** coupled together in the form of a pulse. As one field changes, it generates a new field which extends a bit further on, and so the pulse moves out from one point to the next through space.

We can draw an overly mechanistic, but rather picturesque analogy, if we imagine the electric field lines as a dense radial distribution of strings. When somehow plucked, each individual string distorts to form a kink which travels outward from the source. All of these combine at any instant to yield a three-dimensional expanding pulse.

The **E**- and **B**-fields can more appropriately be considered as two aspects of a single physical phenomenon, *the electromagnetic field*, whose source is a moving charge. The disturbance, once having been generated in the electromagnetic field, is an untethered wave which moves beyond its source and independently of it. Bound together as a single entity, the time-varying electric and magnetic fields regenerate each other in an endless cycle. The electromagnetic waves reaching us from the relatively nearby center of our own galaxy have been on the wing for 30,000 years.

We have not, as yet, considered the direction of propagation of the wave with respect to the constituent fields. Notice however that the high degree of symmetry in Maxwell's equations for free space suggests that the disturbance will propagate in a direction which is symmetric to both **E** and **B**. That would imply that an electromagnetic wave could not be purely longitudinal (i.e. as long as **E** and **B** are not parallel). Let's now replace conjecture with a bit of calculation.

We show in Appendix 1 that Maxwell's equations, for free space, can be manipulated into the form of two extremely concise vector expressions:

$$\nabla^2 \mathbf{E} = \epsilon_0 \mu_0 \frac{\partial^2 \mathbf{E}}{\partial t^2} \qquad \text{[A1.26]}$$

and

$$\nabla^2 \mathbf{B} = \epsilon_0 \mu_0 \frac{\partial^2 \mathbf{B}}{\partial t^2}. \qquad \text{[A1.27]}$$

The Laplacian,* ∇^2, operates on each component of **E** and **B** so that the two vector equations actually represent a total of six scalar equations. Two of these expressions, in Cartesian coordinates, are

$$\frac{\partial^2 E_x}{\partial x^2} + \frac{\partial^2 E_x}{\partial y^2} + \frac{\partial^2 E_x}{\partial z^2} = \epsilon_0 \mu_0 \frac{\partial^2 E_x}{\partial t^2} \qquad (3.22)$$

and

$$\frac{\partial^2 E_y}{\partial x^2} + \frac{\partial^2 E_y}{\partial y^2} + \frac{\partial^2 E_y}{\partial z^2} = \epsilon_0 \mu_0 \frac{\partial^2 E_y}{\partial t^2}, \qquad (3.23)$$

with precisely the same form for E_z, B_x, B_y and B_z. Equations of this sort, which relate the space and time variations of some physical quantity, had been studied long before Maxwell's work and were known to describe wave phenomena. Each and every component of the electromagnetic field (E_x, E_y, E_z, B_x, B_y, B_z) therefore obeys the scalar differential wave equation

$$\frac{\partial^2 \psi}{\partial x^2} + \frac{\partial^2 \psi}{\partial y^2} + \frac{\partial^2 \psi}{\partial z^2} = \frac{1}{v^2} \frac{\partial^2 \psi}{\partial t^2}, \qquad \text{[2.59]}$$

provided that

$$v = 1/\sqrt{\epsilon_0 \mu_0}. \qquad (3.24)$$

---

* In Cartesian coordinates

$$\nabla^2 \mathbf{E} = \hat{\mathbf{i}}\nabla^2 E_x + \hat{\mathbf{j}}\nabla^2 E_y + \hat{\mathbf{k}}\nabla^2 E_z.$$

In order to evaluate v Maxwell made use of the results of electrical experiments performed in 1856 in Leipzig by Wilhelm Weber (1804–91) and Rudolph Kohlrausch (1809–58). Equivalently, since μ_0 is assigned a value of $4\pi \times 10^{-7}$ m kg/C^2 (in MKS) one can determine ϵ_0 directly from simple capacitor measurements. In any event

$$\epsilon_0\mu_0 \approx (8.85 \times 10^{-12}\ \text{s}^2\text{C}^2/\text{m}^3\ \text{kg})\ (4\pi \times 10^{-7}\ \text{m kg/C}^2)$$

or

$$\epsilon_0\mu_0 \approx 11.12 \times 10^{-18}\ \text{s}^2/\text{m}^2.$$

And now the moment of truth—in free space, the predicted speed of all electromagnetic waves would then be

$$v = \frac{1}{\sqrt{\epsilon_0\mu_0}} \approx 3 \times 10^8\ \text{m/s}.$$

This theoretical value was in remarkable agreement with the previously measured speed of light (315,300 km/s) determined by Fizeau. The results of Fizeau's experiments, performed in 1849 using a rotating toothed wheel, were available to Maxwell and led him to comment that:

> This velocity [i.e. his theoretical prediction] is so nearly that of light, that it seems we have strong reason to conclude that light itself (including radiant heat, and other radiations if any) is an electromagnetic disturbance in the form of waves propagated through the electromagnetic field according to electromagnetic laws.

This brilliant analysis was one of the great intellectual triumphs of all times.

It has become customary to designate the speed of light in vacuum by the symbol c, the presently accepted value of which is

$$c = 2.997924562 \times 10^8\ \text{m/s} \pm 1.1\ \text{m/s}.$$

The experimentally verified transverse character of light must now be explained within the context of the electromagnetic theory. To that end, consider the fairly simple case of a plane wave propagating in the positive x-direction. The electric field intensity is a solution of Eq. (A1.26) where $\mathbf{E}$ is constant over each one of an infinite set of planes perpendicular to the x-axis. It is therefore a function only of x and t, i.e. $\mathbf{E} = \mathbf{E}(x, t)$. We now refer back to Maxwell's equations, and in particular to Eq. (3.21) (which is generally read as *the divergence of* $\mathbf{E}$ *equals zero*). Since $\mathbf{E}$ is not a function of either y or z, the equation reduces to

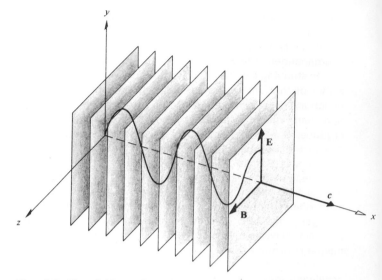

Fig. 3.8 The field configuration in a plane harmonic electromagnetic wave.

$$\frac{\partial E_x}{\partial x} = 0. \tag{3.25}$$

The component of the electric field in the x-direction, i.e. in the direction of propagation, is a constant. It is, accordingly, of no concern to us, since we are interested only in the electromagnetic wave, and not in any unchanging fields which might reside in the same region of space. The $\mathbf{E}$-field associated with the plane wave is then exclusively *transverse*. Without any loss of generality, we shall deal with *plane* or *linearly* polarized waves, where the direction of the vibrating $\mathbf{E}$-vector is fixed. Thus we can orient our coordinate axes so that the electric field is parallel to the y-axis, whereupon

$$\mathbf{E} = \hat{\mathbf{j}}E_y(x, t). \tag{3.26}$$

Returning to Eq. (3.18), it follows that

$$\frac{\partial E_y}{\partial x} = -\frac{\partial B_z}{\partial t} \tag{3.27}$$

and that B_x and B_y are constant, and therefore of no present interest. The time-dependent $\mathbf{B}$-field can only have a component in the z-direction. Clearly then, *in free space, the plane electromagnetic wave is indeed transverse* (Fig. 3.8).

We have not specified the form of the disturbance other than to say that it was a plane wave. Our conclusions are therefore quite general, applying equally well to pulses or continuous waves. We have already pointed out that harmonic functions are of particular interest because any waveform can be expressed in terms of sinusoidal waves using Fourier techniques. We therefore limit the discussion to harmonic waves and write $E_y(x, t)$ as

$$E_y(x, t) = E_{0y} \cos \left[\omega(t - x/c) + \varepsilon \right], \qquad (3.28)$$

the speed of propagation being c. The associated magnetic flux density can be found by directly integrating Eq. (3.27), that is

$$B_z = - \int \frac{\partial E_y}{\partial x} dt.$$

Using Eq. (3.28) we obtain

$$B_z = - \frac{E_{0y}\omega}{c} \int \sin \left[\omega(t - x/c) + \varepsilon \right] dt$$

or

$$B_z(x, t) = \frac{1}{c} E_{0y} \cos \left[\omega(t - x/c) + \varepsilon \right]. \qquad (3.29)$$

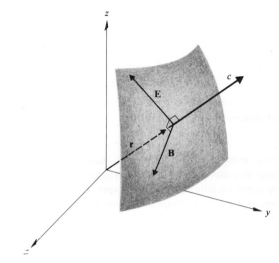

Fig. 3.10 Portion of a spherical wavefront far from the source.

The constant of integration, which represents a time-independent field, has been disregarded. On comparing this result with Eq. (3.28), it is evident that

$$E_y = cB_z. \qquad (3.30)$$

Since E_y and B_z differ only by a scalar, and so have the same time dependence, **E** and **B** are *in phase* at all points in space. Moreover, $\mathbf{E} = \hat{\mathbf{j}}E_y(x, t)$ and $\mathbf{B} = \hat{\mathbf{k}}B_z(x, t)$ are *mutually perpendicular* and their cross product, $\mathbf{E} \times \mathbf{B}$, points in the propagation direction, $\hat{\mathbf{i}}$ (Fig. 3.9).

Plane waves, although of very great importance, are not the only solutions to Maxwell's equations. As we saw in the previous chapter, the differential wave equation allows many solutions, amongst which are cylindrical and spherical waves (Fig. 3.10).

3.3 NONCONDUCTING MEDIA

The response of dielectric or nonconducting materials to electromagnetic fields is of special concern to us in optics. We will, of course, be dealing with transparent dielectrics in the form of lenses, prisms, plates, films, etc. not to mention the surrounding sea of air.

The net effect of introducing a homogeneous, isotropic dielectric into a region of free space is to change ϵ_0 to ϵ and μ_0 to μ in Maxwell's equations. The phase velocity in the medium now becomes

$$v = 1/\sqrt{\epsilon\mu}. \qquad (3.31)$$

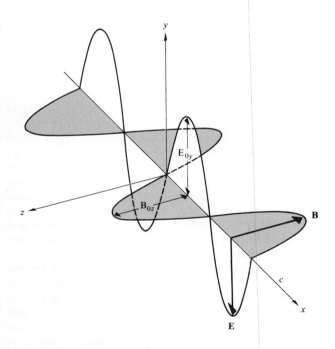

Fig. 3.9 Orthogonal harmonic **E**- and **B**-fields.

The ratio of the speed of an electromagnetic wave in vacuum to that in matter is known as the *absolute index of refraction* *n* and is given by

$$n \equiv \frac{c}{v} = \sqrt{\frac{\epsilon\mu}{\epsilon_0\mu_0}}. \qquad (3.32)$$

In terms of the relative permittivity and relative permeability of the medium, *n* becomes

$$n = \sqrt{K_e K_m}. \qquad (3.33)$$

The great majority of substances, with the exception of ferromagnetic materials, are only weakly magnetic; none is actually nonmagnetic. Even so, K_m generally doesn't deviate from one by any more than a few parts in 10^4 (e.g. for diamond $K_m = 1 - 2.2 \times 10^{-5}$). Setting $K_m = 1$ in the formula for *n* results in an expression known as *Maxwell's relation*, viz.

$$n = \sqrt{K_e}, \qquad (3.34)$$

Table 3.1   Maxwell's relation.

Gases at 0°C and 1 atm		
Substance	$\sqrt{K_e}$	*n*
Air	1.000294	1.000293
Helium	1.000034	1.000036
Hydrogen	1.000131	1.000132
Carbon dioxide	1.00049	1.00045
Liquids at 20°C		
Substance	$\sqrt{K_e}$	*n*
Benzene	1.51	1.501
Water	8.96	1.333
Ethyl alcohol (ethanol)	5.08	1.361
Carbon tetrachloride	4.63	1.461
Carbon disulfide	5.04	1.628
Solids at room temp		
Substance	$\sqrt{K_e}$	*n*
Diamond	4.06	2.419
Amber	1.6	1.55
Fused silica	1.94	1.458
Sodium chloride	2.37	1.50

Values of K_e correspond to the lowest possible frequencies, in some cases as low as 60 Hz, whereas *n* is measured at about 0.5×10^{15} Hz. Sodium D light was used ($\lambda = 589.29$ nm).

wherein K_e is presumed to be the *static dielectric constant*. As indicated in Table 3.1, this relationship seems to work well only for some simple gases. The difficulty arises because K_e, and therefore *n*, are actually *frequency dependent*. The dependence of *n* on the wavelength (or color) of light is a well-known effect called *dispersion*. Indeed, Sir Isaac Newton used prisms to disperse white light into its constituent colors over three hundred years ago and the phenomenon was well known if not well understood even then.

There are two interrelated questions that come to mind at this point: (1) What is the physical basis for the frequency dependence of *n*? and (2) What is the mechanism whereby the phase velocity in the medium is effectively made different from *c*? The answers to both these questions can be found by examining the interaction of an incident electromagnetic wave with the array of atoms constituting a dielectric material.

3.3.1 Dispersion

When a dielectric is subjected to an applied electric field the internal charge distribution distorts under its influence. This corresponds to the generation of electric dipole moments which, in turn, contribute to the total internal field. More simply stated, the external field separates positive and negative charges in the medium (each pair of which is a dipole) and these then contribute an additional field component. The resultant dipole moment per unit volume is called the *electric polarization* **P**. For most materials **P** and **E** are proportional and can satisfactorily be related by

$$(\epsilon - \epsilon_0)\mathbf{E} = \mathbf{P}. \qquad (3.35)$$

The redistribution of charge and the consequent polarization can occur by way of the following mechanisms. There are molecules which have a permanent dipole moment as a result of unequal sharing of valence electrons. These are known as *polar molecules*, of which the nonlinear water molecule is a fairly typical example (Fig. 3.11). Each hydrogen—oxygen bond is polar covalent, with the H-end positive with respect to the O-end. Thermal agitation keeps the molecular dipoles randomly oriented. With the introduction of an electric field, the dipoles align themselves and the dielectric takes on an *orientational polarization*. In the case of *nonpolar molecules and atoms*, the applied field distorts the electron cloud, shifting it relative to the nucleus and thereby producing a dipole moment. In addition to this *electronic polarization*, there is another process which is applicable specifically to molecules, as for example the

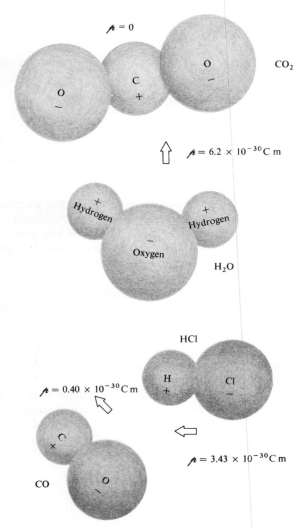

Fig. 3.11 Assorted molecules and their dipole moments.

of inertia. At high driving frequencies ω, polar molecules will be unable to follow the field alternations. Their contributions to **P** will decrease and K_e will drop markedly. The relative permittivity of water is fairly constant at approximately 80, up to about 10^{10} Hz, after which it falls off quite rapidly.

In contrast, electrons have little inertia and can continue to follow the field contributing to $K_e(\omega)$ even at optical frequencies (of about 5×10^{14} Hz). Thus the dependence of n on ω is governed by the interplay of the various electric polarization mechanisms contributing at the particular frequency.

It is possible to derive an analytical expression for $n(\omega)$ in terms of what's happening within the medium on an atomic level. Even though this is actually the domain of quantum mechanics the classical treatment leads to very similar results and in so doing provides a highly useful conceptual model. Indeed that model will be called upon time and time again as we go on to examine reflection, refraction, diffraction and many other phenomena.[†] Imagine that the valence or outermost electrons are bound to their respective atoms or molecules by an elastic restoring force $(-m_e\omega_0^2 x)$ which is proportional to the displacement x of the electrons from equilibrium. The atom then resembles a classical forced oscillator being driven by the alternating

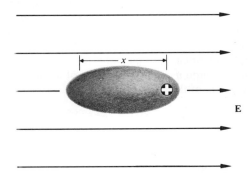

Fig. 3.12 (a) Distortion of the electron cloud in response to an applied **E**-field.

ionic crystal NaCl. In the presence of an electric field, the positive and negative ions undergo a shift with respect to each other. Dipole moments are therefore induced, resulting in what is called *ionic or atomic polarization*.

If the dielectric is subjected to an incident harmonic electromagnetic wave, its internal charge structure will experience time-varying forces and/or torques. These will be proportional to the electric field component of the wave.[*] For polar dielectrics the molecules actually undergo rapid rotations, aligning themselves with the **E**(t) field. But these molecules are relatively large and have appreciable moments

* Forces arising from the magnetic component of the field have the form $\mathbf{F}_M = q\mathbf{v} \times \mathbf{B}$ in comparison to $\mathbf{F}_E = q\mathbf{E}$ for the electric component; but $v \ll c$ and so it follows from Eq. (3.30) that $\mathbf{F}_M$ is generally negligible.

†For more details see e.g. Symon, *Mechanics*, p. 55, or Longhurst, *Geometrical and Physical Optics*, p. 457.

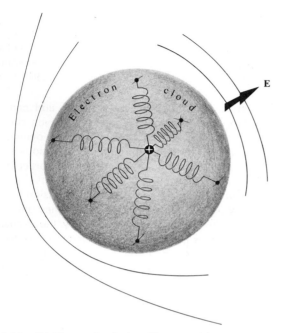

Fig. 3.12   (b) The mechanical oscillator model.

$E(t)$ field which we assume to be applied along the x-direction. Figure 3.12(b) is a mechanical representation of just such an oscillator in an isotropic medium where the negatively charged shell is fastened to a stationary positive nucleus by identical springs. The force (F_E) exerted on an electron of charge q_e by the $E(t)$ field of a harmonic wave of frequency ω is of the form

$$F_E = q_e E(t) = q_e E_0 \cos \omega t.$$

Consequently, Newton's second law provides the equation of motion, i.e. the sum of the forces equals the mass times the acceleration:

$$q_e E_0 \cos \omega t - m_e \omega_0^2 x = m_e \frac{d^2 x}{dt^2}.$$

The constant ω_0 is the *natural frequency* of the oscillator equal to the square root of the ratio of the elastic constant to the mass. It's the oscillatory frequency of the undriven system. To satisfy this expression x will have to be a function whose second derivative isn't very much different from x itself. Furthermore we can anticipate that the electron will oscillate at the same frequency as $E(t)$ and so we "guess" at

the solution

$$x(t) = x_0 \cos \omega t$$

and substitute it back in the equation to evaluate the amplitude x_0. In this way we find that

$$x(t) = \frac{q_e / m_e}{(\omega_0^2 - \omega^2)} E_0 \cos \omega t$$

or

$$x(t) = \frac{q_e / m_e}{(\omega_0^2 - \omega^2)} E(t).$$

Without a driving force (no incident wave) the electron-oscillator will vibrate at its natural or *resonance frequency* ω_0. In the presence of a field whose frequency is less than ω_0, $E(t)$ and $x(t)$ have the same sign, which means that the charge can follow the applied force, i.e. it is in phase with it. However, when $\omega > \omega_0$, the displacement $x(t)$ is in the opposite direction to that of the instantaneous force $q_e E(t)$ and therefore 180° out of phase with it. Keep in mind that we are talking about oscillating dipoles where for $\omega_0 > \omega$ the relative motion of the positive charge is a vibration in the direction of the field. Above resonance the positive charge is 180° out of phase with the field and the dipole is said to lag by π rad.

The dipole moment is equal to the charge q_e times its displacement and if there are N contributing electrons per unit volume the electric polarization, or density of dipole moments, is

$$P = q_e x N.$$

Hence

$$P = \frac{q_e^2 N E / m_e}{(\omega_0^2 - \omega^2)}$$

and from Eq. (3.35)

$$\epsilon = \epsilon_0 + \frac{P(t)}{E(t)} = \epsilon_0 + \frac{q_e^2 N / m_e}{(\omega_0^2 - \omega^2)}.$$

Using the fact that $n^2 = K_e = \epsilon / \epsilon_0$ we can arrive at an expression for n as a function of ω which is known as a *dispersion equation*:

$$n^2(\omega) = 1 + \frac{N q_e^2}{\epsilon_0 m_e} \left(\frac{1}{\omega_0^2 - \omega^2} \right).$$

Thus far we have presumed the existence of only a single

natural frequency ω_0. To account for the observation of more complicated behavior, let's generalize matters by supposing that there are N molecules per unit volume each with f_j oscillators having natural frequencies ω_{0j} where $j = 1, 2, 3 \ldots$. In that case

$$n^2(\omega) = 1 + \frac{Nq_e^2}{\epsilon_0 m_e} \sum_j \left(\frac{f_j}{\omega_{0j}^2 - \omega^2} \right). \qquad (3.36)$$

This is essentially the same result as that arising from the quantum-mechanical treatment, with the exception that some of the terms must be reinterpreted. Accordingly, the quantities ω_{0j} would then be the characteristic frequencies at which an atom may absorb or emit radiant energy. The f_j-terms, which satisfy the requirement that $\Sigma_j f_j = 1$, are weighting factors known as *oscillator strengths*. They reflect the emphasis that should be placed on each one of the modes. Being a measure of the likelihood of occurrence of a given atomic transition, the f_j are also known as *transition probabilities*.

A similar reinterpretation of the f_j-terms is even required classically since agreement with the experimental data demands that they be less than unity. This is obviously contrary to the definition of the f_j which led to Eq. (3.36). One then supposes that a molecule has many oscillatory modes but that each of these has a distinct natural frequency and strength.

Notice that when ω equals any of the characteristic frequencies, n is discontinuous, contrary to actual observation. This is simply the result of having neglected the damping term which should have appeared in the denominator of the sum. Incidentally, the damping, in part, is attributable to energy lost when the forced oscillators (which are, of course, accelerating charges) reradiate electromagnetic energy. In solids, liquids, and gases at high pressure ($\approx 10^3$ atm), the interatomic distances are roughly 10 times less than those of a gas at STP. Atoms and molecules in this relatively close proximity experience strong mutual interactions and a resulting "frictional" force. The effect is a damping of the oscillators and a dissipation of their energy within the substance in the form of heat (molecular motion). This latter process is called *absorption*.

Had we included a damping force proportional to the speed (of the form $\gamma \, dx/dt$) in the equation of motion, the dispersion equation (3.36) would have turned out to have been

$$n^2(\omega) = 1 + \frac{Nq_e^2}{\epsilon_0 m_e} \sum_j \frac{f_j}{\omega_{0j}^2 - \omega^2 + i\gamma_j \omega}. \qquad (3.37)$$

While this expression is fine for rare media such as gases there is yet another complication which must be dealt with if it is to be applied to dense substances. Each atom interacts with the local electric field in which it is immersed. Yet unlike the isolated atoms considered above, those in a dense material will also experience the induced field set up by their brethren. Consequently an atom "sees" in addition to the applied field $E(t)$ another field,* namely $P(t)/3\epsilon_0$. Without going into the details here it can be shown that

$$\frac{n^2 - 1}{n^2 + 2} = \frac{Nq_e^2}{3\epsilon_0 m_e} \sum_j \frac{f_j}{\omega_{0j}^2 - \omega^2 + i\gamma_j \omega}. \qquad (3.38)$$

Thus far we have been considering electron-oscillators exclusively, but the same results would have been applicable for ions bound to fixed atomic sites as well. In that instance m_e would be replaced by the considerably larger ion mass. Thus while electronic polarization is important over the entire optical spectrum the contributions from ionic polarization significantly affect n only in regions of resonance ($\omega_{0j} = \omega$).

The implications of a complex index of refraction will be considered later in Section 4.3.5. At the moment we limit the discussion, for the most part, to situations where absorption is negligible (i.e. $\omega_{0j}^2 - \omega^2 \gg \gamma_j \omega$) and n is real, so that

$$\frac{n^2 - 1}{n^2 + 2} = \frac{Nq_e^2}{3\epsilon_0 m_e} \sum_j \frac{f_j}{\omega_{0j}^2 - \omega^2}. \qquad (3.39)$$

Colorless, transparent gases, liquids and solids have their characteristic frequencies outside of the visible region of the spectrum (which is why they are, in fact, colorless and transparent). In particular, glasses have effective natural frequencies above the visible in the ultraviolet where they become opaque. In cases for which $\omega_{0j}^2 \gg \omega^2$ by comparison ω^2 may be neglected in Eq. (3.39) yielding an essentially constant index of refraction over that frequency region. For example, the important characteristic frequencies for glasses occur at wavelengths of about 100 nm. The middle of the visible range is roughly five times that and there, $\omega_{0j}^2 \gg \omega^2$. Notice that as ω increases toward ω_{0j}, ($\omega_{0j}^2 - \omega^2$) decreases and n *gradually increases with frequency* as is clearly evident in Fig. 3.13. This is called *normal dispersion*. In the ultraviolet region, as ω approaches a natural frequency, the oscillators will begin to resonate. Their amplitudes will increase markedly and this will be accompanied by damping

---

* This result, which applies to isotropic media, is derived in almost any text on electromagnetic theory.

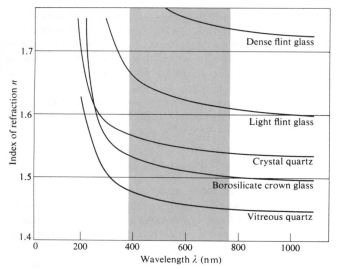

Fig. 3.13 The wavelength dependence of the index of refraction for various materials.

and a strong absorption of energy from the incident wave. When $\omega_{0j} = \omega$ in Eq. (3.38) the damping term obviously becomes dominant. The regions immediately surrounding the various ω_{0j} in Fig. 3.14 are called *absorption bands*. There $dn/d\omega$ is negative and the process is spoken of as *anomalous* (i.e. abnormal) *dispersion*. If white light passes through a glass prism the blue constituent will have a higher index than the red and will therefore be deviated through a larger angle (see Section 5.5.1). In contrast, if we use a liquid-cell prism containing a dye solution having an absorption band in the visible, the spectrum will be altered markedly (Problem 3.6). All substances possess absorption bands somewhere within the electromagnetic frequency spectrum so that the term *anomalous dispersion*, being a carryover from the late eighteen hundreds, is certainly a misnomer.

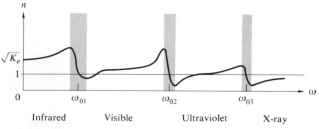

Fig. 3.14 Refractive index versus frequency.

As we have seen, atoms within a molecule can also vibrate about their equilibrium positions. But the nuclei are massive and so the natural oscillatory frequencies will be low, in the infrared. Molecules like H_2O and CO_2 will have resonances in both the infrared and ultraviolet. If water were trapped within a piece of glass during its manufacture, these molecular oscillators would be available and an infrared absorption band would exist. The presence of oxides will also result in infrared absorption. At the even lower frequencies of radio waves, glass will again be transparent. In comparison, a piece of stained glass evidently has a resonance in the visible where it absorbs out a particular range of frequencies, transmitting the complementary color.

As a final point, notice that if the driving frequency is greater than any of the ω_{0j} terms, then $n^2 < 1$ and $n < 1$. Such a situation can occur for example if we beam x-rays onto a glass plate. This is an intriguing result since it leads to $v > c$ in seeming contradiction to special relativity. We will consider this behavior again later on when we discuss the group velocity (Section 7.6).

In partial summary then, over the visible region of the spectrum, electronic polarization is the operative mechanism determining $n(\omega)$. Classically one imagines electron-oscillators vibrating at the frequency of the incident wave. When the wave's frequency is appreciably different from a characteristic or natural frequency, the oscillations are small and there is little absorption. At resonance, however, the oscillator amplitudes are increased and the field does an increased amount of work on the charges. Electromagnetic energy removed from the wave and converted into mechanical energy is then dissipated thermally within the substance and one speaks of an absorption peak or band. The material, although essentially transparent at other frequencies, is fairly opaque to incident radiation at its characteristic frequencies.

3.3.2 The Propagation of Light Through a Dielectric Medium

The process whereby light propagates through a medium at a speed other than c is a fairly complicated one and this section is devoted to making it at least physically reasonable within the context of the simple oscillator model.

Consider an incident or *primary* electromagnetic wave (in vacuum) impinging on a dielectric. As we have seen, it will polarize the medium and drive the electron-oscillators into forced vibration. They, in turn, will reradiate or *scatter*

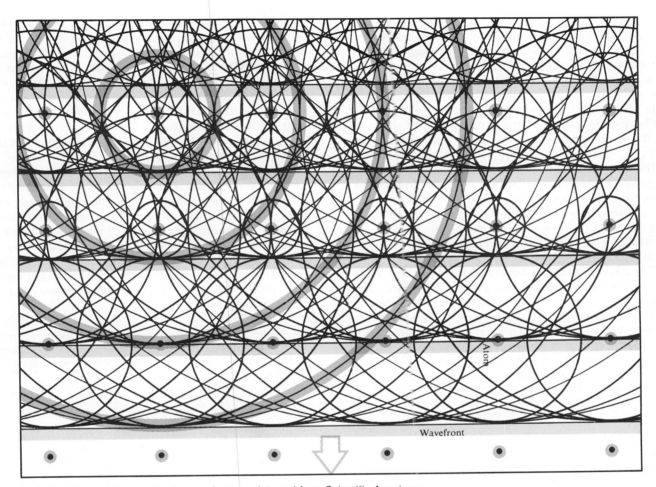

Fig. 3.15 Refracted wave in an ordered array of atoms. Adapted from *Scientific American*

energy in the form of electromagnetic wavelets of the same frequency as the incident wave. In a substance whose atoms or molecules are arranged with some degree of regularity, these wavelets will tend to mutually interfere. That is, they will overlap in certain regions whereupon they will either reinforce or diminish each other to varying degrees. As an example, examine the somewhat overly simplified configuration depicted in Fig. 3.15. There a plane wave, incident on the top, has been scattered into a complicated pattern of wavelets. These in turn have superimposed to form plane wavefronts which we shall refer to as the *secondary* wave. For empirical reasons alone, we can anticipate that the residual primary wave and secondary wave will combine to

yield the only observed disturbance within the medium, namely the *refracted* wave.

Both the primary and secondary electromagnetic waves propagate through the interatomic void with the speed *c*. And yet the medium can certainly possess an index of refraction other than one. The refracted wave may appear to have a phase velocity less than, equal to, or even greater than *c*. The key to this apparent contradiction resides in the phase relationship between the secondary and primary waves.

The classical model predicts that the electron-oscillators will be able to vibrate almost completely in phase with the driving force, i.e. the primary disturbance, only at relatively low frequencies. As the frequency of the electromagnetic field increases, the oscillators will fall behind, lagging in phase by a proportionately larger amount. A detailed analysis leads to the fact that at resonance the phase lag will reach 90°, increasing thereafter to almost 180°, or half a wavelength, at frequencies well above the particular characteristic value.

In addition to these lags there is another effect which must be considered. When the scattered wavelets recombine, the resultant secondary wave* itself lags the oscillators by 90°.

The combined effect of both these mechanisms is that at frequencies below resonance, the secondary wave lags the primary by some amount between approximately 90° to 180°, while at frequencies above resonance, the lag ranges from about 180° to 270°. But a phase lag of $\delta \geq 180°$ is equivalent to a phase lead of $360° - \delta$ [e.g. $\cos(\theta - 270°) = \cos(\theta + 90°)$].

To recapitulate, below resonance the secondary wave lags the primary; above resonance it leads the primary. The resultant or refracted wave will accordingly lead or lag the incident (free space) wave by some amount ε. The process is a progressive one and as the light traverses the medium it is continuously retarded or advanced in phase.

We now wish to show that this is precisely equivalent to a change in the phase velocity. In free space, the *disturbance at some point P* may be written as

$$E_P(t) = E_0 \cos \omega t.$$

If P is now surrounded by a dielectric, there will be a cumulative phase shift ε_P which was built up as the wave moved through the medium to P. The number of wave crests impinging on the dielectric per second must be the same as the number per second propagated into it. That is, the *frequency* must be the same in vacuum as in the dielectric, even though the wavelength and speed may differ. Once again, but this time in the medium, the disturbance at P is

$$E_P(t) = E_0 \cos(\omega t - \varepsilon_P).$$

An observer at P will have to wait a longer time for a given crest to arrive when he is the medium than he would have had to wait in vacuum. In other words, if you imagine two

parallel waves of the same frequency, one in vacuum and one in the material, the vacuum wave will pass P a time ε_P/ω before the other wave. Clearly then, a phase lag of ε_P corresponds to a reduction in speed, $v < c$ and $n > 1$. Similarly, a phase lead yields an increase in speed, $v > c$ and $n < 1$ (Fig. 3.16). The scattering process is a continuous one, and so the cumulative phase shift builds as the light penetrates the medium. That is to say, ε is a function of the length of dielectric traversed; as it must be if v is to be constant (see Problem 3.7).

A rigorous solution to the propagation problem is known as the *Ewald–Oseen extinction theorem*. Although the mathematical formalism, involving integrodifferential equations, is far too complicated to treat here, the results are certainly of interest. It is found that the electron-oscillators generate an electromagnetic wave having essentially two terms. One of these precisely cancels the primary wave within the medium. The other, and only remaining disturbance, propagates through the dielectric at a speed $v = c/n$ as the refracted wave.*

---

* For a discussion of the Ewald–Oseen theorem see *Principles of Optics* by Born and Wolf, Section 2.4.2; this is heavy reading.

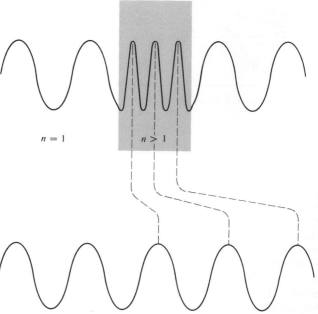

$n = 1$ $n > 1$

Fig. 3.16 The slowing down of a wave in a more dense medium.

---

* This point will be made more plausible when we consider the predictions of the Huygens–Fresnel theory in the diffraction chapter. Most texts on E and M treat the problem of radiation from a sheet of oscillating charges, in which case the 90° phase lag is a natural result.

3.4 ENERGY AND MOMENTUM

3.4.1 Irradiance

One of the most significant properties of the electromagnetic wave is that it transports energy. The light from even the nearest star beyond the sun travels 25 million million miles to reach the earth and yet it still carries enough energy to do work on the electrons within your eye. Any electromagnetic field exists within some region of space and it is therefore quite natural to consider the *radiant energy per unit volume*, i.e. the *energy density u*. For an electric field alone, one can compute the energy density (e.g. between the plates of a capacitor) to be

$$u_E = \frac{\epsilon_0}{2} E^2. \tag{3.40}$$

Similarly, the energy density of the B-field alone (as might be computed within a toroid) is

$$u_B = \frac{1}{2\mu_0} B^2. \tag{3.41}$$

Recall that we derived the relationship $E = cB$ specifically for a plane wave (3.30), nonetheless it is quite general in its applicability. It follows then that

$$u_E = u_B. \tag{3.42}$$

The energy streaming through space in the form of an electromagnetic wave is shared between the constituent electric and magnetic fields. Since

$$u = u_E + u_B,$$

clearly

$$u = \epsilon_0 E^2 \tag{3.43}$$

or equivalently

$$u = \frac{1}{\mu_0} B^2. \tag{3.44}$$

To represent the flow of electromagnetic energy, let S symbolize the transport of energy per unit time (the power) across a unit area. In the MKS system it would then have units of W/m². Figure 3.17 depicts an electromagnetic wave traveling with a speed c through an area A. During a very small interval of time Δt, only the energy contained in the cylindrical volume, $u(c \Delta t A)$, will cross A. Thus

$$S = \frac{uc \Delta t A}{\Delta t A} = uc \tag{3.45}$$

or, using Eq. (3.43)

$$S = \frac{1}{\mu_0} EB. \tag{3.46}$$

We now make the reasonable assumption (for isotropic media) that the energy flows in the direction of propagation of the wave. The corresponding *vector* **S** is then

$$\mathbf{S} = \frac{1}{\mu_0} \mathbf{E} \times \mathbf{B} \tag{3.47}$$

or

$$\mathbf{S} = c^2 \epsilon_0 \mathbf{E} \times \mathbf{B}. \tag{3.48}$$

The magnitude of **S** is the power per unit area crossing a surface whose normal is parallel to **S**. Named after John Henry Poynting (1852–1914), it has come to be known as the *Poynting vector*. Let's now apply these considerations to the case of a harmonic, linearly polarized, plane wave traveling through free space in the direction of **k**:

$$\mathbf{E} = \mathbf{E}_0 \cos (\mathbf{k} \cdot \mathbf{r} - \omega t) \tag{3.49}$$

$$\mathbf{B} = \mathbf{B}_0 \cos (\mathbf{k} \cdot \mathbf{r} - \omega t). \tag{3.50}$$

Using Eq. (3.48) we find

$$\mathbf{S} = c^2 \epsilon_0 \mathbf{E}_0 \times \mathbf{B}_0 \cos^2 (\mathbf{k} \cdot \mathbf{r} - \omega t).$$

It should be evident that **E** × **B** cycles from maxima to minima. At optical frequencies, **S** is an extremely rapidly varying function of time and so its instantaneous value would be an impractical quantity to measure. This rather suggests that we employ an averaging procedure. That is to

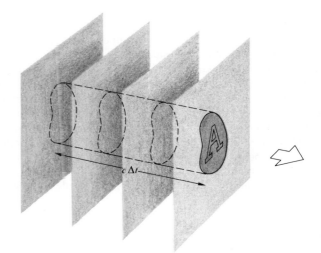

Fig. 3.17 The flow of electromagnetic energy.

say, we absorb the radiant energy during some finite interval of time using, for example, a photocell, a film plate or the retina of a human eye. The time-averaged value of the magnitude of the Poynting vector, symbolized by $\langle S \rangle$, is a measure of the very significant quantity known as the *irradiance*,* *I*. In this case since $\langle \cos^2 (\mathbf{k} \cdot \mathbf{r} - \omega t) \rangle = \frac{1}{2}$ (Problem 3.8),

$$\langle S \rangle = \frac{c^2 \epsilon_0}{2} |\mathbf{E}_0 \times \mathbf{B}_0| \tag{3.51}$$

or

$$I \equiv \langle S \rangle = \frac{c \epsilon_0}{2} E_0^2. \tag{3.52}$$

The irradiance is therefore proportional to the square of the amplitude of the electric field. Two additional alternative ways of saying the same thing are simply

$$I = \frac{c}{\mu_0} \langle B^2 \rangle \tag{3.53}$$

and

$$I = \epsilon_0 c \langle E^2 \rangle. \tag{3.54}$$

Within a linear, homogeneous, isotropic dielectric, the expression for the irradiance becomes

$$I = \epsilon v \langle E^2 \rangle. \tag{3.55}$$

Since, as we have seen, $\mathbf{E}$ is considerably more effective at exerting forces on charges than is $\mathbf{B}$, we shall refer to $\mathbf{E}$ as the *optical field* and use Eq. (3.54) almost exclusively (see Problem 3.3).

The time rate of flow of radiant energy is the power or *radiant flux*, generally expressed in watts. If we divide the radiant flux incident on or exiting from a surface, by the area of the surface we have the *radiant flux density* (W/m²). In the former case, we speak of the *irradiance*, in the latter, the *exitance*; and in either instance the *flux density*.

There are detectors, like the photomultiplier, which serve as *photon counters*. Each quantum of the electromagnetic field, having a frequency v, represents an energy hv (Planck's constant, $h = 6.625 \times 10^{-34}$ J s). If we have a monochromatic beam of frequency v, the quantity I/hv is the average number of photons crossing a unit area (normal to the beam) per unit time, namely the *photon flux density*. Were

---

* In the past physicists generally used the word *intensity* to mean the flow of *energy per unit area per unit time*. Nonetheless, by international, if not universal, agreement that term is now being replaced in optics by the word *irradiance*.

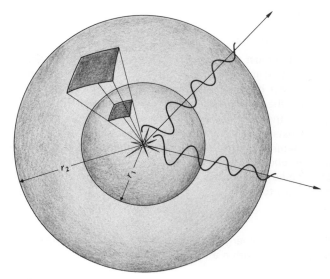

Fig. 3.18 The geometry of the inverse square law.

such a beam to impinge on a counter having an area A, then AI/hv would be the incident *photon flux*, i.e. the average number of photons arriving per unit of time.

We saw earlier that the spherical wave solution of the differential wave equation has an amplitude which varies inversely with r. Let's now examine this same feature within the context of energy conservation. Consider an isotropic point source in free space, emitting energy equally in all directions, i.e. emitting spherical waves. Surround the source with two concentric imaginary spherical surfaces of radii r_1 and r_2, as shown in Fig. 3.18. Let $E_0(r_1)$ and $E_0(r_2)$ represent the amplitudes of the waves over the first and second surfaces, respectively. If energy is to be conserved, the total amount of energy flowing through each surface per second must be equal since there are no other sources or sinks present. Multiplying I by the surface area and taking the square root yields

$$r_1 E_0(r_1) = r_2 E_0(r_2).$$

Inasmuch as r_1 and r_2 were arbitrary, it follows that

$$r E_0(r) = \text{constant}$$

and the amplitude must drop off inversely with r. The irradiance from a point source is proportional to $1/r^2$. This is the well-known *inverse-square law*, which is easily verified using a point source and a photographic exposure meter. Notice that if we envision a beam of photons streaming radially out from the source, the same result clearly obtains.

3.4.2 Momentum

Figure 3.19 depicts a light wave passing by, and interacting with, an electron at some instant in time. The electric field exerts a force $\mathbf{F}_E = q_e\mathbf{E}$ which drives the electron with a velocity $\mathbf{v}_e$. The magnetic field, in turn, exerts a force $\mathbf{F}_M = q_e\mathbf{v}_e \times \mathbf{B}$. Although $\mathbf{F}_E$ changes its direction as $\mathbf{E}$ varies, remaining antiparallel to $\mathbf{E}$ at all times, $\mathbf{F}_M$ is always in the direction of propagation, since $\mathbf{v}_e$ and $\mathbf{B}$ reverse directions simultaneously. The electron undergoes both a rapid transverse oscillation, at the frequency of the wave, and a very slight increase in speed in the direction in which the light propagates. The effect of this small additional component of the velocity, arising from the magnetic field, is quite negligible.

We are interested in determining the amount of linear momentum delivered to the electron by the field. The total force acting on the electron is $\mathbf{F} = \mathbf{F}_E + \mathbf{F}_M$ and its time-averaged value is

$$\langle\mathbf{F}\rangle = q_e\langle\mathbf{v}_e \times \mathbf{B}\rangle$$

or

$$\langle\mathbf{F}\rangle = q_e\langle v_e B\rangle\hat{\mathbf{i}}, \tag{3.56}$$

where $\mathbf{F}_E$ is oscillatory and averages to zero. The magnitudes of $\mathbf{E}$ and $\mathbf{B}$ are related by way of $B = E/c$ and so

$$\langle\mathbf{F}\rangle = \frac{q_e}{c}\langle v_e E\rangle\hat{\mathbf{i}}. \tag{3.57}$$

If W is the instantaneous work done on the electron by the wave, then the rate at which work is being done is

$$\frac{dW}{dt} = \mathbf{v}_e \cdot \mathbf{F} = \mathbf{v}_e \cdot (q_e\mathbf{E} + q_e\mathbf{v}_e \times \mathbf{B}) = q_e v_e E.$$

Thus, on the average

$$\left\langle\frac{dW}{dt}\right\rangle = q_e\langle v_e E\rangle. \tag{3.58}$$

Notice that $\mathbf{v}_e \cdot \langle\mathbf{F}\rangle \approx 0 \neq \langle\mathbf{v}_e \cdot \mathbf{F}\rangle$. Comparing this to Eq. (3.57) and using the fact that the time rate of change of momentum, $d\mathbf{p}/dt$, equals the force, we have

$$\left\langle\frac{d\mathbf{p}}{dt}\right\rangle = \frac{1}{c}\left\langle\frac{dW}{dt}\right\rangle\hat{\mathbf{i}}. \tag{3.59}$$

Over some finite interval in time, during which the electron removes an amount of energy $\mathscr{E}$ from the light wave, it will also take on an amount of momentum equal to $\mathscr{E}/c$, in the direction of propagation. Conservation of momentum obviously requires that the electromagnetic wave be capable of transporting linear momentum.

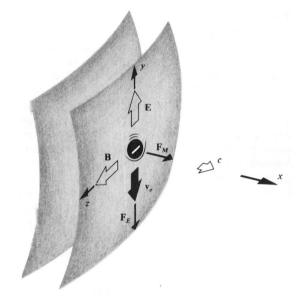

Fig. 3.19 Electromagnetic forces on an electron.

In the photon picture, we envision particle-like quanta each having an energy $\mathscr{E} = h\nu$. We can then expect a photon to carry a momentum $p = \mathscr{E}/c = h/\lambda$. Its vector momentum would be

$$\mathbf{p} = \hbar\mathbf{k} \tag{3.60}$$

where $\mathbf{k}$ is the propagation vector and $\hbar \equiv h/2\pi$. This all fits in rather nicely with special relativity which relates the rest mass m_0, energy and momentum of a particle by way of

$$\mathscr{E} = [(cp)^2 + (m_0c^2)^2]^{\frac{1}{2}}.$$

For a photon $m_0 = 0$ and $\mathscr{E} = cp$.

If an electromagnetic wave impinges on an object and is absorbed or reflected, momentum will be imparted to the electrons within the material and subsequently transmitted to the lattice structure of the object as a whole. The average rate of transfer of momentum per unit area, or equivalently the force per unit area, is known as the *radiation pressure* $\mathscr{P}$. In the special case of normal incidence and complete absorption of the incoming light, it can be shown (Problem 3.18) that

$$\mathscr{P} = I/c. \tag{3.61}$$

For example, the average flux density of electromagnetic energy from the sun impinging normally on a surface just outside of the earth's atmosphere is about 1400 W/m². Assuming complete absorption, the resulting pressure would be 4.7×10^{-6} N/m² or 1.8×10^{-9} ounce/cm² as compared with atmospheric pressure of about 10^5 N/m². Even though it is quite small, radiation pressure is thought to be the agent which deflects the tails of nearby comets, causing them to always point away from the sun (this explanation was given by Kepler in about 1600). It is even possible that there may some day be interplanetary sailships driven by solar radiation pressure. The pressure exerted by light was actually measured as long ago as 1901 by the Russian experimenter Pyotr Nikolaievich Lebedev (1866–1912) and independently by the Americans Edward Leamington Nichols (1854–1937) and Gordon Ferrie Hull (1870–1956).

Light can also transport angular momentum but this will certainly not happen with a linearly polarized wave. Accordingly, we shall defer this rather important discussion to Chapter 8 when circular polarization is examined.

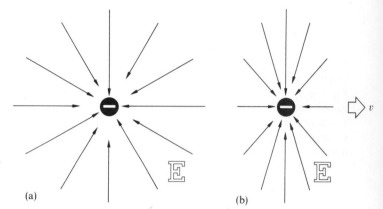

(a)   (b)

Fig. 3.20   (a) Electric field of a stationary electron. (b) Electric field of a moving electron.

3.5 RADIATION

3.5.1 Linearly Accelerating Charges

We have already established, if only in a qualitative way, that accelerating charges are the source of electromagnetic waves. The electric field of a charge at rest can be represented, as in Fig. 3.20, by a uniform, radial distribution of straight *field lines* or *lines of force*. For a charge moving at a constant velocity **v**, the field lines are still radial and straight but they are no longer uniformly distributed. The nonuniformity becomes evident at high speeds and is usually negligible when $v \ll c$. In contrast, Fig. 3.21 shows the field lines associated with an electron accelerating uniformly to the right. The points O_1, O_2, O_3, and O_4 are the positions of the electron after equal time intervals. The field lines are now curved and this, as we shall see, is a most significant difference. As a further contrast Fig. 3.22 depicts the field of an electron at some arbitrary time t_2. Prior to $t = 0$ the particle was always at rest at the point O. The charge was then uniformly accelerated up until time t_1, reaching a speed v, which was maintained constant thereafter. We can anticipate that the surrounding field lines will somehow carry the information that the electron has accelerated. We have ample reason to assume that this "information" will propagate at the speed c. If, for example, $t_2 = 10^{-8}$ s no point beyond 3 m from O would be aware of the fact that the charge had even moved. All of the lines in that region would be uniform, straight and centered on O, as if the charge were still there. At time t_2 the electron is at point O_2 and it is moving with a constant speed v. In the vicinity of O_2 the field lines must then resemble those

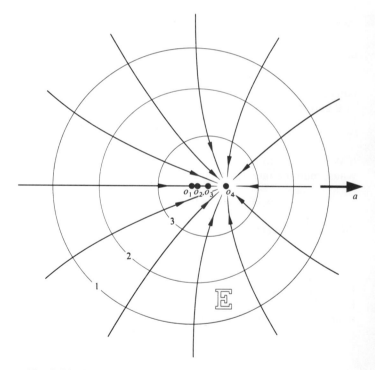

Fig. 3.21   Electric field of a uniformly accelerating electron.

of Fig. 3.20(b). Gauss's law requires that the lines outside of the sphere of radius ct_2 connect to those within the sphere of radius $c(t_2 - t_1)$, since there are no charges between them. It is now apparent that during the interval

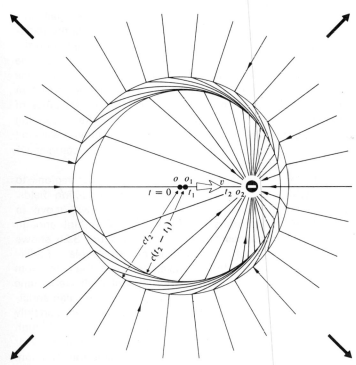

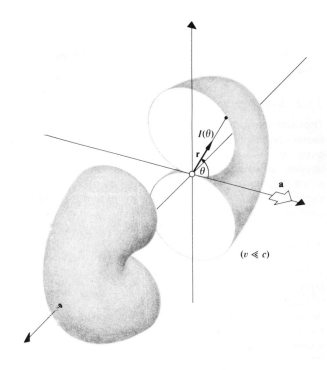

Fig. 3.22 A kink in the **E**-field lines.

while the particle accelerated, the field lines became distorted and a kink appeared. The exact shape of the lines within the region of the kink is of little interest here. What is significant is that there now exists a *transverse component* of the electric field $\mathbf{E}_T$ which propagates outward as a pulse. At some point in space the transverse electric field will be a function of time and it will therefore be accompanied by a magnetic field.

The radial component of the electric field drops off as $1/r^2$, while the transverse component goes as $1/r$. At large distances from the charge the only significant field will be the $\mathbf{E}_T$-component of the pulse which, accordingly, is known as the *radiation field*.* For a positive charge moving slowly ($v \ll c$) the electric and magnetic radiation field can be shown to be proportional to $\mathbf{r} \times (\mathbf{r} \times \mathbf{a})$ and $(\mathbf{a} \times \mathbf{r})$ respectively where $\mathbf{a}$ is the acceleration. For a negative charge

---

* The details of this calculation using J. J. Thomson's method of analyzing the kink can be found in J. R. Tessman and J. T. Finnell, Jr., "Electric Field of an Accelerating Charge", *Am. J. Phys.* **35**, 523 (1967).

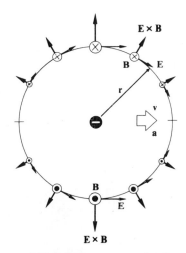

Fig. 3.23 The toroidal radiation pattern of a linearly accelerating charge (split to show cross section).

these reverse as shown in Fig. 3.23. Observe that the irradiance is a function of θ and that $I(0) = I(180°) = 0$ while $I(90°) = I(270°)$ is a maximum.

The energy which is radiated out into the surrounding space by the charge is supplied to it by some external agent. That agent is the one responsible for the accelerating force, which in turn, does work on the charge.

3.5.2 Synchrotron Radiation

The statement that an accelerating charged particle will radiate an electromagnetic wave is the common thread that binds all the varied emission processes. It most certainly applies, equally well, where the acceleration is centripetal rather than linear. There are a number of machines which are cyclical particle accelerators (e.g. betatrons, cyclotrons, synchrotrons). Basically, a high-speed particle is made to interact with a magnetic field which exerts a centripetal force on the charge causing it to move in a circular orbit. Some fraction of the energy imparted to the particle will then be lost in the form of electromagnetic radiation. Even light can be generated in this fashion and indeed the first production of man-made light utilizing this technique took place in an electron synchrotron in 1947. Radiation emitted from charged particles, circulating at relativistic speeds in a magnetic field is known as *synchrotron radiation.* For nonrelativistic particles one speaks of *cyclotron radiation*. If the process were limited to high-energy machines it would be of little present interest to us, and somewhat more than a continuing nuisance to the accelerator people. On the contrary, however, synchrotron radiation, at frequencies including the optical range, does occur in nature. And it is an important source of astronomical information pertinent to our own and other galaxies.

For the sake of brevity and because a complete treat-ment of this material is unwarranted here, we shall only describe its salient features. Figure 3.24 depicts the radiation pattern, or flux-density distribution, at one point on the circular orbit of a charged particle. As v approaches c the radiated beam becomes a narrowing cone-shaped lobe tangent to the trajectory, i.e. in the instantaneous direction of **v**. A distant observer would then see a brief pulse of radiation as the lobe sweeps by, much like a rotating searchlight beam. At relativistic speeds, the charge will emit radiation which is very strongly (but not completely) polarized in the plane of motion.

In the astronomical realm, we can expect regions to exist which are pervaded by magnetic induction fields. Charged particles trapped in these fields will move in circular or helical orbits and if their speeds are high enough they will emit synchrotron radiation. Figure 3.25 shows five photographs of the extragalactic Crab Nebula.* Radiation emanating from the nebula extends over the range from radio frequencies to the extreme ultraviolet. If we assume the source to be trapped circulating charges, we can anticipate strong polarization effects. These are most certainly evident in the first four photographs which were taken through a polarizing filter. The direction of the electric field vector is indicated in each picture. Since in synchrotron radiation, the emitted **E**-field is polarized in the orbital plane, we can conclude that each photo corresponds to a particular uniform magnetic field orientation normal to the orbits and to **E**.

It is believed that a majority of the low-frequency radio waves reaching the earth from outer space have their origin in synchrotron radiation. In 1960, radio astronomers used these long-wavelength emissions to identify the new class of objects known as quasars. In 1955, bursts of polarized radio waves were discovered emanating from Jupiter. Their origin is now attributed to spiraling electrons trapped in radiation belts surrounding the planet.

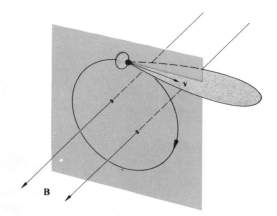

Fig. 3.24 Radiation pattern for an orbiting charge.

---

* The Crab Nebula is believed to be expanding debris left over after the cataclysmic death of a star. From its rate of expansion, astronomers computed the explosion to have taken place in 1050 A.D. This was later corroborated when a study of old Chinese records (the chronicles of the Peiping Observatory) revealed the appearance of an extremely bright star, in the same region of the sky, in the year 1054 A.D.

> In the first year of the period Chihha, the fifth moon, the day Chi-chou [i.e. July 4, 1054], a great star appeared . . . . After more than a year, it gradually became invisible.

There is little doubt that the Crab Nebula is the remnant of that supernova.

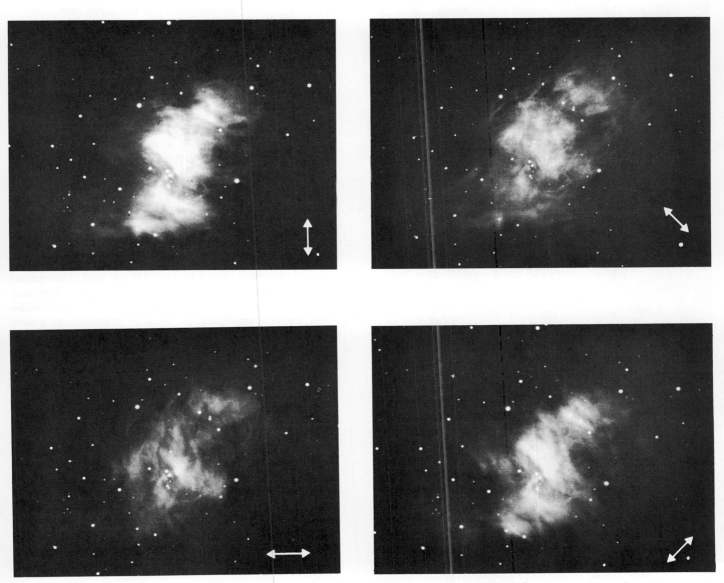

Fig. 3.25(a)   Synchrotron radiation arising from the Crab Nebula. In these photos only light whose **E**-field direction is as indicated was recorded.   [Photos courtesy Mt Wilson and Palomar Observatories.]

Fig. 3.25(b) The Crab Nebula in unpolarized light.

3.5.3   Electric-Dipole Radiation

Both light and ultraviolet radiation arise primarily from the rearrangement of the outermost, or weakly bound, electrons in atoms and molecules. It follows from the quantum-mechanical analysis that the electric dipole moment of the atom is the major source of this radiation. The rate of energy emission from a material system, although a quantum-mechanical process, can be envisioned in terms of the classical oscillating electric dipole. This mechanism is therefore of considerable importance in understanding the manner in which atoms, molecules and even nuclei emit and absorb electromagnetic waves. It will be of particular interest when we study the interaction of light with matter.

We shall again simply make use of the results of what is a lengthy and rather complicated derivation. Figure 3.26 schematically depicts the electric field distribution in the region of an electric dipole. In this simplified configuration, a negative charge is seen oscillating linearly in simple

harmonic motion about an equal stationary positive charge. If the angular frequency of the oscillation is ω, the time-dependent dipole moment $\wp(t)$ has the scalar form

$$\wp = \wp_0 \cos \omega t. \tag{3.62}$$

We should point out that $\wp(t)$ could represent the collective moment of the oscillating charge distribution on the atomic scale or even an oscillating current in a linear TV antenna.

At $t = 0$, $\wp = \wp_0 = qd$ where d is the initial maximum separation between the centers of the two charges (Fig. 3.26a). The dipole moment is, in actuality, a vector in the direction from $-q$ to $+q$. The figure shows a sequence of field line patterns as the displacement, and therefore the dipole moment decreases, then goes to zero and finally reverses direction. When the charges effectively overlap, $\wp = 0$ and the field lines must close on themselves.

Very near the atom, the **E**-field has the form of a static electric dipole. A bit farther out, in the region where the closed loops form, there is no specific wavelength. The detailed treatment shows that the electric field is composed of five different terms and things are obviously complicated. Far from the dipole, in what is called the *wave or radiation zone*, the field configuration takes on a particularly simple form. By then a fixed wavelength has been established; **E** and **B** are transverse, mutually perpendicular, and in phase. Specifically,

$$E = \frac{\wp_0 k^2 \sin \theta}{4\pi\epsilon_0} \frac{\cos (kr - \omega t)}{r} \tag{3.63}$$

and $B = E/c$, where the fields are oriented as in Fig. 3.27. The Poynting vector $\mathbf{S} = \mathbf{E} \times \mathbf{B}/\mu_0$ always points radially outward in the wave zone. There, the **B**-field lines are circles concentric with, and in a plane perpendicular to, the dipole axis. This is understandable since **B** can be considered to arise from the time-varying oscillator current.

The irradiance (radiated radially outward from the source) follows from Eq. (3.52) and is given by

$$I(\theta) = \frac{\wp_0^2 \omega^4}{32\pi^2 c^3 \epsilon_0} \frac{\sin^2 \theta}{r^2} \tag{3.64}$$

again an inverse square law dependence on distance. The angular flux density distribution is toroidal, as in Fig. 3.23.

In the quantum-mechanical treatment of the atom, one finds that energy, rather than being radiated continuously, is emitted in bursts or photons. Each atom undergoing spontaneous emission may be considered, for the moment,

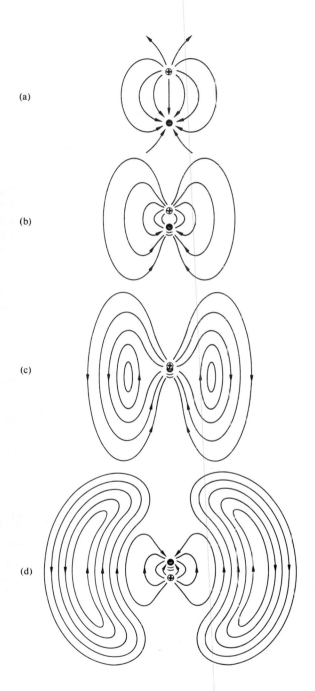

(a)

(b)

(c)

(d)

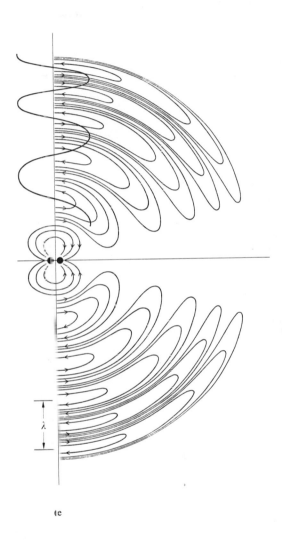

(e)

Fig. 3.26 The **E**-field of an oscillating electric dipole.

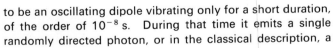

to be an oscillating dipole vibrating only for a short duration, of the order of 10^{-8} s. During that time it emits a single randomly directed photon, or in the classical description, a

single short wave train (or wave packet). We can expect that light emitted from a large assemblage of randomly oriented independent atoms will consist of wave trains in all

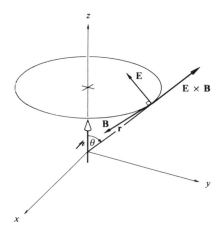

Fig. 3.27 Field orientations for an oscillating electric dipole.

directions. Each one of these will bear no particular consistent phase relation with any of the others, nor will they share a common polarization. This is in marked contrast to the continuous, polarized, extended wave trains generated by sustained current oscillations in a transmitting antenna. Even in this case, however, one does not have truly monochromatic radiation. The simple harmonic functions containing only one frequency are idealizations, at times reasonable ones, but idealizations nonetheless. Prior to switching on even a perfect generator, the radiation will obviously have been zero. Yet a harmonic function has no such limitations on its time dependence and clearly cannot, by itself, represent such a wave. If the generator has been on for a long enough time, the wave it emits will be, at best, nearly monochromatic or *quasimonochromatic*. For many applications, laser light or light passed through a narrow band filter can be adequately represented by a single harmonic function. Even so, since it is not possible to produce monochromatic radiation, the term can only be used loosely and this point must be borne in mind.

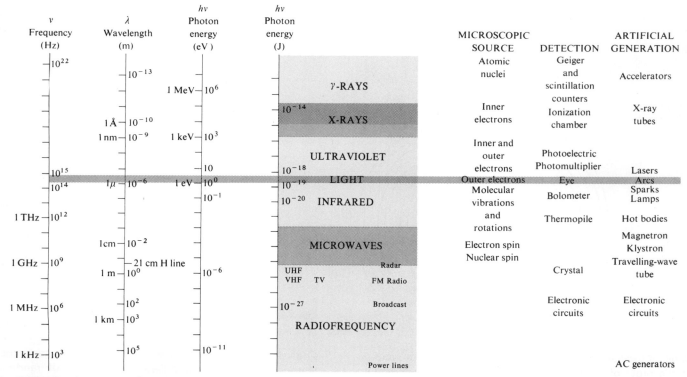

Fig. 3.28 The electromagnetic-photon spectrum.

3.5.4   The Electromagnetic-Photon Spectrum

At the time when Maxwell published the first extensive account of his electromagnetic theory in 1867, the entire known frequency band extended only from the infrared, across the visible, to the ultraviolet.   Although this region is of major concern in optics it is only a small segment of the vast electromagnetic spectrum (see Fig. 3.28).   The following are the main categories (there is actually some overlapping) into which the spectrum is usually divided.

Radiofrequency Waves

In 1887, eight years after Maxwell's death, Heinrich Hertz, then professor of physics at the Technische Hochschule at Karlsruhe, Germany, succeeded in generating and detecting electromagnetic waves.*   His transmitter was essentially an oscillatory discharge across a spark gap (a form of oscillating electric dipole).   As a receiving antenna, he used an open loop of wire with a brass knob on one end and a fine copper point on the other.   A small spark seen across these marked the detection of an incident electromagnetic wave.   Hertz focused the radiation, determined its polarization, reflected and refracted it, caused it to interfere, setting up standing waves and then even measured its wavelength (of the order of a meter).   As he put it:

> I have succeeded in producing distinct rays of electric force, and in carrying out with them the elementary experiments which are commonly performed with light and radiant heat . . . .   We may perhaps further designate them as rays of light of very great wavelength.   The experiments described appear to me, at any rate, eminently adapted to remove any doubt as to the identity of light, radiant heat, and electromagnetic wave motion.

The waves used by Hertz are now classified in the *radiofrequency* range which extends from a few Hz up to about 10^9 Hz (λ, from many kilometers to 0.3 m or so).   These are generally emitted by an assortment of electric circuits.   For example, the 60 Hz alternating current circulating in power lines radiates with a wavelength of 5×10^6 m or about 3×10^3 miles.   There is no upper limit to the theoretical wavelength; one could leisurely swing the proverbial charged pith ball, and in so doing produce a rather long if not very strong wave.   The higher frequency end of the band is used for television and radio broadcasting.

---

* David Hughes may well have been the first person to actually perform this feat, but his experiments of 1879 went unpublished and unnoticed for many years.

At 1 MHz (10^6 Hz) a radiofrequency photon has an energy of 6.62×10^{-28} J or 4×10^{-9} eV, a very small quantity by any measure.   The granular nature of the radiation is generally obscured and only a smooth transfer of energy is apparent.

Microwaves

The microwave region extends from about 10^9 Hz up to about 3×10^{11} Hz.   The corresponding wavelengths go from roughly 30 cm to 1.0 mm.   Radiation capable of penetrating the earth's atmosphere is in the range from less than 1 cm to about 30 m.   Microwaves are therefore of interest in space-vehicle communications as well as radio astronomy.   In particular, neutral hydrogen atoms, distributed over vast regions of space, emit 21 cm (1420 MHz) microwaves.   A good deal of information about the structure of our own and other galaxies has been gleaned from this particular emission.

It is now possible conveniently to check out many theoretical predictions of physical optics using microwave techniques.   Moreover, in many instances this may be the only practical approach.

Photons in the low-frequency end of the microwave spectrum have little energy and one might expect their sources to be electric circuits exclusively.   Emissions of this sort can however arise from atomic transitions if the energy levels involved are quite near each other.   The apparent ground state of the cesium atom is a good example.   It is actually a pair of closely spaced energy levels and transitions between them involve an energy of only 4.14×10^{-5} eV.   The resulting microwave emission has a frequency of 9.1926317×10^9 Hz.   This is the basis for the well-known cesium clock, presently the standard of frequency and time.

Infrared

The infrared region which extends roughly from 3×10^{11} Hz up to about 4×10^{14} Hz was first detected by the renowned astronomer Sir William Herschel (1738–1822) in 1800.   The infrared or IR is often subdivided into four regions: the *near IR*, i.e. near the visible, (780–3000 nm), the *intermediate IR* (3000–6000 nm), the *far IR* (6000–15,000 nm) and the *extreme IR* (15,000 nm–1.0 mm).   This is again a rather loose division and there is no universality in the nomenclature.   Radiant energy at the long wavelength extreme can be generated using either microwave oscillators or incandescent sources, ie. molecular oscillators.   Indeed, any material will radiate and absorb IR via thermal agitation of its constituent molecules.   In addition to the continuous spectra emitted by dense gases, liquids and solids, thermally

Fig. 3.29 Thermograph of one of the authors (E.H.). Note the cool beard.

excited isolated molecules may emit IR in specific narrow ranges. Arising from the vibrations and rotations of these molecules, the emissions are characteristic of the particular chemical bonds involved (e.g. $-C=O$, $-C=C-$, etc.).

Infrared radiant energy is generally measured using a device which responds to the heat generated on absorption of the IR by a blackened surface. There are, for example, thermocouple, pneumatic (e.g. Golay cells), pyroelectric and bolometer detectors. These in turn depend on temperature-dependent variations in induced voltage, gas volume, permanent electric polarization and resistance, respectively. The detector can be coupled by way of a scanning system to a cathode ray tube to produce an instantaneous television-like IR picture (Fig. 3.29). Photographic films sensitive to portions of the IR are also available.

Light

Light corresponds to the electromagnetic radiation in the narrow band of frequencies from about 3.84×10^{14} Hz to roughly 7.69×10^{14} Hz (Table 3.2). It is generally produced by a rearrangement of the outer electrons in atoms and molecules (don't forget synchrotron radiation, which is a different mechanism).*

---

* There is no need here to define light in terms of human physiology. On the contrary, there is some evidence to indicate that this would not be a very good idea, e.g. see T. J. Wang, "Visual Response of the Human Eye to X Radiation," *Am. J. Phys.* **35**, 779 (1967).

In an incandescent material, a hot glowing metal filament or the solar fireball, electrons are randomly accelerated and undergo frequent collisions. The resulting broad emission spectrum is called *thermal radiation*, and it is a major source of light. In contrast, if we fill a tube with some gas and pass an electric discharge through it, the atoms therein will become excited and radiate. The emitted light is characteristic of the particular energy levels of those atoms and it is made up of a series of well-defined frequency bands or lines. Such a device is known as a gas discharge tube. When the gas is the krypton 86 isotope, the lines are particularly narrow (zero nuclear spin, therefore no hyperfine structure). The orange-red line of Kr 86 whose vacuum wavelength is 605.7802105 nm has a width (at half height) of only 0.00047 nm or about 400 MHz. Accordingly it is the present international standard of length (1,650,763.73 wavelengths equals a meter). Observe that electromagnetic radiation serves as the standard for both length and time.

In a flood of bright sunlight where the photon flux density might be 10^{21} photons/m²s, we can generally expect the quantum nature of the energy transport to be thoroughly obscured. However, in very weak beams, since photons in the visible range ($hv \approx 1.6$ eV up to 3.2 eV) are energetic enough to produce effects on a distinctly individual basis, the granularity will become evident. Research on human vision indicates that as few as ten, and possibly even one, light photon may be detectable by the eye.

Table 3.2 *Approximate* frequency and vacuum wave-length ranges for the various colors.

Color	λ_0(nm)	v(THz)*
Red	780–622	384–482
Orange	622–597	482–503
Yellow	597–577	503–520
Green	577–492	520–610
Blue	492–455	610–659
Violet	455–390	659–769

* 1 terahertz (THz) = 10^{12}Hz, 1 nanometer (nm) = 10^{-9} m.

Ultraviolet

Adjacent to light in the spectrum is the ultraviolet region (approximately 8×10^{14} Hz to about 3×10^{17} Hz) discovered by Johann Wilhelm Ritter (1776–1810). Photon energies therein range from roughly 3.2 eV to 1.2×10^3 eV. Ultraviolet, or UV, rays from the sun will thus have more than enough energy to ionize atoms in the upper atmosphere

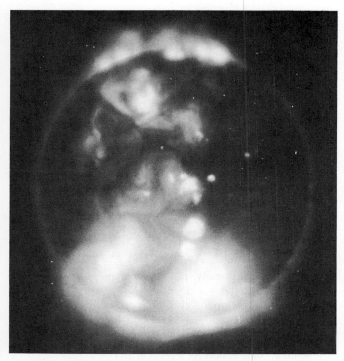

Fig. 3.30 X-ray photo of the sun taken March 1970. The limb of the moon is visible in the southeast corner. [Courtesy Dr. G. Vaiana and NASA.]

and in so doing create the ionosphere. These photon energies are also of the order of the magnitude of many chemical reactions and ultraviolet rays become important in triggering those reactions. Fortunately ozone (O_3) in the atmosphere absorbs out what would otherwise be a lethal stream of solar UV.

Ultraviolet is detectable via fluorescent screens, photographic emulsions and photocells.

X-rays

X-rays were rather fortuitously discovered in 1895 by Wilhelm Conrad Röntgen (1845–1923). They extend in the frequency domain from 3×10^{17} Hz to 5×10^{19} Hz. X-ray photons are emitted by an atom or molecule when the inner, tightly bound, electrons undergo transitions. They are also generated when a high-energy charged particle is made to change its motion (*bremsstrahlung*) and therefore radiate.

At these frequencies photon energies are between 1.2×10^3 eV and 2.1×10^5 eV, high enough so that their interactions with matter can be clearly granular.

In recent times satellites, like the Orbiting Solar Observatory, have recorded extraterrestrial x-ray emissions from the sun and other cosmic sources. Rocket-borne x-ray telescopes incorporating grazing incidence mirrors are being used to probe this new research frontier. Figure 3.30 is a marvelous example of the state of the art. An x-ray spectroheliograph will be aboard the space station Skylab-1 as part of the Post Apollo Program.

Gamma Rays

These are the highest-energy (10^4 eV to about 10^{19} eV), lowest-wavelength electromagnetic radiations. They are emitted by particles undergoing transitions within the atomic nucleus. A single gamma-ray photon carries so much energy that it can be detected with little difficulty. At the same time its wavelength has become so small that it is now extremely difficult to observe any wave-like properties.

We have gone full cycle from the radio frequency wave-like response to gamma-ray particle-like behavior. Somewhere, not far from the (logarithmic) center of the spectrum, is light. As with all electromagnetic radiation, its energy is quantized, but here in particular what we "see" will depend on how we "look".

PROBLEMS

3.1 Consider the plane electromagnetic wave (in MKS) given by the expressions $E_x = 0$, $E_y = 2 \cos [2\pi \times 10^{14} (t - x/c) + \pi/2]$, and $E_z = 0$.

a) What is the frequency, wavelength, direction of motion, amplitude, epoch angle and polarization of the wave?
b) Write an expression for the magnetic flux density.

3.2 Write an expression for the **E**- and **B**-fields which constitute a plane harmonic wave traveling in the $+z$-direction. The wave is linearly polarized with its plane of vibration at 45° to the yz-plane.

3.3* A plane, harmonic, linearly polarized light wave has an electric field intensity given by

$$E_z = E_0 \cos \pi 10^{15}\left(t - \frac{x}{0.65c}\right)$$

while traveling in a piece of glass. Find

a) the frequency of the light,
b) its wavelength,
c) the index of refraction of the glass.

3.4 The relative permittivity of water varies from 88.00 at 0° C. to 55.33 at 100° C. Explain this behavior. Over the same range in temperature, the index of refraction ($\lambda = 589.3$ nm) goes from roughly 1.33 to 1.32. Why is the change in n so much smaller than the corresponding change in K_e?

3.5 Show that for substances of low density, like gases, which have a single resonant frequency ω_0, the index of refraction is given by

$$n \approx 1 + \frac{Nq_e^2}{2\epsilon_0 m_e(\omega_0^2 - \omega^2)}.$$

3.6 Fuchsin is a strong (aniline) dye which, in solution with alcohol, has a deep red color. It appears red because it absorbs out the green component of the spectrum. (As you might expect, the surfaces of crystals of fuchsin reflect green light rather strongly.) Imagine that you have a thin-walled hollow prism filled with this solution. What will the spectrum look like for incident white light? Incidently, anomalous dispersion was first observed in about 1840 by Fox Talbot, and the effect was christened in 1862 by Le Roux. His work was promptly forgotten, only to be rediscovered eight years later by C. Christiansen.

3.7 Imagine that we have a nonabsorbing glass plate of index n and thickness Δy which stands between a source S and an observer P.

a) If the unobstructed wave (without the plate present) is $E_u = E_0 \exp i\omega(t - y/c)$, show that with the plate in place the observer sees a wave

$$E_p = E_0 \exp i\omega[t - (n - 1)\Delta y/c - y/c].$$

b) Show that if either $n \approx 1$ or Δy is very small, then

$$E_p = E_u + \frac{\omega(n - 1)\Delta y}{c} E_u e^{-i\pi/2}.$$

The second term on the right may be envisioned as the field arising from the oscillators in the glass plate.

3.8 The time average of some function $f(t)$ taken over an interval T is given by

$$\langle f(t) \rangle = \frac{1}{T} \int_t^{t+T} f(t')\, dt'$$

where t' is just a dummy variable. If $\tau = 2\pi/\omega$ is the period of a harmonic function, show that

$$\langle \sin^2 (\mathbf{k} \cdot \mathbf{r} - \omega t) \rangle = \tfrac{1}{2},$$

$$\langle \cos^2 (\mathbf{k} \cdot \mathbf{r} - \omega t) \rangle = \tfrac{1}{2}$$

and

$$\langle \sin (\mathbf{k} \cdot \mathbf{r} - \omega t) \cos (\mathbf{k} \cdot \mathbf{r} - \omega t) \rangle = 0$$

when $T = \tau$ and when $T \gg \tau$.

3.9* Consider a linearly polarized plane electromagnetic wave traveling in the $+x$-direction in free space and having as its plane of vibration the xy-plane. Given that its frequency is 10 MHz and its amplitude is $E_0 = 0.08$ V/m,

a) find the period and wavelength of the wave,
b) write an expression for $E(t)$ and $B(t)$,
c) find the flux density, $\langle S \rangle$, of the wave.

3.10 A 1.0 mW laser has a beam diameter of 2 mm. Assuming the divergence of the beam to be negligible, compute its energy density in the vicinity of the laser.

3.11* A cloud of locusts having a density of 100 insects per cubic meter is flying north at a rate of 6 m/min. What is the flux density of locusts, i.e. how many cross an area of 1 m² perpendicular to their flight path per second?

3.12 Imagine that you are standing in the path of an antenna which is radiating plane waves of frequency 100 MHz and flux density 19.88×10^{-2} W/m². Compute the photon flux density, i.e. the number of photons per unit time per unit area. How many photons, on the average, will be found in a cubic meter of this region?

3.13* How many photons per second are emitted from a 100 W yellow light bulb if we assume negligible thermal losses and a quasimonochromatic wavelength of 550 nm? In actuality only about 2.5 percent of the total dissipated power emerges as visible radiation in an ordinary 100 W lamp.

3.14 An ordinary 3.0 V flashlight bulb draws roughly 0.25 A converting about 1.0 percent of the dissipated power into light ($\lambda \approx 550$ nm). If the beam initially has a cross sectional area of 10 cm²,

  a) how many photons are emitted per second?

  b) how many photons occupy each meter of the beam?

  c) what is the flux density of the beam as it leaves the flashlight?

3.15* An isotropic quasimonochromatic point source radiates at a rate of 100 W. What is the flux density at a distance of 1 m? What are the amplitudes of the **E**- and **B**-fields at that point?

3.16 Using energy arguments, show that the amplitude of a cylindrical wave must vary inversely with $\sqrt{r}$. Draw a diagram indicating what's happening.

3.17* What is the momentum of a 10^{19} Hz x-ray photon?

3.18 Consider a normally incident beam of light of irradiance I, which is completely absorbed by some material. Show that the radiation pressure exerted on the material is given by

$$\mathscr{P} = I/c \qquad\qquad [3.61]$$

3.19* Derive an expression for the radiation pressure when the normally incident beam of light is totally reflected. Generalize this result to the case of oblique incidence at an angle θ with the normal.

3.20 A completely absorbing screen receives 300 W of light for 100 s. Compute the total linear momentum transferred to the screen.

3.21* If an ultraviolet photon is to dissociate the oxygen and carbon atoms in the carbon monoxide molecule, it must provide 11 eV of energy. What is the minimum frequency of the appropriate radiation?

3.22 Consider the plight of an astronaut floating in free space with only a 10 W lantern (inexhaustibly supplied with power). How long will it take him to reach a speed of 10 m/s using the radiation as propulsion? His total mass is 100 kg.

The Propagation of Light   4

4.1 INTRODUCTION

We now consider a number of phenomena related to the propagation of light and its interaction with material media. In particular, we shall study the characteristics of light waves as they progress through various substances, crossing interfaces, and being reflected and refracted in the process. For the most part, we shall envision light as a classical electromagnetic wave whose velocity through any medium is dependent upon that material's electric and magnetic properties. It is an intriguing fact that many of the basic principles of optics are predicated on the wave aspects of light and are yet completely independent of the exact nature of that wave. As we shall see, this accounts for the longevity of *Huygens' principle* which has served in turn to describe mechanical ether waves, electromagnetic waves, and now after three hundred years applies to quantum optics.

Suppose, for the moment, that a wave impinges on the interface separating two different media (e.g. a piece of glass in air). As we know from our everyday experiences, a portion of the incident flux density will be diverted back in the form of a *reflected wave*, while the remainder will be transmitted across the boundary as a *refracted wave*. On a submicroscopic scale we envisage an assemblage of atoms which scatter the incident radiant energy. The manner in which these emitted light wavelets superimpose and combine with each other will depend on the spatial distribution of the scattering atoms. As we know from the previous chapter, the scattering process is responsible for the *index of refraction* as well as the resultant *reflected* and *refracted* waves. This atomistic description is quite satisfying conceptually even though it is not a simple matter to treat analytically. It should, however, be kept in mind even when applying macroscopic techniques, as indeed we shall, later on.

We now seek to determine the general principles governing or at least describing the propagation, reflection and refraction of light. In principle it should be possible to trace the progress of radiant energy through any system by applying Maxwell's equations and the associated boundary conditions. In practice, however, this is often an impractical if not an impossible task (see Section 10.1). And so we shall take a somewhat different route, stopping, when appropriate, to verify that our results are in accord with electromagnetic theory.

4.2 THE LAWS OF REFLECTION AND REFRACTION

4.2.1 Huygens' Principle

Recall that a wavefront is a surface over which an optical disturbance has a constant phase. As an illustration, Fig. 4.1 shows a small portion of a spherical wavefront Σ emanating from a monochromatic point source S in a homogeneous medium. Clearly if the radius of the wavefront as shown is r, at some later time t it will simply be $(r + vt)$ where v is the phase velocity of the wave. But suppose instead that the light passes through a nonuniform sheet of glass as in Fig. 4.2 so that the wavefront itself is distorted. How can we determine its new form Σ'? Or for that matter, what will Σ' look like at some later time if it is allowed thereafter to continue unobstructed?

A preliminary step toward the solution of this problem appeared in print in 1690 in the work entitled *Traité de la Lumière* which had been written twelve years earlier by the Dutch physicist Christiaan Huygens. It was there that he enunciated what has since become known as *Huygens' principle*, that *every point on a primary wavefront serves as*

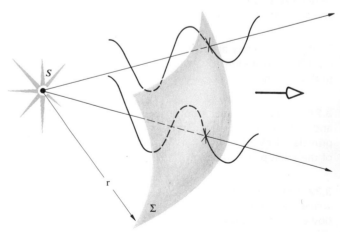

Fig. 4.1 A segment of a spherical wave.

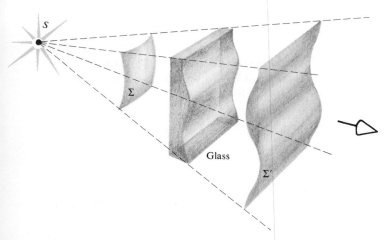

Fig. 4.2 Distortion of a portion of a wavefront on passing through a material of nonuniform thickness.

the source of spherical secondary wavelets such that the primary wavefront at some later time is the envelope of these wavelets. Moreover, the wavelets advance with a speed and frequency equal to that of the primary wave at each point in space. If the medium is homogeneous the wavelets may be constructed with finite radii, whereas if it is inhomogeneous the wavelets will have to have infinitesimal radii. Figure 4.3 should make all of this fairly clear; it shows a view of a wavefront Σ as well as a number of spherical secondary wavelets which, after a time *t*, have propagated out to a radius of *vt*. The envelope of all of these wavelets is then asserted to correspond to the advanced primary wave Σ′. It is easy to visualize the process in terms of mechanical vibrations of an elastic medium. Indeed this is the way that Huygens envisioned it within the context of an all-pervading ether as is evident from this comment by him:

> We have still to consider, in studying the spreading out of these waves, that each particle of matter in which a wave proceeds not only communicates its motion to the next particle to it, which is on the straight line drawn from the luminous point, but that it also necessarily gives a motion to all the others which touch it and which oppose its motion. The result is that around each particle there arises a wave of which this particle is a center.

We can make use of these ideas on two different levels: In treating diffraction theory, a mathematical representation of the wavelets will serve as the basis for a valuable analytical technique. Using it, one is able to trace the progress of a primary wave past all sorts of apertures and obstacles by summing up the wavelet contributions mathematically. In

contrast, Fig. 4.3 in itself represents a graphical application of the essential ideas and as such is known as *Huygens' construction.* For the moment we limit ourselves to the latter approach.

Thus far we have merely stated Huygens' principle without any justification or proof of its validity. As we shall see (Chapter 10) Fresnel in the eighteen hundreds successfully modified Huygens' principle somewhat. A little later on Kirchhoff showed that the *Huygens–Fresnel principle* was a direct consequence of the differential wave equation (2.59), thereby putting it on a firm mathematical base. That there was a need for a bit of reformulation of the principle is evident from Fig. 4.3 where we deceptively only drew hemispherical wavelets. Had we drawn them as spheres, there would have been a *back wave* moving toward the source—something which is not observed. Since this difficulty was taken care of theoretically by Fresnel and Kirchhoff, we shall not be disturbed by it. In fact, we shall merely overlook it completely when applying Huygens' construction.

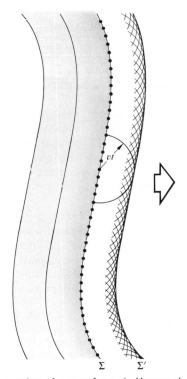

Fig. 4.3 The propagation of a wavefront via Huygens' principle.

Huygens' principle fits in rather nicely with our earlier discussion of the atomic scattering of radiant energy. Each atom of a material substance which interacts with an incident primary wavefront can be regarded as a point source of scattered secondary wavelets. Things are not quite as clear when we consider the principle as applied to the propagation of light through a vacuum. It is helpful however to keep in mind that at any point in empty space on the primary wavefront there exists both a time-varying **E**-field and a time-varying **B**-field. Both of these in turn create new fields which move out from that point. In this sense each point on the wavefront is analogous to a physical scattering center.

4.2.2 Snell's Law and the Law of Reflection

Imagine that we have a monochromatic plane wave incident on the smooth interface separating two different transparent media as in Fig. 4.4. We now wish to determine the wave's behavior using Huygens' construction. Accordingly, each point of the plane wavefront Σ in Fig. 4.5 will be taken as the source of a secondary wavelet.

The number of primary waves arriving at the interface per second corresponds to the frequency of the incident harmonic wave, i.e. v. Clearly then, the reflected and transmitted waves, and therefore all of the secondary wavelets, will have the same frequency—the boundary cannot alter v.* After an elapsed time corresponding to the period $\tau = 1/v$, the wavefronts are as shown in the diagram where Σ has split into Σ_i, Σ_t, and Σ_r. The indices of refraction are such that n_i of the incident medium is less than n_t of the transmitting medium. Thus wavelets in the incident medium travel with a speed $v_i = c/n_i$ while those in the transmitting medium move at $v_t = c/n_t$ where, since $n_t > n_i$, $v_t < v_i$. It might be helpful if you visualized the diagram as a multiple exposure of the wavefront Σ at various times and particularly after an interval τ.

Let's follow the progress of a typical wavelet, for example the one emitted from point b on Σ. After a time t_1 the plane Σ will have moved to Σ', a distance in the incident medium of $v_i t_1$ so that b then corresponds to b'. Presumably, a wavelet will then propagate out from b' into both the incident and transmitting media contributing to the reflected,

---

* This presumes the use of light whose flux density is not so extraordinarily high that the fields are gigantic. With this assumption the medium will behave linearly, as is most often the case. In contrast, observable harmonics can be generated if the fields are made large enough (Section 14.4).

Σ_r, and transmitted, Σ_t, wavefronts. These wavelets are shown here after a time t_2 where $\tau = t_1 + t_2$. The rest of the diagram should be self-explanatory. Figure 4.6 is a somewhat simplified version in which θ_i, θ_r, and θ_t as before are the angles of *incidence, reflection* and *transmission* (or *refraction*) respectively. Notice that

$$\frac{\sin \theta_i}{BD} = \frac{\sin \theta_r}{AC} = \frac{\sin \theta_t}{AE} = \frac{1}{AD}. \tag{4.1}$$

By comparison with Fig. 4.5 it should be evident that

$$BD = v_i t, \qquad AC = v_i t, \qquad AE = v_t t$$

and so substituting into Eq. (4.1) and canceling t, we have

$$\frac{\sin \theta_i}{v_i} = \frac{\sin \theta_r}{v_i} = \frac{\sin \theta_t}{v_t}. \tag{4.2}$$

It then follows from the first two terms that *the angle of incidence equals the angle of reflection*, that is

$$\theta_i = \theta_r. \tag{4.3}$$

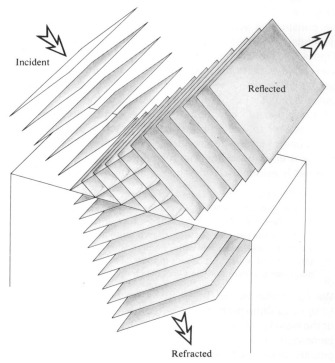

Fig. 4.4 Reflection and transmission of plane waves.

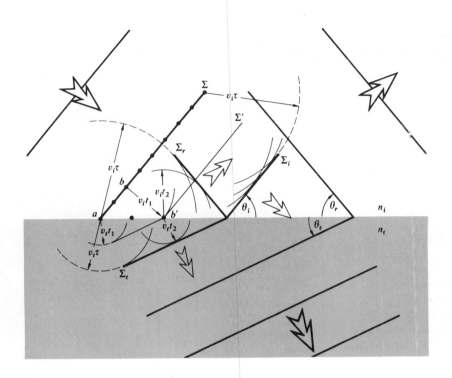

Fig. 4.5 Reflection and transmission at an interface via Huygens' principle.

Known as the *law of reflection*, it first appeared in the book entitled *Catoptrics* which was purported to have been written by Euclid.

The first and last terms of Eq. (4.2) yield

$$\frac{\sin \theta_i}{\sin \theta_t} = \frac{v_i}{v_t} \qquad (4.4)$$

or since $v_i/v_t = n_t/n_i$

$$n_i \sin \theta_i = n_t \sin \theta_t. \qquad (4.5)$$

This is the very important *law of refraction*, the physical consequences of which have been studied, at least on record, for over eighteen hundred years. On the basis of some very fine observations Claudius Ptolemy of Alexandria attempted unsuccessfully to divine the expression. Kepler very nearly succeeded in deriving the law of refraction in his book *Supplements to Vitello* in 1604. Unfortunately he was misled by some erroneous data compiled earlier by Vitello (ca. 1270). The correct relationship seems to have been arrived at independently by Snell* at the University of Leyden and the French mathematician Descartes.† In English-speaking countries Eq. (4.5) is generally referred to as *Snell's law*.

Notice that it can be rewritten in the form

$$\frac{\sin \theta_i}{\sin \theta_t} = n_{ti} \qquad (4.6)$$

where $n_{ti} \equiv n_t/n_i$ is the *ratio of the absolute indices of refraction*. In other words, it is the *relative index of refraction of the two media*. It is evident in Fig. 4.5 where $n_{ti} > 1$, i.e. $n_t > n_i$ and $v_i > v_t$ that $\lambda_i > \lambda_t$ whereas the opposite would be true if $n_{ti} < 1$.

4.2.3 Light Rays

The concept of a light ray is one which will be of sustained interest to us throughout our study of optics. *A ray is a line drawn in space corresponding to the direction of flow of radiant energy.* As such, it is a mathematical device rather than a physical entity. In practice one can produce very narrow *beams* or *pencils* of light (as for example a laser beam) and we might imagine a ray to be the unattainable

---

* This is the most common spelling, although Snel is probably more accurate.

† For a more detailed history, see Max Herzberger, "Optics from Euclid to Huygens," *Appl. Opt.* **5**, 1383 (1966).

limit on the narrowness of such a beam. Bear in mind that in an *isotropic medium*, i.e. one whose properties are the same in all directions, *rays are orthogonal trajectories of the wavefronts*. That is to say, *they are lines normal to the wavefronts at every point of intersection*. Evidently, *in such a medium a ray is parallel to the propagation vector* **k**. As you might suspect, this is not true in *anisotropic* substances whose further consideration we postpone until later (see Section 8.4.1). *Within homogeneous isotropic materials, rays will be straight lines* since by symmetry they cannot bend in any preferred direction, there being none. Moreover, as a result of the fact that the speed of propagation is identical in all directions within a given medium, the spatial separation between two wavefronts, measured along rays, must be the same everywhere.* Points where a single ray intersects a set of wavefronts are called *corresponding points*, as for example A, A', and A'' in Fig. 4.7. *Evidently the separation in time between any two corresponding points on any two sequential wavefronts is identical*. In other words, if wavefront Σ transforms into Σ'' after a time t'', the distance between corresponding points on any and all rays will be traversed in that same time t''. This will obviously be true even if the wavefronts pass from one homogeneous isotropic medium into another. This just means that each point on Σ can be imagined as following the path of a ray to arrive at Σ'' in the time t''.

If a group of rays is such that we can find a surface which is orthogonal to each and every one of them, they are said to form a *normal congruence*. For example, the rays emanating from a point source are perpendicular to a sphere centered at the source and consequently form a normal congruence.

We can now briefly consider an alternative technique to Huygens' principle which will also allow us to follow the progress of light through various isotropic media. The basis for this approach is the *theorem of Malus and Dupin* (introduced in 1808 by E. Malus and modified in 1816 by C. Dupin) according to which *a group of rays will preserve its normal congruence after any number of reflections and refractions* (as in Fig. 4.7). Looking at it from our present vantage point of the wave theory, it is evidently equivalent to the statement that rays remain orthogonal to wavefronts throughout all propagation processes in isotropic media. As shown in Problem 4.9, the theorem can be used to derive the law of

---

* When the material is inhomogeneous or when there is more than one medium involved, it will be the *optical path length* (see Section 4.2.4) between the two wavefronts which is the same.

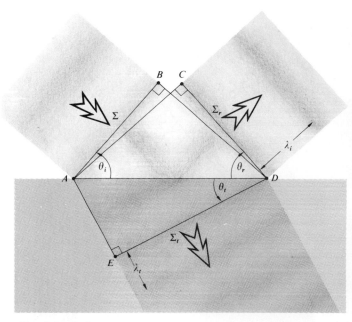

Fig. 4.6   Reflected and transmitted wavefronts at a given instant.

reflection as well as Snell's law. It is often most convenient to carry out a ray trace through an optical system using the laws of reflection and refraction and then reconstruct the wavefronts. The latter can be accomplished in accord with the above considerations of equal transit times between corresponding points and the orthogonality of the rays and wavefronts.

Figure 4.8 depicts the parallel ray formation concomitant with a plane wave where now θ_i, θ_r, and θ_t, which have the

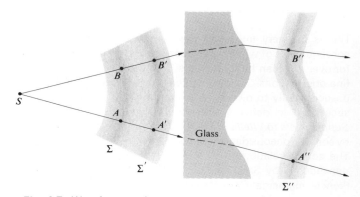

Fig. 4.7   Wavefronts and rays.

exact same meanings as before, are measured from the normal to the interface. The incident ray and the normal determine a plane known as the *plane of incidence*. Because of the symmetry of the situation, we must anticipate that both the reflected and transmitted rays will be undeflected from that plane. In other words, the respective unit propagation vectors $\hat{k}_i$, $\hat{k}_r$, and $\hat{k}_t$ are coplanar.

In summary then, the three basic laws of reflection and refraction are:

1. the incident, reflected and refracted rays all lie in the plane of incidence.
2. $\theta_i = \theta_r$ [4.3]
3. $n_i \sin \theta_i = n_t \sin \theta_t$.   [4.5]

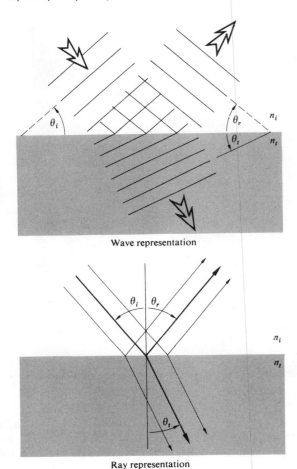

Wave representation

Ray representation

Fig. 4.8 The wave and ray representations of an incident, reflected and transmitted beam.

These are illustrated rather nicely using a narrow light beam in the photographs of Fig. 4.9. In this case as in Fig. 4.10(a) the surface is smooth, or more precisely, any irregularities in it are small compared to a wavelength. A sharply defined or *specularly reflected* beam emerges. In contrast, the *diffuse reflection* of Fig. 4.10(b) occurs when the surface is relatively rough. Despite this, the law of reflection holds exactly over any region which is small enough so as to be considered smooth.

Let $\hat{u}_n$ be a unit vector normal to the interface pointing in the direction from the incident to the transmitting medium (Fig. 4.11). It can be shown (as you will hopefully prove in Problem 4.10) that the first and third basic laws may be combined in the form of a *vector refraction equation*

$$n_i(\hat{k}_i \times \hat{u}_n) = n_t(\hat{k}_t \times \hat{u}_n) \qquad (4.7)$$

or alternatively

$$n_t\hat{k}_t - n_i\hat{k}_i = (n_t \cos \theta_t - n_i \cos \theta_i)\hat{u}_n. \qquad (4.8)$$

It is a bit premature at this point to discuss in detail the processes which on an atomic scale result specifically in the laws of reflection and refraction. As we shall see, scattered wavelets emitted by atoms in the material superimpose, generating a reflected and refracted beam only in the directions of $\hat{k}_r$ and $\hat{k}_t$. Reflection is obviously a surface effect, in fact, it involves only those atoms in a layer about $\lambda/2$ deep. The response is very much like a transmitting antenna array where the spacing (in this case $\sim 10^{-10}$ m) between sources is less than λ (see Section 10.1.3). In contrast, if the incident wavelength is reduced below that of light (~ 500 nm) to the x-ray region ($\sim 10^{-10}$ m) which is of the order of the atomic spacing, there will generally no longer be flux density maxima simply in the directions of θ_r and θ_t as in the case of light. Similarly, if we use some sort of arrangement where the scattering centers have a separation of the order of λ for light (e.g. a diffraction grating), we can observe higher-order reflected and transmitted beams in addition to those (zeroth-order ones) considered above (see Fig. 10.36).

4.2.4 Fermat's Principle

The laws of reflection and refraction and indeed the manner in which light propagates in general, can be viewed from yet another entirely different and intriguing perspective afforded us by something called *Fermat's principle*. The ideas which will unfold presently have had a tremendous influence on the development of physical thought in and beyond the study

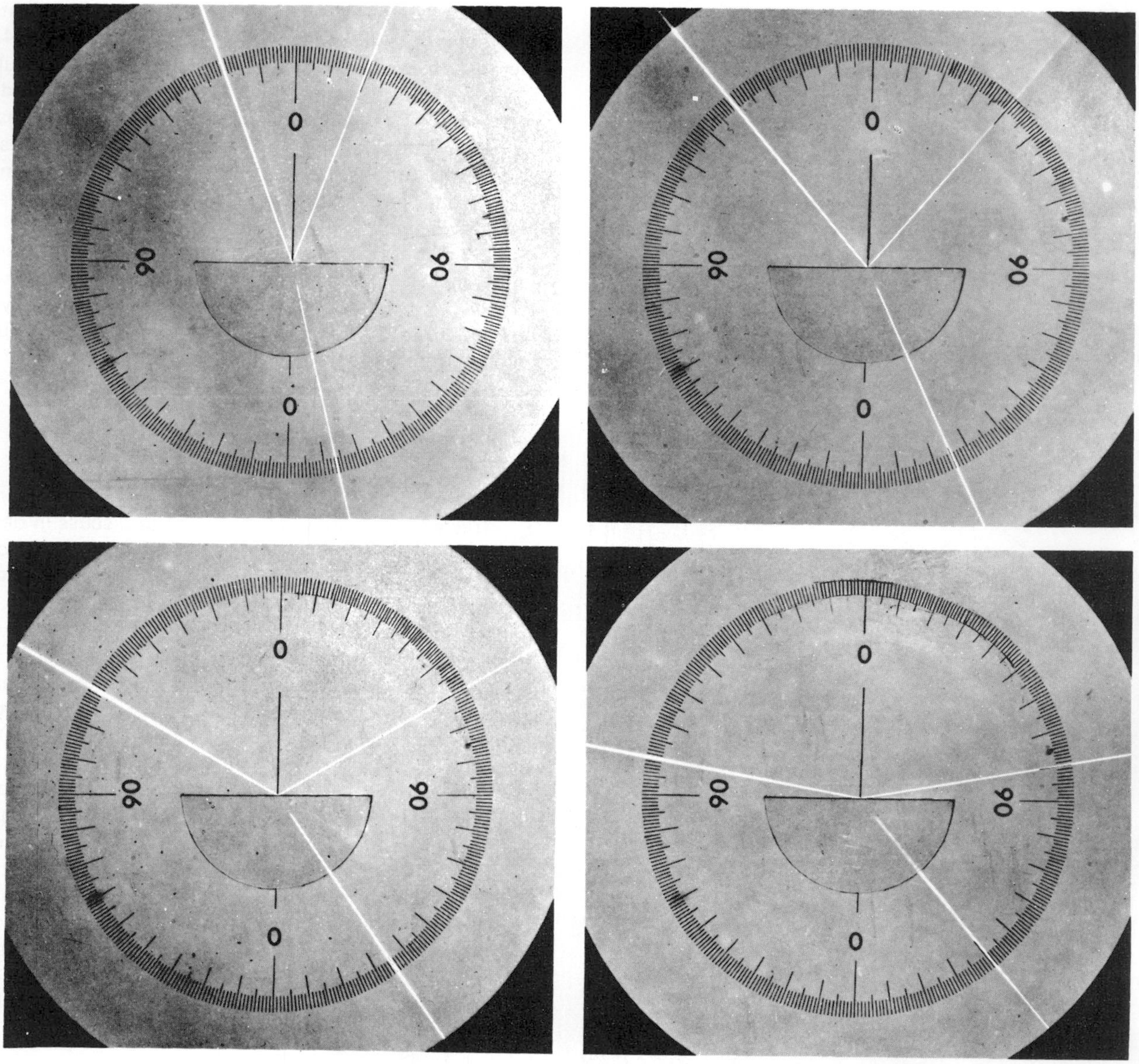

Fig. 4.9 Refraction at various angles of incidence. [Photos courtesy *PSSC College Physics*, D. C. Heath & Co., 1968.]

of classical optics. Apart from its implications in quantum optics (Section 13.6, p. 453) Fermat's principle provides us with an insightful and highly useful way of appreciating and anticipating the behavior of light.

Hero of Alexandria, who lived some time between 150 B.C. and 250 A.D., was the first to set forth what has

since become known as a *variational principle*. In his formulation of the law of reflection, he asserted that *the path actually taken by light in going from some point S to a point P via a reflecting surface was the shortest possible one*. This can be seen rather easily in Fig. 4.12 which depicts a point source *S* emitting a number of rays which are then "reflected"

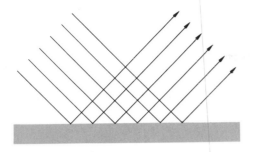

Specular

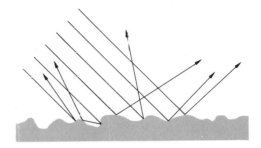

Diffuse

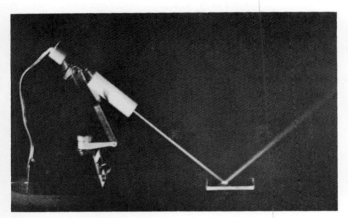

Fig. 4.10 (a) Specular reflection.

Fig. 4.10 (b) Diffuse reflection. [Photos courtesy *PSSC College Physics*, D. C. Heath & Co., 1968.]

toward P. Of course only one of these paths will have any physical reality. If we simply draw the rays as if they emanated from S' (the image of S), none of the distances to P will have been altered, i.e. $SAP = S'AP$, $SBP = S'BP$, etc. But obviously the straight-line path $S'BP$, which corresponds to $\theta_i = \theta_r$, is the shortest possible one. The same kind of reasoning (Problem 4.13) makes it evident that points S, B, and P must lie in what has previously been defined as the plane of incidence. For over fifteen hundred years Hero's curious observation stood alone until in 1657 Fermat propounded his celebrated *principle of least time* which encompassed both reflection and refraction. Quite obviously a beam of light traversing an interface does not take a straight line or *minimum spatial path* between a point in the incident medium and one in the transmitting medium. Fermat consequently reformulated Hero's statement to read: *the actual path between two points taken by a beam of light is the one which is traversed in the least time.* As we shall see in a little while even this form of the statement is some-

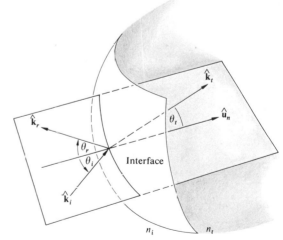

Fig. 4.11 The ray geometry.

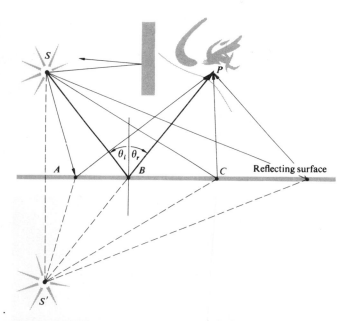

Fig. 4.12 Minimum path from the source S to the observer's eye at P.

what incomplete and a wee bit erroneous at that. For the moment then let us embrace it, but not passionately.

As an example of the application of the principle to the case of refraction refer to Fig. 4.13 where we minimize t, the transit time from S to P, with respect to the variable x. In other words, changing x shifts point O thereby changing the ray from S to P. The smallest transit time will then presumably coincide with the actual path. Hence

$$t = \frac{SO}{v_i} + \frac{OP}{v_t}$$

or

$$t = \frac{(h^2 + x^2)^{1/2}}{v_i} + \frac{[b^2 + (a - x)^2]^{1/2}}{v_t}.$$

To minimize $t(x)$ with respect to variations in x we set $dt/dx = 0$, that is

$$\frac{dt}{dx} = \frac{x}{v_i(h^2 + x^2)^{1/2}} + \frac{-(a - x)}{v_t[b^2 + (a - x)^2]^{1/2}} = 0.$$

Using the diagram this can be rewritten as

$$\frac{\sin \theta_i}{v_i} = \frac{\sin \theta_t}{v_t},$$

which is of course no less than Snell's law (4.4). Thus if a beam of light is to advance from S to P in the least possible time, it must comply with the empirical law of refraction.

Suppose that we have a stratified material composed of m layers each having a different index of refraction as in Fig. 4.14. The transit time from S to P will then be

$$t = \frac{s_1}{v_1} + \frac{s_2}{v_2} + \cdots + \frac{s_m}{v_m}$$

or

$$t = \sum_{i=1}^{m} s_i/v_i,$$

where s_i and v_i are the path length and speed associated with the ith contribution. Thus

$$t = \frac{1}{c} \sum_{i=1}^{m} n_i s_i \qquad (4.9)$$

in which the summation is known as the *optical path length* (O.P.L.) traversed by the ray. This is in contrast to the spatial path length $\Sigma_{i=1}^{m} s_i$. Clearly, for an inhomogeneous medium where n is a function of position, the summation must be changed to an integral.

$$(O.P.L.) = \int_{S}^{P} n(s) \, ds. \qquad (4.10)$$

Inasmuch as $t = (O.P.L.)/c$ we can restate Fermat's principle thus: *light, in going from points S to P, traverses the route*

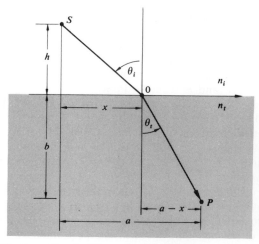

Fig. 4.13 Fermat's principle applied to refraction.

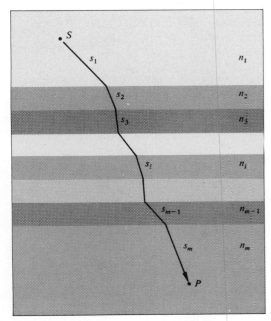

Fig. 4.14 A ray propagating through a layered material.

Fermat's principle in its modern form reads: *A light ray in going from point S to point P must traverse an optical path length which is stationary with respect to variations of that path.* In other words, the O.P.L. for the true trajectory will equal, to a first approximation, that O.P.L. of paths immediately adjacent to it.* And so there will be many curves neighboring the actual one which would take very nearly the same time for the light to traverse. This latter point makes it possible to begin to understand how light manages to be so clever in its meanderings. Suppose that we have a beam of light advancing through a homogeneous isotropic medium so that a ray passes from points S to P. Atoms within the material are driven by the incident disturbance and they reradiate in all directions. Quite generally, wavelets originating in the immediate vicinity of a stationary path will arrive at P by routes which differ only slightly and will therefore reinforce each other (see Section 7.1). Wavelets taking other

---

* The first derivative of the O.P.L. vanishes in its Taylor-series expansion since the path is stationary.

having the smallest optical path length. Accordingly, when light rays from the sun pass through the inhomogeneous atmosphere of the earth, as shown in Fig. 4.15(a), they bend so as to traverse the lower, denser regions as abruptly as possible, thus minimizing the O.P.L. Ergo, one can still see the sun after it has actually passed below the horizon. In the same way, a road viewed at a glancing angle, as in Fig. 4.15(b), will appear to reflect the environs as if it were covered with a sheet of water. The air near the roadway will be warmer and less dense than that farther above it. Rays will bend downward taking the shortest optical path and in so doing they will seem to an observer as if they had reflected from a mirrored surface. The effect is particularly easy to see on long modern highways. The only requirement is that you look at the road at near glancing incidence because the rays bend very gradually.

The original statement of Fermat's *principle of least time* has some serious failings and is, as we shall see, in need of alteration. To that end, recall that if we have a function, say $f(x)$, we can determine the specific value of the variable x which causes $f(x)$ to have a *stationary* value by setting $df/dx = 0$ and solving for x. By a stationary value we mean one where the slope of $f(x)$ versus x is zero or equivalently where the function has a maximum ⌢ , minimum ⌣ or a point of inflection having a horizontal tangent ⤳

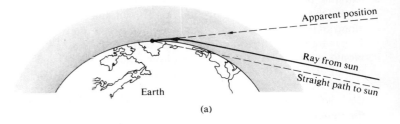

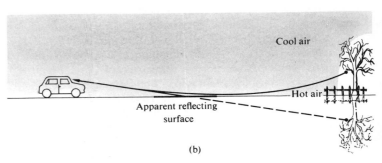

Fig. 4.15 The bending of rays through inhomogeneous media.

paths will arrive at P out of phase and therefore tend to cancel each other. That being the case, energy will effectively propagate along that ray from S to P which satisfies Fermat's principle.

To show that the O.P.L. for a ray need not always be a minimum, examine Fig. 4.16 which depicts a segment of a hollow three-dimensional ellipsoidal mirror. If the source S and the observer P are at the foci of the ellipsoid, then by definition the length SQP will be constant regardless of where on the perimeter Q happens to be. It is also a geometrical property of the ellipse that $\theta_i = \theta_r$ for any location of Q. All optical paths from S to P via a reflection are therefore precisely equal—none is a minimum, and the O.P.L. is clearly stationary with respect to variations. Rays leaving S and striking the mirror will arrive at the focus P. From another viewpoint we can say that radiant energy emitted by S will be scattered by electrons in the mirrored surface such that the wavelets will substantially reinforce each other only at P where they have traveled the same distance and have the same phase. In any case, if there were a plane mirror tangent to the ellipse at Q the exact same path SQP traversed by a ray would then be a relative minimum. At the other extreme, if the mirrored surface conformed to a curve lying within the ellipse, like the dashed one shown, that same ray along SQP would now negotiate a relative maximum O.P.L. This is true even though other unused paths (where $\theta_i \neq \theta_r$) would actually be shorter (i.e. apart from inadmissible curved paths). Thus in all cases the rays travel a stationary O.P.L. in accord with the reformulated Fermat's principle. Note that since the principle speaks only about the path and not the direction along it, a ray going from P to S will trace out the same route as one from S to P. This is the very useful *principle of reversibility*.

Fermat's achievement stimulated a great deal of effort to supersede Newton's laws of mechanics with a similar variational formulation. The work of many men, notably Pierre de Maupertuis (1698–1759) and Leonhard Euler, finally led to the mechanics of Joseph Louis Lagrange (1736–1813) and hence to the *principle of least action* formulated by William Rowan Hamilton (1805–65). The striking similarity between the principles of Fermat and Hamilton played an important role in Schrödinger's development of quantum mechanics. In 1942 Richard Phillips Feynman (b. 1918) showed that quantum mechanics can be fashioned in an alternative way using a variational approach. And so the continuing evolution of variational principles brings us back to optics via the modern formalism of quantum optics (see Chapter 13).

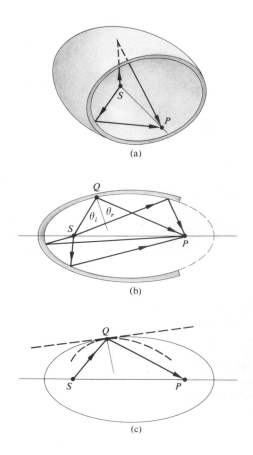

(a)

(b)

(c)

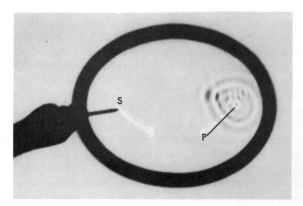

Fig. 4.16 Reflection off an ellipsoidal surface. Observe the reflection of waves using a frying pan filled with water. Even though these are usually circular it's well worth playing with. [Photo courtesy *PSSC College Physics*, D. C. Heath & Co., 1968.]

Fermat's principle is not so much a computational device as it is a concise way of thinking about the propagation of light. It is a statement about the grand scheme of things without any concern for the contributing mechanisms and as such it will yield insights under a myriad of varying circumstances.

4.3 THE ELECTROMAGNETIC APPROACH

Thus far we have been able to deduce the laws of reflection and refraction using three different approaches: *Huygens' principle*, the *theorem of Malus and Dupin* and *Fermat's principle*. Each of these in turn yields a distinctive and valuable point of view of its own. Yet another and even more powerful approach is provided by the electromagnetic theory of light. Unlike the previous techniques which say nothing about the incident, reflected and transmitted radiant flux densities (i.e. I_i, I_r, I_t respectively), the electromagnetic theory treats these within the framework of a far more complete description.

The body of information which forms the subject of optics has accrued over many centuries and as our knowledge of the physical universe becomes more extensive, the concomitant theoretical descriptions must become ever more encompassing. This, quite generally, brings with it an increased complexity. And so, rather than using the formidable mathematical machinery of the quantum theory of light we will most often avail ourselves of the simpler insights of simpler times (e.g. Huygens' and Fermat's principles etc.). Thus even though we are now going to develop still another and more extensive description of reflection and refraction, we will certainly not put aside those earlier methods. In fact, throughout this study we shall use the simplest technique which can yield sufficiently accurate results for our particular purposes.

4.3.1 Waves at an Interface

Suppose that the incident monochromatic light wave is planar so that it has the form

$$\mathbf{E}_i = \mathbf{E}_{0i} \exp [i(\mathbf{k}_i \cdot \mathbf{r} - \omega_i t)] \tag{4.11}$$

or more simply

$$\mathbf{E}_i = \mathbf{E}_{0i} \cos (\mathbf{k}_i \cdot \mathbf{r} - \omega_i t). \tag{4.12}$$

Let's assume that $\mathbf{E}_{0i}$ is constant in time, i.e. the wave is linearly or plane polarized. We'll find in Chapter 8 that any

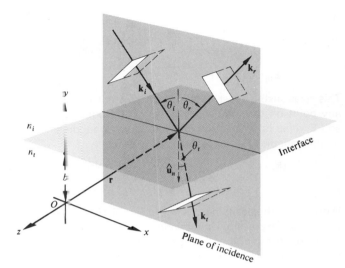

Fig. 4.17 Plane waves incident on the boundary between two dielectric media.

form of light can be represented by two orthogonal linearly polarized waves so that this doesn't actually represent a restriction. Note that just as the origin in time, $t = 0$, is arbitrary so too is the origin O in space where $\mathbf{r} = 0$. Thus making no assumptions about their directions, frequencies, wavelengths, phases or amplitudes, we can write the reflected and transmitted waves as

$$\mathbf{E}_r = \mathbf{E}_{0r} \cos (\mathbf{k}_r \cdot \mathbf{r} - \omega_r t + \varepsilon_r) \tag{4.13}$$

and

$$\mathbf{E}_t = \mathbf{E}_{0t} \cos (\mathbf{k}_t \cdot \mathbf{r} - \omega_t t + \varepsilon_t). \tag{4.14}$$

Here ε_r and ε_t are *phase constants* relative to $\mathbf{E}_i$ which are introduced because the position of the origin is not unique. Figure 4.17 depicts the waves in the vicinity of the planar interface between two dielectric media of indices n_i and n_t.

The laws of electromagnetic theory (Section 3.1) lead to certain requirements which must be met by the fields and these are referred to as the boundary conditions. Specifically, one of these is that the component of the electric field intensity $\mathbf{E}$ which is tangent to the interface must be continuous across it (the same is true for $\mathbf{H}$). In other words, the total tangential component of $\mathbf{E}$ on one side of the surface must equal that on the other (Problem 4.18). Thus since $\hat{\mathbf{u}}_n$ is the unit vector normal to the interface

$$\hat{\mathbf{u}}_r \times \mathbf{E}_i + \hat{\mathbf{u}}_n \times \mathbf{E}_r = \hat{\mathbf{u}}_n \times \mathbf{E}_t \tag{4.15}$$

or

$$\hat{\mathbf{u}}_n \times \mathbf{E}_{0i} \cos(\mathbf{k}_i \cdot \mathbf{r} - \omega_i t) + \hat{\mathbf{u}}_n \times \mathbf{E}_{0r} \cos(\mathbf{k}_r \cdot \mathbf{r} - \omega_r t + \varepsilon_r)$$

$$= \hat{\mathbf{u}}_n \times \mathbf{E}_{0t} \cos(\mathbf{k}_t \cdot \mathbf{r} - \omega_t t + \varepsilon_t). \tag{4.16}$$

This relationship must obtain at any instant in time and at any point on the interface ($y = b$). Consequently, $\mathbf{E}_i$, $\mathbf{E}_r$, and $\mathbf{E}_t$ must have precisely the same functional dependence on the variables t and $\mathbf{r}$ which means that

$$(\mathbf{k}_i \cdot \mathbf{r} - \omega_i t)|_{y=b} = (\mathbf{k}_r \cdot \mathbf{r} - \omega_r t + \varepsilon_r)|_{y=b}$$
$$= (\mathbf{k}_t \cdot \mathbf{r} - \omega_t t + \varepsilon_t)|_{y=b}. \tag{4.17}$$

With this as the case, the cosines in Eq. (4.16) would cancel leaving an expression independent of t and $\mathbf{r}$ as indeed it must be. Inasmuch as this has to be true for all values of time, the coefficients of t must be equal, to wit

$$\omega_i = \omega_r = \omega_t. \tag{4.18}$$

Recall that the electrons within the media are undergoing (linear) forced vibrations at the frequency of the incident wave. Clearly whatever light is scattered has that same frequency. Furthermore

$$(\mathbf{k}_i \cdot \mathbf{r})|_{y=b} = (\mathbf{k}_r \cdot \mathbf{r} + \varepsilon_r)|_{y=b} = (\mathbf{k}_t \cdot \mathbf{r} + \varepsilon_t)|_{y=b}, \tag{4.19}$$

wherein $\mathbf{r}$ terminates on the interface. The values of ε_r and ε_t correspond to a given position of O and thus they allow the relation to be valid regardless of that location. (For example, the origin might be chosen such that $\mathbf{r}$ was perpendicular to $\mathbf{k}_i$ but not $\mathbf{k}_r$ or $\mathbf{k}_t$.) From the first two terms we obtain

$$[(\mathbf{k}_i - \mathbf{k}_r) \cdot \mathbf{r}]_{y=b} = \varepsilon_r. \tag{4.20}$$

Recalling Eq. (2.42), this expression simply says that the endpoint of $\mathbf{r}$ sweeps out a plane (which is of course the interface) perpendicular to the vector $(\mathbf{k}_i - \mathbf{k}_r)$. Phrasing it slightly differently, $(\mathbf{k}_i - \mathbf{k}_r)$ is parallel to $\hat{\mathbf{u}}_n$. Notice however that since the incident and reflected waves are in the same medium $k_i = k_r$. From the fact that $(\mathbf{k}_i - \mathbf{k}_r)$ has no component in the plane of the interface, i.e. $\hat{\mathbf{u}}_n \times (\mathbf{k}_i - \mathbf{k}_r) = 0$, we conclude that

$$k_i \sin \theta_i = k_r \sin \theta_r$$

and hence we have the law of reflection, that is

$$\theta_i = \theta_r.$$

Furthermore since $(\mathbf{k}_i - \mathbf{k}_r)$ is parallel to $\hat{\mathbf{u}}_n$ all three vectors $\mathbf{k}_i$, $\mathbf{k}_r$, and $\hat{\mathbf{u}}_n$ are in the same plane, the plane of incidence.

Again, from Eq. (4.19) we obtain

$$[(\mathbf{k}_i - \mathbf{k}_t) \cdot \mathbf{r}]_{y=b} = \varepsilon_t \tag{4.21}$$

and therefore $(\mathbf{k}_i - \mathbf{k}_t)$ is also normal to the interface. Thus $\mathbf{k}_i$, $\mathbf{k}_r$, $\mathbf{k}_t$, and $\hat{\mathbf{u}}_n$ are all coplanar. As before, the tangential components of $\mathbf{k}_i$ and $\mathbf{k}_t$ must be equal and consequently

$$k_i \sin \theta_i = k_t \sin \theta_t. \tag{4.22}$$

But because $\omega_i = \omega_t$ we can multiply both sides by c/ω_i to get

$$n_i \sin \theta_i = n_t \sin \theta_t$$

which is, of course, Snell's law. Finally, observe that had we chosen the origin O to be in the interface it is evident from Eqs. (4.20) and (4.21) that ε_r and ε_t would both have been zero. That arrangement, although not as instructive, is certainly simpler and consequently we'll use it from here on.

4.3.2 Derivation of the Fresnel Equations

We have just found the relationship which exists amongst the phases of $\mathbf{E}_i(\mathbf{r}, t)$, $\mathbf{E}_r(\mathbf{r}, t)$, and $\mathbf{E}_t(\mathbf{r}, t)$ at the boundary. There is still an interdependence shared by the amplitudes $\mathbf{E}_{0i}$, $\mathbf{E}_{0r}$, and $\mathbf{E}_{0t}$ which can now be evaluated. To that end suppose that a plane monochromatic wave is incident on the planar surface separating two isotropic media. Whatever the polarization of the wave we shall resolve its $\mathbf{E}$- and $\mathbf{B}$-fields into components parallel and perpendicular to the plane of incidence and treat these constituents separately.

Case 1. $\mathbf{E}$ *Perpendicular to the plane of incidence.* We now assume that $\mathbf{E}$ is perpendicular to the plane of incidence and that $\mathbf{B}$ is parallel to it (Fig. 4.18). As you recall $E = vB$ so that

$$\hat{\mathbf{k}} \times \mathbf{E} = v\mathbf{B} \tag{4.23}$$

and of course

$$\hat{\mathbf{k}} \cdot \mathbf{E} = 0, \tag{4.24}$$

i.e. $\mathbf{E}$, $\mathbf{B}$ and the unit propagation vector $\hat{\mathbf{k}}$ form a right-handed system. Again making use of the continuity of the tangential components of the $\mathbf{E}$-field, we have at the boundary at any time and any point

$$\mathbf{E}_{0i} + \mathbf{E}_{0r} = \mathbf{E}_{0t} \tag{4.25}$$

where the cosines cancel. We should mention parenthetically that the field vectors as shown really ought to be envisioned at $y = 0$ (i.e. at the surface) from which they have been displaced for the sake of clarity. Note too that while $\mathbf{E}_r$

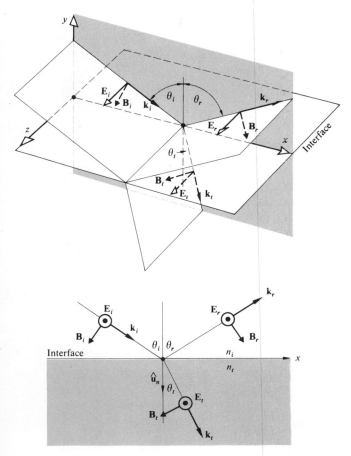

Fig. 4.18 An incoming wave whose **E**-field is normal to the plane of incidence.

and $\mathbf{E}_t$ must be normal to the plane of incidence by symmetry, *we are guessing that they point outward* at the interface when $\mathbf{E}_i$ does. The directions of the **B**-fields then follow from Eq. (4.23).

We will need to invoke another of the boundary conditions in order to get one more equation. The presence of material substances which become electrically polarized by the wave has a definite effect on the field configuration. Thus while the tangential component of **E** is continuous across the boundary, its normal component is not. Instead the normal component of the product $\epsilon\mathbf{E}$ is the same on either side of the interface. Similarly, the normal component of **B** is continuous as is the tangential component of $\mu^{-1}\mathbf{B}$. Here the effect of the two media appears via their permeabili-

ties μ_i and μ_t. This latter boundary condition will be the simplest to use, particularly as applied to reflection from the surface of a conductor.* Thus the continuity of the tangential component of $\mathbf{B}/\mu$ requires that

$$-\frac{\mathbf{B}_i}{\mu_i}\cos\theta_i + \frac{\mathbf{B}_r}{\mu_i}\cos\theta_r = -\frac{\mathbf{B}_t}{\mu_t}\cos\theta_t, \qquad (4.26)$$

where the left and right sides are the total magnitudes of $\mathbf{B}/\mu$ parallel to the interface in the incident and transmitting media, respectively. The positive direction is that of increasing x so that the components of $\mathbf{B}_i$ and $\mathbf{B}_t$ appear with minus signs. From Eq. (4.23) we have

$$\mathbf{B}_i = E_i/v_i \qquad (4.27)$$

$$\mathbf{B}_r = E_r/v_r \qquad (4.28)$$

and

$$\mathbf{B}_t = E_t/v_t. \qquad (4.29)$$

Thus since $v_i = v_r$ and $\theta_i = \theta_r$ Eq. (4.26) can be written as

$$\frac{1}{\mu_i v_i}(E_i - E_r)\cos\theta_i = \frac{1}{\mu_t v_t}E_t\cos\theta_t. \qquad (4.30)$$

Making use of Eqs. (4.12), (4.13) and (4.14) and remembering that the cosines therein equal one at $y = 0$ we obtain

$$\frac{n_i}{\mu_i}(E_{0i} - E_{0r})\cos\theta_i = \frac{n_t}{\mu_t}E_t\cos\theta_t. \qquad (4.31)$$

Combining this with Eq. (4.25) yields

$$\left(\frac{E_{0r}}{E_{0i}}\right)_\perp = \frac{\dfrac{n_i}{\mu_i}\cos\theta_i - \dfrac{n_t}{\mu_t}\cos\theta_t}{\dfrac{n_i}{\mu_i}\cos\theta_i + \dfrac{n_t}{\mu_t}\cos\theta_t} \qquad (4.32)$$

and

$$\left(\frac{E_{0t}}{E_{0i}}\right)_\perp = \frac{2\dfrac{n_i}{\mu_i}\cos\theta_i}{\dfrac{n_i}{\mu_i}\cos\theta_i + \dfrac{n_t}{\mu_t}\cos\theta_t}. \qquad (4.33)$$

---

* In keeping with our intent to use only the E- and B-fields, at least in the early part of this exposition, we have avoided the usual statements in terms of H where

$$\mathbf{H} = \mu^{-1}\mathbf{B} \qquad [A1.14]$$

The $\perp$ subscript serves as a reminder that we are dealing with the case where **E** is perpendicular to the plane of incidence. These two expressions, *which are completely general statements applying to any linear, isotropic, homogeneous media*, are two of what are called the *Fresnel equations*. Quite often one deals with dielectrics for which $\mu_i \approx \mu_t \approx \mu_0$; consequently the most common form of these equations is simply

$$r_\perp \equiv \left(\frac{E_{0r}}{E_{0i}}\right)_\perp = \frac{n_i \cos\theta_i - n_t \cos\theta_t}{n_i \cos\theta_i + n_t \cos\theta_t} \qquad (4.34)$$

and

$$t_\perp \equiv \left(\frac{E_{0t}}{E_{0i}}\right)_\perp = \frac{2n_i \cos\theta_i}{n_i \cos\theta_i + n_t \cos\theta_t}. \qquad (4.35)$$

Here $r_\perp$ denotes the *amplitude reflection coefficient* while $t_\perp$ is the *amplitude transmission coefficient*.

Case 2. **E** *Parallel to the plane of incidence*. A similar pair of equations can be derived when the incoming **E**-field lies in the plane of incidence as shown in Fig. 4.19. Continuity of the tangential components of **E** on either side of the boundary leads to

$$E_{0i} \cos\theta_i - E_{0r} \cos\theta_r = E_{0t} \cos\theta_t. \qquad (4.36)$$

In very much the same way as before, continuity of the tangential components of $\mathbf{B}/\mu$ yields

$$\frac{1}{\mu_i v_i} E_{0i} + \frac{1}{\mu_r v_r} E_{0r} = \frac{1}{\mu_t v_t} E_{0t}. \qquad (4.37)$$

Using the fact that $\mu_i = \mu_r$ and $\theta_i = \theta_r$ these formulas can be combined to give us two more of the *Fresnel equations*:

$$r_\parallel \equiv \left(\frac{E_{0r}}{E_{0i}}\right)_\parallel = \frac{\dfrac{n_t}{\mu_t}\cos\theta_i - \dfrac{n_i}{\mu_i}\cos\theta_t}{\dfrac{n_i}{\mu_i}\cos\theta_t + \dfrac{n_t}{\mu_t}\cos\theta_i} \qquad (4.38)$$

and

$$t_\parallel \equiv \left(\frac{E_{0t}}{E_{0i}}\right)_\parallel = \frac{2\dfrac{n_i}{\mu_i}\cos\theta_i}{\dfrac{n_i}{\mu_i}\cos\theta_t + \dfrac{n_t}{\mu_t}\cos\theta_i}. \qquad (4.39)$$

When both media forming the interface are dielectrics the amplitude coefficients become

$$r_\parallel = \frac{n_t \cos\theta_i - n_i \cos\theta_t}{n_i \cos\theta_t + n_t \cos\theta_i} \qquad (4.40)$$

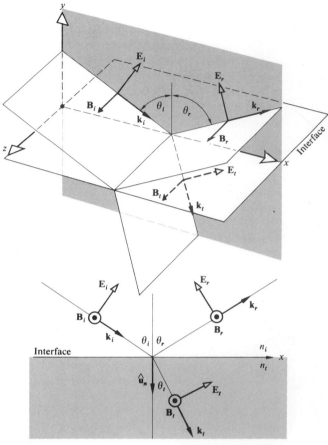

Fig. 4.19 An incoming wave whose **E**-field is in the plane of incidence.

and

$$t_\parallel = \frac{2n_i \cos\theta_i}{n_i \cos\theta_t + n_t \cos\theta_i}. \qquad (4.41)$$

One further notational simplification can be made by availing ourselves of Snell's law whereupon the Fresnel equations for dielectric media become (Problem 4.19)

$$r_\perp = -\frac{\sin(\theta_i - \theta_t)}{\sin(\theta_i + \theta_t)} \qquad (4.42)$$

$$r_\parallel = +\frac{\tan(\theta_i - \theta_t)}{\tan(\theta_i + \theta_t)} \qquad (4.43)$$

$$t_\perp = +\frac{2\sin\theta_t \cos\theta_i}{\sin(\theta_i + \theta_t)} \qquad (4.44)$$

$$t_\parallel = + \frac{2 \sin \theta_t \cos \theta_i}{\sin (\theta_i + \theta_t) \cos (\theta_i - \theta_t)}. \qquad (4.45)$$

A note of caution must be introduced before we move on to examine the considerable significance of the preceding calculation. Bear in mind that the directions (or more precisely the phases) of the fields in Figs. 4.18 and 4.19 were selected rather arbitrarily. For example in Fig. 4.18 we certainly could have assumed that $\mathbf{E}_r$ pointed inward, whereupon $\mathbf{B}_r$ would have had to be reversed as well. Had we done that, the sign of $r_\perp$ would have turned out to be positive, leaving the other amplitude coefficients unchanged. The signs appearing in Eqs. (4.42) through (4.45), in this case +, except for the first, correspond to the particular set of field directions selected. The minus, as we will see, just means that we didn't guess correctly concerning $\mathbf{E}_r$ in Fig. 4.18. Nonetheless be aware that the literature is not standardized and every possible sign variation can be found bearing the title of *Fresnel's equations*—to avoid confusion *they must be related to the specific field directions from which they were derived.*

4.3.3 Interpretation of the Fresnel equations

This section is devoted to an examination of the physical implications of the Fresnel equations. In particular we are interested in determining the fractional amplitudes and flux densities which are reflected and refracted. In addition we shall be concerned with any possible phase shifts which might be incurred in the process.

i) Amplitude Coefficients

Let's now briefly examine the form of the amplitude coefficients over the entire range of θ_i-values. At nearly normal incidence ($\theta_i \approx 0$) the tangents in Eq. (4.43) are essentially equal to sines in which case

$$[r_\parallel]_{\theta_i=0} = [-r_\perp]_{\theta_i=0} = \left[\frac{\sin (\theta_i - \theta_t)}{\sin (\theta_i + \theta_t)}\right]_{\theta_i=0}.$$

We will come back to the physical significance of the minus sign presently. After expanding out the sines and using Snell's law this expression becomes

$$[r_\parallel]_{\theta_i=0} = [-r_\perp]_{\theta_i=0} = \left[\frac{n_t \cos \theta_i - n_i \cos \theta_t}{n_t \cos \theta_i + n_i \cos \theta_t}\right]_{\theta_i=0}. \quad (4.46)$$

which follows as well from Eqs. (4.34) and (4.40). In the limit as θ_i goes to 0, $\cos \theta_i$ and $\cos \theta_t$ both approach one and consequently

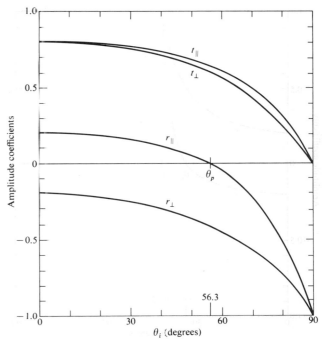

Fig. 4.20 The amplitude coefficients of reflection and transmission as a function of incident angle. These correspond to external reflection $r_t > n_i$ at an air–glass interface ($n_{ti} = 1.5$).

$$[r_\parallel]_{\theta_i=0} = [-r_\perp]_{\theta_i=0} = \frac{n_t - n_i}{n_t + n_i}. \qquad (4.47)$$

Thus, for example, at an air ($n_i = 1$) glass ($n_t = 1.5$) interface at near normal incidence the reflection coefficients equal ± 0.2.

When $n_t > n_i$ it follows from Snell's law that $\theta_i > \theta_t$ and so $r_\perp$ is negative for all values of θ_i (Fig. 4.20). In contrast, $r_\parallel$ starts out positive at $\theta_i = 0$ and decreases gradually until it equals zero when $(\theta_i + \theta_t) = 90°$ since $\tan \pi/2$ is infinite. The particular value of the incident angle for which this occurs is denoted by θ_p and is referred to as the *polarization angle* (see Section 8.6.1). As θ_i increases beyond θ_p, $r_\parallel$ becomes ever more negative until it reaches -1.0 at $90°$.

At normal incidence Eqs. (4.35) and (4.41) lead, rather straightforwardly, to

$$[t_\parallel]_{\theta_i=0} = [t_\perp]_{\theta_i=0} = \frac{2n_i}{n_i + n_t}. \qquad (4.48)$$

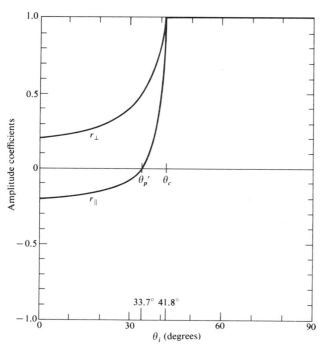

Fig. 4.21 The amplitude coefficients of reflection as a function of incident angle. These correspond to internal reflection $n_t < n_i$ at an air–glass interface ($n_{ti} = 1/1.5$).

It will be shown in Problem 4.21 that the expression

$$t_\perp + (-r_\perp) = 1 \qquad (4.49)$$

holds for all θ_i while

$$t_\parallel + r_\parallel = 1 \qquad (4.50)$$

is true only at normal incidence.

The foregoing discussion, for the most part, was restricted to the case of *external reflection*, i.e. $n_t > n_i$. The opposite situation of *internal reflection* in which the incident medium is the more dense ($n_i > n_t$) is most certainly of interest as well. In that instance $\theta_t > \theta_i$ and $r_\perp$, as described by Eq. (4.42), will always be positive. As shown in Fig. 4.21, $r_\perp$ increases from its initial value (4.47) at $\theta_i = 0$ reaching plus one at what is called the *critical angle*, θ_c. Specifically, θ_c is the special value of the incident angle for which $\theta_t = \pi/2$. Likewise, $r_\parallel$ starts off negatively (4.47) at $\theta_i = 0$ and thereafter increases going to plus one at $\theta_i = \theta_c$,

as is evident from the Fresnel equation (4.40). As before, $r_\parallel$ passes through zero at the *polarization angle* θ'_p. It is left for Problem 4.25 to show that the polarization angles θ'_p and θ_p for internal and external reflection at the interface between the same media are simply the complements of each other. We will return to internal reflection in Section 4.3.4 where it will be shown that $r_\perp$ and $r_\parallel$ are complex quantities for $\theta_i > \theta_c$.

ii) Phase Shifts

It should be evident from Eq. (4.42) that $r_\perp$ is negative regardless of θ_i when $n_t > n_i$. Yet we saw earlier that had we chosen $[\mathbf{E}_r]_\perp$ in Fig. 4.18 to be in the opposite direction, the first Fresnel equation (4.42) would have changed sign causing $r_\perp$ to become a positive quantity. Thus the sign of $r_\perp$ is associated with the relative directions of $[\mathbf{E}_{0i}]_\perp$ and $[\mathbf{E}_{0r}]_\perp$. Bear in mind that a reversal of $[\mathbf{E}_{0r}]_\perp$ is tantamount to introducing a phase shift, $\Delta\varphi_\perp$, of π radians into $[\mathbf{E}_r]_\perp$. Hence at the boundary $[\mathbf{E}_i]_\perp$ and $[\mathbf{E}_r]_\perp$ will be antiparallel and therefore π out of phase with each other as indicated by the negative value of $r_\perp$. When considering components normal to the plane of incidence there is no confusion as to whether two fields are in phase or π radians out of phase; if they are parallel they're in phase; if they're antiparallel they're π out of phase. In summary then, *the component of the electric field normal to the plane of incidence undergoes a phase shift of π radians upon reflection when the incident medium has a lower index than the transmitting medium.* Similarly $t_\perp$ and $t_\parallel$ are always positive and $\Delta\varphi = 0$. *Furthermore, when $n_i > n_t$ no phase shift in the normal component results on reflection, i.e. $\Delta\varphi_\perp = 0$ so long as $\theta_i < \theta_c$.*

Things get a bit less obvious when dealing with $[\mathbf{E}_i]_\parallel$ $[\mathbf{E}_r]_\parallel$ and $[\mathbf{E}_t]_\parallel$. It now becomes necessary to define more explicitly what is meant by *in phase* since the field vectors are coplanar but generally not colinear. The field directions were chosen in Figs. 4.18 and 4.19 such that if you looked down any one of the propagation vectors towards the direction from which the light was coming you would see $\mathbf{E}$, $\mathbf{B}$, and $\mathbf{k}$ to have the same relative orientation whether the ray was incident, reflected or transmitted. We can use this as the required condition in order for two $\mathbf{E}$-fields to be in phase. Equivalently, but more simply, *two fields in the incident plane are in phase if their y-components are parallel, and out of phase if antiparallel.* Notice that when a pair of $\mathbf{E}$-fields are out of phase so too are their associated $\mathbf{B}$-fields and vice versa. With this definition we need only look at the vectors normal to the plane of incidence, whether they be $\mathbf{E}$ or $\mathbf{B}$, to determine the relative phase of the accompanying fields in the

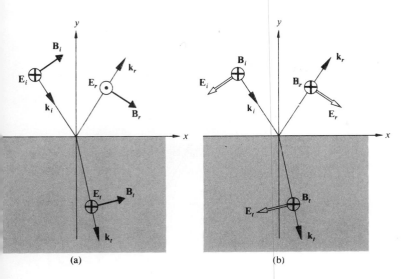

Fig. 4.22 Field orientations and phase shifts.

incident plane. Thus in Fig. 4.22(a) $\mathbf{E}_i$ and $\mathbf{E}_t$ are in phase as are $\mathbf{B}_i$ and $\mathbf{B}_t$ while $\mathbf{E}_i$ and $\mathbf{E}_r$ are out of phase along with $\mathbf{B}_i$ and $\mathbf{B}_r$. Similarly in Fig. 4.22(b) $\mathbf{E}_i$, $\mathbf{E}_r$, and $\mathbf{E}_t$ are in phase as are $\mathbf{B}_i$, $\mathbf{B}_r$, and $\mathbf{B}_t$.

Now, the amplitude reflection coefficient for the parallel component is given by

$$r_{\parallel} = \frac{n_t \cos\theta_i - n_i \cos\theta_t}{n_t \cos\theta_i + n_i \cos\theta_t}$$

which is positive ($\Delta\varphi_{\parallel} = 0$) as long as

$$n_t \cos\theta_i - n_i \cos\theta_t > 0$$

i.e. if

$$\sin\theta_i \cos\theta_i - \cos\theta_t \sin\theta_t > 0$$

or equivalently

$$\sin(\theta_i - \theta_t)\cos(\theta_i + \theta_t) > 0. \qquad (4.51)$$

This will be the case for $n_i < n_t$ if

$$(\theta_i + \theta_t) < \pi/2 \qquad (4.52)$$

and for $n_i > n_t$ when

$$(\theta_i + \theta_t) > \pi/2. \qquad (4.53)$$

And so, when $n_i < n_t$, $[\mathbf{E}_{0r}]_{\parallel}$ and $[\mathbf{E}_{0i}]_{\parallel}$ will be in phase ($\Delta\varphi_{\parallel} = 0$) up until $\theta_i = \theta_p$ and out of phase by π radians thereafter. The transition is not actually discontinuous since $[\mathbf{E}_{0r}]_{\parallel}$ goes to zero at θ_p. In contrast, for internal reflection $r_{\parallel}$ is negative up until θ_p' which means that $\Delta\varphi_{\parallel} = \pi$. From

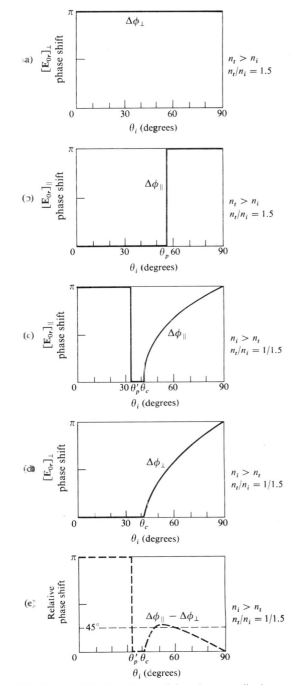

Fig. 4.23 Phase shifts for the parallel and perpendicular components of the **E**-field corresponding to internal and external reflection.

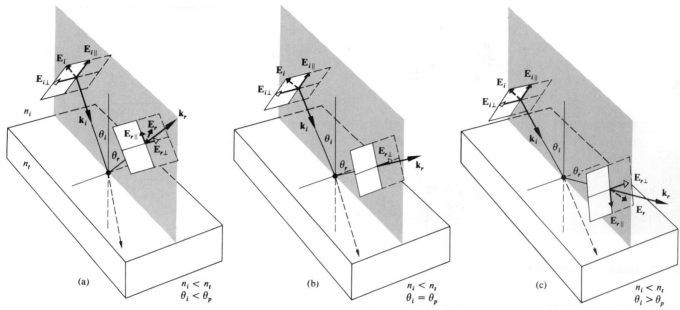

Fig. 4.24 The reflected **E**-field at various angles concomitant with external reflection.

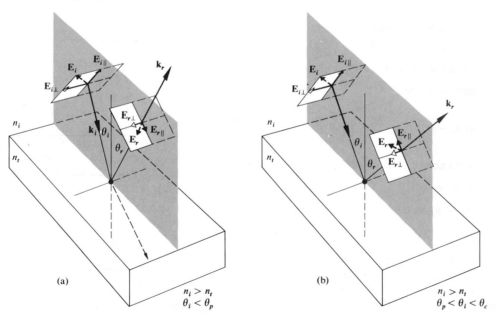

Fig. 4.25 The reflected **E**-field at various angles concomitant with internal reflection.

θ'_p to θ_c, $r_{||}$ is positive and $\Delta\varphi_{||} = 0$. Beyond θ_c, $r_{||}$ becomes complex and $\Delta\varphi_{||}$ gradually increases to π at $\theta_i = 90°$.

Figure 4.23 which summarizes these conclusions will be of continued use to us. The actual functional form of $\Delta\varphi_{||}$ and $\Delta\varphi_{\perp}$ for internal reflection in the region where $\theta_i > \theta_c$ can be found in the literature,* but, the curves depicted here will suffice for our purposes. Figure 4.23(e) is a plot of the relative phase shift between the parallel and perpendicular components, i.e. $\Delta\varphi_{||} - \Delta\varphi_{\perp}$. It is included here because we anticipate its usefulness later on, e.g. when we consider polarization effects. Finally, many of the essential features of this discussion can be illustrated rather unambiguously as in Figs. 4.24 and 4.25. There the amplitudes of the reflected vectors are in accord with those of Figs. 4.20 and 4.21 (recall that this is for an air–glass interface) while the phase shifts agree with those of Fig. 4.23.

One can actually verify many of these conclusions using the simplest experimental facilities, namely two linear polarizers, a piece of glass and a small source such as a flashlight or "high intensity" lamp. By placing one polarizer in front of the source (at 45° to the plane of incidence) you can easily duplicate the conditions of Fig. 4.24. For example, when $\theta_i = \theta_p$ [Fig. 4.24(b)] no light will pass through the second polarizer if its transmission axis is parallel to the plane of incidence. In comparison, at near glancing incidence the reflected beam will vanish when the axes of the two polarizers are almost normal to each other.

iii) Reflectance and Transmittance

Recall that the power per unit area crossing a surface in vacuum whose normal is parallel to **S**, the Poynting vector, is given by

$$\mathbf{S} = c^2\epsilon_0\mathbf{E} \times \mathbf{B}. \qquad [3.48]$$

Furthermore, the radiant flux density (W/m²) or irradiance is then

$$I = \langle S \rangle = \frac{c\epsilon_0}{2}E_0^2. \qquad [3.52]$$

This is the average energy per unit time crossing a unit area normal to **S** (in isotropic media **S** is parallel to **k**). In the case at hand (Fig. 4.26), let I_i, I_r, and I_t be the incident, reflected and transmitted flux densities respectively. Accordingly, the portion of the energy incident *normally on a unit area of the boundary* per second is $I_i \cos\theta_i$. Similarly, $I_r \cos\theta_r$ and $I_t \cos\theta_t$ are the energies per second leaving a

* Born and Wolf, *Principles of Optics*, p. 49.

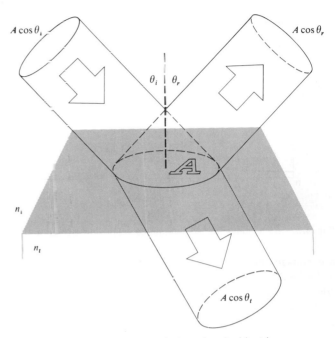

Fig. 4.26 Reflection and transmission of an incident beam.

unit area of the boundary normally on either side. The *reflectance R* is the ratio of the reflected over the incident flux (or power) i.e.

$$R \equiv \frac{I_r \cos\theta_r}{I_i \cos\theta_i} = \frac{I_r}{I_i}, \qquad (4.54)$$

while the *transmittance T* is the ratio of the transmitted over the incident flux and is given by

$$T \equiv \frac{I_t \cos\theta_t}{I_i \cos\theta_i}. \qquad (4.55)$$

The quotient I_r/I_i equals $(v_r\epsilon_r E_{0r}^2/2)/(v_i\epsilon_i E_{0i}^2/2)$ and since the incident and reflected waves are in the same medium $v_r = v_i$, $\epsilon_r = \epsilon_i$, and

$$R = \left(\frac{E_{0r}}{E_{0i}}\right)^2 = r^2. \qquad (4.56)$$

In the same way (assuming $\mu_i = \mu_t = \mu_0$)

$$T = \frac{n_t \cos\theta_t}{n_i \cos\theta_i}\left(\frac{E_{0t}}{E_{0i}}\right)^2 = \left(\frac{n_t \cos\theta_t}{n_i \cos\theta_i}\right)t^2, \qquad (4.57)$$

where use was made of the fact that $\mu_0\epsilon_t = 1/v_t^2$ and $\mu_0 v_t\epsilon_t = n_t/c$. Let's now write an expression representing

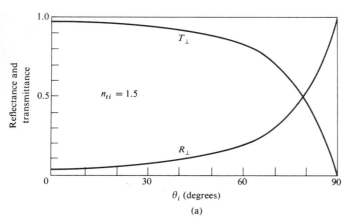

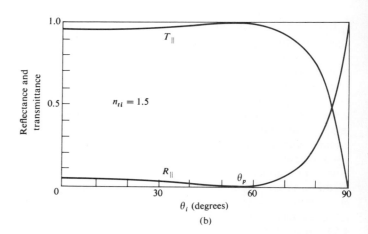

Fig. 4.27   Reflectance and transmittance versus incident angle.

the conservation of energy for the configuration depicted in Fig. 4.26. In other words, the total energy flowing into area A per unit time must equal the energy flowing outward from it per unit time;

$$I_i A \cos \theta_i = I_r A \cos \theta_r + I_t A \cos \theta_t. \qquad (4.58)$$

It is evident here as in Fig. 4.8 that the cross sectional area of the transmitted beam, $A \cos \theta_t$, is larger than that of the incident or reflected beams (which are equal). Multiplying both sides by c this expression becomes

$$n_i E_{0i}^2 \cos \theta_i = n_i E_{0r}^2 \cos \theta_i + n_t E_{0t}^2 \cos \theta_t$$

or

$$1 = \left(\frac{E_{0r}}{E_{0i}}\right)^2 + \left(\frac{n_t \cos \theta_t}{n_i \cos \theta_i}\right)\left(\frac{E_{0t}}{E_{0i}}\right)^2. \qquad (4.59)$$

But this is simply

$$R + T = 1 \qquad (4.60)$$

where there was no absorption. It is convenient to use the component forms, that is

$$R_\perp = r_\perp^2 \qquad (4.61)$$

$$R_\parallel = r_\parallel^2 \qquad (4.62)$$

$$T_\perp = \left(\frac{n_t \cos \theta_t}{n_i \cos \theta_i}\right) t_\perp^2 \qquad (4.63)$$

and

$$T_\parallel = \left(\frac{n_t \cos \theta_t}{n_i \cos \theta_i}\right) t_\parallel^2 \qquad (4.64)$$

which are illustrated in Fig. 4.27. Furthermore, it can be shown (Problem 4.30) that

$$R_\parallel + T_\parallel = 1 \qquad (4.65)$$

and

$$R_\perp + T_\perp = 1. \qquad (4.66)$$

One interesting feature of these curves is easily verified at least qualitatively, and that is that both $R_\parallel$ and $R_\perp$ approach one as $\theta_i \to 90°$. This implies that almost any fairly smooth dielectric surface will actually become mirror-like at glancing incidence. Try looking at a light source using this page as a boundary surface where $\theta_i \approx 90°$. You should be able to see a rather clear image of the source reflected in the paper.

When $\theta_i = 0$ the incident plane becomes undefined and any distinction between the parallel and perpendicular components of R and T vanishes. In this case Eqs. (4.61) through (4.64) along with (4.47) and (4.48) lead to

$$R = R_\parallel = R_\perp = \left(\frac{n_t - n_i}{n_t + n_i}\right)^2 \qquad (4.67)$$

and

$$T = T_\parallel = T_\perp = \frac{4 n_t n_i}{(n_t + n_i)^2}. \qquad (4.68)$$

Thus 4% of the light incident normally on an air–glass interface will be reflected back whether internally, $n_i > n_t$, or externally, $n_i < n_t$ (Problem 4.31). This will obviously become of great concern to anyone who is working with a complicated lens system which might have ten or twenty such air–glass boundaries. Indeed, if you look perpendicularly into a stack of about fifty microscope slides, (cover-glass slides are much thinner and easier to handle in large quantities) most of the light will be reflected. The stack will appear very much like a mirror (Fig. 4.28).

4.3.4 Total Internal Reflection

In the previous section it was evident that something rather interesting was happening in the case of internal reflection ($n_i > n_t$) when θ_i was equal to or greater than θ_c, the so-called *critical angle*. Let's now return to that situation for a somewhat closer look at things. Suppose that we have a source imbedded in an optically dense medium and we allow θ_i to gradually increase as indicated in Fig. 4.29. We know from the preceding section (Fig. 4.21) that $r_\parallel$ and $r_\perp$ increase with increasing θ_i and therefore $t_\parallel$ and $t_\perp$ both decrease. Moreover $\theta_t > \theta_i$ since

$$\sin \theta_i = \frac{n_t}{n_i} \sin \theta_t$$

and $n_i > n_t$ in which case $n_{ti} < 1$. Thus as θ_i becomes larger the transmitted ray gradually approaches tangency with the boundary and as it does so more and more of the available energy appears in the reflected beam. Finally when $\theta_t = 90°$, $\sin \theta_t = 1$ and

$$\sin \theta_c = n_{ti}. \qquad (4.69)$$

As was pointed out earlier, *the critical angle is that special value of θ_i for which $\theta_t = 90°$*. For incident angles greater than or equal to θ_c all of the incoming energy is reflected back into the incident medium in the process known as *total internal reflection*. It should be stressed that the transition

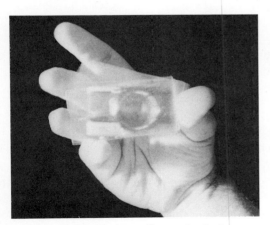

Fig. 4.28 Near normal reflection off a stack of microscope slides. You can see the image of the camera which took the picture. [Photo by E.H.]

from the conditions depicted in Fig. 4.29(a) to those of 4.29(d) takes place without any discontinuities. That is to say, as θ_i becomes larger, the reflected beam grows stronger and stronger while the transmitted beam grows weaker until the latter vanishes, at which point the former carries off all of the energy at $\theta_r = \theta_c$. It's an easy matter to actually observe the diminution of the transmitted beam as θ_i is made larger. Just place a piece of glass, e.g. a microscope slide, on a printed page. At $\theta_i \approx 0$, θ_t is roughly zero and the page as seen through the glass is bright and clear. But if you move your head allowing θ_t (the angle at which you view the interface) to increase, the region of the printed page covered by the glass will appear darker and darker indicating that T has indeed been markedly reduced.

The critical angle for our air–glass interface is roughly 42° (see Table 4.1). Consequently a ray incident normally on the left-hand face of either of the prisms in Fig. 4.30 will have a $\theta_i > 42°$ and therefore be totally internally reflected. This is obviously a convenient way to reflect nearly 100% of the incident light without having to worry about the deterioration which can occur with metallic surfaces.

If we presume that there is no transmitted wave it becomes impossible to satisfy the boundary conditions

Table 4.1 Critical angles.

n_{it}	θ_c (degrees)	θ_c (radians)	n_{it}	θ_c (degrees)	θ_c (radians)
1.30	50.2849	0.8776	1.50	41.8103	0.7297
1.31	49.7612	0.8685	1.51	41.4718	0.7238
1.32	49.2509	0.8596	1.52	41.1395	0.7180
1.33	48.7535	0.8509	1.53	40.8132	0.7123
1.34	48.2682	0.8424	1.54	40.4927	0.7067
1.35	47.7946	0.8342	1.55	40.1778	0.7012
1.36	47.3321	0.8261	1.56	39.8683	0.6958
1.37	46.8803	0.8182	1.57	39.5642	0.6905
1.38	46.4387	0.8105	1.58	39.2652	0.6853
1.39	46.0070	0.8030	1.59	38.9713	0.6802
1.40	45.5847	0.7956	1.60	38.6822	0.6751
1.41	45.1715	0.7884	1.61	38.3978	0.6702
1.42	44.7670	0.7813	1.62	38.1181	0.6653
1.43	44.3709	0.7744	1.63	37.8428	0.6605
1.44	43.9830	0.7676	1.64	37.5719	0.6558
1.45	43.6028	0.7610	1.65	37.3052	0.6511
1.46	43.2302	0.7545	1.66	37.0427	0.6465
1.47	42.8649	0.7481	1.67	36.7842	0.6420
1.48	42.5066	0.7419	1.68	36.5296	0.6376
1.49	42.1552	0.7357	1.69	36.2789	0.6332

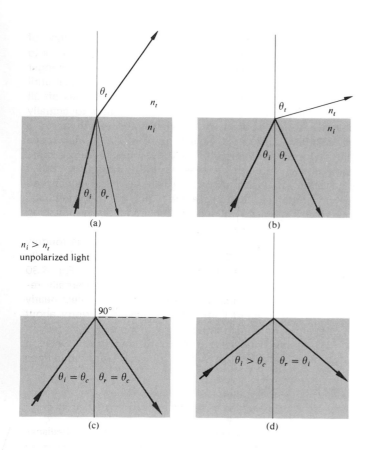

(a)

(b)

$n_i > n_t$
unpolarized light

(c)

(d)

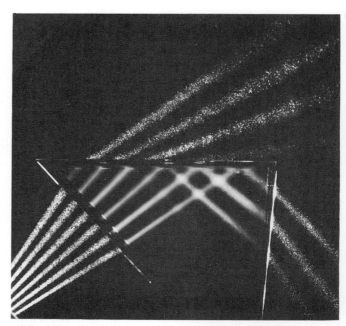

Fig. 4.29 Internal reflection and the critical angle. [Photo courtesy of Educational Service, Inc.]

using only the incident and reflected waves—things are not at all as simple as they might seem. Furthermore we can reformulate Eqs. (4.34) and (4.40) (Problem 4.34) such that

$$r_\perp = \frac{\cos \theta_i - (n_{ti}^2 - \sin^2 \theta_i)^{\frac{1}{2}}}{\cos \theta_i + (n_{ti}^2 - \sin^2 \theta_i)^{\frac{1}{2}}} \qquad (4.70)$$

and

$$r_\| = \frac{n_{ti}^2 \cos \theta_i - (n_{ti}^2 - \sin^2 \theta_i)^{\frac{1}{2}}}{n_{ti}^2 \cos \theta_i + (n_{ti}^2 - \sin^2 \theta_i)^{\frac{1}{2}}}. \qquad (4.71)$$

Clearly then since $\sin \theta_c = n_{ti}$ when $\theta_i > \theta_c$, $\sin \theta_i > n_{ti}$ and both $r_\perp$ and $r_\|$ become complex quantities. Despite this (Problem 4.35) $r_\perp r_\perp^* = r_\| r_\|^* = 1$ and $R = 1$ which means that $I_r = I_i$ and $I_t = 0$. Thus, although there must be a transmitted wave it cannot, on the average, carry energy across the boundary. We shall not perform the complete and rather lengthy computation needed to derive expressions for all of the reflected and transmitted fields but we can get an

appreciation of what's happening in the following way. The wave function for the transmitted electric field is

$$\mathbf{E}_t = \mathbf{E}_{0t} \exp i(\mathbf{k}_t \cdot \mathbf{r} - \omega t)$$

where

$$\mathbf{k}_t \cdot \mathbf{r} = k_{tx} x + k_{ty} y,$$

there being no z-component of $\mathbf{k}$. But

$$k_{tx} = k_t \sin \theta_t$$

and

$$k_{ty} = k_t \cos \theta_t$$

as can be seen from Fig. 4.31. Once again using Snell's law

$$k_t \cos \theta_t = \pm k_t \left(1 - \frac{\sin^2 \theta_i}{n_{ti}^2}\right)^{\frac{1}{2}} \qquad (4.72)$$

or since we are concerned with the case where $\sin \theta_i > n_{ti}$,

$$k_{ty} = \pm i k_t \left(\frac{\sin^2 \theta_i}{n_{ti}^2} - 1\right)^{\frac{1}{2}} \equiv \pm i\beta$$

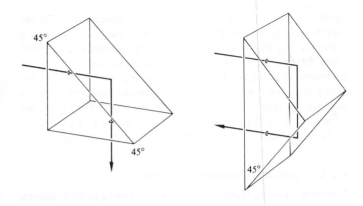

Fig. 4.30 Total internal reflection.

while

$$k_{tx} = \frac{k_t}{n_{ti}} \sin \theta_i.$$

Hence

$$\mathbf{E}_t = \mathbf{E}_{0t} e^{\mp \beta y} e^{i(k_t x \sin \theta_i / n_{ti} - \omega t)}. \qquad (4.73)$$

Neglecting the positive exponential which is physically untenable we have a wave whose amplitude drops off exponentially as it penetrates the less dense medium. The disturbance advances in the x-direction as a so-called *surface* or *evanescent wave*. Notice that the wavefronts or surfaces of constant phase (parallel to the yz-plane) are perpendicular to the surfaces of constant amplitude (parallel to the xy-plane) and as such the wave is *inhomogeneous* (see Section 2.5). Its amplitude decays rapidly in the y-direction, becoming negligible at a distance into the second medium of only a few wavelengths.

If you are still concerned about the conservation of energy, a more extensive treatment would have shown that energy actually circulates back and forth across the interface resulting on the average in a zero net flow through the boundary into the second medium. Yet one puzzling point remains inasmuch as there is still a bit of energy to be accounted for, namely that associated with the evanescent wave which moves along the boundary in the plane of incidence. Since this energy could not have penetrated into the less dense medium under the present circumstances (so long as $\theta_i \geq \theta_c$) we must look elsewhere for its source. Under actual experimental conditions the incident beam would have a finite cross-section and therefore would obviously differ from a true plane wave. This deviation gives rise (via diffraction) to a slight transmission of energy across the interface which is manifest in the evanescent wave.

Incidently, it is clear from Fig. 4.23(c) and (d) that the incident and reflected waves (except at $\theta_i = 90°$) do not differ in phase by π and cannot therefore cancel each other. It follows from the continuity of the tangential component of **E** that there must therefore be an oscillatory field in the less dense medium with a component parallel to the interface having a frequency ω (i.e. the evanescent wave).

The exponential decay of the surface or *boundary wave*, as it is also sometimes called, has quite recently been confirmed experimentally at optical frequencies.[*]

Imagine that a beam of light traveling within a block of glass is internally reflected at a boundary. Presumably if we pressed another piece of glass against the first the air–glass interface could be made to vanish and the beam would then propagate onward undisturbed. Furthermore, we might expect this transition from total to no reflection to occur gradually as the air film thins out. In much the same way, if one holds a drinking glass or a prism you can see the ridges of your fingerprints in a region which, because of total internal reflection, is otherwise mirror like. In more general terms, if the evanescent wave extends with appreciable amplitude across the rare medium into a nearby region occupied by a higher-index material, energy may now flow through the

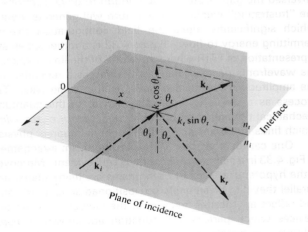

Fig. 4.31 Propagation vectors for internal reflection.

[*] Take a look at the fascinating article "Monomolecular Layers and Light" by K. H. Drexhage, *Sci. Am.* **222**, 108 (1970).

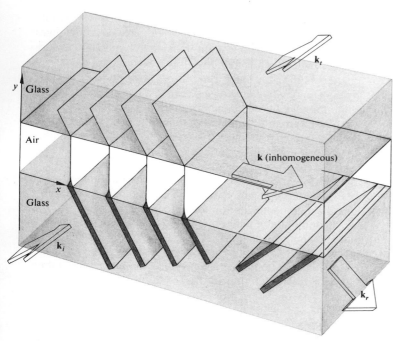

Fig. 4.32 Frustrated total internal reflection.

gap in what is known as *frustrated total internal reflection* (FTIR). In other words, if the evanescent wave, having traversed the gap, is still strong enough to drive electrons in the "frustrating" medium, they in turn will generate a wave which significantly alters the field configuration thereby permitting energy to flow. Figure 4.32 is a rather schematic representation of FTIR where the width of the lines depicting the wavefronts decreases across the gap as a reminder that the amplitude of the field behaves in that same way. The process as a whole is remarkably similar to the quantum-mechanical phenomenon of *barrier penetration* or *tunneling* which finds numerous applications in contemporary physics.

One can demonstrate FTIR with the prism arrangement of Fig. 4.33 in a manner which is fairly self-evident. Moreover, if the hypotenuse faces of both prisms are made planar and parallel they can presumably be positioned so as to transmit and reflect any desired fraction of the incident flux density. Devices which perform this function are known as *beam splitters*. A beam splitter cube can be made rather conveniently by using a thin low index transparent film as a precision spacer. Low-loss reflectors whose transmittance can be controlled by frustrating internal reflection are of considerable practical interest. FTIR can also be observed in

other regions of the electromagnetic spectrum. Three-centimeter microwaves are particularly easy to work with inasmuch as the evanescent wave will extend roughly 10^5 times farther than it would at optical frequencies. One can duplicate the above optical experiments with solid prisms made of paraffin or hollow ones of acrylic plastic filled with kerosene or motor oil. Any one of these would have an index of about 1.5 for 3 cm waves. It then becomes an easy matter to measure directly the dependence of the field amplitude on y.

4.3.5 Optical Properties of Metals

The characteristic feature of conducting media is the presence of a number of free electric charges (free in the sense of being unbound i.e. able to circulate around within the material). For metals these charges are of course electrons and their motion constitutes a current. The current per unit area resulting from the application of a field **E** is related by way of Eq. (A1.15) to the conductivity of the medium σ. For a dielectric there are no free or conduction electrons and $\sigma = 0$ while for actual metals σ is nonzero and finite. In contrast, an idealized "perfect" conductor would have an infinite conductivity. This is equivalent to saying that the

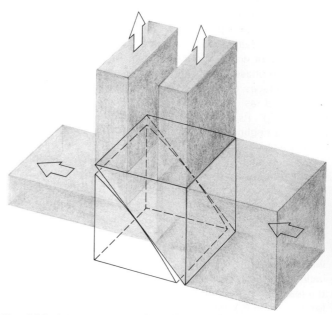

Fig. 4.33 A beam splitter utilizing FTIR.

electrons, driven into oscillation by a harmonic wave, would simply follow the field's alternations. There would be no restoring force, no natural frequencies and no absorption, only reemission. In real metals the conduction electrons undergo collisions with the thermally agitated lattice or with imperfections and in so doing irreversibly convert electromagnetic energy into Joule heat. Evidently the absorption of radiant energy by a material is a function of its conductivity.

i) Waves in a Metal

Visualizing the medium as if it were continuous Maxwell's equations lead to

$$\frac{\partial^2 \mathbf{E}}{\partial x^2} + \frac{\partial^2 \mathbf{E}}{\partial y^2} + \frac{\partial^2 \mathbf{E}}{\partial z^2} = \mu\epsilon\frac{\partial^2 \mathbf{E}}{\partial t^2} + \mu\sigma\frac{\partial \mathbf{E}}{\partial t} \qquad (4.74)$$

which is Eq. (A1.21) in Cartesian coordinates. The last term, $\mu\sigma\,\partial\mathbf{E}/\partial t$, is a first-order time derivative as was the damping force in the oscillator model of Section 3.3.1. The time rate of change of $\mathbf{E}$ generates a voltage, currents circulate and since the material is resistive, light is converted to heat— ergo absorption. It can be shown that this expression reduces to the unattenuated wave equation if the permittivity is reformulated as a complex quantity. This in turn leads to a complex index of refraction which, as we saw earlier (Section 3.3.1), is tantamount to absorption. We then need only substitute the complex index

$$n_c = n_R - in_I \qquad (4.75)$$

(where the real and imaginary indices n_R and n_I are both real numbers) into the corresponding solution for a nonconducting medium. Alternatively, one may utilize the wave equation and appropriate boundary conditions to yield a specific solution. In either event, we can find a simple sinusoidal plane-wave solution applicable within the conductor. Such a wave propagating in the y-direction is ordinarily written as

$$\mathbf{E} = \mathbf{E}_0 \cos{(\omega t - ky)}$$

or as a function of n

$$\mathbf{E} = \mathbf{E}_0 \cos \omega(t - ny/c)$$

but here the refractive index must be taken as complex. Accordingly, writing the wave as an exponential and using Eq. (4.75) it becomes

$$\mathbf{E} = \mathbf{E}_0 e^{(-\omega n_I y/c)} e^{i\omega(t - n_R y/c)} \qquad (4.76)$$

or

$$\mathbf{E} = \mathbf{E}_0 e^{-\omega n_I y/c} \cos \omega(t - n_R y/c). \qquad (4.77)$$

The disturbance advances in the y-direction with a speed c/n_R precisely as if n_R were the more usual index of refraction. As the wave progresses into the conductor, its amplitude, $\mathbf{E}_0 \exp{(-\omega n_I y/c)}$, is exponentially attenuated. Inasmuch as irradiance is proportional to the square of the amplitude, we have

$$I(y) = I_0 e^{-\alpha y}, \qquad (4.78)$$

where $I_0 = I(0)$, i.e. I_0 is the irradiance at $y = 0$ (the interface) and $\alpha \equiv 2\omega n_I/c$ is called the *absorption* or (even better) the *attenuation coefficient*. The flux density will drop by a factor of $e^{-1} = 1/2.7 \approx 1/3$ after the wave has propagated a distance $y = 1/\alpha$ known as the *skin or penetration depth*. For a material to be transparent the penetration depth must be large in comparison to its thickness. The penetration depth for metals, however, is exceedingly small. For example, copper at ultraviolet wavelengths ($\lambda_0 \approx 100$ nm) has a miniscule penetration depth of about 0.6 nm while it is still roughly only 6 nm in the infrared ($\lambda_0 \approx 10,000$ nm). This accounts for the generally observed opacity of metals which nonetheless can become partly transparent when formed into extremely thin films (as e.g. in the case of partially silvered two-way mirrors). The familiar metallic sheen of conductors corresponds to a high reflectance which in turn arises from the fact that the incident wave cannot effectively penetrate the material. Relatively few electrons in the metal "see" the transmitted wave and therefore although each absorbs strongly, little total energy is dissipated by them. Instead, most of the incoming energy reappears as the reflected wave. By far the majority of metals including the less common ones as for example sodium, potassium, cesium, vanadium, niobium, gadolinium, holmium, yttrium, scandium, osmium and many others have a silvery-gray appearance like that of aluminum, tin or steel. They reflect almost all of the incident light regardless of wavelengths and are therefore essentially colorless.

Equation (4.77) is certainly reminiscent of Eq. (4.73) and FTIR. In both cases there is an exponential decay of the amplitude. Moreover a complete analysis would show that the transmitted waves are not strictly transverse, there being a component of the field in the direction of propagation in both instances.

The representation of metal as a continuous medium works fairly well in the low-frequency, long-wavelength domain of the infrared. Yet we certainly might expect that as the wavelength of the incident beam decreased the actual granular nature of matter would have to be reckoned with. Indeed, the continuum model shows large discrepancies

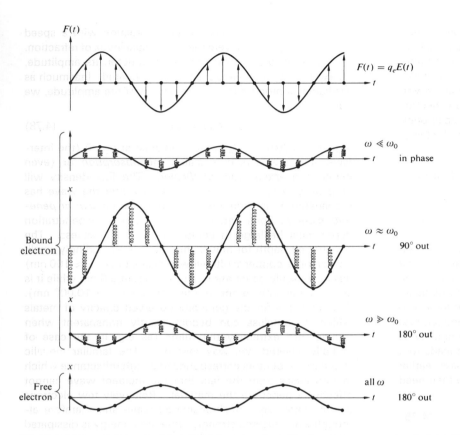

Fig. 4.34 Oscillations of bound and free electrons.

from experimental results at optical frequencies. And so we again turn to the classical atomistic picture initially formulated by Hendrik Lorentz, Paul Karl Ludwig Drude (1863–1906) and others. This simple approach will provide qualitative agreement with the experimental data but the ultimate treatment nonetheless requires quantum theory.

ii) The Dispersion Equation

Envision the conductor as an assemblage of driven, damped oscillators. Some of these correspond to free electrons and will therefore have zero restoring force while others are bound to the atom very much like those in the dielectric media of Section 3.3.1. The conduction electrons are, however, the predominant contributors to the optical properties of metals. Recall that the displacement of a vibrating electron was given (Section 3.3.1) by

$$x(t) = \frac{q_e/m_e}{(\omega_0^2 - \omega^2)} E(t).$$

With no restoring force, $\omega_0 = 0$, the displacement is opposite

in sign to the driving force $q_e E(t)$ and therefore 180° out of phase with it. This is unlike the situation for transparent dielectrics where the resonance frequencies are above the visible and the electrons oscillate in phase with the driving force (Fig. 4.34). Free electrons oscillating out of phase with the incident light will reradiate wavelets which tend to cancel the incoming disturbance. The effect, as we have already seen, is a rapidly decaying refracted wave.

Assuming that the average field experienced by an electron moving about within a conductor is just the applied field $\mathbf{E}(t)$ we can extend the dispersion equation of a rare medium (3.37) to read

$$n^2(\omega) = 1 + \frac{Nq_e^2}{\epsilon_0 m_e}\left[\frac{f_e}{-\omega^2 + i\gamma_e\omega} + \sum_j \frac{f_j}{\omega_{0j}^2 - \omega^2 + i\gamma_j\omega}\right]. \quad (4.79)$$

The first bracketed term is the contribution from the free electrons wherein N is the number of atoms per unit volume. Each of these has f_e conduction electrons which have no natural frequencies. The second term arises from the bound electrons and is identical to Eq. (3.37). It should be noted that if a metal has a particular color it is indicative of the fact that the atoms are partaking of selective absorption by way of the bound electrons. This is in addition to the general absorption characteristic of the free electrons. Recall that a medium which is very strongly absorbing at a given frequency doesn't actually absorb much of the incident light at that frequency but rather *selectively reflects* it. Gold and copper are reddish-yellow because n_I increases with wavelength and the larger values of λ are reflected more strongly. Thus, for example, gold should be fairly opaque to the longer visible wavelengths. Consequently, under white light illumination, a gold foil less than roughly 10^{-6} m thick will indeed transmit predominantly greenish-blue light.

We can get a rough idea of the response of metals to light by making a few gross simplifying assumptions. Accordingly, we neglect the bound electron contribution and furthermore assume that γ_e is also negligible for very large ω, whereupon

$$n^2(\omega) = 1 - \frac{Nq_e^2}{\epsilon_0 m_e \omega^2}. \tag{4.80}$$

The latter presumption is based on the fact that at high frequencies the electrons will undergo a great many oscillations between each collision. Free electrons and positive ions within a metal may be thought of as a plasma whose density oscillates at a natural frequency ω_p, the *plasma frequency*. This in turn can be shown to equal $(Nq_e^2/\epsilon_0 m_e)^{\frac{1}{2}}$ and so

$$n^2(\omega) = 1 - (\omega_p/\omega)^2. \tag{4.81}$$

The plasma frequency serves as a critical value below which the index is complex and the penetrating wave drops off exponentially (4.77) from the boundary; while at frequencies above ω_p, n is real, absorption is small and the conductor is transparent. In the latter circumstance n is less than one as it was for dielectrics at very high frequencies (Section 3.3.1). Hence we can expect metals in general to be fairly transparent to x-rays. Table 4.2 lists the plasma frequencies for some of the alkali metals which are transparent even to ultraviolet.

Most often the index of refraction for a metal will be complex and the impinging wave will suffer absorption in an amount which is frequency dependent. For example, the outer visors on the Apollo space suits are overlayed with a very thin film of gold (Fig. 4.35). The coating reflects about 70% of the incident light and is used under bright conditions such as low and forward sun angles. It was designed to

Fig. 4.35 Edwin Aldrin Jr. at Tranquility Base on the moon. The photographer, Neil Armstrong, is reflected in the gold-coated visor. [Photo courtesy NASA.]

decrease the thermal load on the cooling system by strongly reflecting radiant energy in the infrared while still transmitting adequately in the visible. Inexpensive metal coated sunglasses which are quite similar in principle are also

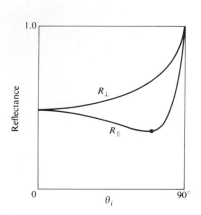

Fig. 4.36 Typical reflectance for a linearly polarized beam of white light incident on an absorbing medium.

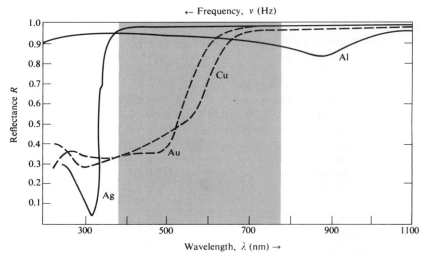

Fig. 4.37 Reflectance versus wavelength for silver, gold, copper and aluminum.

available commercially and they're well worth having just to experiment with.

The ionized upper atmosphere of the earth contains a distribution of free electrons which behave very much like those confined within a metal. The index of refraction of such a medium will be real and less than one for frequencies above ω_p. In July of 1965 the Mariner IV spacecraft made use of this effect to examine the ionosphere of the planet Mars, 216 million kilometers from earth.[*]

Table 4.2 Critical wavelengths and frequencies for some alkali metals.

Metal	λ_p (observed) (in nm)	λ_p (calculated) (in nm)	$\nu_p = c/\lambda_p$ (observed) (in Hz)
Lithium Li	155	155	1.94×10^{15}
Sodium Na	210	209	1.43×10^{15}
Potassium K	315	287	0.95×10^{15}
Rubidium Rb	340	322	0.88×10^{15}

[*] R. Von Eshelman, *Sci. Am.* **220**, 78 (1969).

If we wish to communicate between two distant terrestrial points we might bounce low-frequency waves off the earth's ionosphere. In contrast, in order to speak to someone on the moon we should use high-frequency signals to which the ionosphere would be transparent.

iii) Reflection from a Metal

Imagine that a plane wave initially in air impinges on a conducting surface. The transmitted wave advancing at some angle to the normal will be inhomogeneous. But if the conductivity of the medium is increased, the wavefronts will become aligned with the surfaces of constant amplitude, whereupon $\mathbf{k}_t$ and $\hat{\mathbf{u}}_n$ would approach parallelism. In other words, in a good conductor the transmitted wave propagates in a direction normal to the interface regardless of θ_i.

Let's now compute the reflectance, $R = I_r/I_i$, for the simplest case of normal incidence on a metal. Taking $n_i = 1$ and $n_t = n_c$ (i.e. the complex index) we have, from Eq. (4.47), that

$$R = \left(\frac{n_c - 1}{n_c + 1}\right)\left(\frac{n_c - 1}{n_c + 1}\right)^* \tag{4.82}$$

and therefore since $n_c = n_R - in_I$

$$R = \frac{(n_R - 1)^2 + n_I^2}{(n_R + 1)^2 + n_I^2}. \tag{4.83}$$

If the conductivity of the material goes to zero we have

the case of a dielectric whereupon in principle the index is real ($n_I = 0$) and the attenuation coefficient, α, is zero. Under those circumstances, the index of the transmitting medium n_t is n_R and the reflectance (4.83) becomes identical with that of Eq. (4.67). If instead n_I is large while n_R is comparatively small R in turn becomes large (Problem 4.40). In the unattainable limit where n_c is purely imaginary, one hundred percent of the incident flux density would be reflected ($R = 1$). Notice that it is possible for the reflectance of one metal to be greater than that of another even though its n_I is smaller. For example at $\lambda_0 = 589.3$ nm the parameters associated with solid sodium are roughly $n_R = 0.04$, $n_I = 2.4$, and $R = 0.9$; those for bulk tin are $n_R = 1.5$, $n_I = 5.3$, and $R = 0.8$; while for a gallium single crystal $n_R = 3.7$, $n_I = 5.4$, and $R = 0.7$.

The curves of $R_\parallel$ and $R_\perp$ for oblique incidence shown in Fig. 4.36 are somewhat typical of absorbing media. Thus while R at $\theta_i = 0$ for gold is about 0.5, as opposed to near 0.9 for silver in white light, both have reflectances which are quite similar in shape approaching 1.0 at $\theta_i = 90°$. Just as with dielectrics (Fig. 4.28) $R_\parallel$ drops to a minimum at what is now called the *principal angle of incidence* but here that minimum is nonzero. Figure 4.37 illustrates the spectral reflectance at normal incidence for a number of evaporated metal films under ideal conditions. Observe that while gold transmits fairly well in and below the green region of the spectrum, silver which is highly reflective across the visible becomes transparent in the ultraviolet at about 316 nm.

Phase shifts arising from reflection off a metal occur in both components of the field (i.e. parallel and perpendicular to the plane of incidence). These are generally neither 0 nor π with a notable exception at $\theta_i = 90°$ where, just as with a dielectric, both components shift phase by 180° on reflection.

4.4 FAMILIAR ASPECTS OF THE INTERACTION OF LIGHT AND MATTER

Let's now examine some of the phenomena which paint the everyday world in a marvel of myriad colors.

The mechanism responsible for the yellowish-red hue of gold and copper is, in some respects, similar to the process which causes the sky to appear blue. Putting it rather succinctly (see Section 8.5 for a further discussion of scattering in the atmosphere), the molecules of air have resonances in the ultraviolet and will therefore be driven into larger amplitude oscillations as the frequency of the incident light increases toward the ultraviolet. Consequently they will effectively take energy from and reemit (i.e. scatter)

the blue component of sunlight in all directions, transmitting the complementary red end of the spectrum with little alteration. This is in analogy with the selective reflection or scattering of yellow-red light that takes place at the surface of a gold film and the concomitant transmission of blue-green light. In contradistinction, the characteristic colors of most substances have their origin in the phenomenon of *selective or preferential absorption*. For example, water has a very light green-blue tint by dint of its absorption of red light. That is, the H_2O molecules have a broad resonance in the infrared which extends somewhat into the visible. The absorption isn't very strong so that there is no accentuated reflection of red light at the surface. Rather it is transmitted and gradually absorbed out until at a depth of about 30 m of sea water, red is almost completely removed from sunlight. This same process of selective absorption is responsible for the colors of brown eyes and butterflies, of birds and bees and cabbages and kings. Indeed the great majority of objects in nature appear to have characteristic colors as the result of preferential absorption by pigment molecules. In contrast with most atoms and molecules, which have resonances in the ultraviolet and infrared, the pigment molecules must obviously have resonances in the visible. Yet visible photons have energies of roughly 1.6 eV to 3.2 eV which, as you might expect, are on the low side for ordinary electron excitation and on the high side for excitation via molecular vibration. Despite this, there are atoms where the bound electrons form incomplete shells (e.g. gold) and variations in the configuration of these shells provide a mode for low-energy excitation. In addition to these there is the large group of organic dye molecules which evidently also have resonances in the visible. All such substances, whether natural or synthetic, consist of long chain molecules made up of regularly alternating single and double bonds in what is called a conjugated system. This structure is typified by the carotene molecule $C_{40}H_{56}$ (Fig. 4.38). The carotenoids range in color from yellow to red and are found in carrots, tomatoes, daffodils, dandelions, autumn leaves and people. The chlorophylls are another group of familiar natural pigments but here a portion of the long chain is turned around on itself to form a ring. In any event, conjugated systems of this sort contain a number of particularly mobile electrons known as *pi electrons*. These are not bound to specific atomic sites but instead can range over the relatively large dimensions of the molecular chain or ring. In the phraseology of quantum mechanics we would say that these were long wavelength, low frequency, and therefore low energy, electron states. The energy required to raise a pi electron to an excited state is

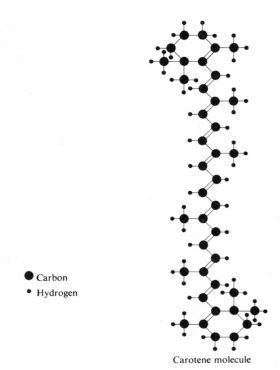

Carbon
Hydrogen

Carotene molecule

Fig. 4.38 The carotene molecule.

accordingly comparatively low corresponding to that of visible photons. In effect, we can imagine the molecule as an oscillator having a resonance frequency in the visible.

The energy levels of an individual atom are precisely defined i.e. the resonances are very sharp. With solids and liquids however the proximity of the atoms results in a broadening of the energy levels into wide bands. In other words, the resonances spread over a broad range of frequencies. Consequently we can expect that a dye will not absorb just a narrow portion of the spectrum; indeed if it did it would reflect most frequencies and appear nearly white.

To see how all of these ideas fit together imagine that we have a piece of blue stained glass or cellophane. White light impinging on the surface (Fig. 4.39) is partially reflected, the remainder being transmitted into the medium. As the beam advances, more and more of the red end of the spectrum is removed from it by absorption until it finally emerges as blue light. It should be clear from the figure that the colors of stained glass windows will be far more vivid when viewed in transmitted rather than reflected light. It is precisely this penetration and gradual preferential absorption which is responsible for the colors of most objects. For example, the

fibers of a sample of white cloth or paper are essentially transparent but when dyed each fiber behaves as if it were a chip of colored glass. The incident light penetrates the paper emerging for the most part as a reflected beam only after undergoing numerous reflections and refractions within the dyed fibers. The exiting light will be colored to the extent that it is minus that frequency component which is absorbed by the dye. This is precisely why a leaf appears green or a banana yellow.

A bottle of ordinary blue ink looks blue in either reflected or transmitted light. But if we paint it on a glass slide and evaporate out the solvent, something rather interesting happens. The concentrated pigment absorbs so effectively that it preferentially reflects at the resonant frequency and we are back to the idea that a strong absorber (large n_I) is a strong reflector. Thus, concentrated bluish ink reflects magenta and transmits blue, while red ink will reflect green and transmit red. Try it with a felt marking pen but you must use reflected light, being careful not to inadvertently inundate the sample with unwanted light from below. (Wipe the ink into a thin layer and then place the slide on a piece of black paper.)

If the range of frequencies being absorbed spreads across the visible, the object will appear black. That is not to say that there is no reflection at all—you obviously can see a reflected image in a piece of black patent leather and a rough black surface reflects as well, only diffusely. If you still have

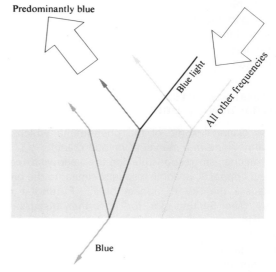

Fig. 4.39 Blue stained glass.

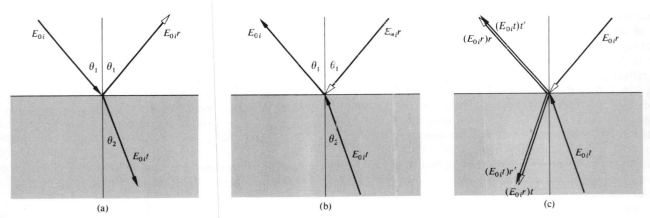

Fig. 4.40 Reflection and refraction via the Stokes treatment.

those red and blue inks, mix them, add some green, and you'll get black.

At the other extreme an object is white under white light illumination when it reflects both effectively and uniformly over the visible spectrum. Although water is essentially transparent, water vapor appears white as does ground glass. The reason is simple enough——if the grain size is small but much larger than the wavelengths involved, light will enter each transparent particle, be reflected and refracted several times and emerge. There will be no distinction made between any of the frequency components and so the reflected light reaching the observer will be white. This is the mechanism accountable for the whiteness of things like sugar, salt, paper, clouds, talcum powder, snow and paint, each grain of which is actually transparent. Similarly clear plastic tape on a thick roll appears to be opaque as does an ordinarily transparent material filled with small air bubbles (e.g. beaten egg white). Even though we usually think of paper, talcum powder, sugar, etc. as each consisting of some sort of opaque white substance, it's an easy matter to dispel that misconception. Cover a printed page with a few of these materials (a sheet of white paper, some grains of sugar or talcum) and illuminate it from behind. You'll have little difficulty in seeing through them. In the case of white paint, one simply suspends colorless transparent particles such as the oxides of zinc, titanium or lead in an equally transparent vehicle, as for example, linseed oil or the newer acrylics. Obviously if the particles and vehicle have the same index of refraction there will not be any reflections at the grain boundaries. The particles will simply disappear into the conglomeration which itself remains clear. In contrast, if the indices are markedly different there will be a good deal of

reflection at all wavelengths (Problem 4.33) and the paint will appear white and opaque [take another look at Eq. (4.67)]. To color paint one need only dye the particles so that they absorb out all frequencies except the desired range.

Carrying the logic in the reverse direction, if we reduce the relative index, n_{ti}, at the grain or fiber boundaries, the particles of material will reflect less, thereby decreasing the overall whiteness of an object. Consequently, a wet white tissue will have a grayish more transparent look. Wet talcum powder loses its sparkling white becoming a dull gray as does wet white cloth. And in the same way a piece of dyed fabric soaked in a clear liquid (e.g. water, gin or benzene) will lose its whitish haze and become very much darker, the colors then being deep and rich like those of a still wet water color painting.

In addition to the above processes that are specifically related to reflection, refraction and absorption, there are many other color-generating mechanisms which we shall explore later on. For example, the scarabeid beetles mantle themselves in the brilliant colors produced by diffraction gratings grown on their wing-cases; while wavelength-dependent interference effects contribute to the color patterns seen on oil slicks, mother-of-pearl, soap bubbles, peacocks and hummingbirds.

4.5 THE STOKES TREATMENT OF REFLECTION AND REFRACTION

A rather elegant and novel way of looking at reflection and transmission at a boundary was developed by the British physicist Sir George Gabriel Stokes (1819–1903). Since we will often make use of his results in the future let's now examine that derivation. Suppose that we have an incident wave of amplitude E_{0i} impinging on the planar interface separating two dielectric media as in Fig. 4.40(a). As we saw earlier in this chapter, since r and t are the fractional amplitudes reflected and transmitted respectively (where $n_i = n_1$ and $n_t = n_2$), then $E_{0r} = rE_{0i}$ and $E_{0t} = tE_{0i}$. Again we are reminded of the fact that Fermat's principle led to the principle of reversibility which implies that the situation depicted in Fig. 4.40(b) where all the ray directions are reversed must also be physically permissible. With the one proviso that there be no energy dissipation (no absorption), a wave's meanderings must be reversible. Equivalently, in the idiom of modern physics one speaks of *time-reversal invariance* i.e. if a process occurs, the reversed process can also occur. Thus if we take a hypothetical motion picture of the wave incident on, reflecting from and transmitting through the interface, the behavior depicted when the film is run backwards must also be physically realizable. Accordingly, examine Fig. 4.40(c) where there are now two incident waves of amplitudes $E_{0i}r$ and $E_{0i}t$. A portion of the wave whose amplitude is $E_{0i}t$ is both reflected and transmitted at the interface. Without making any assumptions let r' and t' be the amplitude reflection and transmission coefficients for a wave incident from below (i.e. $n_i = n_2$, $n_t = n_1$). Consequently, the reflected portion is $E_{0i}tr'$ while that transmitted is $E_{0i}tt'$. Similarly the incoming wave whose amplitude is $E_{0i}r$ splits into segments of amplitude $E_{0i}rr$ and $E_{0i}rt$. If the configuration of Fig. 4.40(c) is to be identical with that of Fig. 4.40(b) then obviously

$$E_{0i}tt' + E_{0i}rr = E_{0i} \qquad (4.84)$$

and

$$E_{0i}rt + E_{0i}tr' = 0. \qquad (4.85)$$

Hence

$$tt' = 1 - r^2 \qquad (4.86)$$

and

$$r' = -r, \qquad (4.87)$$

the latter two equations being known as the Stokes relations. Actually this discussion calls for a bit more caution than is most often granted it. It must be pointed out that *the amplitude coefficients are functions of the incident angles* and therefore Stokes' relations might better be written as

$$t(\theta_1)t'(\theta_2) = 1 - r^2(\theta_1) \qquad (4.88)$$

and

$$r'(\theta_2) = -r(\theta_1), \qquad (4.89)$$

where $n_1 \sin \theta_1 = n_2 \sin \theta_2$. The second equation indicates, by virtue of the minus sign, that *there is a 180° phase difference between the waves internally and externally reflected*. It is most important to keep in mind that here θ_1 and θ_2 are pairs of angles which are related by way of Snell's law. Note as well that we never did say whether n_1 was greater or less than n_2 and so Eqs. (4.88) and (4.89) apply in either case. Let's return for a moment to one of the Fresnel equations, for example

$$r_\perp = -\frac{\sin (\theta_i - \theta_t)}{\sin (\theta_i + \theta_t)}. \qquad [4.42]$$

If a ray enters from above as in Fig. 4.40(a) and we assume $n_2 > n_1$, $r_\perp$ is computed by setting $\theta_i = \theta_1$ and $\theta_t = \theta_2$ (external reflection) the latter arrived at from Snell's law. If on the other hand the wave is incident at that same angle from below (in this instance internal reflection) $\theta_i = \theta_1$ and we again substitute in Eq. (4.42) but here θ_t is not θ_2 as before. The values of $r_\perp$ for internal and external reflection *at the same incident angle* are obviously different. Now then suppose, in this case of internal reflection, that we take $\theta_i = \theta_2$. Then $\theta_t = \theta_1$, the ray directions are the reverse of those in the first situation and Eq. (4.42) yields

$$r'_\perp(\theta_2) = -\frac{\sin (\theta_2 - \theta_1)}{\sin (\theta_2 + \theta_1)}.$$

Although it may be unnecessary we once again point out that this is just the negative of what was determined for $\theta_i = \theta_1$ and external reflection i.e.

$$r'_\perp(\theta_2) = -r_\perp(\theta_1). \qquad (4.90)$$

The use of primed and unprimed symbols to denote the amplitude coefficients should serve as a reminder that we are once more dealing with angles related by Snell's law. In the same way interchanging θ_i and θ_t in Eq. (4.43) leads to

$$r'_\parallel(\theta_2) = -r_\parallel(\theta_1). \qquad (4.91)$$

The 180° phase difference between each pair of components is evident in Fig. 4.23 but do keep in mind that when $\theta_i = \theta_p$, $\theta_t = \theta'_p$ and vice versa (Problem 4.37). Beyond $\theta_i = \theta_c$ there is no transmitted wave, Eq. (4.89) is not applicable and, as we have seen, the phase difference is no longer 180°.

It is common to conclude that both the parallel and perpendicular components of the externally reflected beam change phase by π radians while the internally reflected beam undergoes no phase shift at all. By now, within the particular convention we've established, this should be recognized as obviously incorrect, or at least almost obviously [compare Figs. 4.24(a) and 4.25(a)].

4.6 PHOTONS AND THE LAWS OF REFLECTION AND REFRACTION

Suppose that light consists of a stream of photons and consider one such photon which strikes the interface between two dielectric media at an angle θ_i and is subsequently transmitted across it at an angle θ_t. We know that if this were just one out of billions of such quanta in a narrow laser beam it would obediently conform to Snell's law. To appreciate this behavior let's examine the dynamics associated with the odyssey of our single photon. Recall that

$$\mathbf{p} = \hbar \mathbf{k} \qquad [3.60]$$

and consequently the incident and transmitted momenta are $\mathbf{p}_i = \hbar \mathbf{k}_i$ and $\mathbf{p}_t = \hbar \mathbf{k}_t$ respectively. We assume (without much justification) that while the material in the vicinity of the interface affects the y-component of momentum it leaves the x-component unchanged. Indeed we know experimentally that linear momentum can be transferred to a medium from a light beam (Section 3.4.2). The statement of conservation of the component of momentum parallel to the interface takes the form

$$p_{ix} = p_{tx} \qquad (4.92)$$

or

$$p_i \sin \theta_i = p_t \sin \theta_t.$$

If we use Eq. (3.60) this becomes

$$k_i \sin \theta_i = k_t \sin \theta_t$$

and hence

$$\frac{1}{\lambda_i} \sin \theta_i = \frac{1}{\lambda_t} \sin \theta_t.$$

Multiplying both sides by c/v we have

$$n_i \sin \theta_i = n_t \sin \theta_t,$$

which of course is Snell's law. In exactly the same way if

the photon reflects off the interface instead of being transmitted, Eq. (4.92) leads to

$$k_i \sin \theta_i = k_r \sin \theta_r,$$

and since $\lambda_i = \lambda_r$, $\theta_i = \theta_r$. It is interesting to note that

$$n_{ti} = \frac{p_t}{p_i} \qquad (4.93)$$

and so if $n_{ti} > 1$, $p_t > p_i$. Experiments dating back as far as 1850 to those of Foucault have shown that when $n_{ti} > 1$ the speed of propagation is actually reduced in the transmitting media even though here the momentum apparently increases.*

Do keep in mind that we have been dealing with a very simple representation which leaves much to be desired, for example it says nothing about the atomic structure of the media and nothing about the probability or likelihood of a photon traversing a given path. Even though this treatment is obviously simplistic, it is appealing pedagogically (see Chapter 13).

PROBLEMS

4.1 Calculate the transmission angle for a ray incident in air at 30° on a block of crown-glass ($n_g = 1.52$).

4.2* A ray of yellow light from a sodium discharge lamp falls on the surface of a diamond in air at 45°. If at that frequency $n_d = 2.42$, compute the angular deviation suffered upon transmission.

4.3 Use Huygens' construction to create a wavefront diagram showing the form a spherical wave will have after reflection from a planar surface as in the ripple tank photos of Fig. 4.41. Draw the ray diagram as well.

4.4* Given a water ($n_w = 1.33$) glass ($n_g = 1.50$) interface, compute the transmission angle for a beam incident in the water at 45°. If the transmitted beam is reversed so that it now impinges on the interface, show that $\theta_t = 45°$.

4.5 A beam of 12 cm planar microwaves strikes the surface of a dielectric at 45°. If $n_{ti} = \frac{4}{3}$, compute (a) the wavelength in the transmitting medium and (b) the angle θ_t.

---

* This suggests an increase in the photon's effective mass. See F. R. Tangherlini, "On Snell's Law and the Gravitational Deflection of Light," Am. J. Phys. 36, 1001 (1968). Take a *cautious* look at R. A. Houstoun, "Nature of Light," J. Opt. Soc. Am. 55, 1186 (1965).

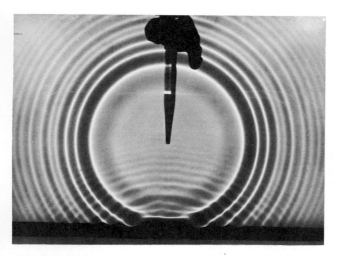

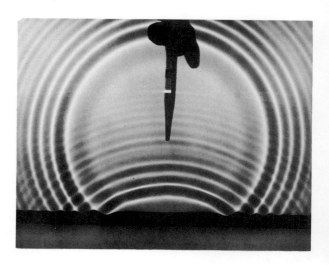

Fig. 4.41   [Photo courtesy *Physics*, Boston, D. C. Heath & Co., 1960.]

4.6* Light of wavelength 600 nm in vacuum enters a block of glass where $n_g = 1.5$. Compute its wavelength in the glass. What color will it appear to someone imbedded in the glass (see Table 3.2)?

4.7 Figure 4.42 shows a bundle of rays entering and emerging from a glass disk (a lens). From the configuration of the rays, determine the shape of the wavefronts at various points. Draw a diagram in profile.

4.8 Make a plot of θ_i versus θ_t for an air–glass boundary where $n_{ga} = 1.5$.

4.9 Making use of the ideas of equal transit times between corresponding points and the orthogonality of rays and wavefronts, derive the law of reflection and Snell's law. The ray diagram of Fig. 4.43 should be helpful.

4.10 Starting with Snell's law prove that the vector refraction equation has the form

$$n_t\hat{\mathbf{k}}_t - n_i\hat{\mathbf{k}}_i = (n_t\cos\theta_t - n_i\cos\theta_i)\hat{\mathbf{u}}_n. \qquad [4.8]$$

4.11 Derive a vector expression equivalent to the law of reflection. As before, let the normal go from the incident to the transmitting medium even though it obviously doesn't really matter.

4.12 In the case of reflection from a planar surface use Fermat's principle to prove that the incident and reflected

rays share a common plane with the normal $\hat{\mathbf{u}}_n$, namely the plane of incidence.

4.13* Derive the law of reflection, $\theta_i = \theta_r$, by using the calculus to minimize the transit time as required by Fermat's principle.

4.14 Show analytically that a beam entering a planar transparent plate, as in Fig. 4.44, emerges parallel to its initial direction. Derive an expression for the lateral displacement of the beam. Incidentally, the fact that the incoming and outgoing rays are parallel would be true even for a stack of plates of different material.

4.15* Show that the two rays which enter the system in Fig. 4.45 parallel to each other emerge from it being parallel.

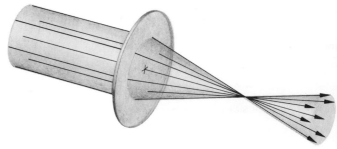

Fig. 4.42

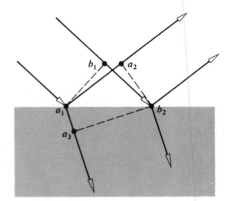

Fig. 4.43

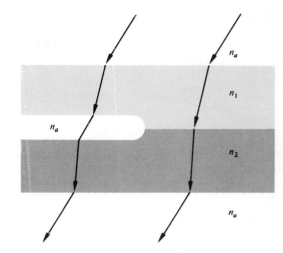

Fig. 4.45

Fig. 4.44

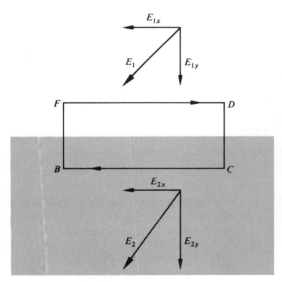

Fig. 4.46

4.16   Discuss the results of Problem 4.14 in the light of Fermat's principle, i.e. how does the relative index n_{21} affect things?   To see the lateral displacement look at a broad source through a thick piece of glass ($\approx \frac{1}{4}$ inch) or a stack (four will do) of microscope slides *held at an angle*. There will be an obvious shift between the region of the source seen directly and the region viewed through the glass.

4.17   Suppose that a light wave which is linearly polarized in the plane of incidence impinges at 30° on a crown glass ($n_g = 1.52$) plate in air. Compute the appropriate amplitude reflection and transmission coefficients at the interface. Compare your results with Fig. 4.20.

4.18 Show that even in the nonstatic case the tangential component of the electric field intensity **E** is continuous across an interface. [*Hint*: using Fig. 4.46 and Eq. (3.5) shrink sides *FB* and *CD* thereby letting the area bounded go to zero.]

4.19 Derive Eqs. (4.42) through (4.45) for $r_\perp$, $r_\parallel$, $t_\perp$, and $t_\parallel$.

4.20 Prove that

$$t_\perp + (-r_\perp) = 1 \qquad [4.49]$$

for all θ_i, first from the boundary conditions and then from the Fresnel equations.

4.21* Verify that

$$t_\perp + (-r_\perp) = 1 \qquad [4.49]$$

for $\theta_i = 30°$ at a crown glass and air interface ($n_{ti} = 1.52$).

4.22* Calculate the critical angle beyond which there is total internal reflection at an air–glass ($n_g = 1.5$) interface. Compare this result with that of Problem 4.8.

4.23 Consider the common mirage associated with an inhomogeneous distribution of air situated above a warm roadway. Envision the bending of the rays as if it were instead a problem in total internal reflection. If an observer, at whose head $n_a = 1.00029$, sees an apprent wet spot at $\theta_i \geq 88.7°$ down the road, find the index of the air immediately above the road.

4.24 Show that $\tan \theta_p = n_t/n_i$ and calculate the polarization angle for external incidence on a plate of crown glass ($n_g = 1.52$) in air.

4.25 Show that the polarization angles for internal and external reflection at a given interface are complementary, i.e. $\theta_p + \theta'_p = 90°$ (see Problem 4.24).

4.26 It is often useful to work with the *azimuthal angle* γ which is defined as the angle between the plane of vibration and the plane of incidence. Thus for linearly polarized light

$$\tan \gamma_i = [E_{0i}]_\perp / [E_{0i}]_\parallel \qquad (4.94)$$

$$\tan \gamma_t = [E_{0t}]_\perp / [E_{0t}]_\parallel \qquad (4.95)$$

and

$$\tan \gamma_r = [E_{0r}]_\perp / [E_{0r}]_\parallel. \qquad (4.96)$$

Figure 4.47 is a plot of γ_r versus θ_i for internal and external reflection at an air–glass interface ($n_{ga} = 1.51$), where $\gamma_i = 45°$. Verify a few of the points on the curves and in addition show that

$$\tan \gamma_r = -\frac{\cos(\theta_i - \theta_t)}{\cos(\theta_i + \theta_t)} \tan \gamma_i. \qquad (4.97)$$

4.27* Making use of the definitions of the azimuthal angles in Problem 4.26 show that

$$R = R_\parallel \cos^2 \gamma_i + R_\perp \sin^2 \gamma_i \qquad (4.98)$$

and

$$T = T_\parallel \cos^2 \gamma_i + T_\perp \sin^2 \gamma_i. \qquad (4.99)$$

4.28 Make a sketch of $R_\perp$ and $R_\parallel$ for $n_i = 1.5$ and $n_t = 1$, i.e. internal reflection.

4.29 Show that

$$T_\parallel = \frac{\sin 2\theta_i \sin 2\theta_t}{\sin^2(\theta_i + \theta_t)\cos^2(\theta_i - \theta_t)} \qquad (4.100)$$

and

$$T_\perp = \frac{\sin 2\theta_i \sin 2\theta_t}{\sin^2(\theta_i + \theta_t)}. \qquad (4.101)$$

4.30* Using the results of Problem 4.29, i.e. Eqs. (4.100) and (4.101), show that

$$R_\parallel + T_\parallel = 1 \qquad [4.65]$$

and

$$R_\perp + T_\perp = 1. \qquad [4.66]$$

4.31 Suppose that we look at a source perpendicularly through a stack of N microscope slides. The source seen through even a dozen slides will be noticeably darker. Assuming negligible absorption show that the total transmittance of the stack is given by

$$T_t = (1 - R)^{2N}$$

and evaluate T_t for three slides in air.

4.32 Making use of the expression

$$I(y) = I_0 e^{-\alpha y} \qquad [4.78]$$

for an absorbing medium we define a quantity called the *unit transmittance* T_1. At normal incidence (4.55) $T = I_t/I_i$ and thus when $y = 1$, $T_1 \equiv I(1)/I_0$. If the total thickness of the slides in the previous problem is d and if they now have a transmittance per unit length T_1, show that

$$T_t = (1 - R)^{2N}(T_1)^d.$$

4.33 Show that at normal incidence on the boundary between two dielectrics as $n_{ti} \to 1$, $R \to 0$, and $T \to 1$. Moreover, prove that as $n_{ti} \to 1$, $R_\parallel \to 0$, $R_\perp \to 0$, $T_\parallel \to 1$, and $T_\perp \to 1$ for all θ_i. Thus as the two media take on more similar indices of refraction, less and less energy is carried off in the reflected wave. It should be obvious that when $n_{ti} = 1$ there will be no interface and no reflection.

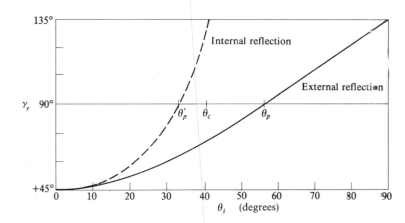

Fig. 4.47

4.34* Derive the expressions for $r_\perp$ and $r_\parallel$ given by Eqs. (4.70) and (4.71).

4.35 Show that when $\theta_i > \theta_c$ at a dielectric interface, $r_\parallel$ and $r_\perp$ are complex and $r_\perp r_\perp^* = r_\parallel r_\parallel^* = 1$.

4.36 Figure 4.48 depicts a ray being multiply reflected by a transparent dielectric plate (the amplitudes of the resulting fragments are indicated). As in Section 4.5 we use the primed coefficient notation because the angles are related by Snell's law.

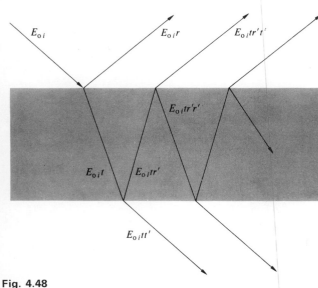

Fig. 4.48

a) Finish labeling the amplitudes of the last four rays.

b) Show, using the Fresnel equations, that

$$t_\parallel t_\parallel' = T_\parallel \tag{4.102}$$

$$t_\perp t_\perp' = T_\perp \tag{4.103}$$

$$r_\parallel^2 = r_\parallel'^2 = R_\parallel \tag{4.104}$$

and

$$r_\perp^2 = r_\perp'^2 = R_\perp. \tag{4.105}$$

4.37* A wave, linearly polarized in the plane of incidence, impinges on the interface between two dielectric media. If $n_i > n_t$ and $\theta_i = \theta_p'$ there is no reflected wave, i.e. $r_\parallel'(\theta_p') = 0$. Using Stokes' technique start from scratch to show that $t_\parallel(\theta_p)t_\parallel'(\theta_p') = 1$, $r_\parallel(\theta_p) = 0$, and $\theta_t = \theta_p$ (see Problem 4.25). How does this compare with Eq. (4.102)?

4.38 Use the Fresnel equations to show that $t_\parallel(\theta_p)t_\parallel'(\theta_p') = 1$ as in the previous problem.

4.39 Figure 4.49 depicts a glass cube surrounded by four glass prisms which are in very close proximity to its sides. Sketch in the paths which will be taken by the two rays shown and discuss a possible application for the device.

4.40 Figure 4.50 is a plot of n_I and n_R versus λ for a common metal. Identify the metal by comparing its characteristics with those considered in the chapter and discuss its optical properties.

4.41 Figure 4.51 shows a prism-coupler arrangement developed at the Bell Telephone Laboratories. Its function is to feed a laser beam into a thin (0.00001 in) transparent film which then serves as a sort of waveguide. One possible application is that of thin film laser beam circuitry—a kind of integrated optics. How do you think the thing works?

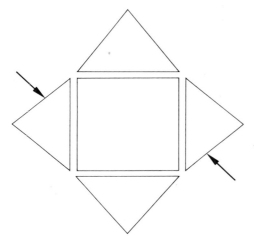

Fig. 4.49

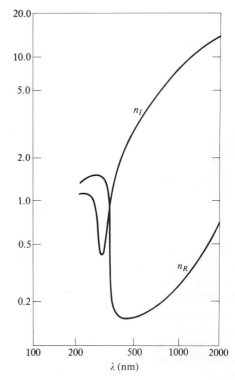

Fig. 4.50

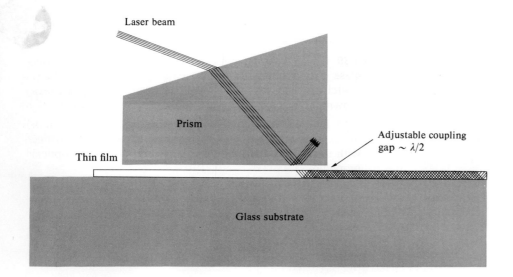

Fig. 4.51

Geometrical Optics— 5
Paraxial Theory

5.1 INTRODUCTORY REMARKS

Suppose that we have an object which is either self-luminous or externally illuminated and imagine its surface as consisting of a large number of point sources. Each of these emits spherical waves, i.e. rays emanate radially in the direction of energy flow, or if you like, in the direction of the Poynting vector (Fig. 4.1). In this case, the rays *diverge* from a given point source *S*, whereas if the spherical wave were collapsing to a point, the rays would of course be *converging*. Generally one deals only with a small portion of a wavefront. *A point from which a portion of a spherical wave diverges, or one towards which the wave segment converges, is known as a focal point of the bundle of rays.*

Now envision the situation where we have a point source in the vicinity of some arrangement of reflecting and refracting surfaces representing an *optical system*. Of the infinity of rays emanating from *S*, generally speaking, only one will pass through an arbitrary point in space. Even so, it is possible to arrange for an infinite number of rays to arrive at a certain point *P*, as in Fig. 5.1. Thus, if for a cone of rays coming from *S* there is a corresponding cone of rays passing through *P*, the system is said to be *stigmatic* for these two points. The energy in the cone (apart from some inadvertent losses due to reflection, scattering and absorption) reaches *P* which is then referred to as a *perfect image* of *S*. The wave could conceivably arrive to form a finite patch of light or *blur spot* about *P*; it would still be an image of *S* but no longer a perfect one.

It follows from the principle of reversibility (Section 4.2.4) that a point source placed at *P* would equally well be imaged at *S* and accordingly these two are spoken of as *conjugate points*. In an *ideal optical system* every point of a three-dimensional region will be perfectly (or stigmatically) imaged in another region; the former being the *object space*, the latter the *image space*.

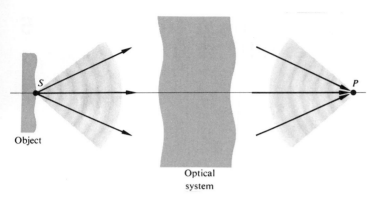

Fig. 5.1 Converging and diverging waves.

Most commonly, the function of an optical device is to collect and reshape a portion of the incident wavefront, often with the ultimate intent of forming an image of an object. Notice that inherent in realizable systems is the limitation of being unable to collect all of the emitted light; the system accepts only a segment of the wavefront. As a result, there will always be an apparent deviation from rectilinear propagation even in homogeneous media—the waves will be *diffracted*. The attainable degree of perfection in the imaging capability of a real optical system will therefore be *diffraction limited* (there will always be a blur spot). As the wavelength of the radiant energy decreases in comparison to the physical dimensions of the optical system, the effects of diffraction become less significant. In the conceptual limit as $\lambda_0 \to 0$ rectilinear propagation obtains in homogeneous media and we have the idealized domain of *geometrical optics.** Behavior which is specifically attributable to the wave nature of light (e.g. interference and diffraction) would no longer be observable. There are many situations in which the great simplicity arising from the approximation of geometrical optics more than compensates for its inaccuracies. In short, *the subject treats the controlled manipulation of wavefronts (or rays) by means of the interpositioning of reflecting and or refracting bodies, neglecting any diffraction effects.*

5.2 LENSES

No doubt the most widely used optical device is the lens and

that notwithstanding the fact that we see the world through a pair of them. Lenses date back to the burning glasses of antiquity and indeed who can say when man first peered through the liquid lens formed by a droplet of water?

As an initial step toward an understanding of what lenses do and how they manage to do it, let's examine what happens when light impinges on the curved surface of a transparent dielectric medium.

5.2.1   Refraction at Aspherical Surfaces

Imagine that we have a point source S whose spherical waves arrive at a boundary between two transparent media as shown in Fig. 5.2. We would like to determine the shape which the interface must have in order that the wave traveling within the second medium converges to a point P, there forming a perfect image of S. Practical reasons for wanting to focus a diverging wave to a point will become evident as we proceed.

The time it takes for each and every portion of a wavefront leaving S to converge at P must be identical. Or in the words of Section 4.2.3, "the distance between corresponding points on any and all rays will be traversed in that same time." Another way to say essentially the same thing from the perspective of Fermat's principle is that if a great many different rays are to go from S to P (i.e. if point A in Fig. 5.3 can be anywhere on the interface) each ray must traverse the same

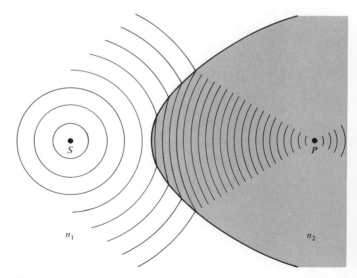

Fig. 5.2 Reshaping a spherical wave at a refracting interface ($n_1 < n_2$).

---

* *Physical optics* deals with situations in which the nonzero wavelength of light must be reckoned with. Analogously, when the de Broglie wavelength of a material object is negligible, we have *classical mechanics*; when it is not, we have the domain of *quantum mechanics* (see Chapter 13).

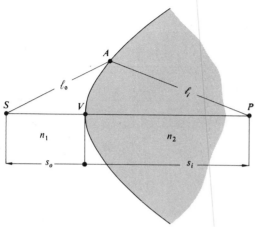

Fig. 5.3 The Cartesian oval.

optical path length. Thus, for example, if S is in a medium of index n_1 and P is in an optically more dense medium of index n_2,

$$\ell_o n_1 + \ell_i n_2 = s_o n_1 + s_i n_2 \qquad (5.1)$$

where s_o and s_i are the *object* and *image distances* measured from the *vertex* or *pole* V respectively. Once we choose s_o and s_i the right-hand side of this equation becomes fixed, and so

$$\ell_o n_1 + \ell_i n_2 = \text{constant.} \qquad (5.2)$$

This is the equation of a *Cartesian oval* whose significance in optics was studied extensively by René Descartes in the early sixteen hundreds (Problem 5.1). Hence, when the boundary between two media has the shape of a Cartesian oval of revolution about the $\overline{SP}$ or *optical axis*, S and P will be conjugate points, i.e. a point source at either location will be perfectly imaged at the other. What's actually occuring physically is rather easy to comprehend. Since $n_2 > n_1$ those regions of the wavefront traveling in the optically more dense medium move slower than those regions traversing the rarer material. Consequently, as the wave begins to pass through the vertex of the oval, the segment immediately about the optical axis is slowed down from c/n_1 to c/n_2. Regions of the same wavefront remote from the axis are still in the first medium traveling with a greater speed, c/n_1. Thus the wavefronts bend and if the boundary is properly configured (in the form of a Cartesian ovoid), the wavefronts will be inverted from diverging to converging spherical segments.

In addition to focusing a spherical wave, we would like to be able to perform a few other reshaping operations using

refracting interfaces; some of these are illustrated in Fig. 5.4. We shall consider these only briefly and more for pedagogical than practical reasons. The surfaces in Fig. 5.4(a) and (b) are ellipsoidal while (c) and (d) are hyperboloidal. Notice that in all cases, the rays either diverge from or converge toward the foci. The arrowheads were omitted to indicate that the rays can go either way. In other words, an incident plane wave will converge to the farthest focus of an ellipsoid just as a spherical wave emitted from that focus will emerge as a plane wave. Furthermore, as you might expect, if we let the point S in Fig. 5.2 move out to infinity the ovoid would gradually metamorphose into an ellipsoid.

Rather than deriving expressions for these surfaces let's just justify the above remarks. To that end, examine Fig. 5.5 which relates back to Fig. 5.4(a). The optical path lengths from any point D on the planar wavefront Σ to the focus F_1 must all be equal to the same constant C, that is

$$(\overline{F_1 A})n_2 + (\overline{AD})n_1 = C$$

or

$$(\overline{F_1 A}) + (\overline{AD})n_{12} = C/n_2. \qquad (5.3)$$

To see that this relationship is indeed satisfied by an ellipsoid of revolution recall that if Σ corresponds to the directrix of the ellipse, $(\overline{F_2 A}) = e(\overline{AD})$ where e is the eccentricity. Thus if $e = n_{12}$, the left-hand side of Eq. (5.3) becomes $(\overline{F_1 A}) + (\overline{F_2 A})$ which is most certainly constant for an ellipse. Here the eccentricity is less than one ($e = n_1/n_2$) but had it been greater than one (i.e. $n_1 > n_2$) the curve would have been a hyperbola instead [compare (a) with (c) and (b) with (d) in Fig. 5.4]. If all of this brings back memories of analytic geometry, you might keep in mind that that subject was originated by Descartes.

The knowledge we have at hand now may be used to construct lenses such that both the object and image points can be in the same medium, which is usually air. The first such device to be considered [Fig. 5.6(a)] is a *double convex hyperbolic* lens which utilizes the response characterized in Fig. 5.4(c). A diverging spherical wave becomes planar after traversing the first hyperbolic surface and then spherically converging on leaving the lens. Alternatively, if the second surface is made planar so that we have a *hyperbolic planar convex* lens as in Fig. 5.6(b), the plane waves within the lens will strike the back surface perpendicularly and emerge unaltered. Another arrangement which will convert diverging spherical waves into plane waves is illustrated in Fig. 5.6(c). This is a *sphero-elliptic convex* lens where F_1 is simultaneously at the center of the spherical

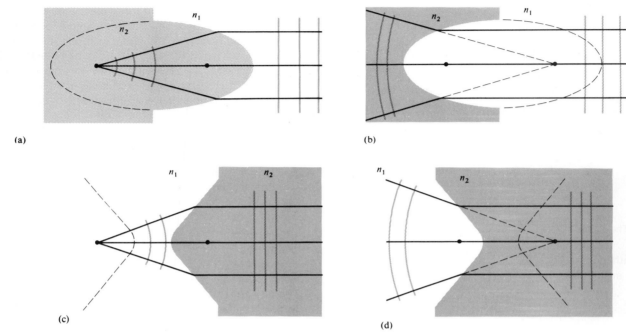

(a)

(b)

(c)

(d)

Fig. 5.4 Ellipsoidal and hyperboloidal refracting surfaces ($n_2 > n_1$).

surface and at the focus of the ellipsoid. Rays from F_1 strike the first surface perpendicularly and are therefore undeviated by it. As in Fig. 5.4(a), the exiting wavefronts are planar. Each of the elements thus far examined have been thicker at their midpoints than they were at their edges and are, for that reason, said to be *convex* (from the Latin *convexus* meaning arched). In contrast, the *planar hyperbolic concave* lens (from the Latin *concavus* meaning hollow; and most easily remembered because it contains the word *cave*) is thinner at the middle than at the edges, as is evident in Fig. 5.6(d). A

number of other arrangements are possible and a few will be considered in the problems (5.2). Note that each of these lenses will work just as well in reverse where the waves shown emerging can instead be thought of as entering from the right.

If a point source is positioned on the optical axis at the point F_1 of the lens in Fig. 5.6(a), rays will *converge* to the conjugate point F_2. A luminous image of the source would appear on a screen placed at F_2, an image which is therefore said to be *real*. On the other hand, in Fig. 5.6(d) the point source is at infinity and the rays emerging from the system this time are *diverging*. They appear to come from a point F_2 but no actual luminous image will appear on a screen at that location. The image here is spoken of as *virtual* just as is the familiar image generated by a plane mirror.

5.2.2 Refraction at Spherical Surfaces

Imagine that we have two pieces of material, one with a concave and the other a convex spherical surface, both having the same radius. It is a unique property of the sphere that such pieces will fit together in intimate contact regardless of their mutual orientation. Thus if we take two roughly spherical objects of suitable curvature, one a grinding tool

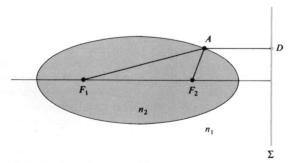

Fig. 5.5 Geometry of an ellipsoid.

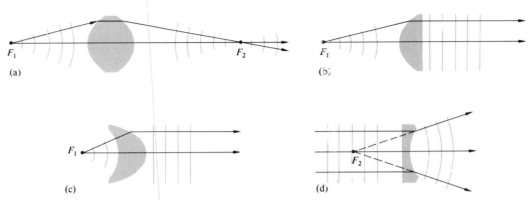

Fig. 5.6 (a) A double hyperbolic lens. (b) A hyperbolic planar convex lens. (c) A sphero-elliptic lens. (d) A planar hyperbolic lens.

and the other a disc of glass, separate them with some abrasive and then randomly move them with respect to each other, we can anticipate that any high spots on either object will wear away. As they wear, both pieces will gradually become ever more spherical (Fig. 5.7). Such surfaces are now commonly generated in batches by automatic grinding and polishing machines. In contrast, high quality aspherical shapes require considerably more effortful working and re-working and accordingly are used, as a rule, only when the concomitant high costs are justifiable. (Molded plastic aspherics are becoming more common in applications not

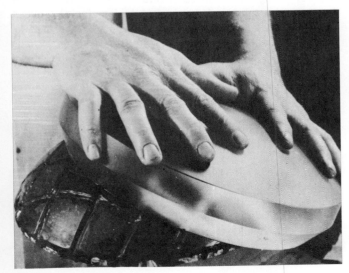

Fig. 5.7 Polishing a spherical lens. [Photo courtesy Optical Society of America.]

requiring great precision.) It should therefore come as no surprise that the vast majority of quality lenses in use today have spherical surfaces. Our intent here is to establish techniques for using such surfaces whereby a great many object points can be satisfactorily imaged simultaneously in light composed of a broad frequency range. Image errors, known as *aberrations*, will occur but it is possible with the present technology to construct high-quality spherical lens systems whose aberrations are so well controlled that image fidelity is limited only by diffraction.

Now that we know why and where we are going, let's move on. Figure 5.8 depicts a wave from the point source S impinging on a spherical interface of radius R centered at C. The ray (SA) will be refracted at the interface toward the local normal $(n_2 > n_1)$ and therefore toward the optical axis. Assume that at some point P it will cross the axis as will all other rays incident at the same angle θ_i (Fig. 5.9). Fermat's principle maintains that the optical path length (O.P.L.) will be stationary, i.e. its derivative with respect to the position variable will be zero. For the ray in question

$$(\text{O.P.L.}) = n_1 \ell_o + n_2 \ell_i. \qquad (5.4)$$

Using the law of cosines in triangles SAC and ACP along with the fact that $\cos \varphi = -\cos(180 - \varphi)$ we get

$$\ell_o = [R^2 + (s_o + R)^2 - 2R(s_o + R)\cos\varphi]^{1/2}$$

and

$$\ell_i = [R^2 + (s_i - R)^2 + 2R(s_i - R)\cos\varphi]^{1/2}.$$

The O.P.L. can be rewritten as

$$(\text{O.P.L.}) = r_1[R^2 + (s_o + R)^2 - 2R(s_o + R)\cos\varphi]^{1/2}$$
$$- n_2[R^2 + (s_i - R)^2 + 2R(s_i - R)\cos\varphi]^{1/2}.$$

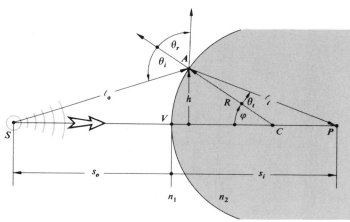

Fig. 5.8 Refraction at a spherical interface.

All of the quantities in the diagram, i.e. s_i, s_o, R etc. are positive numbers and these form the basis of a *sign convention* which is gradually unfolding and to which we shall return time and again (see Table 5.1). Inasmuch as the point A moves at the end of a fixed radius (i.e. $R =$ constant), φ is the position variable and thus setting $d(\text{O.P.L.})/d\varphi = 0$ via Fermat's principle we have

$$\frac{n_1 R(s_o + R)\sin\varphi}{2\ell_o} - \frac{n_2 R(s_i - R)\sin\varphi}{2\ell_i} = 0,$$

from which it follows that

$$\frac{n_1}{\ell_o} + \frac{n_2}{\ell_i} = \frac{1}{R}\left(\frac{n_2 s_i}{\ell_i} - \frac{n_1 s_o}{\ell_o}\right). \quad (5.5)$$

Table 5.1 Sign convention for spherical refracting surfaces and thin lenses* (light entering from the left).

s_o, f_o	+ left of V
x_o	+ left of F_o
s_i, f_i	+ right of V
x_i	+ right of F_i
R	+ if C is right of V
y_o, y_i	+ above optical axis

* This table anticipates the imminent introduction of a few quantities not yet spoken of.

This is the relationship which must hold amongst the parameters for a ray going from S to P by way of refraction at the spherical interface. Although this expression has the attribute of being exact it is rather complicated. We already know that if A is moved to a new location by changing φ the

new ray will not intercept the optical axis at P—this is not a Cartesian oval. The approximations which are used to represent ℓ_o and ℓ_i, and thereby simplify Eq. (5.5), are quite crucial in all that is to follow. Recall that

$$\cos\varphi = 1 - \frac{\varphi^2}{2!} + \frac{\varphi^4}{4!} - \frac{\varphi^6}{6!} + \cdots \quad (5.6)$$

and

$$\sin\varphi = \varphi - \frac{\varphi^3}{3!} + \frac{\varphi^5}{5!} - \frac{\varphi^7}{7!} + \cdots. \quad (5.7)$$

And so if we assume small values of φ, i.e. A close to V, $\cos\varphi \approx 1$. Consequently, the expressions for ℓ_o and ℓ_i yield $\ell_o \approx s_o$, $\ell_i \approx s_i$ and to that approximation

$$\frac{n_1}{s_o} + \frac{n_2}{s_i} = \frac{n_2 - n_1}{R}. \quad (5.8)$$

We could have begun this derivation with Snell's law rather than Fermat's principle (Problem 5.3) in which case small values of φ would have led to $\sin\varphi \approx \varphi$ and Eq. (5.8) once again. This approximation delineates the domain of what is called *first-order theory*—we'll examine *third-order theory* ($\sin\varphi \approx \varphi - \varphi^3/3!$) in the next chapter. Rays which do arrive at shallow angles with respect to the optical axis (such that φ and h are appropriately small) are known as *paraxial rays*. The emerging wavefront segment corresponding to these paraxial rays is essentially spherical and will form a "perfect" image at its center P located at s_i. Notice that Eq. (5.8) is certainly independent of the location of A over a small area about the symmetry axis, namely the *paraxial region*. Gauss in 1841 was the first to give a systematic exposition of the formation of images under the above approximation and the result is variously known as *first-order*, *paraxial*, or *Gaussian optics*. If the optical system is

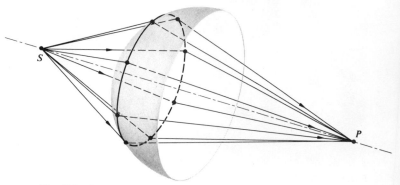

Fig. 5.9 Rays incident at the same angle.

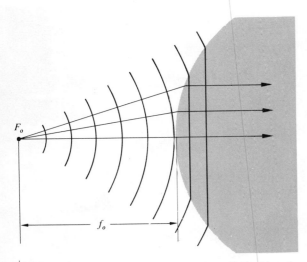

Fig. 5.10 Plane waves propagating beyond a spherical interface—the object focus.

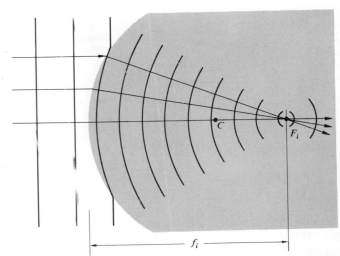

Fig. 5.11 The reshaping of plane into spherical waves at a spherical interface—the image focus.

well corrected, an incident spherical wave will emerge in a form very closely resembling a spherical wave. Consequently as the perfection of the system increases it more closely approaches first-order theory. Deviations from that of paraxial analysis will provide a convenient measure of the quality of an actual optical device.

If the point F_o in Fig. 5.10 is imaged at infinity ($s_i = \infty$) we have

$$\frac{n_1}{s_o} + \frac{n_2}{\infty} = \frac{n_2 - n_1}{R}.$$

That special object distance is defined as the *first focal length* or the *object focal length* $s_o \equiv f_o$, so that

$$f_o = \frac{n_1}{n_2 - n_1} R. \qquad (5.9)$$

The point F_o itself is known as the *first* or *object focus*. Similarly the *second or image focus* is the axial point F_i where the image is formed when $s_o = \infty$, that is

$$\frac{n_1}{\infty} + \frac{n_2}{s_i} = \frac{n_2 - n_1}{R}.$$

Defining the *second or image focal length* f_i as equal to s_i in this special case (Fig. 5.11) we have

$$f_i = \frac{n_2}{n_2 - n_1} R. \qquad (5.10)$$

Recall that an image was virtual when the rays diverged

from it (Fig. 5.12). Analogously, *an object is virtual when the rays converge toward it* (Fig. 5.13). Observe that the virtual object is now on the right-hand side of the vertex and therefore s_o will be taken as a negative quantity. Moreover the surface is concave and its radius will also be negative as required by Eq. (5.9) since f_o would be negative. In the same way the virtual image distance appearing to the left of V is negative.

5.2.3 Thin Lenses

Lenses take a wide range of forms, e.g. there are acoustic and microwave lenses; some of the latter are made of glass or wax in easily recognizable shapes while still others are far more subtle in appearance (Fig. 5.14). In the traditional sense, *a lens is an optical system consisting of two or more refracting interfaces where at least one of these is curved.*

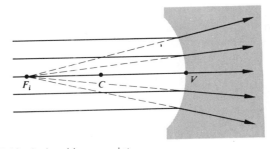

Fig. 5.12 A virtual image point.

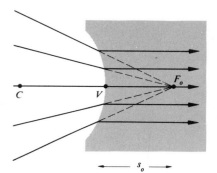

Fig. 5.13 A virtual object point.

Generally the nonplanar surfaces are centered on a common axis. These surfaces are most frequently spherical segments and are often coated with thin dielectric films to control their transmission properties (see Section 9.9). When a lens consists of one element, i.e. when it has only two refracting surfaces, it is a *simple lens*. The presence of more than one element makes it a *compound lens*. A lens is also classified as to whether it is *thin* or *thick*, i.e. whether its thickness is effectively negligible or not. We will limit ourselves, for the most part, to *centered systems* (for which all surfaces are rotationally symmetric about a common axis) of spherical surfaces. Under these restrictions, the simple lens can take the diverse forms shown in Fig. 5.15. Lenses which are variously known as *convex*, *converging* or *positive* are thicker at the center and so tend to decrease the radius of curvature of the wavefronts, i.e. the wave becomes more converging as it traverses the lens. This is, of course, assuming that the index of the lens is greater than that of the media in which it is immersed. *Concave*, *diverging* or *negative* lenses, on the other hand, are thinner at the center and tend to advance that portion of the wavefront causing it to become more diverging than it was upon entry.

i)   *Thin-Lens Equations*

Return for a moment to the discussion of refraction at a single spherical interface where the location of the conjugate points S and P is given by

$$\frac{n_1}{s_o} + \frac{n_2}{s_i} = \frac{n_2 - n_1}{R}. \qquad [5.8]$$

When s_o is large for a fixed $(n_2 - n_1)/R$, s_i is relatively small. As s_o decreases, s_i moves away from the vertex, i.e. both θ_i and θ_t increase until finally $s_o = f_o$ and $s_i = \infty$. At that

Fig. 5.14 A lens for short-wavelength radio waves. The disks serve to refract these waves much as rows of atoms refract light. [Photo courtesy Optical Society of America.]

point, $n_1/s_o = (n_2 - n_1)/R$ so that if s_o gets any smaller, s_i will have to be negative if Eq. (5.8) is to hold. In other words, the image becomes virtual (Fig. 5.16). Let's now locate the conjugate points for the lens of index n_l surrounded by a medium of index n_m as in Fig. 5.17 where we have simply ground another end on the piece in Fig. 5.16(c). This certainly isn't the most general set of circumstances but it is the most common and, even more cogently, it is the simplest.* We know from Eq. (5.8) that the paraxial rays issuing from S at s_{o1} will meet at P' a distance, we now call s_{i1}, from V_1 given by

$$\frac{n_m}{s_{o1}} + \frac{n_l}{s_{i1}} = \frac{n_l - n_m}{R_1}. \qquad (5.11)$$

Thus as far as the second surface is concerned, it "sees" rays coming toward it from P' which serves as its object point a

---

* See Jenkins and White, *Fundamentals of Optics*, p. 57 for a derivation containing three different indices.

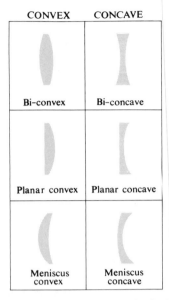

CONVEX   CONCAVE

Bi-convex	Bi-concave
Planar convex	Planar concave
Meniscus convex	Meniscus concave

Fig. 5.15 Cross sections of various centered spherical simple lenses.

distance s_{o2} away. Furthermore, the rays arriving at that second surface are in the medium of index n_l. Thus, the object space for the second interface which contains P' has an index n_l. Note that the rays from P' to that surface are indeed straight lines. Considering the fact that

$$|s_{o2}| = |s_{i1}| + d$$

since s_{o2} is on the left and therefore positive $s_{o2} = |s_{o2}|$ while s_{i1} is also on the left and therefore negative, i.e. $-s_{i1} = |s_{i1}|$, we have

$$s_{o2} = -s_{i1} + d. \qquad (5.12)$$

Thus Eq. (5.8) applied at the second surface yields

$$\frac{n_l}{(-s_{i1} + d)} + \frac{n_m}{s_{i2}} = \frac{n_m - n_l}{R_2}. \qquad (5.13)$$

Here $n_l > n_m$ and $R_2 < 0$ so that the right-hand side is positive. Adding Eqs. (5.11) and (5.13) results in

$$\frac{n_m}{s_{o1}} + \frac{n_m}{s_{i2}} = (n_l - n_m)\left(\frac{1}{R_1} - \frac{1}{R_2}\right) + \frac{n_l d}{(s_{i1} - d)s_{i1}}. \qquad (5.14)$$

If the lens is thin enough ($d \to 0$), the last term on the right is effectively zero. As a further simplification, assume the surrounding medium to be air (i.e. $n_m \approx 1$). Accordingly, we have the very useful *thin-lens equation* often referred to

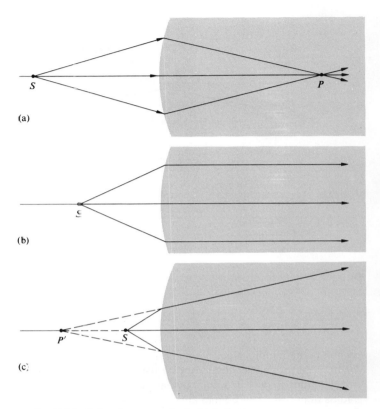

Fig. 5.16 Refraction at a spherical interface.

as the *lensmaker's formula*:

$$\frac{1}{s_o} + \frac{1}{s_i} = (n_l - 1)\left(\frac{1}{R_1} - \frac{1}{R_2}\right), \qquad (5.15)$$

where we let $s_{o1} = s_o$ and $s_{i2} = s_i$. The points V_1 and V_2 tend to coalesce as $d \to 0$ so that s_o and s_i can be measured from either the vertices or the lens center.

Just as in the case of the single spherical surface, if s_o is moved out to infinity, the image distance becomes the focal length f_i or symbolically

$$\lim_{s_o \to \infty} s_i = f_i.$$

Similarly

$$\lim_{s_i \to \infty} s_o = f_o.$$

It is evident from Eq. (5.15) that for a thin lens $f_i = f_o$ and consequently we drop the subscripts altogether. Thus

$$\frac{1}{f} = (n_l - 1)\left(\frac{1}{R_1} - \frac{1}{R_2}\right) \qquad (5.16)$$

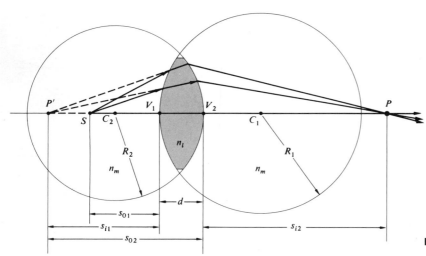

Fig. 5.17 A spherical lens.

and

$$\frac{1}{s_o} + \frac{1}{s_i} = \frac{1}{f},$$ (5.17)

which is the famous *Gaussian lens formula*. As an example of how these expressions might be used, let's compute the focal length in air of a thin planar-convex lens having a radius of curvature of 50 mm and an index of 1.5. With light entering on the planar surface ($R_1 = \infty$, $R_2 = -50$),

$$\frac{1}{f} = (1.5 - 1)\left(\frac{1}{\infty} - \frac{1}{-50}\right),$$

while if instead it arrives at the curved surface ($R_1 = +50$, $R_2 = \infty$),

$$\frac{1}{f} = (1.5 - 1)\left(\frac{1}{+50} - \frac{1}{\infty}\right),$$

and in either case $f = 100$ mm. If an object is alternately placed at distances 600 mm, 200 mm, 150 mm, 100 mm, and 50 mm from the lens on either side, we can find the image points from Eq. (5.17). Hence

$$\frac{1}{600} + \frac{1}{s_i} = \frac{1}{100}$$

and $s_i = 120$ mm. Similarly, the other image distances are 200 mm, 300 mm, ∞, and -100 mm respectively. Interestingly enough, when $s_o = \infty$, $s_i = f$; as s_o decreases s_i increases positively until $s_o = f$ and s_i is negative thereafter. You can qualitatively check this out with a simple convex lens and a small electric light—the "high-intensity" variety that uses auto lamps is probably the most convenient. Standing as far as you can from the source, project a clear image of it onto a white sheet of paper. You should be able to see the lamp quite clearly and not just as a blur. That image distance approximates f. Now move the lens in toward S adjusting s_i to produce a clear image. It will surely increase. As $s_o \rightarrow f$, a clear image of the filament can be projected, but only on an ever-increasingly distant screen. For $s_o < f$ there will just be a blur where the farthest wall intersects the diverging cone of rays—the image is virtual.

ii) Focal Points and Planes

Figure 5.18 summarizes pictorially some of the situations described analytically by Eq. 5.16. Observe that if a lens of index n_l is in a medium of index n_m

$$\frac{1}{f} = (n_{lm} - 1)\left(\frac{1}{R_1} - \frac{1}{R_2}\right).$$ (5.18)

The focal lengths in Fig. 5.18(a) and (b) are equal because the same medium exists on either side of the lens. Since $n_l > n_m$ it follows that $n_{lm} > 1$. In both cases $R_1 > 0$ and $R_2 < 0$ so that each focal length is positive. We have a real object in (a) and a real image in (b). In (c), $n_l < n_m$ and consequently f is negative. In (d) and (e), $n_{lm} > 1$ but $R_1 < 0$ while $R_2 > 0$ and so f is again negative and the object in one case, and the image in the other, are virtual. The last situation shows $n_{lm} < 1$ yielding an $f > 0$.

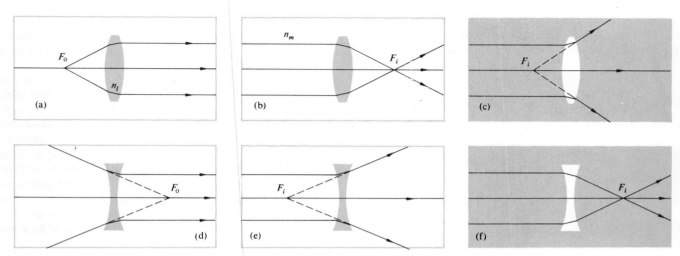

Fig. 5.18 Focal lengths for converging and diverging lenses.

Notice that in each instance it is particularly convenient to draw a ray through the center of the lens which, because it is perpendicular to both surfaces, is undeviated. Suppose however that an off-axis paraxial ray emerges from the lens parallel to its incident direction as in Fig. 5.19. We maintain that all such rays will pass through the point defined as the *optical center* of the lens O. To see this, draw two parallel planes, one on each side tangent to the lens at any pair of points A and B. This can easily be done by selecting A and B such that the radii $\overline{AC_1}$ and $\overline{BC_2}$ are themselves parallel. It is to be shown that the paraxial ray traversing $\overline{AB}$ enters and leaves the lens in the same direction. It is evident from the diagram that triangles AOC_1 and BOC_2 are similar, in the geometrical sense, and therefore their sides are proportional. Hence, $|R_1|(\overline{OC_2}) = |R_2|(\overline{OC_1})$ and since the radii are constant, the location of O is constant, independent of A and B. As we saw earlier (Problem 4.15 and Fig. 4.44), a ray traversing a medium bounded by parallel planes will be displaced laterally, but will suffer no angular deviation. This displacement is proportional to the thickness, which for a thin lens is negligible. *Rays passing through O may, accordingly, be drawn as straight lines.* It is customary when dealing with thin lenses simply to place O midway between the vertices.

Recall that a bundle of parallel paraxial rays incident on a spherical refracting surface comes to a focus at a point on the optical axis (Fig. 5.11). As shown in Fig. 5.20, this implies that several such bundles entering in a narrow cone will be focused on a spherical segment σ, also centered on

C. The undeviated rays normal to the surface, and therefore passing through C, locate the foci on σ. Since the ray cone must indeed be narrow, σ can satisfactorily be represented as a plane normal to the symmetry axis and passing through the image focus. It is known as a *focal plane*. In the same way, limiting ourselves to paraxial theory, a lens will focus all incident parallel bundles of rays[*] onto a surface called the *second or back focal plane* as in Fig. 5.21. Here each point on σ is located by the undeviated ray through O. Similarly, the *first or front focal plane* contains the object focus F_o.

iii) Finite Imagery

Thus far we've dealt with the mathematical abstraction of a single-point source, but now let's suppose there to be a great many such points combining to form a continuous finite object. For the moment, imagine the object to be a segment of a sphere, σ_o, centered on C as in Fig. 5.22. If σ_o is close to the spherical interface, point S will have a virtual image $P(s_i < 0$ and therefore on the left of $V)$. With S farther away, its image will be real ($s_i > 0$ and therefore on the right-hand side). In either case, each point on σ_o has a conjugate point on σ_i lying on a straight line through C. Within the restrictions

---

[*] Perhaps the earliest literary reference to the focal properties of a lens appears in Aristophanes' play, *The Clouds*, which dates back to 423 B.C. In it Strepsiades plots to use a burning glass to focus the sun's rays onto a wax tablet and thereby melt out the record of a gambling debt.

of paraxial theory, these surfaces can be considered as planar. Thus a small planar object normal to the optical axis will be imaged into a small planar region also normal to

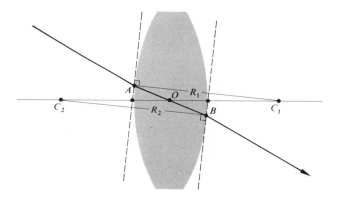

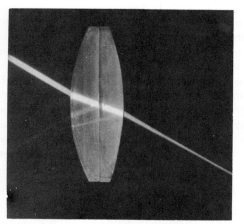

Fig. 5.19 The optical center of a lens. [Photo by E.H.].

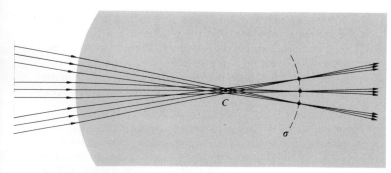

Fig. 5.20 Focusing of several ray bundles.

that axis. It should be noted that if σ_o is moved out to infinity, the cone of rays from each source point will become *collimated*, i.e. parallel, and the image points will lie on the focal plane (Fig. 5.21).

By cutting and polishing the right side of the piece depicted in Fig. 5.22, we can construct a thin lens just as was done in Section (i). Once again, the image (σ_i in Fig. 5.22) formed by the first surface of the lens will serve as the object for the second surface, which in turn will generate a final image. Suppose then that σ_i in Fig. 5.22(a) is the object for the second surface which is assumed to have a negative radius. We already know what will happen next— the situation is identical to Fig. 5.22(b) with the ray directions reversed. The *final image formed by a lens of a small planar object normal to the optical axis will itself be a small plane normal to that axis*.

The location, size and orientation of an image produced by a lens can be determined, particularly simply, using ray diagrams. To find the image of the object in Fig. 5.23, we must locate the image point corresponding to each object point. Since all rays issuing from a source point in a paraxial cone will arrive at the image point, any two such rays will suffice to fix that point. Knowing the positions of the focal points, there are three rays which are especially easy to apply. Two of these make use of the fact that a ray passing through the focal point will emerge from the lens parallel to the optical axis and vice versa; the other is the undeviated ray through O. Incidentally, this technique dates back to the work of Robert Smith as long ago as 1738.

This graphical procedure can be made even simpler by replacing the thin lens with a plane passing through its

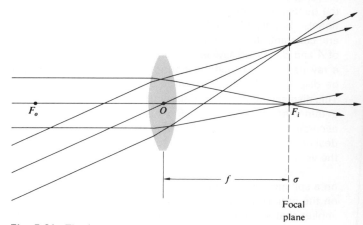

Fig. 5.21 The focal plane of a lens.

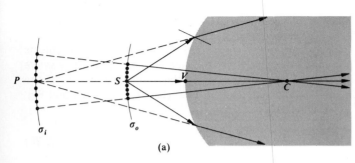

(a)

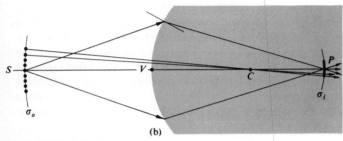

(b)

Fig. 5.22 Finite imagery.

center (Fig. 5.24). Presumably if we were to extend every incoming ray forward a little and every outgoing ray backward a bit, each pair would meet on this plane. Thus the total deviation of any ray can be envisaged as occurring all at once on that plane. This is equivalent to the actual process consisting of two separate angular shifts, one at each interface. (As we will see later, this is tantamount to saying that the two principal planes of a thin lens coincide.)

In accord with convention, transverse distances above the optical axis are taken as positive quantities while those below the axis are given negative numerical values. Therefore in Fig. 5.24 $y_o > 0$ and $y_i < 0$. Here the image is said to be *inverted* whereas if $y_i > 0$ when $y_o > 0$, it is *erect*. Observe that triangles AOF_i and $P_2P_1F_i$ are similar. Ergo

$$\frac{y_o}{|y_i|} = \frac{f}{(s_i - f)},\tag{5.19}$$

Likewise, triangles S_2S_1O and P_2P_1O are similar and

$$\frac{y_o}{|y_i|} = \frac{s_o}{s_i},\tag{5.20}$$

where all quantities other than y_i are positive. Hence

$$\frac{s_o}{s_i} = \frac{f}{(s_i - f)}\tag{5.21}$$

and

$$\frac{1}{f} = \frac{1}{s_o} + \frac{1}{s_i},$$

which is, of course, the Gaussian lens equation (5.17). Furthermore, triangles $S_2S_1F_o$ and BOF_o are similar and accordingly

$$\frac{f}{(s_o - f)} = \frac{|y_i|}{y_o}.\tag{5.22}$$

Using the distances measured from the focal points and combining this with Eq. (5.19) we have

$$x_o x_i = f^2.\tag{5.23}$$

This is the *Newtonian form* of the lens equation, the first statement of which appeared in Newton's *Opticks* in 1704. The signs of x_o and x_i are reckoned with respect to their concomitant foci. By convention x_o is taken to be positive

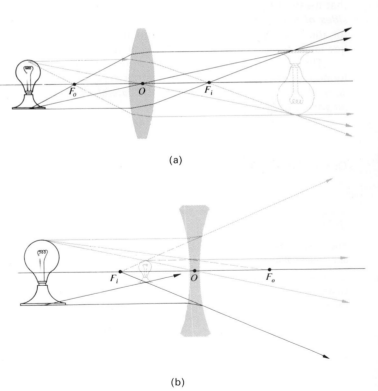

(a)

(b)

Fig. 5.23 (a) A real object and a positive lens. (b) A real object and a negative lens.

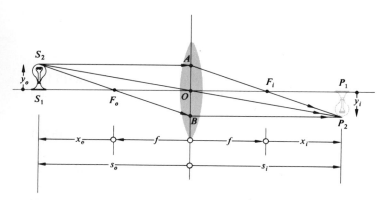

Fig. 5.24 Object and image location for a thin lens.

left of F_o whereas x_i is positive on the right of F_i. To be sure, it is evident from Eq. (5.23) that x_o and x_i have like signs and that means that *the object and image must be on opposite sides of their respective focal points.* This is a good thing for the neophyte to remember when making those hasty freehand ray diagrams for which he is already infamous.

The ratio of the transverse dimensions of the final image formed by any optical system to the corresponding dimension of the object is defined as the *lateral* or *transverse magnification M_T*, that is

$$M_T \equiv \frac{y_i}{y_o}. \tag{5.24}$$

Or from Eq. (5.20)

$$M_T = -\frac{s_i}{s_o}. \tag{5.25}$$

Table 5.2 Meanings associated with the signs of various thin lens and spherical interface parameters.

Quantity	Sign	
	+	−
s_o	Real object	Virtual object
s_i	Real image	Virtual image
f	Converging lens	Diverging lens
y_o	Erect object	Inverted object
y_i	Erect image	Inverted image
M_T	Erect image	Inverted image

Thus *a positive M_T connotes an erect image, while a negative value means the image is inverted* (see Table 5.2). Bear in mind that s_i and s_o are both positive for real objects and images. *Clearly then all such images formed by a single thin lens will be inverted.* The Newtonian expression for the magnification follows from Eqs. (5.19) and (5.22) and Fig. 5.24, whence

$$M_T = -\frac{x_i}{f} = -\frac{f}{x_o}. \tag{5.26}$$

The term magnification is a misnomer since the magnitude of M_T can certainly be less than one in which case the image is smaller than the object. We have $M_T = -1$ when the object and image distances are positive and equal, and that happens (5.17) only when $s_o = s_i = 2f$. This turns out to be (Problem 5.5) the configuration where the object and image are as close together as they can possibly get (namely a distance $4f$ apart). Table 5.3 summarizes a number of image

Table 5.3 Images of real objects formed by thin lenses.

	Convex				
Object		Image			
Location	Type	Location	Orientation	Relative size	
$\infty > s_o > 2f$	Real	$f < s_i < 2f$	Inverted	Minified	
$s_o = 2f$	Real	$s_i = 2f$	Inverted	Same size	
$f < s_o < 2f$	Real	$\infty > s_i > 2f$	Inverted	Magnified	
$s_o = f$		$\pm \infty$			
$s_o < f$	Virtual	$|s_i| > s_o$	Erect	Magnified	

	Concave				
Object		Image			
Location	Type	Location	Orientation	Relative size	
Anywhere	Virtual	$|s_i| < |f|$	Erect	Minified	

configurations resulting from the juxtaposition of a thin lens and a real object.

Presumably the image of a three-dimensional object will itself occupy a three-dimensional region of space. The optical

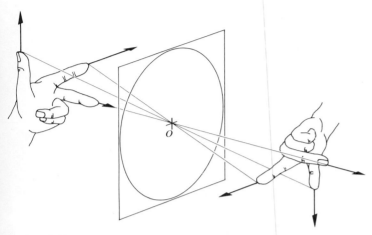

Fig. 5.25 Image orientation for a thin lens.

system can apparently affect both the transverse and longitudinal dimensions of the image. The *longitudinal magnification, M_L,* which relates to the axial direction, is defined as

$$M_L \equiv \frac{dx_i}{dx_o}. \qquad (5.27)$$

This is the ratio of an infinitesimal axial length in the region of the image to the corresponding length in the region of the object. Differentiating Eq. (5.23) leads to

$$M_L = -\frac{f^2}{x_o^2} = -M_T^2 \qquad (5.28)$$

for a thin lens in a single medium. Evidently, $M_L < 0$ which implies that a positive dx_o corresponds to a negative dx_i and vice versa. In other words, a finger pointing towards the lens is imaged pointing away from it (Fig. 5.25).

Form the image of a window on a sheet of paper using a simple convex lens. Assuming a lovely arboreal scene, image the distant trees on the screen. Now move the paper *away* from the lens so that it intersects a different region of the image space. The trees will fade while the nearby window itself comes into view.

iv) Thin-Lens Combinations

It is not our intent here to have the reader develop a proficiency with the subtle intricacies of modern lens design, but rather to bring him to a point where he can begin to appreciate, utilize and adapt those systems already available.

In constructing a new optical system, one generally begins by sketching out a rough arrangement using the quickest approximate calculations. Refinements are then added as the designer goes on to the prodigious and more exact ray-tracing techniques. Nowadays these computations are most often carried out by electronic digital computers. Even so, the simple thin-lens concept provides a highly useful basis for preliminary calculations in a broad range of situations.

No lens is actually a thin lens in the strict sense of having a thickness which approaches zero. Yet many simple lenses, for all practical purposes, function in a fashion equivalent to that of a thin lens. Spectacle lenses which, by the way, have been used at least since the thirteenth century are almost all in this category. When the radii of curvature are large and the lens diameter is small, the thickness will usually be small as well. A lens of this sort would generally have a large focal length compared to which the thickness would be quite small, e.g. many early telescope objectives fit that description perfectly.

We now propose to derive some expressions for parameters associated with thin-lens combinations. The approach here will be fairly simple, leaving the more elaborate traditional treatment for those tenacious enough to pursue the matter into the next chapter.

Suppose we have two thin positive lenses L_1 and L_2 separated by a distance d which is smaller than either focal length as in Fig. 5.26. The resulting image can be located graphically as follows. Overlooking the presence of L_2, the image formed exclusively by L_1 is constructed using rays 1 and 3. As usual, these pass through the lens' object and image foci, namely F_{o1} and F_{i1} respectively. The object is in a normal plane so that two rays determine its top and a perpendicular to the optical axis finds its bottom. Ray 2 is then constructed running backwards from P_1' through O_2. Inserting L_2 has no effect on ray 2 whereas 3 is refracted through the image focus F_{i2} of L_2. The intersection of rays 2 and 3 fixes the image, which in this particular case is real, minified and inverted.

A similar pair of lenses is illustrated in Fig. 5.27 where now the separation has been increased. Once again rays 1 and 3 through F_{i1} and F_{o1} fix the position of the intermediate image generated by L_1 alone. As before, ray 2 is drawn backward from O_2 to P_1' to S_1. The intersection of 2 and 3, as the latter is refracted through F_{i2}, locates the final image. This time it is real and erect. Notice that if the focal

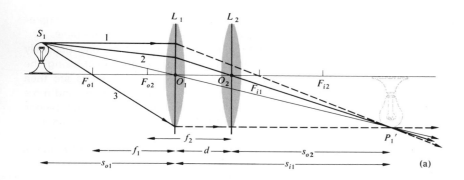

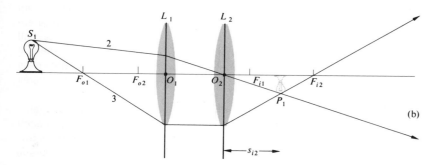

Fig. 5.26 Two thin lenses separated by a distance smaller than either focal length.

length of L_2 is increased with all else constant, the size of the image increases as well.

Analytically we have for L_1

$$\frac{1}{s_{i1}} = \frac{1}{f_1} - \frac{1}{s_{o1}} \qquad (5.29)$$

or

$$s_{i1} = \frac{s_{o1}f_1}{s_{o1} - f_1}. \qquad (5.30)$$

This is positive and the intermediate image is to the right of L_1, when $s_{o1} > f_1$ and $f_1 > 0$. For L_2

$$s_{o2} = d - s_{i1}, \qquad (5.31)$$

and if $d > s_{i1}$ the object for L_2 is real (as in Fig. 5.27) while if $d < s_{i1}$ it is virtual ($s_{o2} < 0$ as in Fig. 5.26). In the former instance the rays approaching L_2 are diverging from P_1', while in the latter they are converging toward it. Furthermore

$$\frac{1}{s_{i2}} = \frac{1}{f_2} - \frac{1}{s_{o2}}$$

or

$$s_{i2} = \frac{s_{o2}f_2}{s_{o2} - f_2}.$$

Using Eq. (5.31) this becomes

$$s_{i2} = \frac{(d - s_{i1})f_2}{(d - s_{i1} - f_2)}. \qquad (5.32)$$

In this same way we could compute the response of any number of thin lenses. It will often be convenient to have a single expression, at least when dealing with only two lenses, and so substituting for s_{i1} from Eq. (5.29) we get

$$s_{i2} = \frac{f_2 d - f_2 s_{o1}f_1/(s_{o1} - f_1)}{d - f_2 - s_{o1}f_1/(s_{o1} - f_1)}. \qquad (5.33)$$

Here s_{o1} and s_{i2} are the object and image distances of the

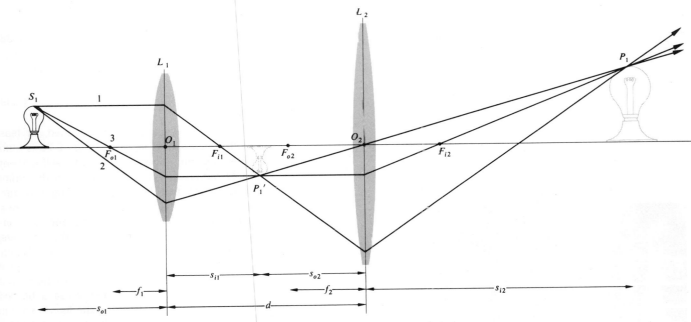

Fig. 5.27 Two thin lenses separated by a distance greater than the sum of their focal lengths.

compound lens, As an example, let's compute the image distance associated with an object placed 50 cm from the first of two positive lenses. These in turn are separated by 20 cm and have focal lengths of 30 cm and 50 cm respectively. By direct substitution (5.33)

$$s_{i2} = \frac{50(20) - 50(50)(30)/(50 - 30)}{20 - 50 - 50(30)/(50 - 30)} = 26.2 \text{ cm}$$

and the image is real. Inasmuch as L_2 "magnifies" the intermediate image formed by L_1, the total transverse magnification of the compound lens is the product of the individual magnifications, that is

$$M_T = M_{T1} M_{T2}.$$

It is left as a problem to show that

$$M_T = \frac{f_1 s_{i2}}{d(s_{o1} - f_1) - s_{o1} f_1}. \tag{5.34}$$

In the above example

$$M_T = \frac{30(26.2)}{20(50 - 30) - 50(30)} = -0.72$$

and just as we should have guessed from Fig. 5.26, the image is minified and inverted.

The distance from the last surface of an optical system to the second focal point of that system as a whole is known as the *back focal length* or b.f.l. Likewise the distance from the vertex of the first surface to the first or object focus is the *front focal length* or f.f.l. Consequently if we let $s_{i2} \to \infty$, s_{o2} approaches f_2, which combined with Eq. (5.31) tells us that $s_{i1} \to d - f_2$. Hence from Eq. (5.29)

$$\frac{1}{s_{o1}}\bigg|_{s_{i2}=\infty} = \frac{1}{f_1} - \frac{1}{(d - f_2)} = \frac{d - (f_1 + f_2)}{f_1(d - f_2)}.$$

But this special value of s_{o1} is the f.f.l.:

$$\text{f.f.l.} = \frac{f_1(d - f_2)}{d - (f_1 + f_2)}. \tag{5.35}$$

In the same way, letting $s_{o1} \to \infty$ in Eq. (5.33), $(s_{o1} - f_1) \to s_{o1}$ and since s_{i2} is then the b.f.l. we have

$$\text{b.f.l.} = \frac{f_2(d - f_1)}{d - (f_1 + f_2)}. \tag{5.36}$$

To see how this works numerically, let's find both the b.f.l.

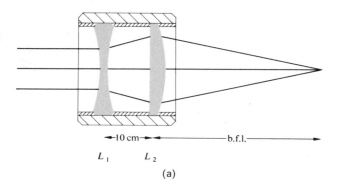

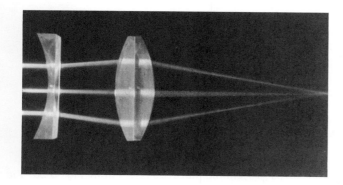

Fig. 5.28 A positive and negative thin-lens combination. [Photo by E.H.]

and f.f.l. for the thin-lens system in Fig. 5.28(a) where $f_1 = -30$ cm and $f_2 = +20$ cm. Then

$$\text{b.f.l.} = \frac{20\,[10 - (-30)]}{10 - (-30 + 20)} = 40 \text{ cm}$$

and similarly f.f.l. = 15 cm. Incidentally, notice that if $d = f_1 + f_2$, plane waves entering the compound lens from either side will emerge as plane waves (Problem 5.18) as in telescopic systems.

Observe that if $d \to 0$, i.e. if the lenses are brought into contact, as e.g. in the case of some achromatic doublets,

$$\text{b.f.l.} = \text{f.f.l.} = \frac{f_2 f_1}{f_2 + f_1}. \qquad (5.37)$$

The resultant thin lens has an *effective focal length f* such

that

$$\frac{1}{f} = \frac{1}{f_1} + \frac{1}{f_2}. \qquad (5.38)$$

This implies that if there are *N* such lenses in contact

$$\frac{1}{f} = \frac{1}{f_1} + \frac{1}{f_2} + \cdots + \frac{1}{f_N}. \qquad (5.39)$$

Many of these conclusions can be verified, at least qualitatively, with a few simple lenses. Figure 5.26 is quite easy to duplicate, and the procedure should be self evident, whereas Fig. 5.27 requires a bit more care. First determine the focal lengths of the two lenses by imaging a distant source. Then hold one of the lenses (L_2) at a fixed distance, *slightly greater than its focal length*, from the plane of observation, i.e. a piece of white paper. Now comes the maneuver that requires some effort if you don't have an optical bench. Move the second lens (L_1) toward the source keeping it reasonably centered. Without any attempts to block out light entering L_2 directly, you will probably see a blurred image of your hand holding L_1. Position the lenses so that the region on the screen corresponding to L_1 is as bright as possible. The scene spread across L_1 (i.e. its image within the image) will become clear and erect as in Fig. 5.27.

5.3 STOPS

5.3.1 Aperture and Field Stops

The intrinsically finite nature of all lenses demands that they will collect only a fraction of the energy emitted by a point source. The physical limitation presented by the periphery of a simple lens therefore determines which rays shall enter the system to ultimately form an image. In that respect, the unobstructed or *clear diameter* of the lens functions as an aperture into which energy flows. Any element, be it the rim of a lens or a separate diaphragm, which determines the amount of light reaching the image is known as the *aperture stop*, abbreviated as A.S. The adjustable leaf diaphragm that is usually located behind the first few elements of a compound camera lens is just such an aperture stop. Evidently it determines the light-gathering capability of the lens as a whole. As shown in Fig. 5.29, highly oblique rays can still enter a system of this sort. These however are usually deliberately restricted in order to control the quality of the image. The element limiting the size or angular breadth of the object which can be imaged by the system is called the *field stop* or F.S.—it determines the field of view

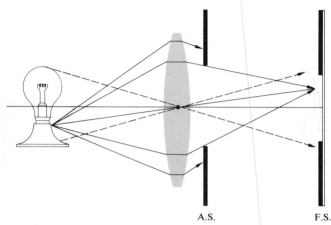

Fig. 5.29 Aperture and field stops.

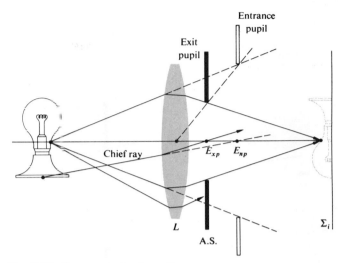

Fig. 5.30 Entrance and exit pupils.

of the instrument. In a camera, it's the edge of the film itself which bounds the image plane and serves as the field stop. Thus, while (Fig. 5.29) the aperture stop controls the number of rays from an object point reaching the conjugate image point, it is the field stop which will or will not obstruct those rays *in toto*. Neither the very top nor the bottom of the object in Fig. 5.29 passes the field stop. Opening the circular aperture stop would cause the system to accept a larger energy cone and in so doing increase the irradiance at each image point. In contrast, opening the field stop would allow the extremities of the object, which were previously blocked, now to be imaged.

5.3.2 Entrance and Exit Pupils

Another concept, quite useful in determining whether or not a given ray will traverse the entire optical system, is the *pupil*. This is simply an *image of the aperture stop*. The *entrance pupil* of a system is the image of the aperture stop as seen from an axial point on the object through those elements preceding the stop. If there are no lenses between the object and the A.S., the latter itself serves as the entrance pupil. To illustrate the point examine Fig. 5.30, which is a lens with a *rear aperture stop*. The image of the aperture stop in L is virtual (see Table 5.3) and magnified. It can be located by sending a few rays out from the edges of the A.S. in the usual way. In contrast, the *exit pupil* is the image of the A.S. as seen from an axial point on the image plane through the interposed lenses, if there are any. In Fig. 5.30 there are no such lenses and so the aperture stop itself serves as the exit pupil. Notice that all of this just means

that the cone of light actually entering the optical system is determined by the entrance pupil while the cone leaving it is controlled by the exit pupil. No rays from the source point proceeding outside of either cone will make it to the image plane.

If you wanted to use a telescope or a monocular as a camera lens, you might attach an external *front aperture stop* to control the amount of incoming light for exposure purposes. Figure 5.31 represents a similar arrangement where

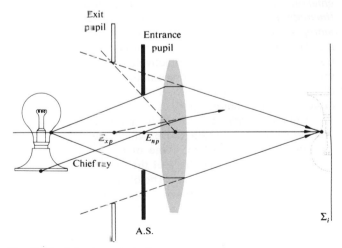

Fig. 5.31 A front aperture stop.

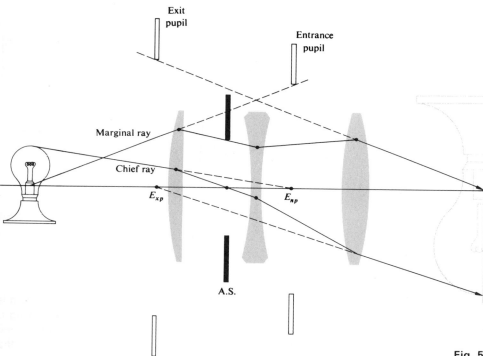

Fig. 5.32 Pupils and stops for a three-lens system.

the entrance and exit pupil locations should be self evident. The last two diagrams included a ray labeled as the *chief ray*. It is defined to be *any ray from an off-axis object point which passes through the center of the aperture stop. The chief ray enters the optical system along a line directed toward the midpoint of the entrance pupil, E_{np}, and leaves the system along a line passing through the center of the exit pupil, E_{xp}.* The chief ray, associated with a conical bundle of rays from a point on the object, effectively behaves as the central ray of the bundle and is representative of it. Chief rays are of particular importance when the aberrations of a lens design are being corrected for.

Figure 5.32 depicts a somewhat more involved situation. The two rays shown are those which are usually traced through an optical system. One of these is the chief ray from a point on the periphery of the object which is to be accommodated by the system. The other is called a *marginal ray* since it goes from the axial object point to the rim or margin of the entrance pupil (or aperture stop).

In a situation where it is not clear which element is the actual aperture stop, each component of the system must be imaged by the remaining elements to its left. *The image*

which subtends the smallest angle at the axial object point is the entrance pupil. The element whose image is the entrance pupil is then the aperture stop of the system for that object point. Problem 5.21 deals with just this kind of calculation.

Notice how the cone of rays, in Fig. 5.33, which can reach the image plane becomes narrower as the object point moves off-axis. The effective aperture stop, which for the axial bundle of rays was the rim of L_1, has been markedly reduced for the off-axis bundle. The result is a gradual fading out of the image at points near its periphery, a process known as *vignetting*.

The locations and sizes of the pupils of an optical system are of considerable practical importance. In visual instruments, the observer's eye is positioned at the center of the exit pupil. The pupil of the eye itself will vary from 2 mm to about 8 mm depending on the general illumination level. Thus a telescope or binocular designed primarily for evening use might have an exit pupil of at least 8 mm (you may have heard the term *night glasses*—they were quite popular on roofs during the Second World War). In contrast, a daylight version will suffice with an exit pupil of 3 or 4 mm. The larger

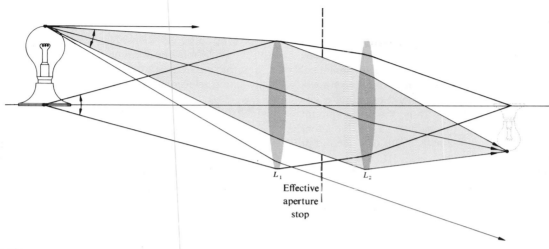

Fig. 5.33 Vignetting.

the exit pupil, the easier it will be to properly align your eye with the instrument. Obviously a telescopic sight for a high-powered rifle should have a large exit pupil located far enough behind the scope so as to avoid injury from recoil.

5.3.3 Relative Aperture and f-number

Suppose we wish to collect the light from an extended source and form an image of it using a lens (or mirror). The amount of energy gathered by the lens (or mirror) from some small region of a distant source will be directly proportional to the area of the lens or more generally to the area of the entrance pupil. A large *clear aperture* will intersect a large cone of rays. Obviously, if the source were a laser having a very narrow beam, this would not necessarily be true. If we neglect losses due to reflections, absorption etc., the incoming energy will be spread across a corresponding region of the image. Thus the energy per unit area per unit time, i.e. the flux density or irradiance, will be inversely proportional to the image area. The entrance pupil area, if circular, varies as the square of its radius and is therefore proportional to the square of its diameter D. Furthermore, the image area will go as the square of its lateral dimension which in turn [Eqs. (5.24) and (5.26)] is proportional to f^2. (Keep in mind that we are talking about an extended object rather than a point source. In the latter case, the image would be confined to a very small area independent of f.) Thus the flux density at the image plane varies as $(D/f)^2$. The ratio D/f is known as the *relative aperture*, while its inverse is said to be the *f-number*

or $f/\#$, that is

$$f/\# \equiv \frac{f}{D} \qquad (5.40)$$

where $f/\#$ should be understood as a single symbol. For example, a lens with a 25 mm aperture and a 50 mm focal length has an f-number of 2 and is usually designated as $f/2$. Figure 5.34 illustrates the point by showing a thin lens behind a variable iris diaphragm operating at either $f/2$ or $f/4$. A smaller f-number clearly permits more light to reach the image plane.

Camera lenses are usually specified by their focal lengths and largest possible apertures, e.g. you might see 50 mm, $f/1.4$ on the barrel of a lens. Since the photographic exposure time is proportional to the square of the f-number, the $f/\#$ is sometimes spoken of as the *speed* of the lens. An $f/1.4$ lens is said to be twice as fast as an $f/2$ lens.

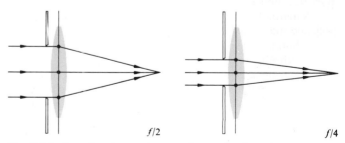

Fig. 5.34 Stopping down a lens to change the f-number.

Usually lens diaphragms have *f-number* markings of 1, 1.4, 2, 2.8, 4, 5.6, 8, 11, 16, 22 etc. The largest relative aperture in this case corresponds to $f/1$ and that's a fast lens—$f/2$ is more typical. Each consecutive diaphragm setting increases the $f/\#$ by a multiplicative factor of $\sqrt{2}$ (within numerical round-off). This corresponds to a decrease in relative aperture by a multiplicative factor of $1/\sqrt{2}$ and therefore a decrease in flux density by *one-half*. Thus, the same amount of light will reach the film if the camera is set for either $f/1.4$ at 1/500th of a second, $f/2$ at 1/250th of a second or $f/2.8$ at 1/125th of a second.

The largest refracting telescope in the world located at the Yerkes Observatory of the University of Chicago, has a 40 inch diameter lens with a focal length of 63 feet and therefore an $f/\#$ of $f/18.9$. The entrance pupil and focal length of a mirror will, in exactly the same way, determine its $f/\#$. Accordingly, the 200 inch diameter mirror of the Mount Palomar telescope with a prime focal length of 666 inches has an $f/\#$ of $f/3.33$.

In precise work where reflection and absorption losses in the lens itself must be taken into consideration, the *T-number* is highly useful. In effect, it is a modified (increased) *f*-number which a given real lens would actually have to have were it to transmit an amount of light corresponding to a particular value of f/D.

5.4 MIRRORS

Mirror systems are finding increasingly more extensive applications, particularly in the ultraviolet and infrared regions of the spectrum. While it is relatively simple to construct a reflecting device which will perform satisfactorily across a broad frequency bandwidth, the same cannot be said of refracting systems. For example, a silicon or germanium lens designed for the infrared will be completely opaque in the visible. As we will see later when we consider their aberrations, mirrors have other attributes which contribute to their usefulness.

A mirror might simply be a piece of black glass or a finely polished metal surface. In the past they were usually made by coating glass with silver, the latter being chosen because of its high efficiency in the UV and IR (see Fig. 4.37), and the former because of its rigidity. In recent times, vacuum-evaporated coatings of aluminum on highly polished substrates have become the accepted standard for quality mirrors. Protective coatings of silicon monoxide or magnesium fluoride are often layered over the aluminum as well. In special applications (e.g. in lasers), where even the small

losses due to metal surfaces cannot be tolerated, mirrors formed of multilayered dielectric films (Section 9.9) are proving to be indispensable.

A whole new generation of lightweight precision mirrors are evolving in anticipation of their use in large-scale orbiting telescopes—the subject is by no means static.

5.4.1 Planar Mirrors

As with all mirror configurations, those which are planar can be either front or back surfaced. The latter is the kind most commonly found in everyday use because it allows the metallic reflecting layer to be completely protected behind glass. In contrast, the majority of mirrors designed for more critical technical usage are front surfaced.

In light of Sections 4.2.2 and 4.2.3, it's a rather easy matter to determine the image characteristics of a planar mirror. Examining the point source and mirror arrangement of Fig. 5.35, we can quickly show that $|s_o| = |s_i|$, i.e. the image P and object S are equidistant from the surface. To wit, $\theta_i = \theta_r$ from the law of reflection; $\theta_i + \theta_r$ is the exterior angle of triangle SPA and is, therefore, equal to the sum of the alternate interior angles, $\angle VSA + \angle VPA$. But $\angle VSA = \theta_i$ and therefore $\angle VSA = \angle VPA$. This makes triangles VAS and VPA congruent, in which case $|s_o| = |s_i|$. (Go back and take another look at Problem 4.3 and Fig. 4.41 for the wave picture of the reflection.)

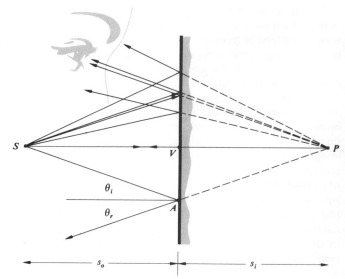

Fig. 5.35 A planar mirror.

We are now faced with the problem of determining a sign convention applicable to mirrors. Whatever we choose, and you should certainly realize that there is a choice, we need only be faithful unto it for all to be well. One obvious dilemma with respect to the convention for lenses is that now the virtual image is to the right of the interface. The observer sees P to be positioned behind the mirror because his eye (or camera) cannot perceive the actual reflection, it merely interpolates the rays backward along straight lines. The rays from P are diverging and no light can be cast upon a screen located at P—the image is certainly virtual. Clearly it is a matter of taste as to whether s_i should be defined as positive or negative in this instance. Since we rather like the idea of virtual object and image distances being negative, we shall define s_o and s_i *as negative when they lie to the right of the vertex V*. This will have the added benefit of yielding a mirror formula identical to the Gaussian lens equation (5.17). Evidently, the same definition of the transverse magnification (5.24) holds, where now as before $M_T = +1$ indicates a *life-sized*, virtual, erect image.

Each point of the extended object of Fig. 5.36, a perpendicular distance s_i from the mirror, is imaged that same distance behind the mirror. In this way, the entire image is built up point by point. This is very much different from the way a lens locates an image. The object in Fig. 5.25 was a left hand and the image formed by the lens was again a left hand; to be sure it might have been distorted

($M_L \neq M_T$), but it was still a left hand. The only evident change was that of a 180° rotation about the optical axis—an effect known as *reversion*. Contrarily, the mirror image of the left hand, determined by dropping perpendiculars from each point, is a right hand (Fig. 5.37). Such an image is sometimes said to be *perverted*, i.e. when looking into a plane mirror, it is the image which is perverted. In deference to the more usual lay connotation of the word, its use in optics is happily waning. The process that converts a right-handed coordinate system in the object space into a left-handed one in the image space is known as *inversion*. Systems with more than one planar mirror can be used to produce either an odd or even number of inversions. In the latter case a right-handed (r-h) object will generate a right-handed image (Fig. 5.38) while in the former instance, the image will be left-handed (l-h).

There are a number of practical devices which utilize rotating planar mirror systems, e.g. choppers, beam deflectors and image rotators. Quite frequently, mirrors are used to amplify and measure the slight rotations of certain laboratory apparatus, e.g. galvanometers, torsion pendulums, current balances, etc. As can be seen in Fig. 5.39, if the mirror rotates through an angle α, the reflected beam or image will move through an angle of 2α.

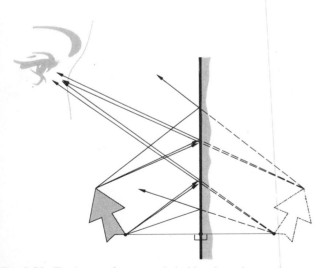

Fig. 5.36 The image of an extended object in a planar mirror.

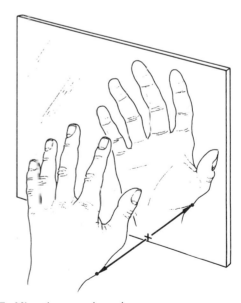

Fig. 5.37 Mirror images—inversion.

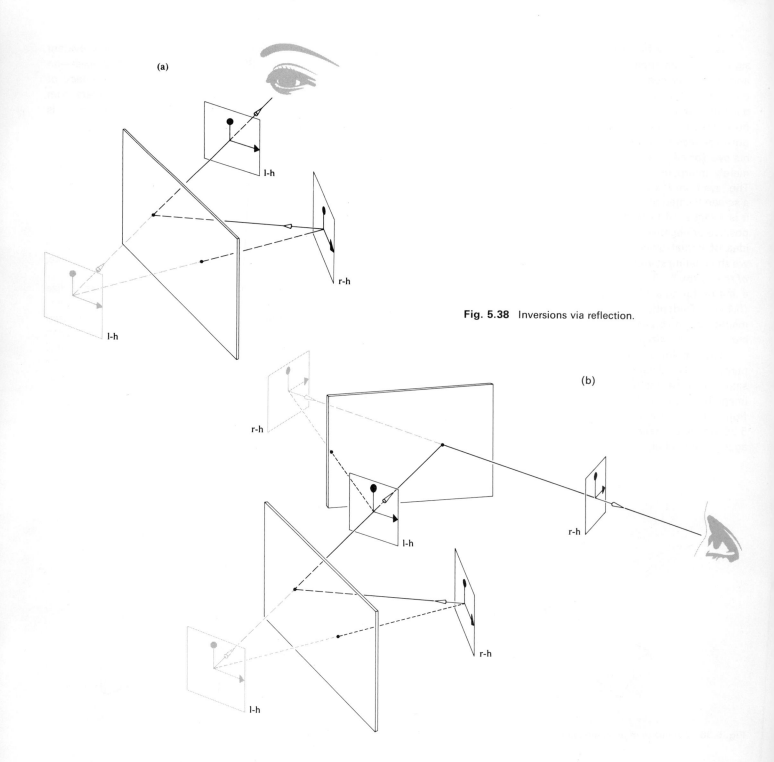

Fig. 5.38 Inversions via reflection.

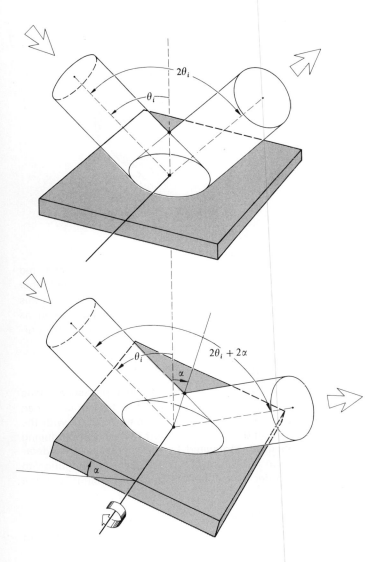

Fig. 5.39 Rotation of a mirror and the concomitant angular displacement of a beam.

5.4.2 Aspherical Mirrors

Curved mirrors which form images very much like those of lenses or curved refracting surfaces have been known since the time of the ancient Greeks. Euclid, who is presumed to have authored the book entitled *Catoptrics*, discusses in it

both concave and convex mirrors.* Fortunately, we developed the conceptual basis for designing such mirrors when we spoke earlier about Fermat's principle as applied to imagery in refracting systems. Suppose then, that we would like to determine the configuration a mirror must have in order that an incident plane wave be reformed upon reflection into a converging spherical wave (Fig. 5.40). If the plane wave is to ultimately converge on some point F, the optical path lengths for all rays must be equal; accordingly, for arbitrary points A_1 and A_2

$$\text{(O.P.L.)} = \overline{W_1 A_1} + \overline{A_1 F} = \overline{W_2 A_2} + \overline{A_2 F}. \quad (5.41)$$

Since the plane Σ is parallel to the incident wavefronts,

$$\overline{W_1 A_1} + \overline{A_1 D_1} = \overline{W_2 A_2} + \overline{A_2 D_2}. \quad (5.42)$$

Equation (5.41) will therefore be satisfied for a surface for which $\overline{A_1 F} = \overline{A_1 D_1}$ and $\overline{A_2 F} = \overline{A_2 D_2}$, or more generally, one where $\overline{AF} = \overline{AD}$ for any point A on the mirror. This same condition was discussed in Section 5.2.1 in which we found $\overline{AF} = e(\overline{AD})$, where e was the eccentricity of a conic section. Here $e = 1$; in other words, the surface is a paraboloid with F as its focus and Σ its directrix. The rays could equally well be reversed, i.e. a point source at the focus of a paraboloid will result in the emission of plane waves from the system. The paraboloidal configuration ranges in its present-day applications from flashlight and auto headlight reflectors to giant radiotelescope antennas (Fig. 5.41), from microwave horns and acoustical dishes to optical telescope mirrors and moon-based communications antennas. The convex paraboloidal mirror is also possible but is less widely in use. Applying what we already know, it should be evident from Fig. 5.42 that an incident parallel bundle of rays will form a virtual image at F when the mirror is convex and a real image when it is concave.

There are several other aspherical mirrors of some interest, namely the ellipsoid ($e < 1$) and hyperboloid ($e > 1$). Both of these produce perfect imagery between a pair of conjugate axial points corresponding to their two foci (Fig. 5.43). As we shall see imminently, the Cassegrainian and Gregorian telescope configurations utilize secondary mirrors which are convex hyperboloidal and ellipsoidal respectively.

It should be noted that all of these devices are readily available commercially. In fact, one can even purchase

---

* *Dioptrics* denotes the optics of refracting elements whereas *catoptrics* connotes the optics of reflecting surfaces.

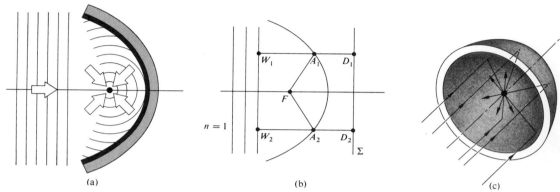

(a)   (b)   (c)

Fig. 5.40 A paraboloidal mirror.

Fig. 5.41 A paraboloidal radio antenna. [Photo courtesy of the Australian News and Information Bureau.]

off-axis elements in addition to the more common centered systems. Thus, in Fig. 5.44, the focused beam can be further processed without obstructing the mirror. Incidentally, this geometry also obtains in the large microwave horn antennas which are playing a significant role in modern communications.

5.4.3 Spherical Mirrors

We are again reminded of the fact that precise aspheric surfaces are considerably more difficult to fabricate than are spherical ones. High costs are commensurate with the increased time and meticulous effort involved in producing quality aspherics. Motivated by these practical considerations, we once more turn to the spherical configuration to determine under what circumstances it might perform adequately.

i) The Paraxial Region

The well-known equation for the circular cross-section of a sphere [Fig. 5.45(a)] is

$$y^2 + (x - R)^2 = R^2, \tag{5.43}$$

where the center C is shifted from the origin O by one radius R. After writing this as

$$y^2 - 2Rx + x^2 = 0$$

we can solve for x:

$$x = R \pm (R^2 - y^2)^{1/2}. \tag{5.44}$$

Let's just concern ourselves with values of x less than R, i.e. we will study a hemisphere, open on the right, corres-

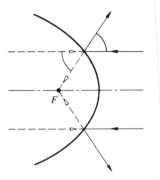

Fig. 5.42 Real and virtual images for a paraboloidal mirror.

ponding to the minus sign in Eq. (5.44). Following expansion in a binomial series, x takes the form

$$x = \frac{y^2}{2R} + \frac{1y^4}{2^2 2! R^3} + \frac{1 \cdot 3 y^6}{2^3 3! R^5} + \cdots . \qquad (5.45)$$

This expression becomes quite meaningful as soon as we realize that the standard equation for a parabola with its vertex at the origin and its focus a distance f to the right [Fig. 5.45(b)] is simply

$$y^2 = 4fx. \qquad (5.46)$$

Thus by comparing these two formulas we see that if $4f = 2R$, i.e. if $f = R/2$, the first contribution in the series can be thought of as being parabolic while the remaining terms represent the deviation therefrom. If that deviation is Δx then

$$\Delta x = \frac{y^4}{8R^3} + \frac{y^6}{16R^5} + \cdots . \qquad (5.47)$$

Evidently this difference will only be appreciable when y is relatively large [Fig. 5.45(c)] in comparison to R. *In the paraxial region, i.e. in the immediate vicinity of the optical axis, these two configurations will be essentially indistinguishable.* Thus if we talk about the paraxial theory of spherical mirrors as a first approximation, we can again embrace the conclusions drawn from our study of the stigmatic imagery of paraboloids. In actual use, however, y will not be so limited and aberrations will appear. Moreover, aspherical surfaces produce perfect images only for pairs of axial points—they too will suffer aberrations.

ii) The Mirror Formula

The paraxial equation which relates conjugate object and image points to the physical parameters of a spherical

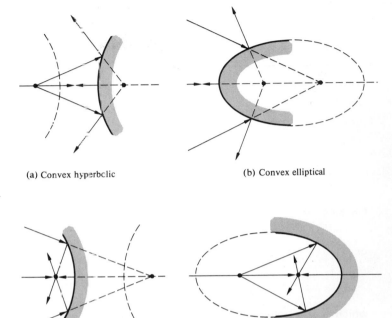

(a) Convex hyperbolic   (b) Convex elliptical

(c) Concave hyperbolic   (d) Concave elliptical

Fig. 5.43 Hyperbolic and elliptical mirrors.

mirror can be derived rather easily with the help of Fig. 5.46. To that end, observe that since $\theta_i = \theta_r$, the $\angle SAP$ is bisected by $\overline{CA}$ which therefore divides the side $\overline{SP}$ of triangle SAP into segments proportional to the remaining two sides, that is

$$\frac{\overline{SC}}{\overline{SA}} = \frac{\overline{CP}}{\overline{PA}}. \qquad (5.48)$$

Furthermore

$$\overline{SC} = s_o - |R| \qquad \text{and} \qquad \overline{CP} = |R| - s_i,$$

where s_o and s_i are on the left and therefore positive. If we use the same sign convention for R as we did when we dealt with refraction, it will be negative here because C is to the left of V, i.e. the surface is concave. Thus $|R| = -R$ and

$$\overline{SC} = s_o + R \qquad \text{while} \qquad \overline{CP} = -(s_i + R).$$

In the paraxial region $\overline{SA} \approx s_o, \overline{PA} \approx s_i$ and so Eq. (5.48) becomes

$$\frac{s_o + R}{s_o} = -\frac{s_i + R}{s_i}$$

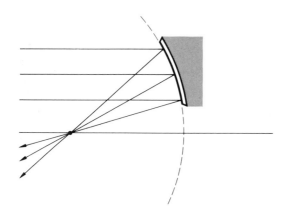

Fig. 5.44 An off-axis parabolic mirror element.

or

$$\frac{1}{s_o} + \frac{1}{s_i} = -\frac{2}{R},$$ (5.49)

which is often referred to as the *mirror formula*. It is equally applicable to concave ($R < 0$) or convex ($R > 0$) mirrors. The primary or object focus is again defined by

$$\lim_{s_i \to \infty} s_o = f_o,$$

while the secondary or image focus corresponds to

$$\lim_{s_o \to \infty} s_i = f_i.$$

Consequently, from Eq. (5.49)

$$\frac{1}{f_o} + \frac{1}{\infty} = \frac{1}{\infty} + \frac{1}{f_i} = -\frac{2}{R},$$

to wit $f_o = f_i = -R/2$, as we knew from Fig. 5.45(c). Thus dropping the subscripts on the focal lengths, we have

$$\frac{1}{s_o} + \frac{1}{s_i} = \frac{1}{f}.$$ (5.50)

Observe that f will be positive for concave mirrors ($R < 0$) and negative for convex mirrors ($R > 0$). In the latter instance the image is formed behind the mirror and is virtual (Fig. 5.47).

iii) Finite Imagery

The remaining mirror properties are so similar to those of lenses and spherical refracting surfaces that we need only mention them briefly without repeating the entire logical development of each item. Accordingly, within the restrictions of paraxial theory, any parallel off-axis bundle of rays will be focused to a point on the *focal plane* passing through F normal to the optical axis. Likewise, a finite planar object perpendicular to the optical axis will be imaged (to a first approximation) in a plane similarly oriented. Essentially we are saying that each object point will have a corresponding image point in the plane. This is certainly true for a plane mirror but it is only approximately the case for other configurations. To be sure, if a spherical mirror is appropriately restricted in its operation, the reflected waves arising from each permitted object point will very closely approximate spherical waves. Under such circumstances good finite images of extended objects can be formed (Fig. 5.48). Just as each image point produced by a thin lens lies along a straight line through the optical center O, each image point for a spherical mirror will lie on a ray passing through both the center of curvature C and the object point. As with the thin lens (Fig. 5.23), the graphical location of the image is quite straightforward. Once more the top of the image is located at the intersection of two rays; e.g. one initially parallel to the axis and passing through F after reflection, the other going straight through C (Fig. 5.49). The ray from any off-axis object point to the vertex makes equal angles with the optical axis on reflection and is therefore particularly convenient to construct as well. So too is the ray which first passes through the focus and after reflection emerges parallel to the axis.

Notice that triangles $S_1 S_2 V$ and $P_1 P_2 V$ are similar and hence their sides are proportional. Taking y_i to be negative as we did in the past, since it is below the axis, $y_i/y_o = -s_i/s_o$ which of course is equal to M_T the *transverse magnification*, identical to that of the lens (5.25).

The only equation which contains information about the structure of the optical element (n, R etc.) is that for f and so, rather understandably, it differs for the thin lens and spherical mirror. The other functional expressions which relate s_o, s_i and f or y_o, y_i and M_T are, however, precisely the same. The only alteration in the previous sign convention appears in Table 5.4 where s_i on the left of V is now taken as positive. The striking similarity between the properties of a concave mirror and a convex lens on one hand and a convex mirror and a concave lens on the other are quite evident on comparing Tables 5.3 and 5.5 which are identical in all respects.

The properties summarized in Table 5.5 can easily be verified empirically. If you don't have a spherical mirror at hand, a fairly crude but functional one can be made by

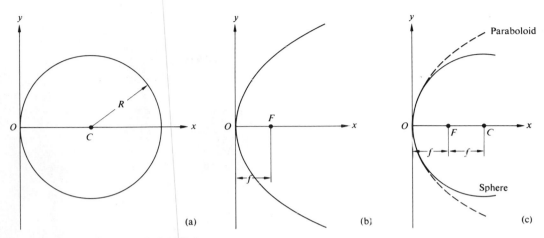

Fig. 5.45 Comparison of a spherical and paraboloidal mirror.

carefully shaping aluminum foil over a spherical form such as the end of a light bulb (in that particular case R and therefore f will be small). A rather nice qualitative experiment involves examining the image of some small object formed by a short focal length concave mirror. As you move it toward the mirror from beyond a distance of $2f = R$, the image will gradually increase until at $s_o = 2f$ it will appear inverted and life-size. Bringing it still closer will cause the image to increase even more until it fills the entire mirror with an un-recognizable blur. As s_o becomes ever smaller, the now erect, magnified image will continue to decrease until the object finally rests on the mirror where the image is again lifesize. If you are not moved by all of this to immediately jump up and make a mirror, you might try examining the image formed by a shiny spoon—either side will be in-teresting.

Table 5.4 Sign convention for spherical mirrors.

Quantity	Sign	
	+	−
s_o	Left of V, real object	Right of V, virtual object
s_i	Left of V, real image	Right of V, virtual image
f	Concave mirror	Convex mirror
R	C right of V, convex	C left of V, concave
y_o	Above axis, erect object	Below axis, inverted object
y_i	Above axis, erect image	Below axis, inverted image

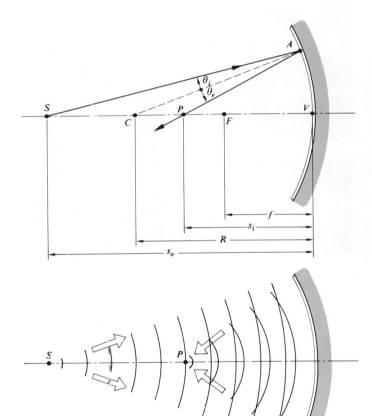

Fig. 5.46 A concave spherical mirror.

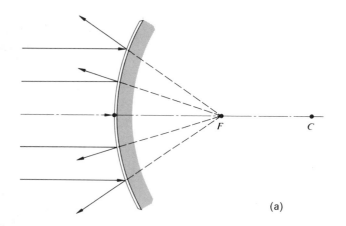

(d)

Fig. 5.47 Focusing of rays via a spherical mirror. [Photos by E.H.]

(a)

(b)

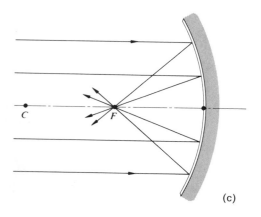

(c)

Table 5.5 Images of real objects formed by spherical mirrors.

Concave				
Object	Image			
Location	Type	Location	Orientation	Relative size
$\infty > s_o > 2f$	Real	$f < s_i < 2f$	Inverted	Minified
$s_o = 2f$	Real	$s_i = 2f$	Inverted	Same size
$f < s_o < 2f$	Real	$\infty > s_i > 2f$	Inverted	Magnified
$s_o = f$		$\pm \infty$		
$s_o < f$	Virtual	$\lvert s_i \rvert > s_o$	Erect	Magnified

Convex				
Object	Image			
Location	Type	Location	Orientation	Relative size
Anywhere	Virtual	$\lvert s_i \rvert < \lvert f \rvert$	Erect	Minified

5.5 PRISMS

Prisms play a great many different roles in optics; there are prism combinations that serve as beam-splitters (Section 4.3.4), polarizing devices (Section 8.4.3) and even interferometers. Despite this diversity, the vast majority of

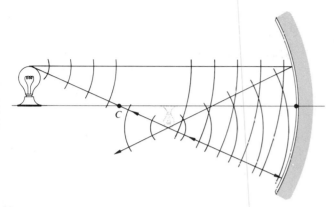

Fig. 5.48 Wavefronts reflected from a concave mirror.

applications make use of only one of two main prism functions. Firstly, a prism can serve as a dispersive device as it does in a variety of spectrum analyzers. That is fo say, it is capable of separating, to some extent, the constituent frequency components in a polychromatic light beam. You might recall that the term *dispersion* was introduced earlier (Section 3.3.1) in connection with the frequency dependence of the index of refraction, $n(\omega)$, for dielectrics. And in fact, the prism provides a highly useful means of measuring $n(\omega)$ over a broad range of frequencies and for a wide variety of materials (including gases and liquids). Its second and most widely utilized function is to affect a change in the orientation of an image or in the direction of propagation of a beam. Prisms are incorporated in a great many optical instruments, often simply to fold the system into a confined space. There are inversion prisms, reversion prisms and prisms that deviate a beam without inversion or reversion and all of this without dispersion.

5.5.1 Dispersing Prisms

Typically, a ray entering a dispersing prism, as in Fig. 5.50, will emerge having been deflected from its original direction by an angle δ known as the *angular deviation*. At the first refraction the ray is deviated through an angle $(\theta_{i1} - \theta_{t1})$ and at the second refraction it is further deflected through $(\theta_{t2} - \theta_{i2})$. The total deviation is then

$$\delta = (\theta_{i1} - \theta_{t1}) + (\theta_{t2} - \theta_{i2}).$$

Since the polygon $ABCD$ contains two right angles, $\angle BCD$ must be the supplement of the *apex angle* α. Now α, being the exterior angle to triangle BCD, is also the sum of the alternate interior angles, that is

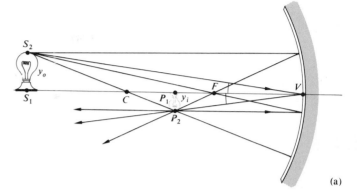

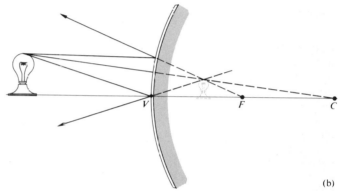

Fig. 5.49 Finite imagery with spherical mirrors.

$$\alpha = \theta_{t1} + \theta_{i2}. \tag{5.51}$$

Thus

$$\delta = \theta_{i1} + \theta_{i2} - \alpha. \tag{5.52}$$

What we would like to do now is write δ as a function of both the angle of incidence for the ray (i.e., θ_{i1}), and the prism angle α; these presumably would be known. It follows from Snell's law, if the prism index is n and it is immersed in air ($n_a \approx 1$), that

$$\theta_{t2} = \sin^{-1}(n \sin \theta_{i2}) = \sin^{-1}[n \sin(\alpha - \theta_{t1})].$$

Upon expanding this expression, replacing $\cos \theta_{t1}$ by $(1 - \sin^2 \theta_{t1})^{1/2}$, and using Snell's law it becomes

$$\theta_{t2} = \sin^{-1}[(\sin \alpha)(n^2 - \sin^2 \theta_{i1})^{1/2} - \sin \theta_{i1} \cos \alpha].$$

The deviation is then

$$\delta = \theta_{i1} + \sin^{-1}[(\sin \alpha)(n^2 - \sin^2 \theta_{i1})^{1/2} - \sin \theta_{i1} \cos \alpha] - \alpha. \tag{5.53}$$

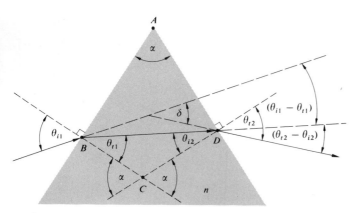

Fig. 5.50   Geometry of a dispersing prism.

Apparently δ increases with n, which is itself a function of frequency, and so we might designate the deviation as $\delta(v)$ or $\delta(\lambda)$. For most transparent dielectrics of practical concern $n(\lambda)$ decreases as the wavelength increases across the visible [refer back to Fig. 3.13 for a plot of $n(\lambda)$ versus λ for various glasses]. Clearly then $\delta(\lambda)$ will be less for red light than it is for blue.

Missionary reports coming out of Asia in the early sixteen hundreds indicated that prisms were well known and highly valued in China because of this ability to generate color. A number of scientists of the era, particularly Marci, Grimaldi and Boyle had made some observations using prisms but it remained for the great Sir Isaac Newton to perform the first definitive studies of dispersion. On February 6, 1672, Newton presented a classic paper to the Royal Society entitled *A New Theory about Light and Colours*. He had concluded that white light consisted of a mixture of various colors and that the process of refraction was color dependent.

Returning to Eq. (5.53), it is evident that the deviation suffered by a monochromatic beam on traversing a given prism (i.e. n and α are fixed) is a function only of the incident angle at the first face, θ_{i1}. A plot of the results of Eq. (5.53) as applied to a typical glass prism is shown in Fig. 5.51. The smallest value of δ is known as the *minimum deviation* δ_m and it is of particular interest for practical reasons. It can be determined analytically by differentiating Eq. (5.53) and then setting $d\delta/d\theta_{i1} = 0$, but a more indirect route will certainly be simpler. Differentiating Eq. (5.52) and setting it equal to zero leads to

$$\frac{d\delta}{d\theta_{i1}} = 1 + \frac{d\theta_{t2}}{d\theta_{i1}} = 0$$

or $d\theta_{t2}/d\theta_{i1} = -1$. Taking the derivative of Snell's law at each interface yields

$$\cos\theta_{i1}\,d\theta_{i1} = n\cos\theta_{t1}\,d\theta_{t1}$$

and

$$\cos\theta_{t2}\,d\theta_{t2} = n\cos\theta_{i2}\,d\theta_{i2}.$$

Note as well on differentiating Eq. (5.51) that $d\theta_{t1} = -d\theta_{i2}$ since $d\alpha = 0$. Dividing the last two equations and substituting for the derivatives, we obtain

$$\frac{\cos\theta_{i1}}{\cos\theta_{t2}} = \frac{\cos\theta_{t1}}{\cos\theta_{i2}}.$$

Making use of Snell's law once again, this can be rewritten as

$$\frac{1 - \sin^2\theta_{i1}}{1 - \sin^2\theta_{t2}} = \frac{n^2 - \sin^2\theta_{i1}}{n^2 - \sin^2\theta_{t2}}.$$

The value of θ_{i1} for which this is true is the one for which $d\delta/d\theta_{i1} = 0$. Inasmuch as $n \neq 1$, it follows that

$$\theta_{i1} = \theta_{t2}$$

and therefore

$$\theta_{t1} = \theta_{i2}.$$

This means that the ray for which the deviation is a minimum traverses the prism symmetrically, i.e. parallel to its base. Incidentally, there is a lovely argument for why θ_{i1} must equal θ_{t2} which is neither as mathematical nor as tedious as the one we have evolved. In brief, suppose a ray undergoes a minimum deviation and $\theta_{i1} \neq \theta_{t2}$. Then if we reverse the

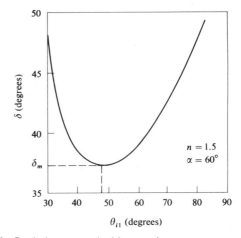

Fig. 5.51   Deviation versus incident angle.

ray, it will retrace the same path and so δ must be unchanged, i.e. $\delta = \delta_m$. But this implies that there are two different incident angles for which the deviation is a minimum and this we know is not true—ergo $\theta_{i1} = \theta_{t2}$.

In the case when $\delta = \delta_m$, it follows from Eqs. (5.51) and (5.52) that $\theta_{i1} = (\delta_m + \alpha)/2$ and $\theta_{t1} = \alpha/2$, whereupon Snell's law at the first interface leads to

$$n = \frac{\sin\left[(\delta_m + \alpha)/2\right]}{\sin \alpha/2}. \tag{5.54}$$

This equation forms the basis of one of the most accurate techniques for determining the refractive index of a transparent substance. Effectively, one fashions a prism out of the material in question and then, measuring α and $\delta_m(\lambda)$, $n(\lambda)$ is computed employing Eq. (5.54) at each wavelength of interest. Hollow prisms whose sides are fabricated of plane-parallel glass can be filled with liquids or gases under high pressure; the glass plates will not result in any deviation of their own.

Figures 5.52 and 5.53 show two examples of *constant deviation dispersing prisms* which are important primarily in spectroscopy. The *Pellin-Broca* is probably the most common of the group. Albeit a single block of glass, it can be envisaged as consisting of two 30°–60°–90° prisms and one 45°–45°–90° prism. Suppose that in the position shown a single monochromatic ray of wavelength λ traverses the component prism *DAE* symmetrically, thereafter to be reflected at 45° from face *AB*. The ray will then traverse prism *CDB* symmetrically having experienced a total deviation of 90°. Effectively, the ray can be thought of as having passed through an ordinary 60° prism (*DAE* combined with *CDB*) at minimum deviation. All other wavelengths present in the beam will emerge at other angles. If the prism is now rotated slightly about an axis normal to the paper, the incoming beam will have a new incident angle. A different wavelength component, say λ_2, will now undergo a minimum deviation which is again 90°—hence the name, *constant deviation*. With a prism of this sort, one can conveniently set up the light source and viewing system at a fixed angle (here 90°) and then simply rotate the prism in order to look at a particular wavelength. The device can be calibrated so that the prism-rotating dial reads directly in wavelength.

5.5.2 Reflecting Prisms

In contrast with the previous section, we now examine *reflecting prisms* in which dispersion is no longer desirable. In the present instance, the beam will be introduced in such a way that at least one internal reflection takes place, for the

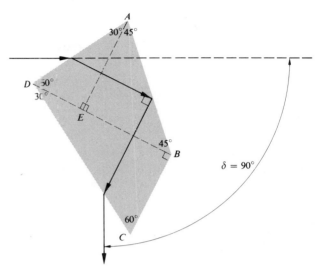

Fig. 5.52 The Pellin–Broca prism.

specific purpose of either changing the direction of propagation, or the orientation of the image, or both.

Let's first establish that it is actually possible to have such an internal reflection without concomitant dispersion. In other words, is δ independent of λ? The prism in Fig. 5.54 is presumed to have as its profile an isosceles triangle—this happens to be a rather common configuration in any event. The ray refracted at the first interface is later reflected from face *FG*. As we saw earlier (Section 4.3.4), this will occur

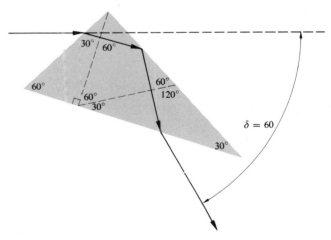

Fig. 5.53 The Abbe prism.

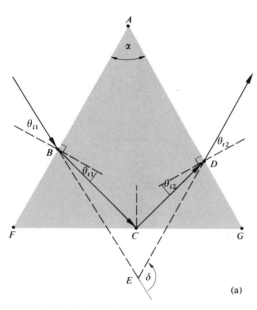

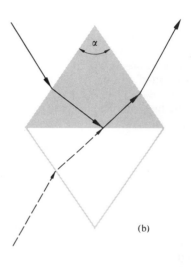

(a)

(b)

Fig. 5.54 Geometry of a reflecting prism.

when the internal incident angle is greater than the critical angle θ_c, defined by

$$\sin \theta_c = n_{ti}. \qquad [4.69]$$

For a glass–air interface, this requires that θ_i be greater than roughly 42°. To avoid any difficulties at smaller angles, let's further suppose the base of our hypothetical prism is silvered as well—certain prisms do in fact require silvered faces. The angle of deviation between the incoming and outgoing rays is

$$\delta = 180° - \angle BED. \qquad (5.55)$$

From the polygon $ABED$ we have that

$$\alpha + \angle ADE + \angle BED + \angle ABE = 360°.$$

Moreover, at the two refracting surfaces

$$\angle ABE = 90° + \theta_{i1}$$

and

$$\angle ADE = 90° + \theta_{t2}.$$

Substituting for $\angle BED$ in Eq. (5.55) we get

$$\delta = \theta_{i1} + \theta_{t2} + \alpha. \qquad (5.56)$$

Since the ray at point C has equal angles of incidence and

reflection $\angle BCF = \angle DCG$. Thus, because the prism is isosceles $\angle BFC = \angle DGC$ and triangles FBC and DGC are similar. It follows that $\angle FBC = \angle CDG$ and finally, therefore, $\theta_{t1} = \theta_{i2}$. From Snell's law we know that this is equivalent to $\theta_{i1} = \theta_{t2}$, whereupon the deviation becomes

$$\delta = 2\theta_{i1} + \alpha, \qquad (5.57)$$

which is certainly independent of both λ and n. The reflection will occur without any color preferences and the prism is said to be *achromatic*. If we unfold the prism, i.e. if we draw its image in the reflecting surface FG, as in Fig. 5.54(b), we see that the prism is equivalent in a sense to a parallelepiped or thick planar plate. The image of the incident ray emerges parallel to itself regardless of wavelength.

A few out of the great many widely used reflecting prisms are shown in the next several figures. These are often made from BSC-2 or C-1 glass (see Table 6.2). For the most part, the illustrations are self explanatory and so the descriptive commentary will be brief.

The *right-angle* prism (Fig. 5.55) deviates rays normal to the incident face by 90°. Notice that the top and bottom of the image have been interchanged, i.e. the arrow has been flipped over while the right and left sides have not. It is therefore an inversion system with the top face acting like a plane mirror. (To see this, imagine that the arrow and

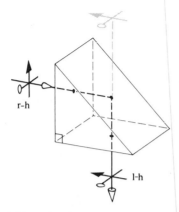

Fig. 5.55 The right-angle prism.

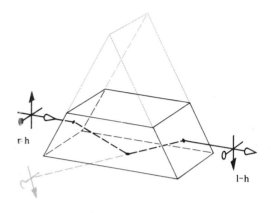

Fig. 5.57 The Dove prism.

lollypop are vectors and take their cross product. The resultant, arrow × lollypop, was initially in the propagation direction but is reversed by the prism.)

The *Porro* prism (Fig. 5.56) is physically the same as the previous one but it is used in a different orientation. After two reflections, the beam is deviated by 180°. Thus, if it enters right-handed, it leaves right-handed.

The *Dove* (Fig. 5.57) is a truncated version (to reduce size and weight) of the right-angle prism, used almost exclusively in collimated light. It has the interesting property (Problem 5.38) of rotating the image twice as fast as it is itself rotated about the longitudinal axis.

The *Amici* (Fig. 5.58) is essentially a truncated right-angle prism with a roof section added on to the hypotenuse face. In its most common usage it has the effect of splitting the image down the middle and interchanging the right and

left portions.* These prisms are expensive because the 90° roof angle must be held to roughly 3 or 4 seconds of arc or a troublesome double image will result. They are often used in simple telescope systems to correct for the reversion introduced by the lenses.

---

* You can see how it actually works by placing two plane mirrors at right angles and looking directly into the combination. If you wink your *right* eye, the image will wink its *right* eye. Incidentally, if your eyes are equally strong, you will see two seams (images of where the mirrors met), one running down the middle of each eye with your nose presumably between them. If either eye is stronger, there will be only one seam down the middle of that eye. If you close it, the seam will jump over to the other eye. This must be tried to be appreciated.

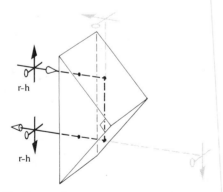

Fig. 5.56 The Porro prism.

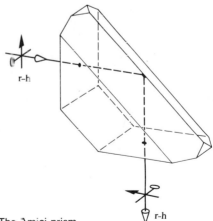

Fig. 5.58 The Amici prism.

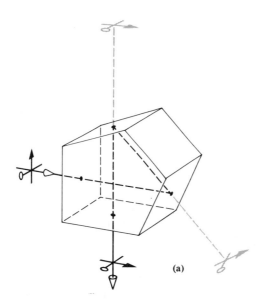

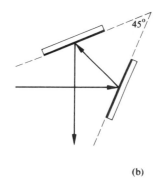

(a)

(b)

Fig. 5.59 The penta prism and its mirror equivalent.

The *Penta* prism (Fig. 5.59) will deviate the beam by 90° without affecting the orientation of the image. Note that two of its surfaces must be silvered. These prisms are often used as end reflectors in small rangefinders.

The *Rhomboid* prism (Fig. 5.60) displaces the line of sight without producing any angular deviation or changes in the orientation of the image.

The *Leman–Springer* prism (Fig. 5.61) also has a 90° roof. Here the line of sight is displaced without being deviated but the emerging image is right-handed and rotated through 180°. The prism can therefore serve to erect images in telescope systems, e.g. gun-sights and the like.

There are many more reflecting prisms which serve specific purposes. For example, if one simply cuts a cube so that the piece removed has three mutually perpendicular faces, it is called a *corner-cube* prism. It has the property of being retrodirective; that is, it will reflect all incoming rays back along their original directions. One hundred of these are sitting in an 18 inch square array 240,000 miles from here, having been placed on the moon during the Apollo 11 flight.*

The most common erecting system consists of two Porro

prisms as illustrated in Fig. 5.62. These are relatively easy to manufacture and are shown here with rounded corners to reduce weight and size. Since there are four reflections, the exiting image will be right-handed. A small slot is often cut in the hypotenuse face to obstruct rays which are internally reflected at glancing angles. Finding these slots after dismantling the family's binoculars is all too often an inexplicable surprise.

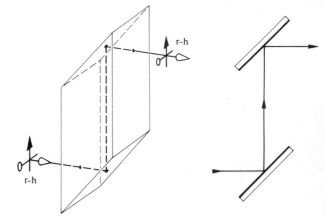

Fig. 5.60 The rhomboid prism and its mirror equivalent.

---

* J. E. Foller and E. J. Wampler, *"The Lunar Laser Reflector," Sci. Am.*, March 1970, p. 38.

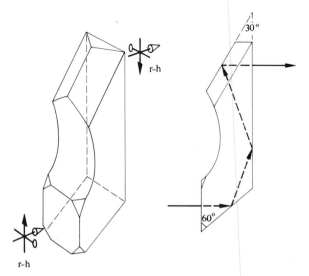

30°

r-h

60°

r-h

Fig. 5.61 The Leman–Springer prism.

5.6 FIBER OPTICS

In recent times, techniques have evolved for efficiently conducting light from one point in space to another via transparent, dielectric fibers. As long as the diameter of these
fibers is large compared to the wavelength of the radiant
energy, the inherent wave nature of the propagation mechanism is of little importance and the process obeys the familiar
laws of geometrical optics. On the other hand, if the diameter
is of the order of λ, the transmission closely resembles the
manner in which microwaves advance along waveguides.
Some of the propagation modes are evident in the photo-
micrographic end views of fibers shown in Fig. 5.63. Here
the wave nature of light must be reckoned with and this
behavior therefore resides in the domain of physical optics.
Although optical waveguides, particularly of the thin film
variety, are of increasing interest this discussion will be
limited to the case of relatively large diameter fibers.

Consider the straight glass cylinder of Fig. 5.64 surrounded by air. Light striking its walls from within will be
totally internally reflected provided that the incident angle
at each reflection is greater than $\theta_c = \sin^{-1} n_a/n_f$, where n_f
is the index of the cylinder or fiber. As we will show, a
meridional ray (i.e. *one which is coplanar with the optical
axis*) might undergo several thousand reflections per foot
as it bounces back and forth along a fiber until it emerges at

the far end (Fig. 5.65). If the fiber has a diameter D and a
length L, the path length ℓ traversed by the ray will be

$$\ell = L/\cos \theta_t,$$

or from Snell's law

$$\ell = n_f L (n_f^2 - \sin^2 \theta_i)^{-1/2}. \qquad (5.58)$$

The number of reflections N is then given by

$$N = \frac{\ell}{D/\sin \theta_t} \pm 1$$

or

$$N = \frac{L \sin \theta_i}{D(n_f^2 - \sin^2 \theta_i)^{1/2}} \pm 1 \qquad (5.59)$$

rounded off to the nearest whole number. The ± 1, which
depends on where the ray strikes the end face, is of no
significance when N is large, as it is in practice. Thus if D
is 50 μ, i.e. 50 microns where $1 \mu = 10^{-6}$ m $= 39.37$
$\times 10^{-6}$ in which is about 2×10^{-3} in, (a hair from the head
of a human is roughly 50 μ in diameter) and if $n_f = 1.6$ and
$\theta_i = 30°$, N turns out to be approximately 2000 reflections
per foot. Fibers are available in diameters from about 2 μ
up to $\frac{1}{4}$ inch or so but are seldom used in sizes much less
than about 10 μ. The large diameter rods are generally
called *light pipes*. Extremely thin glass (or plastic) filaments
are quite flexible as witnessed by the well-known fact that
glass fibers can even be woven into fabric.

The smooth surface of a single fiber must be kept clean
of moisture, dust, oil, etc., if there is to be no leakage of
light (via the mechanism of frustrated total internal reflection).
Similarly if large numbers of fibers are packed in close proximity, light may leak from one fiber to another in what is
known as *cross talk*. For these reasons, it is now customary
to enshroud each fiber in a transparent sheath of lower index
called a *cladding*. This layer need only be thick enough to
provide the desired isolation but for other reasons it generally
occupies about one tenth of the cross sectional area.
Typically a fiber core might have an index (n_f) of 1.62 and
the cladding an index (n_c) of 1.52 although a range of values
is available. A clad fiber is shown in Fig. 5.66. Notice that
there is a maximum value $\theta_{\max}$ of θ_i, for which the internal ray
will impinge at the critical angle, θ_c. Rays incident on the
face at angles greater than $\theta_{\max}$ will strike the interior wall at
angles less than θ_c. They will only be partially reflected at
each such encounter with the core–cladding interface and
will quickly leak out of the fiber. Accordingly, $\theta_{\max}$ defines

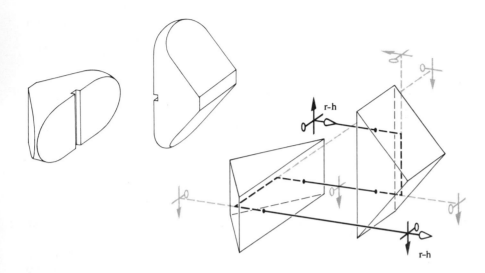

Fig. 5.62 The double Porro prism.

the half-angle of the acceptance cone of the fiber. To determine it we write

$$\sin \theta_c = n_c/n_f = \sin (90 - \theta_t)$$

Thus

$$n_c/n_f = \cos \theta_t$$

or

$$n_c/n_f = (1 - \sin^2 \theta_t)^{1/2}.$$

Making use of Snell's law and rearranging matters, we have

$$\sin \theta_{\max} = \frac{1}{n_o} (n_f^2 - n_c^2)^{1/2}. \qquad (5.60)$$

The quantity $n_o \sin \theta_{\max}$ is defined as the *numerical aperture* or N.A. Its square is a measure of the light gathering power of the system. The term originates in microscopy where the equivalent expression characterizes the corresponding capabilities of the objective lens. It should clearly relate to the *speed* of the system and, in fact,

$$f/\# = \frac{1}{2(\text{N.A.})}.$$

Thus for a fiber

$$\text{N.A.} = (n_f^2 - n_c^2)^{1/2}. \qquad (5.61)$$

The left-hand side of Eq. (5.60) cannot exceed one and in air ($n_o = 1.00028 \approx 1$) that means that the largest value of N.A. is one. In this case, the half-angle $\theta_{\max}$ equals 90°

and the fiber totally internally reflects all light entering its face (Problem 5.39). Fibers with a wide variety of numerical apertures running from about 0.2 up to and including 1.0 are commercially obtainable.

Bundles of free fibers whose ends are bound together (e.g. with epoxy) and ground and polished form flexible light guides. If no attempt is made to align the fibers in an ordered array, they form an *incoherent bundle*. This (unfortunate use of the term which ought not to be confused with coherence theory) just means e.g. that the first fiber in the top row at the entrace face may have its terminus anywhere in the bundle at the exit face. These *flexible light carriers* are, for that reason, relatively easy to make and inexpensive. Their primary function is simply to conduct light from one region to another. Conversely, when the fibers are carefully arranged so that their terminations occupy the same relative positions in both of the bound ends of the bundle, the bundle is said to be *coherent*. Such an arrangement is capable of transmitting images and is consequently known as a *flexible image carrier* (Fig. 5.67). Not all fiber optics arrays are made flexible, for example, fused, rigid, coherent fiber faceplates or *mosaics* are used to replace homogeneous low-resolution sheet glass on cathode-ray tubes, vidicons, image intensifiers, etc. Mosaics consisting of literally millions of fibers with their claddings fused together have mechanical properties almost identical to homogeneous glass. Similarly a sheet of fused *tapered* fibers can either magnify or minify an image depending on whether the light enters the smaller or larger end of the fiber. The compound

Fig. 5.63 Optical waveguide mode patterns seen in the end faces of small-diameter fibers. [Photo courtesy of Narinder S. Kapany.]

eye of an insect like the housefly is effectively a bundle of tapered fiber optical filaments. The rods and cones that make up the human retina may also channel light via total internal reflection. Another common application of mosaics involving imaging is the *field flattener*. If the image formed by a lens system resides on a curved surface, it is often desirable to reshape it into a plane so as, e.g. to match a film plate. A mosaic can be ground and polished on one of its end surfaces to correspond to the contour of the image and on the other to match the detector. Incidentally, there is a naturally occurring fibrous crystal known as *ulexite* which when polished responds surprisingly like a fiber optical mosaic. (Hobby shops often sell it for use in making jewelry.)

If you have never seen the kind of light conduction we've been talking about, try looking down the edges of a stack of microscope slides. Even better are the much thinner

(0.18 mm) cover glass slides. Figure 5.68 shows the way light is conveyed to the upper surface of a stack of a few hundred of these held together by a rubber band.

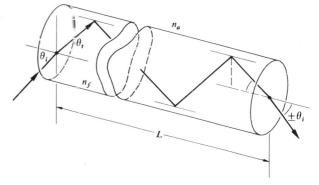

Fig. 5.64 Rays reflected within a dielectric cylinder.

Fig. 5.65 Light emerging from the ends of a loose bundle of glass fibers. [Photo by E.H.]

5.7 OPTICAL SYSTEMS

We have thus far developed paraxial theory to a point where it is now possible to appreciate the underlying principles operative in the majority of practical optical systems. To be sure, the subtleties involved in controlling aberrations are extremely important and still quite beyond this discussion. Even so, for example, one could build a telescope (admittedly not a very good one, but a telescope nonetheless) using the conclusions already drawn from first-order theory.

What better place to begin a discussion of optical instruments than with the most common of all, the eye?

5.7.1 Eyes

For our purposes, we distinguish two main groupings of eyes; those which gather light and form images via a lens system and those which essentially utilize fiber optical bundles for that purpose. Visual lens systems have evolved, independently, in at least three distinct kinds of organisms. Some of the more advanced mollusks (e.g. the octopus), certain spiders (e.g. the Avicularia) and of course the vertebrates, ourselves included, possess eyes with image-forming lenses.

The manner in which these eyes function was, for centuries, a point of great misconception until in 1625 the German Jesuit Christoph Scheiner (1575–1650) performed a classic and irrefutible experiment. He removed the coating on the back of an animal's eye, and peering through its transparent retina from behind, was able to perceive a small inverted image of the scene beyond the eye. At just about the same time Descartes performed similar experiments.

i) *Structure of the Human Eye*

In essence the human eye can be thought of as a positive lens system which casts a real image on a light-sensitive surface. Figure 5.69(a) shows the arrangement of its basic components. The eye is an almost spherical jelly-like mass contained within a tough shell, the *sclera*. Except for the front portion or *cornea* which is transparent, the sclera is white and opaque. Bulging upward from the body of the sphere, the cornea's curved surface (which is slightly flattened, thereby cutting down on spherical aberration) serves as the first and strongest convex element of the lens system. Indeed most of the bending imparted to a bundle of rays takes place at the air–cornea interface. Incidentally, one of the reasons you can't see very well under water ($n_W \approx 1.33$) is that its index is too close to that of the cornea ($n_C \approx 1.376$) to allow for adequate refraction. Light emerging from the cornea passes through a chamber filled with a clear watery fluid called the *aqueous humor* ($n_{ah} \approx 1.336$). A ray which was strongly refracted toward the optical axis at the air–cornea interface will only be slightly redirected at the cornea–aqueous humor interface because of the similarity of their indices. Immersed in the aqueous is a diaphragm known as the *iris* which serves as the aperture stop controlling the amount of light entering the eye by way of the hole

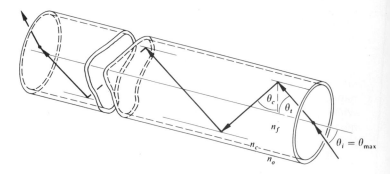

Fig. 5.66 Rays in a clad optical fiber.

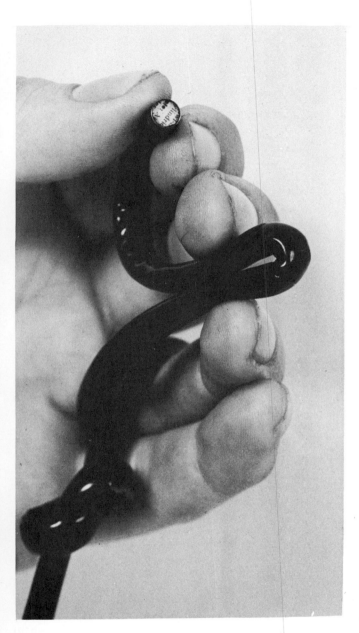

or *pupl*. It is the iris (from the Greek word for rainbow) which gives the eye its characteristic blue, brown, gray, green, or hazel color. Made up of circular and radial muscles, the iris can expand or contract the pupil over a range from about 2 mm in bright light, to roughly 8 mm in darkness. In addit on to this function, it is also linked to the focusing response and will contract to increase image sharpness when doing close work. Immediately behind the iris is the *crystalline lens*. The name, which is somewhat misleading, dates back to about 1000 A.D. and the work of Abû 'Alî al Hasan ibn al Hasan ibn al Haitham alias Alhazen of Cairo who described the eye as partitioned into three regions which were watery, crystalline and glassy, respectively. The lens, which has both the size and shape of a small bean, is a complex layered fibrous mass surrounded by an elastic membrane. In structure it is somewhat like a transparent onion, formed of roughly 22,000 very fine layers. It has some remarkable characteristics which distinguish it from man-made lenses in use today, in addition to the fact that it continues to grow in size. Because of its laminar structure, rays traversing it will follow paths made up of minute, discontinuous segments. The lens as a whole is quite pliable, albeit less so with age. Moreover, its index of refraction varies from about 1.406 at the inner core to approximately 1.336 at the less dense cortex. (Glass lenses with inhomogereous indices are now receiving serious study.) The crystalline lens provides the needed fine-focusing mechanism through changes in its shape, i.e. it has a variable focal length —a feature we'll come back to presently.

Fig. 5.67 A coherent bundle of $10\,\mu$ glass fibers transmitting an image even though knotted and sharply bent. [Photo courtesy of American Cystoscope Makers, Inc.]

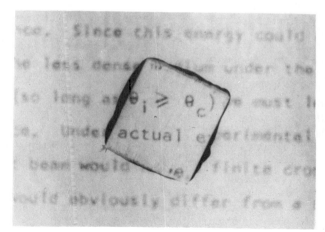

Fig. 5.68 A stack of cover glass slides held together by a rubber band serves as a coherent light guide. [Photo by E.H.]

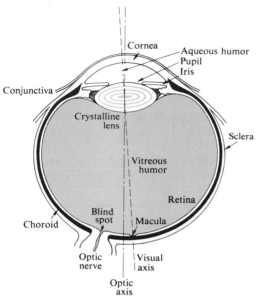
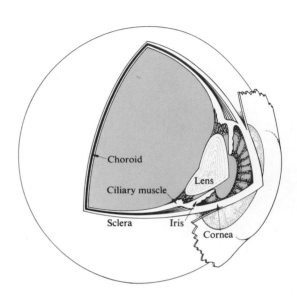

Fig. 5.69 The human eye.

Behind the lens is another chamber filled with a transparent gelatinous substance known as the *vitreous humor* ($n_{vh} \approx 1.337$). As an aside, it should be noted that the vitreous humor contains microscopic particles of cellular debris floating freely about. Their shadows, outlined with diffraction fringes, are easily seen within your own eye by just squinting at a light source or looking at the sky through a pinhole—strange little amoeba-like objects (*muscae volitantes*) will float across the field of view. Incidentally, a marked increase in one's perception of these floaters may be indicative of retinal detachment. While you're at it, squint at the source again (a broad diffuse fluorescent works well). Closing down your lids almost completely, you'll actually be able to see the near circular periphery of your own pupil, beyond which the glare of light will disappear into blackness. If you don't believe it, block and then unblock some of the light; the glare circle will visibly expand and contract, respectively. You are seeing the shadow cast by the iris from the inside! Seeing internal objects like this is known as entoptic perception.

Within the tough sclerotic wall is an inner shell, the choroid It is a dark layer, well supplied with blood vessels and richly pigmented with melanin. The choroid is the absorber of stray light, as is the coat of black paint on the inside of a camera. A thin (about 0.02 in) layer of light receptor cells covers much of the inner surface of the choroid—this is the *retina* (from the Latin *rete* meaning net). The focused beam of light is absorbed via electrochemical reactions in this pinkish multilayered structure. The human eye contains two kinds of photoreceptor cells, the *rods* and *cones* (Fig. 5.70). Roughly 125 million of them are intermingled nonuniformly over the retina. The ensemble of rods in some respects has the characteristics of a high-speed, coarse-grain, black and white film (such as Tri-X). It is exceedingly sensitive, performing in light too dim for the cones to respond to, yet it is unable to distinguish color and the images it relays are not well defined. In contrast, the ensemble of 6 or 7 million cones can be imagined as if it were a separate, but overlapping, low-speed, fine-grain color film. It performs in bright light giving detailed, colored views but is fairly insensitive at low light levels.

The normal wavelength range of human vision is said to be roughly 390 nm to 780 nm (Table 3.2, p. 56). Despite this, studies have extended these limits down to about 310 nm in the ultraviolet and up to roughly 1050 nm in the infrared—indeed people have reported "seeing" X-radiation. The limitation on ultraviolet transmission in the eye is set by the crystalline lens which absorbs in the UV. People who have had a lens removed surgically have greatly improved UV sensitivity.

Fig. 5.70 An electron micrograph of the retina of a salamander (Necturus Maculosus). Two visual cones appear in the foreground and several rods behind them. [Photo from E. R. Lewis, Y. Y. Zeevi and F. S. Werblin, *Brain Research* **15**, 559 (1969).]

The point of exit of the optic nerve from the eye contains no receptors and that area is insensitive to light; accordingly it is known as the *blind spot* (see Fig. 5.71). The optic nerve spreads out over the back of the interior of the eye in the form of the retina.

Just about at the center of the retina is a small depression from 2.5 to 3 mm in diameter known as the yellow spot or *macula*. There is a tiny rod-free region about 0.3 mm in diameter at its center, the *fovea centralis*. (In comparison the image of the full moon on the retina is about 0.2 mm in diameter—Problem 5.40.) Here the cones are thinner and more densely packed than anywhere else in the retina. That region provides the sharpest and most detailed information. For that reason, the eyeball is continuously moving so that light coming from the area on the object of primary interest falls on the fovea. An image is constantly shifted across different receptor cells by these normal eye movements. If

X   **1**   **2**

Fig. 5.71 To verify the existence of the blind spot, close one eye and, at about a distance of 10 inches, look directly at the X—the 2 will disappear. Moving closer will cause the 2 to reappear while the 1 vanishes.

such movements are negated so that the image is kept stationary on a given set of photoreceptors, it would actually tend to fade out. Another fact which indicates the complexity of the sensing system is that the rods are multiply connected to nerve fibers and a single such fiber can be activated by any one of about a hundred rods. By contrast cones in the fovea are individually connected to nerve fibers. The actual perception of a scene is constructed by the eye—brain system in a continuous analysis of the time-varying retinal image. Just think how little trouble the blind spot causes even with one eye closed.

Between the nerve-fiber layer of the retina and the humor is a network of large retinal blood vessels which can be observed entoptically. One way is to close your eye and place a bright *small* source against the lid. You'll "see" a pattern of shadows (*Purkinje Figures*) cast by the blood vessels on the sensitive retinal layer.

ii) Accommodation

The fine focusing or *accommodation* of the human eye is a function performed by the crystalline lens. The lens is suspended in position behind the iris by ligaments which are connected to the *ciliary muscles*. Ordinarily, these muscles are relaxed and in that state they pull back on the network of fine fibers holding the rim of the lens. This draws the pliable lens into a fairly flat configuration increasing its radii which, in turn, increases its focal length (5.16). With the muscles completely relaxed, the light from an object at infinity will be focused on the retina (Fig. 5.72). As the object moves closer to the eye, the ciliary muscles contract, relieving the external tension on the periphery of the lens which then bulges slightly under its own elastic forces. In so doing, the focal length decreases such that s_i is kept constant. As the object comes still closer, the ciliary muscles become more tensely contracted and the lens surfaces take on even smaller radii. The closest point on which the eye can focus is known as the *near point*. In a normal eye it might be about 7 cm for a teenager, 25 cm or so for a young adult, and roughly 100 cm in the middle aged. Visual instruments are designed with this in mind so that the eye need not strain unnecessarily. Clearly, the eye cannot focus on two different objects at once. This will be made obvious if, while looking through a piece of glass, you try to focus on it and the scene beyond at the same time.

Mammals, like man, generally accommodate by varying the lens curvature, but there are other means. Fish, in contrast, move only the lens itself toward or away from the retina, just as the camera lens is moved to focus. Some mollusks

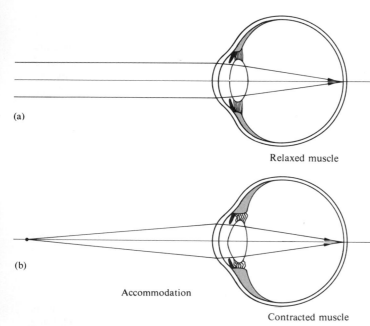

(a)

Relaxed muscle

(b)

Accommodation

Contracted muscle

Fig. 5.72 Accommodation—changes in the lens configuration.

accomplish the same thing by contracting or expanding the whole eye, thus altering the relative distance between lens and retina. For birds of prey, where keeping a rapidly moving object in constant focus over a wide range of distances is a matter of survival, the accommodation mechanism is quite different. They accommodate by greatly changing the curvature of the cornea.

5.7.2 Eyeglasses

Spectacles were probably invented some time in the 13th century, in northern Italy. And polished gem stones were no doubt employed as lorgnettes long before that.

It is customary and quite convenient in physiological optics to speak about the *dioptric power* $\mathscr{D}$ *of a lens, which is simply the reciprocal of the focal length.* When f is in meters, the unit of power is the inverse meter or *diopter.* For example, if a converging lens has a focal length of $+1$ meter, its power is $+1$ diopter; with a focal length of -2 meters (a diverging lens) $\mathscr{D} = -\frac{1}{2}$ diopter; for $f = +10$ cm, $\mathscr{D} = 10$ diopters. Since a thin lens of index n_l in air has a focal length given by

$$\frac{1}{f} = (n_l - 1)\left(\frac{1}{R_1} - \frac{1}{R_2}\right),\qquad [5.16]$$

its power is

$$\mathscr{D} = (n_l - 1)\left(\frac{1}{R_1} - \frac{1}{R_2}\right).\qquad (5.62)$$

You can get a sense of the direction in which we are moving by considering, in rather loose terms, that each surface of a lens bends the incoming rays—the more the bending, the stronger the surface. A convex lens which strongly bends the rays at both surfaces has a short focal length and a large dioptric power. We already know that the focal length for two thin lenses in contact is given by

$$\frac{1}{f} = \frac{1}{f_1} + \frac{1}{f_2}.\qquad [5.38]$$

This means that the combined power is the sum of the individual powers, that is

$$\mathscr{D} = \mathscr{D}_1 + \mathscr{D}_2.$$

Thus a convex lens with $\mathscr{D}_1 = +10$ diopters in contact with a negative lens of $\mathscr{D}_2 = -10$ diopters results in $\mathscr{D} = 0$; the combination behaves like a parallel sheet of glass. Furthermore, we can imagine a lens, e.g. a double convex lens, as being composed of two planar-convex lenses in intimate contact, back to back. The power of each of these follows from Eq. (5.62), thus for the first planar-convex lens $(R_2 = \infty)$

$$\mathscr{D}_1 = \frac{(n_l - 1)}{R_1},\qquad (5.63)$$

while for the second

$$\mathscr{D}_2 = \frac{(n_l - 1)}{-R_2}.\qquad (5.64)$$

These expressions may equally well be defined as giving the *powers of the respective surfaces* of the initial double convex lens. In other words, *the power of any thin lens is equal to the sum of the powers of its surfaces.* Because R_2 for a convex lens is a negative number, both $\mathscr{D}_1$ and $\mathscr{D}_2$ will be positive in that case. The power of a *surface*, defined in this way, is not generally the reciprocal of its focal length, although it is when immersed in air. Relating this terminology to the generally used model for the human eye we note that the power of the crystalline lens *surrounded by air* is about $+19$ diopters. The cornea provides roughly $+43$ of the total $+58.6$ diopters of the intact eye.

A normal eye, despite the connotation of the word, is not really as common an occurrence as one might expect. By the term normal, or its synonym *emmetropic*, we mean an

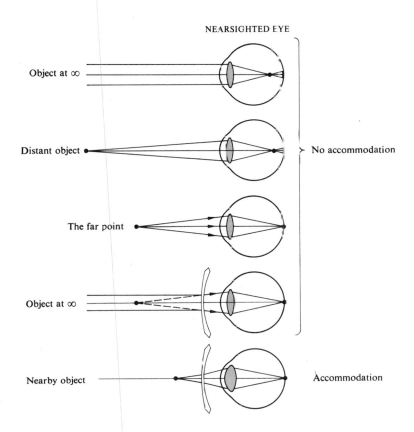

NEARSIGHTED EYE

Object at ∞

Distant object

The far point

Object at ∞

Nearby object

No accommodation

Accommodation

Fig. 5.73 Correction of the nearsighted eye.

eye which is capable of focusing parallel rays on the retina while in a relaxed condition, i.e. one whose second focal point lies on the retina. The most distant point which can be brought to focus, the *far point*, is therefore located at infinity. In contrast, when the second focal point does not lie on the retina the eye is *ametropic* (it suffers e.g. hyperopia, myopia or astigmatism).

i) Nearsightedness—Negative Lenses

Myopia is the condition where parallel rays are brought to focus in front of the retina; the power of the lens system as configured is too large for the anterior–posterior axial length of the eye. This can happen in a number of ways, e.g. the eye may elongate even though its power remains normal; the cornea may experience an increase in curvature; there is also a form which results from abnormal refractivity of the eye's optical media. In any event, images of distant objects fall in front of the retina, the far point is closer in than infinity, and all points beyond it will appear blurred. This is why myopia is often called *nearsightedness*—an eye with this

defect sees nearby objects clearly (Fig. 5.73). To correct the condition, or at least its symptoms, we place an additional lens in front of the eye such that the combined spectacle–eye lens system has its second focal point on the retina. Since the myopic eye can clearly see objects closer than the far point, the spectacle lens must cast relatively nearby images of distant objects. Hence we introduce a negative lens which will diverge the rays a bit. Resist the temptation to suppose that we are merely reducing the power of the system. In point of fact, the power of the lens–eye combination is most often made to equal that of the unaided eye. If you are wearing glasses to correct myopia take them off; the world gets blurry but it doesn't change size. Try casting a real image on a piece of paper using your glasses—it can't be done

Suppose an eye has a far point of 2 m, all would be well if the spectacle lens were to appear to bring more distant objects in closer than 2 m. If the virtual image of an object at infinity is formed by a concave lens at 2 m, the eye will see the object clearly with an unaccommodated lens. Thus using the thin-lens approximation (eyeglasses are generally thin to reduce weight and bulk) we have

$$\frac{1}{f} = \frac{1}{s_o} + \frac{1}{s_i} = \frac{1}{\infty} + \frac{1}{-2},$$ [5.17]

and $f = -2$ m while $\mathscr{D} = -\frac{1}{2}$ diopter. Notice that this calculation overlooks the separation between the spectacle lens and the eye. This is usually made equal to the distance of the first focal point of the eye (≈ 16 mm) from the cornea in order that no magnification of the image over that of the unaided eye occurs. Many people have unequal eyes yet both yield the same magnification. A change in M_T for one and not the other would be a disaster. Placing the correcting lens at the eye's first focal point avoids the problem completely, regardless of the power of that lens [take a look at Eq. (6.8)]. To see this, just draw a ray from the top of some object through that focal point. The ray will enter the eye traversing it parallel to the optic axis, thus establishing the height of the image. Yet, since this ray is unaffected by the presence of the spectacle lens whose center is at the focal point, the image's location may change on insertion of such a lens, but its height and therefore M_T will not (see Eq. 5.24).

ii) Farsightedness—Positive Lenses

Hyperopia (or *hypermetropia*) is the refractive error which causes the second focal point of the unaccommodated eye to lie behind the retina (Fig. 5.74). *Farsightedness*, as you might have guessed it would be called, is often due to a shortening of the anteroposterior axis of the eye—the lens is too close to the retina. To increase the bending of the rays, a positive spectacle lens is placed in front of the eye. The hyperopic eye can and must accommodate to see distant objects distinctly but it will be at its limit to do so for a near point which is much farther away than it would be normally (this we take as 25 cm). It will consequently be unable to see clearly. Suppose that a hyperopic eye has a near point positive power will effectively move a close object out beyond the near point where the eye has adequate acuity, i.e. it will form a distant virtual image which the eye can then see clearly. Suppose that a hyperopic eye has a near point of 125 cm. For an object at $+25$ cm to have its image at $s_i = -125$ cm, so that it can be seen as if through a normal eye, the focal length must be

$$\frac{1}{f} = \frac{1}{(-1.25)} + \frac{1}{0.25} = \frac{1}{0.31}$$

or $f = 0.31$ m and $\mathscr{D} = +3.2$ diopters. This is in accord with

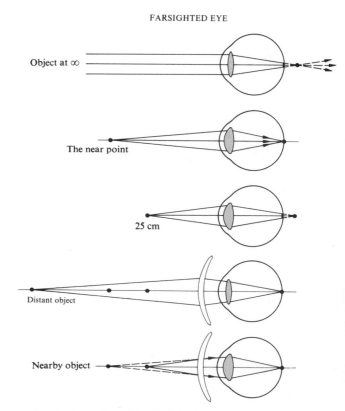

FARSIGHTED EYE

Object at ∞

The near point

25 cm

Distant object

Nearby object

Fig. 5.74 Correction of the farsighted eye.

Table 5.3 where $s_o < f$. These spectacles will cast real images—try it if you're hyperopic.

Very gentle finger pressure on the lids above and below the cornea will temporarily distort it changing your vision from blurred to clear and vice versa.

iii) Astigmatism—Anamorphic Lenses

Another and perhaps the most common eye defect is *astigmatism*. It arises from an uneven curvature of the cornea. In other words, the cornea is asymmetric. Suppose we passed two meridional planes (ones containing the optic axis) through the eye such that the (curvature or) power is maximal on one and minimal on the other. If these planes are perpendicular, the *astigmatism* is *regular* and correctible; if not, it is *irregular* and not easily corrected. Regular astigmatism can take different forms where the eye is emmetropic, myopic or hyperopic in various combinations and degrees on the two perpendicular meridional planes.

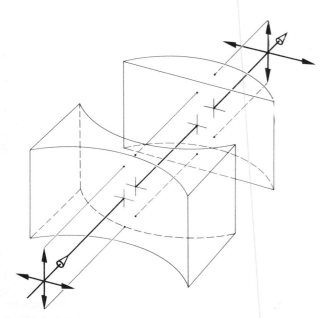

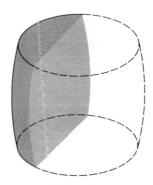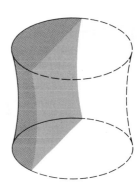

Fig. 5.76 Toric surfaces.

Fig. 5.75 An anamorphic system.

Thus, as a simple example, the columns of a checker board might be well focused while the rows are blurred due to myopia or hyperopia. Obviously these meridional planes need not be horizontal and vertical.

The great astronomer Sir George B. Airy used a concave sphero-cylindrical lens to ameliorate his own myopic astigmatism in 1825. This was probably the first time astigmatism had been corrected. But it was not until the publication in 1862 of a treatise on cylindrical lenses and astigmatism by the Dutchman Franciscus Cornelius Donders (1818–89) that opthalmologists were moved to adopt the method on a large scale.

Any optical system which has a different value of M_T or $\mathscr{D}$ in two principal meridians is said to be *anamorphic*. Thus, for example, if we rebuilt the system depicted in Fig. 5.28, this time using cylindrical lenses (Fig. 5.75), the image would be distorted, having been magnified in only one plane. This is just the sort of distortion needed to correct for astigmatism when a defect exists in only one meridian. An appropriate planar cylindrical spectacle lens, either positive or negative, would restore essentially normal vision. When both perpendicular meridians require correction, the lens may, for example, be *sphero-cylindrical*, or even *toric* as in Fig. (5.76).

Just as an aside, we mention that anamorphic lenses are used in other areas, as for example in the making of wide-screen motion pictures. In that way, an extra-large horizontal field of view is compacted onto the regular film format. When shown through a special lens the distorted picture spreads out again. On occasion a T.V. station will show short excerpts without the special lens—you may have seen the weirdly elongated result.

5.7.3 The Magnifying Glass

An observer can cause an object to appear larger, and therefore examine it in detail, by simply bringing it closer to her eye. As the object is brought nearer and nearer its retinal image increases, remaining in focus until the crystalline lens can no longer provide adequate accommodation. Should the object come closer than this *near point*, the image will blur (Fig. 5.77). A single positive lens can be used, in effect to add refractive power to the eye, so that the object can be brought still closer and yet be in focus. The lens so used is referred to variously as a *magnifying glass*, a *simple magnifier* or a *simple microscope*. In any event, its function is *to provide an image of a nearby object which is larger than that seen by the unaided eye*. Devices of this sort have been around for a long time. In fact, a quartz convex lens ($f \approx 10$ cm), which may have served as a magnifier, was unearthed in 1885 amongst the ruins of the palace of King Sennacherib (705–681 B.C.) of Assyria.

Evidently, it would be desirable for the lens to form a magnified, erect image. Furthermore, the rays entering the normal eye should not be converging. Table 5.3 (p. 112) immediately suggests placing the object within the focal length, i.e. $s_o < f$. The result is shown in Fig. 5.78. Because of the relatively tiny size of the eye's pupil, it will almost certainly always be the aperture stop, and as in Fig. 5.30 (p. 117), it will also be the exit pupil.

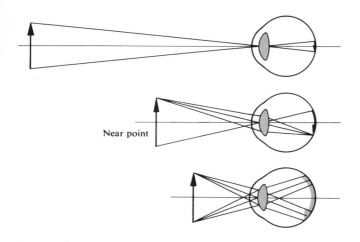

Near point

The *magnifying power, M.P.*, or equivalently the *angular magnification, M_A,* of a visual instrument is defined as *the ratio of the size of the retinal image as seen through the instrument* over *the size of the retinal image as seen by the unaided eye at normal viewing distance.* The latter is generally taken as the distance to the near point, d_o. The ratio of angles α_a and α_u (which are made by chief rays from the top of the object in the instance of the aided and unaided eye, respectively) is equivalent to M.P., that is

$$\text{M.P.} = \frac{\alpha_a}{\alpha_u}. \qquad (5.65)$$

Keeping in mind that we are restricted to the paraxial region $\tan \alpha_a = y_i/L \approx \alpha_a$ and $\tan \alpha_u = y_o/d_o \approx \alpha_u$ and so

$$\text{M.P.} = \frac{y_i d_o}{y_o L},$$

wherein y_i and y_o are above axis and positive. Making d_o and L positive quantities will yield a positive M.P. which is quite reasonable. Using Eqs. (5.24) and (5.25) for M_T along with the thin-lens equation, the expression becomes

$$\text{M.P.} = -\frac{s_i d_o}{s_o L} = \left(1 - \frac{s_i}{f}\right)\frac{d_o}{L}.$$

Inasmuch as the image distance is negative $s_i = -(L - \ell)$ and consequently

$$\text{M.P.} = \frac{d_o}{L}[1 + \mathscr{D}(L - \ell)], \qquad (5.66)$$

$\mathscr{D}$ of course being the power of the magnifier $(1/f)$. There are three situations of particular interest: (1) When $\ell = f$ the magnifying power equals $d_o \mathscr{D}$. (2) When ℓ is effectively zero

$$[\text{M.P.}]_{\ell=0} = d_o\left(\frac{1}{L} + \mathscr{D}\right).$$

In that case the largest value of M.P. corresponds to the smallest value of L which, if vision is to be clear, must equal d_o. Thus

$$[\text{M.P.}]_{\substack{\ell=0 \\ L=d_o}} = d_o \mathscr{D} + 1. \qquad (5.67)$$

Taking $d_o = 0.25$ m for the standard observer we have

$$[\text{M.P.}]_{\substack{\ell=0 \\ L=d_o}} = 0.25\mathscr{D} + 1. \qquad (5.68)$$

As L increases M.P. decreases and similarly as ℓ increases M.P. decreases. If the eye is very far from the lens, the retinal image will indeed be small. (3) This last is perhaps the most common situation. Here we position the object at the focal point $(s_o = f)$ in which case the virtual image is at infinity $(L = \infty)$. Thus from Eq. (5.66)

$$[\text{M.P.}]_{L=\infty} = d_o \mathscr{D} \qquad (5.69)$$

for all practical values of ℓ. Because the rays are parallel, the eye views the scene in a relaxed, unaccommodated configuration, a highly desirable feature. Notice that $M_T = -s_i/s_o$ approaches infinity as $s_o \to f$, while in marked contrast M_A merely decreases by 1 under the same circumstances.

A magnifier with a power of 10 diopters has a focal length $(1/\mathscr{D})$ of 0.1 m and a M.P. equal to 2.5 when $L = \infty$. This is conventionally denoted as 2.5× which means that the retinal image is 2.5 times larger with the object at the focal length of the lens than it would be were the object at the near point of the unaided eye (where the largest clear image is possible). The simplest single-lens magnifiers are limited by aberrations to roughly 2× or 3×. A large field of view generally implies a large lens and that, for practical reasons, most often dictates fairly small curvature on the surfaces. The radii are large, as is f, and therefore M.P. is small. The reading glass, the kind Sherlock Holmes made famous, is a typical example. The watchmaker's eye loupe is frequently a single element lens, also of about 2× or 3×. Figure 5.79 shows a few more complicated magnifiers designed to operate in the range from roughly 10× to 20×. The double lens is quite common in a number of configurations. Although not particularly good, they perform satis-

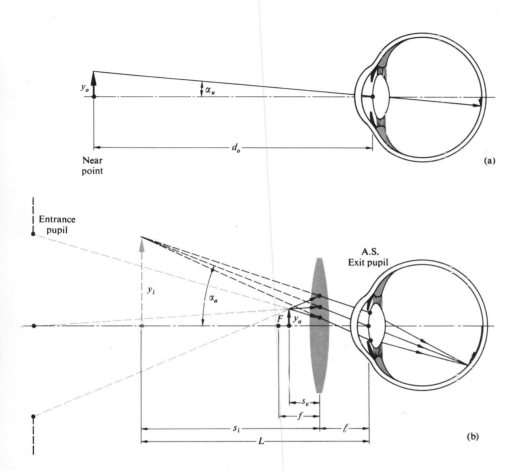

(a)

(b)   **Fig. 5.78** A magnifying glass.

factorily, as for example, in high-powered loupes. The Coddington is essentially a sphere with a slot cut in it to allow an aperture smaller than the pupil of the eye. A clear marble (any small sphere of glass qualifies) will also greatly magnify—but not without a good deal of distortion.

The relative refractive index of a lens and the medium in which it is immersed, n_{lm}, is wavelength dependent. But since the focal length of a simple lens varies with $n_{lm}(\lambda)$, this means that f is a function of wavelength and the constituent colors of white light will focus at different points in space. The resultant defect is known as *chromatic aberration*. In order that the image be free of this coloration, positive and negative lenses made of different glasses are combined to form *achromates* (see Section 6.3.2). Achromatic, cemented, doublet and triplet lenses are comparatively expensive and are usually found in small, highly corrected, high-powered magnifiers.

5.7.4 Eyepieces

The *eyepiece* or *ocular* is a visual optical instrument. Fundamentally a magnifier, its function is to view not an actual object, but the intermediate image of that object as formed by a preceding lens system. In effect, the eye looks into the ocular and the ocular looks into the optical system—be it a spotting scope, compound microscope, telescope or binocular. A single lens could serve the purpose, but poorly. If the retinal image is to be more satisfactory, the ocular cannot generally suffer extensive aberrations. The eyepiece of a special instrument, however, might be designed as part of the complete system so that its lenses can be utilized in the overall scheme to balance out aberrations. Even so, standard eyepieces are used interchangeably on most telescopes and compound microscopes. Moreover, eyepieces are quite difficult to design and the usual, and perhaps most

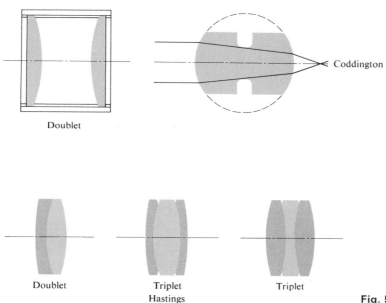

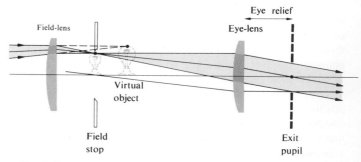

Fig. 5.79   Magnifiers.

fruitful approach is to incorporate or slightly modify one of the existing designs.

The ocular must provide a virtual image (of the intermediate image), most often located at or near infinity, so that it can be comfortably viewed by a normal, relaxed eye. Furthermore, it must position the center of the exit pupil or *eye-point* at which the observer's eye is placed, at some convenient location preferably at least 10 mm or so from the last surface. As before, ocular magnification is the product $d_o\mathscr{D}$ or as is often written M.P. = (250 mm)/f.

The **Huygens** ocular, which dates back over 250 years, is still in wide use today (Fig. 5.80) particularly in microscopy. The lens adjacent to the eye is known as the *eye-lens* while the first lens in the ocular is the *field-lens*. The distance from the eye-lens to the eye-point is known as the *eye relief* and for the Huygens ocular, it's only an uncomfortable 3 mm or so. Notice that this ocular requires the incoming rays to be converging so as to form a virtual object for the eye-lens. Clearly then, the Huygens eyepiece cannot be used as an ordinary magnifier. Its contemporary appeal rests in its low purchase price [see Section 6.3.2(ii)]. Another old standby is the **Ramsden** eyepiece (Fig. 5.81). This time the principal focus is in front of the field-lens and so the intermediate image will appear there in easy access. This is where you would place a *reticle* (or *reticule*) which might contain a set of cross hairs, precision scales, or angularly divided circular grids etc. (When these are formed on a transparent plate, they are often called *graticules*.) Since the reticle and intermediate image are in the same plane, both will be in focus at the same time. Its roughly 12 mm eye relief is an advantage over the previous ocular. The Ramsden is relatively popular and fairly inexpensive (see Problem 6.2). The **Kellner** eyepiece represents a definite increase in image quality although eye relief is between that of the previous two devices. The Kellner is essentially an achromatized Ramsden (Fig. 5.82). It is most commonly used in moderately wide-field telescopic instruments. The **orthoscopic** eyepiece (Fig. 5.83) has a wide field, high

Fig. 5.80   The Huygens eyepiece.

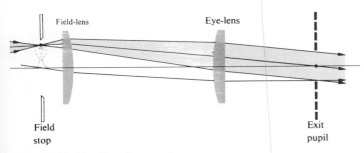

Fig. 5.81   The Ramsden eyepiece.

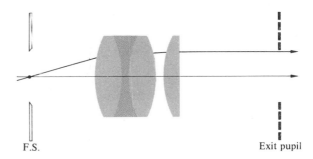

Fig. 5.83   The orthoscopic eyepiece.

magnification and long eye relief (≈ 20 mm). The **symmetrical (Plössl)** eyepiece (Fig. 5.84) has similar characteristics to those of the orthoscopic ocular but is generally somewhat superior to it. The **Erfle** (Fig. 5.85) is probably the most common wide field (roughly $\pm 30°$) eyepiece. It is well corrected for all aberrations and comparatively expensive.*

Although there are many others including variable power *zoom* eyepieces and ones with aspheric surfaces, those discussed above are quite representative. They are the ones you will ordinarily find on telescopes and microscopes and on long lists in the commercial catalogs.

5.7.5   The Compound Microscope

The compound microscope goes the next step beyond the simple magnifier in providing higher angular magnification (greater than about $30\times$) of *nearby* objects. Its invention, which may have occurred as early as 1590, is generally accredited to a Dutch spectacle-maker Zacharias Janssen of Middleburg. Galileo runs a close second, having announced his invention of a compound microscope in 1610.

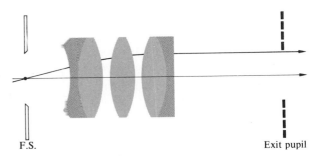

Fig. 5.84   The symmetrical (Plössl) eyepiece.

---

* Detailed designs of these and other oculars can be found in the *Military Standardization Handbook—Optical Design*, MIL-HDBK-141.

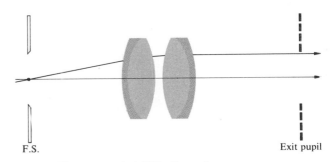

Fig. 5.85   The Erfle eyepiece.

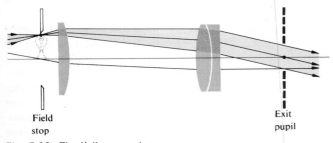

Fig. 5.82   The Kellner eyepiece.

A simple version, which is closer to these earliest devices than it is to a modern laboratory microscope, is depicted in Fig. 5.86. The lens system, here a singlet, closest to the object is referred to as the *objective*. It forms a real, inverted and, usually magnified, image of the object. This image resides in space on the plane of the field stop of the eyepiece. Rays diverging from each point of this image will emerge from the eye-lens (which in this simple case is the eyepiece itself) parallel to each other as in the previous section. The ocular magnifies this intermediate image still further. Thus

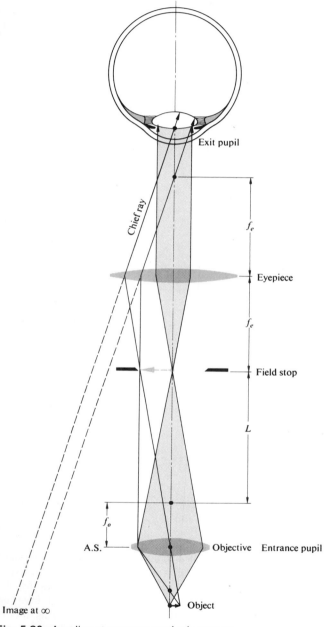

Exit pupil

Chief ray

f_e

Eyepiece

f_e

Field stop

L

f_o

A.S.

Objective   Entrance pupil

Object

Image at ∞

Fig. 5.86 A rudimentary compound microscope.

the magnifying power of the entire system is the product of the transverse linear magnification of the objective, M_{T_o}, and the angular magnification of the eyepiece, M_{Ae}, that is

$$\text{M.P.} = M_{T_o} M_{Ae}. \tag{5.70}$$

Recall that $M_T = -x_i/f$, (5.26), and with this in mind most, but not all, manufacturers design their microscopes such that the distance (corresponding to x_i) from the second focus of the objective to the first focus of the eyepiece is standardized at 160 mm. This distance, known as the *tube length*, is denoted by L in the figure. (Some authors define tube length as the image distance of the objective.) Hence, with the final image at infinity and the standard near point taken as 10 inches or 254 mm

$$\text{M.P.} = \left(-\frac{160}{f_o} \right) \left(\frac{254}{f_e} \right) \tag{5.71}$$

and the image is inverted (M.P. < 0). Accordingly, the barrel of an objective with a focal length f_o of say 32 mm will be engraved with the markings 5× (or ×5) indicating a power of 5. Combined with a 10× eyepiece (f_e = 1 inch) the microscope M.P. would then be 50×.

To maintain the distance relationships between the objective, field stop and ocular, while positioning a focused intermediate image of the object in the first focal plane of the eyepiece, all three elements are moved as a single unit.

The objective itself functions as the aperture stop and entrance pupil. Its image, formed by the eyepiece, is the exit pupil into which the eye is positioned. The field stop, which limits the extent of the largest object that can be viewed is fabricated as part of the ocular. The image of the field stop formed by the optical elements following it is called the *exit window*, and the image formed by the optical elements preceding it is the *entrance window*. The cone angle subtended at the center of the exit pupil by the periphery of the exit window is said to be the *angular field of view in image space*.

A modern microscope objective can be roughly classed as one of essentially three different kinds. It might be designed to work best with the object positioned below a cover glass, with no cover glass (metallurgical instruments) or with the object immersed in a liquid which is in contact with the objective. In some cases, the distinction is not critical and it may be used with or without a cover glass. Four representative objectives are shown in Fig. 5.87 [see Section 6.3.1(i)]. In addition, the ordinary low power

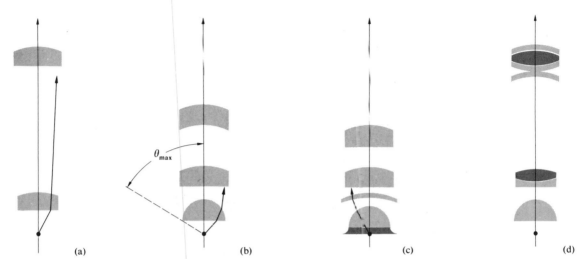

θ_{max}

(a)   (b)   (c)   (d)

Fig. 5.87 Microscope objectives: (a) Lister objective, 10×, N.A. = 0.25, f = 16 mm (two cemented achromates). (b) Amici objective, from 20×, N.A. = 0.5, f = 8 mm to 40×, N.A. = 0.8, f = 4 mm. (c) Oil-immersion objective, 100×, N.A. = 1.3, f = 1.6 mm (see Fig. 6.16). (d) Apochromatic objective, 55 ×, N.A. = 0.95, f = 3.2 (contains two flourite lenses).

(about 5×) cemented doublet achromate is quite common. [Relatively inexpensive medium power (10× or 20×) achromatic objectives, because of their short focal lengths, can also conveniently be used when expanding and spatially filtering laser beams.]

There is one other characteristic quantity of importance which must be mentioned here even if only briefly. The brightness of the image is, in part, dependent on the amount of light gathered in by the objective. The f-number is a useful parameter for describing this quantity, particularly when the object is a distant one (Section 5.3.3). However, for an instrument working at *finite conjugates* (s_i and s_o both finite), the numerical aperture N.A. is more appropriate (Section 5.6). In the present instance

$$\text{N.A.} = n_o \sin \theta_{max} \qquad (5.72)$$

where n_o is the refractive index of the immersing medium (air, oil, water, etc.) adjacent to the objective lens and θ_{max} is the half-angle of the maximum cone of light picked up by that lens [Fig. 5.87(b)]. In other words, θ_{max} is the angle made by a marginal ray with the axis. The numerical aperture ranges from about 0.07 for low-power objectives to 1.4 or so for high-power (100×) ones. Of course, if the object is in the air, the N.A. cannot be greater than 1.0. The N.A. is usually the second number etched in the barrel of the objective. Incidentally, Ernst Abbe (1840–1905), while working in the Carl Zeiss microscope workshop, introduced the concept of the numerical aperture. It was he who recognized that the minimum transverse distance between two object points which can be resolved in the image, i.e. the *resolving power*, varied directly as λ and inversely as the N.A.

5.7.6 The Telescope

It is not at all clear who actually invented the telescope. In point of fact it was probably invented and reinvented many times. Do recall that by the seventeenth century spectacle lenses had been in use in Europe for about three hundred years. During that long span of time, the fortuitous juxtapositioning of two appropriate lens to form a telescope seems almost unavoidable. In any event, it is most likely that a Dutch optician, possibly even the ubiquitous Zacharias Jenssen of microscope fame, first constructed a telescope and in addition had inklings of the value of what he was peering into. The earliest indisputable evidence of the discovery, however, dates to October 2, 1608 when Hans Lippershey petitioned the States-General of Holland for a patent on a device for seeing at a distance (which is what *teleskopos* means in Greek). Incidentally, as you might have guessed, its military possibilities were immediately recognized. His patent was therefore not granted; instead the

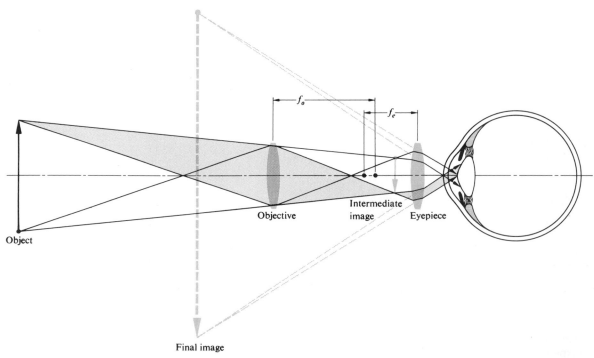

Fig. 5.88 Keplerian astronomical telescope (accommodating eye).

government purchased the rights to the instrument and he received a commission to continue research. Galileo heard of this work and by 1609 he had fashioned a telescope of his own using two lenses and an organ pipe as a tube. It was not long before he constructed a number of greatly improved instruments and began to astound the world with the forth-coming astronomical discoveries for which he is so justly famous.

i) Refracting Telescopes

A simple *astronomical* telescope is shown in Fig. 5.88. Unlike the compound microscope, which it so closely resembles, its primary function is to enlarge the retinal image of a *distant* object. In the illustration, the object is at a finite far distance from the objective so that the real intermediate image is formed just beyond its second focal point. This image will be the object for the next lens system, i.e. the ocular. It follows from Table 5.3 (p. 112) that if the eyepiece is to form a virtual magnified final image (within the range of normal accommodation), the object distance must be less than or equal to the focal length, f_e. In practice, *the position of the intermediate image is fixed and only the eye-piece is moved to focus the instrument*. Notice that *the final image is inverted*, but as long as the scope is used for astro-nomical observations, this is of little consequence especially since most work is photographic.

At very great object distances the incident rays are effectively parallel—the intermediate image resides at the second focus of the objective. Usually the eyepiece is loca-ted so that its first focus overlaps the second focus of the objective in which case rays diverging from a point on the intermediate image will leave the ocular parallel to each other. A normal viewing eye can then focus the rays in a relaxed configuration. Certainly if the eye is nearsighted or farsighted, the ocular can be moved in or out so that the rays diverge or converge a bit to compensate. (If you are astigmatic, you'll have to keep your glasses on when using ordinary visual instruments.) We saw earlier [Section 5.2.3, (iv)] that both the back and front focal lengths of a thin lens combination go to infinity when the two lenses are separated by a distance d equal to the sum of their focal lengths (Fig. 5.89). The astronomical telescope in this configuration of *infinite con-*

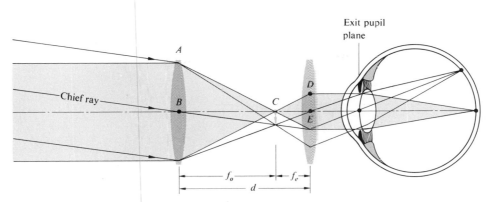

Fig. 5.89 Astronomical telescope—infinite conjugates.

jugates is said to be *afocal*, i.e. without a focal length. As a side note, if you shine a collimated (parallel rays, i.e. plane waves) narrow laser beam into the back end of a scope focused at infinity, it will emerge still collimated, but with an increased cross-section. It is often desirable to have a broad, quasimonochromatic, plane-wave beam, and specific devices of this sort are now available commercially.

The periphery of the objective is the aperture stop and it encompasses the entrance pupil as well, there being no lenses to the left of it. If the telescope is trained directly on some distant galaxy, the visual axis of the eye will presumably be collinear with the central axis of the scope. The entrance pupil of the eye should then coincide in space with the exit pupil of the scope. However, the eye is not immobile. It will move about scanning the entire field of view, which quite often contains many points of interest. In effect, the eye examines different regions of the field by rotating so that rays from a particular area fall on the fovea centralis. The direction established by the chief ray through the center of the entrance pupil to the fovea centralis is the *primary line of sight*. The axial point, fixed in reference to the head, through which the primary line of sight always passes, regardless of the orientation of the eyeball, is called the *sighting intersect*. When it is desirable to have the eye surveying the field, the sighting intersect should be positioned at the center of the telescope's exit pupil. In that case, the primary line of sight will always correspond to a chief ray through the center of the exit pupil, however the eye moves.

Suppose that the margin of the visible object subtends a half-angle of α at the objective (Fig. 5.90). This is essentially the same as the angle α_u which would be subtended at the unaided eye. As in previous sections, the angular magni-

fication is

$$\text{M.P.} = \frac{\alpha_a}{\alpha_u} \qquad [5.65]$$

Here α_u and α_a are measures of the field of view in object and image space, respectively. The first is the half-angle of the actual cone of rays collected while the second relates to the apparent cone of rays. If a ray arrives at the objective with a negative slope, it will enter the eye with a positive slope and vice versa. Thus to make the sign of M.P. positive for erect images, and therefore consistent with previous usage (Fig. 5.78), either α_u or α_a must be taken to be negative—we choose the former because the ray has a negative slope. Observe that the ray passing through the first focus of the objective passes through the second focus of the eyepiece, i.e. F_{o1} and F_{e2} are conjugate points. In the paraxial approximation $\alpha \approx \alpha_u \approx \tan \alpha_u$ and $\alpha_a \approx \tan \alpha_a$. The image fills the region of the field stop and so half its extent equals the distance $\overline{BC} = \overline{DE}$. Thus, from triangles $F_{o1}BC$ and $F_{e2}DE$, the ratio of the tangents yields

$$\text{M.P.} = -\frac{f_o}{f_e}. \qquad (5.73)$$

Another convenient expression for the M.P. comes from considering the transverse magnification of the ocular. Inasmuch as the exit pupil is the image of the objective (Fig. 5.90) we have

$$M_{Te} = -\frac{f_e}{x_o} = -\frac{f_e}{f_o}.$$

Furthermore if D_o is the *diameter of the objective* and D_{ep} is

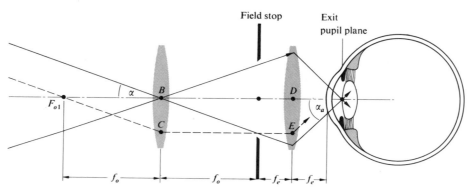

Fig. 5.90 Ray angles for a telescope.

the *diameter of its image, the exit pupil*, then $M_{Te} = D_{ep}/D_o$. These two expressions for M_{Te} compared with Eq. (5.73) yield

$$\text{M.P.} = \frac{D_o}{D_{ep}}. \qquad (5.74)$$

Here D_{ep} is actually a negative quantity since the image is inverted. It is an easy matter to build a simple refracting scope by just holding a long focal length lens in front of one having a short focal length and making sure that $d = f_o + f_e$. But again, well-corrected telescopic instruments generally have multielement objectives, usually doublets or triplets.

To be useful when the orientation of the object is of importance, a scope must contain an additional *erecting system*—such an arrangement is known as a *terrestrial telescope*. Most often a single erecting lens or lens system is located between the ocular and objective with the result that the image is right side up. Figure 5.91 shows one with a cemented doublet objective and a Kellner eyepiece. It will obviously have to have a long draw tube, the picturesque

kind that comes to mind when you think of wooden ships and cannonballs.

For that reason, *binoculars* (binocular telescopes) generally utilize erecting prisms which accomplish the same thing in less space, and also increase the separation of the objectives, thereby enhancing the stereoscopic effect. Most often these are double Porro prisms, as in Fig. 5.92 (notice the involved modified Erfle eyepiece, the wide field stop and achromatic doublet objective). Binoculars customarily bear several numerical markings, e.g. 6 × 30, 7 × 50, or 20 × 50, etc. The initial number is the magnification, here 6×, 7×, or 20×. The second number is the entrance pupil diameter, or equivalently the clear aperture of the objective, expressed in millimeters. It follows from Eq. (5.74) that the exit-pupil diameters will be the second number divided by the first or in this case 5, 7.1 and 2.5, all in millimeters. You can hold the instrument away from your eye and see the bright circular exit pupil surrounded by blackness. To measure it, focus the device at infinity, point it at the sky, and observe the emerging well-defined sharp disk of light

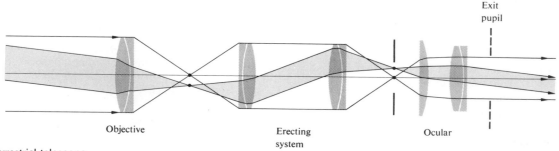

Fig. 5.91 A terrestrial telescope.

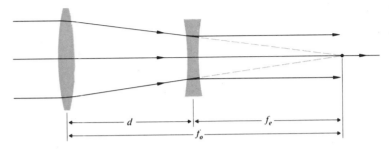

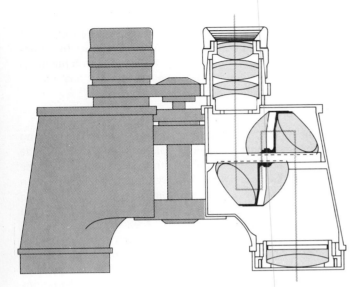

Fig. 5.92 A binocular.

using a piece of paper as a screen. Determine the eye relief while you're at it.

By the way, as long as $d = f_o + f_e$, the scope will be afocal, even if the eyepiece is negative (i.e. $f_e < 0$). The telescope built by Galileo (Fig. 5.93) had just such a negative lens as an eyepiece and therefore formed an erect image [$f_e < 0$ and M.P. > 0 in Eq. (5.73)]. As a telescope the system is now mainly of historical and pedagogical interest although one can still purchase two such scopes mounted side by side to form a Galilean fieldglass. It is quite useful, however, as a laser-beam expander because it has no internal focal points where a high power beam would otherwise ionize the surrounding air.

ii) Reflecting Telescopes

The difficulties inherent in making large lenses become underscored when we note that the largest refracting instrument is the 40-inch Yerkes telescope in Williams Bay, Wisconsin, while the reflector on Palomar Mountain, U.S.A. is 200 inches in diameter and the Soviet Union is constructing a 236 inch reflector at their Crimea Observatory. The problems are evident ones; a lens must be transparent and free of internal bubbles, etc. A front-surfaced mirror obviously need not be so, indeed it need not even be transparent. A lens can only be supported by its rim and may sag under its own weight; a mirror can be supported by its rim and back as well. Furthermore, since there is no refraction, and therefore no

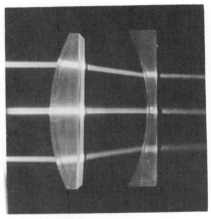

Fig. 5.93 The Galilean telescope. Galileo's first scope had a planar-convex objective (5.6 cm in diameter, $f = 1.7$ m, $R = 93.5$ cm) and a planar-concave eyepiece, both of which he ground himself. It was 3× in contrast to his last scope which was 32×. [Photo by E.H.]

effect on the focal length due to the wavelength dependence of the index mirrors suffer no chromatic aberration. For these and other reasons (e.g. their frequency response) reflectors predominate in large telescopes.

Invented by the Scotsman James Gregory (1638–75) in 1661, the reflecting telescope was first successfully constructed by Newton in 1668, and only became an important research tool in the hands of William Herschel a century later. Figure 5.94 depicts a number of reflector arrangements, each having concave paraboloidal primary mirrors. The 200 inch Hale telescope is so large that a little enclosure, where an observer can sit, is positioned at the prime focus. In the Newtonian version, a plane mirror or prism brings the beam out at right angles to the axis of the scope where it can be photographed, viewed, spectrally analyzed or photoelectrically processed. In the Gregorian arrangement, which is not particularly popular, a concave ellipsoidal secondary

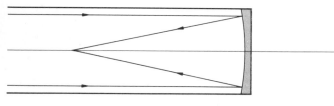

Prime focus (a)

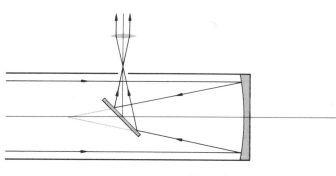

Newtonian (b)

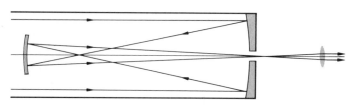

Gregorian (c)

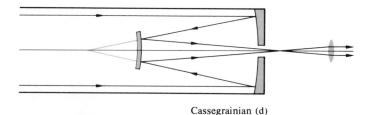

Cassegrainian (d)

Fig. 5.94 Reflecting telescopes.

mirror reinverts the image, returning the beam through a hole in the primary. The Cassegrainian system utilizes a convex hyperboloidal secondary mirror to increase the effective focal length (refer back to Fig. 5.43, p. 125). It functions as if the primary were of the same aperture but had a larger focal length or radius of curvature.

iii) Catadioptric Telescopes

A combination of reflecting (*catoptric*) and refracting (*dioptric*) elements is called a *catadioptric* system. To be sure, the best known of these, although not the first, is the classic *Schmidt optical system*. We must treat it here, even if only too briefly, because it represents the precursor of a new outlook in the design of large-aperture, extended-field reflecting systems. As seen in Fig. 5.95, bundles of parallel rays reflecting off a spherical mirror will form images, of let's say a field of stars, on a spherical image surface, the latter being in practice a curved film plate. The only problem with such a scheme is that although free of other aberrations [see Section 6.3.1 (iv)] we know that rays reflected from the outer regions of the mirror will not arrive at the same focus as those from the paraxial region. In other words, the mirror is a sphere, not a paraboloid and it suffers *spherical aberration* [Fig. 5.95(b)]. If this could be corrected, the system (in theory at least) would be capable of perfect imagery over a wide field of view. Since there is no one central axis, there are, in effect, no off-axis points. Recall that the paraboloid forms perfect images only at axial points, the image deteriorating rapidly off axis. One evening in 1929, while sailing on the Indian ocean (returning from an eclipse expedition to the Philippines), Bernhard Voldemar Schmidt (1879–1935) showed a colleague a sketch of a system he had designed to cope with the spherical aberration of a spherical mirror. He would use a thin glass corrector plate on whose surface would be ground a very shallow toroidal curve [Fig. 5.95(c)]. Light rays traversing the outer regions would be deviated by just the amount needed to be sharply focused on the image sphere. The corrector must overcome one defect without introducing appreciable amounts of other aberrations. This first system was built in 1930 and in 1949 the famous 48 inch Schmidt telescope of the Palomar Observatory was completed. It is a fast (*f*/2.5), wide-field device, ideal for surveying the night sky. A single photograph could encompass a region the size of the bowl of the Big Dipper—this compared with roughly 400 photos by the 200 inch reflector to cover the same area.

Major advances in the design of catadioptric instrumentation have occurred since the introduction of the original

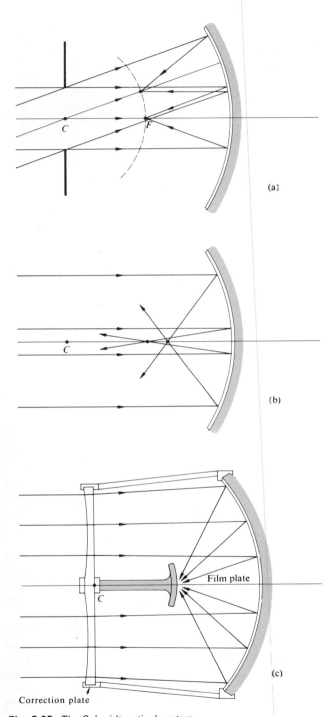

(a)

(b)

Film plate

Correction plate

Fig. 5.95 The Schmidt optical system.

Schmidt system.* There are now catadioptric satellite and missile tracking instruments, meteor cameras, compact commercial telescopes, telephoto objectives and missile-homing guidance systems. Innumerable variations on the theme exist; some replace the correcting plate with concentric meniscus lens arrangements (Bouwers–Maksutov), others use solid thick mirrors. One highly successful approach utilizes a triplet aspheric lens array (Baker).

5.7.7 The Camera

The prototype of the modern photographic camera was a device known as the *camera obscura*, the earliest form of which was simply a dark room containing a small hole in one wall. Light entering the hole cast an inverted image of the sunlit outside scene on an inside screen. The principle was known to Aristotle and his observations were preserved by Arab scholars throughout Europe's long Dark Ages. Alhazen utilized it to examine solar eclipses indirectly, and this over eight hundred years ago. The notebooks of Leonardo da Vinci contain several descriptions of the obscura but the first detailed treatment appears in *Magia naturalis* (Natural Magic) by Giovanni della Porta. He recommended it as a drawing aid, a function to which it was soon quite popularly put. Johannes Kepler, the renowned astronomer, had a portable tent version which he used while surveying in Austria. By the latter part of the sixteen hundreds, small hand-held *camera obscuras* were commonplace. We note parenthetically that the eye of the Nautilus, a little cuttlefish, is literally an open pinhole obscura which simply fills with sea water on immersion.

By replacing the viewing screen with a photosensitive surface, such as a film plate, the obscura becomes a camera in the modern sense of the word. The very first permanent photograph was made in 1826 by Joseph Nicéphore Niépce (1765–1833) using a box camera with a small convex lens, a sensitized pewter plate and a roughly eight hour exposure. It is a roof-top scene, taken from the workroom window of his estate near Châlon-sur-Saône in France. Although blurry and spotty (in its unretouched form), the large slanting roof of a barn, a pigeon-house, and a distant tree are still discernible.

The lensless pinhole camera (Fig. 5.96) is by far the least complicated device for the purpose, and yet it has several endearing and, indeed, remarkable virtues. It can form a very well-defined, practically undistorted, image of

---

* For further reading see e.g. J. J. Villa, "Catadioptric Lenses," *Optical Spectra* (March/April, 1968), p. 57.

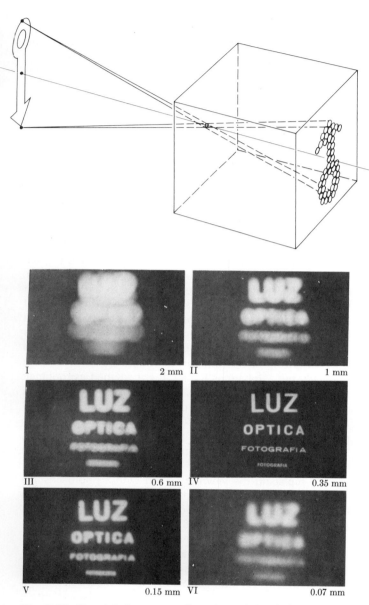

Fig. 5.96 The pinhole camera. Note the variation in image clarity as the hole diameter decreases. [Photos courtesy Dr. N. Joel, UNESCO.]

Fig. 5.97 Pinhole camera photo (Science Building, Adelphi University). Hole diameter 0.5 mm, film plane distance 25 cm, A.S.A. 3000, shutter speed 0.25 s. Note depth of field. [Photo by E.H.]

objects across an extremely wide angular field (due to great depth of focus) and over a large range of distances (great depth of field). If initially the entrance pupil is very large, no image results. As it is decreased in diameter, the image forms and grows ever sharper. After a point, further

reduction in the hole size causes the image to blur again and one quickly finds that the aperture size for maximum sharpness is proportional to its distance from the image plane. (A 0.5 mm diameter hole at 0.25 m from the film plate is convenient and works well.) There is no focusing of the rays at all, and so no defects in that mechanism are responsible for the drop off in clarity. The problem is actually one of diffraction as we shall see later on (Section 10.2.5). In most practical situations, the pinhole camera's one overriding drawback is that it is insufferably slow (roughly f/500). This means that exposure times will generally be far too long, even with the most sensitive films. The obvious exception is a stationary subject such as a building (Fig. 5.97) and there the pinhole camera excels.

Figure 5.98 depicts the essential components of a fairly popular and representative modern camera—the single lens reflex or SLR. Light traversing the first few elements of the lens then passes through an iris diaphragm used, in part, to control the exposure time or equivalently the f-number— it is in effect a variable-aperture stop. On emerging from the lens, light strikes a movable mirror tilted at 45°, then up through the focusing screen to the penta prism and out the finder eyepiece. Pressing the shutter release causes the diaphragm to close down to a pre-set value, the mirror swings up out of the way and the focal-plane shutter opens exposing the film. The shutter closes, the diaphragm opens

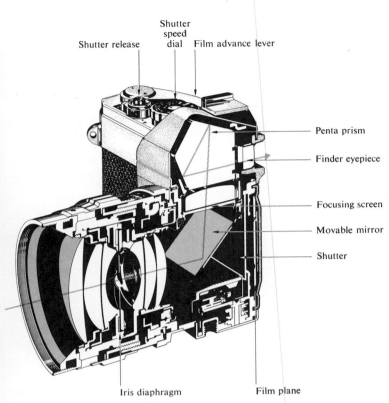

Shutter release

Shutter speed dial

Film advance lever

Penta prism

Finder eyepiece

Focusing screen

Movable mirror

Shutter

Iris diaphragm   Film plane

Fig. 5.98 A single-lens reflex camera.

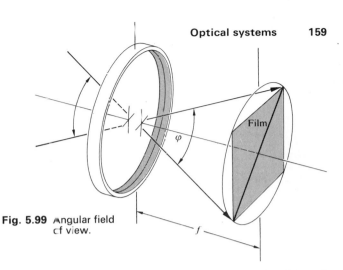

Fig. 5.99 Angular field of view.

full wide and the mirror drops back in place. Nowadays most SLR systems have any one of a number of built-in light-meter arrangements which are automatically coupled to the diaphragm and shutter, but those components were excluded from the diagram for the sake of simplicity.

To focus the camera, the entire lens is moved toward or away from the film plane. Since its focal length is fixed, as s_o varies so too must s_i. The *angular field of view* can loosely be thought of as relating to the fraction of the scene included in the photo. It is furthermore required that the entire photo surface correspond to a region of satisfactory image quality. More precisely, the angle subtended at the lens, by a circle encompassing the film area, is the angular field of view φ (Fig. 5.99). As a rough but reasonable approximation of a common arrangement, take the diagonal distance across the film to equal the focal length. Thus $\varphi/2 \approx \tan^{-1}\frac{1}{2}$, i.e. $\varphi \approx 53°$. If the object comes in from infinity, s_i must increase. The lens is then backed away from the film plate to keep the image in focus and the field of view, as recorded on the film whose periphery is the field

stop, decreases. A *standard* SLR lens has a focal length somewhere in the range from about 50 to 58 mm and a field of view of 40° to 50°. With the film size constant, reducing f results in a wider field angle. Accordingly, *wide-angle* SLR lenses range from $f \approx 40$ mm down to about 6 mm and φ goes from about 50° to a remarkable 220° (the latter being a special purpose lens wherein distortion is unavoidable). The *telephoto* has a long focal length, from roughly 80 mm and upwards. Consequently, its field of view drops off rapidly until it is only a few degrees at $f \approx 1000$ mm.

The standard photographic objective must have a large relative aperture, $1/(f/\#)$, to keep exposure times short. Moreover, the image is required to be flat and undistorted and the lens should have a wide angular field of view as well. All of this is no mean task and it is not surprising that a high quality innovative photographic objective remains particularly difficult to design even with our marvelous, mathematical, electronic *idiot savants*. The evolution of a modern lens still begins with a creative insight which leads to a promising new form. In the past, these were laboriously perfected relying on intuition, experience and, of course, trial and error with a succession of developmental lenses. Today, for the most part, the computer serves this function without the need of numerous prototypes. Many contemporary photographic objectives are variations of well-known successful forms. Figure 5.100 illustrates the general configuration of several important lenses roughly progressing from wide angle to telephoto. Particular specifications are not given because differing variations are quite numerous. The *Aviogon* and Zeiss *Orthometer* are wide angle lenses, while the *Tessar* and *Biotar* are often standard lenses. The *Cooke triplet*, introduced in 1895 by H. Dennis Taylor of Cooke and Sons is still being made (note similarity with the Tessar). Even prior to this (ca. 1840) Josef Max Petzval designed what was then a rapid (portrait) lens for Voight-

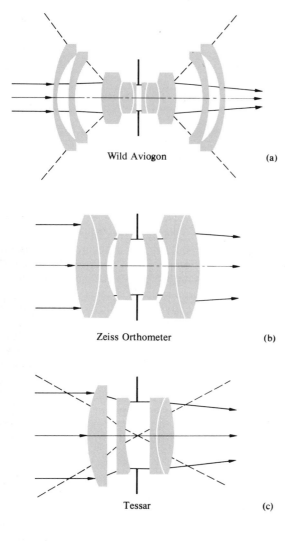

Wild Aviogon   (a)

Zeiss Orthometer   (b)

Tessar   (c)

Double Gauss (Biotar)   (d)

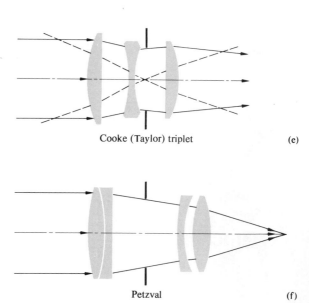

Cooke (Taylor) triplet   (e)

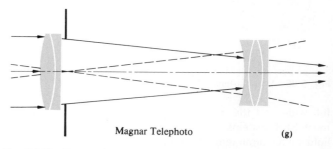

Petzval   (f)

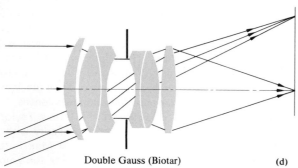

Magnar Telephoto   (g)

Fig. 5.100 Camera lenses.

länder and Son. Its modern offshoots are myriad. In general, a telephoto objective has a positive front grouping and a distant negative rear grouping. It often resembles the Galilean scope except that the lenses are shifted a bit so that the system is not afocal. These are usually rather large and heavy at the longer focal lengths, although calcium fluoride elements have begun to help in both respects. As can be seen in Fig. 5.101, the telephoto has a large effective focal length, e.f.l., i.e. it behaves as if it were a long focal length positive lens located a large distance in front of the focal plane. Thus while the image size is large, the back focal length is conveniently short, allowing it to be handily slipped into a standard camera body.

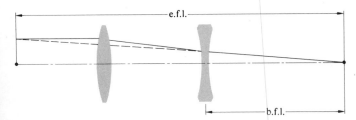

Fig. 5.101 A telephoto lens.

PROBLEMS

5.1 We wish to construct a Cartesian oval such that the conjugate points will be separated by 11 cm when the object is 5 cm from the vertex. If $n_1 = 1$ and $n_2 = \frac{3}{2}$, draw several points on the required surface.

5.2 Diagrammatically construct an ellipto-spheric negative lens showing the form of both rays and wavefronts as they pass through the lens. Do the same for an oval-spheric positive lens.

5.3* Making use of Fig. 5.102, Snell's law, and the fact that in the paraxial region $\alpha = h/s_o$, $\varphi \approx h/R$ and $\beta \approx h/s_i$, derive Eq. (5.8).

5.4 Locate the image of an object placed 1.2 m from the vertex of a gypsy's 20 cm diameter crystal ball ($n = 1.5$). Make a sketch of the thing (not the gypsy, the rays).

5.5 Prove that the minimum separation between conjugate *real* object and image points for a thin positive lens is 4*f*.

5.6 A biconcave lens ($n_l = 1.5$) has radii of 20 cm and 10 cm and an axial thickness of 5 cm. Describe the image of a one inch tall object placed 8 cm from the first vertex.

5.7* Use the thin-lens equation on the previous problem to see how far off it is in determining the final-image location.

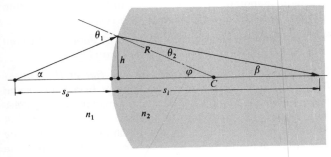

Fig. 5.102

5.8 An object 2 cm high is positioned 5 cm to the right of a positive thin lens of focal length 10 cm. Describe the resulting image completely using *both the Gaussian and Newtonian equations*.

5.9 Make a rough graph of the Gaussian lens equation, i.e. plot s_i versus s_o, using unit intervals of f along each axis. (Get both segments of the curve.)

5.10 What must be the focal length of a thin negative lens in order that it form a virtual image 50 cm away of an ant which is 100 cm away? Given that the ant is to the right of the lens, locate and describe its image.

5.11* Compute the focal length in air of a thin biconvex lens ($n_l = 1.5$) having radii 20 cm and 40 cm. Locate and describe the image of an object 40 cm from the lens.

5.12 Determine the focal length of a planar-concave lens ($n_l = 1.5$) having a radius of curvature of 10 cm. What is its power in diopters?

5.13 Write an expression for the focal length (f_w) of a thin lens immersed in water ($n_w = \frac{4}{3}$) in terms of its focal length when it's in air (f_a).

5.14* A convenient way to measure the focal length of a positive lens makes use of the following fact. If a pair of conjugate object and (real) image points (S and P) are separated by a distance $L > 4f$, there will be two locations of the lens, a distance d apart, for which the same pair of conjugates obtain. Show that

$$f = \frac{L^2 - d^2}{4L}.$$

Note that this avoids measurements made specifically from the vertex which are generally not easy to do.

5.15 An equiconvex thin lens L_1 is cemented in intimate contact with a thin negative lens, L_2, such that the combination has a focal length of 50 cm in air. If their indices are 1.50 and 1.55 respectively and if the focal length of L_2 is −50 cm, determine all of the radii of curvature.

5.16 Verify Eq. (5.34) which gives M_T for a combination of two thin lenses.

5.17 Compute the image location and magnification of an object 30 cm from the front doublet of the thin-lens combination in Fig. 5.103. Do the calculation by finding the effect of each lens separately. Make a sketch of appropriate rays. Compare your value of M_T with that given by Eq. (5.34).

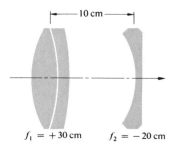

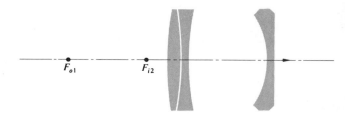

Fig. 5.103

Fig. 5.105

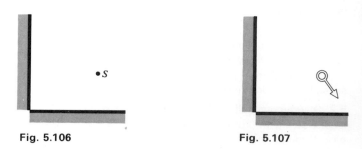

5.18 Draw a ray diagram for the combination of two positive lenses wherein their separation equals the sum of their respective focal lengths. Do the same thing for the case where one of the lenses is negative.

5.19 Redraw the ray diagram for a compound microscope (Fig. 5.86) but this time treat the intermediate image as if it were a real object—this approach should be a bit simpler.

5.20* Redraw the telescope of Fig. 5.89 taking advantage of the fact that the intermediate image can be thought of as a real object (as in the previous problem).

5.21 Consider the case of two positive thin lenses L_1 and L_2 separated by 5 cm. Their diameters are 6 cm and 4 cm respectively and their focal lengths are $f_1 = 9$ cm and $f_2 = 3$ cm. If a diaphragm with a 1 cm dia. hole is located between them, 2 cm from L_2, find (a) the A.S. and (b) the locations and sizes of the pupils for an axial point, S, 12 cm in front of (to the left of) L_1.

5.22 Make a sketch roughly locating the aperture stop and entrance and exit pupils for the lens in Fig. 5.104.

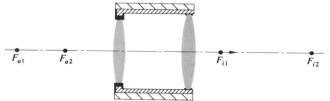

Fig. 5.104

5.23 Make a sketch roughly locating the aperture stop and entrance and exit pupils for the lens in Fig. 5.105 assuming the object point to be beyond (to the left of) F_{o1}.

5.24 Draw a ray diagram locating the images of a point source as formed by a pair of mirrors at 90° (Fig. 5.106).

5.25* Make a sketch of a ray diagram locating the images of the arrow shown in Fig. 5.107.

5.26 Show that Eq. (5.49) for a spherical surface is equally applicable to a plane mirror.

5.27 Locate the image of a paperclip 100 cm away from a convex spherical mirror having a radius of curvature of 80 cm.

5.28* Describe the image you would see standing 5 feet from, and looking directly toward, a 1 ft diameter brass ball hanging in front of a pawn shop.

5.29 The image of a red rose is formed by a concave spherical mirror on a screen 100 cm away. If the rose is 25 cm from the mirror, determine its radius of curvature.

5.30 From the image configuration determine the shape of the mirror hanging on the back wall in Van Eyck's painting of *John Arnolfini and His Wife* (Fig. 5.108).

5.31 Is Venus in Velasquez's painting of *Venus and Cupid* (Fig. 5.109) looking at herself in the mirror?

5.32 The girl in Manet's painting *The Bar at the Folies-Bergère* (Fig. 5.110) is standing in front of a large planar mirror. Reflected in it is her back and a man in evening dress with whom she appears to be talking. It would seem that Manet's intent was to give the uncanny feeling that the viewer is standing where that gentleman must be. From the laws of geometrical optics what is amiss?

Fig. 5.108 *John Arnolfini and His Wife* by Jan Van Eyck—National Gallery, London.

5.33* Suppose you have a concave spherical mirror of focal length 10 cm. At what distance must an object be placed if its image is to be erect and one and a half times as large? What is the radius of curvature of the mirror? Check with Table 5.5.

5.34 Describe the image that would result for a 3 in tall object placed 20 cm from a spherical concave shaving mirror having a radius of curvature of −60 cm.

5.35* Figures 5.111 and 5.112 are taken from an introductory physics book. What's wrong with them?

5.36 Draw a ray diagram for a finite object and a positive lens in each of the configurations treated in Table 5.3.

5.37 Figure 5.113 shows a lens system, an object and the appropriate pupils. Diagrammatically locate the image.

5.38 Referring to the dove prism of Fig. 5.57 rotate it through 90° about an axis along the ray direction. Sketch the new configuration and determine the angle through which the image is rotated.

5.39 Determine the numerical aperture of a single clad optical fiber given that the core has an index of 1.62 and the clad 1.52. When immersed in air, what is its maximum acceptance angle? What would happen to a ray incident at say 45°?

5.40 Using the information on the eye in Section 5.7.2 compute the approximate size (in millimeters) of the image of the moon as cast on the retina. The moon has a diameter of 2160 miles and is roughly 230,000 miles from here although this, of course, varies.

5.41* Figure 5.114 shows an arrangement such that the beam is deviated through a constant angle σ, equal to twice the angle between the plane mirrors, regardless of the angle of incidence. Prove that this is indeed the case.

5.42 An object 20 m from the objective ($f_o = 4$ m) of an astronomical telescope is imaged 30 cm from the eyepiece ($f_e = 60$ cm). Find the total linear magnification of the scope.

5.43* Figure 5.115 which purports to show an erecting lens system, is taken from an old, out-of-print optics text. What's wrong with it?

5.44* If a photo of a moving merry-go-round is perfectly exposed, but blurred, at $\frac{1}{30}$ s and $f/11$, what must the diaphragm setting be if the shutter speed is raised to $\frac{1}{120}$ s in order to "stop" the motion?

5.45 The field of view of a simple two-element astronomical telescope is restricted by the size of the eye lens. Make a ray sketch showing the vignetting which arises.

5.46 A *field lens*, as a rule, is a positive lens placed at (or near) the intermediate image plane in order to collect the rays which would otherwise miss the next lens in the system. In effect, it increases the field of view without changing the power of the system. Redraw the ray diagram of the previous problem to include a field lens. Show that as a consequence the eye relief is reduced somewhat.

5.47* Describe completely the resulting image of a bug which is sitting at the vertex of a thin positive lens. How does this relate directly to the manner in which a field lens works (see previous problem)?

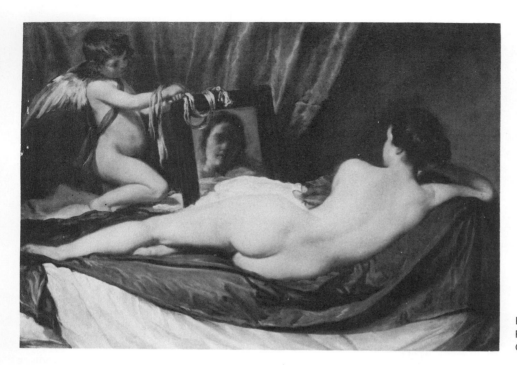

Fig. 5.109 *Venus and Cupid* by Diego Rodríguez de Silva y Velásquez—National Gallery, London.

Fig. 5.110 *The Bar at the Folies Bergères* by Édouard Manet—Courtauld Institute Galleries, London.

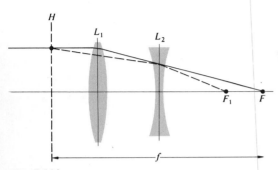

Fig. 5.111

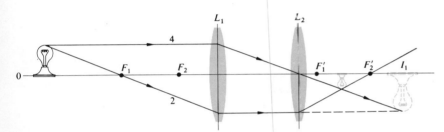

Fig. 5.112

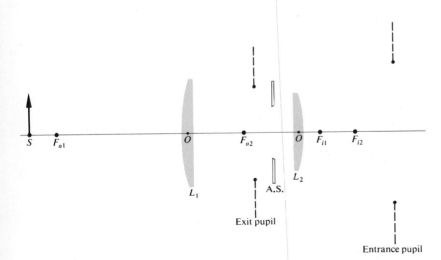

Fig. 5.113

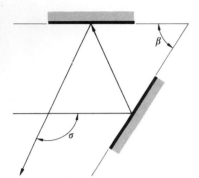

Fig. 5.114

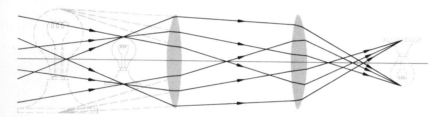

Fig. 5.115

More on Geometrical Optics   6

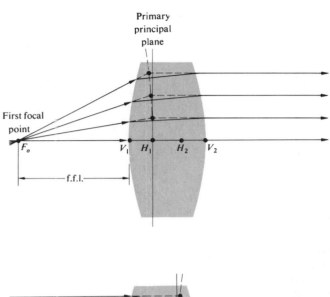

The preceding chapter, for the most part, dealt with paraxial theory as applied to thin spherical lens systems. The two predominant approximations were, rather obviously, that we had *thin* lenses and that first-order theory was sufficient for their analysis. Neither of these assumptions can be maintained throughout the design of a precision optical system but both, taken together, provide the basis for a first rough solution. This chapter will carry things a bit further by examining thick lenses and aberrations; even at that, it is only a beginning. The advent of computerized lens design requires a certain shift in emphasis—there is little need to do what a computer can do better. Moreover, the sheer wealth of existing material developed over centuries demands a bit of judicious pruning to avoid a plethora of pedantry.

6.1 THICK LENSES AND LENS SYSTEMS

Figure 6.1 depicts a thick lens, i.e., one whose thickness is by no means negligible. As we shall see, it could equally well be envisioned more generally as an optical system allowing of the possibility that it consists of a number of simple lenses, not merely one. The first and second focal points, or if you like, the object and image foci, F_o and F_i, can conveniently be measured from the two (outermost) vertices. In that case we have the familiar front and back focal lengths denoted by f.f.l. and b.f.l. When extended, the incident and emerged rays will meet at points, the locus of which forms a curved surface that may or may not reside within the lens. The surface, approximating a plane in the paraxial region, is termed the *principal plane* [see Section 6.3.1(ii)]. Points where the primary and secondary principal planes (as shown in Fig. 6.1) intersect the optical axis are known as the *first* and *second principal points*, H_1 and H_2 respectively. They provide a set of very useful references from which to measure several of the system parameters. We saw earlier (Fig. 5.19, p. 110) that a ray traversing the lens through its optical

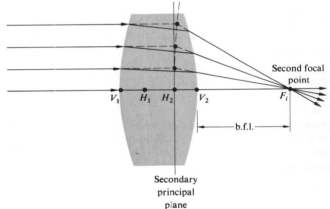

Fig. 6.1 A thick lens.

center emerges parallel to the incident direction. Extending both the incoming and outgoing rays until they cross the optical axis locates what are called the *nodal points*, N_1 and N_2 in Fig. 6.2. When the lens is surrounded on both sides by the same medium, generally air, the nodal and principal points will be coincident. The six points, two focal, two principal and two nodal, constitute the *cardinal points* of the system. As shown in Fig. 6.3 the principal planes can indeed lie completely outside of the lens system. Here although differently configured, each lens in either group has the same power. Observe that in the symmetrical lens the principal planes are, quite reasonably, symmetrically located. In the case of either the planar-concave or planar-convex

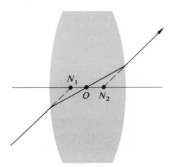

Fig. 6.2 Nodal points.

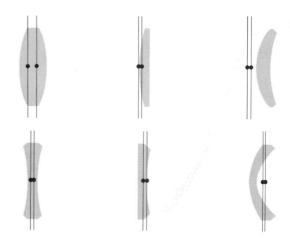

Fig. 6.3 Lens bending.

lens, one principal plane is tangent to the curved surface—as should be expected from the definition (applied to the paraxial region). By way of contrast, the principal points may certainly be external for meniscus lenses. One often speaks of this succession of shapes having the same power as exemplifying *lens bending*. A rule of thumb for ordinary glass lenses in air is that the separation $\overline{H_1 H_2}$ roughly equals one third the lens thickness $\overline{V_1 V_2}$.

The thick lens can be treated as consisting of two spherical refracting surfaces separated by a distance d between their vertices, precisely as was done earlier in Section 5.2.3 where the thin-lens equation was derived. After a great deal of algebraic manipulation,[*] wherein d is not now negligible, one arrives at a very interesting result for the thick lens immersed in air. The expression for the conjugate points once again can be put in the Gaussian form,

$$\frac{1}{s_o} + \frac{1}{s_i} = \frac{1}{f},\tag{6.1}$$

provided that both these object and image distances are measured from the first and second principal planes, respectively. Moreover, the *effective focal length* or simply the *focal length f* is also reckoned with respect to the principal planes and is given by

$$\frac{1}{f} = (n_l - 1)\left[\frac{1}{R_1} - \frac{1}{R_2} + \frac{(n_l - 1)d}{n_l R_1 R_2}\right].\tag{6.2}$$

The principal planes are located at distances of $\overline{V_1 H_1} = h_1$ and $\overline{V_2 H_2} = h_2$ *which are positive when the planes lie to the right of their respective vertices.* Figure 6.4 illustrates the

* For the complete derivation, see Morgan, *Introduction to Geometrical and Physical Optics*, p. 57.

arrangement of the various quantities. The values of h_1 and h_2 are given by

$$h_1 = -\frac{f(n_l - 1)d}{R_2 n_l}\tag{6.3}$$

and

$$h_2 = -\frac{f(n_l - 1)d}{R_1 n_l}.\tag{6.4}$$

In the same way the Newtonian form of the lens equation holds as is evident from the similar triangles in Fig. 6.4. Thus

$$x_i x_o = f^2\tag{6.5}$$

so long as f is given the present interpretation. And from the same triangles

$$M_T = \frac{y_i}{y_o} = -\frac{x_i}{f} = -\frac{f}{x_o}.\tag{6.6}$$

Obviously if $d \rightarrow 0$, Eqs. (6.1), (6.2), and (6.5) transform into the thin-lens expressions (5.17), (5.16), and (5.23). As a numerical example, let's find the image distance for an object positioned 30 cm from the vertex of a double convex lens having radii of 20 cm and 40 cm, a thickness of 1 cm and an index of 1.5. From Eq. (6.2) the focal length (in centimeters) is

$$\frac{1}{f} = (1.5 - 1)\left[\frac{1}{20} - \frac{1}{-40} + \frac{(1.5 - 1)1}{1.5(20)(-40)}\right]$$

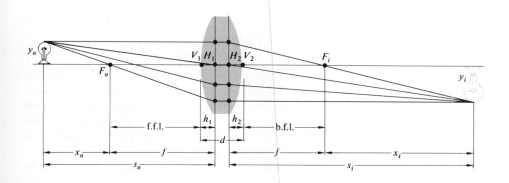

Fig. 6.4 Thick lens geometry.

and so $f = 26.8$ cm. Furthermore

$$h_1 = -\frac{26.8(0.5)1}{-40(1.5)} = +0.22 \text{ cm}$$

and

$$h_2 = -\frac{26.8(0.5)1}{20(1.5)} = -0.44 \text{ cm},$$

which means that H_1 is to the right of V_1 and H_2 is to the left of V_2. Finally, $s_o = 30 + 0.22$, whence

$$\frac{1}{30.2} + \frac{1}{s_i} = \frac{1}{26.8},$$

and $s_i = 238$ cm measured from H_2.

The principal points are conjugate to each other. In other words, since $f = s_o s_i/(s_o + s_i)$, when $s_o = 0$, s_i must be zero because f is finite and thus a point at H_1 is imaged at H_2. Furthermore, an object in the first principal plane ($x_o = -f$) is imaged in the second principal plane ($x_i = -f$) with unit magnification ($M_T = 1$). It is for this reason that these are sometimes spoken of as *unit planes*. Hence any ray directed toward a point on the first principal plane will emerge from the lens as if it originated at the corresponding point (the same distance above or below the axis) on the second principal plane.

Suppose we now have a compound lens consisting of two thick lenses L_1 and L_2 (Fig. 6.5). Let s_{o1}, s_{i1}, f_1 and s_{o2}, s_{i2} and f_2 be the object and image distances and focal lengths for the two lenses, all measured with respect to their own principal planes. We know that the transverse magnification is the product of the magnifications of the individual lenses, that is

$$M_T = \left(-\frac{s_{i1}}{s_{o1}}\right)\left(-\frac{s_{i2}}{s_{o2}}\right) = -\frac{s_i}{s_o}, \qquad (6.7)$$

where s_o and s_i are the object and image distances for the combination as a whole. When s_o is equal to infinity $s_o = s_{o1}$, $s_{i1} = f_1$, $s_{o2} = -(s_{i1} - d)$, and $s_i = f$. Since

$$\frac{1}{s_{o2}} + \frac{1}{s_{i2}} = \frac{1}{f_2}$$

it follows (Problem 6.1), upon substituting into Eq. (6.7), that

$$-\frac{f_1 s_{i2}}{s_{o2}} = f$$

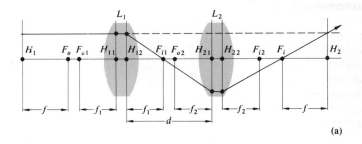

(a)

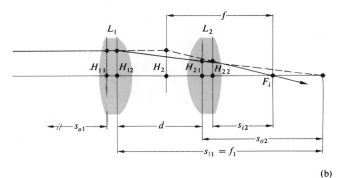

(b)

Fig. 6.5 A compound thick lens.

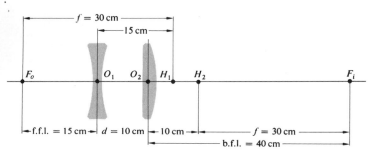

Fig. 6.6 A compound lens.

or

$$f = -\frac{f_1}{s_{o2}}\left(\frac{s_{o2}f_2}{s_{o2} - f_2}\right) = \frac{f_1 f_2}{s_{i1} - d + f_2}.$$

Hence

$$\frac{1}{f} = \frac{1}{f_1} + \frac{1}{f_2} - \frac{d}{f_1 f_2}. \tag{6.8}$$

This is the effective focal length of the combination of two thick lens where all distances are measured from principal planes. The principal planes for the system as a whole are located using the expressions

$$\overline{H_{11}H_1} = \frac{fd}{f_2} \tag{6.9}$$

and

$$\overline{H_{22}H_2} = -\frac{fd}{f_1}, \tag{6.10}$$

which will not be derived here (see Section 6.2.1). We have in effect found an equivalent thick-lens representation of the compound lens. Note that if the component lenses are thin, the pairs of points H_{11}, H_{12} and H_{21}, H_{22} coalesce, whereupon d becomes the center-to-center lens separation as in Section 5.2.3(iv). For example, returning to the thin lenses of Fig. 5.28 where $f_1 = -30$ and $f_2 = 20$, and $d = 10$, as in Fig. 6.6,

$$\frac{1}{f} = \frac{1}{-30} + \frac{1}{20} - \frac{10}{(-30)(20)}$$

and so $f = 30$ cm. We found earlier (p. 116) that b.f.l. = 40 cm and f.f.l. = 15 cm. Moreover since these are thin lenses Eqs. (6.9) and (6.10) can be written as

$$\overline{O_1H_1} = \frac{30(10)}{20} = +15 \text{ cm}$$

and

$$\overline{O_2H_2} = -\frac{30(10)}{-30} = +10 \text{ cm}.$$

Both are positive and therefore the planes lie to the right of O_1 and O_2, respectively. Both computed values agree with the results depicted in the diagram. If light enters from the right the system resembles a telephoto lens which must be placed 15 cm from the film plane and yet has an effective focal length of 30 cm.

The same procedures can be extended to three, four, or more lenses. Thus

$$f = f_1\left(-\frac{s_{i2}}{s_{o2}}\right)\left(-\frac{s_{i3}}{s_{o3}}\right)\cdots \tag{6.11}$$

Equivalently, the first two lenses can be envisioned as combined to form a single thick lens whose principal points and focal length are calculated. It, in turn, is combined with the third lens and so on with each successive element.

6.2 ANALYTICAL RAY TRACING

Ray tracing is unquestionably one of the designer's chief tools. Having formulated an optical system on paper, he can mathematically shine rays through it to evaluate its performance. Any ray, paraxial or otherwise, can be traced through the system exactly. Conceptually it's a simple matter of applying the refraction equation

$$n_i(\hat{k}_i \times \hat{u}_n) = n_t(\hat{k}_t \times \hat{u}_n) \tag{4.7}$$

at the first surface, locating where the transmitted ray then strikes the second surface; applying the equation once again, and so on all the way through. At one time *meridional rays* (those in the plane of the optical axis) were traced almost exclusively because non-meridional or *skew rays* (which do not intersect the axis) are considerably more complicated to deal with mathematically. The distinction is of less importance to a high-speed electronic computer (Fig. 6.7) which simply takes a trifle longer to make the trace. Thus while it would probably take 10 or 15 minutes for a skilled person with a desk calculator to evaluate the trajectory of a single skew ray through a single surface, a computer might require roughly a thousandth of a second for the same job and, equally important, it would be ready for the next calculation with undiminished enthusiasm.

The simplest case which will serve to illustrate the ray-tracing process corresponds to that of a paraxial, meridional

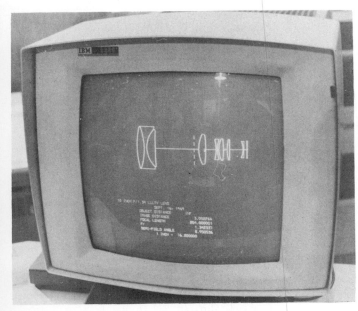

Fig. 6.7 Computer lens display. [Photo by E.H.]

ray traversing a thick spherical lens. In that case, applying Snell's law in Fig. 6.8 at point P_1 yields

$$n_{i1}\theta_{i1} = n_{t1}\theta_{t1}$$

or

$$n_{i1}(\alpha_{i1} + \alpha_1) = n_{t1}(\alpha_{t1} + \alpha_1).$$

Inasmuch as $\alpha_1 = y_1/R_1$ this becomes

$$n_{i1}(\alpha_{i1} + y_1/R_1) = n_{t1}(\alpha_{t1} + y_1/R_1).$$

Rearranging terms, we get

$$n_{t1}\alpha_{t1} = n_{i1}\alpha_{i1} - \left(\frac{n_{t1} - n_{i1}}{R_1}\right)y_1,$$

but as we saw in Section 5.7.2, the power of a single refracting surface is

$$\mathscr{D}_1 = \frac{(n_{t1} - n_{i1})}{R_1}.$$

Hence

$$n_{t1}\alpha_{t1} = n_{i1}\alpha_{i1} - \mathscr{D}_1 y_1; \tag{6.12}$$

this is often called the *refraction equation* pertaining to the first interface. Having undergone refraction at point P_1 the ray advances through the homogeneous medium of the

lens to point P_2 on the second interface. The height of P_2 is expressible as

$$y_2 = y_1 + d_{21}\alpha_{t1}, \tag{6.13}$$

where use was made of the fact that $\tan \alpha_{t1} \approx \alpha_{t1}$. This is known as the *transfer equation* because it allows us to follow the ray from P_1 to P_2. Recall that the angles are positive if the ray has a positive slope. Since we are dealing with the paraxial region $d_{21} \approx \overline{V_2 V_1}$ and y_2 is easily computed. Equations (6.11) and (6.12) are then used successively to trace a ray through the entire system. Of course, these are meridional rays and because of the lenses' symmetry about the optical axis such a ray remains in the same meridional plane throughout its sojourn. The process is two-dimensional; there are two equations and two unknowns α_{t1} and y_2. In contrast, a skew ray would have to be treated in three dimensions.

6.2.1 Matrix Methods

In the beginning of the nineteen-thirties, T. Smith formulated a rather interesting way of handling the ray-tracing equations. The simple linear form of the expressions and the repetitive manner in which they are applied suggested the use of matrices. The processes of refraction and transfer might then be performed mathematically by matrix operators. These initial insights were not widely appreciated for almost thirty years. However, the early nineteen-sixties saw the beginning of a rebirth of interest which is now flourishing.* We shall only outline some of the salient features of the method leaving a more detailed study to the references.

Let's begin by writing the formulas

$$n_{t1}\alpha_{t1} = n_{i1}\alpha_{i1} - \mathscr{D}_1 y_{i1} \tag{6.14}$$

and

$$y_{t1} = 0 + y_{i1}, \tag{6.15}$$

which are not very insightful since we merely replaced y_1 in Eq. (6.12) by the symbol y_{i1} and then let $y_{t1} = y_{i1}$. This last bit of business is for purely cosmetic purposes, as you will see in a moment. In effect it simply says that the height of reference point P_1 above the axis in the incident medium (y_{i1}) equals its height in the transmitting medium (y_{t1})—which is obvious. But now the pair of equations can be

---

* For further reading see K. Hallbach, "Matrix Representation of Gaussian Optics," *Am. J. Phys.* **32**, 90 (1964); W. Brouwer, *Matrix Methods in Optical Instrument Design*; E. L. O'Neill, *Introduction to Statistical Optics*; or A. Nussbaum, *Geometric Optics*.

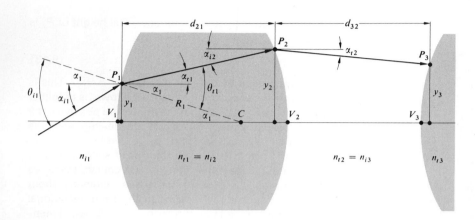

Fig. 6.8 Ray geometry.

recast in matrix form as

$$\begin{bmatrix} n_{t1}\alpha_{t1} \\ y_{t1} \end{bmatrix} = \begin{bmatrix} 1 & -\mathscr{D}_1 \\ 0 & 1 \end{bmatrix} \begin{bmatrix} n_{i1}\alpha_{i1} \\ y_{i1} \end{bmatrix}. \tag{6.16}$$

This could equally well be written as

$$\begin{bmatrix} \alpha_{t1} \\ y_{t1} \end{bmatrix} = \begin{bmatrix} n_{i1}/n_{t1} & -\mathscr{D}_1/n_{t1} \\ 0 & 1 \end{bmatrix} \begin{bmatrix} \alpha_{i1} \\ y_{i1} \end{bmatrix} \tag{6.17}$$

so that the precise form of the 2 × 1 column matrices is actually a matter of preference. In any case, these can be envisioned as rays on either side of P_1, one before and the other after refraction. Accordingly, using $\imath_{t1}$ and $\imath_{i1}$ for the two rays, we can write

$$\imath_{t1} \equiv \begin{bmatrix} n_{t1}\alpha_{t1} \\ y_{t1} \end{bmatrix} \quad \text{and} \quad \imath_{i1} \equiv \begin{bmatrix} n_{i1}\alpha_{i1} \\ y_{i1} \end{bmatrix}. \tag{6.18}$$

The 2 × 2 matrix is the *refraction matrix* denoted as

$$\mathscr{R}_1 \equiv \begin{bmatrix} 1 & -\mathscr{D}_1 \\ 0 & 1 \end{bmatrix}, \tag{6.19}$$

and so Eq. (6.16) can be concisely stated as

$$\imath_{t1} = \mathscr{R}_1 \imath_{i1}, \tag{6.20}$$

which just says that $\mathscr{R}_1$ transforms the ray $\imath_{i1}$ into the ray $\imath_{t1}$ during refraction at the first interface. From Fig. 6.8 we have $n_{i2}\alpha_{i2} = n_{t1}\alpha_{t1}$, that is

$$n_{i2}\alpha_{i2} = n_{t1}\alpha_{t1} + 0 \tag{6.21}$$

and

$$y_{i2} = d_{21}\alpha_{t1} + y_{t1}, \tag{6.22}$$

where $n_{i2} = n_{t1}$, $\alpha_{i2} = \alpha_{t1}$ and use was made of Eq. (6.13) with y_2 rewritten as y_{i2} to make things pretty. And thus

$$\begin{bmatrix} n_{i2}\alpha_{i2} \\ y_{i2} \end{bmatrix} = \begin{bmatrix} 1 & 0 \\ d_{21}/n_{t1} & 1 \end{bmatrix} \begin{bmatrix} n_{t1}\alpha_{t1} \\ y_{t1} \end{bmatrix}. \tag{6.23}$$

The *transfer matrix*

$$\mathscr{T}_{21} \equiv \begin{bmatrix} 1 & 0 \\ d_{21}/n_{t1} & 1 \end{bmatrix} \tag{6.24}$$

takes the transmitted ray at P_1, i.e. $\imath_{t1}$, and transforms it into the incident ray at P_2:

$$\imath_{i2} \equiv \begin{bmatrix} n_{i2}\alpha_{i2} \\ y_{i2} \end{bmatrix}.$$

Hence Eqs. (6.21) and (6.22) become simply

$$\imath_{i2} = \mathscr{T}_{21}\imath_{t1}. \tag{6.25}$$

If we make use of Eq. (6.20) this becomes

$$\imath_{i2} = \mathscr{T}_{21}\mathscr{R}_1\imath_{i1}. \tag{6.26}$$

The 2 × 2 matrix formed by the product of the transfer and refraction matrices $\mathscr{T}_{21}\mathscr{R}_1$ will carry the ray incident at P_1 into the ray incident at P_2. Notice that the determinant of $\mathscr{T}_{21}$, denoted by $|\mathscr{T}_{21}|$, equals 1, i.e., (1) (1) − (0) (d_{21}/n_{t1}) = 1. Similarly $|\mathscr{R}_1| = 1$ and since the determinant of a matrix product equals the product of the individual determinants $|\mathscr{T}_{21}\mathscr{R}_1| = 1$. This provides a quick check on the computations. Carrying the procedure through the second interface (Fig. 6.8) of the lens, which has a refraction matrix $\mathscr{R}_2$, it follows that

$$\boldsymbol{\ell}_{t2} = \mathscr{R}_2 \boldsymbol{\ell}_{i2} \qquad (6.27)$$

or from Eq. (6.26)

$$\boldsymbol{\ell}_{t2} = \mathscr{R}_2(\mathscr{T}_{21}\mathscr{R}_1)\boldsymbol{\ell}_{i1}. \qquad (6.28)$$

The brackets are unnecessary as long as the order of multiplying the 2×2 matrices is first $\mathscr{T}_{21}\mathscr{R}_1$ and then $\mathscr{R}_2\mathscr{T}_{21}\mathscr{R}_1$. The *system matrix* $\mathscr{A}$ is defined as

$$\mathscr{A} \equiv \mathscr{R}_2\mathscr{T}_{21}\mathscr{R}_1 \qquad (6.29)$$

and has the form

$$\mathscr{A} = \begin{bmatrix} a_{11} & a_{12} \\ a_{21} & a_{22} \end{bmatrix}. \qquad (6.30)$$

Since

$$\mathscr{A} = \begin{bmatrix} 1 & -\mathscr{D}_2 \\ 0 & 1 \end{bmatrix}\begin{bmatrix} 1 & 0 \\ d_{21}/n_{t1} & 1 \end{bmatrix}\begin{bmatrix} 1 & -\mathscr{D}_1 \\ 0 & 1 \end{bmatrix}$$

or

$$\mathscr{A} = \begin{bmatrix} 1 & -\mathscr{D}_2 \\ 0 & 1 \end{bmatrix}\begin{bmatrix} 1 & -\mathscr{D}_1 \\ d_{21}/n_{t1} & -\mathscr{D}_1 d_{21}/n_{t1} + 1 \end{bmatrix}.$$

We can write

$$\begin{bmatrix} a_{11} & a_{12} \\ a_{21} & a_{22} \end{bmatrix} = \begin{bmatrix} 1 - \mathscr{D}_2 d_{21}/n_{t1} & -\mathscr{D}_1 + (\mathscr{D}_2\mathscr{D}_1 d_{21}/n_{t1}) - \mathscr{D}_2 \\ d_{21}/n_{t1} & -\mathscr{D}_1 d_{21}/n_{t1} + 1 \end{bmatrix}$$

$$(6.31)$$

and again $|\mathscr{A}| = 1$ (Problem 6.5). The values of each element in $\mathscr{A}$ is expressed in terms of the physical lens parameters such as thickness, index and radii (via $\mathscr{D}$). Thus the cardinal points which are properties of the lens, determined solely by its make-up, should be deducible from $\mathscr{A}$. The system matrix in this case (6.31) transforms an incident ray at the first surface to an emerging ray at the second surface; as a reminder we will write it as $\mathscr{A}_{21}$.

The concept of image formation enters rather directly (Fig. 6.9) after introducing appropriate object and image planes. Consequently, the first operator $\mathscr{T}_{10}$ transfers the reference point from the object, i.e. P_O to P_1. The next operator $\mathscr{A}_{21}$ then carries the ray through the lens, and a final additional transfer $\mathscr{T}_{I2}$ brings it to the image plane, i.e., P_I. Thus the ray at the image point ($\boldsymbol{\ell}_I$) is given by

$$\boldsymbol{\ell}_I = \mathscr{T}_{I2}\mathscr{A}_{21}\mathscr{T}_{10}\boldsymbol{\ell}_O \qquad (6.32)$$

where $\boldsymbol{\ell}_O$ is the ray at P_O. In component form this is

$$\begin{bmatrix} n_I\alpha_I \\ y_I \end{bmatrix} = \begin{bmatrix} 1 & 0 \\ d_{I2}/n_I & 1 \end{bmatrix}\begin{bmatrix} a_{11} & a_{12} \\ a_{21} & a_{22} \end{bmatrix}\begin{bmatrix} 1 & 0 \\ d_{10}/n_O & 1 \end{bmatrix}\begin{bmatrix} n_O\alpha_O \\ y_O \end{bmatrix}.$$

$$(6.33)$$

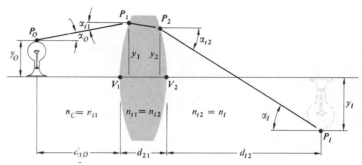

Fig. 6.9 Image geometry.

Notice that $\mathscr{T}_{10}\boldsymbol{\ell}_O = \boldsymbol{\ell}_{i1}$ and that $\mathscr{A}_{21}\boldsymbol{\ell}_{i1} = \boldsymbol{\ell}_{t2}$ and hence $\mathscr{T}_{I2}\boldsymbol{\ell}_{t2} = \boldsymbol{\ell}_I$. The subscripts $O, 1, 2, \ldots, I$ correspond to reference points P_O, P_1, P_2 etc. while subscripts i and t denote on which side of the reference point we are, i.e., whether incident or transmitted. Operation by a refraction matrix will change i to t but not the reference point designation. On the other hand, operation by a transfer matrix obviously does change the latter.

Ordinarily the physical significances of the components of $\mathscr{A}$ are found by expanding out Eq. (6.33), but this is too involved an approach to do here. Instead, let's return to Eq. (6.31) and examine several of the terms. For example,

$$-a_{12} = \mathscr{D}_1 + \mathscr{D}_2 - \mathscr{D}_2\mathscr{D}_1 d_{21}/n_{t1}.$$

If we suppose, for the sake of simplicity, that the lens is in air, then

$$\mathscr{D}_1 = \frac{n_{t1} - 1}{R_1} \quad \text{and} \quad \mathscr{D}_2 = \frac{n_{t1} - 1}{-R_2}$$

as in Eqs. (5.63) and (5.64). Hence

$$-a_{12} = (n_{t1} - 1)\left[\frac{1}{R_1} - \frac{1}{R_2} + \frac{(n_{t1} - 1)d_{21}}{R_1 R_2 n_{t1}}\right].$$

But this is the expression for the focal length of a thick lens (6.2); in other words

$$a_{12} = -1/f. \qquad (6.34)$$

If the imbedding media were different on each side of the lens (Fig. 6.10) this would become

$$a_{12} = -\frac{n_{i1}}{f_o} = -\frac{n_{t2}}{f_i}. \qquad (6.35)$$

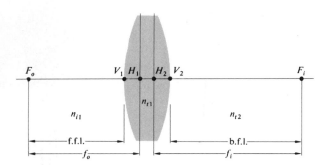

Fig. 6.10 Principal planes and focal lengths.

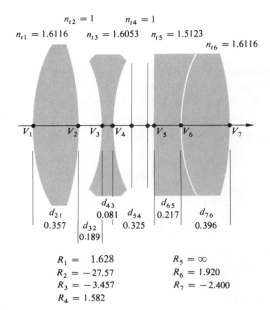

Fig. 6.11 A Tessar.

Similarly it is left as a problem to verify that

$$\overline{V_1 H_1} = \frac{n_{i1}(1 - a_{11})}{-a_{12}} \qquad (6.36)$$

and

$$\overline{V_2 H_2} = \frac{n_{t2}(a_{22} - 1)}{-a_{12}}, \qquad (6.37)$$

which locate the principal points.

As an example of how the technique can be used, let's apply it, at least in principle, to the Tessar lens* shown in Fig. 6.11. The system matrix has the form

$$\mathscr{A}_{71} = \mathscr{R}_7 \mathscr{T}_{76} \mathscr{R}_6 \mathscr{T}_{65} \mathscr{R}_5 \mathscr{T}_{54} \mathscr{R}_4 \mathscr{T}_{43} \mathscr{R}_3 \mathscr{T}_{32} \mathscr{R}_2 \mathscr{T}_{21} \mathscr{R}_1,$$

where

$$\mathscr{T}_{21} = \begin{bmatrix} 1 & 0 \\ \dfrac{0.357}{1.6116} & 1 \end{bmatrix}, \quad \mathscr{T}_{32} = \begin{bmatrix} 1 & 0 \\ \dfrac{0.189}{1} & 1 \end{bmatrix},$$

$$\mathscr{T}_{43} = \begin{bmatrix} 1 & 0 \\ \dfrac{0.081}{1.6053} & 1 \end{bmatrix}, \text{ etc.}$$

---

* We have chosen this particular example primarily because Nussbaum's book *Geometric Optics* contains a simple Fortran computer program specifically written for this lens. It would be almost silly to evaluate the system matrix by hand. Since Fortran is an easily mastered computer language, the program is well worth further study.

Furthermore

$$\mathscr{R}_1 = \begin{bmatrix} 1 & -\dfrac{1.6116 - 1}{1.628} \\ 0 & 1 \end{bmatrix}, \quad \mathscr{R}_2 = \begin{bmatrix} 1 & -\dfrac{1 - 1.6116}{-27.57} \\ 0 & 1 \end{bmatrix},$$

$$\mathscr{R}_3 = \begin{bmatrix} 1 & -\dfrac{1.6053 - 1}{-3.457} \\ 0 & 1 \end{bmatrix}, \text{ etc.}$$

Multiplying out the matrices, in what is obviously a horrendous although conceptually simple calculation, one presumably will get

$$\mathscr{A}_{71} = \begin{bmatrix} 0.848 & -0.198 \\ 1.338 & 0.867 \end{bmatrix}$$

and from that $f = 5.06$, $\overline{V_1 H_1} = 0.77$ and $\overline{V_7 H_2} = -0.67$.

As a last point, it is often convenient to consider a system of thin lenses using the matrix representation and to that end return to Eq. (6.31). It describes the system matrix for a single lens and if we let $d_{21} \to 0$, it corresponds to a thin lens. This is equivalent to making $\mathscr{T}_{21}$ a unit matrix, thus

$$\mathscr{A} = \mathscr{R}_2 \mathscr{R}_1 = \begin{bmatrix} 1 & -(\mathscr{D}_1 + \mathscr{D}_2) \\ 0 & 1 \end{bmatrix}. \qquad (6.38)$$

But as we saw in Section 5.7.2, the power of a thin lens $\mathscr{D}$ is the sum of the powers of its surfaces. Hence

$$\mathscr{A} = \begin{bmatrix} 1 & -\mathscr{D} \\ 0 & 1 \end{bmatrix} = \begin{bmatrix} 1 & -1/f \\ 0 & 1 \end{bmatrix}. \qquad (6.39)$$

In addition, for two thin lenses separated by a distance d, in air, the system matrix is

$$\mathscr{A} = \begin{bmatrix} 1 & -1/f_2 \\ 0 & 1 \end{bmatrix} \begin{bmatrix} 1 & 0 \\ d & 1 \end{bmatrix} \begin{bmatrix} 1 & -1/f_1 \\ 0 & 1 \end{bmatrix}$$

or

$$\mathscr{A} = \begin{bmatrix} 1 - d/f_2 & -1/f_1 + d/f_1 f_2 - 1/f_2 \\ d & -d/f_1 + 1 \end{bmatrix}.$$

Clearly then

$$-a_{12} = \frac{1}{f} = \frac{1}{f_1} + \frac{1}{f_2} - \frac{d}{f_1 f_2}$$

and from Eqs. (6.36) and (6.37)

$$\overline{O_1 H_1} = fd/f_2, \qquad \overline{O_2 H_2} = -fd/f_1,$$

all of which by now should be quite familiar. Note how easy it would be with this approach to find the focal length and principal points for a compound lens composed of three, four or more thin lenses.

6.3 ABERRATIONS

To be sure, we already know that first-order theory is no more than a good approximation—an exact ray trace or even measurements performed on a prototype system would certainly reveal inconsistencies with the corresponding paraxial description. Such departures from the idealized conditions of Gaussian optics are known as *aberrations*. There are two main classifications of these, namely *chromatic aberrations* (which arise from the fact that n is actually a function of frequency or color) and *monochromatic aberrations*. The latter occur even with light which is highly monochromatic, and they, in turn, fall into two subgroupings. There are monochromatic aberrations which deteriorate the image making it unclear, such as *spherical aberration*, *coma* and *astigmatism*. In addition, there are aberrations which

deform the image, as for example, *Petzval field curvature* and *distortion*.

We have known all along that spherical surfaces in general would yield perfect imagery only in the paraxial region. What must now be determined is the kind and extent of the deviations which result simply from using those surfaces with finite apertures. By the judicious manipulation of a system's physical parameters (e.g. the powers, shapes, thicknesses, glass types and separations of the lenses as well as the locations of stops), these aberrations can indeed be minimized. In effect, one cancels out the most undesirable faults by a slight change in the shape of a lens here, or a shift in the position of a stop there (very much like trimming up a circuit with small variable capacitors, coils and pots). When it's all finished, the unwanted deformations of the wavefront incurred as it passes through one surface will hopefully be negated as it traverses some other surfaces further down the line.

Today there are elaborate computer programs for "automatically" doing this kind of analysis. Broadly speaking, you give the computer a quality factor (or merit function) of some sort to aim for, i.e., you essentially tell it how much of each aberration you are willing to tolerate. Then you give it a roughly designed system (e.g. some Tessar configuration) which, in the first approximation, meets the particular requirements. Along with that, you feed in whatever parameters must be held constant, such as a given f-number, focal length, or lens diameter, the field of view, or magnification. The computer will then trace several rays through the system and evaluate the image errors. Having been given leave to vary, say, the curvatures and axial separations of the elements, it will calculate the optimum effect of such changes on the quality factor, make them, and then reevaluate. After perhaps twenty or more iterations, usually taking a matter of minutes, it will have changed the initial configuration so that it now meets the specified limits on aberrations. The final lens design will still be a Tessar, but not the one you started with. The result is, if you will, an *optimum configuration* but probably not *the* optimum. We can be fairly certain that all aberrations cannot be made exactly zero in any real system comprised of spherical surfaces. Moreover, there is no presently known way to determine how close to zero we can actually come. A quality factor is somewhat like a crater-pocked surface in a multidimensional space. The computer will carry the design from one hole to the next until it finds one deep enough to meet the specifications. There it stops and presumably presents us with a perfectly satisfactory configuration. But there is no way to tell if that solution corresponds to the deepest hole without sending the computer out again and again meandering along totally different routes.

We mention all of this so that the reader may appreciate the current state of the art. In a word, it is magnificent, but still incomplete; it is "automatic" but a bit myopic.

6.3.1 Monochromatic Aberrations

The paraxial treatment was based on the assumption that $\sin \varphi$, as in Fig. 5.8, could be represented satisfactorily by φ alone, i.e. the system was restricted to operating in an extremely narrow region about the optical axis. Obviously, if rays from the periphery of a lens are to be included in the formation of an image, the statement that $\sin \varphi \approx \varphi$ is somewhat unsatisfactory. Recall that we also occasionally wrote Snell's law simply as $n_i \theta_i = n_t \theta_t$ which again would be inappropriate. In any event, if the first two terms in the expansion

$$\sin \varphi = \varphi - \frac{\varphi^3}{3!} + \frac{\varphi^5}{5!} - \frac{\varphi^7}{7!} + \cdots \qquad [5.7]$$

are retained as an improved approximation, we have the so-called *third-order theory*. Departures from first-order theory which then result are embodied in the five *primary aberrations* (spherical aberration, coma, astigmatism, field curvature and distortion). These were first studied in detail by Ludwig von Seidel (1821–96) in the eighteen-fifties. Accordingly, they are frequently spoken of as the *Seidel aberrations*. In addition to the first two contributions, the series obviously contains many other terms, smaller to be sure, but still to be reckoned with. Thus, there are most certainly *higher-order aberrations*. The difference between the results of exact ray-tracing and the computed primary aberrations can therefore be thought of as the sum of all contributing higher-order aberrations. We shall restrict this discussion to the primary aberrations exclusively.

i) Sperical Aberration

Let's return for a moment to Section 5.2.2 (p.102) where we computed the conjugate points for a single refracting spherical interface. We found then, for the paraxial region, that

$$\frac{n_1}{s_o} + \frac{n_2}{s_i} = \frac{n_2 - n_1}{R}. \qquad [5.8]$$

If the approximations for ℓ_o and ℓ_i are improved a bit (Problem 6.11) we get the third-order expression:

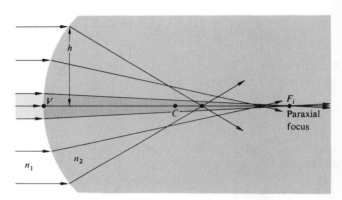

Fig. 6.12 Spherical aberration resulting from refraction at a single interface.

$$\frac{n_1}{s_o} + \frac{n_2}{s_i} = \frac{n_2 - n_1}{R} + h^2 \left[\frac{n_1}{2s_o} \left(\frac{1}{s_o} + \frac{1}{R} \right)^2 + \frac{n_2}{2s_i} \left(\frac{1}{R} - \frac{1}{s_i} \right)^2 \right].$$

$$(6.40)$$

The additional term, which varies approximately as h^2, is clearly a measure of the deviation from first-order theory. As shown in Fig. 6.12 rays striking the surface at greater distances above the axis (h) are focused nearer to the vertex. In brief, spherical aberration or SA corresponds to a dependence of focal length on aperture for nonparaxial rays. Similarly, for a converging lens, as in Fig. 6.13, the marginal rays will, in effect, be bent too much, being focused in front of the paraxial rays. The distance between the axial intersection of a ray and the paraxial focus, F_i, is known as the *longitudinal spherical aberration*, or L·SA, of that ray. In this case, the SA is *positive*. In contrast the marginal rays for a diverging lens will generally intersect the axis behind the paraxial focus and we say that its spherical aberration is therefore *negative*.

If a screen is placed at F_i in Fig. 6.13, the image of a star would appear as a bright central spot on the axis surrounded by a symmetrical halo delineated by the cone of marginal rays. For an extended image, SA would reduce the contrast and degrade the details. The height above the axis where a given ray strikes this screen is called the *transverse* (or *lateral*) *spherical aberration*, T·SA for short. Evidently, SA can be reduced by stopping down the aperture—but that reduces the amount of light entering the system as well. Notice that if the screen is moved to the position labeled Σ_{LC} the image blur will have its smallest diameter. This is known as the *circle of least confusion* and Σ_{LC} is generally the best place to observe the image. If a lens exhibits appreciable SA, it

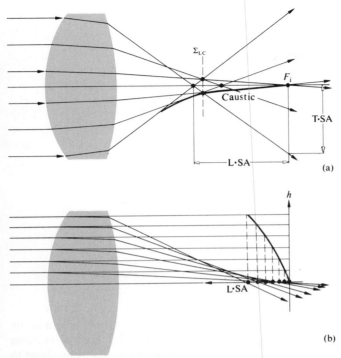

Fig. 6.13 Spherical aberration for a lens. The envelope of the refracted rays is called a caustic. The intersection of the marginal rays and the caustic locates Σ_{LC}.

will have to be refocused after it is stopped down because the position of Σ_{LC} will approach F_i as the aperture decreases.

The amount of spherical aberration, when the aperture and focal length are fixed, varies with both the object distance and the lens shape. For a converging lens, the nonparaxial rays are too strongly bent. Yet if we imagine the lens as roughly resembling two prisms joined at their bases, it is evident that *the incident ray will undergo a minimum deviation when it makes, more or less, the same angle as does*

the emerging ray (Section 5.5.1). A striking example is illustrated in Fig. 6.14 where simply turning the lens around markedly reduces the SA. When the object is at infinity a simple concave or convex lens which has an almost, but not quite, flat rear side will suffer a minimum amount of spherical aberration. In the same way, if the object and image distances are to be equal ($s_o = s_i = 2f$) the lens should be equiconvex to minimize SA. A combination of a converging and a diverging lens (as in an achromatic doublet) can also be utilized to diminish spherical aberration.

Recall that the aspherical lenses of Section 5.2.1 were completely free of spherical aberration for a specific pair of conjugate points. Moreover, Huygens seems to have been the first to discover that two such axial points exist for spherical surfaces as well. These are shown in Fig. 6.15(a) which depicts rays issuing from P and leaving the surface as if they came from P'. It is left for a problem to show that the appropriate locations of P and P' are those indicated in the figure. Just as with the aspherics, lenses can be formed which have this same zero SA for the pair of points P and P'. One simply grinds another surface of radius $\overline{PA}$ centered on P to form either a positive- or negative-meniscus lens. The oil-immersion microscope objective uses this principle to great advantage. The object under study is positioned at P and surrounded by oil of index n_2 as in Fig. 6.16. P and P' are the proper conjugate points for zero SA for the first element while P' and P'' are those for the meniscus lens.

ii) Coma

Coma or *comatic aberration* is an image-degrading, monochromatic, primary aberration associated with an object point even a short distance from the axis. Its origins lie in the fact that the principal "planes" can actually be treated as planes only in the paraxial region. They are, in fact, principal curved surfaces (Fig. 6.1). In the absence of SA a parallel bundle of rays will focus at the axial point F_i, a distance b.f.l. from the rear vertex. Yet the effective focal lengths and therefore the

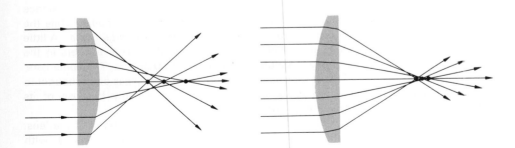

Fig. 6.14 SA for a planar-convex lens.

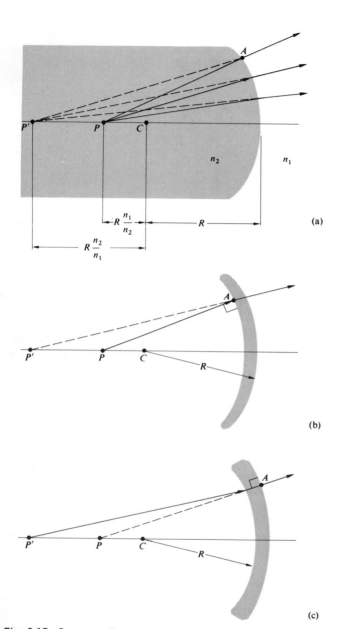

(a)

(b)

(c)

Fig. 6.15 Corresponding axial points for which SA is zero.

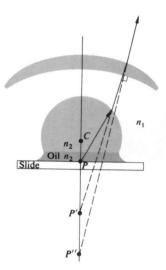

Fig. 6.16 An oil-immersion microscope objective.

transverse magnifications will differ for rays traversing off-axis regions of the lens. When the image point is on the optical axis, this situation is of little consequence, but when the ray bundle is oblique and the image point is off-axis, coma will be

evident. The dependence of M_T on h, the ray height at the lens, is evident in Fig. 6.17. Here meridional rays traversing the extremities of the lens arrive at the image plane closer to the axis than do the rays in the vicinity of the *principal ray* (i.e., the ray which passes through the principal points). In this instance, the least magnification is associated with the marginal rays which would form the smallest image—the coma is said to be negative. By comparison, the coma in Fig. 6.18 is positive because the marginal rays focus further from the axis. Several skew rays are drawn from an extra-axial object point S in Fig. 6.19 to illustrate the formation of the geometrical comatic image of a point. Observe that each circular cone of rays whose endpoints (1-2-3-4-1-2-3-4) form a ring on the lens is imaged in what H. Dennis Taylor called a *comatic circle* on Σ_i. This case corresponds to positive coma and so the larger the ring on the lens, the more distant will be its comatic circle from the axis. When the outer ring is the intersection of marginal rays, the distance from 0 to 1 in the image is the *tangential coma*, while the length from 0 to 3 on Σ_i is termed the *sagittal coma*. A little more than half of the energy in the image appears in the roughly triangular region between 0 and 3. The coma flare, which owes its name to its comet-like tail, is often thought to be the worst of all aberrations, primarily because of its asymmetric configuration.

Like SA, coma is dependent on the shape of the lens. Thus, a strongly concave positive-meniscus lens **)** with

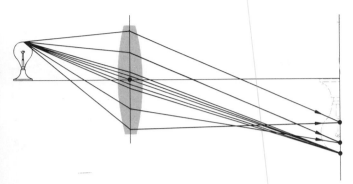

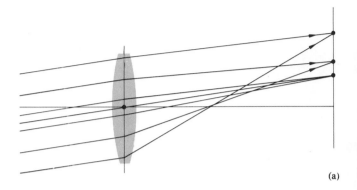

(a)

Fig. 6.17 Negative coma.

the object at infinity will have a large negative coma. Bending the lens so that it becomes planar-convex ❱ , then equi-convex ❘ , convex-planar ❰ and finally convex-meniscus ❰ , will change the coma from negative, to zero, to positive. The fact that it can be made exactly zero for a single lens with a given object distance is quite significant. The particular shape it would then have ($s_o = \infty$) is almost convex-planar and very nearly the configuration for minimum SA.

It is quite important to realize that *a lens which is well corrected for the case where one conjugate point is at infinity ($s_o = \infty$) may not perform satisfactorily when the object is nearby.* One would therefore do well, when using off-the-shelf lenses in a system operating at finite conjugates, to combine two infinite conjugate corrected lenses as in Fig. 6.20. In other words, since it is unlikely that a lens with the desired focal length, which is also corrected for the particular set of finite conjugates, can be obtained ready made, this back-to-back lens approach is an appealing alternative.

Coma can also be negated by using a stop at the proper location, as was discovered in 1812 by William Hyde Wollaston (1766–1828). The order of the list of primary aberrations (SA, coma, astigmatism, Petzval field curvature and distortion) is significant because any one of them, except SA and Petzval curvature, will be affected by the position of a stop, but only if one of the preceding aberrations is also present in the system. Thus while SA is independent of the location along the axis of a stop, coma will not be, as long as SA is present. This can better be appreciated by examining the representation in Fig. 6.21. With the stop at Σ_1, ray 3 is the chief ray, there is SA but no coma; i.e., the ray pairs meet on 3. If the stop is moved to Σ_2, the symmetry is upset, ray 4 be-

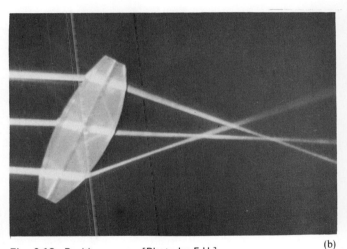

Fig. 6.18 Positive coma. [Photo by E.H.]

(b)

comes the chief ray and the rays on either side of it, such as 3 and 5, meet above it, not on it—there is then positive coma. With the stop at Σ_3, the rays 1 and 3 intersect below the chief ray, 2, and there is negative coma. In this way, controlled amounts of the aberration can be introduced into a compound lens in order to cancel coma in the system as a whole.

The *optical sine theorem* is an important relationship which must be introduced here even if space precludes its formal proof. It was discovered independently in 1873 by Abbe and Helmholtz although a different form of it was given ten years earlier by R. Clausius (of thermodynamics fame). In any event, it states that

$$n_o y_o \sin \alpha_o = n_i y_i \sin \alpha_i, \qquad (6.41)$$

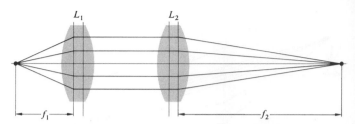

Fig. 6.20 A combination of two infinite conjugate lenses yielding a system operating at finite conjugates.

$$M_T = \frac{y_i}{y_o} \qquad [5.24]$$

must be constant for all rays. Suppose then that we send a marginal and a paraxial ray through the system. The former will comply with Eq. (6.41), the latter with its paraxial version (in which $\sin \alpha_o = \alpha_{op}$, $\sin \alpha_i = \alpha_{ip}$). Since M_T is to be constant over the entire lens, we equate the magnification for both marginal and paraxial rays to get

$$\frac{\sin \alpha_o}{\sin \alpha_i} = \frac{\alpha_{op}}{\alpha_{ip}} = \text{constant}, \qquad (6.42)$$

which is known as the *sine condition*. A necessary criterion for the absence of coma is that the system meet the sine condition. If there is no SA, compliancy with the sine condition will be both necessary and sufficient for zero coma.

It's an easy matter to observe coma. In fact, anyone who has focused sunlight with a simple positive lens has no doubt seen the effects of this aberration. A slight tilt of the lens, so that the nearly collimated rays from the sun make an angle with the optical axis, will cause the focused spot to flare out into the characteristic comet shape.

iii) Astigmatism

When an object point lies an appreciable distance from the optical axis the incident cone of rays will strike the lens asymmetrically, giving rise to a third primary aberration known as *astigmatism*. To facilitate its description, envision the meridional plane (also called the *tangential plane*) containing both the chief ray (i.e. the one passing through the center of the aperture) and the optical axis. The *sagittal plane* is then defined as the plane containing the chief ray which, in addition, is perpendicular to the meridional plane (Fig. 6.22). Unlike the latter which is unbroken from one end of a complicated lens system to the other, the sagittal plane generally changes slope as the chief ray is deviated at

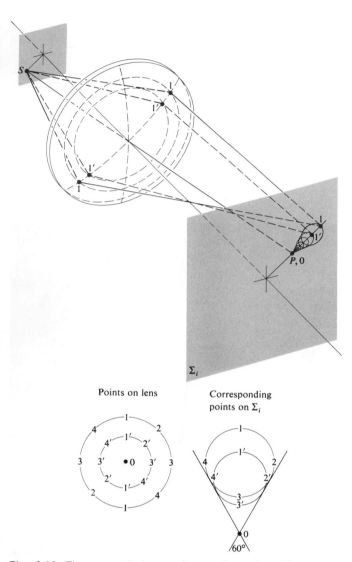

Points on lens   Corresponding points on Σ_i

Fig. 6.19 The geometrical coma image of a point. The central region of the lens forms a point image at the vertex of the cone.

where n_o, y_o, α_o and n_i, y_i, α_i are the index, height and slope angle of a ray in object and image space, respectively at any aperture size* (Fig. 6.9). If coma is to be zero

---

* To be precise, the sine theorem is valid for all values of α_o only in the sagittal plane (from the Latin *sagitta* meaning arrow) which is discussed in the next section.

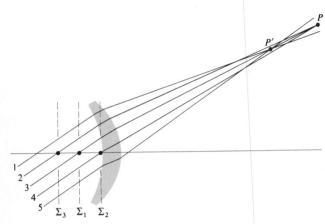

Fig. 6.21 The effect of stop location on coma.

the various elements. Hence to be accurate we should say that there are actually several sagittal planes, one attendant with each region within the system. Nevertheless, all skew rays from the object point lying in a sagittal plane are termed *sagittal rays*.

In the case of an axial object point, the cone of rays is symmetrical with respect to the spherical surfaces of a lens. There is no need to make a distinction between meridional and sagittal planes. The ray configurations in all planes containing the optical axis are identical. In the absence of spherical aberration, all the focal lengths are the same and consequently all rays arrive at a single focus. Contrastingly, the configuration of an oblique, parallel ray bundle will be different in the meridional and sagittal planes. As a result, the focal lengths in these planes will be different as well. In effect, here the meridional rays are tilted more with respect to the lens than are the sagittal rays and they have a shorter focal length. It can be shown,* using Fermat's principle, that the *focal length difference* depends effectively on the power of the lens (as opposed to the shape or index) and the angle at which the rays are inclined. This *astigmatic difference*, as it is often called, increases rapidly as the rays become more oblique, i.e. as the object point moves further off the axis and is, of course, zero on axis.

Having two distinct focal lengths, the incident conical bundle of rays takes on a considerably altered form after refraction (Fig. 6.23). The cross-section of the beam as it leaves the lens is initially circular but it gradually becomes elliptical with the major axis in the sagittal plane until at the

---

* See A. W. Barton, *A Text Book on Light*, p. 124.

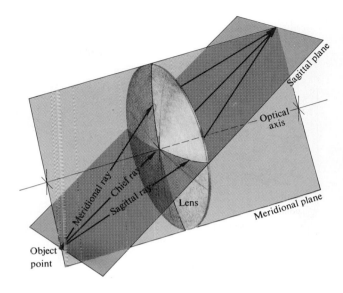

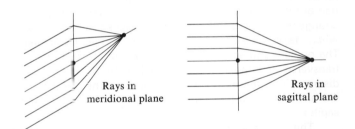

Fig. 6.22 The sagittal and meridional planes.

tangential or meridional focus F_T, the ellipse degenerates into a *line* (at least in third-order theory). All rays from the object point traverse this line which is known as the *primary image*. Beyond this point the beam's cross-section rapidly opens out until it is again circular. At that location the image is a circular blur known as the *circle of least confusion*. Moving further from the lens the beam's cross-section again deforms into a line called the *secondary image*. This time it's in the meridional plane at the *sagittal focus* F_S. Remember that all of this is assuming the absence of SA and coma.

Since the circle of least confusion increases in diameter as the astigmatic difference increases, i.e. as the object moves further off-axis, the image will deteriorate, losing definition around its edges. Observe that the secondary line image will change in orientation with changes in the object position but it will always point toward the optical axis, i.e. it will be

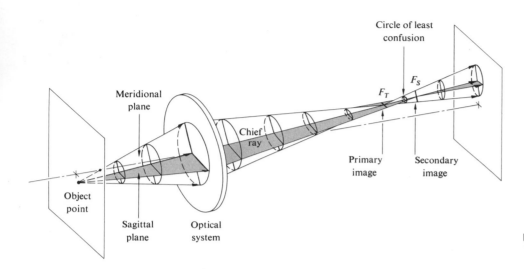

Fig. 6.23 Astigmatism.

radial. Similarly, the primary line image will vary in orientation but it will remain normal to the secondary image. This arrangement causes the interesting effect shown in Fig. 6.24 when the object is made up of radial and tangential elements. The primary and secondary images are, in effect, formed of transverse and radial dashes which increase in size with distance from the axis. In the latter case, the dashes point like arrows toward the center of the image—ergo, the name sagitta.

The existence of the sagittal and tangential foci can be verified directly with a fairly simple arrangement. Place a short focal length positive lens (say about 10 or 20 mm) in the beam of an He–Ne laser. Position another, somewhat longer focal length, positive test lens far enough away so that the now diverging beam fills that lens. A convenient object, to be located between the two lenses, is a piece of ordinary wire screening (or a transparency). Align it so the wires are horizontal (x) and vertical (y). If the test lens is rotated through roughly 45° about the vertical (with the x-, y-, z-axes fixed in the lens), astigmatism should be observable. The meridional is the xz-plane (z being the lens axis, now at about 45° to the laser axis) while the sagittal plane corresponds to the plane of y and the laser axis. As the wire mesh is moved toward the test lens, a point will be reached where the horizontal wires are in focus on a screen beyond the lens, while the vertical wires are not. This is the location of the sagittal focus. Each point on the object is imaged as a short line in the meridional (horizontal) plane, which accounts for the fact that only the horizontal

wires are in focus. Moving the mesh slightly closer to the lens will bring the vertical lines into clarity while the horizontal ones are blurred. This is the tangential focus. Try rotating the mesh about the central laser axis while at either focus.

Note that unlike visual astigmatism which arose from an actual asymmetry in the surfaces of the optical system, the third-order aberration by that same name applies to spherically symmetric lenses.

Mirrors, with the singular exception of the plane mirror, suffer much the same monochromatic aberrations as do lenses. Thus while a paraboloidal mirror is free of SA for an infinitely distant axial object point, its off-axis imagery is quite poor due to astigmatism and coma. This strongly restricts its use to narrow field devices such as searchlights and astronomical telescopes. A concave spherical mirror shows SA, coma and astigmatism. Indeed one could draw a diagram just like Fig. 6.23 with the lens replaced by an obliquely illuminated spherical mirror. Incidentally, such a mirror displays appreciably less SA than would a simple convex lens of the same focal length.

iv) Field Curvature

Suppose that we have an optical system which is free of all of the aberrations thus far considered. There would then be a one-to-one correspondence between points on the object and image surfaces (i.e. stigmatic imagery). We mentioned earlier [Section 5.2.3(iii)] that a planar object normal to the axis will be imaged approximately as a plane only in the

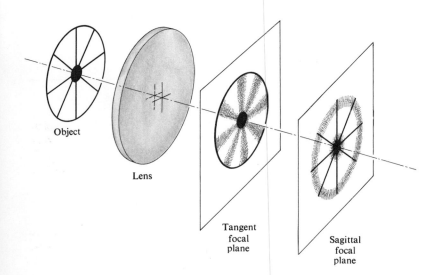

Object

Lens

Tangent
focal
plane

Sagittal
focal
plane

Fig. 6.24 Images in the tangent and sagittal focal planes.

paraxial region.   At finite apertures the resulting curved stigmatic image surface is a manifestation of the primary aberration known as *Petzval field curvature* after the Hungarian mathematician Josef Max Petzval (1807–91). The effect can readily be appreciated by examining Figs. 5.22 (p. 111) and 6.25. A spherical object segment σ_o is imaged by the lens as a spherical segment σ_i, both centered at O. Flattening out σ_o into the plane σ_o' will cause each object point to move toward the lens along the concomitant chief ray, thus forming a paraboloidal *Petzval surface* Σ_P. While the Petzval surface for a positive lens curves *inward* toward the object plane, for a negative lens it curves *outward*, i.e. away from that plane.   Evidently, a suitable combination of positive and negative lenses will negate field curvature. Indeed, the displacement Δx of an image point at height y_i on the Petzval surface from the paraxial image plane is given by

$$\Delta x = \frac{y_i^2}{2} \sum_{j=1}^{m} \frac{1}{n_j f_j},\qquad (6.43)$$

where n_j and f_j are the indices and focal lengths of the m thin lenses forming the system.   This implies that the Petzval surface will be unaltered by changes in the positions or shapes of the lenses, or in the location of the stop, so long as the values of n_j and f_j are fixed.   Notice that for the simple case of two thin lenses ($m = 2$) having any spacing, Δx *can be made zero* provided that

$$\frac{1}{n_1 f_1} + \frac{1}{n_2 f_2} = 0$$

or equivalently

$$n_1 f_1 + n_2 f_2 = 0. \qquad (6.44)$$

This is the so-called *Petzval condition*. As an example of its use, suppose we combine two thin lenses, one positive, the other negative, such that $f_1 = -f_2$ and $n_1 = n_2$. Since

$$\frac{1}{f} = \frac{1}{f_1} + \frac{1}{f_2} - \frac{d}{f_1 f_2}, \qquad [6.8]$$

$$f = \frac{f_1^2}{d},$$

the system can satisfy the Petzval condition, have a flat field, and still have a finite positive focal length.

In visual instruments a certain amount of curvature can be tolerated because the eye can accommodate for it. Clearly in photographic lenses field curvature is most undesirable since it has the effect of rapidly blurring the off-axis image when the film plane is at F_i.   An effective means of nullifying the inward curvature of a positive lens is to place a negative *field flattener* lens near the focal plane. This is often done in projection and photographic objectives when it is not otherwise practicable to meet the Petzval condition (Fig. 6.26).   In this position the flattener will have little effect on other aberrations (take another look at Fig. 6.7).

Astigmatism is intimately related to field curvature.   In the presence of the former aberration, there will be *two* paraboloidal image surfaces, the tangential, Σ_T, and the sagittal, Σ_S (as in Fig. 6.27). These are the loci of all the

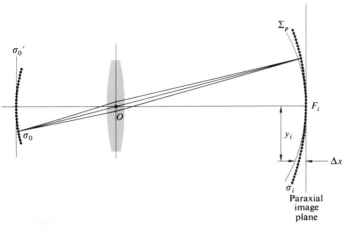

Fig. 6.25 Field curvature.

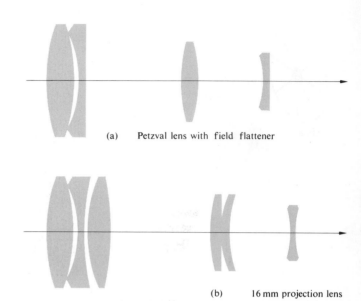

(a)   Petzval lens with field flattener

(b)   16 mm projection lens

Fig. 6.26 The field flattener.

primary and secondary images, respectively, as the object point roams over the object plane. At a given height (y_i), a point on Σ_T always lies three times as far from Σ_P as does the corresponding point on Σ_S and both are on the same side of the Petzval surface (Fig. 6.27). When there is no astigmatism Σ_S and Σ_T coalesce on Σ_P. It is possible to alter the shapes of Σ_S and Σ_T by bending or relocating the lenses or by moving the stop. The configuration of Fig. 6.27(b) is known as an *artificially flattened* field. A stop in front of an inexpensive meniscus box camera lens is usually arranged to produce just this effect. The surface of least confusion, Σ_{LC}, is planar and the image there is tolerable, losing definition at the margins due to the astigmatism. That is to say, although their loci form Σ_{LC}, the circles of least confusion increase in diameter with distance off the axis. Modern good-quality photographic objectives are generally *anastigmats*, i.e. they are designed so that Σ_S and Σ_T cross each other yielding an additional off-axis angle of zero astigmatism. The Cooke Triplet, Tessar, Orthometer and Biotar (Fig. 5.100) are all anastigmats as is the relatively fast Zeiss Sonnar whose residual astigmatism is illustrated graphically in Fig. 6.28. Note the relatively flat field and small amount of astigmatism over most of the film plane.

Let's return briefly to the Schmidt camera of Fig. 5.95 (p. 157) since we are now in a better position to appreciate how it functions. With a stop at the center of curvature of the spherical mirror, all chief rays, which by definition pass through C, are incident normally on the mirror. Moreover each pencil of rays from a distant object point is symmetrical about its chief ray. In effect, each chief ray serves as an

optical axis and so there are no off-axis points and in principle no coma or astigmatism. Instead of attempting to flatten the image surface, curvature is coped with by simply shaping the

v) Distortion

The last of the five primary, monochromatic aberrations is *distortion*. Its origin lies in the fact that the transverse magnification, M_T, may be a function of the off-axis image distance, y_i. Thus, that distance may differ from the one predicted by paraxial theory in which M_T is constant. In the absence of any of the others, this aberration is manifest in a misshaping of the image as a whole, even though each point is sharply focused. Consequently, when processed by an optical system suffering *positive* or *pin-cushion distortion*, a square array deforms as in Fig. 6.29(b). In that instance, each image point is displaced radially outward from the center with the most distant points moving the greatest amount, i.e., M_T increases with y_i. Similarly *negative* or *barrel distortion* corresponds to the situation where M_T decreases with the axial distance and, in effect, each point on the image moves radially inward toward the center [Fig. 6.29(c)]. Distortion can easily be seen by just looking through an aberrant lens at a piece of lined or graph paper. Fairly thin lenses will show essentially no distortion whereas ordinary positive or negative, thick, simple lenses will generally suffer positive or negative distortion, respectively.

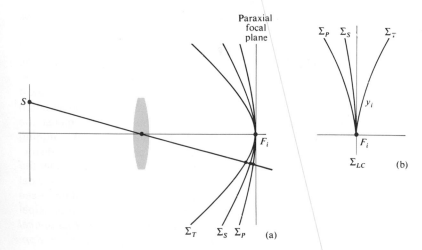

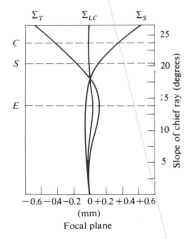

Fig. 6.27 The tangential, sagittal and Petzval image surfaces.

The introduction of a stop into a system of thin lenses is invariably accompanied by distortion, as indicated in Fig. 6.30. One exception to this is when the aperture stop is at the lens so that the chief ray is, in effect, the principal ray (i.e. it passes through the principal points, here coalesced at O). If the stop is in front of a positive lens, as in Fig. 6.30(b), the object distance measured along the chief ray will be greater than it was with the stop at the lens ($S_2A > S_2O$). Thus x_o will be greater and (5.26) M_T will be smaller—ergo, barrel distortion. In other words, M_T for an off-axis point will be less with a front stop in position than it would be without it. The difference is a measure of the aberration which, by the way, exists regardless of the size of the aperture. In the same way a rear stop [Fig. 6.30(c)] decreases x_o along the chief ray, (i.e., $S_2O > S_2B$), thereby increasing M_T and introducing pin-cushion distortion. *Interchanging the object and image thus has the effect of changing the sign of the distortion* for a given lens and stop. The aforementioned stop positions will produce the opposite effect when the lens is negative.

All of this rather suggests using a stop midway between identical lens elements. The distortion from the first lens

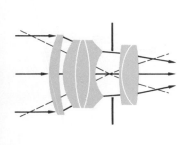

Fig. 6.28 A typical Sonnar. The markings C, S, and E denote the limits of the 35 mm film formate (field stop) i.e. corners, sides and edges. The Sonnar family lies between the double Gauss and the triplet.

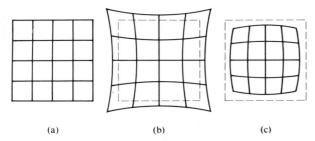

(a)   (b)   (c)

Fig. 6.29 Distortion.

would precisely cancel the contribution from the second. This approach has been used to advantage in the design of a number of photographic lenses (Fig. 5.100). To be sure, if the lens is perfectly symmetrical and operating as in Fig. 6.30(d) the object and image distances will be equal and hence $M_T = 1$. (Incidentally, coma and lateral color would then be identically zero as well.) This applies to (finite conjugate) copy lenses used, for example, to record data. Nonetheless, even when M_T is not one, making the system approximately symmetrical about a stop is a very common practice since it markedly reduces these several aberrations.

Distortion can arise in compound lens systems, as for example in the telephoto arrangement shown in Fig. 6.31. For a distant object point, the margin of the positive achromat serves as the aperture stop. In effect, the arrangement is

like a negative lens with a front stop and so it displays positive or pin-cushion distortion.

Suppose a chief ray enters and emerges from an optical system in the same direction as e.g. in Fig. 6.30(d). The point at which the ray crosses the axis is the optical center of the system; but at the same time, since this is a chief ray, it is also the center of the aperture stop. This is the situation approached in Fig. 6.30(a) with the stop up against the thin lens. In both instances the incoming and outgoing segments of the chief ray are parallel and there is zero distortion, i.e. the system is *orthoscopic*. This also implies that the entrance and exit pupils will correspond to the principal planes (if the system is immersed in a single medium—see Fig. 6.2). Bear in mind that the chief ray is now a principal ray. *A thin-lens system will have zero distortion if its optical center is coincident with the center of the aperture stop.* By the way, in a pinhole camera, the rays connecting conjugate object and image points are straight and pass through the center of the aperture stop. The entering and emerging rays are obviously parallel (being one and the same) and there is no distortion.

6.3.2 Chromatic Aberrations

The five primary or Seidel aberrations have been considered in terms of monochromatic light. To be sure, if the source had a broad spectral bandwidth these aberrations would be

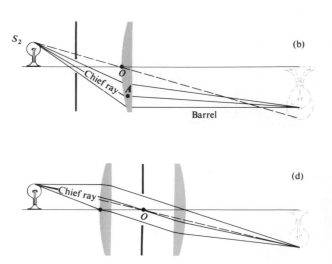

Fig. 6.30 The effect of stop location on distortion.

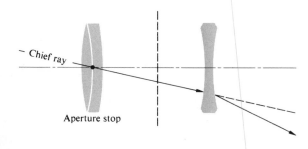

Fig. 6.31 Distortion in a compound lens.

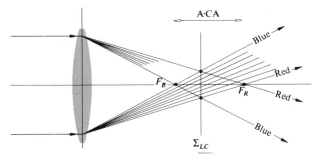

Fig. 6.32 Axial chromatic aberration.

influenced accordingly; but the effects are inconsequential unless the system is quite well corrected. There are, however, *chromatic aberrations* that arise specifically in polychromatic light which, by comparison, are far more significant. The ray-tracing equation (6.12) is a function of the indices of refraction which, in turn, vary with wavelength. Different "colored" rays will traverse a system along different paths, and this is the quintessential feature of chromatic aberration.

Since the thin-lens equation

$$\frac{1}{f} = (n_l - 1)\left(\frac{1}{R_1} - \frac{1}{R_2}\right) \qquad [5.16]$$

is wavelength dependent via $n_l(\lambda)$, the focal length must also vary with λ. In general (Fig. 3.13, p. 42) $n_l(\lambda)$ decreases with wavelength over the visible region and thus $f(\lambda)$ increases with λ. The result is illustrated in Fig. 6.32 where the constituent colors in a collimated beam of white light are focused at different points on the axis. The axial distance between two such focal points spanning a given frequency range (e.g. blue to red) is termed the *axial (or longitudinal) chromatic aberration*, A · CA for short.

It's an easy matter to observe chromatic aberrations, or CA, with a thick, simple converging lens. When illuminated by a polychromatic point source (a candle flame will do) the lens will cast a real image surrounded by a halo. If the plane of observation is then moved nearer the lens, the periphery of the blurred image will become tinged in orange-red. Moving it back away from the lens, beyond the best image, will cause the outlines to become tinted in blue-violet. The location of the circle of least confusion (i.e. the plane Σ_{LC}) corresponds to the position where the best image will appear. Try looking directly through the lens at a source—the coloration will be far more striking.

The image of an off-axis point will be formed of the constituent frequency components, each arriving at a different height above the axis (Fig. 6.33). In essence, the frequency dependence of f causes a frequency dependence of the transverse magnification as well. The vertical distance between two such image points (most often taken to be blue and red) is a measure of the *lateral chromatic aberration* L · CA, or *lateral color*. Consequently, a chromatically aberrant lens illuminated by white light will fill a volume of space with a continuum of more or less overlapping images, varying in size and color. Because the eye is most sensitive to the yellow-green portion of the spectrum, the tendency is to focus the lens for that region. With such a configuration one would see all of the other colored images superimposed and slightly out of focus producing a whitish blur or hazed overlay.

When the blue focus, F_B, is to the left of the red focus, F_R, the A · CA is said to be positive, as it is in Fig. 6.32. Contrarily, a negative lens would generate negative A · CA, with the more strongly deviated blue rays appearing to originate at the right of the red focus. Physically what is happening is that the lens, whether convex or concave, is prismatic in shape, i.e. it becomes either thinner or thicker as the radial distance from the axis increases. As you well know, rays are therefore deviated either toward or away from the axis, respectively. In both cases the rays are bent toward the thicker "base" of the prismatic cross section. But the angular deviation is an increasing function of n and therefore it decreases with λ. Hence blue light is deviated the most and is focused nearest the lens. In other words, for a convex lens the red focus is furthest and to the right; for a concave lens it is furthest and to the left.

i) Thin Achromatic Doublets

All of this rather suggests that a combination of two thin lenses, one positive and one negative, could conceivably result in the precise overlapping of F_R and F_B (Fig. 6.34).

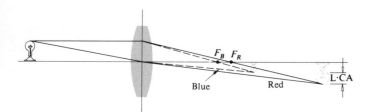

Fig. 6.33 Laterial chromatic aberration.

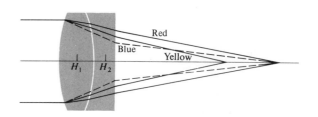

Fig. 6.34 An achromatic doublet.

Such an arrangement is said to be *achromatized* for those two specific wavelengths. Notice that what we would like to do is effectively eliminate the total dispersion (i.e. the fact that each color is deviated by a different amount) and not the total deviation itself. With the two lenses separated by a distance d,

$$\frac{1}{f} = \frac{1}{f_1} + \frac{1}{f_2} - \frac{d}{f_1 f_2}. \qquad [6.8]$$

Rather than carrying around the second term in the thin-lens equation (5.16) let's abbreviate the notation and write $1/f_1 = (n_1 - 1)\rho_1$ and $1/f_2 = (n_2 - 1)\rho_2$ for the two elements. Then

$$\frac{1}{f} = (n_1 - 1)\rho_1 + (n_2 - 1)\rho_2 - d(n_1 - 1)\rho_1(n_2 - 1)\rho_2. \qquad (6.45)$$

This expression will yield the focal length of the doublet for red (f_R) and blue (f_B) light when the appropriate indices are introduced, namely n_{1R}, n_{2R}, n_{1B}, and n_{2B}. But if f_R is to equal f_B, then

$$1/f_R = 1/f_B$$

and

$$(n_{1R} - 1)\rho_1 + (n_{2R} - 1)\rho_2 - d(n_{1R} - 1)\rho_1(n_{2R} - 1)\rho_2$$
$$= (n_{1B} - 1)\rho_1 + (n_{2B} - 1)\rho_2 - d(n_{1B} - 1)\rho_1(n_{2B} - 1)\rho_2. \qquad (6.46)$$

One case of particular importance corresponds to $d = 0$, i.e. the two lenses are in contact. Expanding out Eq. (6.46) with $d = 0$ then leads to

$$\frac{\rho_1}{\rho_2} = -\frac{n_{2B} - n_{2R}}{n_{1B} - n_{1R}}. \qquad (6.47)$$

The focal length of the compound lens (f_Y) can conveniently be specified as that associated with yellow light, roughly midway between the blue and red extremes. For the component lenses in yellow light $1/f_{1Y} = (n_{1Y} - 1)\rho_1$ and $1/f_{2Y} = (n_{2Y} - 1)\rho_2$. Hence

$$\frac{\rho_1}{\rho_2} = \frac{(n_{2Y} - 1)}{(n_{1Y} - 1)} \frac{f_{2Y}}{f_{1Y}}. \qquad (6.48)$$

Equating Eqs. (6.47) and (6.48) leads to

$$\frac{f_{2Y}}{f_{1Y}} = -\frac{(n_{2B} - n_{2R})/(n_{2Y} - 1)}{(n_{1B} - n_{1R})/(n_{1Y} - 1)}. \qquad (6.49)$$

The quantities

$$\frac{n_{2B} - n_{2R}}{n_{2Y} - 1} \quad \text{and} \quad \frac{n_{1B} - n_{1R}}{n_{1Y} - 1}$$

are known as the *dispersive powers* of the two materials forming the lenses. Their reciprocals, V_2 and V_1, are variously known as the *dispersive indices*, *V-numbers*, or *Abbe numbers*. Thus

$$\frac{f_{2Y}}{f_{1Y}} = -\frac{V_1}{V_2}$$

or

$$f_{1Y}V_1 + f_{2Y}V_2 = 0. \qquad (6.50)$$

Since the dispersive powers are positive, so too are the V-numbers. This implies, as we anticipated, that one of the two component lenses must be negative and the other positive if Eq. (6.50) is to obtain, i.e. if f_R is to equal f_B.

At this point we could presumably design an *achromatic doublet* and indeed we presently shall, but a few additional points must be made first. The designation of wavelengths as red, yellow, and blue is far too imprecise for practical application. Instead it is customary to refer to specific spectral lines whose wavelengths are known with great precision. The *Fraunhofer lines*, as they are called, serve as the needed reference markers across the spectrum. Several of these for the visible region are listed in Table 6.1. The lines F, C, and d (i.e. D_3) are most often used (for blue, red, and yellow) and

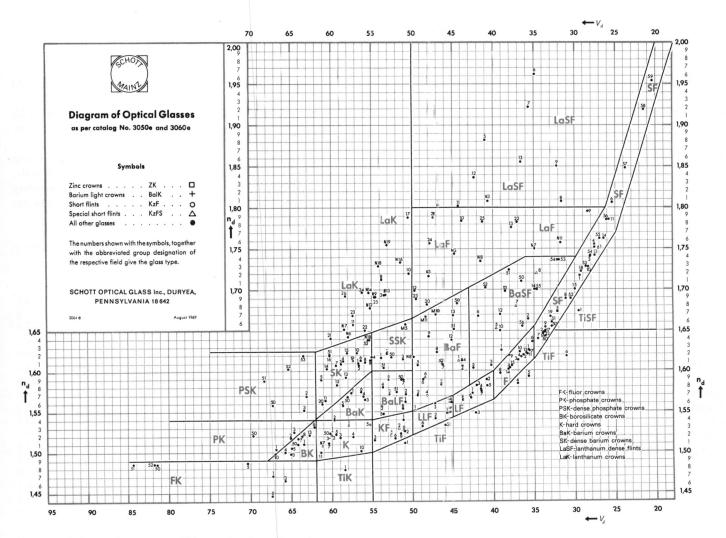

Fig. 6.35 Refractive index versus Abbe number for various glasses.

one generally traces paraxial rays in d-light. Glass manufacturers will usually list their wares in terms of the Abbe number as in Fig. 6.35 which is a plot of the refractive index versus

$$V_d = \frac{n_d - 1}{n_F - n_C}. \quad (6.51)$$

(Take a look at Table 6.2 as well.) Thus Eq. (6.50) might better be written as

$$f_{1d}V_{1d} + f_{2d}V_{2d} = 0, \quad (6.52)$$

where the numerical subscripts pertain to the two glasses used in the doublet and the letter relates to the d-line.

Incidentally, Newton erroneously concluded, on the basis of experiments with the very limited range of materials available at the time, that the dispersive power was constant for all glasses. This is tantamount to saying (6.52) that $f_{1d} = -f_{2d}$, in which case the doublet would have zero power. Newton, accordingly, shifted his efforts from the refracting to the reflecting telescope and this fortunately turned out to be a good move in the long run. The achromat was invented around 1733 by Chester Moor Hall, Esq., but it lay in limbo until it was seemingly reinvented and patented in 1758 by the London optician John Dollond.

Table 6.1   Several strong Fraunhofer lines.

Designation	Wavelength (Å)	Source
C	6562.816 red	H
D_1	5895.923 yellow	Na
D	center of doublet 5892.9	Na
D_2	5889.953 yellow	Na
D_3 or d	5875.618 yellow	He
b_1	5183.618 green	Mg
b_2	5172.699 green	Mg
c	4957.609 green	Fe
F	4861.327 blue	H
f	4340.465 violet	H
g	4226.728 violet	Ca
K	3933.666 violet	Ca

Several forms of the achromatic doublet are shown in Fig. 6.36. Their configurations depend on the glass types selected as well as on the choice of the other aberrations to be controlled. By the way, when purchasing off-the-shelf doublets of unknown origin, be careful not to buy a lens which deliberately includes certain aberrations in order to compensate for errors in the original system from which it came. Perhaps the most commonly encountered doublet is the cemented Fraunhofer achromat. It's formed of a crown* double-convex lens in contact with a concave-planar (or nearly planar) flint lens. The use of a crown front element is quite popular because of its better wear resistance. Since the overall shape is roughly convex-planar, by selecting the proper glasses both spherical aberration and coma can be corrected as well. Suppose then, that we wish to design a Fraunhofer achromat of focal length 50 cm. We can get some idea of roughly how to select glasses by solving Eq. (6.52) simultaneously with the compound-lens equation

$$\frac{1}{f_{1d}} + \frac{1}{f_{2d}} = \frac{1}{f_d}$$

to get

$$\frac{1}{f_{1d}} = \frac{V_{1d}}{f_d(V_{1d} - V_{2d})} \qquad (6.53)$$

---

* Traditionally the glasses in the range $n_d > 1.60$, $V_d > 50$, and $n_d < 1.60$, $V_d > 55$ are known as *crowns* while the others are *flints*. Note the letter designations of Fig. 6.35.

and

$$\frac{1}{f_{2d}} = \frac{V_{2d}}{f_d(V_{2d} - V_{1d})}. \qquad (6.54)$$

Thus, in order to avoid small values of f_{1d} and f_{2d}, which would necessitate strongly curved surfaces on the component lenses, the difference $V_{1d} - V_{2d}$ should be made large (roughly 20 or more is convenient). From Fig. 6.35 (or its equivalent) we select, say, BK1 and F2. These have cataloged indices of $n_C = 1.50763$, $n_d = 1.51009$, $n_F = 1.51566$, and $n_C = 1.61503$, $n_d = 1.62004$, and $n_F = 1.63208$, respectively. Likewise their V-numbers are generally given rather accurately and we needn't compute them. In this instance they are

Table 6.2   Optical glass.

Type number	Name	n_D	V_D
511:635	Borosilicate crown—BSC-1	1.5110	63.5
517:645	Borosilicate crown—BSC-2	1.5170	64.5
513:605	Crown—C	1.5125	60.5
518:596	Crown	1.5180	59.6
523:586	Crown—C-1	1.5230	58.6
529:516	Crown flint—CF-1	1.5286	51.6
541:599	Light barium crown—LBC-1	1.5411	59.9
573:574	Barium crown—LBC-2	1.5725	57.4
574:577	Barium crown	1.5744	57.7
611:588	Dense barium crown—DBC-1	1.6110	58.8
617:550	Dense barium crown—DBC-2	1.6170	55.0
611:572	Dense barium crown—DBC-3	1.6109	57.2
562:510	Light barium flint—LBF-2	1.5616	51.0
588:534	Light barium flint—LBF-1	1.5880	53.4
584:460	Barium flint—BF-1	1.5838	46.0
605:436	Barium flint—BF-2	1.6053	43.6
559:452	Extra light flint—ELF-1	1.5585	45.2
573:425	Light flint—LF-1	1.5725	42.5
580:410	Light flint—LF-2	1.5795	41.0
605:380	Dense flint—DF-1	1.6050	38.0
617:366	Dense flint—DF-2	1.6170	36.6
621:362	Dense flint—DF-3	1.6210	36.2
649:338	Extra dense flint—EDF-1	1.6490	33.8
666:324	Extra dense flint—EDF-5	1.6660	32.4
673:322	Extra dense flint—EDF-2	1.6725	32.2
689:309	Extra dense flint—EDF	1.6890	30.9
720:293	Extra dense flint—EDF-3	1.7200	29.3

From T. Calvert, "Optical Components," *Electromechanical Design* (May 1971).

Type number is given by $(n_D - 1):(10\, V_D)$ where n_D is rounded off to three decimal places—For more data see Smith, *Modern Optical Engineering*, Fig. 7.5.

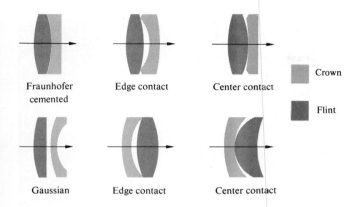

Fraunhofer cemented

Edge contact

Center contact

Crown

Flint

Gaussian

Edge contact

Center contact

$V_{1d} = 63.46$ and $V_{2d} = 36.37$, respectively. The focal lengths, or if you will, the powers of the two lenses, are given by Eqs. (6.53) and (6.54):

$$\mathscr{D}_{1d} = \frac{1}{f_{1d}} = \frac{63.46}{0.50(27.09)}$$

and

$$\mathscr{D}_{2d} = \frac{1}{f_{2d}} = \frac{36.37}{0.50(-27.09)}.$$

Hence $\mathscr{D}_{1d} = 4.685$ diopters and $\mathscr{D}_{2d} = -2.685$ diopters; the sum being 2 diopters which is $1/0.5$, as it should be. For ease of fabrication let the first or positive lens be equi-convex. Consequently its radii R_{11} and R_{12} are equal in magnitude. Hence

$$\rho_1 = \frac{1}{R_{11}} - \frac{1}{R_{12}} = \frac{2}{R_{11}}$$

or equivalently

$$\frac{2}{R_{11}} = \frac{\mathscr{D}_{1d}}{n_{1d} - 1} = \frac{4.685}{0.51009} = 9.185.$$

And so $R_{11} = -R_{12} = 0.2177$ m. Furthermore, having specified that the lenses be in intimate contact we have $R_{12} = R_{21}$, i.e. the second surface of the first lens matches the first surface of the second lens. For the second lens

$$\rho_2 = \frac{1}{R_{21}} - \frac{1}{R_{22}} = \frac{\mathscr{D}_{2d}}{n_{2d} - 1}$$

or

$$\frac{1}{-0.2177} - \frac{1}{R_{22}} = \frac{-2.685}{0.62004}$$

and $R_{22} = -3.819$ m. In summary, the radii of the crown

element are $R_{11} = 21.8$ cm and $R_{12} = -21.8$ cm while the flint has radii of $R_{21} = -21.8$ cm and $R_{22} = -381.9$ cm.

Note that for a thin-lens combination the principal planes coalesce so that achromatizing the focal length corrects both $A \cdot CA$ and $L \cdot CA$. In a thick doublet, however, even though the focal lengths for red and blue are made identical, the different wavelengths may have different principal planes. Consequently, although the magnification is the same for all wavelengths, the focal points may not coincide, i.e. $L \cdot CA$ is corrected for but not $A \cdot CA$.

In the above analysis only the C- and F-rays were brought to a common focus while the d-line was introduced to establish a focal length for the doublet as a whole. It is not possible for *all* wavelengths traversing a doublet achromat to meet at a common focus. The resulting residual chromatism is known as *secondary spectrum*. The elimination of secondary spectrum is particularly troublesome when the design is limited to the glasses presently available. Notwithstanding that, a fluorite (CaF_2) element combined with an appropriate glass element can form a doublet achromatized at three wavelengths and having very little secondary spectrum. More often triplets are used for color correction at three or even four wavelengths. The secondary spectrum of a binocular can easily be observed by looking at a distant white object. Its borders will be slightly haloed in magenta and green—try shifting the focus forward and backward.

ii) Separated Achromatic Doublets

It is also possible to achromatize the focal length of a doublet composed of two widely separated elements of the same glass. Putting it rather succinctly, return to Eq. (6.46) and set $n_{1R} = n_{2R} = n_R$ and $n_{1B} = n_{2B} = n_B$. After a bit of straightforward algebraic manipulation it becomes

$$(n_R - n_B)\left[(\rho_1 + \rho_2) - \rho_1 \rho_2 d(n_B + n_R - 2)\right] = 0$$

or

$$d = \frac{1}{(n_B + n_R - 2)}\left(\frac{1}{\rho_1} + \frac{1}{\rho_2}\right).$$

Again introducing the yellow reference frequency as we did before, namely $1/f_{1Y} = (n_{1Y} - 1)\rho_1$ and $1/f_{2Y} = (n_{2Y} - 1)\rho_2$, ρ_1 and ρ_2 can be replaced. Hence

$$d = \frac{(f_{1Y} + f_{2Y})(n_Y - 1)}{n_B + n_R - 2},$$

where $n_{1Y} = n_{2Y} = n_Y$. Assuming $n_Y = (n_B + n_R)/2$, we have

$$d = \frac{f_{1Y} + f_{2Y}}{2},$$

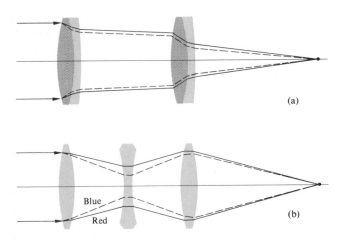

Fig. 6.37 Achromatized lenses.

or in *d*-light

$$d = \frac{f_{1d} + f_{2d}}{2}. \qquad (6.55)$$

This is precisely the form taken by the Huygens ocular (Section 5.7.4). Since the red and blue focal lengths are the same, but the corresponding principal planes for the doublet need not be, the two rays will generally not meet at the same focal point. Thus the ocular's lateral chromatic aberration is well corrected but axial chromatic aberration is not.

In order for a system to be free of both chromatic aberrations, the red and blue rays must emerge parallel to each other (no L·CA) and must intersect the axis at the same point (no A·CA), which means they must overlap. Since this is effectively the case with a thin achromat, it rather implies that multi-element systems, as a rule, should consist of achromatic components in order to keep the red and blue rays from separating (Fig. 6.37). As with all such invocations there are exceptions. The Taylor triplet (Section 5.7.7) is one of these. The two colored rays for which it is achromatized separate within the lens but are recombined and emerge together.

6.3.3 Concluding Remarks

For the practical reason of manufacturing ease, the vast majority of optical systems are limited to lenses having spherical surfaces. There are, to be sure, toric and cylindrical lenses as well as many other aspherics. Indeed, very fine, and as a rule very expensive, devices such as high-altitude

(a)

Fig. 6.38 Photos taken with the Hyac I Panoramic Camera, *f* = 12 inch, *f*/5. The scale in (a) is 1:35,000 while in (b) it is 1:1,200. [Photos courtesy Itek Corporation.]

(b)

reconnaisance cameras and tracking systems may have several aspherical elements. Even so, spherical lenses are here to stay and with them are their inherent aberrations which must satisfactorily be dealt with. As we have seen, the designer (and his faithful electronic companion) must manipulate the system variables (indices, shapes, spacings, stops, etc.) in order to balance out offensive aberrations. This is done to whatever degree and in whatever order is appropriate for the specific optical system. Thus one might tolerate far more distortion and curvature in an ordinary telescope than in a good photographic objective. Likewise, there is little need to worry about chromatic aberration if you want to work exclusively with laser light of almost a single frequency. In any event, this chapter has only touched on the problems (and that more to appreciate than solve them). That they are most certainly amenable to solution is witnessed, e.g. by the remarkable aerial photos of Fig. 6.38 which rather eloquently speak for themselves.

PROBLEMS

6.1* Work out the details leading to Eq. (6.8).

6.2 According to the military handbook MIL-HDBK-141 (23.3.5.3) the Ramsden eyepiece (Fig. 5.81) is made up of two planar-convex lenses of equal focal length f' separated by a distance $2f'/3$. Determine the overall focal length f of the thin-lens combination and locate the principal planes and the position of the field stop.

6.3 Write an expression for the thickness d of a double-convex lens such that its focal length is infinite.

6.4 Suppose we have a positive meniscus lens of radii 6 and 10 and thickness of 3 (any units, as long as you're consistent) and an index of 1.5. Determine its focal length and the locations of its principal points (compare with Fig. 6.3).

6.5* Show that the determinant of the system matrix in Eq. (6.31) is equal to 1.

6.6 Show that Eqs. (6.36) and (6.37) are equivalent to Eqs. (6.3) and (6.4) respectively.

6.7 Show that the planar surface of a concave-planar or convex-planar lens doesn't contribute to the system matrix.

6.8 Compute the system matrix for a thick biconvex lens of index 1.5 having radii of 0.5 and 0.25 and a thickness of 0.3 (in any units you like). Check that $|\mathscr{A}| = 1$.

6.9* For the lens in the previous problem, determine its focal length and the location of the focal points with respect to its vertices V_1 and V_2.

6.10 Referring back to Fig. 6.15, show that when $\overline{P'P} = Rn_2/n_1$ and $\overline{PC} = Rn_1/n_2$ all rays originating at P appear to come from P'.

6.11 Starting with the exact expression given by Eq. (5.5) show that Eq. (6.40) results, rather than Eq. (5.8), when the approximations for ℓ_o and ℓ_i are improved a bit.

6.12 Supposing that Fig. 6.39 is to be imaged by a lens system suffering spherical aberration only, make a sketch of the image.

Fig. 6.39

The Superposition of Waves   7

In succeeding chapters we shall study the phenomena of polarization, interference, and diffraction. These all share a common conceptual basis in that they deal, for the most part, with various aspects of the same process. Stating this in the simplest terms, we are really concerned with what happens when two or more light waves overlap in some region of space. The precise circumstances governing this superposition, of course, determine the final optical disturbance. Amongst other things we are interested in learning how the specific properties of each constituent wave (i.e. amplitude, phase, frequency etc.) influence the ultimate form of the composite disturbance.

Recall that each field component of an electromagnetic wave (E_x, E_y, E_z, B_x, B_y, and B_z) satisfies the scalar three-dimensional differential wave equation,

$$\frac{\partial^2 \psi}{\partial x^2} + \frac{\partial^2 \psi}{\partial y^2} + \frac{\partial^2 \psi}{\partial z^2} = \frac{1}{v^2}\frac{\partial^2 \psi}{\partial t^2}. \qquad [2.59]$$

A rather significant feature of this expression is that it is *linear*, i.e. $\psi(\mathbf{r}, t)$ and its derivatives appear only to the first power. Consequently, if $\psi_1(\mathbf{r}, t), \psi_2(\mathbf{r}, t), \ldots, \psi_n(\mathbf{r}, t)$ are each individually solutions of Eq. (2.59) *any linear combination* of these will, in turn, be a solution. Thus

$$\psi(\mathbf{r}, t) = \sum_{i=1}^{n} C_i \psi_i(\mathbf{r}, t) \qquad (7.1)$$

satisfies the wave equation, where the coefficients C_i are simply arbitrary constants. Known as the *principle of superposition*, this property suggests that the resultant disturbance at any point in a medium is the algebraic sum of the separate constituent waves (Fig. 7.1). At this time we are only interested in linear systems where the superposition principle is actually applicable. Do keep in mind however that large-amplitude waves, whether sound waves or waves on a string, can generate a nonlinear response. The focused beam of a high-intensity laser (where the electric field might

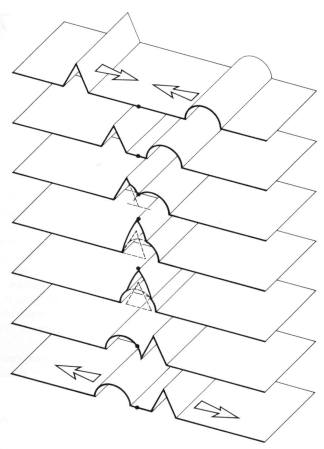

Fig. 7.1 The superposition of two disturbances.

be as high as 10^{10} V/cm) is easily capable of eliciting non-linear effects (see Chapter 14). By comparison the electric field associated with sunlight here on earth has an amplitude of only about 10 V/cm.

There are many instances in which we need not be concerned with the vector nature of light and for the present we will restrict ourselves to such cases. For example, if the light waves all propagate along the same line and share a common constant plane of vibration, they could each be described in terms of one electric-field component. These would all be either parallel or antiparallel at any instant and could thus be treated as scalars. A good deal more will be said about this point as we progress; for now, let's represent the optical disturbance as a scalar function $E(\mathbf{r}, t)$ which is a solution of Eq. (2.59). This approach leads to a simple

scalar theory which is highly useful as long as we are careful about applying it.

THE ADDITION OF WAVES OF THE SAME FREQUENCY

7.1 THE ALGEBRAIC METHOD

Recall that we can write a solution of the differential wave equation in the form

$$E(x, t) = E_0 \sin [\omega t - (kx + \varepsilon)] \qquad (7.2)$$

in which E_0 is the amplitude of the harmonic disturbance propagating along the positive x-axis. Alternatively, let

$$\alpha(x, \varepsilon) = -(kx + \varepsilon) \qquad (7.3)$$

so that

$$E(x, t) = E_0 \sin [\omega t + \alpha(x, \varepsilon)]. \qquad (7.4)$$

Suppose then that we have two such waves

$$E_1 = E_{01} \sin (\omega t + \alpha_1) \qquad (7.5)$$

and

$$E_2 = E_{02} \sin (\omega t + \alpha_2) \qquad (7.6)$$

each with the same frequency and speed, overlapping in space. The resultant disturbance is the linear superposition of these waves. Thus

$$E = E_1 + E_2$$

or, on expanding Eqs. (7.5) and (7.6)

$$E = E_{01}(\sin \omega t \cos \alpha_1 + \cos \omega t \sin \alpha_1)$$
$$+ E_{02}(\sin \omega t \cos \alpha_2 + \cos \omega t \sin \alpha_2).$$

Hence

$$E = (E_{01} \cos \alpha_1 + E_{02} \cos \alpha_2) \sin \omega t$$
$$+ (E_{01} \sin \alpha_1 + E_{02} \sin \alpha_2) \cos \omega t.$$

Since the bracketed terms are constant in time, let

$$E_0 \cos \alpha = E_{01} \cos \alpha_1 + E_{02} \cos \alpha_2 \qquad (7.7)$$

and

$$E_0 \sin \alpha = E_{01} \sin \alpha_1 + E_{02} \sin \alpha_2. \qquad (7.8)$$

This is not an obvious substitution but it will be legitimate as long as we can solve for E_0 and α. To that end, square and add Eqs. (7.7) and (7.8) to get

$$E_0^2 = E_{01}^2 + E_{02}^2 + 2E_{01}E_{02} \cos (\alpha_2 - \alpha_1) \qquad (7.9)$$

and divide Eq. (7.8) by (7.7) to get

$$\tan \alpha = \frac{E_{01} \sin \alpha_1 + E_{02} \sin \alpha_2}{E_{01} \cos \alpha_1 + E_{02} \cos \alpha_2}. \qquad (7.10)$$

The total disturbance then becomes

$$E = E_0 \cos \alpha \sin \omega t + E_0 \sin \alpha \cos \omega t$$

or

$$E = E_0 \sin (\omega t + \alpha). \qquad (7.11)$$

The composite wave (7.11) is harmonic and of the same frequency as the constituents although its amplitude and phase are different. The flux density of a light wave is proportional to its amplitude squared by way of Eq. (3.52). Thus it follows from Eq. (7.9) that the resultant flux density is not simply the sum of the component flux densities—there is an additional contribution $2E_{01}E_{02} \cos (\alpha_2 - \alpha_1)$ known as the *interference term*. The crucial factor is the difference in phase between the two *interfering* waves E_1 and E_2, $\delta \equiv (\alpha_2 - \alpha_1)$. When $\delta = 0, \pm 2\pi, \pm 4\pi, \ldots$ the resultant amplitude is a maximum while $\delta = \pm \pi, \pm 3\pi, \ldots$ yields a minimum. In the former case, the waves are said to be in phase; crest overlaps crest. In the latter instance the waves are 180° out of phase and trough overlaps crest as shown in Fig. 7.2. Realize that the *phase difference* may arise from a difference in path length traversed by the two waves as well as a difference in epoch angle, i.e.

$$\delta = (kx_1 + \varepsilon_1) - (kx_2 + \varepsilon_2) \qquad (7.12)$$

or

$$\delta = \frac{2\pi}{\lambda}(x_1 - x_2) + (\varepsilon_1 - \varepsilon_2). \qquad (7.13)$$

Here x_1 and x_2 are the distances from the sources of the two waves to the point of observation and λ is the wavelength in the pervading medium. If the waves are initially in phase at their respective emitters, then $\varepsilon_1 = \varepsilon_2$, and

$$\delta = \frac{2\pi}{\lambda}(x_1 - x_2). \qquad (7.14)$$

This would also apply to the case where two disturbances from the same source traveled different routes before arriving at the point of observation. Since $n = c/v = \lambda_0/\lambda$

$$\delta = \frac{2\pi}{\lambda_0}n(x_1 - x_2). \qquad (7.15)$$

The quantity $n(x_1 - x_2)$ is known as the *optical path difference* and will be represented by the abbreviation O.P.D. or by the symbol Λ. Bear in mind that it is possible, in more complicated situations, for each wave to travel through a number of different thicknesses of different media (see

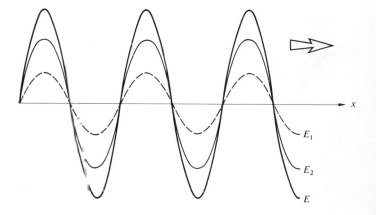

$$E = E_1 + E_2$$

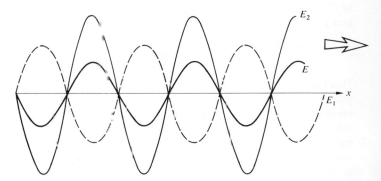

Fig. 7.2 The superposition of two harmonic waves in and out of phase.

Problem 7.4). Notice too that $\Lambda/\lambda_0 = (x_1 - x_2)/\lambda$ is the number of waves in the medium corresponding to the path difference. Since each wavelength is associated with a 2π radian phase change, $\delta = 2\pi(x_1 - x_2)/\lambda$, or more succinctly

$$\delta = k_0\Lambda, \qquad (7.16)$$

k_0 being the propagation number in vacuum, i.e. $2\pi/\lambda_0$.

Waves for which $\varepsilon_1 - \varepsilon_2$ is constant, regardless of its value, are said to be *coherent*; a situation we shall presume obtains throughout most of this discussion.

There is one special case which is of some interest and that is the superposition of the waves

$$E_1 = E_0 \sin [\omega t - k(x + \Delta x)]$$

and

$$E_2 = E_{02} \sin (\omega t - kx),$$

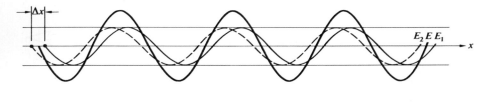

$$E = E_1 + E_2 \qquad\qquad E_2 \text{ leads } E_1 \text{ by } k\,\Delta x$$

Fig. 7.3 Waves out of phase by $k\Delta x$.

where in particular $E_{01} = E_{02}$ and $\alpha_2 - \alpha_1 = k\,\Delta x$. It is left to Problem 7.5 to show that in this case Eqs. (7.9), (7.10), and (7.11) lead to a resultant wave of

$$E = 2E_{01} \cos\left(\frac{k\,\Delta x}{2}\right) \sin\left[\omega t - k\left(x + \frac{\Delta x}{2}\right)\right]. \quad (7.17)$$

This brings out rather clearly the dominant role played by the path-length difference, Δx, especially when the waves are emitted in phase ($\varepsilon_1 = \varepsilon_2$). There are many practical instances where one arranges just these conditions as will be seen later. If $\Delta x \ll \lambda$ the resultant has an amplitude of very nearly $2E_{02}$; while if $\Delta x = \lambda/2$ it is zero. The former situation is referred to as *constructive interference* while the latter is *destructive interference* (see Fig. 7.3).

By repeated applications of the procedure used to arrive at Eq. (7.11) we can show that the *superposition of any number of coherent harmonic waves having a given frequency and traveling in the same direction leads to a harmonic wave of that same frequency* (Fig. 7.4). We happen to have chosen to represent the two waves above in

terms of sine functions but the same results would prevail had we used cosine functions. In general then, the sum of N such waves,

$$E = \sum_{i=1}^{N} E_{0i} \cos\left(\alpha_i \pm \omega t\right)$$

is given by

$$E = E_0 \cos\left(\alpha \pm \omega t\right) \quad (7.18)$$

where

$$E_0^2 = \sum_{i=1}^{N} E_{0i}^2 + 2\sum_{j>i}^{N}\sum_{i=1}^{N} E_{0i}E_{0j} \cos\left(\alpha_i - \alpha_j\right) \quad (7.19)$$

and

$$\tan\alpha = \frac{\displaystyle\sum_{i=1}^{N} E_{0i} \sin\alpha_i}{\displaystyle\sum_{i=1}^{N} E_{0i} \cos\alpha_i}. \quad (7.20)$$

Pause for a moment and satisfy yourself that these relations are indeed true. Imagine that we have a very large number of independent sources (N) where the phase angles, α_i, are

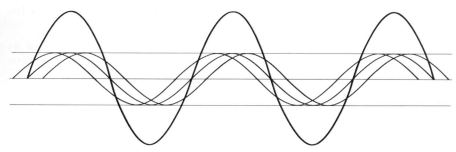

Fig. 7.4 The superposition of three harmonic waves yields a harmonic wave.

now *completely random*. In other words, the epoch angle for each source may have any value between 0 and 2π in a way totally unrelated to any other source, or the emitters may be randomly located, or both. Although applicable to light sources this arrangement can also be visualized with microwave generators or even violins. The results to be obtained apply whether all of the amplitudes are equal or not, but the former case is simpler to appreciate. Accordingly, let each amplitude in Eq. (7.19) be E_{01}. If N is sufficiently large, $\cos(\alpha_i - \alpha_j)$ will take on both positive and negative values with the same likelihood and the second term in Eq. (7.19) will approach zero. Hence

$$E_0^2 = NE_{01}^2. \tag{7.21}$$

The resultant flux density arising from N sources having random phases is given by N times the flux density of any one source. In other words, *it is determined by the sum of the individual flux densities.* For example, the light emanating from a thermal source (as distinct from a laser) is composed of radiation from a large number of atomic emitters. The waves generated by these microscopic sources have random phases and thus their individual flux densities combine in the manner considered above to form the total flux density.

Furthermore, atomic emitters vary in phase rapidly and randomly and therefore so too does the resultant total wave from the source. Two or more separate thermal sources (discharge lamps, flashlights, light bulbs, etc.) would be incoherent by virtue of these rapid variations in δ. Since flux density is proportional to the time average of E_0^2, generally taken over a relatively long interval, and since the α's are functions of time via the epoch angles, $\cos[\alpha_i(t) - \alpha_j(t)]$ would again average to zero.

At the other extreme, if the sources are coherent and in phase at the point of observation, i.e. $\alpha_i = \alpha_j$, Eq. (7.19) will become

$$E_0^2 = \sum_{i=1}^{N} E_{0i}^2 + 2 \sum_{j>i}^{N} \sum_{i=1}^{N} E_{0i}E_{0j}$$

or equivalently

$$E_0^2 = \left(\sum_{i=1}^{N} E_{0i} \right)^2. \tag{7.22}$$

Again supposing that each amplitude is E_{01}, we get

$$E_0^2 = (NE_{01})^2 = N^2 E_{01}^2. \tag{7.23}$$

In this case of in-phase coherent sources, we have a situation where the amplitudes are added first and then squared to determine the resulting flux density. The superposition of

coherent waves generally has the effect of altering the spatial distribution of the energy but not the total amount present. If there are regions where the flux density is greater than the sum of the individual flux densities there will be regions where it is less than that sum.

7.2 THE COMPLEX METHOD

It is often mathematically convenient to make use of the complex representation of trigonometric functions when dealing with the superposition of harmonic disturbances. The wave

$$E_1 = E_{01} \cos(kx \pm \omega t + \varepsilon_1)$$

or

$$E_1 = E_{01} \cos(\alpha_1 \mp \omega t)$$

can then be written as

$$E_1 = E_{01}e^{i(\alpha_1 \mp \omega t)}, \tag{7.24}$$

if we remember that we are only interested in the real part (see Section 2.4). Suppose that there are N such overlapping waves having the same frequency and traveling in the *positive x-direction*. The resultant wave is given by

$$E = E_0 e^{i(\alpha + \omega t)}$$

which is equivalent to Eq. (7.18) or upon summation of the component waves

$$E = \left[\sum_{j=1}^{N} E_{0j}e^{i\alpha_j} \right] e^{+i\omega t}. \tag{7.25}$$

The quantity

$$E_0 e^{i\alpha} = \sum_{j=1}^{N} E_{0j}e^{i\alpha_j} \tag{7.26}$$

is known as the *complex amplitude* of the composite wave and is simply the sum of the complex amplitudes of the constituents. Since

$$E_0^2 = (E_0 e^{i\alpha})(E_0 e^{i\alpha})^*, \tag{7.27}$$

we can always compute the resultant irradiance from Eqs. (7.26) and (7.27). For example, if $N = 2$,

$$E_0^2 = (E_{01}e^{i\alpha_1} + E_{02}e^{i\alpha_2})(E_{01}e^{-i\alpha_1} + E_{02}e^{-i\alpha_2}),$$

whence

$$E_0^2 = E_{01}^2 + E_{02}^2 + E_{01}E_{02}[e^{i(\alpha_1 - \alpha_2)} + e^{-i(\alpha_1 - \alpha_2)}]$$

or

$$E_0^2 = E_{01}^2 + E_{02}^2 + 2E_{01}E_{02} \cos(\alpha_1 - \alpha_2),$$

which is identical to Eq. (7.9).

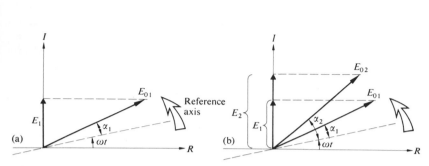

Fig. 7.5 Phasor addition.

7.3 PHASOR ADDITION

The summation described in Eq. (7.26) can be represented graphically as an addition of vectors in the complex plane (recall the Argand diagram of Fig. 2.9). In the parlance of electrical engineering the complex amplitude is known as a *phasor* and it is specified by its magnitude and phase, often written simply in the form $E_0 \angle \alpha$. The method of phasor addition to be developed now can be employed without any appreciation of its relationship to the complex-number formalism. For simplicity's sake, we will for the most part circumvent the use of that interpretation in what is to follow. Imagine then that we have a disturbance described by

$$E_1 = E_{01} \sin (\omega t + \alpha_1).$$

In Fig. 7.5(a) we represent the wave by a vector of length E_{01} rotating counterclockwise at a rate ω such that its projection on the vertical axis is $E_{01} \sin (\omega t + \alpha_1)$. Were we concerned with cosine waves we would take the projection on the horizontal axis. Incidentally, the rotating vector is, of course, a phasor $E_{01} \angle \alpha_1$ and the R and I designations signify the real and imaginary axes. Similarly, a second wave

$$E_2 = E_{02} \sin (\omega t + \alpha_2)$$

is depicted along with E_1 in Fig. 7.5(b). Their algebraic sum, $E = E_1 + E_2$, is the projection on the I-axis of the resultant phasor determined by the vector addition of the component phasors, as in Fig. 7.5(c). The law of cosines applied to the triangle of sides E_{01}, E_{02}, and E_0 yields

$$E_0^2 = E_{01}^2 + E_{02}^2 + 2E_{01}E_{02} \cos (\alpha_2 - \alpha_1),$$

where use was made of the fact that $\cos [\pi - (\alpha_2 - \alpha_1)] = -\cos (\alpha_2 - \alpha_1)$. This is identical to Eq. (7.9), as it must be. Using the same diagram observe that $\tan \alpha$ is given by Eq.

(7.10) as well. We are usually concerned with finding E_0 rather than $E(t)$ and since E_0 is unaffected by the constant revolving of all of the phasors, it will often be convenient to set $t = 0$ and thus eliminate that rotation.

Some rather elegant schemes like the *vibration curve* and the *Cornu spiral* (Chapter 10) will be predicated on the technique of phasor addition. Moreover, it is a pictorial approach and that often helps to develop insights. As a final example let's briefly examine the wave resulting from the addition of

$$E_1 = 5 \sin \omega t$$

$$E_2 = 10 \sin (\omega t + 45°)$$

$$E_3 = \sin (\omega t - 15°)$$

$$E_4 = 10 \sin (\omega t + 120°)$$

and

$$E_5 = 8 \sin (\omega t + 180°).$$

where ω is in degrees per second. The appropriate phasors $5 \angle 0°$, $10 \angle 45°$, $1 \angle -15°$, $10 \angle 120°$, and $8 \angle 180°$ are plotted in Fig. 7.6. Notice that each phase angle, whether positive or negative, is referenced to the horizontal. One need only read off $E_0 \angle \alpha$ with a scale and protractor to get $E = E_0 \sin (\omega t + \alpha)$. It is evident that this technique offers a tremendous advantage in speed and simplicity if not in accuracy.

7.4 STANDING WAVES

We saw in Chapter 2 that the general solution of the differential wave equation consisted of a sum of two traveling waves,

$$\psi(x, t) = C_1 f(x - vt) + C_2 g(x + vt). \qquad [2.12]$$

In particular let us choose to examine *two harmonic waves of* the same frequency propagating in opposite directions. A situation of practical concern arises when the incident wave is reflected backward off some sort of mirror; a rigid wall will do for sound waves or a conducting sheet for electromagnetic waves. Imagine then that an incoming wave traveling to the left,

$$E_I = E_{0I} \sin (kx + \omega t + \varepsilon_I) \qquad (7.28)$$

strikes a mirror at $x = 0$ and is reflected to the right in the form

$$E_R = E_{0R} \sin (kx - \omega t + \varepsilon_R). \qquad (7.29)$$

The composite wave in the region to the right of the mirror is $E = E_I + E_R$. We could perform the indicated summation and arrive at a general solution* much like that of Section 7.1. There are, however, some valuable physical insights to be gained by taking a slightly more restricted approach.

The epoch angle ε_I may be set to zero by merely starting our clock at a time when $E_I = E_{0I} \sin kx$. There are some qualifications determined by the physical setup which must be met by the mathematical solution and these are known formally as *boundary conditions*. For example, if we were talking about a rope having one end tied to a wall at $x = 0$, that point must always have a zero displacement. The two

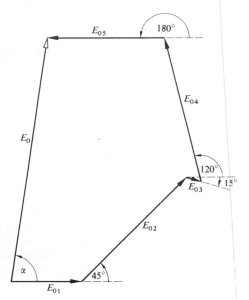

Fig. 7.6 The sum of E_1, E_2, E_3, E_4, and E_5.

---

* See for example J. M. Pearson, *A Theory of Waves*.

overlapping waves, one incident and the other reflected, would have to add in such a way as to yield a zero resultant wave at $x = 0$. Similarly, at the boundary of a perfectly conducting sheet the resultant electromagnetic wave must have a zero electric-field component parallel to the surface. Assuming $E_{0I} = E_{0R}$, the boundary conditions require that at $x = 0$, $E = 0$ and since $\varepsilon_I = 0$ it follows from Eqs. (7.28) and (7.29) that $\varepsilon_R = 0$. The composite disturbance is then

$$E = E_{0I}[\sin (kx + \omega t) + \sin (kx - \omega t)].$$

Applying the identity

$$\sin \alpha + \sin \beta = 2 \sin \tfrac{1}{2}(\alpha + \beta) \cos \tfrac{1}{2}(\alpha - \beta)$$

we obtain

$$E(x, t) = 2E_{0I} \sin kx \cos \omega t. \qquad (7.30)$$

This is the equation of a *standing* or *stationary wave* as opposed to a traveling wave. Its profile does not move through space; it is clearly not of the form $f(x \pm vt)$. At any point $x = x'$ the amplitude is a constant equal to $2E_{0I} \sin kx'$ and $E(x', t)$ varies harmonically as $\cos \omega t$. At certain points, namely $x = 0, \lambda/2, \lambda, 3\lambda/2, \ldots$, the disturbance will be zero at all times. These are known as *nodes* or *nodal points* (Fig. 7.7). Halfway between each adjacent node, that is at $x = \lambda/4, 3\lambda/4, 5\lambda/4, \ldots$, the amplitude has a maximum value of $\pm 2E_{0I}$ and these points are known as the *antinodes*. The disturbance $E(x, t)$ will be zero at all values of x whenever $\cos \omega t = 0$, i.e. when $t = (2m + 1)\tau/4$ where $m = 0, 1, 2, 3, \ldots$ and τ is the period of the component waves.

If the reflection off the mirror is not perfect, as is often the case, the composite wave will contain a traveling component along with the stationary wave. Under such conditions there will be a net transfer of energy in contrast to the pure standing wave for which there is none.

It was by measuring the distances between the nodes of standing waves that Hertz was able to determine the wavelength of the radiation in his historic experiments (see Section 3.5.4). A few years later in 1890 Otto Wiener first demonstrated the existence of standing light waves. The arrangement he used is depicted in Fig. 7.8. It shows a normally incident parallel beam of quasimonochromatic light reflecting off a front silvered mirror. A thin, transparent photographic film, less than $\lambda/20$ thick deposited on a glass plate, was inclined to the mirror at an angle of about 10^{-3} radians. In that way the film plate cut across the pattern of standing plane waves. After developing the emulsion it was found to be blackened along a series of equidistant parallel

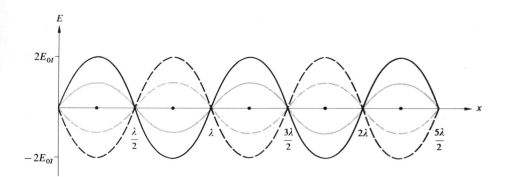

Fig. 7.7 A standing wave.

bands. These corresponded to the regions where the photographic layer had intersected the antinodal planes. Quite significantly, there was no blackening of the emulsion at the mirror's surface. It can be shown that the nodes and antinodes of the magnetic field component of an electromagnetic standing wave alternate with those of the electric field (Problem 7.8). We might suspect as much from the fact that at $t = (2m + 1)\tau/4$, $E = 0$ for all values of x and so to conserve energy it follows that $B \neq 0$. In agreement with theory, Hertz had previously (1888) determined the existence of a nodal point of the electric field at the surface of his reflector. Accordingly, Wiener could conclude that the blackened regions were associated with antinodes of the **E**-field. *Thus it is the electric field which triggers the photochemical process.* In a very similar way Drude and Nernst showed that the **E**-field is responsible for fluorescence. These observations are all quite understandable since the force exerted on an electron by the **B**-field component of an electromagnetic wave is generally negligible in comparison to that of the **E**-field. *It is for these reasons that the electric field is referred to as the optic disturbance or light field.*

THE ADDITION OF WAVES OF DIFFERENT FREQUENCY

Thus far the analysis has been restricted to the superposition of waves all having the same frequency. Yet one never actually has disturbances, of any kind, that are strictly monochromatic. It will be far more realistic, as we shall see, to speak of quasimonochromatic light which is composed of a narrow range of frequencies. The study of such light will lead us to the important concepts of bandwidth and coherence time.

The ability to effectively modulate light (Section 8.11.3) makes it possible to couple electronic and optical systems in a way which will most certainly have far-reaching effects on

the entire technology in the decades to come. Moreover, with the advent of electro-optical techniques, light is beginning to play a new and significant role as a carrier of information. This section is devoted to developing some of the mathematical ideas needed to appreciate this new emphasis.

7.5 BEATS

Consider now the composite disturbance arising from the combination of the waves

$$E_1 = E_{01} \cos (k_1 x - \omega_1 t)$$

and

$$E_2 = E_{01} \cos (k_2 x - \omega_2 t),$$

which have equal amplitudes and zero epoch angles. The net wave

$$E = E_{01}[\cos (k_1 x - \omega_1 t) + \cos (k_2 x - \omega_2 t)]$$

can be reformulated as

$$E = 2E_{01} \cos \tfrac{1}{2}[(k_1 + k_2)x - (\omega_1 + \omega_2)t]$$
$$\times \cos \tfrac{1}{2}[(k_1 - k_2)x - (\omega_1 - \omega_2)t],$$

using the identity

$$\cos \alpha + \cos \beta = 2 \cos \tfrac{1}{2}(\alpha + \beta) \cos \tfrac{1}{2}(\alpha - \beta).$$

We now define the quantities $\bar{\omega}$ and $\bar{k}$, which are the *average angular frequency* and *average propagation number*, respectively. Similarly the quantities ω_m and k_m are designated as the *modulation frequency and modulation propagation number*. Accordingly, let

$$\bar{\omega} \equiv \tfrac{1}{2}(\omega_1 + \omega_2) \qquad \omega_m \equiv \tfrac{1}{2}(\omega_1 - \omega_2) \qquad (7.31)$$

and

$$\bar{k} \equiv \tfrac{1}{2}(k_1 + k_2) \qquad k_m \equiv \tfrac{1}{2}(k_1 - k_2); \qquad (7.32)$$

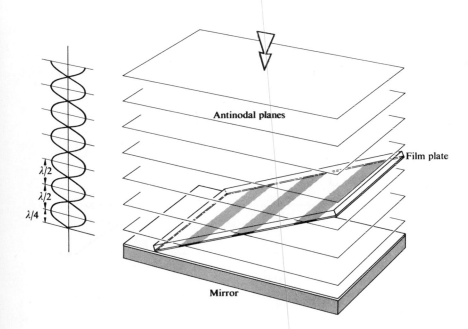

Fig. 7.8 Wiener's experiment.

thus

$$E = 2E_{01} \cos (k_m x - \omega_m t) \cos (\bar{k} x - \bar{\omega} t). \qquad (7.33)$$

The total disturbance may be regarded as a traveling wave of frequency $\bar{\omega}$ having a time-varying or modulated amplitude $E_0(x, t)$ such that

$$E(x, t) = E_0(x, t) \cos (\bar{k} x - \bar{\omega} t), \qquad (7.34)$$

where

$$E_0(x, t) = 2E_{01} \cos (k_m x - \omega_m t). \qquad (7.35)$$

In applications of interest here, ω_1 and ω_2 will always be rather large. In addition, if they are comparable to each other, $\omega_1 \approx \omega_2$, then $\bar{\omega} \gg \omega_m$ and $E_0(x, t)$ would change slowly whereas $E(x, t)$ would vary quite rapidly (Fig. 7.9). The irradiance is proportional to

$$E_0^2(x, t) = 4E_{01}^2 \cos^2 (k_m x - \omega_m t)$$

or

$$E_0^2(x, t) = 2E_{01}^2 [1 + \cos 2(k_m x - \omega_m t)].$$

Notice that $E_0^2(x, t)$ oscillates about a value of $2E_{01}^2$ with a frequency of $2\omega_m$ or simply $(\omega_1 - \omega_2)$ which is known as the *beat frequency*. In other words, the modulation frequency, corresponding to the envelope of the curve, is half of the beat frequency. It might seem from Fig. 7.9(b) that because the wave form between two consecutive nodes repeats itself that that distance should be the wavelength of the envelope, but this is not generally the case.

Beats were first observed using light in 1955 by Forrester, Gudmundsen and Johnson.* In order to obtain two waves of slightly different frequency they used the Zeeman effect. When the atoms of a discharge lamp, in this case mercury, are subjected to a magnetic field, their energy levels split. As a result the emitted light contains two frequency components v_1 and v_2 which differ in proportion to the magnitude of the applied field. When these components are recombined at the surface of a photoelectric mixing tube the beat frequency, $v_1 - v_2$, is generated. Specifically, the field was adjusted so that $v_1 - v_2 = 10^{10}$ Hz which conveniently corresponds to a 3 cm microwave signal. The recorded photoelectric current had the same form as the $E_0^2(x)$ curve of Fig. 7.9(d).

The advent of the laser has since made the observation of beats using light considerably easier. Even a beat frequency of a few Hz out of 10^{14} Hz can be seen as a variation in phototube current. The observation of beats now represents a particularly sensitive and fairly simple means of detecting small frequency differences. For example, a modern version of the famous Michelson–Morley experiment that beats two infrared laser beams will be considered in Section 9.10.3. The ring laser (Section 9.10.6), functioning

---

* A. T. Forrester, R. A. Gudmundsen, and P. O. Johnson, "Photoelectric Mixing of Incoherent Light," *Phys. Rev.* **99**, 1691 (1955).

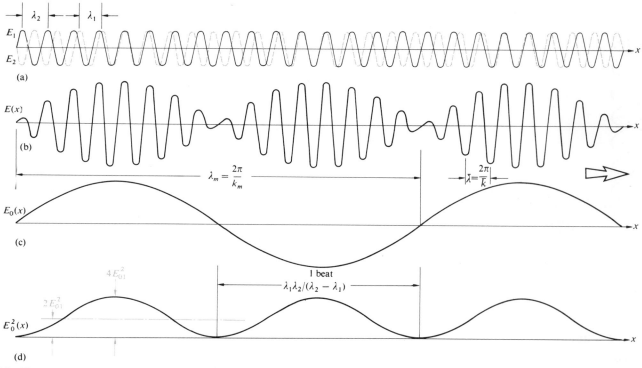

Fig. 7.9 The superposition of two harmonic waves of different frequency.

as a gyroscope, utilizes beats to measure frequency differences induced as a result of the rotation of the system. The Doppler effect, which accounts for the frequency shift when light is reflected off a moving surface, provides another series of applications of beats. By scattering light off a target, whether solid, liquid or even gaseous, and then beating the original and reflected waves, we get a precise measure of the target speed. In very much the same way on an atomic scale, laser light will shift in phase upon interacting with sound waves moving in a material (this phenomenon is called Brillouin scattering). Thus $2\omega_m$ becomes a measure of the speed of sound in the medium.

7.6 GROUP VELOCITY

The disturbance examined in the previous section,

$$E(x, t) = E_0(x, t) \cos(\bar{k}x - \bar{\omega}t),   [7.34]$$

consists of a high-frequency ($\bar{\omega}$) *carrier wave, amplitude modulated* by a cosine function. Suppose, for a moment,

that the wave in Fig. 7.9(b) were not modulated, i.e. $E_0 = $ constant. Each small peak in the carrier would travel to the right with the usual phase velocity. In other words

$$v = -\frac{(\partial \varphi/\partial t)_x}{(\partial \varphi/\partial x)_t}.   [2.32]$$

From Eq. (7.34) the phase is given by $\varphi = (\bar{k}x - \bar{\omega}t)$ and hence

$$v = \bar{\omega}/\bar{k}.   (7.36)$$

Clearly, this is the phase velocity whether the carrier is modulated or not. In the former case the peaks simply change amplitude periodically as they stream along.

Evidently, there is another motion to be concerned with and that is the propagation of the modulation envelope. Return to Fig. 7.9(a) and suppose that the constituent waves, $E_1(x, t)$ and $E_2(x, t)$, advance with the same speed, $v_1 = v_2$. Imagine if you will the two harmonic functions having different wavelengths and frequencies drawn on separate sheets of clear plastic. When these are overlaid in some way [as in Fig. 7.9(a)] the resultant is a stationary beat pattern. If the sheets are both moved to the right at the same speed so as to resemble traveling waves, the beats will obviously move with that same speed. The rate at which the modulation envelope advances is known as the *group velocity* or symbolically as v_g. In this instance the group velocity equals the phase velocity of the carrier (the average speed, $\bar{\omega}/\bar{k}$). In other words, $v_g = v = v_1 = v_2$. This applies specifically to *nondispersive media* in which the phase velocity is independent of wavelength so that the two waves could have the same speed. For a more generally applicable solution examine the expression for the modulation envelope:

$$E_0(x, t) = 2E_{01} \cos (k_m x - \omega_m t). \qquad [7.35]$$

The speed with which that wave moves is again given by Eq. (2.32) where now we can forget the carrier wave. The modulation therefore advances at a rate dependent on the phase of the envelope $(k_m x - \omega_m t)$ and so

$$v_g = \frac{\omega_m}{k_m}$$

or

$$v_g = \frac{\omega_1 - \omega_2}{k_1 - k_2} = \frac{\Delta \omega}{\Delta k}.$$

Realize however that ω may be dependent on λ or equivalently on k. The particular function $\omega = \omega(k)$ is called a *dispersion relation*. When the frequency range $\Delta\omega$, centered about $\bar{\omega}$, is small $\Delta\omega/\Delta k$ is approximately equal to the derivative of the dispersion relation, that is

$$v_g = \frac{d\omega}{dk}. \qquad (7.37)$$

The modulation or signal propagates at a speed v_g which may be greater, equal to, or less than v, the phase velocity of the carrier. Equation (7.37) is quite general and will be true, as well, for any group of overlapping waves as long as their frequency range is narrow.

Since $\omega = kv$, Eq. (7.37) yields

$$v_g = v + k\frac{dv}{dk}. \qquad (7.38)$$

As a consequence, in nondispersive media where v is independent of λ, $dv/dk = 0$ and $v_g = v$. Specifically in vacuum $\omega = kc$, $v = c$, and $v_g = c$. In dispersive media ($v_1 \neq v_2$ as in Fig. 7.10) where $n(k)$ is known, $\omega = kc/n$ and it is useful to reformulate v_g as

$$v_g = \frac{c}{n} - \frac{kc}{n^2}\frac{dn}{dk},$$

or

$$v_g = v\left(1 - \frac{k}{n}\frac{dn}{dk}\right). \qquad (7.39)$$

For optical media, in regions of normal dispersion, the refractive index increases with frequency ($dn/dk > 0$) and as a result $v_g < v$. Clearly, one should also define a *group index of refraction*

$$n_g \equiv c/v_g \qquad (7.40)$$

which must be carefully distinguished from n. A. A. Michelson in 1885 measured n_g in carbon disulfide using pulses of white light and got 1.758 in comparison to $n = 1.635$.

The special theory of relativity makes it quite clear that there are no circumstances under which a signal can propagate at a speed greater than c. Yet we have already seen that under certain circumstances (Section 3.3.1) the phase velocity can exceed c. The contradiction is only an apparent one arising from the fact that while a monochromatic wave can indeed have a speed in excess of c, it cannot convey information. In contrast, a signal in the form of any modulated wave will propagate at the group velocity which is always less than c in normally dispersive media.*

7.7 ANHARMONIC PERIODIC WAVES—FOURIER ANALYSIS

Figure 7.11 depicts a disturbance which arises from the superposition of two harmonic functions having different amplitudes and wavelengths. Notice that something rather curious has taken place—the composite disturbance is *anharmonic*, i.e. it is not sinusoidal. As we have already said, and will certainly say again, *purely sinusoidal* waves have no actual physical existence. This fact emphasizes the practical significance of anharmonic disturbances and is the motivation for our present concern with them. Returning to Fig. 7.11, one gets the feeling that by using a number of sinusoidal functions whose amplitudes, wavelengths, and relative phases have been judiciously selected, it would be possible to synthesize some rather interesting wave profiles. An exceptionally beautiful mathematical technique for doing precisely this was devised by the French physicist Jean

---

* In regions of anomalous dispersion (Section 3.3.1) where $dn/dk < 0$, v_g may be greater than c. Here, however, the signal propagates at yet a different speed, known as the *signal velocity* v_s. Thus $v_s = v_g$ except in a resonance absorption band. In all cases v_s corresponds to the velocity of transfer of energy and never exceeds c.

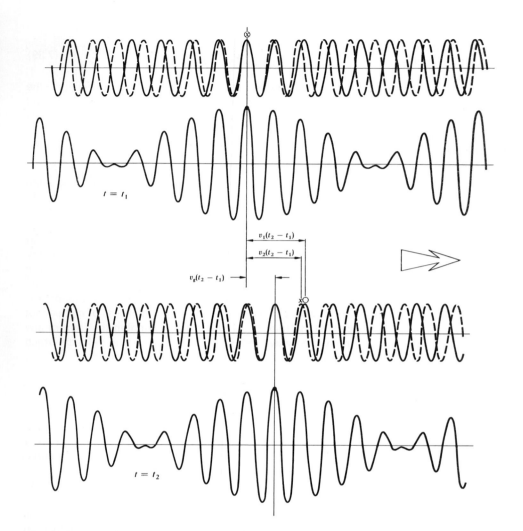

Fig. 7.10 Group and phase velocities.

Baptiste Joseph, Baron de Fourier (1768–1830). This theory is predicated on what has come to be known as *Fourier's theorem*, which states that *a function f(x), having a spatial period λ, can be synthesized by a sum of harmonic functions whose wavelengths are integral submultiples of λ (that is, λ, λ/2, λ/3, etc.).* This Fourier-series representation has the mathematical form

$$f(x) = C_0 + C_1 \cos\left(\frac{2\pi}{\lambda}x + \varepsilon_1\right) + C_2 \cos\left(\frac{2\pi}{\lambda/2}x + \varepsilon_2\right) + \cdots$$

(7.41)

where the C-values are constants and, of course, the profile $f(x)$ may correspond to a traveling wave $f(x - vt)$. To get

some sense of how this scheme works, observe that while C_0 by itself is obviously a poor substitute for the original function, it will be appropriate at those few points where it crosses the $f(x)$ curve. In the same way, adding on the next term improves things a bit since the function

$$[C_0 + C_1 \cos (2\pi x/\lambda + \varepsilon_1)]$$

will be chosen so as to cross the $f(x)$ curve even more frequently. If the synthesized function [the right-hand side of Eq. (7.41)] is constituted of an infinite number of terms, selected in order that they intersect the anharmonic function at an infinite number of points, the series will presumably be identical to $f(x)$.

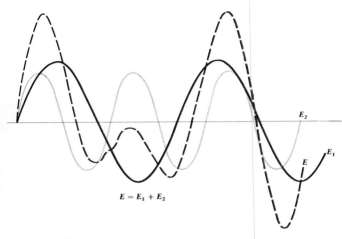

Fig. 7.11 The superposition of two harmonic waves of different frequency.

It is usually more convenient to reformulate Eq. (7.41) by making use of the trigonometric identity

$$C_m \cos(mkx + \varepsilon_m) = A_m \cos mkx + B_m \sin mkx,$$

where $k = 2\pi/\lambda$, λ being the wavelength of $f(x)$, $A_m = C_m \cos \varepsilon_m$ and $B_m = -C_m \sin \varepsilon_m$. Thus

$$f(x) = \frac{A_0}{2} + \sum_{m=1}^{\infty} A_m \cos mkx + \sum_{m=1}^{\infty} B_m \sin mkx. \quad (7.42)$$

The first term is written as $A_0/2$ because of the mathematical simplification it will lead to later on. The process of determining the coefficients A_0, A_m, and B_m for a specific periodic function $f(x)$ is referred to as *Fourier analysis*. We'll spend a moment now deriving a set of equations for these coefficients which can be used henceforth. To that end, integrate both sides of Eq. (7.42) over any spatial interval equal to λ, e.g. from 0 to λ or $-\lambda/2$ to $+\lambda/2$ or more generally from x' to $x' + \lambda$. Since over any such interval

$$\int_0^\lambda \sin mkx \, dx = \int_0^\lambda \cos mkx \, dx = 0$$

there is only one nonzero term to be evaluated, namely

$$\int_0^\lambda f(x) \, dx = \int_0^\lambda \frac{A_0}{2} \, dx = A_0 \frac{\lambda}{2},$$

and thus

$$A_0 = \frac{2}{\lambda} \int_0^\lambda f(x) \, dx. \quad (7.43)$$

In order to find A_m and B_m we will make use of the *orthogonality of sinusoidal functions* (Problem 7.16), i.e. the fact that

$$\int_0^\lambda \sin akx \cos bkx \, dx = 0 \quad (7.44)$$

$$\int_0^\lambda \cos akx \cos bkx \, dx = \frac{\lambda}{2} \delta_{ab} \quad (7.45)$$

$$\int_0^\lambda \sin akx \sin bkx \, dx = \frac{\lambda}{2} \delta_{ab}, \quad (7.46)$$

where a and b are nonzero positive integers and δ_{ab}, known as the *Kronecker delta*, is a shorthand notation equal to zero when $a \neq b$ and 1 when $a = b$. To find A_m we now multiply both sides of Eq. (7.42) by $\cos \ell kx$, ℓ being a positive integer, and then integrate over a spatial period. Only one term is nonvanishing and that is the single contribution in the second sum which corresponds to $\ell = m$, in which case

$$\int_0^\lambda f(x) \cos mkx \, dx = \int_0^\lambda A_m \cos^2 mkx \, dx = \frac{\lambda}{2} A_m.$$

Thus

$$A_m = \frac{2}{\lambda} \int_0^\lambda f(x) \cos mkx \, dx. \quad (7.47)$$

This expression can be used to evaluate A_m for *all values of m including m = 0* as is evident from a comparison of Eqs. (7.43) and (7.47). Similarly, multiplying Eq. (7.42) by $\sin \ell kx$ and integrating leads to

$$B_m = \frac{2}{\lambda} \int_0^\lambda f(x) \sin mkx \, dx. \quad (7.48)$$

In summary then, a periodic function $f(x)$ can be represented as a Fourier series

$$f(x) = \frac{A_0}{2} + \sum_{m=1}^{\infty} A_m \cos mkx + \sum_{m=1}^{\infty} B_m \sin mkx \quad [7.42]$$

where, knowing $f(x)$, the coefficients are computed using

$$A_m = \frac{2}{\lambda} \int_0^\lambda f(x) \cos mkx \, dx \quad [7.47]$$

and

$$B_m = \frac{2}{\lambda} \int_0^\lambda f(x) \sin mkx \, dx. \quad [7.48]$$

Be aware that there are some mathematical subtleties related to the convergence of the series and the number of singularities in $f(x)$, but we need not be concerned with these matters here.

There are certain symmetry conditions that are well worth recognizing because they lead to some effort-saving computational shortcuts. Thus if a function $f(x)$ is *even*, i.e. if $f(-x) = f(x)$, or equivalently if it is symmetric about $x = 0$, its Fourier series will contain only cosine terms ($B_m = 0$ for all m) which are themselves even functions. Likewise *odd* functions which are not symmetric about $x = 0$, i.e. ones where $f(-x) = -f(x)$, will have series expansions containing only sine functions ($A_m = 0$ for all m). In either case, one need not bother to calculate both sets of coefficients. This is particularly helpful when the location of the origin ($x = 0$) is arbitrary and we can choose it so as to make life as simple as possible. Nonetheless, keep in mind that many common functions are neither odd nor even (for example, e^x).

As an example of the technique, let's compute the Fourier series which corresponds to a square wave. We select the location of the origin as shown in Fig. 7.12 and so

$$f(x) = \begin{cases} +1 \text{ when } 0 < x < \lambda/2 \\ -1 \text{ when } \lambda/2 < x < \lambda. \end{cases}$$

Since $f(x)$ is odd $A_m = 0$, while

$$B_m = \frac{2}{\lambda} \int_0^{\lambda/2} (+1) \sin mkx\, dx + \frac{2}{\lambda} \int_{\lambda/2}^{\lambda} (-1) \sin mkx\, dx$$

thus

$$B_m = \frac{1}{m\pi} \left[-\cos mkx \right]_0^{\lambda/2} + \frac{1}{m\pi} \left[\cos mkx \right]_{\lambda/2}^{\lambda}.$$

Remembering that $k = 2\pi/\lambda$, this becomes

$$B_m = \frac{2}{m\pi}(1 - \cos m\pi).$$

The Fourier coefficients are therefore

$$B_1 = \frac{4}{\pi}, B_2 = 0, B_3 = \frac{4}{3\pi}, B_4 = 0, B_5 = \frac{4}{5\pi}, \cdots$$

and the required series is simply

$$f(x) = \frac{4}{\pi}(\sin kx + \tfrac{1}{3}\sin 3kx + \tfrac{1}{5}\sin 5kx + \cdots). \quad (7.49)$$

Figure 7.13 is a plot of a few partial sums of the series as the number of terms increases. We could pass over to the time domain to find $f(t)$ by just changing kx to ωt. Suppose that

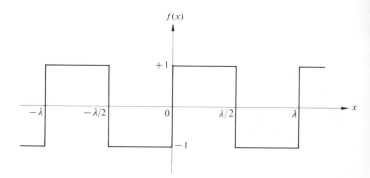

Fig. 7.12 A periodic square wave.

we have three ordinary electronic oscillators whose output voltages vary sinusoidally and are controllable in both frequency and amplitude. If these are connected in series with their frequencies set at ω, 3ω and 5ω and the total signal is examined on an oscilloscope, we could synthesize any of these curves, e.g. Fig. 7.13(d). Similarly we might simultaneously strike three keys on an appropriately tuned piano with just the correct force on each to create a chord, or composite sound wave, having Fig. 7.13(c) as its profile. Curiously enough, the human ear–brain audio system is capable of Fourier-analyzing a simple composite wave into its harmonic constituents—presumably there are people who could even name each note in the chord.

Earlier we postponed any detailed consideration of anharmonic periodic functions, such as those of Fig. 2.6, and restricted our analysis to purely sinusoidal waves. We now have a cogent rationale for having done so. From now on we can envision this kind of disturbance as a superposition of harmonic constituents of different frequencies whose individual behavior can be studied separately. Accordingly, we may write

$$f(x \pm vt) = \frac{A_0}{2} + \sum_{m=1}^{\infty} A_m \cos mk(x \pm vt)$$
$$+ \sum_{m=1}^{\infty} B_m \sin mk(x \pm vt) \qquad (7.50)$$

or equivalently

$$f(x \pm vt) = \sum_{m=0}^{\infty} C_m \cos [mk(x \pm vt) + \varepsilon_m] \qquad (7.51)$$

for any such *anharmonic periodic wave*.

As a last example let's now analyze the square wave of Fig. 7.14 into its Fourier components. We notice that with

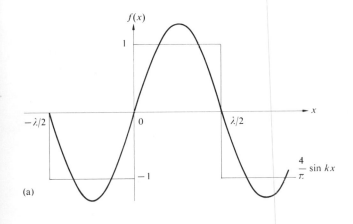

(a) $\dfrac{4}{\pi}\sin kx$

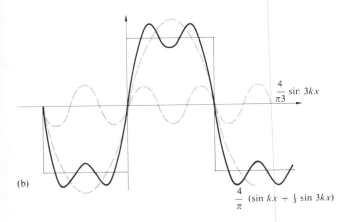

(b) $\dfrac{4}{\pi}\left(\sin kx + \tfrac{1}{3}\sin 3kx\right)$

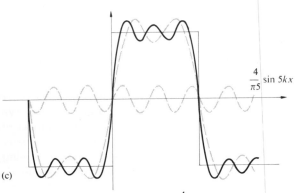

(c) $\dfrac{4}{\pi}\left(\sin kx + \tfrac{1}{3}\sin 3kx + \tfrac{1}{5}\sin 5kx\right)$

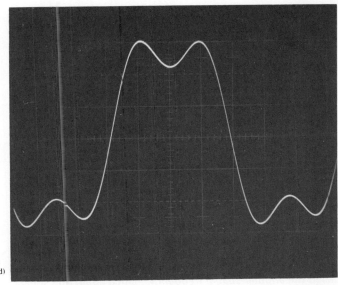

(d)

Fig. 7.13 Synthesis of a periodic square wave. [Photo by E.H.]

the origin chosen as shown, the function is even and all of the B_m-terms are zero. The appropriate Fourier coefficients (Problem 7.17) are then

$$A_0 = \frac{4}{a} \quad \text{and} \quad A_m = \frac{4}{a}\left(\frac{\sin m2\pi/a}{m2\pi/a}\right). \quad (7.52)$$

Unlike the previous function we now have a nonzero value of A_0. You might have already noticed that $A_0/2$ is actually the *mean value* of $f(x)$ and since the curve lies completely above the axis it will clearly not be zero.

The expression $(\sin u)/u$ arises so frequently in optics that it is given the special name sinc u and its values are listed in Table 1 (p. 513). Since the limit of sinc u as u goes to zero is 1, A_m can represent all the coefficients if we let $m = 0, 1, 2, \ldots$.

The form we are using is rather general inasmuch as the width of the square peak, $2(\lambda/a)$, can be any fraction of the total wavelength, depending on a. The Fourier series is then

$$f(x) = \frac{2}{a} + \sum_{m=1}^{\infty} \frac{4}{a}\left(\frac{\sin m2\pi/a}{m2\pi/a}\right)\cos mkx. \quad (7.53)$$

If we were synthesizing the corresponding function of time, $f(t)$, having a square peak of width $2(\tau/a)$, the same expression (7.53) would apply where kx was simply replaced by ωt. Here ω is the *angular temporal frequency* of the periodic function $f(t)$ and is known as the *fundamental*. It is the

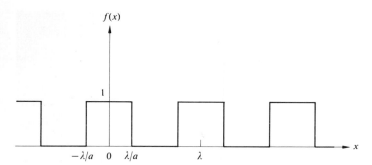

Fig. 7.14 A periodic anharmonic wave.

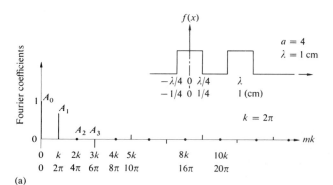

(a)

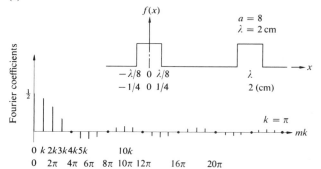

(b)

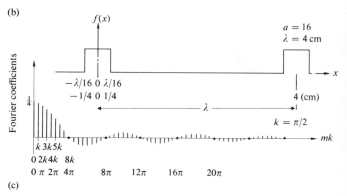

(c)

Fig. 7.15 The square pulse as a limiting case. The negative coefficients correspond to a phase shift of π radians.

lowest frequency of the cosine term and arises when $m = 1$. Frequencies of 2ω, 3ω, 4ω, etc. are known as *harmonics* of the fundamental and are associated, of course, with $m = 2$, 3, 4, etc. In much the same way, since λ is the *spatial period*, $\varkappa \equiv 1/\lambda$ is the *spatial frequency* and $k = 2\pi\varkappa$ might be called the *angular spatial frequency*. Once again one speaks of the harmonics of frequency $2k$, $3k$, $4k$, etc. where these are spatial alternations. Evidently, the dimensions of $\varkappa$ are cycles per unit length (e.g. cycles per mm or possibly just cm^{-1}) while those of k are radians per unit length.

Suppose that we set $a = 4$, or in other words, we cause the square peak to have a width of $\lambda/2$. In that instance

$$f(x) = \frac{1}{2} + \frac{2}{\pi}(\cos kx - \tfrac{1}{3}\cos 3kx + \tfrac{1}{5}\cos 5kx - \cdots). \quad (7.54)$$

As a matter of fact if the graph of the function $f(x)$ is such that a horizontal line could divide it into equally shaped segments, above and below that line, the Fourier series will consist of only odd harmonics. Were we to plot the curve representing the partial sum of the terms through $m = 9$, it would quite closely resemble the square wave. In contrast, if the width of the peak is reduced the number of terms in the series needed to produce the same general resemblance to $f(x)$ will be increased. This can be appreciated by examining the ratio

$$\frac{A_m}{A_1} = \frac{\sin m2\pi/a}{m\sin 2\pi/a}. \quad (7.55)$$

Observe that for $a = 4$, the ninth term, i.e. $m = 9$, is fairly small, $A_9 \approx 10\%\, A_1$. In comparison, for a peak one hundred times narrower viz. $a = 400$, $A_9 \approx 99\%\, A_1$. Similarly, while it takes terms through $m = 4$ to duplicate the curve of Fig. 7.13(b) when $a = 4$, it will take up to $m = 8$ to produce roughly the equivalent profile when $a = 8$. Making

the peak narrower has the effect of introducing higher-order harmonics which in turn have smaller wavelengths. We might guess then that it is not so much the total number of terms in the series which is of prime importance but rather the relative dimensions of the smallest features being reproduced and the corresponding wavelengths available.[*]

---

[*] Evidently one is not going to be able to build a castle of blocks unless the blocks are a good deal smaller than the castle.

If there are fine details in the profile the series must contain comparatively short-wavelength (or in the time domain, short-period) contributions.

7.8 NONPERIODIC WAVES—FOURIER INTEGRALS

Return to Fig. 7.14 and imagine that we keep the width of the square peak constant while λ is made to increase without limit. Apparently as λ approaches infinity the resulting function will no longer appear periodic. We would then have one single square pulse, the adjacent peaks having moved off to infinity. This suggests a possible way of generalizing the method of Fourier series to include nonperiodic functions. As we shall see these are of the greatest practical interest in physics, particularly in optics and quantum mechanics.

To see how this can be accomplished, let's initially set $a = 4$ and choose some value of λ: anything will do, say $\lambda = 1$ cm. The peak then has a width of $\frac{1}{2}$ cm, i.e. $2(\lambda/a)$ centered at $x = 0$, as illustrated in Fig. 7.15(a). The importance of each particular frequency, mk, can be appreciated by examining the value of the corresponding Fourier coefficient, in this case A_m. The coefficients may be thought of as weighting factors which appropriately emphasize the various harmonics. Figure 7.15(a) contains a plot of a number of values of A_m (where $m = 0, 1, 2, \ldots$) versus mk for the foregoing square wave—such a curve is known as the *spatial frequency spectrum*. We can regard A_m as a function, $A(mk)$, of mk which may be nonzero only at values of $m = 0, 1, 2, \ldots$. If the quantity a is now made equal to 8 while λ is increased to 2 cm, the peak width will be completely unaffected. The only alteration is a doubling of the space between peaks. Yet a very interesting change in the spatial frequency spectrum is evident in Fig. 7.15(b). Note that the density of components along the mk-axis has increased markedly. Nonetheless, $A(mk)$ is still zero when $mk = 4\pi, 8\pi, 12\pi, \ldots$ but since k is now π rather than 2π, there will be more terms between these zero points. Finally let $a = 16$ while λ is increased to 4 cm. Again the individual peaks are unaltered in shape but the terms in the frequency spectrum are now even more densely packed. In effect, the pulse as compared to λ is getting smaller and smaller, thereby requiring higher frequencies to synthesize it. Observe that the envelope of the curve which was barely discernible in Fig. 7.15(a) is quite evident in Fig. 7.15(c). In fact, the envelope is identical in each case except for a scale factor. It is determined only by the shape of the original signal and will be quite different for other configurations. We can conclude that as λ increases and the function takes on the

appearance of a single square pulse, the space between each of the $A(mk)$ contributions in the spectrum will decrease. The discrete spectral lines, while decreasing in amplitude, will gradually merge, becoming individually unresolvable. In other words, in the limit as λ approaches ∞ the spectral lines will become infinitely close to each other. As k becomes extremely small, m must consequently become exceedingly large if mk is to be at all appreciable. And so, changing notation we replace mk, the angular frequency of the harmonics, by k_m. Although comprised of discrete terms, in the limit k_m will be transformed into k, i.e. a continuous frequency distribution. The function $A(k_m)$ in the limit will become the envelope shown in Fig. 7.15. It is obviously no longer meaningful to talk about the fundamental frequency and its harmonics. The pulse being synthesized, $f(x)$, has no apparent fundamental frequency.

Recall that an integral is actually the limit of a sum as the number of elements goes to infinity and their size tends to zero. Thus it should not be surprising that the *Fourier series* must be replaced by the so-called *Fourier integral* as λ goes to infinity. That integral, which we state here without proof, is

$$f(x) = \frac{1}{\pi}\left[\int_0^\infty A(k)\cos kx\,dk + \int_0^\infty B(k)\sin kx\,dk\right] \quad (7.56)$$

provided that

$$A(k) = \int_{-\infty}^\infty f(x)\cos kx\,dx \quad \text{and} \quad B(k) = \int_{-\infty}^\infty f(x)\sin kx\,dx. \quad (7.57)$$

The similarity with the series representation should be obvious. The quantities $A(k)$ and $B(k)$ are interpreted as the amplitudes of the sine and cosine contributions in the range of angular spatial frequency between k and $k + dk$. They are generally spoken of as the *Fourier cosine and sine transforms*, respectively. In the foregoing example of a square pulse, it is the cosine transform, $A(k)$, which will be found to correspond to the envelope of Fig. 7.15.

7.9 PULSES AND WAVE PACKETS

Let's now determine the Fourier-integral representation of the square pulse of Fig. 7.16 which is described by the function

$$f(x) = \begin{cases} E_0 & \text{when } |x| \le L/2 \\ 0 & \text{when } |x| > L/2. \end{cases}$$

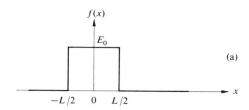

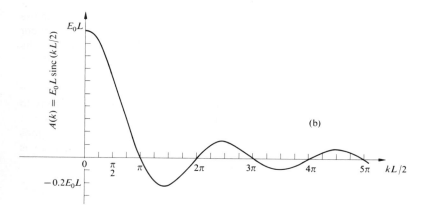

Fig. 7.16 The square pulse and its transform.

Since $f(x)$ is an even function, the sine transform, $B(k)$, will be found to be zero (7.57) while

$$A(k) = \int_{-\infty}^{\infty} f(x) \cos kx \, dx = \int_{-L/2}^{+L/2} E_0 \cos kx \, dx.$$

Hence

$$A(k) = \frac{E_0}{k} \sin kx \Big|_{-L/2}^{+L/2} = \frac{2E_0}{k} \sin kL/2.$$

And so, multiplying numerator and denominator by L and rearranging terms, we have

$$A(k) = E_0 L \frac{\sin kL/2}{kL/2}$$

or equivalently

$$A(k) = E_0 L \text{ sinc } (kL/2). \tag{7.58}$$

The Fourier transform of the square pulse is plotted in Fig. 7.16(b) and should be compared with the envelope of Fig. 7.15. Realize that as L increases, the spacing between successive zeroes of $A(k)$ decreases and vice versa. Moreover, when $k = 0$ it follows from Eq. (7.58) that $A(0) = E_0 L$. We will spend a good deal more time looking specifically at

Fourier transforms and their applications to optics in Chapter 11. Accordingly, we defer some rather interesting observations until then.

It is a simple matter to write out the integral representation of $f(x)$ using Eq. (7.56):

$$f(x) = \frac{1}{\pi} \int_0^{\infty} E_0 L \text{ sinc } (kL/2) \cos kx \, dk. \tag{7.59}$$

An evaluation of this integral is left for Problem 7.18.

In the past when we talked about monochromatic waves, we pointed out that they were in fact fictitious, at least physically. There will always have been some point in time when the generator, however perfect, was turned on. Figure 7.17 depicts a somewhat idealized harmonic pulse corresponding to the function

$$E(x) = \begin{cases} E_0 \cos k_p x & \text{when } -L \leq x \leq L \\ 0 & \text{when } |x| > L. \end{cases}$$

We chose to work in the space domain but could certainly have envisioned the disturbance as a function of time. Effectively we are examining the spatial profile of the wave $E(x - vt)$ at $t = 0$ rather than the temporal profile at $x = 0$.

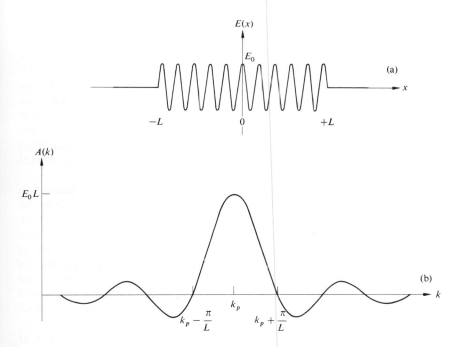

Fig. 7.17 A finite cosine wave train and its transform.

The spatial frequency k_p is that of the harmonic region of the pulse itself. Proceeding with the analysis, we note that $E(x)$ is an even function, consequently $B(k) = 0$ and

$$A(k) = \int_{-L}^{+L} E_0 \cos k_p x \cos kx \, dx.$$

This is identical to

$$A(k) = \int_{-L}^{+L} E_0 \tfrac{1}{2}[\cos(k_p + k)x + \cos(k_p - k)x] \, dx,$$

which integrates to

$$A(k) = E_0 L \left[\frac{\sin(k_p + k)L}{(k_p + k)L} + \frac{\sin(k_p - k)L}{(k_p - k)L} \right]$$

or, if you like,

$$A(k) = E_0 L[\text{sinc}\,(k_p + k)L + \text{sinc}\,(k_p - k)L]. \quad (7.60)$$

When there are many waves in the train ($\lambda_p \ll L$), $k_p L \gg 2\pi$. Thus $(k_p + k)L \gg 2\pi$ and therefore sinc $(k_p + k)L$ is down to fairly small values. Contrastingly, when $k_p = k$ the second sinc function in the brackets has its maximum value of 1. Consequently we may neglect the first sinc in this particular case and write the transforms as

$$A(k) = E_0 L \,\text{sinc}\,(k_p - k)L \quad (7.61)$$

[Fig. 7.17(b)]. Despite the fact that the wave train is very long, since it is not infinitely long it must be synthesized out of a continuous range of spatial frequencies. And so it can be thought of as the composite of an infinite ensemble of harmonic waves. In that context one speaks of such pulses as *wave packets* or *wave groups*. As we might have expected, the dominant contribution is associated with $k = k_p$. Had the analysis been carried out in the time domain the same results would obtain where the transform was centered about the temporal angular frequency ω_p. Quite clearly as the wave train becomes infinitely long (i.e. $L \to \infty$), its frequency spectrum shrinks and the curve of Fig. 7.17(b) closes down to a single tall spike at k_p (or ω_p). This obviously is the limiting case of the idealized monochromatic wave.

Since we may think of $A(k)$ as the amplitude of the contributions to $E(x)$ in the range k to $k + dk$, $A^2(k)$ must be related to the energy of the wave in that range (Problem 7.19). We'll come back to this point in Chapter 11 when we consider the *power spectrum*. For the moment, merely observe [Fig. 7.17(b)] that most of the energy is carried in the spatial frequency range from $k_p - \pi/L$ to $k_p + \pi/L$ extending between the minima on either side of the central peak. Increasing the length of the wave train causes the energy of the wave to become concentrated in an ever-narrowing range of k about k_p.

The wave packet in the time domain, that is

$$E(t) = \begin{cases} E_0 \cos \omega_p t & \text{when } -T \le t \le T \\ 0 & \text{when } |t| > T \end{cases}$$

has the transform

$$A(\omega) = E_0 T \operatorname{sinc}(\omega_p - \omega)T, \qquad (7.62)$$

where ω and k are related by the phase velocity. The frequency spectrum, except for the notational change from k to ω and L to T, is identical to that of Fig. 7.17(b). For the particular wave packet being studied the range of frequencies (ω or k) comprising the transform is certainly not finite. Yet if we were to speak of the *width* of the transform ($\Delta\omega$ or Δk), Fig. 7.17(b) rather suggests that we use $\Delta k = 2\pi/L$ or $\Delta\omega = 2\pi/T$. In contrast, the spatial or temporal extent of the pulse is quite unambiguous at $\Delta x = 2L$ or $\Delta t = 2T$, respectively. The product of the width of the packet in what might be called *k-space* and its width in *x-space* is $\Delta k\,\Delta x = 4\pi$ or analogously $\Delta\omega\,\Delta t = 4\pi$. One speaks of the quantities Δk and $\Delta\omega$ as the *frequency bandwidths*. Had we used a differently shaped pulse the bandwidth–pulse length product might certainly have been somewhat different. The ambiguity arises because we have not yet chosen one of the alternative possibilities for specifying $\Delta\omega$ and Δk. For example, rather than using the first minima of $A(k)$ (there are transforms which have no such minima, such as the Gaussian function of Section 11.2), we could have let Δk be the width of $A^2(k)$ at a point where the curve had dropped to $\frac{1}{2}$ or possibly $1/e$ of its maximum value. In any event, it will suffice for the time being to observe that

$$\Delta\nu \sim 1/\Delta t, \qquad (7.63)$$

that is, the frequency bandwidth is of the order of magnitude of the reciprocal of the temporal extent of the pulse (Problem 7.20). If the wave packet has a narrow bandwidth, it will extend over a large region of space and time. Accordingly, a radio tuned to receive a bandwidth of $\Delta\nu$ will be capable of detecting pulses of duration no shorter than $\Delta t \sim 1/\Delta\nu$.

These considerations are of profound importance in quantum mechanics where wave packets describe particles and Eq. (7.63) is akin to the Heisenberg uncertainty principle.

7.10 OPTICAL BANDWIDTHS

Suppose that we examine the light emitted by what is loosely termed a monochromatic source, e.g. a sodium discharge lamp. When the beam is passed through some sort of spectrum analyzer we will be able to observe all of its various frequency components. Typically we would find that there were a number of fairly narrow frequency ranges which contained most of the energy and that these were separated by much larger regions of darkness. Each such brightly colored band is known as a *spectral line*. There are devices where the light enters by way of a slit and each line would actually be a colored image of that slit. Other analyzers will represent the frequency distribution on the screen of an oscilloscope. In any event, the individual spectral lines are never infinitely sharp. They always consist of a band of frequencies, however small (Fig. 7.18).

The electron transitions responsible for the generation of light have a duration of the order of 10^{-8} s. Because the emitted wave trains are finite there will be a spread in the frequencies present known as the *natural linewidth* [see Section 11.3.4(ii)]. Moreover, since the atoms are in random thermal motion the frequency spectrum will be altered by the Doppler effect. In addition, the atoms suffer collisions which interrupt the wave trains and again tend to broaden the frequency distribution. The total effect of all of these mechanisms is that each spectral line has a bandwidth $\Delta\nu$ rather than one single frequency. The time Δt which satisfies Eq. (7.63) is referred to as the *coherence time* and the length Δx given by

$$\Delta x = c\,\Delta t \qquad (7.64)$$

is the *coherence length*.

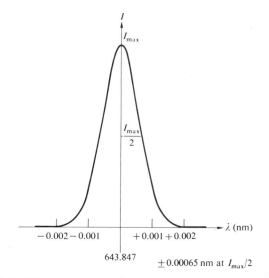

Fig. 7.18 The cadmium red ($\bar{\lambda} = 643.847$ nm) spectral line from a low pressure lamp.

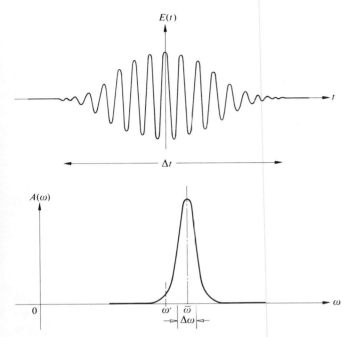

Fig. 7.19 A cosinusoidal wave train modulated by a Gaussian envelope along with its Gaussian transform.

Because of the quantized nature of the radiation process, light is emitted in the form of individual photons which for our present purposes can be represented by finite wave trains. Suppose that we have a quasimonochromatic thermal (nonlaser) light beam composed of N of these wave packets bearing a random phase relation to one another. The configuration of each wave packet, as drawn in Fig. 7.19, is assumed such that the square of its Fourier transform $A(\omega)$ resembles an irradiance versus frequency distribution often observed for spectral lines. Imagine then that we only look at one and the same harmonic frequency component in each packet, e.g. the one corresponding to ω'. Remember that each such component is an infinitely long, constant-amplitude wave. If every wave packet is assumed to be identical in form, the amplitude of the Fourier component associated with ω' will be the same for each. Since these have a random relative phase distribution it follows from Eq. (7.21) that the resultant will be a harmonic wave of frequency ω' having an amplitude proportional to $N^{\frac{1}{2}}$. The equivalent will be true, of course, for every frequency within the range constituting the wave packets. In other words, there is the same amount of energy present at each frequency in the composite wave as there is in the sum of the separate constituent wave trains at that frequency. As a result of the randomness of the wave trains, the individual harmonic components of the resultant wave will not have the same relative phases as they did in each packet. Thus the profile of the resultant will differ from that of the separate wave packets even though the amplitude of each frequency component present in the resultant is simply $N^{\frac{1}{2}}$ times its amplitude in any one packet. The observed spectral line corresponds to the power spectrum of the resultant beam, to be sure, but it also corresponds to the power spectrum of an individual packet. Ordinarily there will be a tremendous number of arbitrarily overlapping wave groups so that the envelope of the resultant will rarely, if ever, be zero. If the source is quasimonochromatic, i.e. if the bandwidth is small compared to the mean frequency $\bar{v}$, we can envision the resultant as being "almost" sinusoidal. In summary then, the composite wave can be pictured as in Fig. 7.20. We might imagine the frequency and amplitude to be randomly varying; the former over a range Δv centered at $\bar{v}$. Accordingly, the *frequency stability* defined as $\Delta v / \bar{v}$ is a useful measure of spectral purity. Even a coherence time as short as 10^{-9} s corresponds to roughly a few million wavelengths of the rapidly oscillating carrier ($\bar{v}$) so that any amplitude or frequency variations will occur quite slowly in comparison. Equivalently we can introduce a time-varying phase factor such that the disturbance can be written as

$$E(t) = E_0(t) \cos [\varepsilon(t) - 2\pi \bar{v} t], \qquad (7.65)$$

where the separation between wave crests changes in time.

The average duration of a wave packet is Δt and so two points on the wave of Fig. 7.20 separated by more than Δt must lie on different contributing wave trains. These points would thus be completely uncorrelated in phase. In other words, if we determine the electric field of the composite wave as it passes by an idealized detector we could predict its phase fairly accurately for times much less than Δt later, but not at all for times greater than Δt. In Chapter 12 we will consider the *degree of coherence* which applies over the region between these extremes as well.

White light has a frequency range from 0.4×10^{15} Hz to about 0.7×10^{15} Hz, that is, a bandwidth of about 0.3×10^{15} Hz. The coherence time is then roughly 3×10^{-15} s which corresponds (7.64) to wave trains having a spatial extent only a few wavelengths long. *Accordingly, white light may be envisaged as a random succession of very short pulses.* Were we to synthesize white light, we would have to superimpose a broad continuous range of

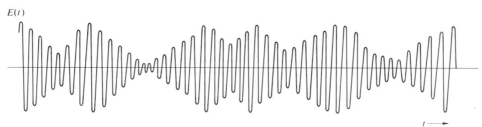

$E(t)$

Fig. 7.20 A quasimonochromatic light wave.

harmonic constituents in order to produce the very short wave packets. Inversely, we can pass white light through a Fourier analyzer, such as a diffraction grating or a prism, and in so doing actually generate those components.

The available bandwidth in the visible spectrum (≈ 300 THz) is so broad that it represents something of a "wonderland" for the communications engineer. For example, a typical television channel occupies a range of 4 MHz in the electromagnetic spectrum ($\Delta \nu$ is determined by the duration of the pulses needed to control the scanning electron beam). Thus the visible region could carry roughly 75 million television channels. Needless to say this is an area of active research (see Section 8.11).

Ordinary discharge lamps have relatively large bandwidths leading to coherence lengths only of the order of several millimeters. In contrast, the spectral lines emitted by low-pressure isotope lamps such as Hg^{198} ($\lambda_{air} = 546.078$ nm) or the international standard Kr^{86} ($\lambda_{air} = 605.616$ nm) have bandwidths of roughly 1000 MHz. The corresponding coherence lengths are of the order of 1 m and coherence times are about 1 ns. The frequency stability is about one part per million—these sources are certainly quasimonochromatic.

By far the most spectacular of all present-day sources is the laser. Under optimum conditions, where temperature variations and vibrations were meticulously suppressed, a laser was actually operated at quite close to its theoretical limit of frequency constancy. A short-term frequency stability of about 8 parts per 10^{14} was attained* with a He−Ne continuous gas laser at $\lambda = 1153$ nm. That corresponds to a remarkably narrow bandwidth of about 20 Hz. More commonly, frequency stabilities of several parts per 10^9 are

not very difficult to come by. There are commercially available CO_2 lasers which provide a short-term ($\sim 10^{-1}$ s) $\Delta \nu / \bar{\nu}$ ratio of 10^{-9} and a long-term ($\sim 10^3$ s) value of 10^{-8}.

PROBLEMS

7.1 Determine the resultant of the superposition of the parallel waves $E_1 = E_{01} \sin (\omega t + \varepsilon_1)$ and $E_2 = E_{02} \sin (\omega t + \varepsilon_2)$ when $\omega = 120\pi$, $E_{01} = 6$, $E_{02} = 8$, $\varepsilon_1 = 0$, and $\varepsilon_2 = \pi/2$. Plot each function and the resultant.

7.2* Show that the *optical path*, defined as the sum of the products of the various indices times the thicknesses of media traversed by a beam, that is, $\Sigma_i n_i x_i$, is equivalent to the length of the path in vacuum which would take the same time for that beam to negotiate.

7.3 a) How many wavelengths of $\lambda_0 = 500$ nm light will span a 1 m gap in vacuum?
b) How many waves span the gap when a 5 cm thick glass plate ($n = 1.5$) is inserted in the path?
c) Determine the O.P.D. between the two situations.
d) Verify that Λ/λ_0 corresponds to the difference between the solutions to (a) and (b) above.

7.4* Determine the optical path difference for the two waves A and B both having vacuum wavelengths of 500 nm depicted in Fig. 7.21; the glass ($n = 1.52$) tank is filled with water ($n = 1.33$). If the waves started out in phase and all of the above numbers are exact, find their relative phase difference at the finishing line.

7.5* Using Eqs. (7.9), (7.10), and (7.11) show that the resultant of the two waves

$$E_1 = E_{01} \sin [\omega t - k(x + \Delta x)]$$

and

$$E_2 = E_{01} \sin (\omega t - kx)$$

---

* T. S. Jaseja, A. Javan, and C. H. Townes, "Frequency Stability of Helium−Neon Masers and Measurements of Length," *Phys. Rev. Letters* **10**, 165 (1963).

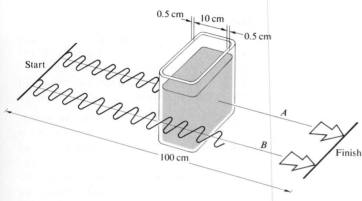

Fig. 7.21

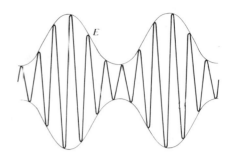

Fig. 7.22

is

$$E = 2E_{01} \cos\left(\frac{k \, \Delta x}{2}\right) \sin\left[\omega t - k\left(x + \frac{\Delta x}{2}\right)\right]. \quad [7.17]$$

7.6 Add the two waves of Problem 7.5 directly to find Eq. (7.17).

7.7 Use the complex representation to find the resultant $E = E_1 + E_2$, where

$$E_1 = E_0 \cos (kx + \omega t) \quad \text{and} \quad E_2 = -E_0 \cos (kx - \omega t).$$

Describe the composite wave.

7.8 The electric field of a standing electromagnetic plane wave is given by

$$E(x, t) = 2E_0 \sin kx \cos \omega t; \quad [7.30]$$

derive an expression for $B(x, t)$. (You might want to take another look at Section 3.2.) Make a sketch of the standing wave.

7.9* Imagine that we strike two tuning forks, one of frequency 340 Hz, the other of 342 Hz. What will we hear?

7.10 Figure 7.22 shows a carrier of frequency ω_c being amplitude modulated by a sine wave of frequency ω_m, that is

$$E = E_0(1 + \alpha \cos \omega_m t) \cos \omega_c t.$$

Show that this is equivalent to the superposition of three waves of frequencies ω_c, $\omega_c + \omega_m$, and $\omega_c - \omega_m$. When a number of modulating frequencies are present, we write E as a Fourier series and sum over all values of ω_m. The terms $\omega_c + \omega_m$ constitute what is called the *upper sideband* while all of the $\omega_c - \omega_m$ terms form the *lower sideband*. What bandwidth would you need in order to transmit the complete audible range?

7.11 Given the dispersion relation $\omega = ak^2$, compute both the phase and group velocities.

7.12 The speed of propagation of a surface wave in a liquid of depth much greater than λ is given by

$$v = \sqrt{\frac{g\lambda}{2\pi} + \frac{2\pi\Upsilon}{\rho\lambda}},$$

where

g = acceleration of gravity
λ = wavelength
ρ = density
Υ = surface tension

Compute the group velocity of a pulse in the long wavelength limit (these are called *gravity waves*).

7.13* Show that the group velocity can be written as

$$v_g = v - \lambda \frac{dv}{d\lambda}.$$

7.14 Show that the group velocity can be written as

$$v_g = \frac{c}{n + \omega(dn/d\omega)}.$$

7.15 Using the dispersion equation

$$n^2(\omega) = 1 + \frac{Nq_e^2}{\epsilon_0 m_e} \sum_j \left(\frac{f_j}{\omega_{0j}^2 - \omega^2}\right) \quad [3.36]$$

show that the group velocity is given by

$$v_g = \frac{c}{1 + Nq_e^2/\epsilon_0 m_e \omega^2 2}$$

for high-frequency electromagnetic waves, e.g. x-rays.

Keep in mind that since f_j are the weighting factors, $\Sigma_j f_j = 1$. What is the phase velocity? Show that $v v_g \approx c^2$.

7.16 Show that

$$\int_0^\lambda \sin akx \cos bkx \, dx = 0 \qquad [7.44]$$

$$\int_0^\lambda \cos akx \cos bkx \, dx = \frac{\lambda}{2}\delta_{ab} \qquad [7.45]$$

$$\int_0^\lambda \sin akx \sin bkx \, dx = \frac{\lambda}{2}\delta_{ab}, \qquad [7.46]$$

where $a \neq 0$, $b \neq 0$, and a and b are positive integers.

7.17 Compute the Fourier series components for the periodic function shown in Fig. 7.14.

7.18 Change the upper limit of Eq. (7.59) from ∞ to a and evaluate the integral. Leave the answer in terms of the so-called *sine integral*:

$$\text{Si}(z) = \int_0^z \text{sinc } w \, dw,$$

which is a function whose values are commonly tabulated.

7.19 Write an expression for the transform $A(\omega)$ of the harmonic pulse of Fig. 7.23. Check that sinc u is 50 % or greater for values of u *roughly* less than $\pi/2$. With that in mind show that $\Delta\nu \, \Delta t \sim 1$ where $\Delta\nu$ is the bandwidth of the transform at half its maximum amplitude. Verify that $\Delta\nu \, \Delta t \sim 1$ at half the maximum irradiance as well. The intent here is to get some sense of the kind of approximations used in the discussion.

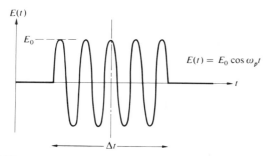

$$E(t) = E_0 \cos \omega_p t$$

Fig. 7.23

7.20 Derive an expression for the coherence length (in vacuum) of a wave train which has a frequency bandwidth $\Delta\nu$; express your answer in terms of the *linewidth* $\Delta\lambda_0$ and the mean wavelength $\bar{\lambda}_0$ of the train.

7.21 Consider a photon in the visible region of the spectrum emitted during an atomic transition of about 10^{-8} s. How long is the wave packet? Keeping in mind the results of the previous problem (i.e. if you've done it), estimate the linewidth of the packet ($\bar{\lambda}_0 = 500$ nm). What can you say about its monochromaticity as indicated by the frequency stability?

7.22 The first[*] experiment directly measuring the bandwidth of a laser (in this case a continuous-wave $Pb_{0.88}Sn_{0.12}Te$ diode laser) has been successfully carried out. The laser, operating at $\lambda_0 = 10{,}600$ nm, was heterodyned with a CO_2 laser and bandwidths as narrow as 54 kHz were observed. Compute the corresponding frequency stability and coherence length for the lead-tin-telluride laser.

7.23[*] A magnetic-field technique for stabilizing a He–Ne laser to 2 parts in 10^{10} has recently been patented. At 632.8 nm, what would be the coherence length of a laser with such a frequency stability?

7.24 Imagine that we chop a continuous laser beam (assumed to be monochromatic at $\lambda_0 = 632.8$ nm) into 0.1 ns pulses using some sort of shutter. Compute the resultant linewidth $\Delta\lambda$, bandwidth and coherence length. Find the bandwidth and linewidth which would result if we could chop at 10^{15} Hz.

7.25[*] Suppose that we have a filter which has a pass band of 1.0 Å centered at 600 nm and we illuminate it with sunlight. Compute the coherence length of the emerging wave.

---

[*] D. Hinkley and C. Freed, *Phys. Rev. Letters* **23**, 277 (1969).

Polarization 8

8.1 THE NATURE OF POLARIZED LIGHT

We have already established that light may be treated as a transverse electromagnetic wave. Thus far we have considered only *linearly* polarized or *plane*-polarized light, that is, light for which the orientation of the electric field is constant although its magnitude and sign varies in time (Fig. 3.9). The electric field or optical disturbance therefore resides in what is known as the *plane of vibration*. That fixed plane contains both **E** and **k**, the electric field vector and the propagation vector in the direction of motion. Imagine now that we have two harmonic, linearly polarized light waves of the same frequency, moving through the same region of space, in the same direction. If their electric field vectors are collinear, the superimposing disturbances will simply combine to form a resultant linearly polarized wave. Its amplitude and phase will be examined in detail, under a diversity of conditions, in the next chapter when we consider the phenomenon of interference. In contradistinction, if the two light waves are such that their respective electric field directions are mutually perpendicular, the resultant wave may or may not be linearly polarized. Exactly what form that light will take (i.e. its *state of polarization*), how we can observe it, produce it, change it and make use of it will be the concern of this chapter.

8.1.1 Linear Polarization

We can represent the two orthogonal optical disturbances which were considered above in the form

$$\mathbf{E}_x(z, t) = \hat{\mathbf{i}}E_{0x} \cos(kz - \omega t) \tag{8.1}$$

and

$$\mathbf{E}_y(z, t) = \hat{\mathbf{j}}E_{0y} \cos(kz - \omega t + \varepsilon) \tag{8.2}$$

where ε is the relative phase difference between the waves, both of which are traveling in the z-direction. The resultant optical disturbance is then simply

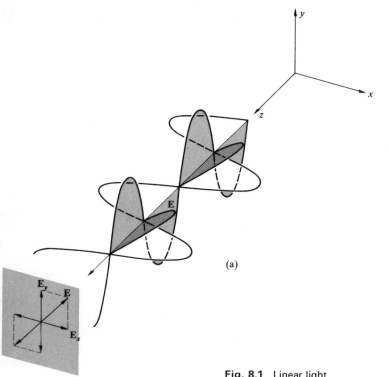

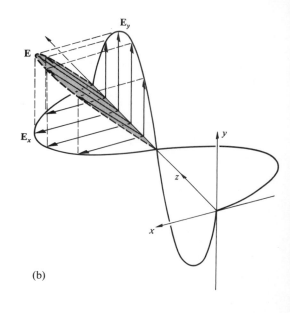

(a)

(b)

Fig. 8.1   Linear light.

$$\mathbf{E}(z, t) = \mathbf{E}_x(z, t) + \mathbf{E}_y(z, t). \qquad (8.3)$$

If ε is zero or an integral multiple of $\pm 2\pi$, the waves are said to be *in phase*. In that particular case Eq. (8.3) becomes

$$\mathbf{E} = (\hat{\mathbf{i}} E_{0x} + \hat{\mathbf{j}} E_{0y}) \cos (kz - \omega t). \qquad (8.4)$$

The resultant wave therefore has a fixed amplitude equal to $(\hat{\mathbf{i}} E_{0x} + \hat{\mathbf{j}} E_{0y})$, i.e. it too is linearly polarized as shown in Fig. 8.1. This process can equally well be carried out in reverse, that is, we can resolve any plane-polarized wave into two orthogonal components.

Suppose now that ε is an odd integer multiple of $\pm \pi$. The two waves are said to be 180° out of phase and

$$\mathbf{E} = (\hat{\mathbf{i}} E_{0x} - \hat{\mathbf{j}} E_{0y}) \cos (kz - \omega t). \qquad (8.5)$$

This wave is again linearly polarized but the plane of vibration has been rotated (and not necessarily by 90°) from that of the previous condition, as indicated in Fig. 8.2.

8.1.2   Circular Polarization

Another special case of particular interest arises when both constituent waves have equal amplitudes, i.e. $E_{0x} = E_{0y} = E_0$ and in addition, their relative phase difference $\varepsilon = -\pi/2 + 2m\pi$ where $m = 0, \pm 1, \pm 2, \ldots$. Accordingly

$$\mathbf{E}_x(z, t) = \hat{\mathbf{i}} E_0 \cos (kz - \omega t) \qquad (8.6)$$

and

$$\mathbf{E}_y(z, t) = \hat{\mathbf{j}} E_0 \sin (kz - \omega t). \qquad (8.7)$$

The consequent wave is given by

$$\mathbf{E} = E_0 [\hat{\mathbf{i}} \cos (kz - \omega t) + \hat{\mathbf{j}} \sin (kz - \omega t)] \qquad (8.8)$$

(Fig. 8.3). Notice that now the scalar amplitude of $\mathbf{E}$, which is equal to E_0, is a constant. But the direction of $\mathbf{E}$ is time varying and it is not restricted as before to a single plane. Figure 8.4 depicts what is happening at some arbitrary point z_0 on the axis. At $t = 0$ $\mathbf{E}$ lies along the reference axis in Fig. 8.4(a) and so

$$\mathbf{E}_x = \hat{\mathbf{i}} E_0 \cos kz_0 \qquad \text{and} \qquad \mathbf{E}_y = \hat{\mathbf{j}} E_0 \sin kz_0.$$

At a later time, $t = kz_0/\omega$, $\mathbf{E}_x = \hat{\mathbf{i}} E_0$, $\mathbf{E}_y = 0$, and $\mathbf{E}$ is along the x-axis. The resultant electric field vector $\mathbf{E}$ is rotating *clockwise* at an angular frequency of ω as seen by an ob-

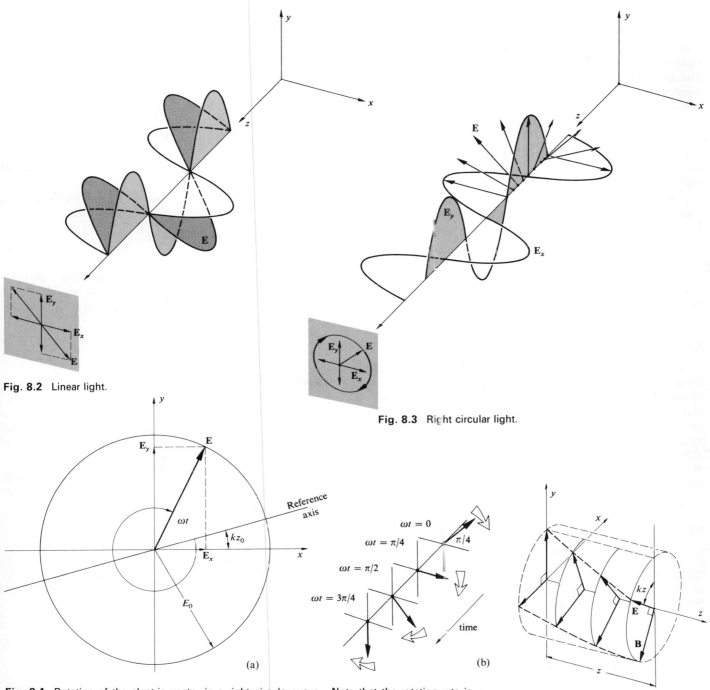

Fig. 8.2 Linear light.

Fig. 8.3 Right circular light.

Fig. 8.4 Rotation of the electric vector in a right circular wave. Note that the rotation rate is ω and $kz = \pi/4$.

Fig. 8.5 Right circular light.

server toward whom the wave is moving (i.e. looking back at the source). Such a wave is said to be *right-circularly polarized* (Fig. 8.5) and one generally simply refers to it as *right-circular light*. The **E**-vector makes one complete rotation as the wave advances through one wavelength. In comparison, if $\varepsilon = \pi/2, 5\pi/2, 9\pi/2$ etc., i.e. $\varepsilon = \pi/2 + 2m\pi$ where $m = 0, \pm 1, \pm 2, \pm 3, \ldots$, then

$$\mathbf{E} = E_0[\hat{\mathbf{i}} \cos (kz - \omega t) - \hat{\mathbf{j}} \sin (kz - \omega t)]. \quad (8.9)$$

the amplitude is unaffected, but **E** now rotates *counter-clockwise* and the wave is referred to as *left-circularly polarized*.

A linearly polarized wave can be synthesized from two oppositely polarized circular waves of equal amplitude. In particular, if we add the right-circular wave of Eq. (8.8) to the left-circular wave of Eq. (8.9) we get

$$\mathbf{E} = 2E_0\hat{\mathbf{i}} \cos (kz - \omega t), \quad (8.10)$$

which has a constant amplitude vector of $2E_0\hat{\mathbf{i}}$ and is therefore linearly polarized.

8.1.3 Elliptical Polarization

As far as the mathematical description is concerned, both linear and circular light may be considered to be special cases of *elliptically polarized*, or more simply, *elliptical light*. By that we mean that, in general, the resultant electric field vector **E** will both rotate and change its magnitude as well. In such cases the endpoint of **E** will trace out an ellipse, in a fixed plane perpendicular to **k**, as the wave sweeps by. We can better see this by actually writing an expression for the curve traversed by the tip of **E**. To that end recall that

$$E_x = E_{0x} \cos (kz - \omega t) \quad (8.11)$$

and

$$E_y = E_{0y} \cos (kz - \omega t + \varepsilon). \quad (8.12)$$

The equation of the curve we are looking for should neither be a function of position nor time, i.e. we should be able to get rid of the $(kz - \omega t)$ dependence. Expand the expression for E_y into

$$E_y/E_{0y} = \cos (kz - \omega t) \cos \varepsilon - \sin (kz - \omega t) \sin \varepsilon$$

and combine it with E_x/E_{0x} to yield

$$\frac{E_y}{E_{0y}} - \frac{E_x}{E_{0x}} \cos \varepsilon = - \sin (kz - \omega t) \sin \varepsilon. \quad (8.13)$$

It follows from Eq. (8.11) that

$$\sin (kz - \omega t) = [1 - (E_x/E_{0x})^2]^{1/2}$$

and so Eq. (8.13) leads to

$$\left(\frac{E_y}{E_{0y}} - \frac{E_x}{E_{0x}} \cos \varepsilon\right)^2 = \left[1 - \left(\frac{E_x}{E_{0x}}\right)^2\right] \sin^2 \varepsilon.$$

Finally then, on rearranging terms we have

$$\left(\frac{E_y}{E_{0y}}\right)^2 + \left(\frac{E_x}{E_{0x}}\right)^2 - 2\left(\frac{E_x}{E_{0x}}\right)\left(\frac{E_y}{E_{0y}}\right) \cos \varepsilon = \sin^2 \varepsilon. \quad (8.14)$$

This is the equation of an ellipse making an angle α with the (E_x, E_y)-coordinate system (Fig. 8.6) such that

$$\tan 2\alpha = \frac{2E_{0x}E_{0y} \cos \varepsilon}{E_{0x}^2 - E_{0y}^2}. \quad (8.15)$$

Equation (8.14) might be a bit more recognizable if the principal axes of the ellipse were aligned with the coordinate axes, i.e. $\alpha = 0$ or equivalently $\varepsilon = \pm\pi/2, \pm 3\pi/2, \pm 5\pi/2, \ldots$, in which cases we have the familiar form

$$\frac{E_y^2}{E_{0y}^2} + \frac{E_x^2}{E_{0x}^2} = 1. \quad (8.16)$$

Furthermore, if $E_{0y} = E_{0x} = E_0$ this reduces to

$$E_y^2 + E_x^2 = E_0^2$$

which, in agreement with our previous results, is a circle. If ε is an even multiple of π, Eq. (8.14) results in

$$E_y = \frac{E_{0y}}{E_{0x}} E_x \quad (8.18)$$

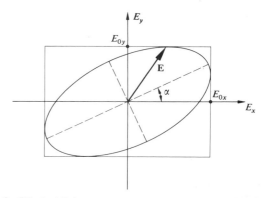

Fig. 8.6 Elliptical light.

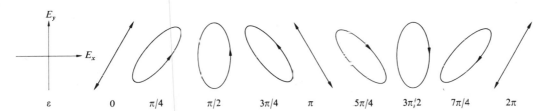

Fig. 8.7 Various polarization configurations corresponding to specific values of ε. Here E_x leads E_y by ε. The light would be circular when $\varepsilon = \pi/2$ or $3\pi/2$ if $E_{0x} = E_{0y}$.

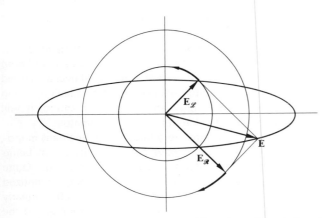

Fig. 8.8 Elliptical light as the superposition of an $\mathscr{R}$- and $\mathscr{L}$-state.

and similarly for odd multiples of π

$$E_y = -\frac{E_{0y}}{E_{0x}} E_x. \qquad (8.19)$$

These are both straight lines having slopes of $\pm E_{0y}/E_{0x}$, i.e. we have linear light. Figure 8.7 diagrammatically summarizes most of these conclusions.

We are now in a position to refer to a particular light wave in terms of its specific *state of polarization*. We shall say that linearly or plane polarized light is in a $\mathscr{P}$-state, while right or left circular light is in an $\mathscr{R}$- or $\mathscr{L}$-state, respectively. Similarly, the condition of elliptic polarization corresponds to an $\mathscr{E}$-state. We've already seen that a $\mathscr{P}$-state can be represented as a superposition of $\mathscr{R}$- and $\mathscr{L}$-states, and the same is true for an $\mathscr{E}$-state. In this case, as shown in Fig. 8.8, the amplitudes of the two circular waves are different. (An analytical treatment is left for Problem 8.3.)

8.1.4 Natural Light

An ordinary light source consists of a very large number of randomly oriented atomic emitters. Each excited atom ra-

diates a polarized wave train for roughly 10^{-8} s. All of the emissions having the same frequency will combine to form a single resultant polarized wave which persists for no longer than 10^{-8} s. New wave trains are constantly emitted and the overall polarization changes in a completely unpredictable fashion (see Section 8.9). If these changes take place at so rapid a rate as to render any single resultant polarization state indiscernible, the wave is referred to as *natural light*. It is also known as *unpolarized light*, but this is a bit of a misnomer since in actuality the light is composed of a rapidly varying succession of the different polarization states.

We can mathematically represent natural light in terms of two arbitrary, *incoherent*, orthogonal, linearly polarized waves of equal amplitude (i.e. waves for which the relative phase difference varies rapidly and randomly).

Keep in mind that an idealized monochromatic plane wave must be depicted as an infinite wave train. If this disturbance is resolved into two orthogonal components perpendicular to the direction of propagation they, in turn, must have the same frequency, be infinite in extent and therefore mutually coherent (i.e. $\varepsilon = $ constant). In other words, a perfectly monochromatic plane wave is always polarized. In fact Eqs. (8.1) and (8.2) are just the Cartesian components of a transverse ($E_z = 0$) harmonic plane wave.

Generally light, whether natural in origin or artificial, is neither completely polarized nor unpolarized; both cases are extremes. More often, the electric field vector varies in a way which is neither totally regular nor totally irregular and one refers to such an optical disturbance as being *partially polarized*. One useful way of describing this behavior is to envision it as the result of the superposition of specific amounts of natural and polarized light.

8.1.5 Angular Momentum and the Photon Picture

We have already seen that an electromagnetic wave impinging on an object can impart both energy and linear momentum to that body (Section 3.4.2). Moreover, if the incident plane wave is circularly polarized we can expect electrons within the material to be set into circular motion in response to the force generated by the rotating **E**-field. Alternatively we might picture the field as being composed of two orthogonal $\mathscr{P}$-states which are 90° out of phase. These simultaneously drive the electron in two perpendicular directions with a $\pi/2$ phase difference. The resulting motion is again circular. In effect the torque exerted by the **B**-field averages to zero over an orbit and the **E**-field drives the electron with an angular velocity ω equal to the frequency of the electromagnetic wave. Angular momentum will thus be imparted by the wave to the substance in which the electrons are imbedded and to which they are bound. We can treat the problem rather simply without actually going into the details of the dynamics. The power delivered to the system is the energy transferred per unit time, $d\mathscr{E}/dt$. Furthermore, the power generated by a torque Γ acting on a rotating body is just $\omega\Gamma$ (which is analogous to vF for linear motion) and so

$$\frac{d\mathscr{E}}{dt} = \omega\Gamma. \tag{8.20}$$

Since the torque is equal to the time rate of change of the angular momentum L, it follows that on the average

$$\frac{d\mathscr{E}}{dt} = \omega\frac{dL}{dt}. \tag{8.21}$$

A charge which absorbs a quantity of energy $\mathscr{E}$ from the incident circular wave will simultaneously absorb an amount of angular momentum L such that

$$L = \frac{\mathscr{E}}{\omega}. \tag{8.22}$$

If the incident wave is in an $\mathscr{R}$-state its **E**-vector rotates clockwise, looking toward the source. This is the direction in which a positive charge in the absorbing medium would rotate and the angular momentum vector is therefore taken to point in the direction opposite to the propagation direction* as shown in Fig. 8.9.

According to the quantum-mechanical description, an electromagnetic wave transfers energy in quantized packets or photons such that $\mathscr{E} = h\nu$. Thus $\mathscr{E} = \hbar\omega$ ($\hbar \equiv h/2\pi$), and the *intrinsic* or *spin* angular momentum of a photon is either

---

* This choice of terminology is admittedly a bit awkward. Yet its use in optics is fairly well established and persists even though it is completely antithetic to the more reasonable convention adopted in elementary particle physics.

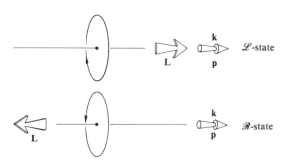

Fig. 8.9 Angular momentum of a photon.

$-\hbar$ or $+\hbar$, where the signs indicate right- or left-handedness, respectively. Notice that *the angular momentum of a photon is completely independent of its energy*. Whenever a charged particle emits or absorbs electromagnetic radiation, along with changes in its energy and linear momentum, it will undergo a change of $\pm\hbar$ in its angular momentum.*

The energy transferred to a target by an incident monochromatic electromagnetic wave can be envisaged as being transported in the form of a stream of identical photons. Quite obviously we can anticipate a corresponding quantized transport of angular momentum. A purely left-circularly polarized plane wave will impart angular momentum to the target as if all of the constituent photons in the beam had their spins aligned in the direction of propagation. Changing the light to right circular reverses the spin orientation of the photons as well as the torque exerted by them on the target. Using an extremely sensitive torsion pendulum, Richard A. Beth (b. 1906) in 1935 was actually able to perform such measurements.[†]

Thus far we've had no difficulty in describing purely right- and left-circular light in the photon picture; but what is linearly or elliptically polarized light? Classically, light in a $\mathscr{P}$-state can be synthesized by the coherent superposition

---

*As a rather important yet simple example, consider the hydrogen atom. It is composed of a proton and an electron, each having a spin of $\hbar/2$. The atom has slightly more energy when the spins of both particles are in the same direction. It is possible however, that once in a very long time, roughly 10^7 years, one of the spins will flip over and be antiparallel to the other. The change in angular momentum of the atom is then $\hbar$ and this is imparted to an emitted photon which carries off the slight excess in energy as well. This is the origin of the 21 cm microwave emission which is so very significant in radio astronomy.

[†]Richard A. Beth, "Mechanical Detection and Measurement of the Angular Momentum of Light," *Phys. Rev.* **50**, 115 (1936).

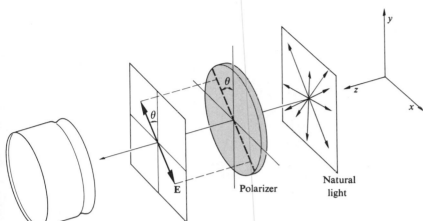

Fig. 3.10 A linear polarizer.

of equal amounts of light in $\mathscr{R}$- and $\mathscr{L}$-states (with an appropriate phase difference). Any single photon whose angular momentum is somehow measured will be found to have its spin totally either parallel or antiparallel to **k**. A beam of linear light will interact with matter as if it were composed, at that instant, of equal numbers of right- and left-handed photons. There is a subtle point that has to be made here. We cannot say that the beam is actually made up of precisely equal amounts of well-defined right- and left-handed photons; the photons are all identical. Rather, each individual photon exists simultaneously in both possible spin states with equal likelihood. On measuring the angular momentum of the constituent photons, $-\hbar$ would result equally as often as $+\hbar$. This is all we can observe. We are not privy to what the photon is doing prior to the measurement (if indeed it exists *prior* to the measurement). As a whole, the beam will therefore impart no total angular momentum to a target.

In contrast, if each photon does not occupy both spin states with the same probability, one angular momentum, say $+\hbar$, will be measured to occur somewhat more often than the other, $-\hbar$. In this instance, a net positive angular momentum will therefore be imparted to the target. The result *en masse* is elliptically polarized light, i.e. a superposition of unequal amounts of $\mathscr{R}$- and $\mathscr{L}$-light bearing a particular phase relationship.

8.2 POLARIZERS

Now that we have some idea of what polarized light is, the next logical step is to develop an understanding of the techniques used to generate it, change it, and in general manipulate it to fit our needs. An optical device whose input is natural light and whose output is some form of polarized light is quite reasonably known as a *polarizer*. For example, recall that one possible representation of unpolarized light is the superposition of two equal-amplitude, incoherent, orthogonal $\mathscr{P}$-states. An instrument which separates these two components, discarding one and passing on the other, is known as a *linear polarizer*. Depending on the form of the output we could also have *circular* or *elliptical polarizers*. And all of these devices vary in effectiveness down to what might be called leaky or *partial* polarizers.

Polarizers take on many different configurations as we shall see, but they are all based on one of four fundamental physical mechanisms: *dichroism* or selective absorption, *reflection, scattering* and *birefringence* or double refraction. There is however, one underlying property which they all share and that is simply that *there must be some form of asymmetry associated with the process*. This is certainly understandable since the polarizer must somehow select a particular polarization state and discard all others. In truth, the asymmetry may be a subtle one related to the incident or viewing angle but more usually it is an obvious anisotropy in the material of the polarizer itself.

8.2.1 Malus's Law

One matter needs to be settled before we go on, and that is: how do we determine experimentally whether or not a device is actually a linear polarizer?

By definition, if natural light is incident on an ideal linear polarizer as in Fig. 8.10, only light in a $\mathscr{P}$-state will be

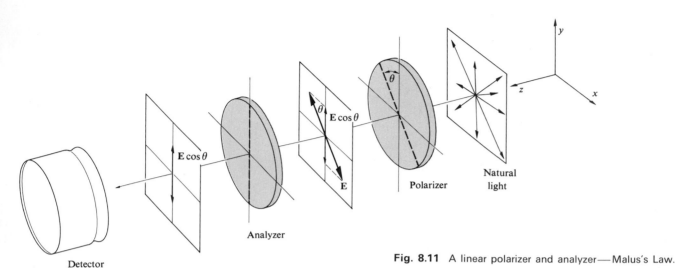

Fig. 8.11 A linear polarizer and analyzer—Malus's Law.

transmitted. That $\mathscr{P}$-state will have an orientation parallel to a specific direction which we will call the *transmission axis* of the polarizer. In other words, only the component of the optical field parallel to the transmission axis will pass through the device essentially unaffected. If the polarizer in Fig. 8.10 is rotated about the z-axis the reading of the detector (e.g. a photocell) will be unchanged because of the complete symmetry of unpolarized light. Keep in mind that we are most certainly dealing with waves but because of the very high frequency of light our detector will, for practical reasons, measure only the incident irradiance. Since the irradiance is proportional to the square of the amplitude of the electric field [Eq. (3.52)] we need only concern ourselves with that amplitude.

Now suppose that we introduce a second identical ideal polarizer, or *analyzer*, whose transmission axis is vertical (Fig. 8.11). If the amplitude of the electric field transmitted by the polarizer is E_0 only its component, $E_0 \cos \theta$, parallel to the transmission axis of the analyzer will be passed on to the detector (assuming no absorption). According to Eq. (3.52) the irradiance reaching the detector is then given by

$$I(\theta) = \frac{c\epsilon_0}{2} E_0^2 \cos^2 \theta. \tag{8.23}$$

The maximum irradiance, $I(0) = c\epsilon_0 E_0^2 / 2$, occurs when the angle θ between the transmission axes of the analyzer and polarizer is zero. Equation (8.23) can accordingly be rewritten as

$$I(\theta) = I(0) \cos^2 \theta. \tag{8.24}$$

This is known as *Malus's law*, having first been published in 1809 by Étienne Malus, military engineer and captain in the army of Napoleon.

Observe that $I(90°) = 0$. This arises from the fact that the electric field that has passed through the polarizer is perpendicular to the transmission axis of the analyzer (the two devices so arranged are said to be *crossed*). The field is therefore parallel to what is called the *extinction axis* of the analyzer and hence obviously has no component along the transmission axis.

Evidently, we can use the set-up of Fig. 8.11 along with Malus's law to determine whether or not a particular device is in fact a linear polarizer.

8.3 DICHROISM

In its broadest sense the term *dichroism* refers to the selective absorption of one of the two orthogonal $\mathscr{P}$-state components of an incident beam. The dichroic polarizer itself is physically anisotropic, producing a strong asymmetric or preferential absorption of one field component while being essentially transparent to the other.

8.3.1 The Wire-Grid Polarizer

The simplest device of this sort is a grid of parallel conducting wires as shown in Fig. 8.12. Imagine that an unpolarized electromagnetic wave impinges on the grid from the right.

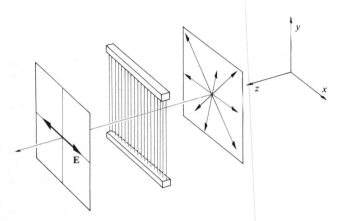

Fig. 8.12 A wire-grid polarizer.

The electric field can be resolved into the usual two orthogonal components; in this case, one chosen to be parallel to the wires and the other perpendicular to them. The y-component of the field drives the conduction electrons along the length of each wire, thus generating a current. The electrons in turn collide with lattice atoms imparting energy to them and thereby heating the wires (joule heat). In this manner energy is transferred from the field to the grid. In addition, electrons accelerating along the y-axis radiate in both the forward and backward directions. As should be expected the incident wave tends to be canceled by the wave reradiated in the forward direction, resulting in little or no transmission of the y-component of the field. The radiation propagating in the backward direction simply appears as a reflected wave. In contrast, the electrons are not free to move very far in the x-direction and the corresponding field component of the wave is essentially unaltered as it propagates through the grid. *The transmission axis of the grid is accordingly perpendicular to the wires.* It is an extremely common error to naïvely assume that the y-component of the field somehow slips through the spaces between the wires.

One can easily confirm our conclusions using microwaves and a grid made of ordinary electrical wire. It is not so easy a matter however to fabricate a grid which will polarize light, but it has been done! In 1960, George R. Bird and Maxfield Parrish Jr.* constructed a grid having an incredible 2160 wires per mm. Their feat was accomplished

---

* G. R. Bird and M. Parrish, Jr., "The Wire Grid as a Near-Infrared Polarizer," *J. Opt. Soc. Am.* **50**, 886 (1960).

by evaporating a stream of gold (or at other times aluminum) atoms at nearly grazing incidence onto a plastic diffraction grating replica (see Section 10.2.7). The metal accumulated along the edges of each step in the grating to form thin microscopic "wires" whose width and spacing were less than one wavelength across.

Although the wire grid is useful particularly in the infrared, we mention it here more for pedagogical than practical reasons. The underlying principle on which it is based is shared by other, more common, dichroic polarizers.

8.3.2 Dichroic Crystals

There are certain materials which are inherently dichroic because of an anisotropy in their respective crystalline structures. Probably the best known of these is the naturally occurring mineral *tourmaline*, a semiprecious stone often used in jewelry. Actually there are several tourmalines which are boron silicates of differing chemical composition [e.g. $NaFe_3B_3Al_6Si_6O_{27}(OH)_4$]. For this substance there is a specific direction within the crystal known as the principal or *optic* axis which is determined by its atomic configuration. The electric-field component of an incident light wave which is perpendicular to the principal axis is strongly absorbed by the sample. The thicker the crystal the more complete the absorption (Fig. 8.13). A plate cut from a tourmaline crystal parallel to its principal axis and several millimeters thick will accordingly serve as a linear polarizer. In this instance the crystal's principal axis becomes the polarizer's transmission axis. But the usefulness of tourmaline is rather limited by the fact that its crystals are comparatively small. Moreover, even the transmitted light suffers a certain amount of absorption. To complicate matters, this undesirable absorption is strongly wavelength dependent and the specimen will therefore be colored. Holding a tourmaline crystal up to natural white light, it might appear green (they come in other colors as well) when viewed normal to the principal axis and nearly black when viewed along that axis, where all the **E**-fields are perpendicular to it (ergo the term dichroic, meaning two colors).

There are several other substances which display similar characteristics. A crystal of the mineral hypersthene, a ferromagnesian silicate, might look green under white light polarized in one direction and pink for a different polarization direction.

We can get a qualitative picture of the mechanism which gives rise to crystal dichroism by considering the microscopic structure of the sample (you might want to take another look

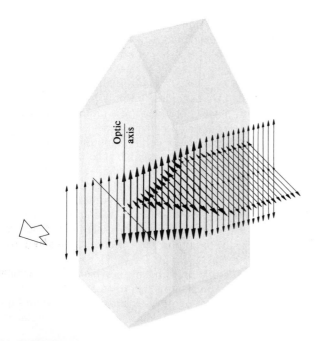

Fig. 8.13 A dichroic crystal. The naturally occurring ridges evident in the photo of the tourmaline crystals correspond to the optic axis. [Photo by E.H.]

at Section 3.3.1). The atoms within a crystal are strongly bound together by short-range forces to form a periodic lattice. The electrons, which are responsible for the optical properties, can be envisioned as elastically tied to their respective equilibrium positions. Electrons associated with a given atom are also under the influence of the surrounding nearby atoms, which themselves may not be symmetrically distributed. As a result, the elastic binding forces on the electrons will be different in different directions. Accordingly, their response to the harmonic electric field of an incident electromagnetic wave will vary with the direction of **E**. If in addition to being anisotropic the material is absorbing, a detailed analysis would have to include an orientation-dependent conductivity. Currents will exist and energy from the wave will be converted into joule heat. The attenuation, in addition to varying in direction, may be frequency dependent as well. This means that if the incoming white light is in a $\mathscr{P}$-state the crystal will appear colored and the color will depend on the orientation of **E**. Substances which display two or even three different colors are said to be dichroic or trichroic respectively.*

8.3.3 Polaroid

In 1928 Edwin Herbert Land, then a 19-year-old undergraduate at Harvard College, invented the first dichroic sheet polarizer known commercially as *polaroid J-sheet*. It incorporated a synthetic dichroic substance called *herapathite* or *quinine sulfate periodide*.[†] Land's own retrospective account of his early work is rather informative and makes fascinating reading. It is particularly interesting to follow the sometimes whimsical origins of what is now, no doubt, the most widely used group of polarizers. The following is an excerpt from Land's remarks:

> In the literature there are a few pertinent high spots in the development of polarizers, particularly the work of William Bird Herapath, a physician in Bristol, England, whose pupil, a Mr. Phelps, had found that when he dropped iodine into the urine of a dog that had been fed quinine, little scintillating green crystals formed in the reaction liquid. Phelps went to his teacher,

---

* More will be said about these processes later on when we consider birefringence. Suffice it to say now that for crystals classified as *uniaxial* there are two distinct directions and therefore two colors may be displayed by *absorbing* specimens. In *biaxial* crystals there are three distinct directions and the possibility of three colors.

† E. H. Land, "Some Aspects of the Development of Sheet Polarizers," *J. Opt. Soc. Am.* **41**, 957 (1951).

and Herapeth then did something which I [Land] think was curious under the circumstances; he looked at the crystals under a microscope and noticed that in some places they were light where they overlapped and in some places they were dark. He was shrewd enough to recognize that here was a remarkable phenomenon, a new polarizing material [now known as herapathite]. . . .

Herapath's work caught the attention of Sir David Brewster, who was working in those happy days on the kaleidoscope. . . . Brewster, who invented the kaleidoscope, wrote a book about it, and in that book he mentioned that he would like to use herapathite crystals for the eyepiece. When I was reading this book, back in 1926 and 1927, I came across his reference to these remarkable crystals, and that started my interest in herapathite.

Land's initial approach to creating a new form of linear polarizer was to grind herapathite into millions of submicroscopic crystals which fortunately were naturally needle shaped. Their small size lessened the problem of the scattering of light. In his earliest experiments the crystals were aligned nearly parallel to each other using magnetic or electric fields. Later he found that they would be mechanically aligned when a viscous colloidal suspension of the herapathite needles was extruded through a long narrow slit. The resulting J-sheet was *effectively* a large flat dichroic crystal. The individual submicroscopic crystals still scattered light a bit and as a result J-sheet was somewhat hazy. In 1938 Land invented *H-sheet* which is now probably the most widely used linear polarizer. It does not contain dichroic crystals but is instead a molecular analog of the wire grid. A sheet of clear polyvinyl alcohol is heated and stretched in a given direction, its long hydrocarbon molecules becoming aligned in the process. The sheet is then dipped into an ink solution rich in iodine. The iodine impregnates the plastic and attaches to the straight long-chain polymeric molecules effectively forming a chain of its own. The conduction electrons associated with the iodine can move along the chains as if they were long thin wires. The component of **E** in an incident wave which is parallel to the molecules drives the electrons, does work on them, and is strongly absorbed. The transmission axis of the polarizer is therefore perpendicular to the direction in which the film was stretched.

Each separate miniscule dichroic entity is known as a *dichromophore*. In *H*-sheet the dichromophores are of molecular dimensions and so scattering represents no problem. *H*-sheet is a very effective polarizer across the entire visible spectrum but is somewhat less so at the blue end.

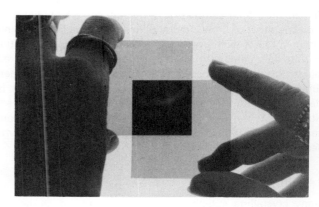

Fig. 8.14 A pair of crossed polaroids. Each polaroid appears gray because it absorbs roughly half the incident light. [Photo by E.H.]

When viewing a bright white light through a pair of crossed *H*-sheet polaroids, as in Fig. 8.14, the *extinction* color will be a deep blue as a result of this leakage. *HN*-50 would be the designation of a hypothetical, ideal *H*-sheet having a *neutral color* (*N*) and transmitting 50% of the incident natural light while absorbing the other 50%, which is the undesired polarization component. In practice however, about 4% of the incoming light will be reflected back at each surface (antireflection coatings are not generally used) leaving 92%. Half of this is presumably absorbed and thus we might contemplate an *HN*-46 polaroid. Actually, large quantities of *HN*-38, *HN*-32, and *HN*-22, each differing by the amount of iodine present, are produced commercially and are readily available (Problem 8.4).

Many other forms of polaroid have been developed.* *K*-sheet, which is humidity and heat resistant, has as its dichromophore the straight-chain hydrocarbon polyvinylene. A combination of the ingredients of *H*- and *K*-sheets leads to *HR*-sheet, a near-infrared polarizer.

Polaroid vectograph is a commercially available material designed to be incorporated in a process for making three-dimensional photographs. It can, however, be used to produce some rather thought-provoking if not mystifying demonstrations. Vectograph film is a water-clear plastic laminate of two sheets of polyvinyl alcohol arranged so that their stretch directions are at right angles to each other. In this form there are no conduction electrons available and the film is not a polarizer. Using an iodine solution, imagine

---

* See *Polarized Light: Production and Use* by Shurcliff or its more readable little brother *Polarized Light* by Shurcliff and Ballard.

that we draw an X on one side of the film and a Y overlapping it on the other. Under natural illumination the light passing through the X will be in a $\mathscr{P}$-state perpendicular to the $\mathscr{P}$-state light coming from the Y. In other words, the painted regions form two crossed polarizers. Both will be seen superimposed on each other. Now, if the vectograph is viewed through a linear polarizer which can be rotated, either the X, the Y, or both will be seen. Obviously more imaginative drawings can be made (one need only remember to make the one on the far side backwards).

8.4 BIREFRINGENCE

Many crystalline substances (i.e. solids whose atoms are arranged in some sort of regular repetitive array) are *optically anisotropic*. In other words, their optical properties are not the same in all directions within any given sample. The dichroic crystals of the previous section are but one special subgroup. We saw there that if the crystal's lattice atoms were not completely symmetrically arrayed, the binding forces on the electrons would be anisotropic. Earlier, in Fig. 3.12(b) we represented the isotropic oscillator using the simple mechanical model of a spherical charged shell bound by identical springs to a fixed point. This was a fitting representation for *optically isotropic* substances (amorphous solids like glass and plastic are usually, but not always, isotropic). Figure 8.15 again shows a charged shell, this time bound by springs of differing stiffness, i.e. having different spring constants. An electron which is displaced from equilibrium along a direction parallel to one set of "springs" will evidently oscillate with a different characteristic frequency than it would were it displaced in some other direction. As we have pointed out previously (Section 3.3.2) light propagates through a transparent substance by exciting the electrons within the medium. The electrons are driven by the **E**-field and they reradiate; these secondary wavelets recombine and the resultant refracted wave moves on. The speed of the wave, and therefore the index of refraction, is determined by the difference between the frequency of the **E**-field and the natural or characteristic frequency of the electrons. *An anisotropy in the binding force will therefore be manifest in an anisotropy in the refractive index.* For example, if $\mathscr{P}$-state light were to move through some hypothetical crystal so that it encountered electrons which could be represented by Fig. 8.15, its speed would be governed by the orientation of **E**. If **E** were parallel to the stiff springs, i.e. in a direction of strong binding, here along the x-axis, the electron's natural frequency would be high

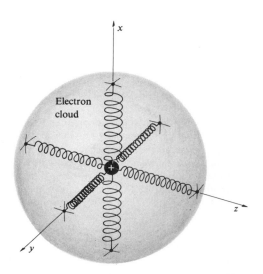

Fig. 8.15 Mechanical model depicting a negatively charged shell bound to a positive nucleus by pairs of springs having different stiffness.

(proportional to the square root of the spring constant). In contrast, with **E** along the y-axis, where the binding force is weaker, the natural frequency would be somewhat lower. Keeping in mind our earlier discussion of dispersion and the $n(\omega)$ curve of Fig. 3.14, the appropriate indices of refraction might look like those of Fig. 8.16. A material of this sort which displays two different indices of refraction is said to be *birefringent*.* If the crystal is such that the frequency of the incident light appears in the vicinity of ω_d, in Fig. 8.16, it resides in the absorption band of $n_y(\omega)$. A crystal so illuminated will be strongly absorbing for one polarization direction (y) and transparent for the other (x). Clearly, a birefringent material which absorbs one of the orthogonal $\mathscr{P}$-states, passing on the other, is in fact *dichroic*. Further, suppose that the crystal symmetry is such that the binding forces in the y- and z-directions are identical, i.e. each of these springs has the same natural frequency and they are equally lossy. The x-axis now defines the direction of the *optic axis*. Inasmuch as a crystal can be represented by an array of these oriented anisotropic charged oscillators, *the optic axis is actually a direction and not merely a single line.*

---

* The word *refringence* used to be used instead of our present day term *refraction*. It comes from the latin *refractus* by way of an etymological route beginning with *frangere* meaning to break.

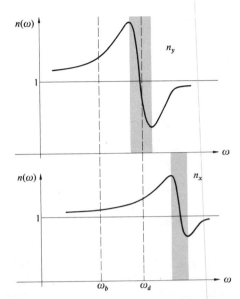

Fig. 8.16 Refractive index versus frequency along two axes in a crystal. Regions where $dn/d\omega < 0$ correspond to absorption bands.

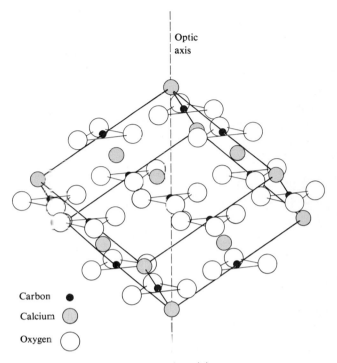

Carbon ●
Calcium ◯ (shaded)
Oxygen ◯

Fig. 8.17 Arrangement of atoms in calcite.

The model works rather nicely for dichroic crystals since if light were to propagate along the optic axis (**E** in the yz-plane) it would be strongly absorbed, while if it moved normal to that axis, it would emerge linearly polarized.

Often the characteristic frequencies of birefringent crystals are above the optical range and they appear colorless. This is represented by Fig. 8.16 where the incident light is now considered to have frequencies in the region of ω_b. Two different indices are apparent but absorption for either polarization is negligible. Equation (3.36) shows that $n(\omega)$ varies inversely with the natural frequency. This means that a large effective spring constant, i.e. strong binding, corresponds to a low polarizability, a low dielectric constant and a low refractive index.

We will construct, if only pictorially, a linear polarizer utilizing birefringence by causing the two orthogonal $\mathscr{P}$-states to follow different paths and thus actually separate. But even more fascinating things can be done with birefringent crystals as we shall see later.

8.4.1 Calcite

Let's now spend a moment relating the above ideas to an actual and somewhat typical birefringent crystal, viz. calcite. Calcite or calcium carbonate ($CaCO_3$) is a rather common naturally occurring substance. Both marble and limestone are made up of many small calcite crystals bonded together. Of particular interest are the beautiful large single crystals which, although becoming rare, can still be found.

Figure 8.17 shows the distribution of carbon, calcium and oxygen within the calcite structure; while Fig. 8.18 is a view from above looking down along what has, in anticipation, been labeled the optic axis in Fig. 8.17. Each CO_3 group forms a triangular cluster whose plane is perpendicular to the optic axis. Notice that if we rotate Fig. 8.18 about a line normal to and passing through the center of any one of the carbonate groups, the same exact configuration of atoms would appear three times during each revolution. The direction we have designated as the optic axis corresponds to a rather special crystallographic orientation, in that it is an axis of 3-*fold symmetry*. The large birefringence displayed by calcite arises from the fact that the carbonate groups are all in planes normal to the optic axis. The behavior of their electrons, or rather the mutual interaction of the induced oxygen dipoles, is markedly different when **E** is either in or normal to those planes (Problem 8.13). In any event the asymmetry is clear enough.

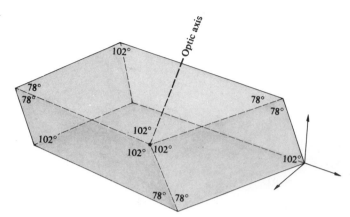

Fig. 8.19 Calcite cleavage form.

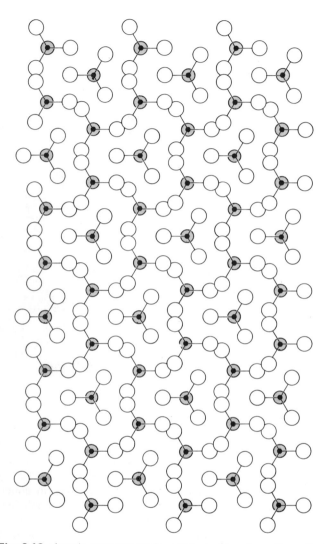

Fig. 8.18 Atomic arrangement for calcite looking down the optical axis.

Calcite samples can readily be split, forming smooth surfaces known as *cleavage planes*. The crystal is essentially made to come apart between specific planes of atoms where the interatomic bonding is relatively weak. All cleavage planes in calcite (Fig. 8.18) are normal to three different directions. As a crystal grows, atoms are added layer upon layer following the same pattern. But more raw material may be available to the growth process on one side than another resulting in a crystal having an externally complicated shape. Even so, the cleavage planes are dependent on the atomic configuration and if one cuts a sample so that each surface is a cleavage plane, its form will be related to the basic arrangement of its atoms. Such a specimen is referred to as a *cleavage form*. In the case of calcite it is a rhombohedron, with each face a parallelogram whose angles are 78° 5′ and 101° 55′ (Fig. 8.19). Note that there are only two *blunt corners* where the surface planes meet to form three obtuse angles. A line passing through the vertex of either of the blunt corners, so oriented that it makes equal angles with each face (45.5°) and each edge (63.8°), is clearly an axis of 3-fold symmetry. (This would be a bit more obvious if we cut the rhomb to have edges of equal length.) Evidently such a line must correspond to the optic axis. Whatever the natural shape of a particular calcite specimen, you need only find a *blunt corner* and you have the optic axis.

In 1669 Erasmus Bartholinus (1625–92), doctor of medicine and professor of mathematics at the University of Copenhagen (and incidently, Römer's father-in-law), came upon a new and remarkable optical phenomenon in calcite which he called *double refraction*. Calcite had been discovered not long before near Eskifjordur in Iceland and was then known as *Iceland spar*. In the words of Bartholinus:*

> Greatly prized by all men is the diamond, and many are the joys which similar treasures bring, such as precious stones and pearls . . . but he, who, on the other hand, prefers the knowledge of unusual phenomena to these delights, he will, I hope, have

---

* W. F. Magie, *A Source Book in Physics*.

Fig. 8.20 Double image formed by a calcite crystal (not cleavage form). [Photo by E.H.]

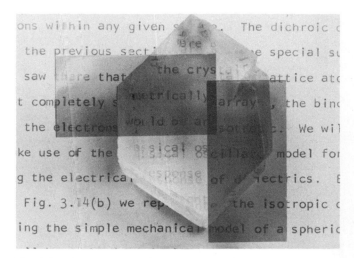

Fig. 8.21 A calcite crystal (blunt corner on the bottom). The transmission axes of the two polarizers are parallel to their short edges. Where the image is doubled the lower undeflected one is the ordinary image. Take a long look, there's a lot in this one. [Photo by E.H.]

no less joy in a new sort of body, namely, a transparent crystal, recently brought to us from Iceland, which perhaps is one of the greatest wonders that nature has produced. . . .

As my investigation of this crystal proceeded there showed itself a wonderful and extraordinary phenomenon: objects which are looked at through the crystal do not show, as in the case of other transparent bodies, a single refracted image, but they appear double.

The double image referred to by Bartholinus is quite evident in the photograph of Fig. 8.20. If we send a narrow beam of natural light into a calcite crystal normal to a cleavage plane it will split and emerge as two parallel beams. To see the same effect quite simply, we need only place a black dot on a piece of paper and then cover it with a calcite rhomb. The image will now consist of two gray dots (black where they overlap). Rotating the crystal will cause one of the dots to remain stationary while the other appears to move in a circle about it, following the motion of the crystal. The rays forming the fixed dot, which is the one invariably closer to the upper blunt corner, behave as if they had merely passed through a plate of glass. In accord with a suggestion made by Bartholinus they are known as the *ordinary* or *o-rays*. The rays coming from the other dot, which behave in such an unusual fashion, are known as the *extraordinary* or *e-rays*. If the crystal is now examined through an analyzer, it will be found that the ordinary and extraordinary images are linearly polarized (Fig. 8.21). Moreover, the two emerging $\mathscr{P}$-states are orthogonal.

Any number of planes can be drawn through the rhomb so as to contain the optic axis and these are all called *principal planes*. More specifically, if the principal plane is also normal to a pair of opposite surfaces of the cleavage form, it slices the crystal across a *principal section*. There are evidently three of these passing through any one point; each is a parallelogram having angles of 109° and 71°. Figure 8.22 is a diagrammatic representation of an initially unpolarized beam traversing a principal section of a calcite rhomb. The filled-in circles and/or arrows drawn along each ray indicate that the *o*-ray has its electric field vector normal to the principal section, while the field of the *e*-ray is parallel to the principal section.

To simplify matters a bit, let **E** in the incident plane wave be linearly polarized perpendicular to the optic axis as shown in Fig. 8.23. The wave strikes the surface of the crystal, thereupon driving electrons into oscillation and they in turn reradiate secondary wavelets. The wavelets superimpose and recombine to form the refracted wave and the process is repeated over and over again until the wave emerges from the crystal. This represents a cogent physical argument for applying the ideas of Huygens' principle. Huygens himself, although without benefit of electromagnetic theory, used his construction to successfully explain many aspects of double refraction in calcite as long ago as 1690. It should be made clear from the outset however that his treatment is rather

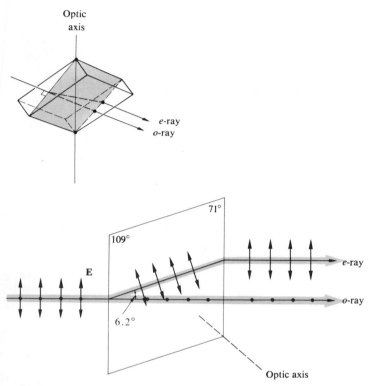

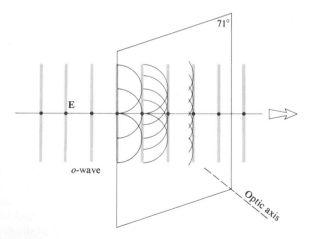

Fig. 8.22 A light beam with two orthogonal field components traversing a calcite principal section.

Fig. 8.23 An incident plane wave polarized perpendicular to the principal section.

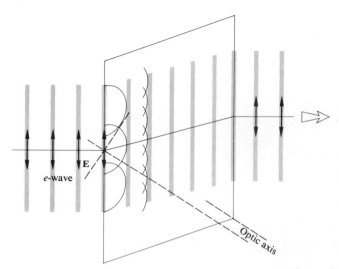

Fig. 8.24 An incident plane wave polarized parallel to the principal section.

incomplete,* in which form it is appealingly, although deceptively, simple.

Inasmuch as the **E**-field is perpendicular to the optic axis, one assumes that every point on the wavefront (which initially corresponds to the surface) acts as a source of spherical wavelets, all of which are in phase. Presumably, as long as the *field of the wavelets is everywhere normal to the optic axis*, they will expand into the crystal in all directions with a speed $v_\perp$, as they would in an isotropic medium. (Keep in mind that the speed is a function of frequency.) Since the *o*-wave displays no anomalous behavior, this assumption seems a reasonable one. The envelope of the wavelets is essentially a portion of a plane wave which in turn serves as a distribution of secondary point sources. The process continues and the wave moves straight across the crystal.

In contrast, consider the incident wave of Fig. 8.24 whose **E**-field is parallel to the principal section. Notice that **E** now has a component normal to the optic axis as well as a component parallel to it. Since the medium is birefringent, light of a given frequency polarized parallel to the optic axis propagates with a speed $v_\parallel$ where $v_\parallel \neq v_\perp$. In particular for calcite and sodium yellow light ($\lambda = 589$ nm), $1.486v_\parallel = 1.658v_\perp = c$. What kind of Huygens' wavelets can we expect now? At the risk of oversimplifying matters

---

* A. Sommerfeld, *Optics* p. 148.

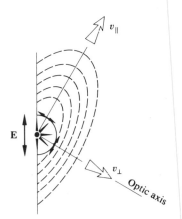

Fig. 8.25 Wavelets within calcite.

we represent each e-wavelet, for the moment at least, as a small sphere (Fig. 8.25). But $v_{\parallel} > v_{\perp}$ so that the wavelet will elongate in all directions normal to the optic axis. We therefore speculate, as Huygens did, that the secondary wavelets associated with the e-wave are ellipsoids of revolution about the optic axis. The envelope of all of the ellipsoidal wavelets is essentially a portion of a plane wave parallel to the incident wave. This plane wave will, however, evidently undergo a sidewise displacement in traversing the crystal. The beam moves in a direction parallel to the lines connecting the origin of each wavelet and the point of tangency with the planar envelope. It is known as the *ray direction and corresponds to the direction in which energy propagates.* This is an instance in which the direction of the ray is not normal to the wavefront.

Now then, if the incident beam is natural light both situations depicted in Figs. 8.23 and 8.24 will exist simultaneously with the result that the beam will split into two orthogonal linearly polarized beams (Fig. 8.22). You can actually see the two diverging beams within a crystal by using a properly oriented narrow laser beam (**E** neither normal nor parallel to the principal plane, which is usually the case). Light will scatter off internal flaws making its path fairly visible.

The electromagnetic description of what is happening is rather complicated but well worth examining at this point, even if only superficially. Recall from Chapter 3 that the incident **E**-field will polarize the dielectric, i.e. it will shift the distribution of charges thereby creating electric dipoles. The field within the dielectric is thus altered by the inclusion of an induced field and one is led to introduce a new quantity, the *displacement* **D** (see Appendix 1). In isotropic media **D**

is related to **E** by a scalar quantity and the two are therefore always parallel. In anisotropic crystals **D** and **E** are related by a tensor and are most certainly not always parallel. If we now apply Maxwell's equations to the problem of a wave moving through such a medium, we find that the fields vibrating within the wavefront are **D** and **B** and not, as before, **E** and **B**. In other words, the propagation vector **k**, which is normal to the surfaces of constant phase, is now perpendicular to **D** rather than **E**. In fact, **D**, **E**, and **k** are all coplanar. Clearly then, the *ray direction* corresponds to the direction of the Poynting vector $\mathbf{S} = c^2 \epsilon_0 \mathbf{E} \times \mathbf{B}$ which is generally different from that of **k**. Because of the manner in which the atoms are distributed **E** and **D** will however be collinear when they are both either parallel or perpendicular to the optic axis.* This means that the o-wavelet will encounter an effectively isotropic medium and thus be spherical, having **S** and **k** collinear. In contrast the e-wavelets will have **S** and **k**, or equivalently **E** and **D**, parallel only in directions along or normal to the optic axis. At all other points on the wavelet it is **D** which is tangent to the ellipsoid and therefore it is always **D** which ends up in the envelope or composite planar wavefront within the crystal (Fig. 8.26).

8.4.2 Birefringent Crystals

Cubic crystals like sodium chloride, i.e. common salt, have their atoms arranged in a relatively simple and highly symmetric form. (There are *four* 3-fold symmetry axes each running from one corner to an opposite corner, unlike calcite which has one such axis.) Light emanating from a point source within such a crystal will propagate uniformly in all directions as a spherical wave. As with amorphous solids, there will be no preferred directions in the material. It will have a single index of refraction and be *optically isotropic* (Fig 8.27). In that case all of the springs in the oscillator model will evidently be identical.

Crystals belonging to the *hexagonal, tetragonal*, and *trigonal* systems have their atoms so arranged that light propagating in some general direction will encounter an asymmetric structure. Such substances are optically anisotropic and birefringent. For them, the optic axis corresponds

---

* In the oscillator model the general case corresponds to the situation in which **E** is not parallel to any of the spring directions. The field will drive the charge, but its resultant motion will not be in the direction of **E** because of the anisotropy of the binding forces. The charge will be displaced most, for a given force component, in the direction of weakest restraint. The induced field will thus not have the same orientation as **E**.

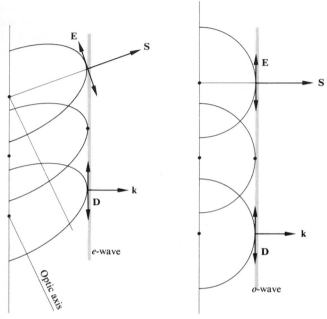

Fig. 8.26 Orientations of the **E**-, **D**-, **S**- and **k**-vectors.

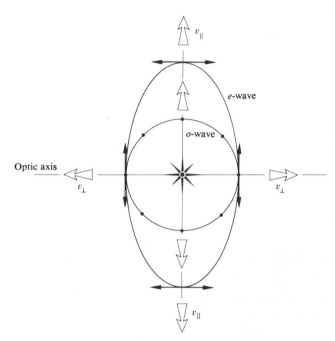

Fig. 8.28 Wavelets in a negative uniaxial crystal.

to a direction about which the atoms are arranged symmetrically. Crystals like these, for which there is only one such direction, are known as *uniaxial*. A point source of natural light imbedded within one of these specimens gives rise to spherical *o*-wavelets and ellipsoidal *e*-wavelets. It

is the orientation of the field with respect to the optic axis which determines the speeds with which these wavelets expand. The **E**-*field of the o-wave is everywhere normal to the optic axis* and so it moves at a speed $v_\perp$ in all directions. Similarly the *e*-wave has a speed $v_\perp$ only in the direction of the optic axis (Fig. 8.25) along which it is always tangent to the *o*-wave. Normal to this direction, **E** *is parallel to the optic axis* and that portion of the wavelet expands at a speed $v_\parallel$ (Fig. 8.28). Uniaxial materials have two principal indices of refraction, $n_o \equiv c/v_\perp$ and $n_e \equiv c/v_\parallel$ (Problem 8.16).

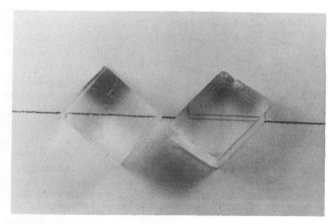

Fig. 8.27 Images in sodium chloride and calcite single crystals. [Photo by E.H.]

Table 8.1 Refractive indices of some uniaxial birefringent crystals ($\lambda = 589.3$ nm).

Crystal	n_o	n_e
Tourmaline	1.669	1.638
Calcite	1.6584	1.4864
Quartz	1.5443	1.5534
Sodium Nitrate	1.5854	1.3369
Ice	1.309	1.313
Rutile (TiO$_2$)	2.616	2.903

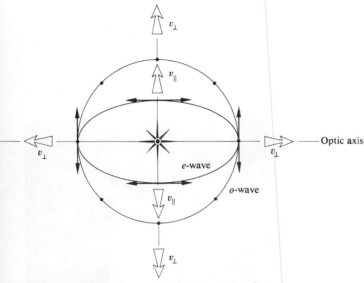

Fig. 8.29 Wavelets in a positive uniaxial crystal.

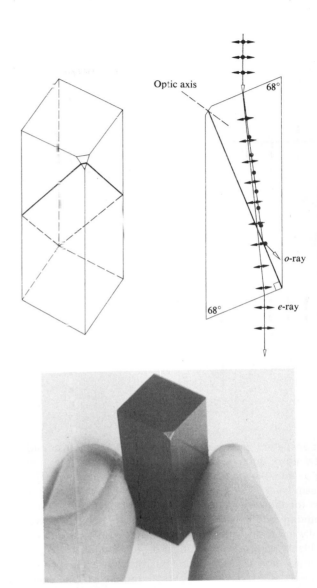

Fig. 8.30 The Nicol prism. (The little flat on the blunt corner locates the optic axis.) [Photo by E.H.]

The difference $\Delta n = (n_e - n_o)$ is a measure of the birefringence. In calcite $v_\parallel > v_\perp$, $(n_e - n_o)$ is -0.172 and it is said to be *negative uniaxial*. In comparison, there are crystals, e.g. quartz (crystallized silicon dioxide) and ice, for which $v_\perp > v_\parallel$. Consequently, the ellipsoidal *e*-wavelets are enclosed within the spherical *o*-wavelets as shown in Fig. 8.29 (quartz is optically active and therefore actually a bit more complicated). In that case, $(n_e - n_o)$ is positive and the crystal is said to be *positive uniaxial*.

The remaining crystallographic systems, namely *orthorhombic*, *monoclinic*, and *triclinic*, have two optic axes and are therefore said to be *biaxial*. Such substances, e.g. mica $[KH_2Al_3(SO_4)_3]$, have three different principal indices of refractions. Each set of springs in the oscillator model would then be different. The birefringence of biaxial crystals is measured as the numerical difference between the largest and smallest of these indices.

8.4.3 Birefringent Polarizers

It will now be a rather easy matter, at least conceptually, to make some sort of linear birefringent polarizer. Any number of schemes for separating the *o*- and *e*-waves have been employed, all of them of course relying on the fact that $n_e \neq n_o$.

The most renowned birefringent polarizer was introduced in 1828 by the Scottish physicist William Nicol (1768–1851). The *Nicol prism*, as it is called, is now mainly

of historical interest, it having long been superseded by other more effective polarizers. Putting it rather succinctly, the device is made by first grinding and polishing the ends (from 71° to 68°; see Fig. 8.23) of a suitably long, narrow calcite rhombohedron; then, after cutting the rhomb diagonally, the two pieces are polished and cemented back together with canada balsam (Fig. 8.30). The balsam cement

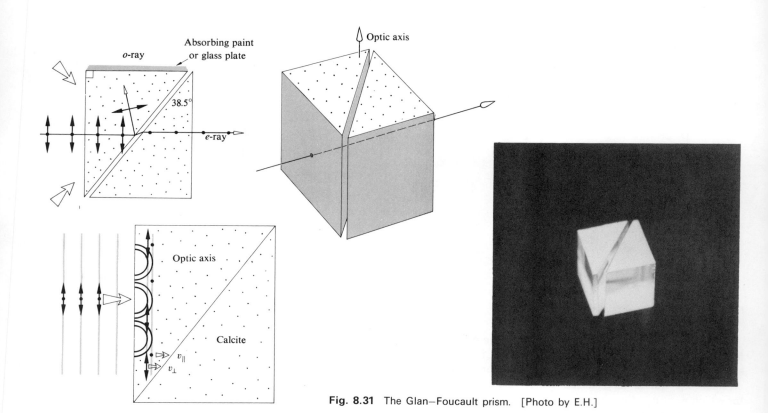

Fig. 8.31 The Glan—Foucault prism.   [Photo by E.H.]

is transparent and has an index of 1.55 almost midway between n_e and n_o. The incident beam enters the "prism", the o- and e-rays are refracted, they separate and strike the balsam layer. The critical angle at the calcite–balsam interface for the o-ray is about 69° (Problem 8.18). The o-ray (entering within a narrow cone of roughly 28°) will be totally internally reflected and thereafter absorbed by a layer of black paint on the sides of the rhomb. The e-ray emerges, laterally displaced, but otherwise essentially unscathed, at least in the optical region of the spectrum (canada balsam absorbs in the ultraviolet).

The *Glan–Foucault polarizer* (Fig. 8.31) is constructed of nothing other than calcite, which is transparent from roughly 5000 nm in the infrared to about 230 nm in the ultraviolet. It therefore can be used over a broad spectral range. The incoming ray strikes the surface normally and **E** can be resolved into components which are either completely parallel or perpendicular to the optic axis. The two rays traverse the first calcite section without any deviation. (We'll come back to this point later on when we talk about

retarders.) Notice that if the angle of incidence on the calcite–air interface is θ, one need only arrange things so that $n_e < 1/\sin \theta < n_o$ in order for the o-ray, and not the e-ray, to be totally internally reflected. If the two prisms are now cemented together (glycerine or mineral oil are used in the ultraviolet) and the interface angle is changed appropriately, the device is known as a Glan–Thompson polarizer. Its field of view is roughly 30° in comparison to about 10° for the Glan–Foucault or *Glan–Air* as it is often called. The latter, however, has the advantage of being able to handle the considerably higher power levels often encountered with lasers. For example, while the maximum irradiance for a *Glan–Thompson* could be about 1 W/cm² (continuous wave as opposed to pulsed), a typical Glan–Air might have an upper limit of 100 W/cm² (continuous wave). The difference is, of course, due to deterioration of the interface cement (and the absorbing paint if it's used).

The *Wollaston prism* is actually a polarizing beamsplitter because it passes both orthogonally polarized components. It can be made of calcite or quartz in the form in-

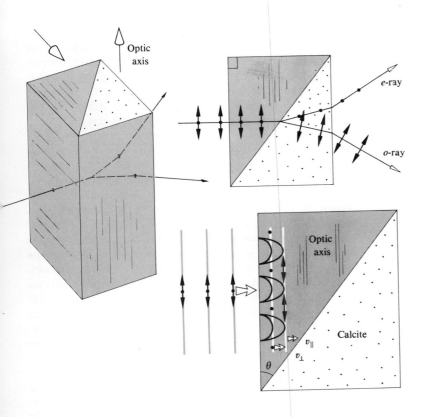

Fig. 8.32 The Wollaston prism.

dicated in Fig. 8.32. Observe that the two component rays separate at the diagonal interface. There, the *e*-ray becomes an *o*-ray changing its index accordingly. In calcite $n_e < n_o$ and the emerging *o*-ray is bent towards the normal. Similarly, the *o*-ray, whose field is initially perpendicular to the optic axis, becomes an *e*-ray in the right-hand section. This time, in calcite the *e*-ray is bent away from the normal to the interface (see Problem 8.19). The deviation angle between the two emerging beams is determined by the prism's wedge angle, θ. Prisms providing deviations ranging from about 15° to roughly 45° are available commercially. They can be purchased cemented, e.g. with castor oil or glycerine, or not cemented at all, i.e. optically contacted, depending on the frequency and power requirements.

8.5 SCATTERING AND POLARIZATION

8.5.1 An Introduction to Scattering

The interaction of light and matter must obviously be one of our prevailing concerns. We can begin to understand a great many apparently unrelated phenomena in terms of differing aspects of the same recurring atomic processes, and so we again return to the electron. When an electromagnetic wave impinges on an atom or molecule it interacts with the bound electron cloud, imparting energy to the atom. We can picture the effect as if the lowest energy or ground state of the atom were set into vibration. The oscillatory frequency of the electron cloud is equal to the driving frequency v, i.e. the frequency of the harmonic **E**-field of the light wave. The amplitude of the oscillation will be large only when v is in the vicinity of the resonant frequency of the atom. In fact, at resonance we can employ the simple description of the atom as first being in its ground state; upon absorbing a photon (having the resonating frequency), it makes the transition to an excited state. In dense media, the atom will most likely return to its ground state having dissipated its excess energy in the form of heat. In rarefied gases the atom will generally make the downward transition by emitting a photon, an effect known as *resonance radiation*.

At frequencies below or above resonance, the electrons vibrating with respect to the nucleus may be regarded as

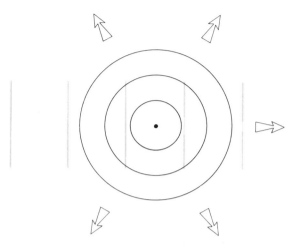

Fig. 8.33 Scattering of a spherical wavelet.

oscillating electric dipoles and as such they will reradiate electromagnetic energy at a frequency which coincides with that of the incident light. This nonresonant emission propagates out in the dipole radiation pattern of Figs. 3.24 and 3.30. *The removal of energy from an incident wave and the subsequent reemission of some portion of that energy is known as scattering* (Fig. 8.33). It is the underlying physical mechanism operative in reflection, refraction and diffraction; the scattering process is fundamental indeed.

In addition to the electron-oscillators which generally have resonances in the ultraviolet, there are atomic-oscillators which correspond to the vibration of the constituent atoms within a molecule. Because of their large masses atomic-oscillators usually have resonances in the infrared. Moreover, they have relatively small vibrational amplitudes and are therefore of little present concern.

The amplitude of an oscillator and therefore the amount of energy removed from the incident wave increases as the frequency of the wave approaches a natural frequency of the atom. For low-density gases where interatomic interactions are negligible, absorption will be insignificant and the reradiated or scattered wave will carry off increasingly more energy as the driving frequency approaches a resonance. This results in some rather interesting effects when the atom's natural frequencies are in the ultraviolet and the incident wave is in the visible region. In that case, as the frequency of the incoming light increases, more and more of it will be scattered. As an example, imagine yourself out of doors on a bright clear morning. The sky is a brilliant blue and you are surrounded, even inundated with blue light. Sunlight streaming into the atmosphere from one direction is scattered in all directions by the air molecules. Without an atmosphere, the daytime sky would be as black as the void of space; a point well made in the Apollo lunar photographs (Fig. 8.34). An observer would then only see light that shone directly at him. With an atmosphere, the red end of the spectrum will be, for the most part, undeviated while the blue or high-frequency end will be substantially scattered. This high-frequency scattered light will reach the observer from many directions and the entire sky will appear bright and blue (Fig. 8.35). When the sun is very low in the sky, its rays pass through a great thickness of air. The blues and violets are scattered sideways out of the beam much more strongly than are the yellows and reds which continue to propagate along a line of sight from the sun to form the earth's familiar fiery sunsets.

Lord Rayleigh was the first to work out the dependence of the scattered flux density on frequency. In accord with Eq. (3.64), which describes the radiation pattern for an oscillating dipole, *the scattered flux density is directly proportional to the fourth power of the driving frequency.* The scattering of light by objects which are small in comparison to the wavelength is known as *Rayleigh scattering.* The molecules of dense transparent media, be they gaseous, liquid or solid, will similarly scatter predominantly bluish light, if only feebly. The effect is quite weak particularly in liquids and solids because the oscillators are arrayed in a more orderly fashion and the reemitted wavelets tend to reinforce each other only in the forward direction, canceling sideways scattering.*

The smoke rising from the end of a lighted cigarette is made up of particles which are smaller than the wavelength of light, and as such, it appears blue when seen against a dark background. In contrast, exhaled smoke contains relatively large water droplets and appears white. Each droplet is larger than the constituent wavelengths of light and thus contains so many oscillators as to be able to sustain the ordinary processes of reflection and refraction. These effects are not preferential to any one frequency component in the incident white light. The light reflected and refracted several times by a droplet and then finally returned to the observer is therefore also white. This accounts for the whiteness of small grains of salt and sugar, fog, clouds, paper, powders,

---

* Recall that you can see the two beams passing through a birefringent calcite crystal only if the sample contains enough flaws to act as scattering centers.

Fig. 8.34 A half-earth hanging in the black moon sky. [Photo courtesy N.A.S.A.]

ground glass, and, more ominously, the typical pallid, polluted city sky.

Particles which are approximately the size of a wave-length (remember that atoms are roughly a fraction of a nanometer across) scatter light in a very distinctive way. A large distribution of such equally sized particles can give rise to a whole range of transmitted colors. In 1883 the volcanic island Krakatoa, located in the Sunda Strait west of Java, blew apart in a fantastic conflagration. Great quantities of fine volcanic dust were spewed high into the atmosphere

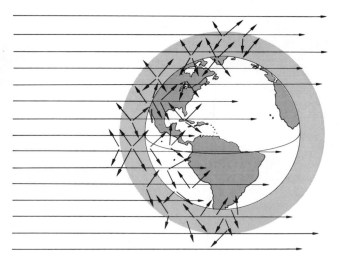

Fig. 8.35 Scattering of sky light.

and drifted over vast regions of the earth. For a few years afterwards the sun and moon repeatedly appeared green or blue and sunrises and sunsets were abnormally colored.

Gustav Mie (1868–1957) in 1908 published a rigorous solution of the scattering problem for homogeneous spherical particles of any size. Although complicated, his solution has great practical value, particularly when applied to studying colloidal and metallic suspensions, interstellar particles, fog, clouds, and the solar corona, to mention only a few.

8.5.2 Polarization by Scattering

Imagine that we have a linearly polarized plane wave incident on an air molecule, as pictured in Fig. 8.36. The orientation of the electric field of the scattered radiation, i.e. E_s, follows the dipole pattern (Section 3.5.3) such that E_s, the Poynting vector **S**, and the oscillating dipole are all co-planar (Fig. 3.27). The vibrations induced in the atom are parallel to the **E**-field of the incoming light wave and so are perpendicular to the propagation direction. Observe once again that the dipole does not radiate in the direction of its axis. Now if the incident wave is unpolarized, it can be represented by two orthogonal, incoherent $\mathscr{P}$-states, in which case the scattered light (Fig. 8.37) is equivalent to a super-position of the conditions shown in Figs. 8.36(a) and (b). Evidently, the scattered light in the forward direction is completely unpolarized; off that axis it is partially polarized, becoming increasingly more polarized as the angle increases. When the direction of observation is normal to the primary beam, the light is completely linearly polarized.

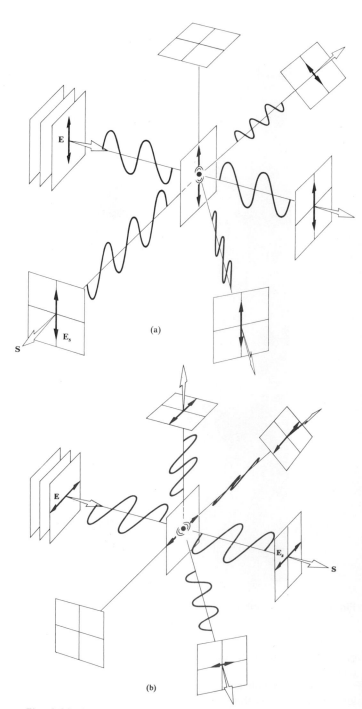

Fig. 8.36 Scattering of polarized light by a molecule.

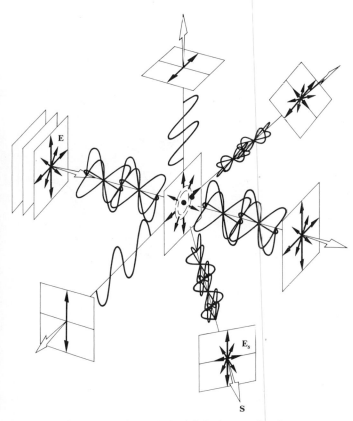

Fig. 8.37 Scattering of unpolarized light by a molecule.

the operative mechanism is Rayleigh scattering. Accordingly, the scattered light will also be partially polarized as anticipated.

Using very much the same ideas Charles Glover Barkla (1877–1944) in 1906 established the transverse wave nature of x-ray radiation by showing that it could be polarized in certain directions as a result of scattering off matter.

You can easily verify these conclusions if you happen to have a piece of polaroid. Locate the sun and then examine a region of the sky at roughly 90° to the solar rays. You'll find that portion of the sky to be quite clearly partially polarized normal to the rays (see Fig. 8.38). It's not completely polarized mainly because of molecular anisotropies, the presence of large particles in the air, and the depolarizing effects of multiple scattering. The latter condition can be illustrated by placing a piece of waxed paper between crossed polaroids (Fig. 8.39). Because the light undergoes a good deal of scattering and multiple reflections within the waxed paper, a given oscillator may "see" the superposition of many essentially unrelated **E**-fields. The resulting emission is almost completely depolarized.

As a final experiment, put a few drops of milk in a glass of water and illuminate it (perpendicular to its axis) using a bright flashlight. The solution will appear bluish white in scattered light and yellowish in direct light indicating that

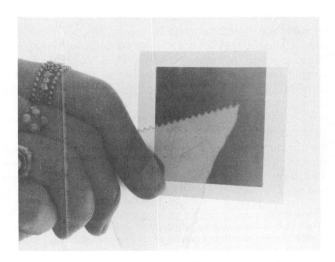

8.6 POLARIZATION BY REFLECTION

One of the most common sources of polarized light is the ubiquitous process of reflection from dielectric media. The glare spread across a window pane, a sheet of paper, or a balding head, the sheen on the surface of a telephone, a billiard ball or a book jacket are all generally partially polarized.

The effect was first studied by Étienne Malus in 1808. The Paris Academy had offered a prize for a mathematical theory of double refraction and Malus accordingly undertook a study of the problem. He was standing at the window of his house in the Rue d'Enfer one evening examining a calcite crystal. The sun was setting and its image reflected toward him from the windows of the Luxembourg Palace not far away. He held up the crystal and looked through it at the sun's reflection. To his astonishment, he saw one of the double images disappear as he rotated the calcite. After the sun had set he continued into the night verifying his observations using candle light reflected from the surfaces of water and glass.* The significance of birefringence and the actual nature of polarized light was becoming clear for the first time. At that time no satisfactory explanation of polarization existed within the context of the wave theory. Within the next thirteen years the work of many men, principally Thomas Young and Augustin Fresnel finally produced the representation of light as some sort of transverse vibration. (Keep in mind that all of this predates the electromagnetic theory of light by roughly forty years.)

The electron-oscillator model leads to a remarkably simple picture of what's happening when light is polarized on reflection. Unfortunately it's not a very complete description since it does not account for the behavior of magnetic nonconducting materials. Nonetheless consider an incoming plane wave linearly polarized so that its **E**-field is perpendicular to the plane of incidence (Fig. 8.40). The wave is refracted at the interface, entering the medium at some transmission angle θ_t. Its electric field drives the bound electrons, in this case normal to the plane of incidence, and they in turn reradiate. A portion of that reemitted energy appears in the form of a reflected wave. It should be clear then from the geometry and the dipole radiation pattern that both the reflected and refracted waves must also be in

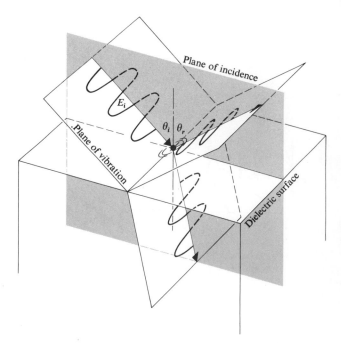

Fig. 8.40 A wave reflecting and refracting at an interface.

$\mathscr{P}$-states normal to the incident plane.* In contradistinction, if the incoming **E**-field is in the incident plane the electron-oscillators near the surface will vibrate under the influence of the refracted wave as shown diagrammatically in Fig. 8.41. Observe that a rather interesting thing is happening to the reflected wave. Its flux density is now relatively low because the reflected ray direction makes a small angle θ with the dipole axis. If we could arrange things so that $\theta = 0$, or equivalently $\theta_r + \theta_t = 90°$, the reflected wave would vanish entirely. *Under those circumstances, for an incoming unpolarized wave made up of two incoherent orthogonal $\mathscr{P}$-states, only the component polarized normal to the incident plane and therefore parallel to the surface will be reflected.* The particular angle of incidence for which this situation occurs is designated by θ_p and referred to as the *polarization angle* or *Brewster's angle*, whereupon $\theta_p + \theta_t = 90°$. Hence, from Snell's law

$$n_i \sin \theta_p = n_t \sin \theta_t$$

---

* Try it with a candle flame and a piece of glass. Hold the glass at $\theta_p \approx 56°$ for the most pronounced effect. At near glancing incidence both of the images will be bright and neither will vanish as you rotate the crystal—Malus apparently lucked in at a good angle to the palace window.

---

* The angle of reflection is determined by the scattering array as discussed in Section 10.2.7. It is found there that the scattered wavelets in general combine constructively in only one direction yielding a reflected ray at an angle equal to that of the incident ray.

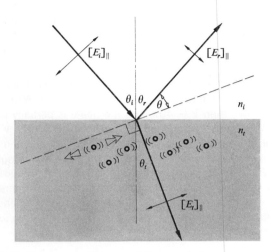

Fig. 8.41 Electron-oscillators and Brewster's Law.

and the fact that $\theta_t = 90° - \theta_p$, it follows that

$$n_i \sin \theta_p = n_t \cos \theta_p$$

and

$$\tan \theta_p = n_t/n_i. \qquad (8.25)$$

This is known as *Brewster's law* after the man who discovered it empirically, Sir David Brewster (1781–1868), professor of physics at St. Andrews University and, of course, inventor of the kaleidoscope.

When the incident beam is in air $n_i = 1$ and if the transmitting medium is glass, in which case $n_t \approx 1.5$, the polarization angle is $\approx 56°$. Similarly if an unpolarized beam strikes the surface of a pond ($n_t \approx 1.33$ for H_2O) at an angle of 53°, the reflected beam will be completely polarized with its **E**-field perpendicular to the plane of incidence or, if you like, parallel to the water's surface. This then suggests a rather handy way to locate the transmission axis of an unmarked polarizer, one just needs a piece of glass or a pond.

The problem encountered in utilizing this phenomenon to construct an effective polarizer lies in the fact that the reflected beam, although completely polarized, is weak; while the transmitted beam, although strong, is only partially polarized. One scheme, illustrated in Fig. 8.42, is often referred to as a *pile-of-plates polarizer*. It was invented by Dominique F. J. Arago in the year 1812. Devices of this kind can be fabricated with glass plates in the visible, silver chloride plates in the infrared and quartz or vycor in the ultraviolet. It's an easy matter to construct a crude arrangement of this sort with a dozen or so microscope slides. (The

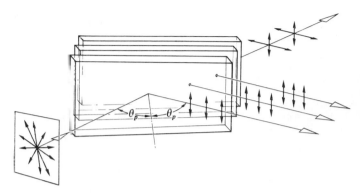

Fig. 8.42 The pile-of-plates polarizer.

beautiful colors that may appear when the slides are in contact are discussed in the next chapter.)

8.6.1 An Application of the Fresnel Equations

In Chapter 4 we obtained a set of formulas known as the Fresnel equations which describe the effects of an incoming electromagnetic plane wave falling on the interface between two different dielectric media. These equations relate the reflected and transmitted field amplitudes to the incident amplitude by way of the angles of incidence θ_i and transmission θ_t. For linear light having its **E**-field parallel to the plane of incidence, we defined the *amplitude reflection coefficient* as $r_{||} \equiv [E_{0r}/E_{0i}]_{||}$, i.e. the ratio of the reflected to incident electric field *amplitudes*. Similarly when the electric field is normal to the incident plane, we have $r_{\perp} \equiv [E_{0r}/E_{0i}]_{\perp}$. The corresponding irradiance ratio (the incident and reflected beams have the same cross-sectional area) is known as the *reflectance* and, since irradiance is proportional to the square of the amplitude of the field,

$$R_{||} = r_{||}^2 = [E_{0r}/E_{0i}]_{||}^2 \quad \text{and} \quad R_{\perp} = r_{\perp}^2 = [E_{0r}/E_{0i}]_{\perp}^2.$$

Squaring the appropriate Fresnel equations yields

$$R_{||} = \frac{\tan^2 (\theta_i - \theta_t)}{\tan^2 (\theta_i + \theta_t)} \qquad (8.26)$$

and

$$R_{\perp} = \frac{\sin^2 (\theta_i - \theta_t)}{\sin^2 (\theta_i + \theta_t)}. \qquad (8.27)$$

Observe that while $R_{\perp}$ can never be zero, $R_{||}$ is indeed zero when the denominator is infinite, i.e. when $\theta_i + \theta_t = 90°$. The reflectance, for linear light with **E** parallel to the plane of incidence, thereupon vanishes; $E_{r||} = 0$ and the beam is

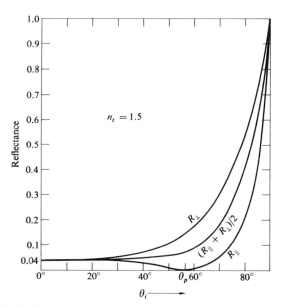

Fig. 8.43 Reflectance versus incident angle.

completely transmitted. This is of course the essence of Brewster's law.

If the incoming light is unpolarized, we can represent it by two now familiar orthogonal, incoherent, equal amplitude, $\mathscr{P}$-states. Incidentally, the fact that they are equal in amplitude means that the amount of energy in one of these two polarization states is the same as that in the other, i.e. $I_{i\parallel} = I_{i\perp} = I_i/2$, which is quite reasonable. Thus

$$I_{r\parallel} = I_{r\parallel} I_i/2 I_{i\parallel} = R_{\parallel} I_i/2$$

and in the same way $I_{r\perp} = R_{\perp} I_i/2$. The reflectance in natural light, $R = I_r/I_i$, is therefore given by

$$R = \frac{I_{r\parallel} + I_{r\perp}}{I_i} = \tfrac{1}{2}(R_{\parallel} + R_{\perp}). \tag{8.28}$$

Figure 8.43 is a plot of Eqs. (8.26), (8.27), and (8.28) for the particular case when $n_i = 1$ and $n_t = 1.5$. The middle curve, which corresponds to incident natural light, shows that only about 7.5% of the incoming light is reflected when $\theta_i = \theta_p$. The transmitted light is then evidently partially polarized. When $\theta_i \neq \theta_p$ both the transmitted and reflected waves are partially polarized.

It is very often desirable to make use of the concept of

the *degree of polarization V*, defined most generally as

$$V = \frac{I_p}{I_p + I_u}, \tag{8.29}$$

in which I_p and I_u are the constituent flux densities of polarized and unpolarized light. For example, if $I_p = 4\,\text{W/m}^2$ and $I_u = 6\,\text{W/m}^2$, then $V = 40\%$ and the beam is partially polarized. With unpolarized light $I_p = 0$ and obviously $V = 0$, while in the opposite extreme if $I_u = 0$, $V = 1$ and the light is completely polarized; thus $0 \le V \le 1$. One frequently deals with partially polarized linear quasimonochromatic light. In that case if we rotate an analyzer in the beam, there will be an orientation in which the transmitted irradiance is maximum (I_{max}) and perpendicular to this, a direction where it is a minimum (I_{min}). Clearly $I_p = I_{\text{max}} - I_{\text{min}}$ and so

$$V = \frac{I_{\text{max}} - I_{\text{min}}}{I_{\text{max}} + I_{\text{min}}}. \tag{8.30}$$

Note that V is actually a property of the beam which may obviously be partially or even completely polarized before encountering any sort of polarizer.

8.7 RETARDERS

We shall now consider a class of optical elements known as *retarders* which serve to change the polarization of an incident wave. In principle the operation of a retarder is quite simple. One of the two constituent coherent $\mathscr{P}$-states is somehow caused to lag in phase behind the other by a predetermined amount. Upon emerging from the retarder, the relative phase of the two components is different than it was initially and thus the polarization state is different as well. Indeed, once having developed the concept of the retarder, we will be able to convert any given polarization state into any other and in so doing create circular and elliptic polarizers as well.

8.7.1 Wave Plates and Rhombs

Recall that a plane monochromatic wave, incident on a uniaxial crystal such as calcite, was generally divided in two, emerging as an ordinary and an extraordinary beam. In contrast, we can cut and polish a calcite crystal so that its optic axis will be normal to both the front and back surfaces (Fig. 8.44). A normally incident plane wave can only have its **E**-field perpendicular to the optic axis. The secondary spherical and ellipsoidal wavelets will be tangent to each other in the direction of the optic axis. The o- and e-waves, which are envelopes of these wavelets, will be coincident and

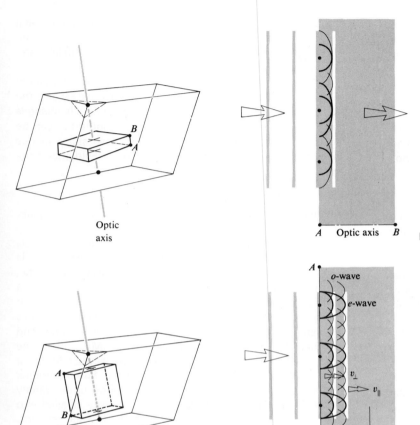

Fig. 8.44 A calcite plate cut perpendicular to the optic axis.

Fig. 8.45 A calcite plate cut parallel to the optic axis.

a single undeflected plane wave will pass through the crystal; there are no relative phase shifts and no double images.*

Now suppose that the direction of the optic axis is arranged to be parallel to the front and back surfaces as shown in Fig. 8.45. If the **E**-field of an incident monochromatic plane wave has components parallel and perpendicular to the optic axis, two separate plane waves will

propagate through the crystal. Since $v_{\parallel} > v_{\perp}$, $n_o > n_e$ and the e-wave will move across the specimen more rapidly than the o-wave. After traversing a plate of thickness d the resultant electromagnetic wave is the superposition of the e- and o-waves which now have a relative phase difference of $\Delta\varphi$. Keep in mind that these are harmonic waves of the same frequency whose **E**-fields are orthogonal. Now then, the relative optical path difference is given by

$$\Lambda = d(|n_o - n_e|) \tag{8.31}$$

and since $\Delta\varphi = k_0 \Lambda$

$$\Delta\varphi = \frac{2\pi}{\lambda_0} d(|n_o - n_e|) \tag{8.32}$$

---

* If you have a calcite rhomb, find the blunt corner and orient the crystal until you are looking along the direction of the optic axis through one of the faces. The two images will converge until they completely overlap.

where λ_0, as always, is the wavelength in vacuum (the form containing the absolute value of the index difference is the most general statement). The state of polarization of the emergent light evidently depends on the amplitudes of the incoming orthogonal field components and of course on $\Delta\varphi$.

The Full-wave Plate

If $\Delta\varphi$ is equal to 2π, the *relative retardation* is one wavelength; the *e*- and *o*-waves are back in phase and there is no observable effect on the polarization of the incident monochromatic beam. When the *relative retardation* $\Delta\varphi$, which is also known as the *retardance*, is 360° the device is called a *full-wave plate*. (This does not mean that $d = \lambda$.) In general the quantity $|n_o - n_e|$ in Eq. (8.32) changes little over the optical range so that $\Delta\varphi$ varies effectively as $1/\lambda_0$. Evidently a *full-wave plate* can only function in the manner discussed for a particular wavelength and retarders of this sort are thus said to be *chromatic*. If such a device is placed at some arbitrary orientation between crossed linear polarizers, all of the light entering it, and in this case let it be white light, will be linear. Only the one wavelength which satisfies Eq. (8.32) will pass through the retarder unaffected, thereafter to be absorbed in the analyzer. All other wavelengths will undergo some retardance and will accordingly emerge from the wave plate as various forms of elliptical light. Some portion of this light will proceed through the analyzer, finally emerging as the complementary color to that which was extinguished. It is a common error to assume that a full-wave plate behaves as if it were isotropic at all frequencies; it obviously doesn't.

Recall that in calcite, the wave whose **E**-field vibrations are parallel to the optic axis travels fastest, i.e. $v_\parallel > v_\perp$. The direction of the optic axis in a *negative* uniaxial retarder is therefore often referred to as the *fast axis* while the direction perpendicular to it is the *slow axis*. For *positive* uniaxial crystals like quartz these principal axes are reversed, the slow axis now corresponding to the optic axis.

The Half-wave Plate

A retardation plate which introduces a relative phase difference of π radians or 180° between the *o*- and *e*-waves is known as a *half-wave plate*. Suppose that the plane of vibration of an incoming beam of linear light makes some arbitrary angle θ with the fast axis as shown in Fig. 8.46. In a negative material the *e*-wave will have a higher speed (same v) and a longer wavelength than the *o*-wave. On emerging from the plate there will be a relative phase shift of $\lambda_0/2$ (i.e. 2π radians/2) with the effect that **E** will have rotated

through 2θ. Going back to Fig. 8.7 it should be evident that a half-wave plate will similarly flip elliptical light. In addition, it will invert the handedness of circular or elliptical light, changing right to left and vice versa.

As the *e*- and *o*-waves progress through any retardation plate, their relative phase difference $\Delta\varphi$ increases and the state of polarization of the wave therefore gradually changes from one point in the plate to the next. Figure 8.7 can be envisioned as a sampling of a few of these states at one instant in time taken at different locations. Evidently if the thickness of the material is such that

$$d(|n_o - n_e|) = (2m + 1)\lambda_0/2,$$

where $m = 0, 1, 2, \ldots$, it will function as a half-wave plate ($\Delta\varphi = \pi, 3\pi, 5\pi$, etc.).

Calcite, although its behavior is simple to visualize, is not actually often used to make retardation plates. It is quite brittle and difficult to handle in thin slices but more than that its birefringence, the difference between n_e and n_o, is a bit too large for convenience. On the other hand, quartz with its much smaller birefringence is frequently used, but it has no natural cleavage planes and must be cut, ground, and polished and is therefore rather expensive. Most often, one uses the biaxial crystal mica. There are several forms of mica which serve the purpose admirably, e.g. fluophlogopite, biotite or muscovite. The most commonly occurring variety is the pale brown muscovite. It is very easily cleaved into strong flexible, and exceedingly thin, large-area sections. Moreover, its two principal axes are almost exactly parallel to the cleavage planes. Along those axes the indices are about 1.599 and 1.594 for sodium light and although these numbers vary slightly from one sample to the next, their difference is fairly constant. The minimum thickness of a mica half-wave plate is about 60 microns.

Retarders are also made from sheets of polyvinyl alcohol which have been stretched so as to align their long-chain organic molecules. Because of the evident anisotropy, electrons in the material do not experience the same binding forces along and perpendicular to the direction of these molecules. Substances of this sort are therefore permanently birefringent even though they are not crystalline.

You can make a rather nice half-wave plate by just attaching a strip of ordinary (glossy) cellophane tape over the surface of a microscope slide. The fast axis, i.e. the vibration direction of the faster of the two waves, corresponds to the transverse direction across the tape's width while the slow axis is along its length. During its manufacture cellophane is formed into sheets and in the process its

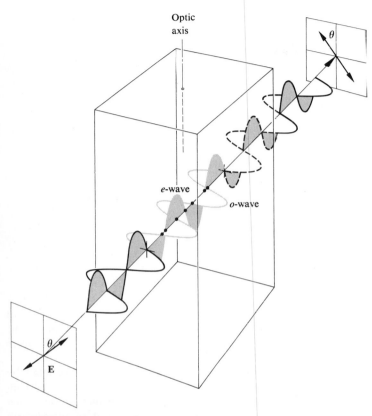

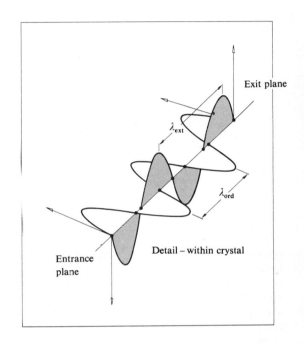

Fig. 8.46 A half-wave plate.

molecules become aligned leaving it birefringent. If you put your half-wave plate between crossed linear polarizers it will show no effect when its principal axes coincide with those of the polarizers. If, however, it is set at 45° with respect to the polarizer, the **E**-field emerging from the tape will be flipped 90° and will thus be parallel to the transmission axis of the analyzer. Light will pass through the region covered by the tape as if it were a hole cut in the black background of the crossed polarizers (Fig. 8.47). A piece of cellophane wrapping (e.g. from certain cigarette packs) will generally also function as a half-wave plate. See if you can determine the orientation of each of its principal axes using the tape retarder and crossed polaroids. (Notice the fine parallel ridges on the sheet cellophane.)

The Quarter-wave Plate

The quarter-wave plate is an optical element which introduces a relative phase shift of $\Delta\varphi = \pi/2$ between the con-

Fig. 8.47 A hand holding a piece of Scotch tape stuck to a microscope slide between two crossed polaroids. [Photo by E.H.]

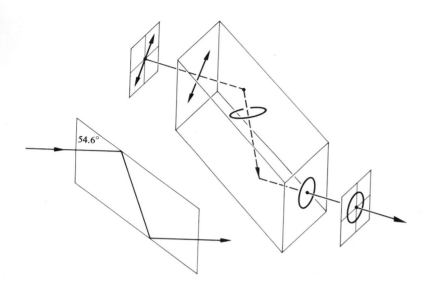

54.6°

Fig. 8.48 The Fresnel rhomb.

stituent orthogonal *o*- and *e*-components of a wave. It
follows once again from Fig. 8.7 that a phase shift of 90° will
convert linear to elliptical light and vice versa. It should be
apparent that linear light incident parallel to either principal
axis will be unaffected by any sort of retardation plate. You
can't have a *relative* phase difference without having two
components. With incident *natural* light, the two constituent
$\mathscr{P}$-states are incoherent, i.e. their relative phase difference
changes randomly and rapidly. The introduction of an
additional constant phase shift by any form of retarder will
therefore still result in a random phase difference and thus
have no noticeable effect. When linear light at 45° to either
principal axis is incident on a quarter-wave plate, its *o*- and
e-components have equal amplitudes. Under these special
circumstances a 90° phase shift converts the wave into
circular light. Similarly, an incoming circular beam will
emerge linearly polarized.

Quarter-wave plates are also usually made of quartz,
mica or organic polymeric plastic. In any case, the thickness
of the birefringent material must satisfy the expression
$d(|n_o - n_e|) = (4m + 1)\lambda_0/4$. You can make a crude
quarter-wave plate using household plastic food-wrap, the
thin stretchy stuff that comes on rolls. Like cellophane, it has
ridges running in the long direction which coincides with a
principal axis. Overlap about a half dozen layers, being
careful to keep the ridges parallel. Position the plastic at 45°
to the axes of a polarizer and examine it through a rotating
analyzer. Keep adding one layer at a time until the irradiance

stays roughly constant as the analyzer turns; at that point
you will have circular light and a quarter-wave plate. This
is easier said than done in white light but it's well worth
trying.

Commercial wave plates are generally designated by
their *linear retardation* which might be, for example, 140 nm
for a quarter-wave plate. This simply means that the device
has a 90° retardance only for green light of wavelength
560 nm (i.e. 4×140). The linear retardation is usually not
given quite that precisely; something like 140 ± 20 nm is
more realistic. The retardation of a wave plate can be
increased or decreased from its specified value by tilting it
somewhat. If the plate is rotated about its fast axis, the
retardation will increase while a rotation about the slow
axis has the opposite effect. In this way a wave plate can be
tuned to a specific frequency in a region about its nominal
value.

The Fresnel Rhomb

We saw in Chapter 4 that the process of total internal reflection introduced a relative phase difference between the two
orthogonal field components. In other words, the components parallel and perpendicular to the plane of incidence
were shifted in phase with respect to each other. In glass,
($n = 1.51$) a shift of 45° accompanies internal reflection at
the particular incident angle of 54.6° [Fig. 4.23(e)]. The
Fresnel rhomb shown in Fig. 8.48 utilizes this effect by
causing the beam to be internally reflected twice, thereby

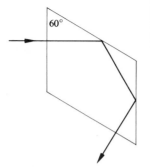

Fig. 8.49 The Mooney rhomb.

imparting a 90° relative phase shift to its components. If the incoming plane wave is linearly polarized at 45° to the plane of incidence, the field components $[E_i]_{\parallel}$ and $[E_i]_{\perp}$ will initially be equal. After the first reflection the wave within the glass will be elliptically polarized. After the second reflection it will be circular. Since the retardance is almost independent of frequency over a large range, the rhomb is essentially an *achromatic* 90° retarder. The Mooney rhomb ($n = 1.65$) shown in Fig. 8.49 is similar in principle, although its operating characteristics are different in some respects.

8.7.2 Compensators

A compensator is an optical device which is capable of impressing a controllable retardance on a wave. Unlike a wave plate where $\Delta\varphi$ is fixed, the relative phase difference arising from a compensator can be varied continuously. Of the many different kinds of compensators we shall only consider two of those that are used most widely. The Babinet compensator, depicted in Fig. 8.50, consists of two independent calcite, or more usually quartz, wedges whose optic axes are indicated by the lines and dots in the figure. A ray passing vertically downward through the device at some arbitrary point will traverse a thickness of d_1 in the upper wedge and d_2 in the lower one. The relative phase difference imparted to the wave by the first crystal is $2\pi d_1(|n_o - n_e|)/\lambda_0$ while that of the second crystal is $-2\pi d_2(|n_o - n_e|)/\lambda_0$. As in the Wollaston prism, which this system closely resembles but which has larger angles and is much thicker, the o and e-rays in the upper wedge become the e and o-rays, respectively, in the bottom wedge. The compensator is thin (the wedge angle is typically about 2.5°) and thus the separation of the rays is negligible. The total phase difference is then

$$\Delta\varphi = \frac{2\pi}{\lambda_0}(d_1 - d_2)(|n_o - n_e|). \qquad (8.33)$$

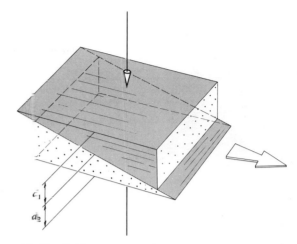

Fig. 8.50 The Babinet compensator.

If the compensator is made of calcite, the e-wave leads the o-wave in the upper wedge and therefore if $d_1 > d_2$, $\Delta\varphi$ corresponds to the total angle by which the e-component leads the o-component. The converse is true for a quartz compensator, i.e., if $d_1 > d_2$, $\Delta\varphi$ is the angle by which the o-wave leads the e-wave. At the center, where $d_1 = d_2$, the effect of one wedge is exactly canceled by the other and $\Delta\varphi = 0$ for all wavelengths. The retardation will vary from point to point over the surface, being constant in narrow regions running the width of the compensator along which the wedge thicknesses are themselves constant. If light enters by way of a slit parallel to one of these regions and if we then move either wedge horizontally with a micrometer screw, we can get any desired $\Delta\varphi$ to emerge.

When the Babinet is positioned at 45° between crossed polarizers a series of parallel equally spaced dark extinction fringes will appear across the width of the compensator. These mark the positions where the device acts as if it were a full-wave plate. In white light the fringes will be colored with the exception of the black central band ($\Delta\varphi = 0$). The retardance of an unknown plate can be found by placing it on the compensator and examining the fringe shift it produces.

The Babinet can be modified to produce a uniform retardation over its surface by merely rotating the top wedge 180° about the vertical so that its thin edge rests on the thin edge of the lower wedge. This configuration will, however, slightly deviate the beam. Another variation of the Babinet which has the advantage of producing a uniform retardance over its surface and no beam deviation is the *Soleil com-*

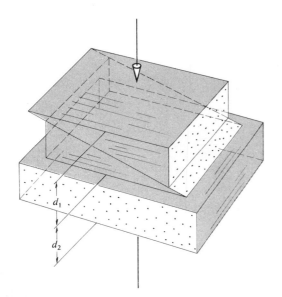

Fig. 8.51 The Soleil compensator.

pensator shown in Fig. 8.51. Generally made of quartz (although MgF_2 and CdS have been used in the infrared), it consists of two wedges and one plane-parallel slab whose optic axes are oriented as indicated. The quantity d_1 now corresponds to the total thickness of both wedges, which is constant for any setting of the positioning micrometer screw.

8.8 CIRCULAR POLARIZERS

Earlier we concluded that linear light whose **E**-field is at 45° to the principal axes of a quarter-wave plate will emerge from that plate circularly polarized. Any series combination of an appropriately oriented linear polarizer and a 90° retarder will therefore perform as a *circular polarizer*. The two elements function completely independently and while one might be birefringent, the other could be of the reflection type. The handedness of the emergent circular light depends on whether the transmission axis of the linear polarizer is at +45° or −45° to the fast axis of the retarder. Either circular state, $\mathscr{L}$ or $\mathscr{R}$, can be generated quite easily. In fact, if the linear polarizer is situated between two retarders, one oriented at plus and the other at minus 45°, the combination will be "ambidextrous". In short, it will yield an $\mathscr{R}$-state for light entering from one side and an $\mathscr{L}$-state when the input is on the other side.

CP-HN is the commercial designation for a popular one-piece circular polarizer. It is a laminate of an HN polaroid and a stretched polyvinyl alcohol 90° retarder. The *input side* of such an arrangement is evidently the face of the linear polarizer. If the beam is incident on the *output side*, i.e. on the retarder, it will thereafter pass through the H-sheet and can only emerge linearly polarized.

A circular polarizer can be used as an analyzer to determine the handedness of a wave which is already known to be circular. To see how this might be done, imagine that we have the four elements labeled A, B, C, and D in Fig. 8.52.

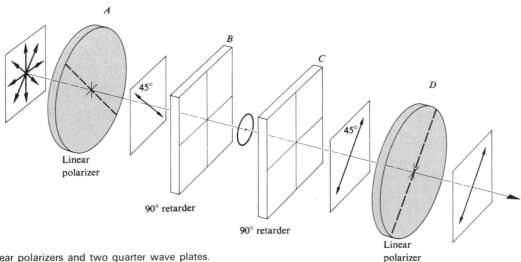

Fig. 8.52 Two linear polarizers and two quarter wave plates.

The first two, A and B, taken together form a circular polarizer as do C and D. The precise handedness of these polarizers is unimportant now as long as they are both the same; which is tantamount here to saying that the fast axes of the retarders are parallel. Linear light coming from A receives a 90° retardance from B at which point it is then circular. In passing through C another 90° retardance is added on, resulting once more in a linearly polarized wave. In effect, B and C together form a half-wave plate which merely flips the linear light from A through a spatial angle of 2θ, in this case 90°. Since the linear wave from C is parallel to the transmission axis of D it passes through it and out of the system. In this simple process we've actually proved something which is rather subtle. If the circular polarizers $A + B$ and $C + D$ are both left-handed, we've shown that *left-circular light entering a left-circular polarizer from the output side will be transmitted.* Furthermore, it should be apparent, at least after some thought, that right-circular light will produce a $\mathcal{P}$-state perpendicular to the transmission axis of D and so will be absorbed. The converse is true as well, i.e. *of the two circular forms, only light in an $\mathcal{R}$-state will pass through a right-circular polarizer having entered from the output side.*

8.9 POLARIZATION OF POLYCHROMATIC LIGHT

8.9.1 Bandwidth and Coherence Time of a Polychromatic Wave

We are again reminded of the fact that by its very nature purely monochromatic light, which is of course not a physical reality, must be polarized. The two orthogonal components of such a wave have the same frequency and each has a constant amplitude. If the amplitude of either sinusoidal component varied, it would be equivalent to the presence of other additional frequencies in the Fourier-analyzed spectrum. Moreover the two components have a constant relative phase difference, i.e. they are coherent. A monochromatic disturbance is an infinite wave train whose properties have been fixed for all time; whether it be in an $\mathcal{R}$-, $\mathcal{L}$-, $\mathcal{P}$- or $\mathcal{E}$-state, the wave is completely polarized.

Actual light sources are polychromatic, that is to say, they emit radiant energy having a range of frequencies. Let's now examine what happens on a submicroscopic scale, paying particular attention to the polarization state of the emitted wave. Envision an electron-oscillator which, having been excited into vibration (possibly by a collision), thereupon radiates. Depending on its precise motion, the oscillator will emit some form of polarized light. As in Section 7.2.6 we

picture the radiant energy from a single atom as a wave train having a finite spatial extent Δx. Assume for the moment that its polarization state is essentially constant for a duration of the order of the coherence time Δt (which, as you recall, corresponds to the temporal extent of the wave train, i.e. $\Delta x/c$). A typical source generally consists of a large collection of such radiating atoms. And these we can envision as oscillating with different phases at some dominant frequency $\bar{\nu}$. Suppose then that we examine the light coming from a very small region of the source, such that the emitted rays arriving at a point of observation are essentially parallel. During a time which is short in comparison with the average coherence time the amplitudes and phases of the wave trains from the individual atoms will be essentially constant This means that if we were to look toward the source in some direction we would, at least for an instant, "see" a coherent superposition of the waves emitted in that direction. In other words we would "see" a resultant wave having a given polarization state. That state would only last for an interval less than the coherence time before it changed, but even so it would correspond to a great many oscillations at the frequency $\bar{\nu}$. Clearly, if the bandwidth $\Delta\nu$ is broad, the coherence time ($\Delta t \sim 1/\Delta\nu$) will be small and any polarization state will be short lived. Evidently *the concepts of polarization and coherence are related in a fundamental way.*

Now consider a wave whose bandwidth is very small in comparison with its mean frequency, i.e. a quasimonochromatic wave. It can be represented by two orthogonal harmonic $\mathcal{P}$-states as in Eqs. (8.1) and (8.2) but here the amplitudes and epoch angles are functions of time. Furthermore, the frequency and propagation number correspond to the mean values of the spectrum present in the wave, namely $\bar{\omega}$ and $\bar{k}$. Thus

$$\mathbf{E}_x(t) = \hat{\mathbf{i}} E_{0x}(t) \cos[\bar{k}z - \bar{\omega}t + \varepsilon_x(t)] \qquad (8.34a)$$

and

$$\mathbf{E}_y(t) = \hat{\mathbf{j}} E_{0y}(t) \cos[\bar{k}z - \bar{\omega}t + \varepsilon_y(t)]. \qquad (8.34b)$$

The polarization state, and accordingly $E_{0x}(t)$, $E_{0y}(t)$, $\varepsilon_x(t)$, $\varepsilon_y(t)$, will vary slowly, remaining essentially constant over a large number of oscillations. Keep in mind that the narrow bandwidth implies a relatively large coherence time. If we watch the wave during a much longer interval, the amplitudes and epoch angles will vary somehow, each either independently or in some correlated fashion. If the variations are completely uncorrelated, the polarization state will remain constant only for an interval small compared to the

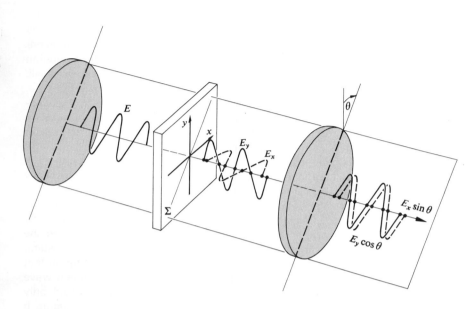

Fig. 8.53 The origin of interference colors.

coherence time. In other words, the ellipse describing the polarization state may change shape, orientation and handedness. Since, speaking practically, no existing detector could discern any one particular state lasting for so short a time, we would conclude that the wave was unpolarized. Antithetically, if the ratio $E_{0x}(t)/E_{0y}(t)$ were constant even though both terms varied, and if $\varepsilon = \varepsilon_y(t) - \varepsilon_x(t)$ were constant as well, the wave would be polarized. Here the necessity for correlation amongst these different functions is quite obvious. Yet we can actually impress these conditions on the wave by merely passing it through a polarizer, thereby removing any undesired constituents. The time interval over which the wave would thereafter maintain its polarization state is no longer dependent on the bandwidth because the wave's components have been appropriately correlated. The light could be polychromatic (even white) and yet completely polarized. It will behave very much like the idealized monochromatic waves treated in Section 8.1. Between these two extremes of completely polarized and unpolarized light is the condition of partial polarization. In fact it can be shown that any quasimonochromatic wave can be represented as the sum of a polarized and an unpolarized wave where the two are independent and either may be zero.

8.9.2 Interference Colors

Insert a crumpled sheet of cellophane between two polaroids illuminated by white light. The resulting pattern will be a profusion of multicolored regions which vary in hue as either polaroid rotates. These *interference colors*, as they are generally called, arise from the wavelength dependence of the retardation. The usual variegated nature of the patterns is due to local variations in thickness, or birefringence, or both.

The appearance of interference colors is quite common and can easily be observed in any number of substances. For example the effect can be seen using a piece of multi-layered mica, a chip of ice, a stretched plastic bag, or finely crushed particles of an ordinary white (quartz) pebble. To appreciate how this phenomenon occurs, examine Fig. 8.53. A narrow beam of monochromatic linear light is schematically shown passing through some small region of a birefringent plate Σ. Over that area the birefringence and thickness are both assumed constant. The transmitted light is most generally elliptical. Equivalently, we envision the light emerging from Σ as composed of two orthogonal linear waves (i.e. the x- and y-components of the total $\mathbf{E}$-field) which have a relative phase difference $\Delta\varphi$ determined by Eq. (8.32). Only the components of these two disturbances which are in the direction of the transmission axis of the analyzer will pass through it and on to the observer. Now these components, which also have a phase difference of $\Delta\varphi$, are coplanar and can thus interfere. When $\Delta\varphi = \pi$, 3π, 5π, etc. they are completely out of phase and cancel each other. When $\Delta\varphi = 0$, 2π, 4π, etc. the waves are in phase and reinforce each other. Suppose then that the retardance arising

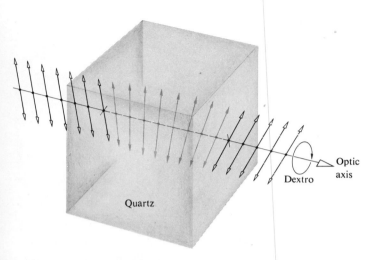

Fig. 8.54 Optical activity displayed by quartz.

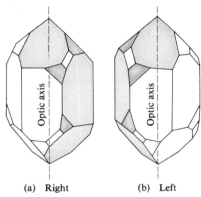

(a) Right   (b) Left

Fig. 8.55 Right- and left-handed quartz crystals.

8.10 OPTICAL ACTIVITY

As is often the case, the manner in which light interacts with material substances can yield a great deal of valuable information about their molecular structures. The process to be examined next, although of specific interest in the study of optics, has had and is continuing to have far-reaching effects in the sciences of chemistry and biology.

In 1811, the French physicist Dominique F. J. Arago first observed the rather fascinating phenomenon now known as *optical activity*. It was then that he discovered that the plane of vibration of a beam of linear light underwent a continuous rotation as it propagated along the optic axis of a quartz plate (Fig. 8.54). At about the same time Jean Baptiste Biot (1774–1862) saw this same effect while using both the vaporous and liquid forms of various natural substances like turpentine. Any such material which causes the **E**-field of an incident linear plane wave to appear to rotate is said to be *optically active*. Moreover, as Biot found, one must distinguish between right- and left-handed rotation. If while looking in the direction of the source, the plane of vibration appears to have revolved clockwise, the substance is referred to as *dextrorotatory* or *d-rotatory* (from the Latin *dextro* meaning right). Alternatively, if **E** appears to have been displaced counterclockwise, the material is *levorotatory* or *l-rotatory* (from the Latin *levo* meaning left).

In 1822, the English astronomer Sir John F. W. Herschel (1792–1871), recognized that *d*-rotatory and *l*-rotatory behavior in quartz actually corresponded to two different crystallographic structures. Although the molecules are identical (SiO_2), crystal quartz can be either right- or left-handed depending on the arrangement of those molecules. As shown in Fig. 8.55 the external appearances of these two

at some point P_1 on Σ for blue light ($\lambda_0 = 435$ nm) is 4π. In that case blue will be strongly transmitted. It follows from Eq. (8.32) that $\lambda_0 \Delta\varphi = 2\pi d(|n_o - n_e|)$ is essentially a constant determined by the thickness and the birefringence. At the point in question therefore, $\lambda_0 \Delta\varphi = 1740\pi$ for all wavelengths. If we now change to incident yellow light ($\lambda_0 = 580$ nm), $\Delta\varphi \approx 3\pi$ and the light from P_1 would be completely canceled. Under white-light illumination that particular point on Σ will seem as if it had removed yellow completely, passing on all the other colors, but none as strongly as blue. Another way of saying this is that the blue light emerging from the region about P_1 is linear ($\Delta\varphi = 4\pi$) and parallel to the analyzer's transmission axis. In contrast, the yellow light is linear ($\Delta\varphi = 3\pi$) and along the extinction axis; the other colors are elliptical. The region about P_1 behaves like a half-wave plate for yellow and a full-wave plate for blue. If the analyzer were now rotated 90° the yellow would be transmitted and the blue extinguished. By definition two colors are said to be complementary when their combination yields white light. Thus on rotating the analyzer through 90° it will alternately transmit or absorb complementary colors. In much the same way there might be a point P_2 somewhere else on Σ where $\Delta\varphi = 4\pi$ for red ($\lambda_0 = 650$ nm). Then, $\lambda_0 \Delta\varphi = 2600\pi$, whereupon green light ($\lambda_0 = 520$ nm) will have a retardance of 5π and be extinguished. Clearly then if the retardance varies from one region to the next over the specimen, so too will the color of the light transmitted by the analyzer.

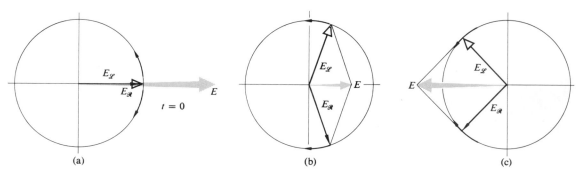

Fig. 8.56 The superposition of an $\mathscr{R}$- and an $\mathscr{L}$-state at $z = 0$.

forms are the same in all respects except that one is the mirror image of the other; they are said to be *enantiomorphs* of each other.   All transparent enantiomorphic substances are optically active.   Furthermore, molten quartz or *fused* quartz, neither of which are crystalline, are not optically active. Evidently, in quartz optical activity is associated with the structural distribution of the molecules as a whole.   There are many substances both organic and inorganic (e.g. benzil and $NaBrO_3$ respectively) which, like quartz, only exhibit optical activity when in crystal form.   In contrast many naturally occurring organic compounds such as sugar, tartaric acid and turpentine are optically active in solution or in the liquid state.   Here the *rotatory power*, as it is often referred to, is evidently an attribute of the individual molecules.   There are also more complicated substances for which optical activity is associated with both the molecules themselves and their arrangement within the various crystals.   This is evidenced, for example, by rubidium tartrate.   A *d*-rotatory solution of that compound will change to *l-rotatory* when crystallized.

In 1825, Fresnel, without addressing himself to the actual mechanism involved, proposed a simple phenomenological description of optical activity.   Since the incident linear wave can be represented as a superposition of $\mathscr{R}$- and $\mathscr{L}$-states, he suggested that these two forms of circular light propagate at different speeds.   An active material shows *circular birefringence*, i.e. it possesses two indices of refraction, one for $\mathscr{R}$-states ($n_{\mathscr{R}}$) and one for $\mathscr{L}$-states ($n_{\mathscr{L}}$). In traversing an optically active specimen the two circular waves would get out of phase and the resultant linear wave would appear to have rotated.   We can see how this is possible analytically by returning to Eqs. (8.8) and (8.9) which described monochromatic right- and left-circular light propagating in the *z*-direction.   It was seen in Eq. (8.10)

that the sum of these two waves is indeed linearly polarized. We now alter these expressions slightly in order to remove the factor of two in the amplitude of Eq. (8.10) in which case

$$\mathbf{E}_{\mathscr{R}} = \frac{E_0}{2}[\hat{\mathbf{i}} \cos (k_{\mathscr{R}}z - \omega t) + \hat{\mathbf{j}} \sin (k_{\mathscr{R}}z - \omega t)] \quad (8.35a)$$

and

$$\mathbf{E}_{\mathscr{L}} = \frac{E_0}{2}[\hat{\mathbf{i}} \cos (k_{\mathscr{L}}z - \omega t) - \hat{\mathbf{j}} \sin (k_{\mathscr{L}}z - \omega t)] \quad (8.35b)$$

represent the right- and left-handed constituent waves. Since ω is constant $k_{\mathscr{R}} = k_0 n_{\mathscr{R}}$ and $k_{\mathscr{L}} = k_0 n_{\mathscr{L}}$. The resultant disturbance is given by $\mathbf{E} = \mathbf{E}_{\mathscr{R}} + \mathbf{E}_{\mathscr{L}}$ and, after a bit of trigonometric manipulation, it becomes

$$\mathbf{E} = E_0 \cos [(k_{\mathscr{R}} + k_{\mathscr{L}})z/2 - \omega t] [\hat{\mathbf{i}} \cos (k_{\mathscr{R}} - k_{\mathscr{L}})z/2 \\ + \hat{\mathbf{j}} \sin (k_{\mathscr{R}} - k_{\mathscr{L}})z/2]. \quad (8.36)$$

At the position where the wave enters the medium ($z = 0$) it is linearly polarized along the *x*-axis as shown in Fig. 8.56, i.e.

$$\mathbf{E} = E_0\hat{\mathbf{i}} \cos \omega t. \quad (8.37)$$

Notice that at any point along the path, both components have the same time dependence and are therefore in phase. This just means that anywhere along the *z*-axis the resultant is linearly polarized (Fig. 8.57), although its orientation is certainly a function of *z*.   Moreover if $n_{\mathscr{R}} > n_{\mathscr{L}}$ or equivalently $k_{\mathscr{R}} > k_{\mathscr{L}}$, $\mathbf{E}$ will rotate counterclockwise while if $k_{\mathscr{L}} > k_{\mathscr{R}}$ the rotation is clockwise (looking toward the source).   Traditionally the angle β through which $\mathbf{E}$ rotates is defined to be positive when it is clockwise.   Keeping this sign convention in mind, it should be clear from Eq. (8.36) that the field at point *z* makes an angle of $\beta = -(k_{\mathscr{R}} - k_{\mathscr{L}})z/2$ with respect

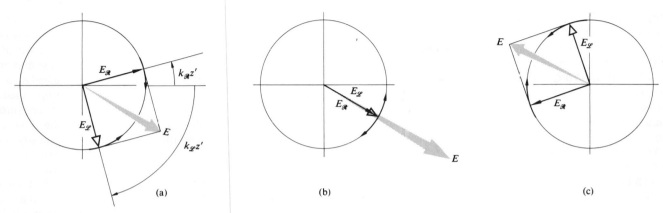

Fig. 8.57 The superposition of an $\mathcal{R}$- and an $\mathcal{L}$-state at $z = z'$ $(k_{\mathcal{L}} > k_{\mathcal{R}})$.

to its original orientation. If the medium has a thickness d the angle through which the plane of vibration rotates is then

$$\beta = \frac{\pi d}{\lambda_0}(n_{\mathcal{L}} - n_{\mathcal{R}}) \qquad (8.38)$$

where $n_{\mathcal{L}} > n_{\mathcal{R}}$ is d-rotatory and $n_{\mathcal{R}} > n_{\mathcal{L}}$ is l-rotatory (Fig. 8.58).

Fresnel was actually able to separate the constituent $\mathcal{R}$- and $\mathcal{L}$-states of a linear beam using the composite

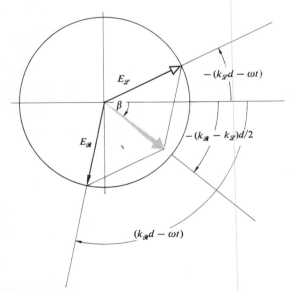

Fig. 8.58 The superposition of an $\mathcal{R}$- and an $\mathcal{L}$-state at $z = d$ $(k_{\mathcal{L}} > k_{\mathcal{R}}, n_{\mathcal{L}} > n_{\mathcal{R}}, \lambda_{\mathcal{L}} < \lambda_{\mathcal{R}}$ and $v_{\mathcal{L}} < v_{\mathcal{R}})$.

prism of Fig. 8.59. It consists of a number of right- and left-handed quartz segments cut with their optic axes as shown. The $\mathcal{R}$-state propagates more rapidly in the first prism than in the second and is thus refracted toward the normal to the oblique boundary. The opposite is true for the $\mathcal{L}$-state and the two circular waves increase in angular separation at each interface.

In sodium light the *specific rotatory power*, which is defined as β/d, is found to be 21.7°/mm for quartz. Thus it follows that $|n_{\mathcal{L}} - n_{\mathcal{R}}| = 7.1 \times 10^{-5}$ for light propagating along the optic axis. In that particular direction ordinary double refraction, of course, vanishes. However, with the incident light propagating normal to the optic axis (as is frequently the case in polarizing prisms, wave plates and compensators) quartz behaves as would any optically inactive, positive, uniaxial crystal. There are other birefringent optically active crystals, both uniaxial and biaxial, such as cinnabar, HgS $(n_o = 2.854, n_e = 3.201)$, which has a rotatory power of 32.5°/mm. Contrastingly, the substance

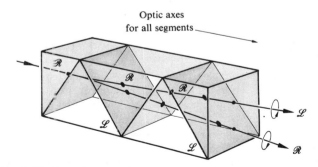

Fig. 8.59 Fresnel composite prism.

$NaClO_3$ is optically active (3.1°/mm) but not birefringent. The rotatory power of liquids in comparison is so relatively small that it is usually specified in terms of 10 cm path lengths; for example, in the case of turpentine ($C_{10}H_6$) it is only −37°/10 cm (10°C with $\lambda_0 = 589.3$ nm). The rotatory power of solutions varies with concentration. This fact is particularly helpful in determining the amount of sugar present in, for example, a urine sample or a commercial sugar syrup.

You can observe optical activity rather easily using colorless corn syrup, the kind available in any grocery store. You won't need much of it since β/d is roughly +30°/inch. Put about an inch of syrup in a glass container between crossed polaroids and illuminate it with a flashlight. The beautiful colors that appear as the analyzer is rotated arise from the fact that β is a function of λ_0, an effect known as *rotatory dispersion*. Using a filter to get roughly monochromatic light, you can readily determine the rotatory power of the syrup.[*]

The first great scientific contribution made by Louis Pasteur (1822−95) came in 1848 and was associated with his doctoral research. He showed that racemic acid, which is an optically inactive form of tartaric acid, is actually composed of a mixture containing equal quantities of right- and left-handed constituents. Substances of this sort which have the same molecular formulas but differ somehow in structure are called *isomers*. He was able to crystallize racemic acid and then separate the two different types of mirror image crystals (enantiomorphs) that resulted. On dissolving these separately in water they formed *d*-rotatory and *l*-rotatory solutions. This implied the existence of molecules that, although chemically the same, were themselves mirror images of each other; such molecules are now known as *optical stereoisomers*. These ideas were the basis for the development of the stereochemistry of organic and inorganic compounds, where one is concerned with the three-dimensional spatial distribution of atoms within a given molecule.

8.10.1 A Useful Model

The phenomenon of optical activity is an extremely complicated one which, although it can be treated in terms of classical electromagnetic theory, actually requires a quantum-mechanical solution.[*] Despite this we will consider a simplified model which will yield a qualitative, yet plausible, description of the process. Recall that we represented an optically isotropic medium by a homogeneous distribution of isotropic electron-oscillators which vibrated parallel to the **E**-field of an incident wave. An optically anisotropic medium similarly was depicted as a distribution of anisotropic oscillators which vibrated at some angle to the driving **E**-field.. We now imagine that the electrons in optically active substances are constrained to move along twisting paths which, for simplicity, are assumed to be helical. In other words such a molecule is pictured much as if it were a conducting helix. The silicon and oxygen atoms in a quartz crystal are known to be arranged in either right- or left-handed spirals about the optic axis as indicated in Fig. 8.60. In the present representation this crystal would correspond to a parallel array of helices. In comparison, an active sugar solution would be analogous to a distribution of randomly oriented helices, each having the same handedness.[†]

In quartz we might anticipate that the incoming wave would interact differently with the specimen depending on whether it "saw" right- or left-handed helices. And thus we could expect different indices for the $\mathscr{R}$- and $\mathscr{L}$-components of the wave. The detailed treatment of the process which leads to circular birefringence in crystals is by no means simple, but at least the necessary asymmetry is evident. How then can a random array of helices, corresponding to a solution, produce optical activity? Let us examine one such molecule in this simplified representation; for example, one whose axis happens to be parallel to the harmonic **E**-field of the electromagnetic wave. That field will drive charges up and down along the length of the molecule, effectively producing a time-varying electric-dipole moment $p(t)$, parallel to the axis. In addition, we now have a current associated with the spiraling motion of the electrons. This in turn generates an oscillating magnetic

---

[*] A gelatin filter would be fine but a piece of colored cellophane will also do nicely. Just remember that the cellophane will act as a wave plate (see Section 8.7.1), so don't put it between the polaroids unless you align its principal axes appropriately.

[*] The review article "Optical Activity and Molecular Dissymmetry," by S. F. Mason, *Contemp. Phys.* **9**, 239 (1968) contains a fairly extensive list of references for further reading.

[†] In addition to these solid and liquid states, there is a third classification of substances which promises to be rather useful because of its remarkable optical properties. It is known as the *mesomorphic* or *liquid crystal* state. Liquid crystals are organic compounds which can flow and yet maintain their characteristic molecular orientations. In particular *cholesteric* liquid crystals have a helical structure and therefore exhibit extremely large rotatory powers; of the order of 40,000°/mm. The pitch of the screw-like molecular arrangement is considerably smaller than that of quartz.

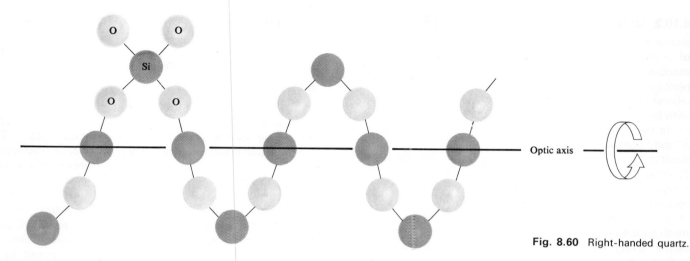

Fig. 8.60 Right-handed quartz.

dipole moment $m(t)$ which is also along the helix axis (Fig. 8.61). In contrast, if the molecule were parallel to the **B**-field of the wave there would be a time-varying flux and thus an induced electron current circulating around the molecule. This would again yield oscillating axial electric and magnetic dipole moments.   In either case *$p(t)$ and $m(t)$ will be parallel or antiparallel to each other depending on the sense of the particular molecular helix.*   Clearly, energy has been removed from the field and both oscillating dipoles will scatter, i.e. reradiate, electromagnetic waves.   The electric field E_p emitted in a given direction by an electric dipole is perpendicular to the electric field E_m emitted by a magnetic dipole. Accordingly, the sum of these, which is the resultant field E_s scattered by a helix, will not be parallel to the incident field E_i along the direction of propagation (the same is of course true for the magnetic fields).   The plane of vibration of the resultant transmitted light ($E_s + E_i$) will thus be rotated in a direction determined by the sense of the helix.   The amount of the rotation will vary with the orientation of each molecule, but it will always be in the same direction for helices of the same sense.

   Although this discussion of optically active molecules as helical conductors is admittedly superficial, the analogy is well worth keeping in mind. In fact, if we direct a linear 3 cm microwave beam onto a box filled with a large number of identical copper helices (e.g. 1 cm long by 0.5 cm in diameter and insulated from each other) the transmitted wave will, indeed, undergo a rotation of its plane of vibration.*

---

* I. Tinoco and M. P. Freeman, "The Optical Activity of Oriented Copper Helices," *J. Phys. Chem.* **61**, 1196 (1957).

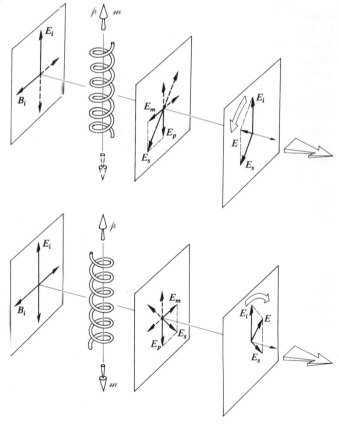

Fig. 8.61 Optical activity via right- and left-handed helices.

8.10.2 Optically Active Biological Substances

Before moving on to other things we should mention a few of what are probably the most fascinating observations associated with optical activity, namely those in the field of biology. Whenever organic molecules are synthesized in the laboratory, an equal number of *d*- and *l*-isomers are produced with the effect that the compound is optically inactive. One might then expect that if they exist at all, equal amounts of *d*- and *l*-optical stereoisomers will be found in natural organic substances. This is by no means the case. Natural sugar (sucrose $C_{12}H_{22}O_{11}$), no matter where it is grown, whether extracted from sugar cane or sugar beets, is always *d*-rotatory. Moreover the simple sugar dextrose or *d*-glucose ($C_6H_{12}O_6$), which as its name implies is *d*-rotatory, is the most important carbohydrate in human metabolism. Evidently, living things can somehow distinguish between optical isomers.

All proteins are fabricated of compounds known as *amino acids*. These in turn are combinations of carbon, hydrogen, oxygen, and nitrogen. There are twenty-odd amino acids and all of them (with the exception of the simplest one, glycine, which is not enantiomorphic) are generally *l*-rotatory. This means that if we break up a protein molecule, whether it comes from an egg or an eggplant, a beetle or a Beatle, the constituent amino acids will be *l*-rotatory. One important exception is the group of antibiotics such as penicillin which do contain some dextro amino acids. In fact, this may well account for the toxic effect penicillin has on bacteria.

It is intriguing to speculate about the possible origins of life on this and other planets. For example, did life on earth originally consist of both mirror image forms? To date, five amino acids have been found in a meteorite which fell in Victoria, Australia on September 28, 1969. Indications of the presence of amino acids have also been observed in lunar samples. Study of the meteorite has revealed the existence of roughly equal amounts of the optically right- and left-handed forms of four of the amino acids. This is in marked contrast to the overwhelming predominance of the left-handed form found in terrestrial rocks. The implications are many and marvelous.*

8.11 INDUCED OPTICAL EFFECTS—OPTICAL MODULATORS

There are a number of different physical effects involving polarized light which all share the single common feature of

_____

* See *Physics Today*, Feb. 1971, p. 17 for additional discussion and references for further reading.

somehow being externally induced. In these instances one exerts an external influence (e.g. a mechanical force, a magnetic or electric field) on the optical medium thereby changing the manner in which it transmits light.

8.11.1 Photoelasticity

In 1816 Sir David Brewster discovered that normally transparent isotropic substances could be made optically anisotropic by the application of mechanical stress. The phenomenon is variously known as *mechanical birefringence*, *photoelasticity* or *stress birefringence*. Under compression or tension the material takes on the properties of a negative or positive uniaxial crystal, respectively. In either case the effective optic axis is in the direction of the stress and the induced birefringence is proportional to the stress. Clearly then, if the stress is not uniform over the sample neither will be the birefringence nor the retardance imposed on a transmitted wave [Eq. (8.32)].

Photoelasticity serves as the basis of a technique for studying the stresses in both transparent and opaque mechanical structures (Fig. 8.62). Improperly annealed or carelessly mounted glass, whether serving as an automobile windshield or a telescope lens, will develop internal stresses that can easily be detected. Information concerning the surface strain on opaque objects can be gotten by bonding photoelastic coatings to the parts under study. More commonly a transparent scale model of the part is made out of a *stress-optically sensitive* material such as epoxy, glyptol or modified polyester resins. The model is then subjected to the forces which the actual component would experience in use. Since the birefringence varies from point to point over the surface of the model, when placed between crossed polarizers, a complicated variegated fringe pattern will reveal the internal stresses. Examine almost any piece of clear plastic or even a block of unflavored gelatin between two polaroids; try stressing it further and watch the pattern change accordingly (Fig. 8.63).

The retardance at any point on the sample is proportional to the *principal stress difference*, i.e. ($\sigma_1 - \sigma_2$) where the sigmas are the orthogonal principal stresses. For example if the sample were a plate under vertical tension, σ_1 would be the maximum principal stress in the vertical direction and σ_2 would be the minimum principal stress, in this case zero, horizontally. In more complicated situations, the principal stresses, as well as their differences, will vary from one region to the next. Under white-light illumination, the loci of all points on the specimen for which ($\sigma_1 - \sigma_2$) is constant are known as *isochromatic regions* and each such

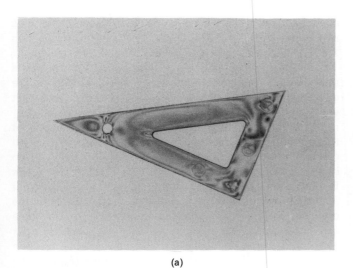

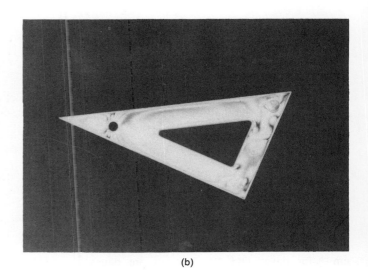

(a)   (b)

Fig. 8.62 A clear plastic triangle between polaroids. [Photo by E.H.]

region corresponds to a particular color. Superimposed on these colored fringes will be a separate system of black bands. At any point where the **E**-field of the incident linear light is parallel to either local principal stress axis, the wave will pass through the sample unaffected regardless of wavelength. With crossed polarizers, that light will be absorbed by the analyzer yielding a black region known as an *isoclinic* band (Problem 8.27). In addition to having the attribute of simply being a beautiful thing to look at, the fringes also

provide both a qualitative map of the stress pattern and a basis for quantitative calculations as well.

8.11.2 The Faraday Effect

Michael Faraday, in 1845, discovered that the manner in which light propagated through a material medium could be influenced by the application of an external magnetic field. In particular, he found that the plane of vibration of linear light incident on a piece of glass rotated when a strong magnetic

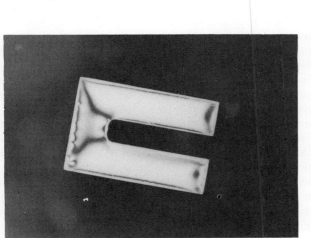

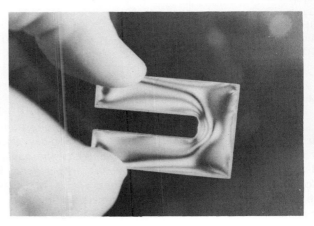

(a)   (b)

Fig. 8.63 A stressed piece of clear plastic between polaroids. [Photo by E.H.]

field was applied in the propagation direction. The *Faraday or magneto-optic effect* was one of the earliest indications of the interrelationship between electromagnetism and light. Although it is reminiscent of *optical activity* there is, as we shall see, an important distinction between these two effects.

The angle β (measured in minutes of arc) through which the plane of vibration rotates is given by the empirically determined expression

$$\beta = \mathcal{V} B d \qquad (8.39)$$

in which B is the static magnetic flux density (usually in gauss), d is the length of medium traversed (in cm) and $\mathcal{V}$ is a factor of proportionality known as the *Verdet constant*. The Verdet constant for a particular medium varies with both frequency (dropping off rapidly as v decreases) and temperature. It is roughly of the order of 10^{-5} min of arc gauss^{-1} cm^{-1} for gases and 10^{-2} min of arc gauss^{-1} cm^{-1} for solids and liquids (see Table 8.2). You can get a better feeling for the meaning of these numbers by imagining that we have, for example, a 1 cm long sample of H_2O in the moderately large field of 10^4 gauss (the earth's field is about one half gauss). In that particular case, a rotation of 2°11' would result since $\mathcal{V} = 0.0131$.

Table 8.2 Verdet constants for some selected substances.

Material	Temperature (°C)	$\mathcal{V}$ (min of arc gauss^{-1} cm^{-1})
Light flint glass	18	0.0317
Water	20	0.0131
NaCl	16	0.0359
Quartz	20	0.0166
$NH_4Fe(SO_4)_2.12H_2O$	26	−0.00058
Air*	0	6.27×10^{-6}
CO_2*	0	9.39×10^{-6}

* $\lambda = 578$ nm and 760 mm Hg.
More extensive listings are given in the usual handbooks.

By convention, *a positive Verdet constant corresponds to a (diamagnetic) material for which the Faraday effect is l-rotatory when the light moves parallel to the applied B-field and d-rotatory when it propagates antiparallel to B.* Note that no such reversal of handedness occurs in the case of natural optical activity. For a convenient mnemonic, imagine the **B**-field to be generated by a solenoidal coil wound about the sample. The plane of vibration, when $\mathcal{V}$ is positive, rotates in the same direction as the current in the

coil regardless of the beam's propagation direction along its axis. The effect can, accordingly, be amplified by reflecting the light back and forth a few times through the sample.

The theoretical treatment of the Faraday effect involves the quantum-mechanical theory of dispersion including the effects of **B** on the atomic or molecular energy levels. It will suffice here to merely outline the limited classical argument for nonmagnetic materials. Suppose the incident light to be circular and monochromatic. An elastically bound electron will take on a steady-state circular orbit being driven by the rotating **E**-field of the wave (the effect of the wave's **B**-field is negligible). The introduction of a large constant applied magnetic field perpendicular to the plane of the orbit will result in a radial force F_M on the electron. That force can either point toward or away from the circle's center depending on the handedness of the light and the direction of the constant **B**-field. The total radial force (F_M plus the elastic restoring force) can therefore have two different values and so too can the radius of the orbit. Consequently, for a given magnetic field there will be two possible values of the electric dipole moment, the polarization, and the permittivity, and finally also two values of the index of refraction, $n_\mathscr{R}$ and $n_\mathscr{L}$. The discussion can then proceed in precisely the same fashion as that of Fresnel's treatment of optical activity. As before, one speaks of two normal modes of propagation of electromagnetic waves through the medium, the $\mathscr{R}$- and $\mathscr{L}$-states.

For ferromagnetic substances things are somewhat more complicated. In the case of a magnetized material β is proportional to the component of the magnetization in the direction of propagation rather than the component of the applied dc field.

There are a number of practical applications of the Faraday effect. It can be used to analyze mixtures of hydrocarbons, since each constituent has a characteristic magnetic rotation. Moreover, when utilized in spectroscopic studies it yields information about the properties of energy states above the ground level. In recent times the Faraday effect has been put to even more exciting and promising uses. Since the advent of the laser in the early nineteen-sixties, a tremendous effort has been put forth in an attempt to utilize the enormous potential of laser light as a communications medium (see Section 7.2.6). An essential component of any such system is the *modulator*, whose function it is to impress information on the beam. Such a device must have the capability of somehow varying the light wave at high speeds and in a controlled fashion. It might, for example, alter the wave's amplitude, polarization, propaga-

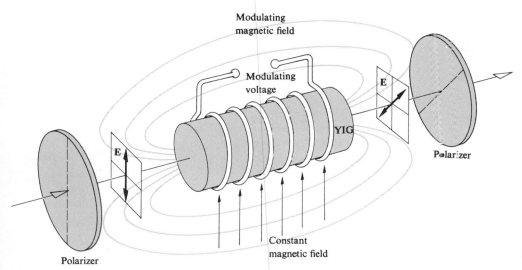

Fig. 8.64 A Faraday effect modulator.

tion direction, phase or frequency in a manner related to the signal which is to be transmitted. The Faraday effect provides one possible basis for such a modulator. Clearly if a device of this sort is to function efficiently, each unit length of the medium must absorb as little light as possible while imparting as large a rotation to the beam as is possible. A number of rather exotic ferromagnetic materials have been studied to this end. An infrared modulator of this sort has been constructed by R. C. LeCraw. It utilizes the synthetic magnetic crystal yttrium–iron garnet (YIG) to which has been added a quantity of gallium. YIG has a structure similar to that of natural gem garnets. The device is depicted schematically in Fig. 8.64. A linear infrared laser beam enters the crystal from the left. A transverse dc magnetic field saturates the magnetization of the YIG crystal in that direction. The total magnetization vector (arising from the constant field and the field of the coil) can vary in direction, being tilted towards the axis of the crystal by an amount proportional to the modulating current in the coil. Since the Faraday rotation depends on the axial component of the magnetization, the coil current controls β. The analyzer then converts this polarization modulation to amplitude modulation by way of Malus's law [Eq. (8.24)]. In short, the signal to be transmitted is introduced across the coil as a modulating voltage and the emerging laser beam carries that information in the form of amplitude variations.

There are actually several other magneto-optic effects. We shall only consider two of these, and rather succinctly at that. The *Voigt* and *Cotton–Mouton effects* both arise when

a constant magnetic field is applied to a transparent medium perpendicular to the direction of propagation of the incident light beam. The former occurs in vapors while the latter, which is considerably stronger, occurs in liquids. In either case the medium displays birefringence similar to a uniaxial crystal whose optic axis is in the direction of the dc magnetic field, i.e. normal to the light beam [Eq. (8.32)]. The two indices of refraction now correspond to the situations where the plane of vibration of the wave is either normal or parallel to the constant magnetic field. Their difference Δn (i.e. the birefringence) is proportional to the square of the applied magnetic field. It arises in liquids from an aligning of the optically and magnetically anisotropic molecules of the medium with that field. If the incoming light propagates at some angle to the static field other than 0 or $\pi/2$, the Faraday and Cotton–Mouton effects occur concurrently, with the former generally being the much larger of the two. The Cotton–Mouton is the magnetic analog of the Kerr electro-optic effect to be considered next.

8.11.3 The Kerr and Pockels Effects

The first electro-optic effect was discovered by the Scottish physicist John Kerr (1824–1907) in 1875. He found that an isotropic transparent substance becomes birefringent when placed in an electric field **E**. The medium takes on the characteristics of a uniaxial crystal whose optic axis corresponds to the direction of the applied field. The two indices, $n_{\parallel}$ and $n_{\perp}$, are associated with the two orientations of the plane of vibration of the wave, namely, parallel and per-

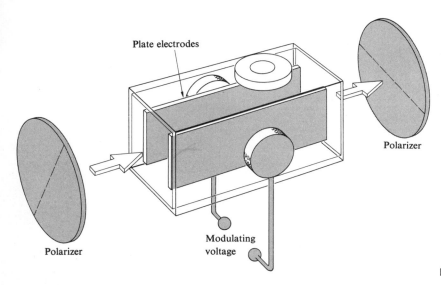

Plate electrodes

Polarizer

Polarizer

Modulating
voltage

Fig. 8.65 Kerr cell.

pendicular to the applied electric field respectively. Their difference, Δn, is the birefringence and it is found to be

$$\Delta n = \lambda_0 K E^2 \qquad (8.40)$$

where K is the *Kerr constant*. When K is positive, as it most often is, Δn, which can be thought of as $n_e - n_o$, is positive and the substance behaves like a positive uniaxial crystal. Values of the Kerr constant (Table 8.3) are usually listed in electrostatic units so that one must remember to enter E in Eq. (8.40) in statvolts per cm (one statvolt ≈ 300 V). Observe that, as with the Cotton–Mouton effect, *the Kerr effect is proportional to the square of the field and is often referred to as the quadratic electro-optic effect*. The phenomenon in liquids is attributed to a partial alignment of anisotropic molecules by the **E**-field. In solids the situation is considerably more complicated.

Table 8.3 Kerr constants for some selected liquids.

Substance		K (in units of 10^{-7} cm statvolt^{-2})
Benzene	C_6H_6	0.6
Carbon disulfide	CS_2	3.2
Chloroform	$CHCl_3$	-3.5
Water	H_2O	4.7
Nitrotoluene	$C_5H_7NO_2$	123
Nitrobenzene	$C_6H_5NO_2$	220

(20°C, $\lambda_0 = 589.3$ nm)

Figure 8.65 depicts an arrangement known as a Kerr shutter or optical modulator. It consists of a glass cell containing two electrodes which is filled with a polar liquid. This *Kerr cell*, as it is called, is positioned between crossed linear polarizers whose transmission axes are at $\pm 45°$ to the applied **E**-field. With zero voltage across the plates no light will be transmitted; the shutter is closed. The application of a modulating voltage generates a field causing the cell to function as a variable wave plate and thus opening the shutter proportionately. The great value of such a device lies in the fact that it can respond effectively to frequencies roughly as high as 10^{10} Hz. Kerr cells, usually containing nitrobenzene or carbon disulfide, have been used for a number of years in a variety of applications. They serve as shutters in high-speed photography and as light-beam choppers to replace rotating toothed wheels. As such, they have been utilized in measurements of the speed of light. Kerr cells are extensively used as Q-switches (see Chapter 14) in pulsed laser systems.

If the plates comprising the electrodes have an effective length of ℓ cm and are separated by a distance d, the retardation is given by

$$\Delta\varphi = 2\pi K\ell V^2/d^2, \qquad (8.41)$$

in which V is the applied voltage. Thus a nitrobenzene cell where d is one cm and ℓ is several cm will require a rather large voltage of roughly 3×10^4 V in order to respond as a half-wave plate. This is a characteristic quantity known as the *half-wave voltage*, $V_{\lambda/2}$. Another drawback is that nitrobenzene is both poisonous and explosive. Transparent solid substances like the mixed crystal potassium tantalate niobate ($KTa_{0.65}Nb_{0.35}O_3$), KTN for short, or barium titanate ($BaTiO_3$) which show a Kerr effect are therefore of interest as electro-optical modulators.

There is yet another very important electro-optical effect known as the *Pockels effect* after the German physicist Friedrich Carl Alwin Pockels (1865–1913) who studied it extensively in 1893. It is a linear electro-optical effect inasmuch as the induced birefringence is proportional to the first power of the applied **E**-field and therefore the applied voltage. The Pockels effect exists only in certain crystals which lack a center of symmetry; in other words, crystals having no central point through which every atom can be reflected into an identical atom. There are thirty-two crystal symmetry classes and of these twenty may show the Pockels effect. Incidentally, these same twenty classes are also piezoelectric. Thus, many crystals and all liquids are excluded from displaying a linear electro-optic effect.

The first practical Pockels cell, which could perform as a shutter or modulator, had to wait until the nineteen-forties for the development of suitable crystals. The operating principle for such a device is one we've already discussed. In brief, the birefringence is varied electronically by means of a controlled applied electric field. The retardance can be altered as desired, thereby changing the state of polarization of the incident linear wave. In this way, the system functions as a polarization modulator. Early devices were made of ammonium dihydrogen phosphate ($NH_4H_2PO_4$) or ADP, and potassium dihydrogen phosphate (KH_2PO_4) known as KDP; both are still widely in use. A great improvement was provided by the introduction of single crystals of potassium dideuterium phosphate (KD_2PO_4) or KD*P which yields the same retardation with voltages less than half of those needed for KDP. Tremendous effort has gone into research on electro-optical crystals. The development of these materials is continuously adding exotic names to the jargon of the new technology, such as lithium tantalate, rubidium dihydrogen phosphate, lithium niobate, barium titanate, and barium sodium niobate to mention only a few.

A *Pockels cell* is simply an appropriate noncentrosymmetric, oriented, single crystal immersed in a controllable electric field. Typically such devices can be operated at fairly low voltages (roughly 5 to 10 times less than an equivalent Kerr cell); they are linear and of course there is no problem with toxic liquids. The response time of KDP is quite short, typically less than 10 ns, and it can modulate a light beam at up to about 25 GHz (i.e. 25×10^9 Hz). There are two common cell configurations referred to as *transverse*

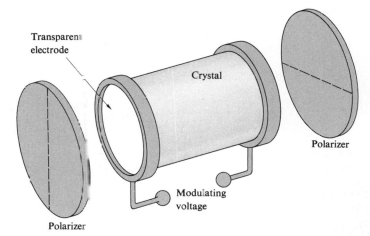

Fig. 8.66 Pockels cell.

and *longitudinal* depending on whether the applied **E**-field is perpendicular or parallel to the direction of propagation, respectively. The longitudinal type is illustrated, in its most basic form, in Fig. 8.66. Since the beam traverses the electrodes, these are usually made of transparent metaloxide coatings (e.g. SnO, InO or CdO), thin metal films, grids or rings. The crystal itself is generally uniaxial in the absence of an applied field and it is aligned such that its optic axis is along the beam's propagation direction. For such an arrangement the retardance is given by

$$\Delta\varphi = 2\pi n_o^3 r_{63} V / \lambda_0 \qquad (8.42)$$

where r_{63} is the *electro-optic constant* in m/V, n_o is the *ordinary* index of refraction, V is the potential difference in volts and λ_0 is the vacuum wavelength in meters.* Since the crystals are anisotropic, their properties vary in different directions and they must be described by a group of terms referred to collectively as the second-rank electro-optic tensor r_{ij}. Fortunately, we need only concern ourselves here with one of its components, namely r_{63}, values of which are given in Table 8.4. The half-wave voltage corresponds to a value of $\Delta\varphi = \pi$ in which case

$$\Delta\varphi = \pi \frac{V}{V_{\lambda/2}} \qquad (8.43)$$

---

* This expression, along with the appropriate one for the transverse mode, is derived rather nicely in A. Yariv, *Quantum Electronics*. Even so, the treatment is sophisticated and not recommended for casual reading.

and from Eq. (8.42)

$$V_{\lambda/2} = \frac{\lambda_0}{2n_o^3 r_{63}}. \tag{8.44}$$

Table 8.4 Electro-optic constants (room temperature, $\lambda_0 = 546.1$ nm).

Material	r_{63} (units of 10^{-12} m/V)	(approx.) n_o	$V_{\lambda/2}$ (in kV)
ADP ($NH_4H_2PO_4$)	8.5	1.52	9.2
KDP (KH_2PO_4)	10.6	1.51	7.6
KDA (KH_2AsO_4)	~13.0	1.57	~6.2
KD*P (KD_2PO_4)	~23.3	1.52	~3.4

For example, for KDP, $r_{63} = 10.6 \times 10^{-12}$ m/V, $n_o = 1.51$ and we obtain $V_{\lambda/2} \approx 7.6 \times 10^3$ V at $\lambda_0 = 546.1$ nm.

Pockels cells have been used as ultra-fast shutters, Q-switches for lasers and dc to 30 GHz light modulators. They are also being applied in a wide range of electro-optical systems, e.g. data processing and display techniques.*

8.12 A MATHEMATICAL DESCRIPTION OF POLARIZATION

Thus far we have considered polarized light in terms of the electric-field component of the wave. The most general representation was, of course, that of elliptical light. There we envisioned the end-point of the vector **E** continuously sweeping along the path of an ellipse having a particular shape—the circle and line being special cases. The period over which the ellipse was traversed equaled that of the light wave, i.e. roughly 10^{-15} s, and was thus far too short to be detected. In contrast, measurements made in practice are generally averages over comparatively long time intervals. Clearly, it would be advantageous to formulate an alternative description of polarization in terms of convenient observables, namely irradiances. Our motivations are far more than the ever-present combination of esthetics and pedagogy. The formalism to be considered has far-reaching significance in other areas of study, e.g. particle physics (the photon is after all an elementary particle) and quantum mechanics. It serves

---

* The reader interested in light modulation in general should consult D. F. Nelson, "The Modulation of Laser Light," *Scientific American* (June 1968). For some of the practical details see R. S. Ploss, "A Review of Electro-Optics Materials, Methods and Uses," *Optical Spectra* (Jan./Feb. 1969) or R. Goldstein, "Pockels Cell Primer," *Laser Focus Magazine* (Feb. 1968) both of which contain useful bibliographies.

in some respects to link the classical and quantum-mechanical pictures. But even more demanding of our present attention are the considerable practical advantages to be gleaned from this alternative description. We shall evolve an elegant procedure for predicting the effects of complex systems of polarizing elements on the ultimate state of an emergent wave. The mathematics, written in the compressed form of matrices, will only require the simplest manipulation of those matrices. The complicated logic associated with phase retardations, relative orientations, etc. for a tandem series of wave plates and polarizers is almost all built in. One need only select appropriate matrices from a chart and drop them into the mathematical mill.

8.12.1 The Stokes Parameters

The modern representation of polarized light actually had its origins in 1852 in the work of G. G. Stokes. He introduced four quantities which are functions only of observables of the electromagnetic wave and are now known as the *Stokes parameters*.* The polarization state of a beam of light (either natural, totally or partially polarized) can be described in terms of these quantities. We will first define the parameters operationally and then relate them to electromagnetic theory. Imagine that we have a set of four filters each of which, under *natural* illumination, will transmit half of the incident light, the other half being discarded. The choice is not a unique one and a number of equivalent possibilities exist. Suppose then that the first filter is simply isotropic, passing all states equally, while the second and third are linear polarizers whose transmission axes are horizontal and at $+45°$ (diagonal along the first and third quadrants) respectively. The last filter is a circular polarizer opaque to $\mathscr{L}$-states. Each one of these four filters is positioned alone in the path of the beam under investigation and the transmitted irradiances I_0, I_1, I_2, I_3 are measured with a meter of a type insensitive to polarization (not all of them are). The operational definition of the Stokes parameters is then given by the relations

$$\mathscr{S}_0 = 2I_0 \tag{8.45a}$$

$$\mathscr{S}_1 = 2I_1 - 2I_0 \tag{8.45b}$$

$$\mathscr{S}_2 = 2I_2 - 2I_0 \tag{8.45c}$$

$$\mathscr{S}_3 = 2I_3 - 2I_0. \tag{8.45d}$$

---

* Much of the material in this section is treated more extensively in Shurcliff's *Polarized Light: Production and Use*, which is something of a classic on the subject. You might also look at M. J. Walker, "Matrix Calculus and the Stokes Parameters of Polarized Radiation," *Am. J. Phys.* **22**, 170 (1954).

Notice that S_0 is simply the incident irradiance while S_1, S_2, and S_3 specify the state of polarization. Thus S_1 reflects a tendency for the polarization to more nearly resemble either a horizontal $\mathscr{P}$-state (whereupon $S_1 > 0$) or a vertical one (in which case $S_1 < 0$). When the beam displays no preferential orientation with respect to these axes ($S_1 = 0$) it may be elliptical at $\pm 45°$, circular, or unpolarized. Similarly S_2 implies a tendency for the light to more closely resemble a $\mathscr{P}$-state oriented either in the direction of $+45°$ (when $S_2 > 0$) or in the direction of $-45°$ (when $S_2 < 0$) or neither ($S_2 = 0$). In quite the same way S_3 reveals a tendency of the beam toward right-handedness ($S_3 > 0$), left-handedness ($S_3 < 0$) or neither ($S_3 = 0$).

Now recall the expressions for quasimonochromatic light,

$$\mathbf{E}_x(t) = \hat{\mathbf{i}} E_{0x}(t) \cos\left[(\bar{k}z - \bar{\omega}t) + \varepsilon_x(t)\right] \quad [8.34(a)]$$

and

$$\mathbf{E}_y(t) = \hat{\mathbf{j}} E_{0y}(t) \cos\left[(\bar{k}z - \bar{\omega}t) + \varepsilon_y(t)\right] \quad [8.34(b)]$$

where $\mathbf{E}(t) = \mathbf{E}_x(t) + \mathbf{E}_y(t)$. Using these in a fairly straightforward way, the Stokes parameters can be recast* as

$$S_0 = \langle E_{0x}^2 \rangle + \langle E_{0y}^2 \rangle \tag{8.46a}$$

$$S_1 = \langle E_{0x}^2 \rangle - \langle E_{0y}^2 \rangle \tag{8.46b}$$

$$S_2 = \langle 2E_{0x}E_{0y} \cos \varepsilon \rangle \tag{8.46c}$$

$$S_3 = \langle 2E_{0x}E_{0y} \sin \varepsilon \rangle. \tag{8.46d}$$

Here $\varepsilon = \varepsilon_y - \varepsilon_x$ and we've dropped the constant $\epsilon_0 c/2$ so that the parameters are now *proportional* to irradiances. For the hypothetical case of perfectly monochromatic light $E_{0x}(t)$, $E_{0y}(t)$, and $\varepsilon(t)$ are time independent and one need only drop the $\langle \ \rangle$ brackets in Eq. (8.46) to get the applicable Stokes parameters. Interestingly enough, these same results can be gotten by time averaging Eq. (8.14), which is the general equation for elliptical light.†

If the beam is unpolarized $\langle E_{0x}^2 \rangle = \langle E_{0y}^2 \rangle$; neither average to zero because the amplitude squared is always positive. In that case $S_0 = \langle E_{0x}^2 \rangle + \langle E_{0y}^2 \rangle$ but $S_1 = S_2 = S_3 = 0$. The latter two parameters go to zero since both $\cos \varepsilon$ and $\sin \varepsilon$ average to zero independently of the amplitudes. It is very often convenient to *normalize* the Stokes parameters by dividing each one by the value of S_0. This has

---

* For the details see E. Hecht, "Note on an Operational Definition of the Stokes Parameters," *Am. J. Phys.* **38**, 1156 (1970).

† E. Collett, "The Description of Polarization in Classical Physics," *Am. J. Phys.* **36**, 713 (1968).

the effect of using an incident beam of unit irradiance. The set of parameters (S_0, S_1, S_2, S_3) for *natural light* in the normalized representation is then $(1, 0, 0, 0)$. If the light is horizontally polarized it has no vertical component and the normalized parameters are $(1, 1, 0, 0)$. Similarly for vertically polarized light we have $(1, -1, 0, 0)$. Representations of a few other polarization states are listed in Table 8.5 (the parameters are displayed vertically for reasons to be discussed later). Notice that for completely polarized light it follows from Eq. (8.46) that

$$S_0^2 = S_1^2 + S_2^2 + S_3^2. \tag{8.47}$$

Table 8.5 Stokes and Jones vectors for some polarization states.

State of Polarization	Stokes vectors	Jones vectors
Horizontal $\mathscr{P}$-state	$\begin{bmatrix} 1 \\ 1 \\ 0 \\ 0 \end{bmatrix}$	$\begin{bmatrix} 1 \\ 0 \end{bmatrix}$
Vertical $\mathscr{P}$-state	$\begin{bmatrix} 1 \\ -1 \\ 0 \\ 0 \end{bmatrix}$	$\begin{bmatrix} 0 \\ 1 \end{bmatrix}$
$\mathscr{P}$-state at $+45°$	$\begin{bmatrix} 1 \\ 0 \\ 1 \\ 0 \end{bmatrix}$	$\dfrac{1}{\sqrt{2}} \begin{bmatrix} 1 \\ 1 \end{bmatrix}$
$\mathscr{P}$-state at $-45°$	$\begin{bmatrix} 1 \\ 0 \\ -1 \\ 0 \end{bmatrix}$	$\dfrac{1}{\sqrt{2}} \begin{bmatrix} 1 \\ -1 \end{bmatrix}$
$\mathscr{R}$-state	$\begin{bmatrix} 1 \\ 0 \\ 0 \\ 1 \end{bmatrix}$	$\dfrac{1}{\sqrt{2}} \begin{bmatrix} 1 \\ -i \end{bmatrix}$
$\mathscr{L}$-state	$\begin{bmatrix} 1 \\ 0 \\ 0 \\ -1 \end{bmatrix}$	$\dfrac{1}{\sqrt{2}} \begin{bmatrix} 1 \\ i \end{bmatrix}$

Moreover, for partially polarized light it can be shown that the degree of polarization (8.29) is given by

$$V = (S_1^2 + S_2^2 + S_3^2)^{1/2}/S_0. \tag{8.48}$$

Imagine now that we have two quasimonochromatic waves described by (S_0', S_1', S_2', S_3') and $(S_0'', S_1'', S_2'', S_3'')$

which are superimposed in some region of space. As long as the waves are *incoherent*, any one of the Stokes parameters of the resultant will be the sum of the corresponding parameters of the constituents (all of which are proportional to irradiance). In other words the set of parameters describing the resultant is $(S_0' + S_0'', S_1' + S_1'', S_2' + S_2'', S_3' + S_3'')$. For example, if a unit-flux density, vertical $\mathscr{P}$-state $(1, -1, 0, 0)$ is added to an *incoherent* $\mathscr{L}$-state (see Table 8.5) of flux density two $(2, 0, 0, -2)$ the composite wave has parameters $(3, -1, 0, -2)$. It is an ellipse of flux density 3, more nearly vertical than horizontal $(S_1 < 0)$, left-handed $(S_3 < 0)$ and having a degree of polarization of $\sqrt{5}/3$.

The set of Stokes parameters for a given wave can be envisaged as a *vector*, where we have already seen how two such (incoherent) *vectors* add.* Indeed, it will not be the usual kind of three-dimensional vector but this sort of representation is rather widely used in physics to great advantage. More specifically the parameters (S_0, S_1, S_2, S_3) are arranged in the form of what is called a *column vector*,

$$S = \begin{bmatrix} S_0 \\ S_1 \\ S_2 \\ S_3 \end{bmatrix}. \qquad (8.49)$$

8.12.2 The Jones Vectors

Another representation of polarized light which complements that of the Stokes parameters was invented in 1941 by the American physicist R. Clark Jones. The technique he evolved has the advantages of being applicable to coherent beams and at the same time being extremely concise. Yet unlike the previous formalism it is *only applicable to polarized waves*. In that case it would seem that the most natural way to represent the beam would be in terms of the electric vector itself. Written in column form this *Jones vector* is

$$\mathbf{E} = \begin{bmatrix} E_x(t) \\ E_y(t) \end{bmatrix} \qquad (8.50)$$

where $E_x(t)$ and $E_y(t)$ are the instantaneous scalar components of **E**. Obviously knowing **E** we know everything about the polarization state. And, if we preserve the phase information, we will be able to handle coherent waves. With

---

* The detailed requirements for a collection of objects to form a vector space and themselves be vectors in such a space are discussed in e.g. Davis, *Introduction to Vector Analysis*.

this in mind rewrite Eq. (8.50) as

$$\mathbf{E} = \begin{bmatrix} E_{0x}e^{i\varphi_x} \\ E_{0y}e^{i\varphi_y} \end{bmatrix} \qquad (8.51)$$

where φ_x and φ_y are the appropriate phases. Horizontal and vertical $\mathscr{P}$-states are thus given by

$$\mathbf{E}_h = \begin{bmatrix} E_{0x}e^{i\varphi_x} \\ 0 \end{bmatrix} \quad \text{and} \quad \mathbf{E}_v = \begin{bmatrix} 0 \\ E_{0y}e^{i\varphi_y} \end{bmatrix} \qquad (8.52)$$

respectively. The sum of two coherent beams, as with the Stokes vectors, is formed by a sum of the corresponding components. Since $\mathbf{E} = \mathbf{E}_h + \mathbf{E}_v$ when for example $E_{0x} = E_{0y}$ and $\varphi_x = \varphi_y$, **E** is given by

$$\mathbf{E} = \begin{bmatrix} E_{0x}e^{i\varphi_x} \\ E_{0x}e^{i\varphi_x} \end{bmatrix} \qquad (8.53)$$

or after factoring, by

$$\mathbf{E} = E_{0x}e^{i\varphi_x} \begin{bmatrix} 1 \\ 1 \end{bmatrix} \qquad (8.54)$$

which is a $\mathscr{P}$-state at $+45°$. This is the case since the amplitudes are equal and the phase difference is zero. There are many applications in which it is not necessary to know the exact amplitudes and phases. In such instances we can *normalize* the irradiance to unity, thereby forfeiting some information but gaining much simpler expressions. This is done by dividing both elements in the vector by the same scalar (real or complex) quantity such that the sum of the squares of the components is one. For example, dividing both terms of Eq. (8.53) by $\sqrt{2}\, E_{0x}e^{i\varphi_x}$ leads to

$$\mathbf{E}_{45} = \frac{1}{\sqrt{2}} \begin{bmatrix} 1 \\ 1 \end{bmatrix}. \qquad (8.55)$$

Similarly, in normalized form

$$\mathbf{E}_h = \begin{bmatrix} 1 \\ 0 \end{bmatrix} \quad \text{and} \quad \mathbf{E}_v = \begin{bmatrix} 0 \\ 1 \end{bmatrix}. \qquad (8.56)$$

Right-circular light has $E_{0x} = E_{0y}$ and the y-component leading the x-component by 90°. Since we are using the form $(kz - \omega t)$ we will have to add $-\pi/2$ to φ_y, thus

$$\mathbf{E}_{\mathscr{R}} = \begin{bmatrix} E_{0x}e^{i\varphi_x} \\ E_{0x}e^{i(\varphi_x - \pi/2)} \end{bmatrix}.$$

Dividing both components by $E_{0x}e^{i\varphi_x}$ we have

$$\begin{bmatrix} 1 \\ e^{-i\pi/2} \end{bmatrix} = \begin{bmatrix} 1 \\ -i \end{bmatrix};$$

hence the normalized Jones vector is[†]

$$\mathbf{E}_{\mathscr{R}} = \frac{1}{\sqrt{2}}\begin{bmatrix} 1 \\ -i \end{bmatrix} \quad \text{and similarly} \quad \mathbf{E}_{\mathscr{L}} = \frac{1}{\sqrt{2}}\begin{bmatrix} 1 \\ i \end{bmatrix}. \quad (8.57)$$

The sum $\mathbf{E}_{\mathscr{R}} + \mathbf{E}_{\mathscr{L}}$ is

$$\frac{1}{\sqrt{2}}\begin{bmatrix} 1+1 \\ -i+i \end{bmatrix} = \frac{2}{\sqrt{2}}\begin{bmatrix} 1 \\ 0 \end{bmatrix}.$$

This is a horizontal $\mathscr{P}$-state having an amplitude twice that of either component; a result in agreement with our earlier calculation of Eq. (8.10). The Jones vector for elliptical light can be gotten by the same procedure used to arrive at $\mathbf{E}_{\mathscr{R}}$ and $\mathbf{E}_{\mathscr{L}}$ where now E_{0x} may not be equal to E_{0y} and the phase difference need not be 90°. In essence, for vertical and horizontal $\mathscr{E}$-states we have only to stretch out the circular form into an ellipse by multiplying either component by a scalar. Thus

$$\frac{1}{\sqrt{5}}\begin{bmatrix} 2 \\ -i \end{bmatrix} \quad (8.58)$$

describes one possible form of horizontal, right-handed, elliptical light.

Two vectors **A** and **B** are said to be orthogonal when $\mathbf{A} \cdot \mathbf{B} = 0$; similarly two complex vectors are orthogonal when $\mathbf{A} \cdot \mathbf{B}^* = 0$. One refers to polarization states as being *orthogonal* when their Jones vectors are orthogonal. For example

$$\mathbf{E}_{\mathscr{R}} \cdot \mathbf{E}_{\mathscr{L}}^* = \tfrac{1}{2}[(1)(1)^* + (-i)(i)^*] = 0$$

or

$$\mathbf{E}_h \cdot \mathbf{E}_v^* = [(1)(0)^* + (0)(1)^*] = 0$$

where taking the complex conjugates of real numbers obviously leaves them unaltered. Any polarization state will have a corresponding orthogonal state. Notice that

$$\mathbf{E}_{\mathscr{R}} \cdot \mathbf{E}_{\mathscr{R}}^* = \mathbf{E}_{\mathscr{L}} \cdot \mathbf{E}_{\mathscr{L}}^* = 1$$

and

$$\mathbf{E}_{\mathscr{R}} \cdot \mathbf{E}_{\mathscr{L}}^* = \mathbf{E}_{\mathscr{L}} \cdot \mathbf{E}_{\mathscr{R}}^* = 0.$$

Such vectors form an *orthonormal set* as do $\mathbf{E}_h$ and $\mathbf{E}_v$. And, as we have seen, any polarization state can be described by

---

[†] Had we used $(\omega t - kz)$ for the phase, the terms in $\mathbf{E}_{\mathscr{R}}$ would have been interchanged. The present notation although possibly a bit more difficult to keep straight (e.g. $-\pi/2$ for a phase lead) is more often used in modern works. Be wary when consulting references (e.g. Shurcliff).

a linear combination of the vectors of either one of these orthonormal sets. These same ideas are of considerable importance in quantum mechanics where one deals with orthonormal wave functions.

8.12.3 The Jones and Mueller Matrices

Suppose that we have a polarized incident beam represented by its Jones vector $\mathbf{E}_i$ which passes through an optical element, emerging as a new vector $\mathbf{E}_t$ corresponding to the transmitted wave. The optical element has transformed $\mathbf{E}_i$ into $\mathbf{E}_t$, a process which can be described mathematically using a 2×2 matrix. Recall that a matrix is just an array of numbers which has prescribed addition and multiplication operations. Let $\mathscr{A}$ represent the transformation matrix of the optical element in question. Then

$$\mathbf{E}_t = \mathscr{A}\mathbf{E}_i \quad (8.59)$$

where

$$\mathscr{A} = \begin{bmatrix} a_{11} & a_{12} \\ a_{21} & a_{22} \end{bmatrix} \quad (8.60)$$

and the column vectors are to be treated like any other matrices. As a reminder we write Eq. (8.59) as

$$\begin{bmatrix} E_{tx} \\ E_{ty} \end{bmatrix} = \begin{bmatrix} a_{11} & a_{12} \\ a_{21} & a_{22} \end{bmatrix} \begin{bmatrix} E_{ix} \\ E_{iy} \end{bmatrix} \quad (8.61)$$

which, upon expanding, yields

$$E_{tx} = a_{11}E_{ix} + a_{12}E_{iy},$$
$$E_{ty} = a_{21}E_{ix} + a_{22}E_{iy}.$$

Table 8.6 contains a brief listing of Jones matrices for various optical elements. To appreciate how these are used let's examine a few applications. Suppose that $\mathbf{E}_i$ represents a $\mathscr{P}$-state at $+45°$ which passes through a quarter-wave plate whose fast axis is vertical (i.e. in the y-direction). The polarization state of the emergent wave is found as follows, where we drop the constant-amplitude factors for convenience:

$$\begin{bmatrix} 1 & 0 \\ 0 & -i \end{bmatrix} \begin{bmatrix} 1 \\ 1 \end{bmatrix} = \begin{bmatrix} E_{tx} \\ E_{ty} \end{bmatrix}$$

and thus

$$\mathbf{E}_t = \begin{bmatrix} 1 \\ -i \end{bmatrix}.$$

Table 8.6 Jones and Mueller matrices.

Linear optical element	Jones matrix	Mueller matrix
Horizontal linear polarizer $\leftrightarrow$	$\begin{bmatrix} 1 & 0 \\ 0 & 0 \end{bmatrix}$	$\frac{1}{2}\begin{bmatrix} 1 & 1 & 0 & 0 \\ 1 & 1 & 0 & 0 \\ 0 & 0 & 0 & 0 \\ 0 & 0 & 0 & 0 \end{bmatrix}$
Vertical linear polarizer $\updownarrow$	$\begin{bmatrix} 0 & 0 \\ 0 & 1 \end{bmatrix}$	$\frac{1}{2}\begin{bmatrix} 1 & -1 & 0 & 0 \\ -1 & 1 & 0 & 0 \\ 0 & 0 & 0 & 0 \\ 0 & 0 & 0 & 0 \end{bmatrix}$
Linear polarizer at $+45°$ $\nearrow$	$\frac{1}{2}\begin{bmatrix} 1 & 1 \\ 1 & 1 \end{bmatrix}$	$\frac{1}{2}\begin{bmatrix} 1 & 0 & 1 & 0 \\ 0 & 0 & 0 & 0 \\ 1 & 0 & 1 & 0 \\ 0 & 0 & 0 & 0 \end{bmatrix}$
Linear polarizer at $-45°$ $\searrow$	$\frac{1}{2}\begin{bmatrix} 1 & -1 \\ -1 & 1 \end{bmatrix}$	$\frac{1}{2}\begin{bmatrix} 1 & 0 & -1 & 0 \\ 0 & 0 & 0 & 0 \\ -1 & 0 & 1 & 0 \\ 0 & 0 & 0 & 0 \end{bmatrix}$
Quarter-wave plate fast axis vertical	$e^{i\pi/4}\begin{bmatrix} 1 & 0 \\ 0 & -i \end{bmatrix}$	$\begin{bmatrix} 1 & 0 & 0 & 0 \\ 0 & 1 & 0 & 0 \\ 0 & 0 & 0 & -1 \\ 0 & 0 & 1 & 0 \end{bmatrix}$
Quarter-wave plate fast axis horizontal	$e^{i\pi/4}\begin{bmatrix} 1 & 0 \\ 0 & i \end{bmatrix}$	$\begin{bmatrix} 1 & 0 & 0 & 0 \\ 0 & 1 & 0 & 0 \\ 0 & 0 & 0 & 1 \\ 0 & 0 & -1 & 0 \end{bmatrix}$
Homogeneous circular polarizer right $\circlearrowright$	$\frac{1}{2}\begin{bmatrix} 1 & i \\ -i & 1 \end{bmatrix}$	$\frac{1}{2}\begin{bmatrix} 1 & 0 & 0 & 1 \\ 0 & 0 & 0 & 0 \\ 0 & 0 & 0 & 0 \\ 1 & 0 & 0 & 1 \end{bmatrix}$
Homogeneous circular polarizer left $\circlearrowleft$	$\frac{1}{2}\begin{bmatrix} 1 & -i \\ i & 1 \end{bmatrix}$	$\frac{1}{2}\begin{bmatrix} 1 & 0 & 0 & -1 \\ 0 & 0 & 0 & 0 \\ 0 & 0 & 0 & 0 \\ -1 & 0 & 0 & 1 \end{bmatrix}$

The beam, as you well know, is right-circular. If the wave passes through a series of optical elements represented by the matrices $\mathscr{A}_1, \mathscr{A}_2, \ldots, \mathscr{A}_n$ then

$$\mathbf{E}_t = \mathscr{A}_n \cdots \mathscr{A}_2 \mathscr{A}_1 \mathbf{E}_i.$$

The matrices do not commute; they must be applied in the proper order. The wave leaving the first optical element in the series is $\mathscr{A}_1 \mathbf{E}_i$; after passing through the second element it becomes $\mathscr{A}_2 \mathscr{A}_1 \mathbf{E}_i$ and so on. To illustrate the process, return

to the wave considered above, i.e. a $\mathscr{P}$-state at $+45°$, but now have it pass through two quarter-wave plates both with their fast axes vertical. Thus, again discarding the amplitude factors,

$$E_t = \begin{bmatrix} 1 & 0 \\ 0 & -i \end{bmatrix}\begin{bmatrix} 1 & 0 \\ 0 & -i \end{bmatrix}\begin{bmatrix} 1 \\ 1 \end{bmatrix}$$

whereupon

$$E_t = \begin{bmatrix} 1 & 0 \\ 0 & -i \end{bmatrix}\begin{bmatrix} 1 \\ -i \end{bmatrix}$$

and finally

$$E_t = \begin{bmatrix} 1 \\ -1 \end{bmatrix}.$$

The transmitted beam is a $\mathscr{P}$-state at $-45°$, having essentially been flipped through 90° by a half-wave plate. When the same series of optical elements is being used to examine various states it becomes desirable to replace the product $\mathscr{A}_n \cdots \mathscr{A}_2 \mathscr{A}_1$ by the single 2×2 *system matrix* gotten by carrying out the multiplication (the order in which it is calculated should be $\mathscr{A}_2 \mathscr{A}_1$, then $\mathscr{A}_3 \mathscr{A}_2 \mathscr{A}_1$, etc.).

In 1943 Hans Mueller, then a professor of physics at the Massachusetts Institute of Technology, devised a matrix method for dealing with the Stokes vectors. Recall that the Stokes vectors have the attribute of being applicable to both polarized and partially polarized light. The Mueller method shares this quality and thus serves to complement the Jones method. The latter however can easily deal with coherent waves while the former cannot. The Mueller, 4×4, matrices are applied in much the same way as are the Jones matrices. There is therefore little need to discuss the method at length and a few simple examples augmented by Table 8.6 should suffice. Imagine then that we pass a unit-irradiance unpolarized wave through a linear horizontal polarizer, the Stokes vector of the emerging wave $\mathcal{S}_t$ is

$$\mathcal{S}_t = \frac{1}{2}\begin{bmatrix} 1 & 1 & 0 & 0 \\ 1 & 1 & 0 & 0 \\ 0 & 0 & 0 & 0 \\ 0 & 0 & 0 & 0 \end{bmatrix}\begin{bmatrix} 1 \\ 0 \\ 0 \\ 0 \end{bmatrix} = \begin{bmatrix} \frac{1}{2} \\ \frac{1}{2} \\ 0 \\ 0 \end{bmatrix}.$$

The transmitted wave has an irradiance of $\frac{1}{2}$ $(\mathcal{S}_0 = \frac{1}{2})$ and is linearly polarized horizontally $(\mathcal{S}_1 > 0)$. As a final example, suppose we have a partially polarized elliptical wave whose Stokes parameters have been determined to be, say, $(4, 2, 0, 3)$. Its irradiance is 4; it is more nearly horizontal than vertical $(\mathcal{S}_1 > 0)$; it is right-handed $(\mathcal{S}_3 > 0)$, and has a degree of polarization of 90%. Since none of the parameters can be larger than $\mathcal{S}_0$, a value of $\mathcal{S}_3 = 3$ is fairly large, indicating that the ellipse resembles a circle. If the wave is now made to traverse a quarter-wave plate with a vertical fast axis, then

$$\mathcal{S}_t = \begin{bmatrix} 1 & 0 & 0 & 0 \\ 0 & 1 & 0 & 0 \\ 0 & 0 & 0 & -1 \\ 0 & 0 & 1 & 0 \end{bmatrix} \begin{bmatrix} 4 \\ 2 \\ 0 \\ 3 \end{bmatrix}$$

and thus

$$\mathcal{S}_t = \begin{bmatrix} 4 \\ 2 \\ -3 \\ 0 \end{bmatrix}.$$

The emergent wave has the same irradiance and degree of polarization but is now partially linearly polarized.

We have only touched on a few of the more important aspects of the matrix methods. The full extent of the subject goes far beyond these introductory remarks.*

PROBLEMS

8.1 Describe completely the state of polarization of each of the following waves:

a) $\mathbf{E} = \hat{\mathbf{i}}E_0 \cos(kz - \omega t) - \hat{\mathbf{j}}E_0 \cos(kz - \omega t)$
b) $\mathbf{E} = \hat{\mathbf{i}}E_0 \sin 2\pi(z/\lambda - vt) - \hat{\mathbf{j}}E_0 \sin 2\pi(z/\lambda - vt)$
c) $\mathbf{E} = \hat{\mathbf{i}}E_0 \sin(\omega t - kz) + \hat{\mathbf{j}}E_0 \sin(\omega t - kz + \pi/4)$
d) $\mathbf{E} = \hat{\mathbf{i}}E_0 \cos(\omega t - kz) + \hat{\mathbf{j}}E_0 \cos(\omega t - kz + \pi/2)$.

8.2 Consider the disturbance given by the expression $\mathbf{E}(z, t) = [\hat{\mathbf{i}} \cos \omega t + \hat{\mathbf{j}} \cos(\omega t - \pi/2)]E_0 \sin kz$. What kind of wave is it? Draw a rough sketch showing its main features.

8.3 Analytically, show that the superposition of an $\mathscr{R}$- and an $\mathscr{L}$-state having different amplitudes will yield an $\mathscr{E}$-state as is shown in Fig. 8.8. What must ε be to duplicate that figure?

8.4 If light which is initially natural and of flux density I_i passes through two sheets of HN-32 whose transmission axes are parallel, what will be the flux density of the emerging beam?

---

*One can weave a more elaborate and mathematically satisfying development in terms of something called the coherence matrix. For further, but more advanced, reading see O'Neill, *Introduction to Statistical Optics*.

8.5* What will be the irradiance of the emerging beam if the analyzer of the previous problem is rotated 30°?

8.6* Suppose that we have a pair of crossed polarizers with transmission axes vertical and horizontal. The beam emerging from the first polarizer has flux density I_1 and of course no light passes through the analyzer, i.e. $I_2 = 0$. Now insert a perfect linear polarizer (HN-50) with its transmission axis at 45° to the vertical between the two elements—compute I_2. Think about the motion of the electrons that are radiating in each polarizer.

8.7 Suppose that an ideal polarizer is rotated at a rate ω between a similar pair of stationary crossed polarizers. Show that the emergent flux density will be modulated at four times the rotational frequency. In other words, show that

$$I = \frac{I_1}{8}(1 - \cos 4\omega t)$$

where I_1 is the flux density emerging from the first polarizer and I is the final flux density.

8.8 Figure 8.67 shows a ray traversing a calcite crystal at nearly normal incidence, bouncing off a mirror and then going through the crystal again. Will the observer see a double image of the spot on Σ?

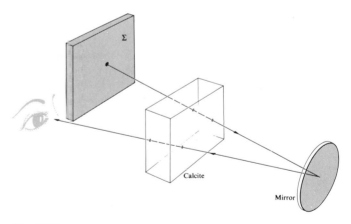

Fig. 8.67

8.9* A pencil mark on a sheet of paper is covered by a calcite crystal. With illumination from above, isn't the light impinging on the paper already polarized having passed through the crystal? Why then do we see two images? Test your solution by polarizing the light from a flashlight and then reflecting it off a sheet of paper. Try specular reflection off glass; is the reflected light polarized?

8.10 Discuss in detail what you see in Fig. 8.68. The crystal in the photograph is calcite and it has a blunt corner at the upper left. The two polaroids have their transmission axes parallel to their *short* edges.

8.11 The calcite crystal in Fig. 8.69 is shown in three different orientations. Its blunt corner is on the left in (a), the lower left in (b), and the bottom in (c). The polaroid's transmission axis is horizontal. Explain each photo, particularly (b).

8.12* Imagine that we have a transmitter of microwaves which radiates a linearly polarized wave whose **E**-field is known to be parallel to the dipole direction. We wish to reflect the entire beam off the surface of a pond having an index of refraction of 9.0. Find the necessary orientation of the beam and its incident angle.

8.13 In discussing calcite we pointed out that its large birefringence arises from the fact that the carbonate groups lie in parallel planes (normal to the optic axis). Show in a sketch and explain why the polarization of the group will be less when **E** is perpendicular to the CO_3 plane than when **E** is parallel to it. What does this mean with respect to $v_\perp$ and $v_\parallel$, i.e. the wave's speeds when **E** is linearly polarized perpendicular or parallel to the optic axis?

8.14 A ray of yellow light is incident on a calcite plate at 50°. The plate is cut so that the optic axis is parallel to the front face and perpendicular to the plane of incidence. Find the angular separation between the two emerging rays.

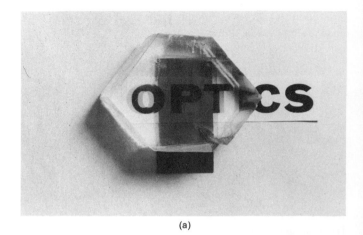

(a)

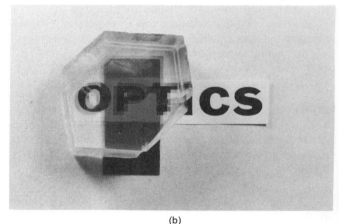

(b)

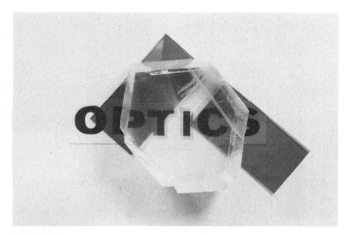

Fig. 8.68

(c)

Fig. 8.69

8.15* A beam of light is incident normally on a quartz plate whose optic axis is perpendicular to the beam. If $\lambda_0 = 589.3$ nm compute the wavelengths of the ordinary and extraordinary waves. What are their frequencies?

8.16 A beam of light enters a calcite prism from the left as shown in Fig. 8.70. There are three possible orientations of the optic axis of particular interest and these correspond to the x-, y-, and z-directions. Imagine then that we have three such prisms. In each case sketch the entering and emerging beams showing the state of polarization. How can any one of these be used to determine n_o and n_e?

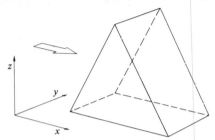

Fig. 8.70

8.17 The electric-field vector of an incident $\mathscr{P}$-state makes an angle of $+30°$ with the horizontal fast axis of a quarter-wave plate. Describe, in detail, the state of polarization of the emergent wave.

8.18 Compute the critical angle for the ordinary ray, i.e. the angle for total internal reflection at the calcite–balsam layer of a Nicol prism.

8.19* Draw a quartz Wallaston prism showing all pertinent rays and their polarization states.

8.20 The prism shown in Fig. 8.71 is known as a *Rochon polarizer*. Sketch all of the pertinent rays assuming:

a) that it is made of calcite
b) that it is made of quartz
c) Why might such a device be more useful than a dichroic polarizer when functioning with high flux density laser light?
d) What valuable feature of the Rochon is lacking in the Wollaston polarizer?

8.21* An $\mathscr{L}$-state traverses an eighth-wave plate having a horizontal fast axis. What is its polarization state on emerging?

8.22* Figure 8.72 shows two polaroid linear polarizers and between these a microscope slide to which is attached a piece of cellophane tape. Explain what you see.

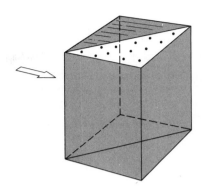

Fig. 8.71

8.23 A Babinet compensator is positioned at 45° between crossed linear polarizers and is being illuminated with sodium light. When a thin sheet of mica (indices 1.599 and 1.594) is placed on the compensator, the black bands all shift by $\frac{1}{4}$ of the space separating them. Compute the retardance of the sheet and its thickness.

8.24 Imagine that we have unpolarized room light incident almost normally on the glass surface of a radar screen. A portion of it would be specularly reflected back toward the viewer and would thus tend to obscure the display. Suppose now that we cover the screen with a right-circular polarizer as shown in Fig. 8.73. Trace the incident and reflected beams indicating their polarization states. What happens to the reflected beam?

8.25 Is it possible for a beam to consist of two orthogonal incoherent $\mathscr{P}$-states and not be natural light? Explain. How might you arrange to have such a beam?

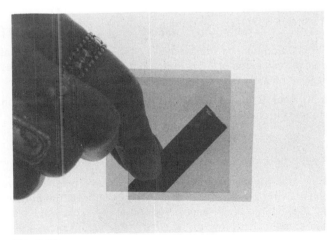

Fig. 8.72

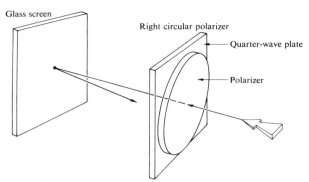

Glass screen

Right circular polarizer

Quarter-wave plate

Polarizer

Fig. 8.73

8.26* The specific rotatory power for sucrose dissolved in water at 20°C ($\lambda_0 = 589.3$ nm) is $+66.45°$ per 10 cm of path traversed through a solution containing 1 g of active substance (sugar) per cm^3 of solution. A vertical $\mathscr{P}$-state (sodium light) enters at one end of a 1 m tube containing 1000 cm^3 of solution of which 10 g is sucrose. At what orientation will the $\mathscr{P}$-state emerge?

8.27 On examining a piece of stressed photoelastic material between crossed linear polarizers one would see a set of colored bands (isochromatics) and superimposed on these a set of dark bands (isoclinics). How might we remove the isoclinics leaving only the isochromatics? Explain your solution. Incidently, the proper arrangement is independent of the orientation of the photoelastic sample.

8.28* Consider a Kerr cell whose plates are separated by a distance d. Let ℓ be the effective length of those plates (slightly different from the actual length because of fringing of the field). Show that

$$\Delta\varphi = 2\pi K\ell V^2/d^2. \qquad [8.41]$$

8.29 Compute the half-wave voltage for a longitudinal Pockels cell made of ADA (ammonium dihydrogen arsenate) at $\lambda_0 \approx 550$ nm where $r_{63} = 5.5 \times 10^{-12}$ and $n_0 = 1.58$.

8.30 Find a Jones vector $\mathbf{E}_2$ representing a polarization state orthogonal to

$$\mathbf{E}_1 = \begin{bmatrix} 1 \\ -2i \end{bmatrix}.$$

Sketch both of these.

8.31 Use Table 8.6 to derive a Mueller matrix for a half-wave plate having a vertical fast axis. Utilize your result to convert an $\mathscr{R}$-state into an $\mathscr{L}$-state. Verify that the same wave plate will convert an $\mathscr{L}$- to an $\mathscr{R}$-state. Advancing or retarding the relative phase by $\pi/2$ should have the same effect. Check

this by deriving the matrix for a horizontal fast axis half-wave plate.

8.32 Construct one possible Mueller matrix for a right circular polarizer made out of a linear polarizer and a quarter-wave plate. Such a device is obviously an inhomogeneous two-element train and will differ from the *homogeneous* circular polarizer of Table 8.6. Test that your matrix will convert natural light to an $\mathscr{R}$-state. Show that it will pass $\mathscr{R}$-states as will the homogeneous matrix. Your matrix should convert $\mathscr{L}$-states incident on the input side to $\mathscr{R}$-states while the homogeneous polarizer will totally absorb them. Verify this.

8.33* If the Pockels cell modulator shown in Fig. 8.66 is illuminated by light of irradiance I_i it will transmit a beam of irradiance I_t such that

$$I_t = I_i \sin^2 (\Delta\varphi/2).$$

Make a plot of I_t/I_i versus applied voltage. What is the significance of the voltage which corresponds to maximum transmission? What is the lowest voltage above zero which will cause I_t to be zero for ADP ($\lambda_0 = 546.1$ nm)? How can things be rearranged to yield a maximum value of I_t/I_i for zero voltage? In this new configuration what irradiance results when $V = V_{\lambda/2}$?

8.34 Construct a Jones matrix for an isotropic plate of absorbing material having an amplitude transmission coefficient of t. It might sometimes be desirable to keep track of the phase, since even if $t = 1$ such a plate is still an isotropic phase retarder. What is the Jones matrix for a region of vacuum? What is it for a perfect absorber?

8.35 Construct a Mueller matrix for an isotropic plate of absorbing material having an amplitude transmission coefficient of t. What Mueller matrix will completely depolarize any wave without affecting its irradiance (it has no physical counterpart)?

8.36 Keeping Eq. (8.29) in mind, write an expression for the unpolarized flux-density component (I_u) of a partially polarized beam in terms of the Stokes parameters. To check your result, add an unpolarized Stokes vector of flux density 4 to an $\mathscr{R}$-state of flux density 1. Then see if you get $I_u = 4$ for the resultant wave.

Interference

9

The intricate color patterns shimmering across an oil slick on a wet asphalt pavement result from one of the more common manifestations of the phenomenon of interference.[*] On a macroscopic scale we might consider the related problem of the interaction of surface ripples in a pool of water. Our everyday experience with this kind of situation allows us to envision a complex distribution of disturbances (as shown e.g. in Fig. 9.1). There might be regions where two (or more) waves have overlapped, partially or even completely canceling each other. Still other regions might exist in the pattern, where the resultant troughs and crests were even more pronounced than those of any of the constituent waves. After being superimposed, the individual waves separate and continue on, completely unaffected by their previous encounter.

Phenomena arising from optical interference would, of course, be quite difficult to interpret in terms of a purely corpuscular model. The wave theory of the electromagnetic nature of light, however, provides a natural basis from which to proceed. Recall that the expression describing the optical disturbance is a second-order homogeneous linear partial differential equation (3.22). As we have seen, it therefore obeys the important *principle of superposition*. Accordingly, the resultant electric-field intensity E, at a point in space where two or more light waves overlap, is equal to the *vector sum* of the individual constituent disturbances. Briefly then, *optical interference may be termed an interaction of two or more light waves yielding a resultant irradiance which deviates from the sum of the component irradiances.*

Out of the multitude of optical systems which produce

---

[*] The layer of water on the asphalt allows the oil film to assume the shape of a smooth planar surface. The black asphalt absorbs the transmitted light preventing back reflection which would tend to obscure the fringes.

Fig. 9.1 Water waves from two point sources in a ripple tank.

interference, we will choose a few of the more important to examine. Interferometric devices will be divided, for the sake of discussion, into two groups, *wavefront splitting* and *amplitude splitting*. In the first instance, portions of the primary wavefront are used either directly as sources to emit secondary waves, or, in conjunction with optical devices, to produce virtual sources of secondary waves. These secondary waves are then brought together, thereupon to interfere. In the latter case of amplitude splitting, the primary wave itself is divided into two segments which travel different paths before recombining and interfering.

9.1 GENERAL CONSIDERATIONS

We have already examined the problem of the superposition of two scalar waves (Section 7.1) and in many respects those results will again be applicable. But light is, of course, a vector phenomenon; the electric and magnetic fields are vector fields. And an appreciation of this fact is fundamental to any kind of intuitive understanding of optics. Needless to say, there are many situations in which the particular optical system is so configured that the vector nature of light is of little practical significance. We will therefore derive the basic interference equations within the context of the vector model, thereafter delineating the conditions under which the scalar treatment is applicable.

In accordance with the principle of superposition, the electric-field intensity **E**, at a point in space, arising from the separate fields $\mathbf{E}_1$, $\mathbf{E}_2$, . . . of various contributing sources is given by

$$\mathbf{E} = \mathbf{E}_1 + \mathbf{E}_2 + \cdots .$$

Once again we point out that the optical disturbance, or

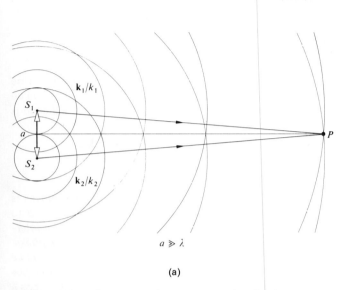

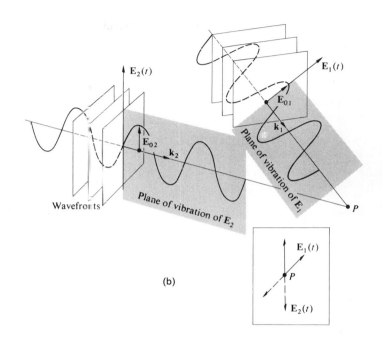

Fig. 9.2 Waves from two point sources overlapping in space.

light field **E**, varies in time at an exceedingly rapid rate, roughly

$$4.3 \times 10^{14}\ \text{Hz} \qquad \text{to} \qquad 7.5 \times 10^{14}\ \text{Hz},$$

making the actual field an impractical quantity to detect. On the other hand, the irradiance I can be measured directly using a wide variety of sensors (e.g. photocells, bolometers, photographic emulsions or eyes). Indeed then, if we are to study interference, we had best approach the problem by way of the irradiance.

Much of the analysis to follow can be performed without specifying the particular shape of the wavefronts and the results are therefore quite general in their applicability (Problem 9.1). For the sake of simplicity, however, consider two point sources S_1 and S_2 emitting monochromatic waves of the same frequency, in a homogeneous medium. Furthermore, let their separation a be much greater than λ. Locate the point of observation P, far enough away from the sources so that at P the wavefronts will be planes (Fig. 9.2). For the moment, we will only consider linearly polarized waves of the form

$$\mathbf{E}_1(\mathbf{r}, t) = \mathbf{E}_{01} \cos(\mathbf{k}_1 \cdot \mathbf{r} - \omega t + \varepsilon_1)$$

and

$$\mathbf{E}_2(\mathbf{r}, t) = \mathbf{E}_{02} \cos(\mathbf{k}_2 \cdot \mathbf{r} - \omega t + \varepsilon_2).$$

We saw in Chapter 3 that the irradiance at P is given by

$$I = \epsilon v \langle \mathbf{E}^2 \rangle.$$

Inasmuch as we will be concerned only with relative irradiances within the same medium we will, for the time being at least, simply neglect the constants and set

$$I = \langle \mathbf{E}^2 \rangle.$$

What is meant by $\langle \mathbf{E}^2 \rangle$ is of course the time average of the magnitude of the electric field intensity squared or $\langle \mathbf{E} \cdot \mathbf{E} \rangle$. Accordingly

$$\mathbf{E}^2 = \mathbf{E} \cdot \mathbf{E},$$

where now

$$\mathbf{E}^2 = (\mathbf{E}_1 + \mathbf{E}_2) \cdot (\mathbf{E}_1 + \mathbf{E}_2)$$

and thus

$$\mathbf{E}^2 = \mathbf{E}_1^2 + \mathbf{E}_2^2 + 2\mathbf{E}_1 \cdot \mathbf{E}_2.$$

Taking the time average of both sides, the irradiance becomes

$$I = I_1 + I_2 + I_{12} \tag{9.1}$$

provided that

$$I_1 = \langle \mathbf{E}_1^2 \rangle,$$

$$I_2 = \langle \mathbf{E}_2^2 \rangle$$

and

$$I_{12} = 2 \langle \mathbf{E}_1 \cdot \mathbf{E}_2 \rangle.$$

The latter expression is known as the *interference term*. To evaluate it in this specific instance, we form

$$\mathbf{E}_1 \cdot \mathbf{E}_2 = \mathbf{E}_{01} \cdot \mathbf{E}_{02} \cos (\mathbf{k}_1 \cdot \mathbf{r} - \omega t + \varepsilon_1)$$
$$\times \cos (\mathbf{k}_2 \cdot \mathbf{r} - \omega t + \varepsilon_2)$$

or equivalently

$$\mathbf{E}_1 \cdot \mathbf{E}_2 = \mathbf{E}_{01} \cdot \mathbf{E}_{02} [\cos (\mathbf{k}_1 \cdot \mathbf{r} + \varepsilon_1)$$
$$\times \cos \omega t + \sin (\mathbf{k}_1 \cdot \mathbf{r} + \varepsilon_1) \sin \omega t]$$
$$\times [\cos (\mathbf{k}_2 \cdot \mathbf{r} + \varepsilon_2) \cos \omega t \qquad (9.2)$$
$$+ \sin (\mathbf{k}_2 \cdot \mathbf{r} + \varepsilon_2) \sin \omega t].$$

Recall that the time average of some function $f(t)$, taken over an interval T, is

$$\langle f(t) \rangle = \frac{1}{T} \int_t^{t+T} f(t')\, dt'. \qquad (9.3)$$

The period τ of the harmonic functions is $2\pi/\omega$ and for our present concern $T \gg \tau$. In that case the $1/T$ coefficient in front of the integral has a dominant effect. After multiplying out and averaging, Eq. (9.2) becomes

$$\langle \mathbf{E}_1 \cdot \mathbf{E}_2 \rangle = \tfrac{1}{2}\mathbf{E}_{01} \cdot \mathbf{E}_{02} \cos (\mathbf{k}_1 \cdot \mathbf{r} + \varepsilon_1 - \mathbf{k}_2 \cdot \mathbf{r} - \varepsilon_2),$$

where use was made of the fact that $\langle \cos^2 \omega t \rangle = \tfrac{1}{2}$, $\langle \sin^2 \omega t \rangle = \tfrac{1}{2}$ and $\langle \cos \omega t \sin \omega t \rangle = 0$. The interference term is then

$$I_{12} = \mathbf{E}_{01} \cdot \mathbf{E}_{02} \cos \delta \qquad (9.4)$$

and δ, equal to $(\mathbf{k}_1 \cdot \mathbf{r} - \mathbf{k}_2 \cdot \mathbf{r} + \varepsilon_1 - \varepsilon_2)$, is the *phase difference* arising from a combined path-length and epoch-angle difference. Notice that if $\mathbf{E}_{01}$ and $\mathbf{E}_{02}$ (and therefore $\mathbf{E}_1$ and $\mathbf{E}_2$) are perpendicular, $I_{12} = 0$ and $I = I_1 + I_2$. Two such orthogonal $\mathscr{P}$-states will combine to yield an $\mathscr{R}$-, $\mathscr{L}$-, $\mathscr{P}$- or $\mathscr{E}$-state but the flux-density distribution will be unaltered.

By far the most commonly occurring situation in the work to follow corresponds to $\mathbf{E}_{01}$ parallel to $\mathbf{E}_{02}$. In that case, the irradiance reduces to the value found in the scalar treatment of Section 7.1. Under those conditions

$$I_{12} = E_{01}E_{02} \cos \delta.$$

This can be written in a more convenient way by noticing that

$$I_1 = \langle \mathbf{E}_1^2 \rangle = \frac{E_{01}^2}{2}$$

and

$$I_2 = \langle \mathbf{E}_2^2 \rangle = \frac{E_{02}^2}{2}.$$

The interference term becomes

$$I_{12} = 2\sqrt{I_1 I_2} \cos \delta,$$

whereupon the total irradiance is

$$I = I_1 + I_2 + 2\sqrt{I_1 I_2} \cos \delta. \qquad (9.5)$$

At various points in space, the resultant irradiance can be greater, less than, or equal to $I_1 + I_2$ depending on the value of I_{12}, i.e. depending on δ. A maximum in the irradiance is obtained when $\cos \delta = 1$, so that

$$I_{max} = I_1 + I_2 + 2\sqrt{I_1 I_2}$$

when

$$\delta = 0, \pm 2\pi, \pm 4\pi, \dots .$$

In this case the phase difference between the two waves is an integer multiple of 2π, and the disturbances are said to be *in phase*. One speaks of this as *total constructive interference*. When $0 < \cos \delta < 1$ the waves are *out of phase*, $I_1 + I_2 < I < I_{max}$ and the result is known as *constructive interference*. At $\delta = \pi/2$, $\cos \delta = 0$, the optical disturbances are said to be 90° out of phase and $I = I_1 + I_2$. For $0 > \cos \delta > -1$ we have the condition of *destructive* interference, $I_1 + I_2 > I > I_{min}$. The minimum in the irradiance results when the waves are 180° out of phase, troughs overlap crests, $\cos \delta = -1$, and

$$I_{min} = I_1 + I_2 - 2\sqrt{I_1 I_2}.$$

This, of course, occurs when $\delta = \pm \pi, \pm 3\pi, \pm 5\pi, \dots$, and it is referred to as *total destructive interference*.

Another somewhat special, yet very important case arises when the amplitudes of both waves reaching P in Fig. 9.2 are equal, i.e. $\mathbf{E}_{01} = \mathbf{E}_{02}$. Since the irradiance contributions from both sources are then equal, let $I_1 = I_2 = I_0$. Equation (9.5) can now be written as

$$I = 2I_0(1 + \cos \delta) = 4I_0 \cos^2 \frac{\delta}{2} \qquad (9.6)$$

from which it follows that $I_{min} = 0$ and $I_{max} = 4I_0$.

Equation (9.5) holds equally well for the spherical waves emitted by S_1 and S_2. Such waves can be expressed as

$$\mathbf{E}_1(r_1, t) = \mathbf{E}_{01}(r_1) \exp [i(kr_1 - \omega t + \varepsilon_1)]$$

and

$$\mathbf{E}_2(r_2, t) = \mathbf{E}_{02}(r_2) \exp [i(kr_2 - \omega t + \varepsilon_2)].$$

The terms r_1 and r_2 are the radii of the spherical wavefronts overlapping at P, i.e. they specify the distances from the sources to P. In this case

$$\delta = k(r_1 - r_2) + (\varepsilon_1 - \varepsilon_2).$$

The flux density in the region surrounding S_1 and S_2 will certainly vary from point to point as $(r_1 - r_2)$ varies. Nonetheless, from the principle of conservation of energy, we expect the spatial average of I to remain constant and equal to the average of $I_1 + I_2$. The space average of I_{12} must therefore be zero, a property verified by Eq. (9.4) since the average of the cosine term is, in fact, zero (for further discussion of this point see Problem 9.2).

Equation (9.6) will be applicable when the separation between S_1 and S_2 is small in comparison with r_1 and r_2 and when, in addition, the interference region is also small in the same sense. Under these circumstances $\mathbf{E}_{01}$ and $\mathbf{E}_{02}$ may be considered independent of position, i.e. constant over the small region examined. If the emitting sources are of equal strength $E_{01} = E_{02}$, $I_1 = I_2 = I_0$ and we have

$$I = 4I_0 \cos^2 \tfrac{1}{2}[k(r_1 - r_2) + (\varepsilon_1 - \varepsilon_2)].$$

Irradiance maxima occur when

$$\delta = 2\pi m$$

provided that $m = 0, \pm 1, \pm 2, \ldots$. Similarly minima, for which $I = 0$, arise when

$$\delta = \pi(2m + 1).$$

These expressions can be rewritten such that maximum irradiance occurs when

$$(r_1 - r_2) = [2\pi m + (\varepsilon_2 - \varepsilon_1)]/k \qquad (9.7a)$$

and minimum when

$$(r_1 - r_2) = [\pi(2m + 1) + (\varepsilon_2 - \varepsilon_1)]/k. \qquad (9.7b)$$

Either one of these equations defines a family of surfaces, each of which is a hyperboloid of revolution. The vertices of the hyperboloids are separated by distances equal to the right-hand sides of Eqs. (9.7). The foci are located at S_1 and S_2. If the waves are in phase at the emitter $\varepsilon_1 - \varepsilon_2 = 0$, and Eqs. (9.7) simplify to

$$(r_1 - r_2) = 2\pi m/k = m\lambda$$

$$(r_1 - r_2) = \pi(2m + 1)/k = (m + 1/2)\lambda$$

for maximum and minimum irradiance, respectively. Figure 9.3(a) shows a few of the surfaces over which there are irradiance maxima. The dark and light zones which would be seen on a screen placed in the region of interference are known as *interference fringes* [Fig. 9.3(b)].

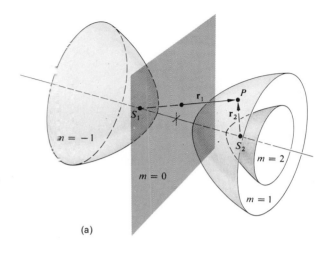

(a)

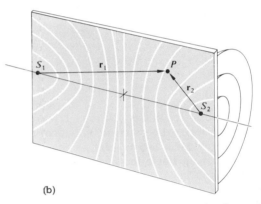

(b)

Fig. 9.3 Hyperboloidal surfaces of maximum irradiance for two point sources

9.2 CONDITIONS FOR INTERFERENCE

If the interference pattern corresponding to Eqs. (9.7) is to be observable, the phase difference $(\varepsilon_1 - \varepsilon_2)$ between the two sources must remain fairly constant in time. Such sources are said to be coherent.[*] Two overlapping beams arising from separate emitters will interfere, but the resultant pattern will not be sustained long enough to be readily observable. A typical source contains a large number of excited atoms, each capable of radiating a wave train for about 10^{-8} s. Two distinct sources could therefore maintain

---

[*] Chapter 10 is devoted to the study of coherence and so here we'll merely touch on those aspects which are immediately pertinent.

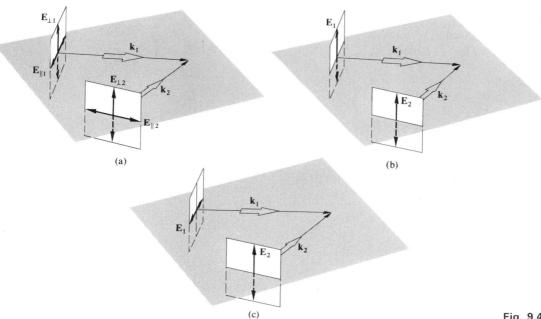

(a)

(b)

(c)

Fig. 9.4 Interference of polarized light.

their relative phases for, at best, 10^{-8} s. The resulting interference pattern would be constant in space for only that
brief duration, before it would vary as the phase shifted,
thereupon remaining stable for another short while, only to
change again and so on. It would therefore be futile to
attempt to see or photograph the interference pattern
resulting from two light bulbs. Two separate lasers have
been used to generate interference patterns, but we'll come
back to that later. The most common means of overcoming
this problem, as we shall see, is to make one source serve
to produce two coherent secondary sources.

If two beams are to interfere to produce a stable pattern,
they must have very nearly the same frequency. A significant
frequency difference would result in a rapidly varying, time
dependent, phase difference which in turn would cause
I_{12} to average to zero during the detection interval (see
Section 7.2.1).

The clearest patterns will exist when the interfering
waves have equal or very nearly equal amplitudes. The
central regions of the dark and light fringes will then correspond to complete destructive and constructive interference,
respectively, yielding maximum contrast.

In the previous section, we assumed that the two overlapping optical disturbance vectors were linearly polarized

and parallel. Nonetheless, the formulae of Section 9.1 apply
as well to more complicated situations; indeed the treatment
is applicable regardless of the polarization state of the waves.
To appreciate this, recall that any polarization state can be
synthesized out of two orthogonal $\mathscr{P}$-states. For natural
(unpolarized) light these $\mathscr{P}$-states are mutually incoherent,
but that represents no particular difficulty.

Suppose that every wave has its propagation vector
in the same plane so that we can label the constituent
orthogonal $\mathscr{P}$-states with respect to that plane, e.g. $\mathbf{E}_{\parallel}$ and
$\mathbf{E}_{\perp}$, which are parallel and perpendicular to the plane, respectively [Fig. 9.4(a)]. Thus any plane wave, whether
polarized or not, can be written in the form $(\mathbf{E}_{\parallel} + \mathbf{E}_{\perp})$.
Imagine then, that the waves $(\mathbf{E}_{\parallel 1} + \mathbf{E}_{\perp 1})$ and $(\mathbf{E}_{\parallel 2} + \mathbf{E}_{\perp 2})$
emitted from two identical coherent sources superimpose in
some region of space. The resulting flux-density distribution would consist of two independent, precisely over-
lapping, interference patterns, $\langle (\mathbf{E}_{\parallel 1} + \mathbf{E}_{\parallel 2})^2 \rangle$ and
$\langle (\mathbf{E}_{\perp 1} + \mathbf{E}_{\perp 2})^2 \rangle$. Therefore, while we derived the equations
of the previous section specifically for linear light, they are
applicable, as well, for any polarization state including
natural light.

Notice that even though $\mathbf{E}_{\perp 1}$ and $\mathbf{E}_{\perp 2}$ are always parallel
to each other $\mathbf{E}_{\parallel 1}$ and $\mathbf{E}_{\parallel 2}$, which are in the reference plane,

need not be. They will only be parallel when the two beams are themselves parallel (i.e. $\mathbf{k}_1 = \mathbf{k}_2$). The inherent vector nature of the interference process as manifest in the dot-product representation (9.4) of I_{12} cannot therefore be ignored. As we shall see, there are a great many practical situations in which the beams approach being parallel and for these the scalar theory will do rather nicely. Even so, Figs. 9.4(b) and (c) are included as an urge to caution. They depict the imminent overlapping of two coherent linearly polarized waves. In Fig. 9.4(b) the optical vectors are parallel, even though the beams aren't, and interference would nonetheless result. In Fig. 9.4(c) the optical vectors are perpendicular and $I_{12} = 0$, which would be the case here even if the beams were parallel.

Fresnel and Arago made an extensive study of the conditions under which the interference of polarized light occurs and their conclusions summarize some of the above considerations. The *Fresnel-Arago laws* are as follows:

1. Two orthogonal, coherent $\mathscr{P}$-states cannot interfere in the sense that $I_{12} = 0$ and no fringes result.

2. Two parallel, coherent $\mathscr{P}$-states will interfere in the same way as will natural light.

3. The two constituent orthogonal $\mathscr{P}$-states of natural light cannot interfere to form a readily observable fringe pattern even if rotated into alignment. This last point is understandable since these $\mathscr{P}$-states are incoherent.

9.3 WAVEFRONT-SPLITTING INTERFEROMETERS

Return for a moment to Fig. (9.3) where the equation

$$(r_1 - r_2) = m\lambda$$

determined the surfaces of maximum irradiance. Since the wavelength λ for light is very small, a large number of surfaces corresponding to the lower values of m will exist close to, and on either side of, the plane $m = 0$. A number of fairly straight parallel fringes will therefore appear on a screen placed perpendicular to that ($m = 0$) plane and in the vicinity of it, and for this case the approximation $r_1 \approx r_2$ will hold. If S_1 and S_2 are then displaced normal to the $\overline{S_1 S_2}$ line, the fringes will merely be displaced parallel to themselves. Practically, two narrow slits will therefore increase the irradiance leaving the central region of the two-point source pattern otherwise essentially unchanged.

Consider a hypothetical monochromatic plane wave illuminating a long narrow slit. From that primary slit a cylindrical wave will emerge and suppose that this wave, in turn, falls on two parallel, narrow, closely spaced slits

S_1 and S_2. This is shown in a three-dimensional view in Fig. 9.5(a). When symmetry exists, the segments of the primary wavefront arriving at the two slits will be exactly in phase, and the slits will constitute two coherent secondary sources. We expect that wherever the two waves coming from S_1 and S_2 overlap, interference will occur (provided that the optical path difference is less than the coherence length, $c \, \Delta t$).

Consider the construction shown in Fig. 9.5(c). In a realistic physical situation the distance between each of the screens would be very large in comparison with the distance a between the two slits, and all the fringes would be fairly close to the center O of the screen. The path difference between the rays along $\overline{S_1 P}$ and $\overline{S_2 P}$ can be obtained, to a good approximation, by dropping a perpendicular from S_2 onto $\overline{S_1 P}$ This path difference is given by

$$(\overline{S_1 B}) = (\overline{S_1 P}) - (\overline{S_2 P}) \tag{9.8}$$

or

$$(\overline{S_1 B}) = r_1 - r_2.$$

Continuing with this approximation (Problem 9.4) the path difference can be expressed as

$$r_1 - r_2 = a\theta \tag{9.9}$$

since $\theta \approx \sin \theta$.

Notice that

$$\theta = \frac{y}{s} \tag{9.10}$$

and so

$$r_1 - r_2 = \frac{a}{s} y.$$

In accordance with Section 9.1, *constructive* interference will occur when

$$r_1 - r_2 = m\lambda. \tag{9.11}$$

Thus, from the last two relations we obtain

$$y_m = \frac{s}{a} m\lambda. \tag{9.12}$$

This gives the position of the mth bright fringe on the screen if we count the maximum at 0 as the zeroth fringe. The angular position of the fringe is obtained by substituting the last expression into Eq. (9.10); thus

$$\theta_m = \frac{m\lambda}{a}. \tag{9.13}$$

This relationship can be obtained directly by inspecting Fig. 9.5(c). For the mth-order interference, m whole wavelengths

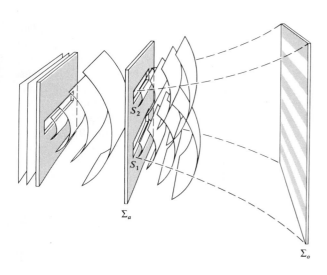

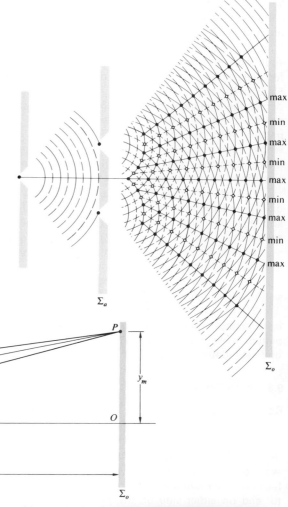

Fig. 9.5 Young's experiment. [Photo courtesy M. Cagnet, M. Francon, and J. C. Thrierr: *Atlas optischer Erscheinungen*, Berlin—Heidelberg—New York: Springer, 1962.]

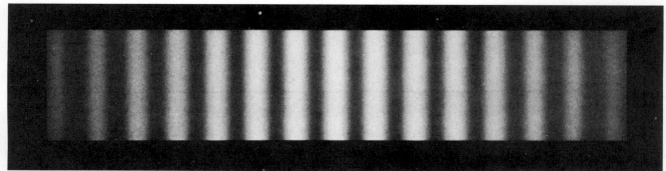

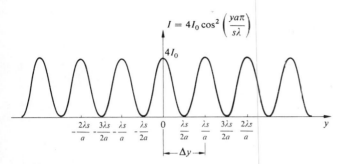

$$I = 4I_0 \cos^2\left(\frac{ya\pi}{s\lambda}\right)$$

Fig. 9.6 Irradiance versus distance.

should fit within the distance $r_1 - r_2$. Therefore, from the triangle $S_1 S_2 B$,

$$\theta_m = m\lambda/a.$$

The spacing of the fringes on the screen can be gotten readily from Eq. (9.12). The difference in the positions of two consecutive maxima is

$$y_{m+1} - y_m = \frac{s}{a}(m+1)\lambda - \frac{s}{a}m\lambda$$

or

$$\Delta y = \frac{s}{a}\lambda. \qquad (9.14)$$

Since this pattern is equivalent to that obtained for two overlapping spherical waves (at least in the $r_1 \approx r_2$ region), we can apply Eq. (9.6). Using the phase difference

$$\delta = k(r_1 - r_2).$$

Equation (9.6) can be rewritten as

$$I = 4I_0 \cos^2\frac{k(r_1 - r_2)}{2},$$

provided, of course, that the two beams are coherent and have equal irradiances I_0. With

$$r_1 - r_2 = ya/s$$

the resultant irradiance becomes

$$I = 4I_0 \cos^2\frac{ya\pi}{s\lambda}.$$

As shown in Fig. 9.6 consecutive maxima are separated by the Δy given in Eq. (9.14).*

---

* Modifications of this pattern arising as a result of increasing the width of either the primary S, or secondary-source slits will be considered in later chapters (10 and 12). In the former case, fringe contrast will be used as a measure of the degree of coherence (Section 12.1). In the latter, diffraction effects become significant.

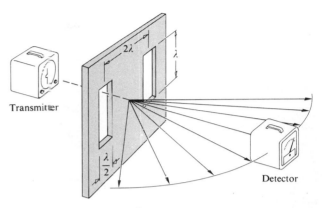

Fig. 9.7 A microwave interferometer.

A direct visual observation of the fringe pattern can be made by punching two small pinholes in a thin card. The holes should be approximately the size of the type symbol for a period on this page and the separation between their centers about three radii. A street lamp, car headlight, or traffic signal at night, located a few hundred feet away will serve as a plane wave source. The card should be positioned directly in front of, and *very close up to the eye*. The fringes will appear perpendicular to the line of centers. The pattern is much more readily seen using slits as discussed in Section 10.2.2, but you should give the pinholes a try.

Microwaves, because of their long wavelength, also offer an easy way to observe double-slit interference. Two slits (e.g. $\lambda/2$ wide by λ long, separated by 2λ) cut in a piece of sheet metal or foil will serve quite well as secondary sources (Fig. 9.7).

The interferometric configuration discussed above, with either point or slit sources, is known as *Young's experiment*. The same physical and mathematical considerations apply directly to a number of other wavefront-splitting interferometers. Most common among these are Fresnel's double mirror, Fresnel's double prism and Lloyd's mirror.

Fresnel's double mirror consists of two plane front silvered mirrors inclined to each other at a very small angle, as shown in Fig. 9.8. One portion of the cylindrical wavefront coming from slit S is reflected from the first mirror, while another portion of the wavefront is reflected from the second mirror. An interference field exists in space in the region where the two reflected waves are superimposed on each other. The images (S_1 and S_2) of the slit S in the two mirrors can be considered as separate coherent sources, placed at a distance a apart. It follows from the laws of reflection, as

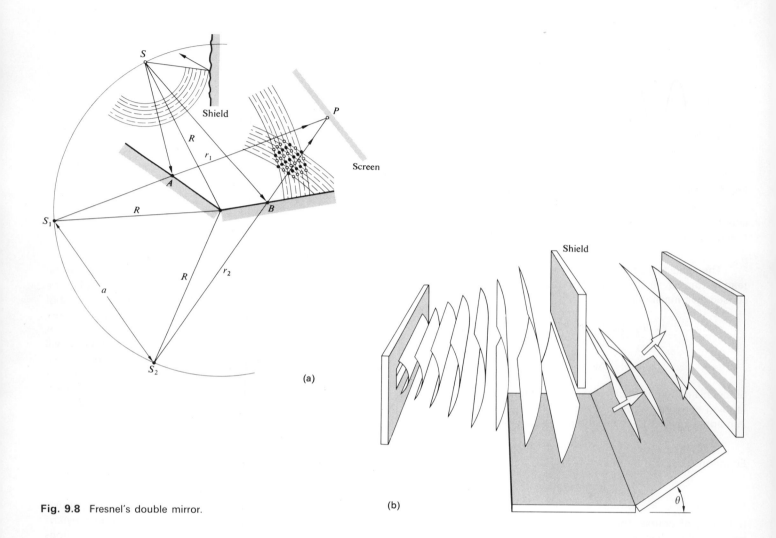

Fig. 9.8 Fresnel's double mirror.

(a)

(b)

illustrated in Fig. 9.8(a), that $\overline{SA} = \overline{S_1 A}$, $\overline{SB} = \overline{S_2 B}$, so that $\overline{SA} + \overline{AP} = r_1$ and $\overline{SB} + \overline{BP} = r_2$. The optical path-length difference between the two rays is then simply $r_1 - r_2$. The various maxima occur at $r_1 - r_2 = m\lambda$ as was the case for Young's interferometer. Again, the separation of the fringes is given by

$$\Delta y = \frac{s}{a}\lambda$$

where s is the distance between the plane of the two virtual sources (S_1, S_2) and the screen. The arrangement in Fig. 9.8 has again been deliberately exaggerated to make the geometry somewhat clearer. Notice that the angle θ between the

mirrors must be quite small if the electric field vectors for each of the two beams are to be parallel, or nearly so. Let $\mathbf{E}_1$ and $\mathbf{E}_2$ represent the light waves emitted from the coherent virtual sources S_1 and S_2. At any instant in time at the point P in space, each of these vectors can be resolved into components, parallel and perpendicular to the plane of the figure. With $\mathbf{k}_1$ and $\mathbf{k}_2$ parallel to AP and BP respectively, it should be apparent that the components of $\mathbf{E}_1$ and $\mathbf{E}_2$ in the plane of the figure will approach being parallel only for small θ.

The Fresnel double prism or biprism consists of two thin prisms joined at their bases as shown in Fig. 9.9. A single cylindrical wavefront impinges on both prisms. The top

(a)

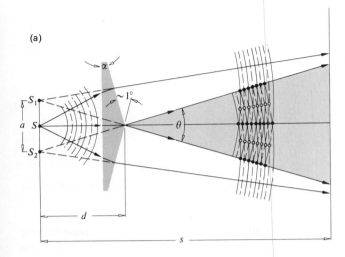

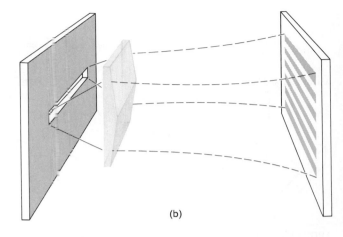

Fig. 9.9 Fresnel's biprism.

portion of the wavefront is refracted downward, while the lower segment is refracted upward. In the region of super-position, interference occurs. Here, again, two virtual sources S_1 and S_2 exist, separated by a distance a which can be expressed in terms of the prism angle α (Problem 9.5) where $s \gg a$. The expression for the separation of the fringes is the same as before.

The last wavefront-splitting interferometer which we will consider is Lloyd's mirror. It is shown in Fig. 9.10 and con-

sists of a flat piece of either dielectric or metal which serves as a mirror from which is reflected a portion of the cylindrical wavefront coming from slit S. Another portion of the wave-front proceeds directly from the slit to the screen. For the separation a, between the two coherent sources, we take the distance between the actual slit and its image S_1 in the mirror. The spacing of the fringes is once again given by $(s/a)\lambda$. The distinguishing feature of this device is that at glancing incidence ($\theta_i \approx \pi/2$) the reflected beam undergoes a

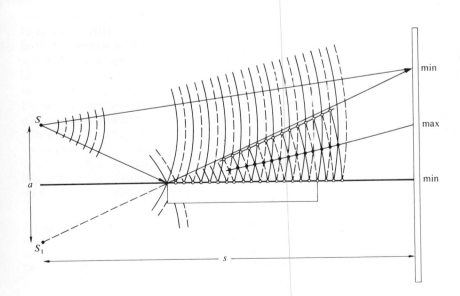

Fig. 9.10 Lloyd's mirror.

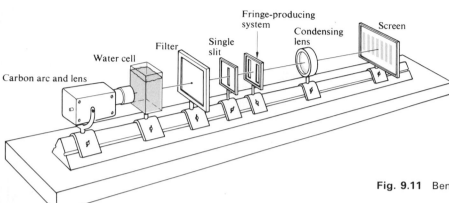

Fig. 9.11 Bench set up to study wavefront splitting arrangements.

180° phase shift (recall that the amplitude reflection coefficients are then both equal to −1). With an additional phase shift of $\pm\pi$

$$\delta = k(r_1 - r_2) \pm \pi$$

and the irradiance becomes

$$I = 4I_0 \sin^2 \left(\frac{\pi a y}{s \lambda} \right).$$

The fringe pattern for Lloyd's mirror is complementary to that of Young's interferometer; the maxima of one pattern exist at values of y which correspond to minima in the other pattern. The top edge of the mirror is equivalent to $y = 0$ and will be the center of a dark fringe rather than a bright one as in Young's device. The lower half of the pattern will be obstructed by the presence of the mirror itself. Consider then, what would happen if a thin sheet of transparent material were placed in the path of the rays traveling directly to the screen. The transparent sheet would have the effect of increasing the number of wavelengths in each direct ray. The entire pattern would accordingly move upward to where the reflected rays would travel a bit farther before interfering. Because of the obvious inherent simplicity of this device, it has found use over a very wide region of the electromagnetic spectrum. The actual reflecting surfaces have varied from crystals for x-rays, to ordinary glass for light, wire screening for microwaves, to a lake or even the earth's ionosphere for radio waves.*

---

* For a discussion of the effects of having a finite slit width and a finite frequency bandwidth see R. N. Wolfe and F. C. Eisen, "Irradiance Distribution in a Lloyd Mirror Interference Pattern," *J. Opt. Soc. Am.* **38**, 706 (1948).

All of the above interferometers can be demonstrated quite readily. The necessary parts, mounted on a single optical bench, are shown diagrammatically in Fig. 9.11. The source of light should be a strong one; if a laser is not available, a discharge lamp or a carbon arc followed by a water cell, to cool things down a bit, will do nicely. The light is then not monochromatic but the fringes, which will be colored, can still be observed. A satisfactory approximation to monochromatic light can be obtained with a filter placed in front of the arc. A low-power He–Ne laser is perhaps the easiest source to work with and you won't need a water cell or filter.

9.4 AMPLITUDE-SPLITTING INTERFEROMETERS

Suppose that a light wave were incident on a half-silvered mirror.* Part of the wave would be transmitted and part of it would be reflected. Both the transmitted and the reflected waves would, of course, have lower amplitudes than the original one. One might say figuratively that the amplitude had been "split". If the two separate waves could somehow be brought together again at a detector, interference would result as long as the original coherence between the two had not been destroyed. If the path lengths differ by a distance greater than that of the wave train (i.e. the coherence length), the portions reunited at the detector will correspond to different wave trains. No unique phase relationship will exist between them in that case and the fringe pattern will be

---

* A *half silvered mirror* is one which is semitransparent because the metallic coating is too thin to be opaque. You can look through it and at the same time you can see your reflection in it. *Beam splitters*, as devices of this kind are called, can also be made of thin stretched plastic films known as *pellicles* or even uncoated glass plate.

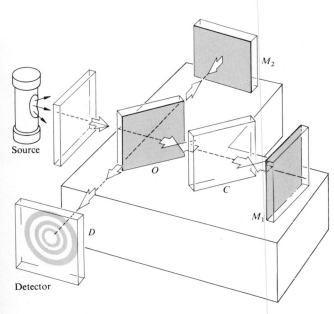

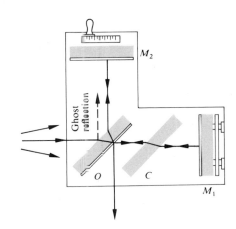

Fig. 9.12 The Michelson interferometer.

unstable to the point of being unobservable. We will get back to these ideas when we consider coherence theory in more detail. For the moment we restrict ourselves to those cases where the path difference is less than the coherence length.

There are a great number of amplitude-splitting interferometers of which we shall discuss only a very few. By far the best known, and historically the most important, is the Michelson interferometer. Its configuration is illustrated in Fig. 9.12. An extended source (which may, for example, be a diffusing ground-glass plate illuminated by a discharge lamp) emits a wave, part of which travels to the right. The beam splitter at O divides the wave into two, one segment traveling to the right and one up into the background. The two waves are reflected by mirrors M_1 and M_2 and return to the beam splitter. Part of the wave coming from M_2 passes through the beam splitter going downward and part of the wave coming from M_1 is deflected by the beam splitter toward the detector. Thus the two waves are united and interference can be expected.

Notice that one beam passes through O three times while the other only traverses it once. Consequently, each beam will pass through equal thicknesses of glass only when a *compensator plate* C is inserted in the arm OM_1. The compensator is an exact duplicate of the beam splitter with

the exception of any possible silvering or thin film coating on the beam splitter. It is positioned at an angle of 45° so that O and C are parallel to each other. With the compensator in place any optical path difference arises from the actual path difference. In addition, because of the dispersion of the beam splitter, the optical path is a function of λ. Accordingly, for quantitative work, the interferometer without the compensator plate can only be used with a quasimonochromatic source. The inclusion of a compensator negates the effect of dispersion so that even a source with a very broad bandwidth will then generate discernible fringes.

To understand how fringes are formed, refer to the construction shown in Fig. 9.13 where the physical components are represented more as mathematical surfaces. An observer at the position of the detector will simultaneously see both mirrors M_1 and M_2 along with the source Σ in the beam splitter. Accordingly, we can redraw the interferometer as if all the elements were in a straight line. Here M_1' corresponds to the image of mirror M_1 in the beam splitter and Σ has been swung over in line with O and M_2. The positions of these elements in the diagram depend on their relative distances from O (e.g. M_1' can be in front of, behind, or coincident with M_2 and can even pass through it). The surfaces Σ_1 and Σ_2 are the images of the source Σ in mirrors M_1 and M_2, respectively. Now consider a single point S on the source

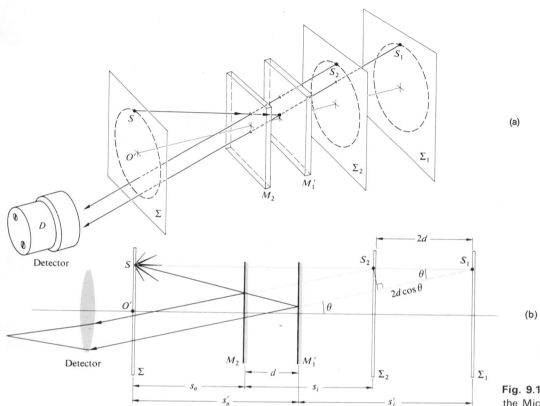

Fig. 9.13 A conceptual rearrangement of the Michelson interferometer.

emitting light in all directions; let's follow the course of one emerging ray. In actuality a wave from S will be split at O and its segments will thereafter be reflected by M_1 and M_2. In our schematic diagram we represent this by reflecting the ray off both M_2 and M_1'. To an observer at D the two reflected rays will appear to have come from the image points S_1 and S_2 [note that all rays shown in Fig. 9.14(a) and (b) share a common plane of incidence]. For all practical purposes S_1 and S_2 are coherent point sources and we can anticipate a flux-density distribution obeying Eq. (9.5). As can be seen from the figure, the optical path difference for these rays is very nearly $2d\cos\theta$ which represents a phase difference of $k_0 2d\cos\theta$. There is an additional phase term arising from the fact that the wave traversing the arm OM_2 is internally reflected in the beam splitter, whereas the OM_1-wave is externally reflected at O. If the beam splitter is simply an uncoated glass plate, the relative phase shift resulting from the two reflections will (Section 4.5) be π radians. *Destructive*, rather than constructive, interference

will then exist when

$$2d\cos\theta_m = m\lambda_0 \qquad (9.15)$$

where m is an integer. If this condition is fulfilled for the point S, then it will be equally well fulfilled for any point on Σ which lies on the circle of radius $O'S$ where O' is located on the axis of the detector. As illustrated in Fig. 9.14, an observer will see a circular fringe system concentric with the central axis of her eyes' lens. Because of the small aperture of the eye the observer will not be able to see the entire pattern without the use of a large lens near the beam splitter to collect most of the emergent light.

The dependence of θ_m on λ_0 in Eq. (9.15) requires that if we use a source containing a number of frequency components (e.g. a mercury discharge lamp) each such component will generate a fringe system of its own. Note too, that since $2d\cos\theta_m$ must be less than the coherence length of the source, it follows that laser light will be particularly easy to use in demonstrating the interferometer (see Section

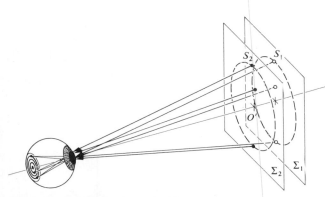

Fig. 9.14 Formation of circular fringes.

9.6). This point would be made strikingly evident were we to compare the fringes produced by laser light with those generated by "white" light from an ordinary tungsten bulb or a candle. In the latter case, the path difference must be very nearly zero if we are to see any fringes at all while in the former instance a difference of 10 cm has little noticeable effect.

    An interference pattern in quasimonochromatic light typically consists of a large number of alternatively bright and dark rings. A particular ring corresponds to a fixed *order m*. As M_2 is moved toward M_1', d decreases and according to Eq. (9.15) $\cos \theta_m$ increases while θ_m therefore decreases. The rings shrink toward the center with the highest-order one disappearing whenever d decreases by $\lambda_0/2$. Each remaining ring broadens as more and more fringes vanish at the center, until only a few fill the whole screen. By the time $d = 0$ has been reached, the central fringe will have spread out filling the entire field of view. With a phase shift of π resulting from reflection off the beam splitter, the whole screen will then be an interference minimum (lack of perfection in the optical elements can render this unobservable). Moving M_2 still farther causes the fringes to reappear at the center and move outward.

    Notice that a central dark fringe for which $\theta_m = 0$ in Eq. (9.15) can be represented by

$$2d = m_0 \lambda_0. \qquad (9.16)$$

(Keep in mind that this is a special case. The central region might correspond to neither a maximum nor a minimum.) Even if d is 10 cm, which is fairly modest in laser light, and $\lambda_0 = 500$ nm, m_0 will be quite large, namely 400,000. At a fixed value of d, successive dark rings will satisfy the expressions

$$2d \cos \theta_1 = (m_0 - 1)\lambda_0$$

$$2d \cos \theta_2 = (m_0 - 2)\lambda_0$$

$$\vdots$$

$$2d \cos \theta_p = (m_0 - p)\lambda_0. \qquad (9.17)$$

The angular position of any ring, e.g. the pth ring, is gotten by combining Eqs. (9.16) and (9.17) to yield

$$2d(1 - \cos \theta_p) = p\lambda_0. \qquad (9.18)$$

Since $\theta_m \equiv \theta_p$, both are just the half-angle subtended at the detector by the particular ring, and since $m = m_0 - p$, Eq. (9.18) is equivalent to Eq. (9.15). The new form is somewhat more convenient since (using the same example as above) with $d = 10$ cm, the sixth dark ring can be specified by stating that $p = 6$ or, in terms of the *order* of the pth ring, that $m = 399,994$. If θ_p is small

$$\cos \theta_p = 1 - \frac{\theta_p^2}{2}$$

and Eq. (9.13) yields

$$\theta_p = \left(\frac{p\lambda_0}{d}\right)^{\frac{1}{2}} \qquad (9.19)$$

for the angular radius of the pth fringe.

    The construction of Fig. 9.13 represents one possible configuration, the one where we consider only pairs of parallel emerging rays. Since these rays do not actually meet, they cannot form an image without a condensing lens of some sort. Indeed that lens is most often provided by the observer's eye focused at infinity. The resulting *fringes of equal inclination* ($\theta_m = $ const.) located at infinity, are also sometimes referred to as *Haidinger fringes* after the Austrian physicist Wilhelm Karl Haidinger (1795–1871). A comparison of Figs. 9.14(b) and 9.3(a), both showing two coherent point sources, rather suggests that in addition to these (virtual) fringes at infinity there might also be (real) fringes formed by converging rays. These fringes do in fact exist. Hence, if you illuminate the interferometer with a *broad source* and shield out all extraneous light, you can easily see the projected pattern on a screen in a darkened room (see Section 9.6). The fringes will appear in the space in front of the interferometer (i.e. where the detector is shown) and their size will increase with increasing distance from the beam splitter. We will consider the (real) fringes arising from point source illumination a little later on.

When the mirrors of the interferometer are inclined with respect to each other making a small angle, i.e. when M_1 and M_2 are not quite perpendicular, *Fizeau fringes* are observed. The resultant wedge-shaped air film between M_2 and M_1' creates a pattern (which will be discussed in Section 9.5.2) of straight parallel fringes. The interfering rays appear to diverge from a point behind the mirrors. The eye would have to focus on this point in order to make these *localized fringes* observable. It can be shown analytically[*] that by appropriately adjusting the orientation of the mirrors M_1 and M_2 fringes can be produced which are straight, circular, elliptical, parabolic or hyperbolic—this holds as well for the real and virtual fringes.

It is apparent that the Michelson interferometer can be used to make extremely accurate length measurements. As the moveable mirror is displaced by $\lambda_0/2$, each fringe will move to the position previously occupied by an adjacent fringe. Using a microscope arrangement, one need only count the number of fringes N, or portions thereof, which have moved past a reference point in order to determine the distance traveled by the mirror Δd, i.e.

$$\Delta d = N \frac{\lambda_0}{2}$$

and, of course, nowadays this can be done electronically fairly easily. Michelson used the method to measure the number of wavelengths of the red cadmium line corresponding to the standard meter in Sèvres near Paris.[†]

The Michelson interferometer can be used along with a few polaroid filters to verify the Fresnel–Arago laws. A polarizer inserted in each arm will allow the optical path-length difference to remain fairly constant, while the vector field directions of the two beams are easily changed.

A *microwave Michelson interferometer* can be constructed with sheet-metal mirrors and a chicken-wire beam splitter. With the detector located at the central fringe, it can easily measure shifts from maxima to minima as one of the mirrors is moved, thereby determining λ. A few sheets of plywood, plastic or glass inserted in one arm will change the central fringe. Counting the number of fringe shifts yields a value for the index of refraction and from that we can compute the dielectric constant of the material.

The *Mach–Zehnder* interferometer is yet another amplitude-splitting device. As shown in Fig. 9.15, it consists

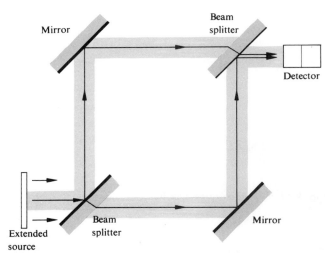

Fig. 9.15 The Mach–Zehnder interferometer.

of two beam splitters and two totally reflecting mirrors. The two waves within the apparatus travel along separate paths. A difference between the optical paths can be introduced by a slight tilt of one of the beam splitters. Since the two paths are separated, the interferometer is relatively difficult to align. For the very same reason, however, the interferometer finds myriad applications. It has even been used, in a rather altered yet conceptually similar form, to obtain electron interference fringes.[*]

Interposing an object in one beam will alter the optical path-length difference, thereby changing the fringe pattern. A common application of the device is to observe the density variations in gas-flow patterns within research chambers, e.g. wind tunnels, shock tubes, etc. One beam passes through the optically flat windows of the test chamber, while the other beam traverses appropriate compensator plates. The beam within the chamber will propagate through regions having a spatially varying index of refraction. The resulting distortions in the wavefront generate the fringe contours. A particularly nice application is shown in Fig. 9.16 which is a photograph of the magnetic compression device known as Scylla IV. It's used to study controlled thermonuclear reactions at the Los Alamos Scientific Laboratory. In this application the Mach–Zehnder interferometer has the form of a parallelogram as illustrated in Fig. 9.17. The two ruby laser *interferograms*, as these photos are called, show (Fig. 9.18) the background pattern without a plasma in the tube

---

[*] See e.g. Valasek, *Optics*, p. 135.

[†] A discussion of the procedure he used to avoid counting the 3,106,327 fringes directly can be found in Strong, *Concepts of Classical Optics*, p. 238 or Williams, *Applications of Interferometry*, p. 51.

---

[*] L. Marton, J. Arol Simpson, and J. A. Suddeth, *Rev. Sci. Instr.* **25**, 1099, (1954) and *Phys. Rev.* **90**, 490, (1953).

Fig. 9.16 Scylla IV.

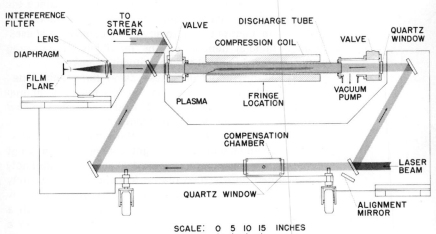

Fig. 9.17 Schematic of Scylla IV.

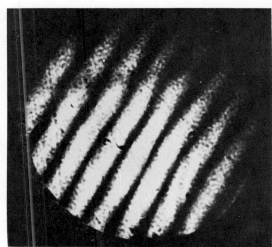

Fig. 9.18 Interferogram sans plasma.

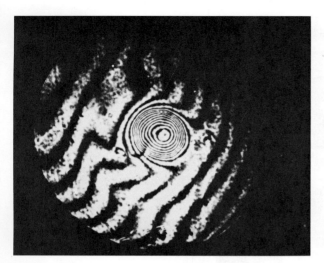

Fig. 9.19 Interferogram with plasma. [Photos courtesy Los Alamos Scientific Laboratory.]

and the density contours within the plasma during a reaction (Fig. 9.19).

Another amplitude-splitting device which contrasts with the previous instrument in many respects is one which we shall refer to as the *Sagnac interferometer*. It is very easy to

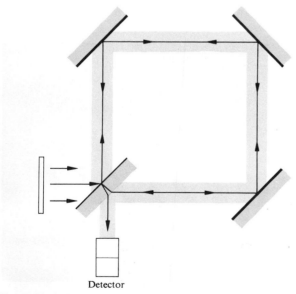

Fig. 9.20 A Sagnac interferometer.

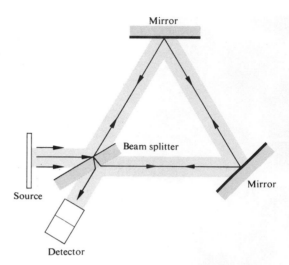

Fig. 9.21 Another variation of the Sagnac interferometer.

align, quite stable and yet rather difficult to make practical use of. One very interesting application is discussed in the last section of this chapter where we consider its use as a gyroscope. A form of the Sagnac interferometer is shown in Fig. 9.20 and another in Fig. 9.21; still others are possible. Notice that the main feature of the device is that there are two identical but oppositely directed paths taken by the beams and that both form closed loops before they are united to produce interference. A slight deliberate shift in the orientation of one of the mirrors will produce a path-length difference and a resulting fringe pattern. Since the beams are superimposed and therefore inseparable, the interferometer cannot be put to any of the conventional uses. These in general depend on the possibility of imposing variations on only one of the constituent beams.

We will now consider these same three amplitude-splitting interferometers, together with a fourth additional one where now in each case we use a point source. The new interferometer is the *Pohl fringe-producing system*; its configuration is illustrated in Fig. 9.22. It is simply a thin transparent film illuminated by the light coming from a point source. In this case, the fringes are real and can accordingly be intercepted on a screen placed anywhere in the vicinity of the interferometer without a condensing-lens system. A convenient light source is a mercury lamp covered with a shield having a small hole ($\approx \frac{1}{4}$ inch diameter) in it. For a thin film use a piece of ordinary mica taped to a dark-colored book cover which serves as an opaque backing. If you

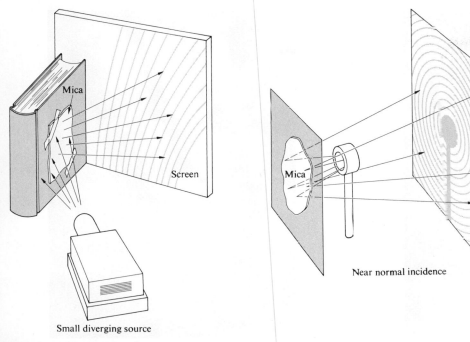

Fig. 9.22 The Pohl interferometer.

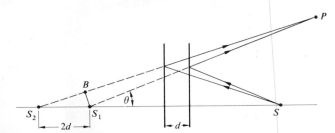

Fig. 9.23 Point-source illumination of parallel surfaces.

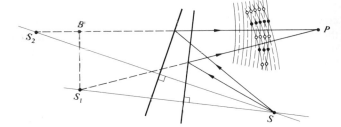

Fig. 9.24 Point-source illumination of inclined surfaces.

have a laser, its remarkable coherence length and high flux density will allow you to perform this same experiment with almost anything smooth and transparent. Expand the beam to about an inch or 2 in diameter by passing it through a lens (a focal length of 50 to 100 mm will do). Then just reflect the beam off the surface of a glass plate (e.g. a microscope slide) and the fringes will be evident within the illuminated disk wherever it strikes a screen.

The underlying physical principle involved with point-source illumination for all four devices can be appreciated with the help of a construction, variations of which are shown in Fig. 9.23 and 9.24.* The two vertical lines in Fig. 9.23, or the inclined ones in Fig. 9.24, represent either the positions of the mirrors or the two sides of the thin sheet in the Pohl interferometer. Let's assume that point P in the surrounding medium is a point at which there is constructive interference. A screen placed at that point would intercept this maximum as well as a whole fringe pattern, without any

---

* This discussion and part of Section 9.6 are treated at length in A. Zajac, H. Sadowski, and S. Licht, "The Real Fringes in the Sagnac and the Michelson Interferometers," *Am. J. Phys.* **29**, 669 (1961).

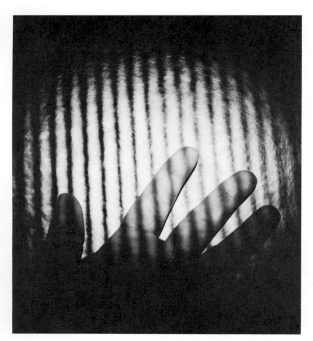

Fig. 9.25 Real Michelson fringes using He–Ne laser light. [Photo by E.H.]

.condensing system. The coherent virtual sources emitting the interfering beams are mirror images S_1 and S_2 of the actual point source S. It should be noted that this kind of real fringe pattern can be observed with both the Michelson and Sagnac interferometers, although it is not the one most often demonstrated (Figs. 9.25 and 9.26). If either device is illuminated with an expanded laser beam, a real fringe pattern will be generated directly by the emerging waves. This is an extremely simple and beautiful demonstration.

9.5 DIELECTRIC FILMS—DOUBLE-BEAM INTERFERENCE

Interference effects are observable in transparent materials, the thicknesses of which vary over a very broad range. A range extending from films less than the length of a light wave (e.g. for green light λ_0 equals about $\frac{1}{150}$ the thickness of this printed page) to plates several centimeters thick. A layer of some material is referred to as a *thin film* for a given wavelength of electromagnetic radiation, when its thickness is of the order of that wavelength. Prior to the early 1940s the interference phenomena associated with thin dielectric films, although well known, had fairly limited practical appli-

(AT 5 METERS)

(AT 2 METERS)

Fig. 9.26 Real Sagnac fringes using a white-light point source.

cability. The rather spectacular color displays arising from oil slicks and soap films, however pleasing esthetically and theoretically, were practically mainly beautiful curiosities.

The advent of suitable vacuum deposition techniques in the 1930s brought with it the ability to produce precisely

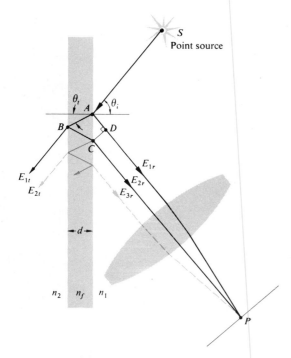

Fig. 9.27 Fringes of equal inclination.

controlled coatings on a commercial scale and with that, in turn, a rebirth of interest. During the Second World War, both sides were finding the enemy with a variety of coated optical devices and by the 1960s multilayered coatings were in widespread use.

9.5.1   Fringes of Equal Inclination

Initially, consider the simple case of a transparent parallel plate of dielectric material having a thickness d. One possible condition resulting from the reflection of two separate incident rays was depicted earlier in Fig. 9.23 in conjunction with our discussion of the Pohl interferometer. Another possibility is illustrated in Fig. 9.27. Suppose that the film is nonabsorbing and that the amplitude-reflection coefficients at the interfaces are so low that only the first two reflected beams E_{1r} and E_{2r} (both having undergone only one reflection) need be considered. In practice the amplitudes of the higher-order reflected beams (E_{3r}, etc.) generally decrease very rapidly, as can be shown for the air–water and air–glass interfaces (Problem 9.10). For the moment, consider S to be a monochromatic point source. The film serves as an amplitude-splitting device, so that E_{1r} and E_{2r}

may be considered as arising from two coherent virtual sources lying behind the film. The reflected rays are parallel on leaving the film and can be brought together at a point P on the focal plane of a telescope objective or on the retina of the eye when focused at infinity. From Fig. 9.27 the optical path-length difference for the first two reflected beams is given by

$$\Lambda = n_f[(\overline{AB}) + (\overline{BC})] - n_1(\overline{AD})$$

and since $(\overline{AB}) = (\overline{BC}) = d/\cos\theta_t$,

$$\Lambda = \frac{2n_f d}{\cos\theta_t} - n_1(\overline{AD}).$$

Now, to find an expression for $(\overline{AD})$, write

$$(\overline{AD}) = (\overline{AC})\sin\theta_i;$$

if we make use of Snell's law this becomes

$$(\overline{AD}) = (\overline{AC})\frac{n_f}{n_1}\sin\theta_t,$$

where

$$(\overline{AC}) = 2d\tan\theta_t. \tag{9.20}$$

The expression for Λ now becomes

$$\Lambda = \frac{2n_f d}{\cos\theta_t}(1 - \sin^2\theta_t)$$

or finally

$$\Lambda = 2n_f d\cos\theta_t. \tag{9.21}$$

The corresponding phase difference associated with the optical path-length difference is then just the product of the free-space propagation number and Λ, i.e., $k_0\Lambda$. If the film is immersed in a single medium, the index of refraction can simply be written as $n_1 = n_2 = n$. Realize, of course, that n may be less than n_f, as in the case of a soap film in air; or greater than n_f, as with an air film between two sheets of glass. In either case there will be an additional phase shift arising from the reflections themselves. Recall that, regardless of the polarization of the incoming light, the two beams, one internally and one externally reflected, will experience a *relative phase shift* of π radians. Accordingly

$$\delta = k_0\Lambda \pm \pi$$

and more explicitly

$$\delta = \frac{4\pi n_f}{\lambda_0}d\cos\theta_t \pm \pi \tag{9.22}$$

or

$$\delta = \frac{4\pi d}{\lambda_0}(n_f^2 - n^2\sin^2\theta_i)^{\frac{1}{2}} \pm \pi. \tag{9.23}$$

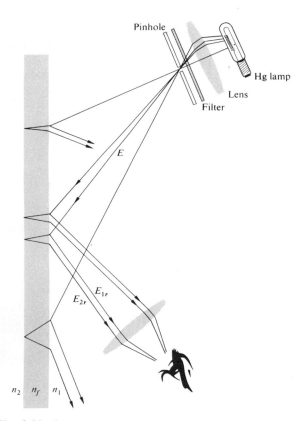

Fig. 9.28 Fringes seen on a small portion of the film.

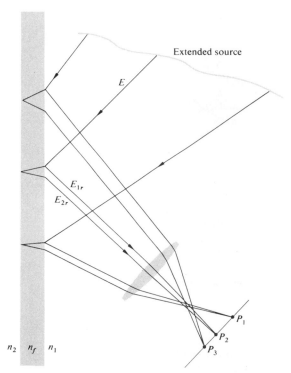

Fig. 9.29 Fringes seen on a large region of the film.

The sign of the phase shift is immaterial so that we will choose the negative sign in order to make the equations a bit simpler in form.   In reflected light an interference maximum, a bright spot, appears at P when $\delta = 2m\pi$, i.e. an even multiple of π. In that case Eq. (9.22) can be rearranged to yield

$$\text{(maxima)}\quad d\cos\theta_t = (2m+1)\frac{\lambda_f}{4},\quad m = 0, 1, 2, \ldots,$$

$$(9.24)$$

where use was made of the fact that $\lambda_f = \lambda_0/n_f$. This also corresponds to minima in the transmitted light.   Interference minima in reflected light (maxima in transmitted) result when $\delta = (2m \pm 1)\pi$, i.e. odd multiples of π. For such cases Eq. (9.22) yields

$$\text{(minima)}\qquad d\cos\theta_t = 2m\frac{\lambda_f}{4}.\qquad (9.25)$$

The appearance of odd and even multiples of $\lambda_f/4$ in Eqs. (9.24) and (9.25) is rather significant, as will be seen later.

We could, of course, have a situation where $n_1 > n_f > n_2$ or where $n_1 < n_f < n_2$ as with a fluoride film deposited on an optical element of glass immersed in air.   The π phase shift would then not be present and the above equations would simply be modified appropriately.

If the lens used to focus the rays has a small aperture, interference fringes will appear on a small portion of the film. Only the rays leaving the point source which are reflected directly into the lens will be seen (Fig. 9.28).   For an extended source, light will reach the lens from various directions and the fringe pattern will spread out over a large area of the film (Fig. 9.29).

The angle θ_i or equivalently θ_t, determined by the position of P, will in turn control δ.   The fringes appearing at points P_1 and P_2 in Fig. 9.30 are, accordingly, known as *fringes of equal inclination* (Problem 9.26 discusses some easy ways to see these fringes).   Keep in mind that each source point on the extended source is incoherent with respect to the others.

Notice that as the film becomes thicker, the separation $(\overline{AC})$ between E_{1r} and E_{2r} also increases since

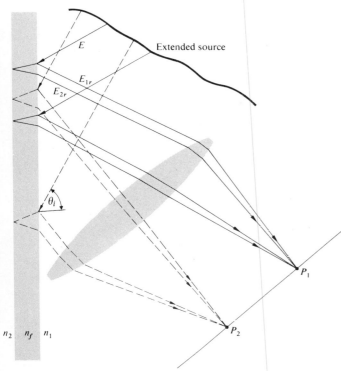

Fig. 9.30 All rays inclined at the same angle arrive at the same point.

$$(\overline{AC}) = 2d \tan \theta_t. \qquad [9.20]$$

When only one of the two rays is able to enter the pupil of the eye, the interference pattern will disappear. The larger lens of a telescope can then be used to gather in both rays, once again making the pattern visible. The separation can also be reduced by reducing θ_t and therefore θ_i, i.e. by viewing the film at nearly normal incidence. The equal-inclination fringes which are seen in this manner for thick plates are known as *Haidinger fringes*. With an extended source they consist of a series of concentric circular bands centered on the perpendicular drawn from the eye to the film (Fig. 9.31). As the observer moves, the interference pattern follows along.

9.5.2 Fringes of Equal Thickness

A whole class of interference fringes exists, for which the optical thickness, $n_f d$, is the dominant parameter rather than θ_i. These are referred to as *fringes of equal thickness*. Under white-light illumination the iridescence of soap bubbles, oil slicks (a few wavelengths thick), and even oxidized metal

surfaces, are all the result of variations in film thickness. Interference bands of this kind are analogous to the constant-height contour lines of a topographical map. Each fringe is the locus of all points in the film for which the optical thickness is a constant. In general, n_f does not vary, so that the fringes actually correspond to regions of constant film thickness. As such, they can be quite useful in determining the surface features of optical elements; lenses, prisms, etc. For example, a surface to be examined may be put into contact with an *optical flat*.* The air in the space between the two generates a thin-film interference pattern. If the test surface is flat, a series of straight, equally spaced bands indicates a wedge-shaped air film, usually resulting from dust between the flats. Two pieces of plate glass separated at one end by a strip of paper will form a satisfactory wedge with which to observe these bands.

When viewed at nearly normal incidence in the manner illustrated in Fig. 9.32, the contours arising from a nonuniform film are called *Fizeau fringes*. For a thin wedge of small angle α the optical path length difference between two reflected rays may be approximated by Eq. (9.21), where d is the thickness at a particular point, i.e.

$$d = x\alpha.$$

For small values of θ_i the condition for an interference maximum becomes

$$(m + \tfrac{1}{2})\lambda_0 = 2n_f d_m$$

or

$$(m + \tfrac{1}{2})\lambda_0 = 2\alpha x_m n_f.$$

Since $n_f = \lambda_0 / \lambda_f$, x_m may be written as

$$x_m = \left(\frac{m + 1/2}{2\alpha} \right) \lambda_f.$$

Maxima occur at distances from the apex given by $\lambda_f/4\alpha$, $3\lambda_f/4\alpha$, etc. and consecutive fringes are separated by a distance Δx, given by

$$\Delta x = \lambda_f/2\alpha.$$

Notice that the difference in film thickness between adjacent maxima is simply $\lambda_f/2$. Since the beam reflected from the

---

* A surface is said to be optically flat when it deviates by not more than about $\lambda/4$ from a perfect plane. In the past, the best flats were made of clear fused quartz. Now glass—ceramic materials (e.g. CER-VIT) having extremely small thermal coefficients of expansion (about one sixth that of quartz) are available. Individual flats of $\lambda/200$ or a bit better can be made.

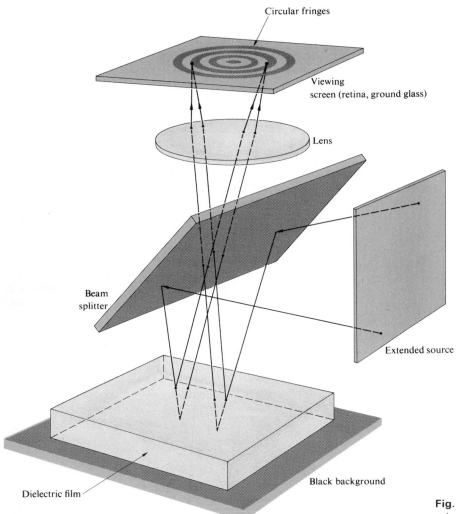

Circular fringes

Viewing
screen (retina, ground glass)

Lens

Beam
splitter

Extended source

Dielectric film

Black background

Fig. 9.31 Circular Haidinger fringes centered on the lens
axis.

lower surface traverses the film twice ($\theta_i \approx \theta_t \approx 0$), adjacent
maxima differ in optical path length by λ_f. Note too, that
the film thickness at the various maxima is given by

$$d_m = (m + \tfrac{1}{2})\frac{\lambda_f}{2},$$

which is an odd multiple of a quarter wavelength. Traversing
the film twice yields a phase shift of π which, when added to
the shift of π resulting from reflection, puts the two rays
back in phase.

Figure 9.33 is a photograph of a soap film held vertically
so that it settles into a wedge shape under the influence of
gravity. When illuminated with white light the bands are of
various colors. The black region at the top is a portion where
the film is less than $\lambda_f/4$ thick. Twice this, plus an additional
shift of $\lambda_f/2$ due to the reflection is less than a whole wavelength. The reflected rays are therefore out of phase. As
the thickness decreases still further, the total phase difference
approaches π. The irradiance at the observer goes to a
minimum (Eq. 9.5) and the film appears black in reflected
light.*

---

* The relative phase shift of π between internal and external reflection
is required if the reflected flux density is to go to zero smoothly, as
the film gets thinner and finally disappears.

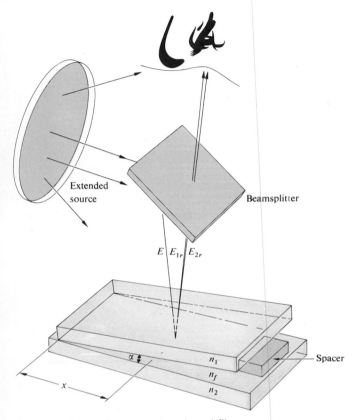

Fig. 9.32 Fringes from a wedge-shaped film.

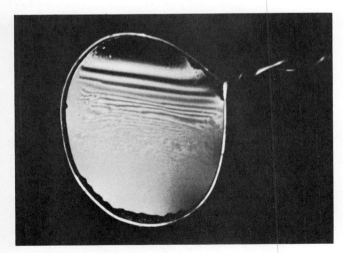

Fig. 9.33 A wedge-shaped film made of liquid dishwashing soap. [Photo by E.H.]

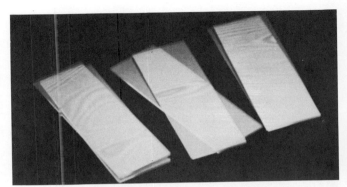

Fig. 9.34 Fringes in an air film between two microscope slides. [Photo by E.H.]

Press two well-cleaned microscope slides together. The enclosed air film will usually not be uniform. In ordinary room light a series of irregular, colored bands (fringes of equal thickness) will be clearly visible across the surface (Fig. 9.34). The thin glass slides distort under pressure and the fringes move and change accordingly. Indeed, if the two pieces of glass are forced together at a point as might be done by pressing on them with a sharp pencil a series of concentric nearly circular, fringes are formed about that point (Fig. 9.35). Known as Newton's rings,* this pattern is more precisely examined using the arrangement of Fig. 9.36. Here a lens is placed on an optical flat and illuminated at normal incidence with quasimonochromatic light. The amount of uniformity in the concentric circular pattern is a measure of the degree of perfection in the shape of the lens. With R as the radius of curvature of the convex lens, the relation between the distance x and the film thickness d is given by

$$x^2 = R^2 - (R - d)^2,$$

or more simply by

$$x^2 = 2Rd - d^2$$

Since $R \gg d$ this becomes

$$x^2 = 2Rd.$$

---

* Robert Hooke (1635–1703) and Isaac Newton both independently studied a whole range of thin-film phenomena from soap bubbles to the air film between lenses. Quoting from Newton's *Opticks*:

> I took two Object-glasses, the one a Planoconvex for a fourteen Foot Telescope, and the other a large double Convex for one of about fifty Foot; and upon this, laying the other with its plane side downwards, I pressed them slowly together to make the Colours successively emerge in the middle of the Circles.

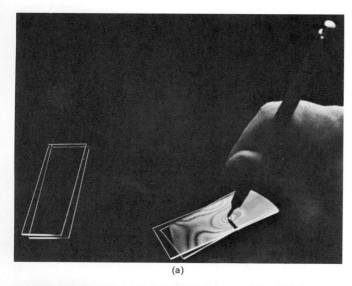

(a)

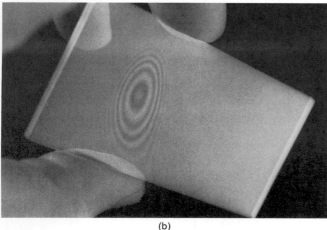

(b)

Fig. 9.35 Newton's rings with two microscope slides.
[Photos by E.H.]

We again approximate by assuming that we need only
examine the first two reflected beams E_{1r} and E_{2r}. The
mth-order interference maximum will occur in the thin film
when its thickness is in accord with the relationship

$$2n_f d_m = (m + 1/2)\lambda_0.$$

The radius of the mth bright ring is therefore found by combining the last two expressions to yield

$$x_m = [(m + \tfrac{1}{2})\lambda_f R]^{1/2}. \tag{9.26}$$

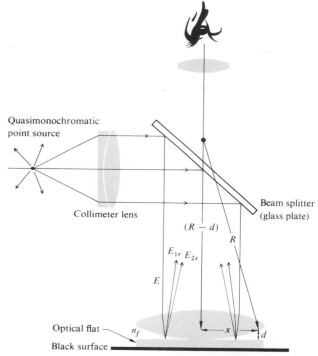

Fig. 9.36 A standard set-up to observe Newton's rings.

Similarly, the radius of the mth dark ring is

$$x_m = (m\lambda_f R)^{1/2}. \tag{9.27}$$

If the two pieces of glass are in good contact (no dust), the
central fringe at that point ($x_0 = 0$) will clearly be the zeroth-
order minimum, an understandable result since d goes to
zero at that point. In transmitted light, the observed pattern
will be the complement of the reflected one discussed above,
so that the center will now appear bright.

Newton's rings, which are Fizeau fringes, can be distinguished from the circular pattern of Haidinger's fringes
by the manner in which the diameters of the rings vary with
the order m. The central region in the Haidinger pattern
corresponds to the maximum value of m (Problem 9.11)
while just the opposite applies to Newton's rings.

An optical shop, in the business of making lenses, will
have a set of precision spherical test plates or gauges. A
designer may then specify the surface accuracy of a new lens
in terms of the number and regularity of the Newton rings
which will be seen with a particular test gauge. We should
mention that the use of test plates in the manufacture of high-

quality lenses is giving way to far more sophisticated techniques involving laser interferometers (Section 9.10.5).

9.6 TYPES AND LOCALIZATION OF INTERFERENCE FRINGES

Often it is important to know where the fringes produced in a given interferometric system will be located, i.e. in what region we need focus our detector (eye, camera, telescope). In general, the problem of locating fringes is characteristic of a given interferometer, i.e. it has to be solved for each individual device.

Fringes can be classified in two categories as firstly either *real* or *virtual* and then as either *nonlocalized* or *localized*. Real fringes are those which can be seen on a screen without the use of an additional focusing system. The rays forming these fringes converge to the point of observation, all by themselves. Virtual fringes cannot be projected onto a screen without a focusing system. In this case the rays obviously do not converge.

Nonlocalized fringes are real and exist everywhere within an extended (three-dimensional) region of space. The pattern is quite literally nonlocalized in that it is not restricted to some small region. Young's experiment, as illustrated in Fig. 9.5, fills the space beyond the secondary sources with a whole array of real fringes. Nonlocalized fringes of this sort are generally produced by small sources, i.e. point or line sources, be they real or virtual. In contrast, localized fringes are clearly observable only over a particular surface. The pattern is quite literally localized, whether near a thin film or at infinity. This type of fringe will always result from the use of extended sources but can be generated with a point source as well.

The Pohl interferometer (Fig. 9.21) is particularly useful in illustrating these principles since with a point source it will produce both real nonlocalized and virtual localized fringes. The real nonlocalized fringes (Fig. 9.37 upper half) can be intercepted on a screen almost anywhere in front of the mica film.

For the nonconverging rays, realize that since the aperture of the eye is quite small it will intercept only those rays which are directed almost exactly at it. For this small pencil of rays, the eye, at a particular position, sees either a bright or dark spot but not much more. To perceive an extended fringe pattern formed by parallel rays of the type shown on the bottom in Fig. 9.37, a large lens will have to be used to gather in light entering at other orientations. In practice, however, the source is usually somewhat extended

and fringes can generally be seen by looking into the film with the eye focused at infinity. These virtual fringes are localized at infinity and are equivalent to the *equal-inclination fringes* of Section 9.5. Similarly, if the mirrors M_1 and M_2 in the Michelson interferometer are parallel, the usual circular, virtual, equal-inclination fringes localized at infinity will be seen. We can imagine a thin air film between the surfaces of the mirrors M_2 and M_1' acting to generate these fringes. As with the configuration of Fig. 9.37 for the Pohl device, real nonlocalized fringes will also be present.

The geometry of the fringe pattern seen in reflected light from a transparent wedge of small angle α is shown in Fig. 9.38. The fringe location P will be determined by the direction of incidence of the incoming light. Newton's rings have this same kind of localization as do the Michelson, Sagnac and other interferometers for which the equivalent interference system consists of two reflecting planes inclined slightly to each other. Figures 9.25 and 9.26 are photographs of the real, equal-thickness fringes taken from the Sagnac and Michelson devices under point-source illumination. The wedge set-up of the Mach–Zehnder interferometer is distinctive in that the resulting virtual fringes can be localized on any plane within the region generally occupied by the test chamber by rotating the mirrors (Fig. 9.39).

9.7 MULTIPLE-BEAM INTERFERENCE

Thus far we have examined a number of situations in which two coherent beams have been combined under a diversity of conditions to produce interference patterns. There are, however, other circumstances under which a much larger number of mutually coherent waves are made to interfere. In fact, if the amplitude-reflection coefficients, the r's, for the parallel plate illustrated in Fig. 9.27 are not small, as was previously the case, the higher-order reflected waves E_{3r}, E_{4r}, ... become quite significant. A glass plate, slightly silvered on both sides so that the r's approach unity, will generate a large number of multiply internally reflected rays. For the moment, we will only consider situations where the film, substrate, and surrounding medium are transparent dielectrics. This avoids the more complicated phase changes resulting from metal-coated surfaces.

To begin the analysis as simply as possible, let the film be nonabsorbing and let $n_1 = n_2$. The notation will be in accord with that of Section 4.5, i.e. the amplitude-transmission coefficients are represented by t, the fraction of the amplitude of a wave transmitted on entering into the film, and t', the fraction transmitted when a wave leaves the film.

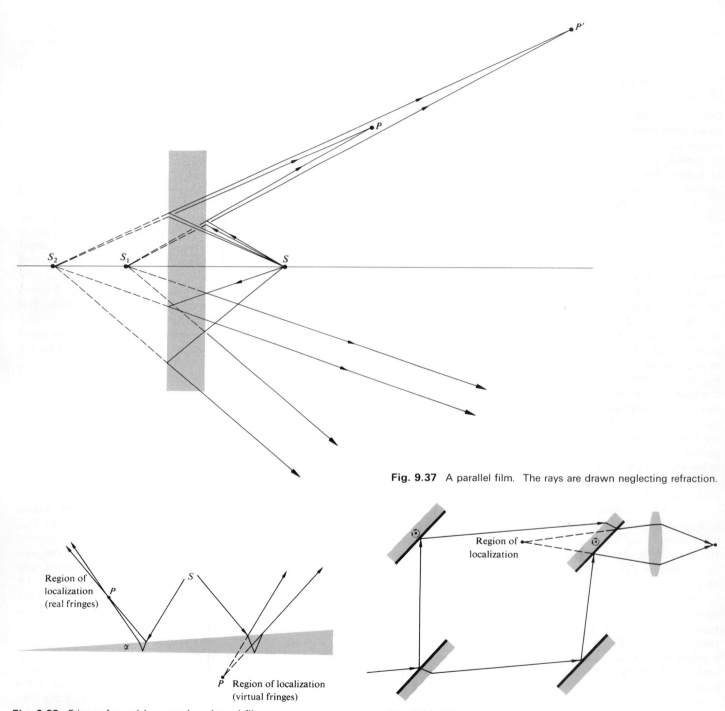

Fig. 9.37 A parallel film. The rays are drawn neglecting refraction.

Fig. 9.38 Fringes formed by a wedge-shaped film.

Fig. 9.39 Fringes in the Mach–Zehnder interferometer.

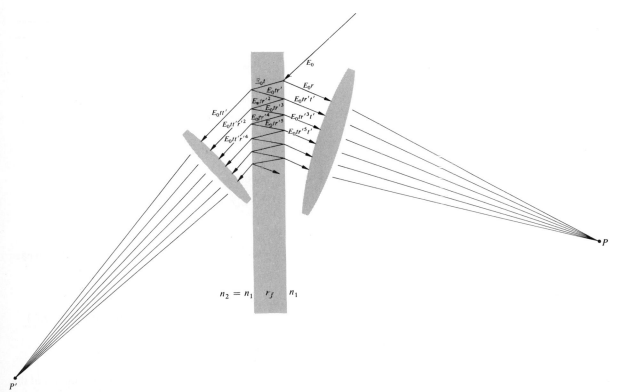

Fig. 9.40 Multiple-beam interference from a parallel film.

Keep in mind that the rays are actually lines drawn perpendicular to the wavefronts and therefore are also perpendicular to the optical fields $\mathbf{E}_{1r}$, $\mathbf{E}_{2r}$, etc. Since the rays will remain nearly parallel, the scalar theory will suffice as long as we are careful to account for any possible phase shifts. As shown in Fig. 9.40, the scalar amplitudes of the reflected waves $\mathbf{E}_{1r}$, $\mathbf{E}_{2r}$, $\mathbf{E}_{3r}$, ..., are respectively $E_0 r$, $E_0 t r' t'$, $E_0 t r'^3 t'$, ..., where E_0 is the amplitude of the initial incoming wave and $r = -r'$ via Eq. (4.89). The minus sign indicates a phase shift which we will consider later. Similarly, the transmitted waves $\mathbf{E}_{1t}$, $\mathbf{E}_{2t}$, $\mathbf{E}_{3t}$, ... will have amplitudes $E_0 t t'$, $E_0 t r'^2 t'$, $E_0 t r'^4 t'$, ... Consider the set of parallel reflected rays. Each ray bears a fixed phase relationship to all of the other reflected rays. The phase differences arise from a combination of optical path-length differences and phase shifts occurring at the various reflections. Nonetheless, the waves are mutually coherent and if they are collected and brought to focus at a point P by a lens, they will all interfere. The resultant irradiance expression has a particularly simple form for two special cases.

The difference in optical path length between adjacent rays is given by

$$\Lambda = 2 n_f d \cos \theta_t. \qquad [9.21]$$

All of the waves except for the first, $\mathbf{E}_{1r}$, undergo an odd number of reflections *within* the film. It follows from Fig. 4.23 that at each internal reflection the component of the field parallel to the plane of incidence changes phase by either 0 or π depending on the internal incident angle, $\theta_i < \theta_c$. The component of the field perpendicular to the plane of incidence suffers no change in phase on internal reflection when $\theta_i < \theta_c$. Clearly then, no relative change in phase amongst these waves results from an odd number of such reflections (Fig. 9.41). As the *first special case*, if $\Lambda = m\lambda$, the 2nd, 3rd, 4th, etc. waves will all be in phase at P. The wave $\mathbf{E}_{1r}$, however, because of its reflection at the top surface of the film, will be out of phase by 180° with respect to all the other waves. The phase shift is embodied in the fact that $r = -r'$ and r' occurs only in odd powers. The sum of the scalar amplitudes, i.e. the total *reflected amplitude*

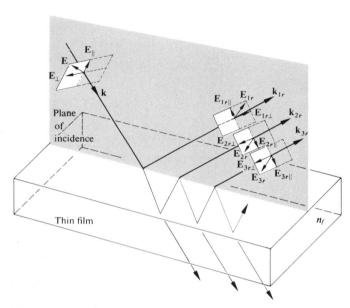

Fig. 9.41 Phase shifts arising purely from the reflections (internal $\theta_i < \theta_p'$).

at point P, is then

$$E_{0r} = E_0 r - (E_0 trt' + E_0 tr^3 t' + E_0 tr^5 t' + \cdots)$$

or

$$E_{0r} = E_0 r - E_0 trt'(1 + r^2 + r^4 + \cdots),$$

where since $\Lambda = m\lambda$, we've just replaced r' by $-r$. The geometric series in parentheses converges to the finite sum $1/(1 - r^2)$ as long as $r^2 < 1$ so that

$$E_{0r} = E_0 r - \frac{E_0 trt'}{(1 - r^2)}.$$

It was shown in Section 4.5 when considering Stokes' treatment of the principle of reversibility (Eq. 4.86) that $tt' = 1 - r^2$, and so it follows that

$$E_{0r} = 0.$$

Thus when $\Lambda = m\lambda$ the 2nd, 3rd, 4th, etc. waves exactly cancel the first reflected wave as is shown in Fig. 9.42. In this case no light is reflected; all of the incoming energy is transmitted. The *second special case* arises when $\Lambda = (m + \frac{1}{2})\lambda$. Now the first and second rays are in phase while all other adjacent waves are $\lambda/2$ out of phase, i.e. the 2nd is out of phase with the 3rd, the 3rd is out of phase with

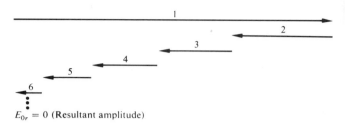

$E_{0r} = 0$ (Resultant amplitude)

Fig. 9.42 Phasor diagram.

the 4th etc. The resultant *scalar amplitude* is then

$$E_{0r} = E_0 r + E_0 trt' - E_0 tr^3 t' + E_0 tr^5 t' - \cdots$$

or

$$E_{0r} = E_0 r + E_0 rtt'(1 - r^2 + r^4 - \cdots).$$

The series in parentheses is equal to $1/(1 + r^2)$, in which case

$$E_{0r} = E_0 r \left[1 + \frac{tt'}{(1 + r^2)} \right].$$

Again, $tt' = 1 - r^2$, therefore, as is illustrated in Fig. 9.43

$$E_{0r} = \frac{2r}{(1 + r^2)} E_0.$$

Since this particular arrangement results in the addition of the 1st and 2nd waves, which have relatively large amplitudes, it should yield a large reflected flux density. Since the irradiance is proportional to $E_{0r}^2/2$, via Eq. (3.50)

$$I_r = \frac{4r^2}{(1 + r^2)^2} \left(\frac{E_0^2}{2} \right) \tag{9.28}$$

That this is in fact the maximum, $(I_r)_{max}$ will be shown later.

We will now consider the problem of multiple-beam interference in a more general fashion, making use of the complex representation. Again let $n_1 = n_2$, thereby avoiding

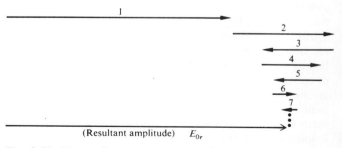

(Resultant amplitude) E_{0r}

Fig. 9.43 Phasor diagram.

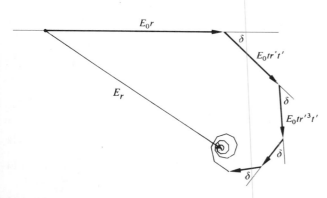

Fig. 9.44 Phasor diagram.

the need to introduce different reflection and transmission coefficients at each interface. The optical fields at point P are given by

$$E_{1r} = E_0 r e^{i\omega t}$$

$$E_{2r} = E_0 t r' t' e^{i(\omega t - \delta)}$$

$$E_{3r} = E_0 t r'^3 t' e^{i(\omega t - 2\delta)}$$

$$\vdots$$

$$E_{Nr} = E_0 t r'^{(2N-3)} t' e^{i[\omega t - (N-1)\delta]},$$

where $E_0 e^{i\omega t}$ is the incident wave.

The terms $\delta, 2\delta, \ldots, (N-1)\delta$ are the contributions to the phase arising from an optical path-length difference between adjacent rays ($\delta = k_0 \Lambda$). There is an additional phase contribution arising from the optical distance traversed in reaching point P, but this is common to each ray and has been omitted. The relative phase shift undergone by the 1st ray as a result of the reflection is embodied in the quantity r'. The resultant *reflected scalar wave* is then

$$E_r = E_{1r} + E_{2r} + E_{3r} + \cdots + E_{Nr},$$

or upon substitution (Fig. 9.44)

$$E_r = E_0 r e^{i\omega t} + E_0 t r' t' e^{i(\omega t - \delta)} + \cdots + E_0 t r'^{(2N-3)} t' e^{i[\omega t - (N-1)\delta]}.$$

This can be rewritten as

$$E_r = E_0 e^{i\omega t} \{ r + r' t t' e^{-i\delta} [1 + (r'^2 e^{-i\delta})$$

$$+ (r'^2 e^{-i\delta})^2 + \cdots + (r'^2 e^{-i\delta})^{N-2}] \}.$$

If $r'^2 e^{-i\delta} < 1$, and if the number of terms in the series approaches infinity, the series converges. The resultant wave becomes

$$E_r = E_0 e^{i\omega t} \left[r + \frac{r' t t' e^{-i\delta}}{1 - r'^2 e^{-i\delta}} \right]. \tag{9.29}$$

In the case of zero absorption, no energy being taken out of the waves, we can use the relations $r = -r'$ and $tt' = 1 - r^2$ to rewrite Eq. (9.29) as

$$E_r = E_0 e^{i\omega t} \left[\frac{r(1 - e^{-i\delta})}{1 - r^2 e^{-i\delta}} \right].$$

The reflected flux density at P is then $I_r = E_r E_r^* / 2$, that is

$$I_r = \frac{E_0^2 r^2 (1 - e^{-i\delta})(1 - e^{+i\delta})}{2(1 - r^2 e^{-i\delta})(1 - r^2 e^{+i\delta})},$$

which can be transformed into

$$I_r = I_i \frac{2r^2(1 - \cos\delta)}{(1 + r^4) - 2r^2 \cos\delta}. \tag{9.30}$$

The symbol $I_i = E_0^2 / 2$ represents the incident flux density since, of course, E_0 was the amplitude of the incident wave. Similarly, the amplitudes of the transmitted waves given by

$$E_{1t} = E_0 t t' e^{i\omega t}$$

$$E_{2t} = E_0 t t' r'^2 e^{i(\omega t - \delta)}$$

$$E_{3t} = E_0 t t' r'^4 e^{i(\omega t - 2\delta)}$$

$$\vdots$$

$$E_{Nt} = E_0 t t' r'^{2(N-1)} e^{i[\omega t - (N-1)\delta]}$$

add to yield

$$E_t = E_0 e^{i\omega t} \left[\frac{tt'}{1 - r^2 e^{-i\delta}} \right]. \tag{9.31}$$

Multiplying this by its complex conjugate we obtain (Problem 9.18) the irradiance of the transmitted beam

$$I_t = \frac{I_i (tt')^2}{(1 + r^4) - 2r^2 \cos\delta}. \tag{9.32}$$

If we use trigonometric identity $\cos\delta = 1 - 2\sin^2(\delta/2)$, Eqs. (9.30) and (9.32) become

$$I_r = I_i \frac{[2r/(1 - r^2)]^2 \sin^2(\delta/2)}{1 + [2r/(1 - r^2)]^2 \sin^2(\delta/2)}. \tag{9.33}$$

and

$$I_t = I_i \frac{1}{1 + [2r/(1 - r^2)]^2 \sin^2(\delta/2)}, \tag{9.34}$$

where energy is not absorbed, i.e., $tt' + r^2 = 1$. If indeed none of the incident energy is absorbed, the flux density of the incoming wave should exactly equal the sum of the flux density reflected off the film and total transmitted flux

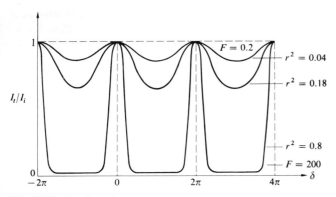

Fig. 9.45 Airy function.

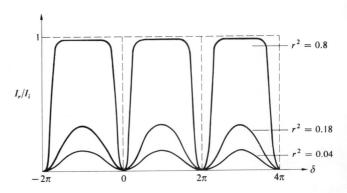

Fig. 9.46 One minus the Airy function.

density emerging from the film. It follows from Eqs. (9.33) and (9.34), that this is indeed the case, that is

$$I_i = I_r + I_t. \tag{9.35}$$

This will not be true, however, if the dielectric film is coated with a thin layer of semitransparent metal. Surface currents induced in the metal will dissipate a portion of the incident electromagnetic energy (see Section 4.3.5).

Consider the transmitted waves as described by Eq. (9.32). A maximum will exist when the denominator is as small as possible, i.e. when $\cos\delta = 1$, in which case $\delta = 2\pi m$ and

$$(I_t)_{\max} = I_i.$$

Under these conditions Eq. (9.30) indicates that

$$(I_r)_{\min} = 0$$

as was to be expected from Eq. (9.35). Again, from Eq. (9.32) it is clear that a minimum transmitted flux density will exist when the denominator is a maximum, i.e. when $\cos\delta = -1$. In that case $\delta = (2m + 1)\pi$ and

$$(I_t)_{\min} = I_i \frac{(1 - r^2)^2}{(1 + r^2)^2}. \tag{9.36}$$

The corresponding maximum in the reflected flux density is

$$(I_r)_{\max} = I_i \frac{4r^2}{(1 + r^2)^2}. \tag{9.37}$$

Notice that the constant-inclination fringe pattern has its maxima when $\delta = (2m + 1)\pi$ or

$$\frac{4\pi n_f}{\lambda_0} d \cos\theta_t = (2m + 1)\pi,$$

which is the same result arrived at previously, Eq. (9.24), by only using the first two reflected waves. Note too, that Eq. (9.37) verifies that Eq. (9.28) was indeed a maximum.

The form of Eqs. (9.33) and (9.34) suggests that we introduce a new quantity known as the *coefficient of finesse* F, such that

$$F \equiv \left(\frac{2r}{1 - r^2}\right)^2 \tag{9.38}$$

whereupon these equations can be written as

$$\frac{I_r}{I_i} = \frac{F \sin^2(\delta/2)}{1 + F \sin^2(\delta/2)} \tag{9.39}$$

and

$$\frac{I_t}{I_i} = \frac{1}{1 + F \sin^2(\delta/2)}. \tag{9.40}$$

The term $[1 + F \sin^2(\delta/2)]^{-1} \equiv \mathscr{A}(\theta)$ is known as the *Airy function*. It represents the transmitted flux density distribution, and is plotted in Fig. 9.45. The complementary function $[1 - \mathscr{A}(\theta)]$, i.e. Eq. (9.39), is plotted as well, in Fig. 9.46. When $\delta/2 = m\pi$ the Airy function is equal to unity for all values of F and therefore r. When r approaches one, the transmitted flux density is very small, except within the sharp spikes centered about the points $\delta/2 = m\pi$. Multiple-beam interference has resulted in a redistribution of the energy density in comparison to the sinusoidal two-beam pattern (of which the curves corresponding to a small reflectance are reminiscent). This effect will be further demonstrated when we consider the diffraction grating. At that time we will clearly see this same peaking effect, resulting from increasing the number of coherent sources contributing to the interference pattern. Remember that the Airy function is, in fact, a function of θ_t or θ_i by way of its dependence on δ, as follows from Eqs. (9.22) and (9.23), ergo the notation $\mathscr{A}(\theta)$. Each spike in the flux-density curve corresponds to a particular δ and therefore a particular θ_i. For a plane-parallel plate, the fringes, in transmitted light, will consist of a series of narrow bright rings on an almost completely dark background. In reflected light, the fringes will be narrow and dark, on an almost uniformly bright background.

Constant-thickness fringes can also be made sharp and narrow by lightly silver coating the relevant reflecting surfaces to produce multiple-beam interference. This procedure has a number of practical applications, one of which will be discussed in Section 9.10, when we consider the use of multiple-beam Fizeau fringes to examine surface topography.

9.8 THE FABRY–PEROT INTERFEROMETER

9.8.1 Modus Operandi

The multiple-beam interferometer, first constructed by Charles Fabry and Alfred Perot in the late eighteen hundreds, is of considerable importance in modern optics. Its particular value arises from the fact that besides being a spectroscopic device of extremely high resolving power, it also serves as the basic laser resonant cavity. In principle, the device consists of two plane, parallel, highly reflecting surfaces separated by some distance d. This is the simplest configuration and as we shall see, other forms are also widely in use. In practice, two semisilvered or aluminized glass optical flats form the reflecting boundary surfaces. The enclosed air gap generally varies from several millimeters to several centimeters when the apparatus is used interferometrically, and often to considerably greater lengths when serving as a laser resonant cavity. If the gap can be mechanically varied by moving one of the mirrors, it's referred to as an interferometer. When the mirrors are held fixed and adjusted for parallelism by screwing down on some sort of spacer (invar or quartz are commonly used), it's said to be an *etalon* (although it is, of course, still an interferometer in the broad sense). Indeed, if the two surfaces of a single quartz plate are appropriately polished and silvered, it too will serve as an etalon; the gap need not be air. The unsilvered sides of the plates are often made to have a slight wedge shape (a few minutes of arc) to reduce the interference pattern arising from reflections off these sides. The etalon in Fig. 9.47 is shown illuminated by a broad source which might be a mercury arc or a Helium–Neon laser beam spread out in diameter to several centimeters. This can be done rather nicely by sending the beam into the back end of a telescope

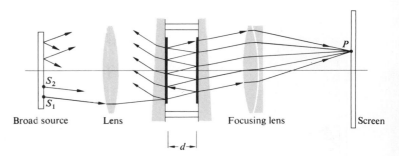

Fig. 9.47 Fabry–Perot etalon.

focused at infinity. The light can then be made diffuse by passing it through a sheet of ground glass. Only one ray emitted from some point S_1 on the source is traced through the etalon. Entering by way of the partially silvered plate, it is multiply reflected within the gap. The transmitted rays are collected by a lens and brought to a focus on a screen, where they interfere to form either a bright or dark spot. Consider this particular plane of incidence which contains all of the reflected rays. Any other ray emitted from a different point S_2, parallel to the original ray and in that plane of incidence, will form a spot at the same point P on the screen. As we shall see, the discussion of the previous section is again applicable, so that Eq. (9.32) determines the transmitted flux density I_t. The multiple waves generated in the cavity, arriving at P from either S_1 or S_2, are coherent amongst themselves. But the rays arising from S_1 are completely incoherent with respect to those from S_2, so that there is no sustained mutual interference. The contribution to the irradiance I_t at P is just the sum of the two irradiance contributions.

All of the rays incident on the gap at a given angle will result in a single circular fringe of uniform irradiance (Fig. 9.48). With a broad diffuse source, the interference bands will be narrow concentric rings, corresponding to the multiple-beam transmission pattern.

The fringe system can be observed visually, by looking directly into the etalon, while focusing at infinity. The job of the focusing lens, which is no longer needed, is done by the eye. At large values of d, the rings will be close together and a telescope might be needed to magnify the pattern. A relatively inexpensive monocular will serve the same purpose and will allow for photographing the fringes localized at infinity. As might be expected from the considerations of Section 9.6, it is possible to produce real nonlocalized fringes using a bright point source.

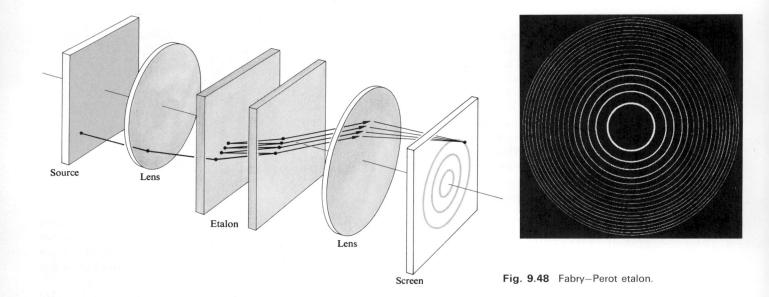

Fig. 9.48 Fabry–Perot etalon.

The partially transparent metal films which are often used to increase the reflectance ($R = r^2$), will absorb a fraction A of the flux density; this fraction is referred to as the absorptance.

The expression

$$tt' + r^2 = 1$$

or

$$T + R = 1, \qquad [4.60]$$

where T is the transmittance, must now be rewritten as

$$T + R + A = 1. \qquad (9.41)$$

One further complication is introduced by the metallic films and that is, an additional phase shift $\phi(\theta_i)$, which can differ from either zero or π. The phase difference between two successively transmitted waves is then

$$\delta = \frac{4\pi n_f}{\lambda_0} d \cos \theta_t + 2\phi. \qquad (9.42)$$

For the conditions under consideration, θ_i is small and ϕ may be considered to be constant. In general, d is so large, and λ_0 so small, that ϕ can be neglected. We can now express Eq. (9.32) as

$$\frac{I_t}{I_i} = \frac{T^2}{1 + R^2 - 2R \cos \delta},$$

or equivalently

$$\frac{I_t}{I_i} = \left(\frac{T}{1 - R} \right)^2 \frac{1}{1 + [4R/(1 - R)]^2 \sin^2 (\delta/2)}. \qquad (9.43)$$

Making use of Eq. (9.41) and the definition of the Airy function, we obtain

$$\frac{I_t}{I_i} = \left[1 - \frac{A}{(1 - R)} \right]^2 \mathscr{A}(\theta) \qquad (9.44)$$

as compared with the equation for zero absorption

$$\frac{I_t}{I_i} = \mathscr{A}(\theta). \qquad [9.40]$$

Inasmuch as the absorbed portion A is never zero, the transmitted flux-density maxima $(I_t)_{max}$ will always be somewhat less than I_i. [Recall that for $(I_t)_{max}$ $\mathscr{A}(\theta) = 1$.]

Accordingly, the *peak transmission* is defined as $(I_t/I_i)_{max}$:

$$\frac{(I_t)_{max}}{I_i} = \left[1 - \frac{A}{(1 - R)} \right]^2. \qquad (9.45)$$

A silver film 50 nm thick would be approaching its maximum value of R, e.g. about 0.94, while T and A might be respectively 0.01 and 0.05. In this case, the peak transmission will be down to $\frac{1}{36}$. The relative irradiance of the fringe

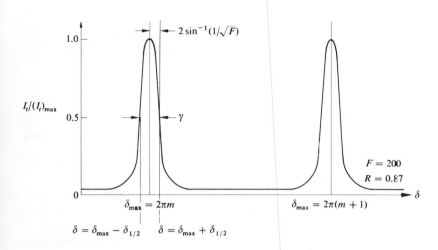

Fig. 9.49 Fabry–Perot fringes.

pattern will still be determined by the Airy function, since

$$\frac{I_t}{(I_t)_{\text{max}}} = \mathscr{A}(\theta). \qquad (9.46)$$

A measure of the sharpness of the fringes, i.e. how rapidly the irradiance drops off on either side of the maximum, is given by the half-width γ. Shown in Fig. 9.49, γ is the width of the peak, in radians, when $I_t = (I_t)_{\text{max}}/2$.

Peaks in the transmission occur at specific values of the phase difference $\delta_{\text{max}} = 2\pi m$. Accordingly, the irradiance will drop to half its maximum value, i.e. $\mathscr{A}(\theta) = \frac{1}{2}$ whenever $\delta = \delta_{\text{max}} \pm \delta_{1/2}$. Inasmuch as

$$\mathscr{A}(\theta) = [1 + F \sin^2(\delta/2)]^{-1},$$

then when

$$[1 + F \sin^2(\delta_{1/2}/2)]^{-1} = \frac{1}{2}$$

it follows that

$$\delta_{1/2} = 2 \sin^{-1}(1/\sqrt{F}).$$

Since F is generally rather large, $\sin^{-1}(1/\sqrt{F}) \approx 1/\sqrt{F}$ and therefore the half-width, $\gamma = 2\delta_{1/2}$, becomes

$$\gamma = 4/\sqrt{F}. \qquad (9.47)$$

Recall that $F = 4R/(1 - R)^2$, so that the larger R is, the sharper will be the transmission peaks.

Another quantity of particular interest is the ratio of the separation of adjacent maxima to the half-width. Known as the *finesse*, $\mathscr{F} \equiv 2\pi/\gamma$ or, from Eq. (9.47),

$$\mathscr{F} = \frac{\pi\sqrt{F}}{2}. \qquad (9.48)$$

Over the visible spectrum, the finesse of most ordinary Fabry–Perot instruments is about 30. The physical limitation on $\mathscr{F}$ is set by deviations in the mirrors from plane parallelism. Keep in mind that as the finesse increases, the half-width decreases, but so too does the peak transmission. Incidently, finesse of about 1000 is attainable with curved-mirror systems using dielectric thin-film coatings.*

9.8.2 Fabry–Perot Spectroscopy

The Fabry–Perot interferometer is frequently used to examine the detailed structure of spectral lines. We will not attempt a complete treatment of interference spectroscopy, but rather will define the relevant terminology, briefly outlining appropriate derivations.[†]

As has been seen, a hypothetical purely monochromatic light wave generates a particular circular fringe system. But δ is a function of λ_0, so that if the source were made up of two such monochromatic components, two superimposed ring systems would result. When the individual fringes partially overlap, a certain amount of ambiguity exists in deciding when the two systems are individually discernible, i.e. when they are said to be *resolved*. Lord Rayleigh's[‡]

---

* The paper "Multiple Beam Interferometry" by H. D. Polster, *Appl. Opt.* **8**, 522 (1969) should be of interest.

[†] A more complete treatment can be found in Born and Wolf, *Principles of Optics*, or W. E. Williams, *Applications of Interferometry*, to name only two.

[‡] The criterion will be reconsidered with respect to diffraction in the next chapter (see Fig. 10.40).

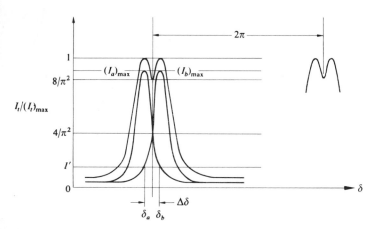

Fig. 9.50 Overlapping fringes.

criterion for resolving two equal irradiance overlapping slit images is well accepted, even if somewhat arbitrarily in the present application. Using it, however, will allow a comparison with prism or grating instruments. The essential feature of this criterion, is that the fringes are *just resolvable* when the combined irradiance of both fringes at the center, or saddle point, of the resultant broad fringe is $8/\pi^2$ times the maximum irradiance. This just means that one would see a broad bright fringe with a grey central region. To be a bit more analytic about it, examine Fig. 9.50, keeping in mind the previous derivation of the half-width. Consider the case when both constituent fringes have equal irradiances, $(I_a)_{max} = (I_b)_{max}$. The peaks in the resultant, occurring at $\delta = \delta_a$ and $\delta = \delta_b$, will have equal irradiances,

$$(I_t)_{max} = (I_a)_{max} + I'. \tag{9.49}$$

At the saddle point, the irradiance, $(8/\pi^2)\,(I_t)_{max}$ is the sum of the two constituent irradiances, so that, recalling Eq. (9.46),

$$(8/\pi^2)\frac{(I_t)_{max}}{(I_a)_{max}} = [\mathscr{A}(\theta)]_{\delta=\delta_a+\Delta\delta/2} + [\mathscr{A}(\theta)]_{\delta=\delta_b+\Delta\delta/2}. \tag{9.50}$$

Using $(I_t)_{max}$ given by Eq. (9.49), along with the fact that

$$\frac{I'}{(I_a)_{max}} = [\mathscr{A}(\theta)]_{\delta=\delta_a+\Delta\delta}$$

Eq. (9.50) can be solved for $\Delta\delta$. For large values of F,

$$(\Delta\delta) \approx \frac{4.2}{\sqrt{F}}. \tag{9.51}$$

This then represents the smallest phase increment, $(\Delta\delta)_{min}$,

separating two resolvable fringes. It can be related to equivalent minimum increments in wavelength $(\Delta\lambda_0)_{min}$, frequency $(\Delta v)_{min}$, and wave number $(\Delta\varkappa)_{min}$. From Eq. (9.42), for $\delta = 2\pi m$, we have

$$m\lambda_0 = 2n_f d \cos\theta_t + \frac{\phi\lambda_0}{\pi}. \tag{9.52}$$

Dropping the term $\phi\lambda_0/\pi$, which is clearly negligible, and then differentiating, yields

$$m(\Delta\lambda_0) + \lambda_0(\Delta m) = 0$$

or

$$\frac{\lambda_0}{(\Delta\lambda_0)} = -\frac{m}{(\Delta m)}.$$

The minus will be omitted, since it only means that the order increases when λ_0 decreases. When δ changes by 2π, m changes by 1 and so

$$\frac{2\pi}{(\Delta\delta)} = \frac{1}{(\Delta m)}.$$

Now then

$$\frac{\lambda_0}{(\Delta\lambda_0)} = \frac{2\pi m}{(\Delta\delta)}. \tag{9.53}$$

The ratio of λ_0 to the least resolvable wavelength difference $(\Delta\lambda_0)_{min}$ is known as the *chromatic resolving power* $\mathscr{R}$, of any spectroscope. And so at nearly normal incidence

$$\mathscr{R} \equiv \frac{\lambda_0}{(\Delta\lambda_0)_{min}} \approx \mathscr{F}\frac{2n_f d}{\lambda_0} \tag{9.54}$$

or

$$\mathscr{R} \approx \mathscr{F} m.$$

For a wavelength of 500 nm, $n_f d = 10$ mm, and $R = 90\%$, the resolving power is well over a million, a range only recently achieved by the finest diffraction gratings. It follows as well, in this example, that $(\Delta\lambda_0)_{min}$ is less than a millionth of λ_0. In terms of frequency, the *minimum resolvable bandwidth* is

$$(\Delta v)_{min} = \frac{c}{\mathscr{F}2n_f d} \tag{9.55}$$

inasmuch as $|\Delta v| = |c\Delta\lambda_0/\lambda_0^2|$.

As the two components present in the source become increasingly different in wavelength, the peaks shown overlapping in Fig. 9.50 separate. As the wavelength difference increases, the mth-order fringe for one wavelength λ_0 will approach the $(m + 1)$th-order for the other wavelength

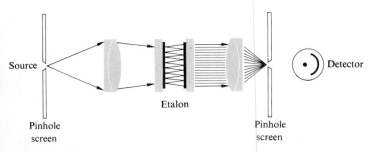

Source

Pinhole
screen

Etalon

Pinhole
screen

Detector

Fig. 9.51 Central spot scanning.

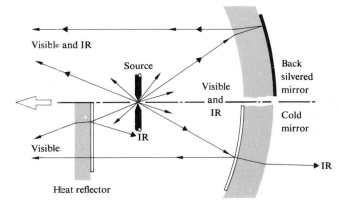

Visible and IR

Source

Visible
and
IR

Back
silvered
mirror

Cold
mirror

IR

Visible

Heat reflector

Fig. 9.52 A composite drawing showing an ordinary system on top and a coated one on the bottom.

$(\lambda_0 - \Delta\lambda_0)$. The particular wavelength difference at which overlapping takes place, $(\Delta\lambda_0)_{\text{fsr}}$, is known as the *free spectral range*. From Eq. (9.53), a change in δ of 2π corresponds to $(\Delta\lambda_0)_{\text{fsr}} = \lambda_0/m$, or at near normal incidence,

$$(\Delta\lambda_0)_{\text{fsr}} \approx \lambda_0^2/2n_f d, \qquad (9.56)$$

and similarly

$$(\Delta\nu)_{\text{fsr}} \approx c/2n_f d. \qquad (9.57)$$

Continuing the above example, i.e. $\lambda_0 = 500$ nm and $n_f d = 10$ mm, $(\Delta\lambda_0)_{\text{fsr}} = 0.0125$ nm. Clearly, if we attempt to increase the resolving power by merely increasing d, the *free spectral range* will decrease, bringing with it the resulting confusion from the overlapping of orders. What is needed is that $(\Delta\lambda_0)_{\text{min}}$ *be as small as possible* and $(\Delta\lambda_0)_{\text{fsr}}$ *as large as possible*. But lo and behold

$$\frac{(\Delta\lambda)_{\text{fsr}}}{(\Delta\lambda)_{\text{min}}} = \mathscr{F}. \qquad (9.58)$$

This result should not be too surprising in view of the original definition of $\mathscr{F}$.

Both the applications and configurations of the Fabry–Perot interferometer are numerous indeed. Etalons have been arranged in series with other etalons, as well as with grating and prism spectroscopes and multilayer dielectric films have been used to replace the metallic mirror coatings.

Scanning techniques are now widely in use. These take advantage of the superior linearity of photoelectric detectors over photographic plates, to obtain more reliable flux-density measurements. The basic set-up for *central-spot scanning* is illustrated in Fig. 9.51. Scanning is accomplished by varying δ, either by changing n_f or d rather than $\cos\theta_t$. In some arrangements, n_f is smoothly varied by altering the air pressure within the etalon. Alternatively, mechanically vibrating one mirror with a displacement of $\lambda/2$ will be enough to scan the free spectral range, corresponding as it does to $\Delta\delta = 2\pi$.

A popular technique for accomplishing this utilizes a piezoelectric mirror mount. This kind of material will change its length, and therefore d, as a voltage is applied to it. The voltage profile determines the mirror motion.

Rather than photographically recording irradiance over a large region in space, at a single point in time, this method records irradiance over a large region in time, at a single point in space.

The actual configuration of the etalon itself has also undergone some significant variations. Pierre Connes in 1956 first described the *spherical-mirror Fabry–Perot interferometer*. Since then curved-mirror systems have become prominent as laser cavities and in addition are finding increasing use as spectrum analyzers.

9.9 APPLICATIONS OF SINGLE AND MULTILAYER FILMS

The optical uses to which coatings of thin dielectric films have been put in recent times are many indeed. Coatings to eliminate unwanted reflections off a diversity of surfaces, from showcase glass to quality camera lenses, are now commonplace. Multilayer, nonabsorbing beam splitters and *dichroic* mirrors (color selective beam splitters which transmit and reflect particular wavelengths) can be purchased commercially. Figure 9.52 is a segmented diagram illustrating the use of a *cold mirror* in combination with a *heat reflector* to channel infrared radiation to the rear of a motion-picture projector. The intense unwanted infrared (IR) radiation emitted by the source is removed from the beam to avoid heating problems at the photographic film. The top half of

Fig. 9.52 is an ordinary back silvered mirror shown for comparison. Solar cells, which are one of the prime power-supply systems for space vehicles, and even the astronauts' helmets and visors, are shielded with similar heat control coverings. Multilayer broad and narrow band-pass filters, ones which transmit only over a specific spectral range, can be made to span the region from infrared to ultraviolet. In the visible, for example, they play an important role in splitting up the image in color television cameras, while in the IR, they're used in missile guidance systems, CO_2 lasers and satellite horizon sensors. The applications of thin-film devices are manifold, as are their structures, which extend from the simplest single coatings to intricate arrangements of 100 or more layers.

The treatment of multilayer film theory used here will deal with the *total* electric and magnetic fields and their boundary conditions in the various regions. This is a far more practical approach for many-layered systems than is the multiple-wave technique used earlier.[*]

9.9.1 Mathematical Treatment

Consider the linearly polarized wave shown in Fig. 9.53, impinging on a thin dielectric film between two semi-infinite transparent media. In practice, this might correspond to a dielectric layer a fraction of a wavelength thick, deposited on the surface of a lens, a mirror, or a prism. One point must be made clear at the outset: each wave E_{rI}, E'_{rII}, E_{tII}, etc. represents the resultant of all possible waves traveling in that direction, at that point in the medium. The summation process is, therefore, built in. As discussed in Section 4.3.2 the boundary conditions require that the tangential components of both the electric (E) and magnetic ($H = B/\mu$) fields be continuous across the boundaries (i.e. equal on both sides). At boundary I

$$E_I = E_{iI} + E_{rI} = E_{tI} + E'_{rII} \tag{9.59}$$

and

$$H_I = \sqrt{\frac{\epsilon_0}{\mu_0}}(E_{iI} - E_{rI})n_0 \cos\theta_{iI} = \sqrt{\frac{\epsilon_0}{\mu_0}}(E_{tI} - E'_{rII})n_1 \cos\theta_{iII}, \tag{9.60}$$

where use is made of the fact that E and H in nonmagnetic media are related by way of the index of refraction and the

[*] For a very readable nonmathematical discussion, see P. Bowmeister and G. Pincus, "Optical Interference Coatings," *Sci. Amer.* **223**, 59 (December 1970).

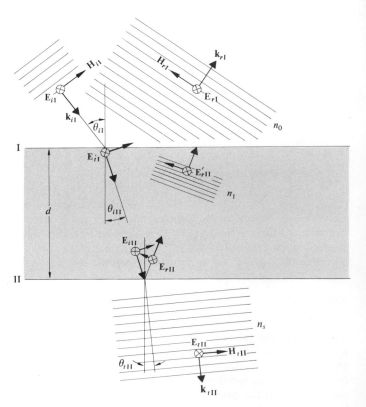

Fig. 9.53 Fields at the boundaries.

unit propagation vector:

$$H = \sqrt{\frac{\epsilon_0}{\mu_0}}n\hat{k} \times E.$$

At boundary II

$$E_{II} = E_{iII} + E_{rII} = E_{tII} \tag{9.61}$$

and

$$H_{II} = \sqrt{\frac{\epsilon_0}{\mu_0}}(E_{iII} - E_{rII})n_1 \cos\theta_{iII}$$

$$= \sqrt{\frac{\epsilon_0}{\mu_0}}E_{tII}n_s \cos\theta_{tII}, \tag{9.62}$$

the substrate having an index n_s. In accord with Eq. (9.21), a wave which traverses the film once undergoes a shift in phase of $k_0(2n_1 d \cos\theta_{iII})/2$ which will be denoted by $k_0 h$ so that

$$E_{iII} = E_{tI}e^{-ik_0 h} \tag{9.63}$$

and

$$E_{rII} = E'_{rII}e^{+ik_0 h} \tag{9.64}$$

Equations (9.61) and (9.62) can now be written as

$$E_{II} = E_{tI}e^{-ik_0h} + E'_{rII}e^{+ik_0h} \tag{9.65}$$

and

$$H_{II} = (E_{tI}e^{-ik_0h} - E'_{rII}e^{+ik_0h})\sqrt{\frac{\epsilon_0}{\mu_0}}\, n_1 \cos\theta_{iII}. \tag{9.66}$$

These last two equations can be solved for E_{tI} and E'_{rII} which, when substituted into Eqs. (9.59) and (9.60) yield

$$E_I = E_{II}\cos k_0 h + H_{II}(i\sin k_0 h)/\Upsilon_1 \tag{9.67}$$

and

$$H_I = E_{II}\Upsilon_1\, i\sin k_0 h + H_{II}\cos k_0 h, \tag{9.68}$$

where

$$\Upsilon_1 \equiv \sqrt{\frac{\epsilon_0}{\mu_0}}\, n_1 \cos\theta_{iII}.$$

The above calculations carried out for the case when **E** is in the plane of incidence result in similar equations, provided that now

$$\Upsilon_1 \equiv \sqrt{\frac{\epsilon_0}{\mu_0}}\, n_1/\cos\theta_{iII}.$$

In matrix notation, the above linear relations take the form

$$\begin{bmatrix} E_I \\ H_I \end{bmatrix} = \begin{bmatrix} \cos k_0 h & (i\sin k_0 h)/\Upsilon_1 \\ \Upsilon_1 i\sin k_0 h & \cos k_0 h \end{bmatrix} \begin{bmatrix} E_{II} \\ H_{II} \end{bmatrix} \tag{9.69}$$

or

$$\begin{bmatrix} E_I \\ H_I \end{bmatrix} = \mathcal{M}_I \begin{bmatrix} E_{II} \\ H_{II} \end{bmatrix}. \tag{9.70}$$

The *characteristic matrix* $\mathcal{M}_I$ relates the fields at the two adjacent boundaries. It follows therefore, that if two overlaying films were deposited on the substrate, there would be three boundaries or interfaces, and now

$$\begin{bmatrix} E_{II} \\ H_{II} \end{bmatrix} = \mathcal{M}_{II} \begin{bmatrix} E_{III} \\ H_{III} \end{bmatrix}. \tag{9.71}$$

Multiplying both sides of this expression by $\mathcal{M}_I$, we obtain

$$\begin{bmatrix} E_I \\ H_I \end{bmatrix} = \mathcal{M}_I \mathcal{M}_{II} \begin{bmatrix} E_{III} \\ H_{III} \end{bmatrix}. \tag{9.72}$$

In general, if p is the number of layers, each with a particular value of n and h, then the first and the last boundaries are related by

$$\begin{bmatrix} E_I \\ H_I \end{bmatrix} = \mathcal{M}_I \mathcal{M}_{II} \cdots \mathcal{M}_p \begin{bmatrix} E_{(p+1)} \\ H_{(p+1)} \end{bmatrix}. \tag{9.73}$$

The characteristic matrix of the entire system is the resultant of the product (in the proper sequence) of the individual two-by-two matrices, that is

$$\mathcal{M} = \mathcal{M}_I \mathcal{M}_{II} \cdots \mathcal{M}_p = \begin{bmatrix} m_{11} & m_{12} \\ m_{21} & m_{22} \end{bmatrix}. \tag{9.74}$$

To see how all of this fits together, we will derive expressions for the amplitude coefficients of reflection and transmission using the above scheme. By reformulating Eq. (9.70) in terms of the boundary conditions [(9.59), (9.60) and (9.62)] and setting

$$\Upsilon_0 = \sqrt{\frac{\epsilon_0}{\mu_0}}\, n_0 \cos\theta_{iI}$$

and

$$\Upsilon_s = \sqrt{\frac{\epsilon_0}{\mu_0}}\, n_s \cos\theta_{tII},$$

we obtain

$$\begin{bmatrix} (E_{iI} + E_{rI}) \\ (E_{iI} - E_{rI})\Upsilon_0 \end{bmatrix} = \mathcal{M}_I \begin{bmatrix} E_{tII} \\ E_{tII}\Upsilon_s \end{bmatrix}.$$

On expanding the matrices, the last relation becomes

$$1 + r = m_{11}t + m_{12}\Upsilon_s t$$

and

$$(1 - r)\Upsilon_0 = m_{21}t + m_{22}\Upsilon_s t$$

in as much as

$$r = E_{rI}/E_{iI} \qquad \text{and} \qquad t = E_{tII}/E_{iI}.$$

Consequently,

$$r = \frac{\Upsilon_0 m_{11} + \Upsilon_0\Upsilon_s m_{12} - m_{21} - \Upsilon_s m_{22}}{\Upsilon_0 m_{11} + \Upsilon_0\Upsilon_s m_{12} + m_{21} + \Upsilon_s m_{22}} \tag{9.75}$$

and

$$t = \frac{2\Upsilon_0}{\Upsilon_0 m_{11} + \Upsilon_0\Upsilon_s m_{12} + m_{21} + \Upsilon_s m_{22}}. \tag{9.76}$$

To find either r or t for any configuration of films, we need only compute the characteristic matrices for each film, multiply them, and then substitute the resulting matrix elements into the above equations.

9.9.2 Antireflection Coatings

Now consider the extremely important case of normal incidence, that is

$$\theta_{iI} = \theta_{iII} = \theta_{tII} = 0,$$

which in addition to being the simplest, is quite frequently approximated in practical situations. If we put a subscript on r to indicate the number of layers present, the reflection coefficient for a single film becomes

$$r_1 = \frac{n_1(n_0 - n_s)\cos k_0 h + i(n_0 n_s - n_1^2)\sin k_0 h}{n_1(n_0 + n_s)\cos k_0 h + i(n_0 n_s + n_1^2)\sin k_0 h}. \quad (9.77)$$

Multiplying r_1 by its complex conjugate leads to the reflectance

$$R_1 = \frac{n_1^2(n_0 - n_s)^2 \cos^2 k_0 h + (n_0 n_s - n_1^2)^2 \sin^2 k_0 h}{n_1^2(n_0 + n_s)^2 \cos^2 k_0 h + (n_0 n_s + n_1^2)^2 \sin^2 k_0 h}. \quad (9.78)$$

This formula becomes particularly simple when $k_0 h = \frac{1}{2}\pi$, which is equivalent to saying that the optical thickness h of the film is an odd multiple of $\frac{1}{4}\lambda_0$. In this case $d = \frac{1}{4}\lambda_f$, and

$$R_1 = \frac{(n_0 n_s - n_1^2)^2}{(n_0 n_s + n_1^2)^2} \quad (9.79)$$

which, quite remarkably, will equal zero when

$$n_1^2 = n_0 n_s. \quad (9.80)$$

Generally, d is chosen so that h equals $\frac{1}{4}\lambda_0$ in the yellow-green portion of the visible spectrum, where the eye is most sensitive. Cryolite ($n = 1.35$), a sodium aluminum fluoride compound, and magnesium fluoride ($n = 1.38$) are common low-index films. Since MgF_2 is by far the more durable, it is most frequently used. On a glass substrate, ($n_s \approx 1.5$), both these films would have indices which are still somewhat too large to satisfy Eq. (9.80). Nonetheless, a single $\frac{1}{4}\lambda_0$ layer of MgF_2, will reduce the reflectance of glass from about 4% to a bit more than 1%, over the visible spectrum. It is now common practice to apply antireflection coatings to the elements of optical instruments. On camera lenses, such coatings produce a decrease in the haziness caused by stray internally scattered light, as well as a marked increase in image brightness. At wavelengths on either side of the central yellow-green region, R increases and the lens surface will appear blue-red in reflected light.

For a double layer, quarter-wavelength antireflection coating

$$\mathscr{M} = \mathscr{M}_I \mathscr{M}_{II}$$

or more specifically

$$\mathscr{M} = \begin{bmatrix} 0 & i/\Upsilon_1 \\ i\Upsilon_1 & 0 \end{bmatrix}\begin{bmatrix} 0 & i/\Upsilon_2 \\ i\Upsilon_2 & 0 \end{bmatrix}. \quad (9.81)$$

At normal incidence this becomes

$$\mathscr{M} = \begin{bmatrix} -n_2/n_1 & 0 \\ 0 & -n_1/n_2 \end{bmatrix}. \quad (9.82)$$

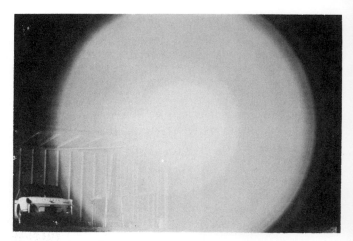

Fig. 9.54 Lens elements coated with a single layer of MgF_2.

Substituting the appropriate matrix elements into Eq. (9.15) yields r_2 which, when squared, leads to the reflectance

$$R_2 = \left[\frac{n_2^2 n_0 - n_s n_1^2}{n_2^2 n_0 + n_s n_1^2}\right]^2. \quad (9.83)$$

In order that R_2 be exactly zero at a particular wavelength, we need

$$\left(\frac{n_2}{n_1}\right)^2 = \frac{n_s}{n_0}. \quad (9.84)$$

This kind of film is referred to as a *double-quarter, single-minimum* coating. When n_1 and n_2 are as small as possible, the reflectance will have its single broadest minimum equal to zero at the chosen frequency. It should be clear from Eq. (9.84) that $n_2 > n_1$; accordingly, it is now a common practice to designate a (glass)−(high index)−(low index)−(air) system as $gHLa$. Zirconium dioxide ($n = 2.1$), titanium dioxide ($n = 2.40$), and zinc sulfide ($n = 2.32$) are commonly used for H-layers while magnesium fluoride ($n = 1.38$) and cerium fluoride ($n = 1.63$) often serve as L-layers.

Other double- and triple-layer schemes can be designed to satisfy specific requirements on spectral response, incident angle, cost, etc. Fig. 9.54 is a scene photographed through a fifteen-element zoom lens, with a 150 W lamp pointing directly into the camera. The lens elements were covered with a single layer of MgF_2. For Fig. 9.55 a triple-layer antireflection coating was used. The improved contrast and glare reduction is apparent.

Fig. 9.55 Lens elements coated with a multi-layer film structure. [Photos courtesy Optical Coating Laboratory, Inc., Santa Rosa, California.]

9.9.3 Multilayer Periodic Systems

The simplest kind of periodic system is the *quarter-wave stack* which, not unreasonably, is made up of a number of quarter-wave layers. The periodic structure of alternately high- and low-index materials, illustrated in Fig. 9.56, is designated by

$$g(HL)^3a.$$

Figure 9.57 illustrates the general form of a portion of the spectral reflectance for a few multilayer filters. The width of the high-reflectance central zone increases with increasing values of the index ratio n_H/n_L while its height increases with the number of layers. Note that the maximum reflectance of a periodic structure such as $g(HL)^ma$ can be increased further by adding another H-layer such that it now has the form $g(HL)^mHa$. Very high reflectance mirror surfaces can be produced using this arrangement.

The small peak on the short-wavelength side of the central zone can be decreased by adding an eighth-wave low-index film to both ends of the stack, in which case the whole arrangement will be denoted by

$$g(0.5L)(HL)^mH(0.5L)a.$$

This has the effect of increasing the short-wavelength high-frequency transmittance and is therefore known as a *high-pass filter*. Similarly, the structure

$$g(0.5H)L(HL)^m(0.5H)a$$

merely corresponds to the case when the end H-layers are

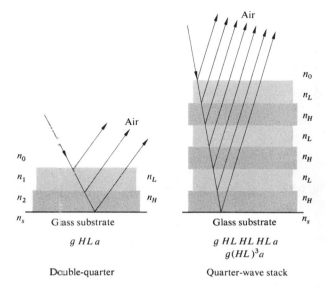

Fig. 9.56 A periodic structure.

$\lambda_0/8$ thick. It has a higher transmittance at the long-wavelength, low-frequency range and serves as a *low-pass filter*.

At nonnormal incidence, up to about 30°, there is quite frequently little degradation in the response of thin-film coatings. In general, the effect of increasing the incident angle is a shift in the whole reflectance curve down to slightly shorter wavelengths. This kind of behavior is evidenced by several naturally occurring periodic structures, e.g. peacock and hummingbird feathers, butterflies' wings, as well as the backs of several varieties of beetles.

The last multilayer system to be considered is the *interference*, or more precisely the *Fabry–Perot*, filter. If the separation between the plates of an etalon is of the order of λ, the transmission peaks will be widely separated in wavelength. It will then be possible to block all of the peaks but one, by using absorbing filters of colored glass or gelatin. The transmitted light then corresponds to a single sharp peak and the etalon serves as a narrow band-pass filter. Such devices can be fabricated by depositing a semitransparent metal film onto a glass support, followed by a MgF_2 spacer and another metal coating.

All-dielectric, essentially nonabsorbing Fabry–Perot filters have an analogous structure, two possible examples of which are

$$g\ HLH\ LL\ HLH\ a$$

and

$$g\ HLHL\ HH\ LHLH\ a.$$

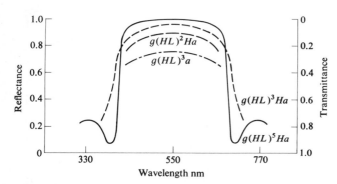

Fig. 9.57 Reflectance and transmittance for several periodic structures.

The characteristic matrix for the first of these is

$$\mathcal{M} = \mathcal{M}_H\,\mathcal{M}_L\,\mathcal{M}_H\,\mathcal{M}_L\,\mathcal{M}_H\,\mathcal{M}_L\,\mathcal{M}_H$$

but from Eq. (9.82)

$$\mathcal{M}_L\,\mathcal{M}_L = \begin{bmatrix} -1 & 0 \\ 0 & -1 \end{bmatrix},$$

or

$$\mathcal{M}_L\,\mathcal{M}_L = -\mathcal{I},$$

where $\mathcal{I}$ is the unity matrix. The central double layer, corresponding to the Fabry–Perot cavity, is a half-wavelength thick ($d = \frac{1}{2}\lambda_f$). It therefore has no effect on the reflectance *at the particular wavelength under consideration.* Thus, it is said to be an absentee layer, and as a consequence

$$\mathcal{M} = -\,\mathcal{M}_H\,\mathcal{M}_L\,\mathcal{M}_H\,\mathcal{M}_H\,\mathcal{M}_L\,\mathcal{M}_H$$

The same conditions prevail again at the center, and will obviously finally result in

$$\mathcal{M} = \begin{bmatrix} 1 & 0 \\ 0 & 1 \end{bmatrix}$$

At the special frequency for which the filter was designed, r, according to Eq. (9.75) at normal incidence reduces to

$$r = \frac{n_0 - n_s}{n_0 + n_s},$$

the value for the uncoated substrate. In particular, for glass ($n_s = 1.5$), in air ($n_0 = 1$) the theoretical peak transmission is 96% (neglecting reflections from the back surface of the substrate, as well as losses in both the blocking filter and the films themselves).

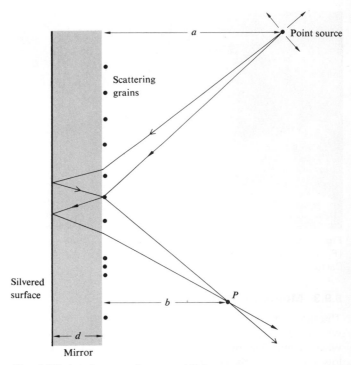

Fig. 9.58 Interference of scattered light.

9.10 APPLICATIONS OF INTERFEROMETRY

There are a great many physical applications utilizing the principles of interferometry. Some of these are now only of historical or pedagogical significance, while others are presently being used extensively. The advent of the laser and the resultant availability of highly coherent quasimonochromatic light has made it particularly easy to create new interferometer configurations.

9.10.1 Scattered-light Interference

Probably the earliest recorded study of interference fringes arising from scattered light is to be found in Sir Isaac Newton's *Optiks* (1704, Book Two, Part IV). Our present interest in this phenomenon is twofold. Firstly, it provides an extremely easy way to see some rather beautiful colored interference fringes. Secondly, it is the basis for a remarkably simple and highly useful interferometer.

To see the fringes, lightly rub a thin layer of ordinary talcum powder onto the surface of any common back-silvered mirror (dew will do as well). Neither the thickness

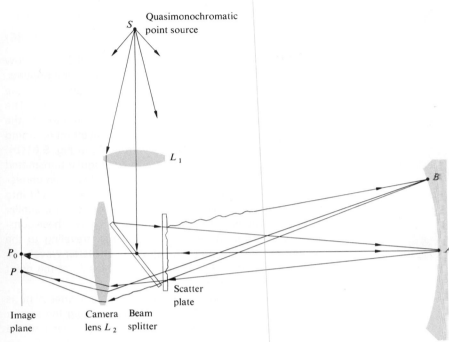

Quasimonochromatic point source

S

L_1

B

P_0

A

Scatter
plate

P

Image
plane

Camera
lens L_2

Beam
splitter

Test mirror

Fig. 9.59 Scatter plate set up.   Adapted from R. M. Scott, *Appl. Opt.* **8**, 531 (1969).

nor the uniformity of the coating is particularly important. The use of a bright point source, however, is crucial. A satisfactory source can be made by taping a heavy piece of cardboard having a hole about $\frac{1}{4}$ inch in diameter over a good flashlight.   Initially, stand back from the mirror about three or four feet; the fringes will be too fine and closely spaced to see if you stand very much nearer. Hold the flashlight alongside your cheek and illuminate the mirror so that you can see the brightest reflection of the bulb in it. The fringes will then be clearly seen as a number of alternately bright and dark bands.

Two coherent rays leaving the point source are shown in Fig. 9.58 arriving at point P after traveling different routes. One ray is reflected from the mirror and then scattered by a single transparent talcum grain toward P. The second ray is first scattered downward by the grain after which it crosses the mirror and is reflected back toward P. The resulting optical path-length difference determines the interference at P. At normal incidence, the pattern is a series of concentric rings of radius*

___

* For more of the details, see A. J. deWitte, "Interference in Scattered Light," *Am. J. Phys.* **35**, 301 (1967).

$$\rho \approx \left[\frac{nm\lambda a^2 b^2}{d(a^2 - b^2)} \right]^{\frac{1}{2}}.$$

Now consider a related device which is very useful in testing optical systems. Known as a *scatter plate*, it generally consists of a slightly rough-surfaced, transparent sheet. In an arrangement such as the one shown in Fig. 9.59, it serves as an amplitude-splitting element. In this application it must have a center of symmetry, i.e. each scattering site is required to have a duplicate, symmetrically located about a central point.

In the system under consideration, a point source of quasimonochromatic light S is imaged, by means of lens L_1 on the surface, at point A of the mirror being tested.   A portion of the light coming from the source is scattered by the scatter plate and thereafter illuminates the entire surface of the mirror. The mirror, in turn, reflects light back to the scatter plate. This wave, as well as the light forming the image of the pinhole at point A, passes through the scatter plate again, and finally reaches the image plane (either on a screen or in a camera).   Fringes are formed on this latter plane. The interference process, which is manifest in the formation of these fringes, occurs because each point in the

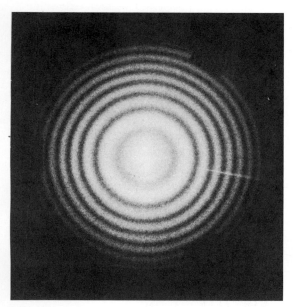

Fig. 9.60   Fringes in scattered light.

final image plane is illuminated by light arriving via two dis-similar routes; one originating at *A* and the other at some point *B* which reflects scattered light. Indeed, as strange as it may look at first sight, well-defined fringes do result as may be seen from Fig. 9.60.

Examining the passage of light through the system in a bit more detail, consider the light initially incident on the scatter plate and assume that the wave is planar as shown in Fig. 9.61. After it passes through the scatter plate, the incident plane wavefront E_i will be distorted into a transmitted wavefront E_T. We envision this wave, in turn, split into a

series of Fourier components consisting of plane waves, i.e.

$$E_T = E_1 + E_2 + \cdots . \qquad (9.85)$$

Two of these constituents are shown in Fig. 9.61(a). Now suppose we attach a specific meaning to these components, namely E_1 is taken to represent the light traveling to the point *A* in Fig. 9.59 and E_2 that traveling towards *B*. The analysis of the stages that follow could be continued in the same way. And so, let the portion of the wavefront returning from *A* be represented by the wavefront E_A in Fig. 9.61(b). The scatter plate will transform it into an irregular transmitted wave denoted by E_{AT} in the same figure. This again corresponds to a complicated configuration but it can be split into Fourier components consisting of plane waves in a similar way as in the above case. In Fig. 9.61(b), two of these component wavefronts have been drawn, one traveling to the left and the other inclined at an angle θ. The latter wavefront, which is denoted by $E_{A\theta}$, is focused by lens L_2 at the point *P* on the screen (Fig. 9.59).

The wavefront returning from *B* to the scatter plate is denoted by E_B in Fig. 9.61(c). Upon traversing the scatter plate, it will be reshaped into the wave E_{BT}. One of the Fourier components of this wavefront, denoted by $E_{B\theta}$, is inclined at the angle θ and will therefore be focused at the same point *P* on the screen.

Some of the waves arriving at *P* will be coherent in the sense that interference occurs. To obtain the resultant ir-radiance I_P, first add the amplitudes of all the waves arriving at *P*, i.e. E_P and then square and time average E_P.

In the discussion above, only two point sources at the mirror were considered. Actually, of course, the whole sur-face of the mirror is illuminated by the ongoing light, and every point of it will serve as a secondary source of returning

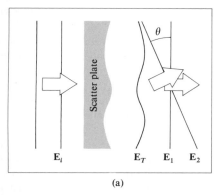
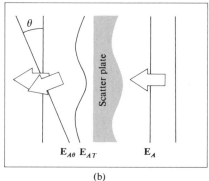
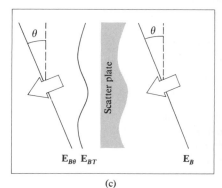

(a)   (b)   (c)

Fig. 9.61   Wavefronts passing through the scatter plate.

waves. All the waves will be deformed by the scatter plate and these, in turn, can be split into plane wave components. In each series of component waves, there will be one inclined at an angle θ and all of these will be focused at the same point P on the screen. The resultant amplitude will then have the form

$$\mathbf{E}_P = \mathbf{E}_{A\theta} + \mathbf{E}_{B\theta} + \cdots.$$

The light reaching the image plane may be envisioned as being made up in part of two optical fields of special interest. One of these results from light which was scattered only on its passage through the plate toward the mirror and the other from light which was scattered only on the way toward the image plane. The former broadly illuminates the test mirror and ultimately results in an image of it on the screen. The latter, which was initially focused to the region about A, scatters a diffuse blur across the screen. The point A is chosen so that the small area in the vicinity of it is free of aberrations. In that case, the wave reflected from it serves as a reference with which to compare the wavefront corresponding to the entire mirror surface. The interference pattern will show, as a series of contour fringes, any deviations from perfection in the mirror surface.[*]

9.10.2 Thin-film Measurements by Multiple-beam Interferometry

Return to Fig. 9.32 and now suppose that the wedge has a step in it. Figure 9.62 illustrates the fringe pattern that might be seen under these circumstances. If the wedge angle is the same for each surface, i.e. if the top surfaces are parallel, the fringes will be equally spaced.

When the separation of the fringes is b, and the shift is a, then the height of the step is given by

$$t = \frac{a}{b}\frac{\lambda_f}{2}.$$

If one of the boundaries of the film is an optical flat and the other boundary is a crystal surface or some other surface examined for flatness, then these Fizeau fringes are contours of the surface under examination.

* For further discussion of the scatter plate, the reader might consult the rather succinct papers by J. M. Burch *Nature* 171, 889 (1953) and *J. Opt. Soc. Am.* 52, 600 (1962). Reference should be made to J. Strong, *Concepts of Classical Optics*, p. 383. Also see R. M. Scott, "Scatter Plate Interferometry," *Appl. Opt.* 8, 531 (1969) and J. B. Houston, Jr., "How to Make and Use a Scatterplate Interferometer," *Optical Spectra* (June 1970), p. 32.

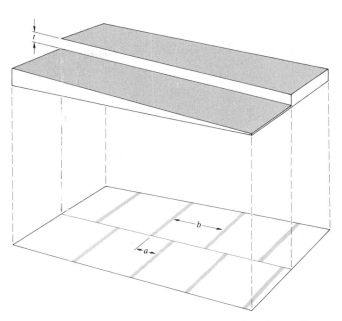

Fig. 9.62 Fringes arising from a stepped wedge-shaped film.

An actual optical system for measuring the thickness of a thin film deposited on a glass substrate, is shown in Fig. 9.63. The film whose thickness is to be determined is coated with an opaque layer of silver, about 70 nm thick, which accurately contours the undersurface. The opposing silvered surfaces generate a sharp multiple-wave Fizeau pattern. The upper plate is tilted slightly to create an air film in the form of Fig. 9.62, so that the same arrangement of fringes is now observed, Fig. 9.64. Film thicknesses of about 2.0 nm can readily be determined in this manner. Such methods yield a resolution in depth comparable to the lateral resolution of an electron microscope. Tolansky, using the multiple-beam techniques which he invented, has measured height changes of 1×10^{-8} inches, very nearly the size of a single atom.

9.10.3 The Michelson–Morley Experiment

The Michelson interferometer has, over the years since 1881, found innumerable applications, most of which are now mainly of historical interest. One of the most significant of these was its use in the Michelson–Morley experiment.

During the last century there was a common belief amongst scientists that there existed a medium, the *luminiferous* (light-carrying) *ether*, which permeated all matter, pervaded all space, was massless and neither solid, liquid, nor

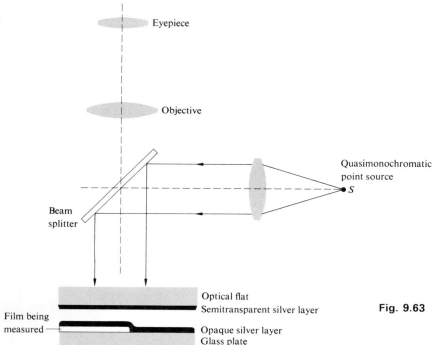

Eyepiece

Objective

Beam
splitter

Quasimonochromatic
point source

• S

Optical flat
Semitransparent silver layer

Film being
measured

Opaque silver layer
Glass plate

Fig. 9.63 Arrangement for measuring film thickness.

gas.   As James Maxwell wrote in the *Encyclopaedia Britannica*:

> Ethers were invented for the planets to swim in, to constitute electric atmospheres and magnetic affluvia, to convey sensations from one part of our bodies to another, and so on, until all space had been filled three or four times over with ethers . . . The only ether which has survived is that which was invented by Huygens to explain the propagation of light.

It was well established that light was a wave, and so it was only natural to have a medium in which the disturbance propagated.   Assuming that much, the nature of the ether had to match terrestrial and astronomical observations. At the time, there was no denying the actual existence of ether; the debate centered on its physical properties.   Was the ether stationary in space, thereby providing a reference frame from which to measure the absolute motion of all other objects? Or, was it dragged along by the planets as they moved through space?   If the ether were stationary an observer on the earth would be able to detect an ether wind streaming over its surface, as it moved in orbit. A. A. Michelson, later joined by E. W. Morley, set out to measure the effects of the ether

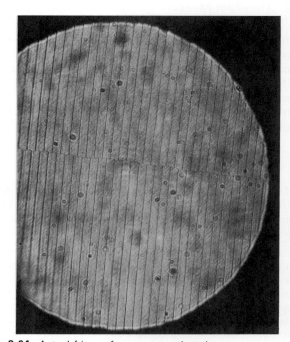

Fig. 9.64 Actual fringes from a stepped wedge.

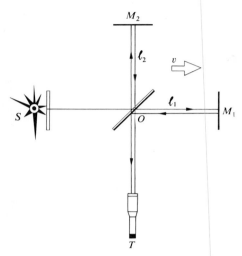

Fig. 9.65 The Michelson—Morley experiment.

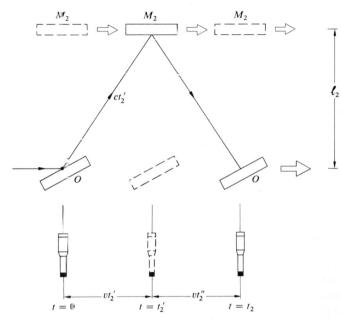

Fig. 9.66 The Michelson—Morley experiment.

wind using his interferometer, which was designed specifically for that purpose. It was oriented as shown in Fig. 9.65, with the arm OM_1 parallel to the velocity **v** of the earth through space. The basic reasoning of the Michelson—Morley approach, derived from purely classical laws of physics, was as follows: when the beam of light travels to the right, its relative speed with respect to the moving interferometer is $c - v$, it is moving against the ether wind, and the time to travel the length OM_1 is

$$t_1' = \frac{\ell_1}{c - v}.$$

For the return trip, M_1O, the beam travels with the ether wind and

$$t_1'' = \frac{\ell_1}{c + v}.$$

The total time, $t_1' + t_1''$, to traverse OM_1O is

$$t_1 = \frac{\ell_1}{c - v} + \frac{\ell_1}{c + v},$$

which can be written as

$$t_1 = \frac{2\ell_1}{c} \beta^2,$$

where

$$\beta = \frac{1}{\sqrt{1 - v^2/c^2}}.$$

The time of travel towards the second mirror can be obtained with the help of Fig. 9.66. From the right triangle, where t_2' is the transit time to cover OM_2,

$$c^2 t_2'^2 = v^2 t_2'^2 + \ell_2^2$$

from which it follows that

$$t_2' = \frac{\ell_2}{c} \beta.$$

But this is also the time t_2'' that it takes the beam of light to return from M_2 to O and so, since $t_2 = t_2' + t_2''$,

$$t_2 = \frac{2\ell_2}{c} \beta.$$

Notice that even when $\ell_1 = \ell_2 = \ell$, $t_1 \neq t_2$ and

$$t_1 - t_2 = \frac{2\ell}{c} (\beta^2 - \beta).$$

Using the binomial expansion with $c \gg v$,

$$\beta^2 = (1 - v^2/c^2)^{-1} = 1 + v^2/c^2$$

and

$$\beta = (1 - v^2/c^2)^{-\frac{1}{2}}$$

or

$$\beta = 1 + \tfrac{1}{2} v^2/c^2.$$

We find that with $\Delta t = t_1 - t_2$

$$\Delta t = \frac{\ell}{c}\left(\frac{v}{c}\right)^2.$$

A time difference Δt in the two paths corresponds to a difference in the number of wavelengths fitting between OM_1O and OM_2O:

$$\Delta N = \Delta t/\tau \qquad \text{or} \qquad \Delta N = \nu\Delta t,$$

where τ is the period and ν the frequency. This is also the number of pairs of fringes (i.e. a maximum and a minimum) which would shift past the telescope crosshairs if a time difference Δt were somehow introduced during the observation. Suppose that the earth were stationary in space and then started moving with a speed v, such that $\Delta N = \frac{1}{2}$. Furthermore, suppose the observer sets the crosshairs initially at the center of a bright fringe. As the earth begins to move, the bright fringe would sweep by, and the crosshairs would shift to the center of the adjacent dark fringe. We cannot, of course, stop the world but we can rotate the interferometer. If the instrument is rotated 90° the new transit time difference can be gotten by just interchanging the 1 and 2 subscripts and is therefore equal to $-\Delta t$. This means that if the observer were to rotate the interferometer 90°, a time difference of $2\,\Delta t$ would be introduced for which, in that example, $\Delta N = 1$, and the crosshairs would end up on the next bright fringe.

This is essentially what Michelson and Morley did do. Their apparatus was multimirrored to make the path length as large as possible, $\ell_1 \approx \ell_2 \approx 11.0$ m. It rested on a massive stone which floated on a trough filled with mercury (Fig. 9.67). Each man took turns slowly pushing the stone, while continuously observing the fringe pattern. Assuming a velocity equal to the earth's orbital speed of about 30 km/s and $\lambda_0 = 550$ nm, the fringe shift on rotation would be

$$\Delta N = \frac{2\ell}{\lambda}\left(\frac{v}{c}\right)^2$$

or

$$\Delta N = 0.4.$$

They made thousands of observations at different hours of the earth's daily cycle and on different days during its yearly orbit. And yet, even though they could have detected a shift of a minute fraction of a fringe, they saw none whatever. There was no ether wind; Michelson and Morley had sounded the prelude to special relativity.

Ten years later, Michelson interferometrically tested the possibility that the ether was being dragged along with the

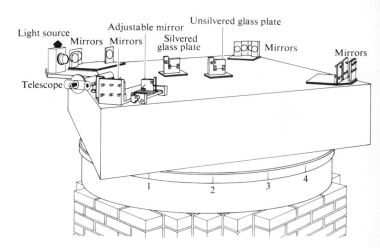

Fig. 9.67 The Michelson—Morley experiment.

earth. His results showed that this too was not true, and the ether theory was doomed.

A modern version of the Michelson—Morley experiment,[*] shown here in Fig. 9.68, compared the frequencies of two infrared lasers. (Recall that in Section 7.2.1, we considered the application of lasers to the problem of generating beats.) The combined beam reaching the photomultiplier, being the resultant of two coplanar harmonic waves, was *amplitude modulated* by a relatively slow variation. These *beats* had a frequency equal to the difference between those of the two constituent laser beams. The precise frequency of the mode in which each laser operated, was governed by the length of the particular resonant cavity, and the speed of light therein. If both lasers, functioning at about 3×10^{14} Hz, were rotated 90°, the *ether wind* would affect the speed of light in the cavities and therefore the frequency difference between them. A relative change in ν of 3 MHz would be expected from the ether wind hypothesis, due to the earth's orbital velocity. No change in the beat frequency, to within an accuracy of 3 kHz or $\frac{1}{1000}$ of that predicted, was detected.

9.10.4 The Twyman—Green Interferometer

The Twyman—Green is essentially a variation of the Michelson interferometer. It's an instrument of very great importance in the domain of modern optical testing. Amongst its

---

* T. S. Jaseja, A. Javan, J. Murray, and C. H. Townes, "Test of Special Relativity or of the Isotropy of Space by Use of Infrared Masers," *Phys. Rev.* **133**, A1221 (1964).

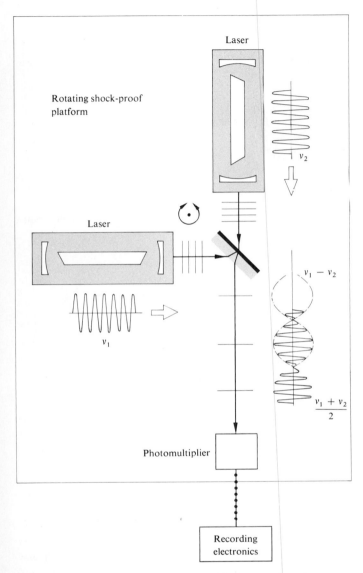

Fig. 9.68 A variation of the Michelson–Morley experiment.

times. These tend to minimize unwanted vibration effects. Laser versions of the Twyman–Green are among the most effective testing tools in optics. As shown in the figure, the device is set up to examine a lens. The spherical mirror M_2 has its center of curvature coincident with the focal point of the lens. If the lens being tested is free of aberrations, the emerging reflected light returning to the beam splitter will again be a plane wave. If however, astigmatism, coma, or spherical aberration deform the wavefront, a fringe pattern clearly manifesting these distortions will be seen and can be photographed. When M_2 is replaced by a plane mirror, a number of other elements, e.g. prisms, optical flats, etc. can be tested equally well. The optician interpreting the fringe pattern can then mark the surface for further polishing to correct high or low spots. In the fabrication of the very finest optical systems, telescopes, high altitude cameras, etc., the interferograms may even be scanned electronically and the resulting data computer analyzed. Computer-controlled plotters can then automatically produce surface contour maps or perspective "three-dimensional" drawings of the distorted wavefront generated by the element being tested. These procedures can be used throughout the fabrication process to ensure the highest-quality optical instruments. Complex systems with wavefront aberrations in the fractional-wavelength range are the result of what might be called the *new technology*.*

9.11 THE ROTATING SAGNAC INTERFEROMETER

The application of the Sagnac interferometer to the task of measuring the rotational speed of a system has become of interest in recent times. In particular the *ring laser*, which is essentially a Sagnac interferometer containing a laser in one or more of its arms, was designed specifically for that purpose. The first ring laser gyroscope was introduced in 1963 and work is continuing on various devices of this sort (Fig. 9.70). The initial experiments which gave impetus to these efforts were performed by Sagnac in 1911. At that time he rotated the entire interferometer, mirrors, source, and detector, about a perpendicular axis passing through its center (Fig. 9.71). Recall, from Section 9.4, that two overlapping beams traverse the interferometer, one clockwise, the other counterclockwise. The rotation effectively shortens the path taken by one beam in comparison to that of the other. In the interferometer the result is a fringe shift proportional to the angular

distinguishing physical characteristics are (as illustrated in Fig. 9.69) a quasimonochromatic point source and lens L_1 to provide a source of incoming *plane waves*, and a lens L_2 which permits all of the light from the aperture to enter the eye so that the entire field can be seen, i.e., any portion of M_1 and M_2. A continuous laser serves as a superior source in that it provides both the convenience of long path-length differences and, in addition, short photographic exposure

*Take a look at R. Berggren, "Analysis of Interferograms," *Optical Spectra*, (Dec. 1970), p. 22.

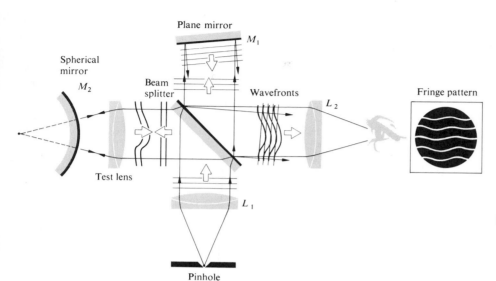

Spherical mirror

M_2

Plane mirror

M_1

Beam splitter

Wavefronts

L_2

Fringe pattern

Test lens

L_1

Pinhole

Fig. 9.69 The Twyman–Green interferometer.

Fig. 9.70 A ring laser gyro. [Photo courtesy Autonetics, a Division of North American Rockwell Corp.]

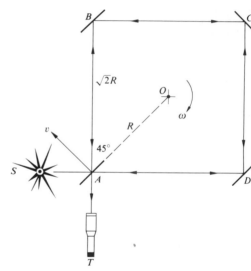

Fig. 9.71 The rotating Sagnac interferometer. Originally it was 1 m × 1 m with $\omega = 120$ rev/min.

speed of rotation ω. In the ring laser, it is a frequency difference between the two beams which is proportional to ω.

Consider the arrangement depicted in Fig. (9.71). The corner A (and every other corner) moves with a linear speed $v = R\omega$, where R is half the diagonal of the square. Using classical reasoning we find that the time of travel of light along AB is

$$t_{AB} = \frac{R\sqrt{2}}{v/\sqrt{2} + c}$$

or

$$t_{AB} = \frac{2R}{\omega R + \sqrt{2} c}.$$

The time of travel of the light from A to D is

$$t_{AD} = \frac{2R}{\sqrt{2} c - \omega R}.$$

The total time for counterclockwise and clockwise travel is given respectively by

$$t_\circlearrowleft = \frac{8R}{\sqrt{2} c - \omega R}$$

and

$$t_\circlearrowright = \frac{8R}{\sqrt{2} c + \omega R}.$$

For $\omega R \ll c$ the difference between these two intervals is

$$\Delta t = t_\circlearrowleft - t_\circlearrowright$$

or, using the binomial series

$$\Delta t = \frac{8R^2 \omega}{c^2}.$$

This can be expressed in terms of the area $A = 2R^2$ of the square formed by the beams of light as

$$\Delta t = \frac{4A\omega}{c^2}.$$

Let the period of the monochromatic light used be $\tau = \lambda/c$; then the fractional displacement of the fringes, given by $\Delta N = \Delta t/\tau$, is

$$\Delta N = \frac{4A\omega}{c\lambda},$$

a result which has been verified experimentally. In particular, Michelson and Gale* used this method to determine the angular velocity of the earth.

The preceding classical treatment is obviously lacking inasmuch as it assumes speeds in excess of c which is contrary to the dictates of special relativity. Furthermore, it would appear that since the system is accelerating, general relativity would prevail. In fact, all of these formalisms yield the same results.

PROBLEMS

9.1 Returning to Section 9.1 let

$$\mathbf{E}_1(\mathbf{r}, t) = \mathbf{E}_1(\mathbf{r}) e^{-i\omega t}$$

---

* Michelson and Gale, *Astrophys. J.* **61**, 140 (1925).

and

$$\mathbf{E}_2(\mathbf{r}, t) = \mathbf{E}_2(\mathbf{r}) e^{-i\omega t},$$

where the wavefront shapes are not explicitly specified and $\mathbf{E}_1$ and $\mathbf{E}_2$ are complex vectors depending on space and epoch angle. Show that the interference term is then given by

$$I_{12} = \tfrac{1}{2}(\mathbf{E}_1 \cdot \mathbf{E}_2^* + \mathbf{E}_1^* \cdot \mathbf{E}_2). \qquad (9.86)$$

You will have to evaluate terms of the form

$$\langle \mathbf{E}_1 \cdot \mathbf{E}_2 e^{-2i\omega t} \rangle = \frac{\mathbf{E}_1 \cdot \mathbf{E}_2}{T} \int_t^{t+T} e^{-2i\omega t'} \, dt'$$

for $T \gg \tau$ (take another look at Problem 3.8). Show that Eq. (9.86) leads to Eq. (9.4) for plane waves.

9.2 In Section 9.1 we considered the spatial distribution of energy for two point sources. We mentioned that for the case where the separation $a \gg \lambda$, I_{12} spatially averages to zero. Why is this true? What happens when a is much less than λ?

9.3 Will we get an interference pattern in Young's experiment (Fig. 9.5) if we replace the source slit S by a single long-filament light bulb? What would occur if we replaced the slits S_1 and S_2 by these same bulbs?

9.4 To examine the conditions under which the approximations of Eq. (9.9) are valid:

a) Apply the law of cosines to triangle $S_1 S_2 P$ in Fig. 9.5(c) to get

$$\frac{r_2}{r_1} = \left[1 - 2\left(\frac{a}{r_1}\right) \sin \theta + \left(\frac{a}{r_1}\right)^2 \right]^{1/2}.$$

b) Expand this in a Maclaurin series yielding

$$r_2 = r_1 - a \sin \theta + \frac{a^2}{2r_1} \cos^2 \theta + \cdots.$$

c) In light of Eq. (9.6), show that if $(r_1 - r_2)$ is to equal $a \sin \theta$ it is required that $r_1 \gg a^2/\lambda$.

9.5* Show that a for the Fresnel biprism of Fig. 9.9 is given by $a = 2d(n - 1)\alpha$.

9.6* In the Fresnel double mirror $s = 2$ m, $\lambda_0 = 589$ nm, and the separation of the fringes was found to be 0.5 mm. What is the angle of inclination of the mirrors if the perpendicular distance of the actual point source to the intersection of the two mirrors is 1 m?

9.7* The Fresnel biprism is used to obtain fringes from a point source which is placed 2 m from the screen and the

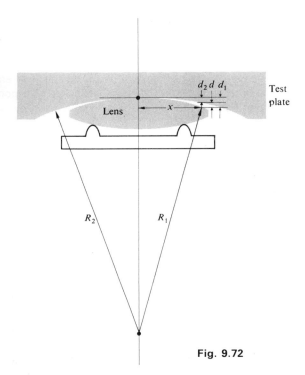

Fig. 9.72

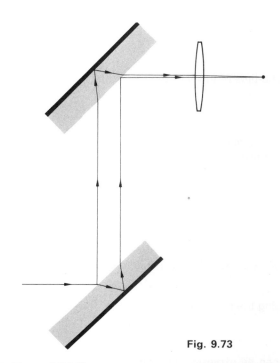

Fig. 9.73

prism is midway between the source and the screen. Let the light have the wavelength $\lambda_0 = 500$ nm, and the index of refraction of the glass be $n = 1.5$. What is the prism angle if the separation of the fringes is 0.5 mm?

9.8 What is the general expression for the separation of the fringes of a Fresnel biprism of index n immersed in a medium having an index of refraction n'?

9.9 Using Lloyd's mirror, x-ray fringes were observed, the spacing of which was found to be 0.0025 cm. The wavelength used was 8.33 Å. If the source–screen distance is 3 m, how high above the mirror plane was the point source of x-rays placed?

9.10* If the plate in Fig. 9.27 is glass in air, show that the amplitudes of E_{1r}, E_{2r}, and E_{3r} are respectively $0.2E_{0i}$, $0.192 E_{0i}$ and $0.008 E_{0i}$ where E_{0i} is the incident amplitude. Make use of the Fresnel coefficients at normal incidence assuming no absorption. You might repeat the calculation for a water film in air.

9.11 Consider the circular pattern of Haidinger's fringes resulting from a film of thickness 2 mm and index of refraction 1.5. For monochromatic illumination of $\lambda_0 = 600$ nm, find the order of the central fringe ($\theta_t = 0$). Will it be bright or dark?

9.12 Figure 9.72 illustrates a set-up used for testing lenses. Show that

$$d = x^2(R_2 - R_1)/2R_1R_2$$

when d_1 and d_2 are negligible in comparison with $2R_1$ and $2R_2$, respectively. (Recall the theorem from plane geometry which relates the products of the segments of intersecting chords.) Prove that the radius of the mth dark fringe is then

$$x_m = [R_1R_2m\lambda_f/(R_1 - R_2)]^{1/2}.$$

How does this relate to Eq. (9.27)?

9.13 Draw the configuration which you would use to see Newton's rings in a Twyman–Green interferometer.

9.14* Newton rings are observed with quasimonochromatic light of wavelength 500 nm. If the 20th bright ring has a radius of 1 cm, what is the radius of curvature of the lens forming one part of the interfering system?

9.15 Fringes are observed when a parallel beam of light of wavelength 500 nm is incident perpendicularly onto a wedge-shaped film of index of refraction 1.5. What is the angle of the wedge if the fringe separation is $\frac{1}{3}$ cm?

9.16* A form of the Jamin interferometer is illustrated in Fig. 9.73. How does it work? To what use might it be put?

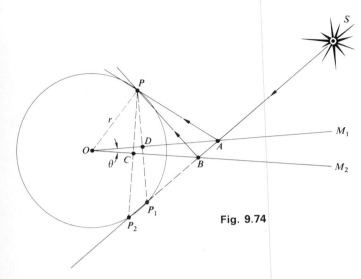

Fig. 9.74

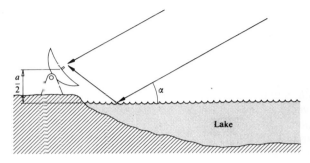

Fig. 9.75

9.17* In Searle's method of localization of fringes [*Phil. Mag.* **37**, 361 (1946)] (Fig. 9.74) the two film boundary planes $\overline{OM}_1$ and $\overline{OM}_2$ reflect a beam coming from the source. Show that the path difference for the two rays is $2r \sin \theta$ and discuss the usefulness of this construction.

9.18 Starting with Eq. (9.31) for the transmitted wave, compute the flux density, i.e. Eq. (9.32).

9.19* Satisfy yourself of the fact that a film of thickness $\lambda_f/4$ and index n_1 will always reduce the reflectance of the substrate on which it is deposited as long as $n_s > n_1 > n_0$. Consider the simplest case of normal incidence and $n_0 = 1$.

Show that this is equivalent to saying that the waves reflected back from the two interfaces cancel.

9.20 Verify that the reflectance of a substrate can be increased by coating it with a $\lambda_f/4$, high-index layer i.e. $n_1 > n_s$. Show that the reflected waves interfere constructively. The quarter-wave stack $g(HL)^m Ha$ can be thought of as a series of such structures.

9.21 Imagine that we have an antenna at the edge of a lake picking up a signal from a distant radio star (Fig. 9.75) which is just coming up above the horizon. Write expressions for δ and for the angular position of the star when the antenna detects its first maximum.

9.22 Determine the refractive index and thickness of a film to be deposited on a glass surface ($n_g = 1.54$) such that no normally incident light of wavelength 540 nm is reflected.

9.23* Figure 9.76 is a Newton's ring fringe pattern generated by a lens and an optical flat under $\lambda_0 = 600$ nm illumination.

Draw a possible contour map of the lens surface. Is it possible, with this figure alone, to determine whether the lens surface is concave or convex? Explain.

9.24 Consider the interference pattern of the Michelson interferometer as arising from two equal flux density beams. Using Equation (9.6) compute the half-width. What is the separation, in δ, between adjacent maxima? What then is the finesse?

9.25 Given that the mirrors of a Fabry–Perot interferometer have an amplitude reflection coefficient of $r = 0.8944$, find

a) the coefficient of finesse,

b) the half-width,

Fig. 9.76

c) the finesse, and,

d) the *contrast factor* defined by

$$C \equiv \frac{(I_t/I_i)_{max}}{(I_t/I_i)_{min}}.$$

9.26 Illuminate a microscope slide (or even better a thin cover glass slide). Colored fringes can easily be seen with an ordinary fluorescent lamp serving as a broad source, or a mercury street light as a point source. Describe the fringes. Now rotate the glass. Does the pattern change? Duplicate the conditions shown in Figs. 9.28 and 9.29. Try it again with a sheet of plastic food wrap stretched across the top of a cup.

9.27*In order to get some feeling for the need for antireflection coatings, consider the following somewhat over-simplified problem. If there is normal incidence, no absorption and $n = 1.5$, what fraction of the incident light will be transmitted through a twelve-element lens system?

9.28* Discuss the influence of the linewidth $\Delta\lambda$ of the source on the sharpness of fringes in an interferometer. Show that the fringes will disappear when $\Delta\lambda \approx \lambda/m$ where m, the order of interference, is a large number. Determine the order for which the fringes first vanish with the red cadmium line of Fig. 7.18. Take a look at Section 12.2 for further study.

9.29 To fill in some of the details in the derivation of the smallest phase increment separating two resolvable Fabry–Perot fringes, that is

$$(\Delta\delta) \approx 4.2/\sqrt{F}, \qquad\qquad [9.51]$$

satisfy yourself that

$$[\mathscr{A}(\theta)]_{\delta = \delta_a \pm \Delta\delta/2} = [\mathscr{A}(\theta)]_{\delta = \Delta\delta/2}.$$

Show that Eq. (9.50) can be rewritten as

$$2[\mathscr{A}(\theta)]_{\delta = \Delta\delta/2} = 0.81\{1 + [\mathscr{A}(\theta)]_{\delta = \Delta\delta}\}.$$

When F is large γ is small, and $\sin(\Delta\delta) \approx \Delta\delta$. Prove that Eq. (9.51) then follows.

9.30 A stream of electrons, each having an energy of 0.5 eV, impinges on a pair of extremely thin slits separated by 10^{-2} mm. What is the distance between adjacent minima on a screen 20 m behind the slits? ($m_e = 9.108 \times 10^{-31}$ kg, 1 eV $= 1.602 \times 10^{-19}$ J).

Diffraction 10

10.1 PRELIMINARY CONSIDERATIONS

An opaque body placed midway between a screen and a point source casts an intricate shadow made up of bright and dark regions quite unlike anything one might expect from the tenets of geometrical optics (Fig. 10.1).* The work of Francesco Grimaldi in the sixteen hundreds was the first published detailed study of this *deviation of light from rectilinear propagation*, something he called *"diffractio"*. *The effect is a general characteristic of wave phenomena occurring whenever a portion of a wavefront, be it sound, a matter wave or light, is obstructed in some way.* If in the course of encountering an obstacle, either transparent or opaque, a region of the wavefront is altered in amplitude or phase, diffraction will occur.† The various segments of the wavefront which propagate beyond the obstacle interfere to cause the particular energy-density distribution referred to as the diffraction pattern. There is no significant physical distinction between *interference* and *diffraction*. It has however become somewhat customary, if not always appropriate, to speak of interference when considering the superposition of only a few waves and diffraction when

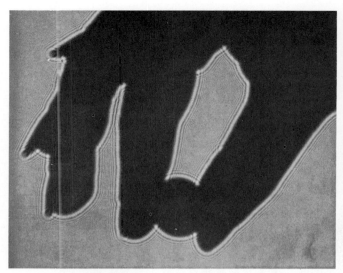

Fig. 10.1 The shadow of a hand holding a dime, cast directly on 4 × 5 Polaroid A.S.A. 3000 film using a He—Ne beam and no lenses. [Photo by E.H.]

treating a large number of waves. Even so, one refers to multiple-beam interference in one context and diffraction from a grating in another.

We might mention parenthetically that the wave theory, although the most natural, is not the only means for dealing with certain diffraction phenomena. For example, diffraction from a grating (Section 10.2.7) can be analyzed using a corpuscular quantum approach.* For our purposes however, the classical wave theory which provides the simplest effective formalism will more than suffice throughout this chapter.

It should be emphasized that optical instruments make use of only a portion of the complete incident wavefront. Diffraction effects are accordingly of great significance in the detailed understanding of devices containing lenses, stops, source slits, mirrors, etc. If all defects in a lens system were removed, the ultimate sharpness of an image would be limited by diffraction.

As an initial approach to the problem, let's reconsider Huygens' principle (Section 4.2.1). Accordingly, each point on a wavefront can be envisaged as a source of secondary spherical wavelets. The progress through space of the wavefront or any portion thereof can then presumably be determined. At any particular time, the shape of the wave-

---

* The effect is easily seen but you need a fairly strong source. A high-intensity lamp shining through a small hole works well. If you look at the shadow pattern arising from a pencil under point-source illumination, you will see an unusual bright region bordering the edge and even a faintly illuminated band down the middle of the shadow. Take a close look at the shadow cast by your hand in direct sunlight.

† Diffraction associated with transparent obstacles is not usually considered, although if you have ever driven an automobile at night with a few rain droplets on your eye glasses you are no doubt quite familiar with the effect. If you have not, put a droplet of water or saliva on a glassplate, hold it very close to your eye, and look directly through it at a point source. You'll see bright and dark fringes.

---

* W. Duane, *Proc. Nat. Acad. Sci.* **9**, 158 (1923).

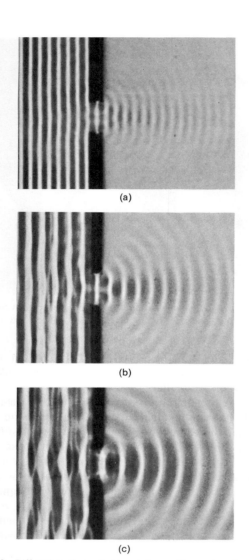

(a)

(b)

(c)

Fig. 10.2 Diffraction through an aperture with varying λ as seen in a ripple tank. [Photo courtesy PSSC *Physics*, D. C. Heath, Boston, 1960.]

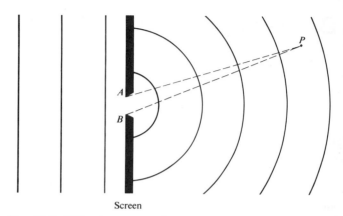

Screen

Fig. 10.3 Diffraction at a small aperture.

front is supposed to be the envelope of the secondary wavelets (Fig. 4.3). The technique however ignores most of each secondary wavelet, retaining only that portion common to the envelope. As a result of this inadequacy, Huygens' principle is unable to account for the diffraction process. That this is indeed the case is borne out by everyday experience. Sound waves (e.g. $\nu = 500$ Hz, $\lambda \approx 68$ cm) easily "bend" around large objects like telephone poles and

trees which on the contrary cast fairly distinct shadows when illuminated by light. Yet Huygens' principle is independent of any wavelength considerations and would predict the same wavefront configurations in both situations. The difficulty was resolved by Fresnel with his addition of the concept of interference. The corresponding *Huygens–Fresnel* principle states that *every unobstructed point of a wavefront, at a given instant in time, serves as a source of spherical secondary wavelets (of the same frequency as the primary wave). The amplitude of the optical field at any point beyond is the superposition of all of these wavelets (considering their amplitudes and relative phases).* Applying these ideas on the very simplest qualitative level, refer to the ripple tank photographs of Fig. 10.2 and the illustration in Fig. 10.3. If each unobstructed point on the incoming plane wave acts as a coherent secondary source, the maximum optical path-length difference amongst these will be $\Lambda_{max} = |\overline{AP} - \overline{BP}|$, corresponding to a source point at each edge of the aperture. But Λ_{max} is less than or equal to $\overline{AB}$, the latter being the case when P is on the screen. When $\lambda \gg \overline{AB}$, as in Fig. 10.3, it follows that $\lambda \gg \Lambda_{max}$ and since the waves were initially in phase, they must all interfere constructively (to varying degrees) wherever P happens to be [see Fig. 10.2(c)]. The antithetic situation occurs when $\lambda \ll \overline{AB}$ as in Fig. 10.2(a). Now the area where $\lambda \gg \Lambda_{max}$ is limited to a small region extending out directly in front of the aperture and it is only there that all of the wavelets will interfere constructively. Beyond this zone some of the wavelets can interfere destructively and the "shadow" begins. Keep in mind that the idealized *geometrical shadow* corresponds to $\lambda \to 0$.

The Huygens—Fresnel principle has some shortcomings (which we will examine later) in addition to the fact that the whole thing at this point is rather hypothetical. Gustav Kirchhoff developed a more rigorous theory based directly on the solution of the differential wave equation. Kirchhoff, although a contemporary of Maxwell, did his work prior to Hertz's demonstration (and the resulting popularization) of the propagation of electromagnetic waves in 1887. Accordingly, Kirchhoff employed the older elastic-solid theory of light. His refined analysis lent credence to the assumptions of Fresnel and led to an even more precise formulation of Huygens' principle as an exact consequence of the wave equation. Even so, the Kirchhoff theory is itself an approximation which is valid for sufficiently small wavelengths, i.e. when the diffracting apertures have dimensions which are large in comparison to λ. The difficulty arises from the fact that we require the solution of a partial differential equation which meets the boundary conditions imposed by the obstruction. This kind of rigorous solution is only obtainable in a very few special cases. Kirchhoff's theory however works fairly well even though it deals only with scalar waves and is insensitive to the fact that light is a transverse vector field.*

It should be stressed that the problem of determining an exact solution for a particular diffracting configuration is amongst the most troublesome to be dealt with in optics. The first such solution, utilizing the electromagnetic theory of light, was published by Arnold Johannes Wilhelm Sommerfeld (1868–1951) in 1896. But there the problem was physically somewhat unrealistic in that it involved an infinitely thin yet opaque, perfectly conducting, plane screen. The result was nonetheless extremely valuable, providing a good deal of insight into the fundamental processes involved.

Rigorous solutions of this sort do not exist even today for many of the configurations of practical interest. We will therefore, out of necessity, rely on the approximate treatments of Huygens—Fresnel and Kirchhoff. In recent times, microwave techniques have been employed to conveniently study features of the diffraction field which might otherwise be almost impossible to examine optically. The Kirchhoff theory has held up remarkably well under this kind of

scrutiny.* In many cases, the simpler Huygens—Fresnel treatment will prove quite adequate for our purposes.

10.1.1 Opaque Obstructions

Diffraction may be envisioned as arising from the interaction of electromagnetic waves with some sort of physical obstruction. We would do well therefore to briefly re-examine the processes involved, i.e. what actually takes place within the material of the opaque object?

One possible description is that a screen may be considered to be a continuum, i.e. its microscopic structure may be neglected. For a nonabsorbing metal sheet (no joule heating, therefore infinite conductivity) we can write Maxwell's equations in the metal, and in the surrounding medium, and then match the two at the boundaries. Precise solutions can thus be obtained for very simple configurations (A. Sommerfeld). The reflected and diffracted waves then result from the current distribution within the sheet.

Now, examining the screen on a submicroscopic scale, imagine the electron cloud of each atom set into vibration by the electric field of the incident radiation. The classical model, which speaks of electron-oscillators vibrating and reemitting at the source frequency (Section 3.3.1) serves quite well so that we need not be concerned with the quantum-mechanical description. The amplitude and phase of a particular oscillator within the screen is determined by the local electric field surrounding it. This in turn is a superposition of the incident field and the fields of all of the other vibrating electrons. A large opaque screen with no apertures, be it made of black paper or aluminum foil, has one obvious effect: there is no optical field in the region beyond it. Electrons near the illuminated surface are driven into oscillation by the impinging light. They emit radiant energy which is ultimately reflected backward, or absorbed by the material in the form of heat, or both. In any case, the incident primary wave and the electron-oscillator fields superimpose in such a way as to yield zero light at any point beyond the screen. This might seem a remarkably special balance, but it actually is not. If the primary wave were not canceled completely it would propagate deeper into the material of the screen, exciting more electrons to radiate. This in turn would further weaken the primary wave until it ultimately vanished (if the screen were thick enough). Even an opaque material like silver, in the form of a sufficiently thin sheet, is transparent (recall the half-silvered mirror).

---

* A vectorial formulation of the scalar Kirchhoff theory is discussed in J. D. Jackson, *Classical Electrodynamics*, p. 283. Also see Sommerfeld, *Optics*, p. 325. You might as well take a look at B. B. Baker and E. T. Copson, *The Mathematical Theory of Huygens' Principle*, as a general reference to diffraction. None of these texts is easy reading.

* C. L. Andrews, *Am. J. Phys.* **19**, 250 (1951); S. Silver, *J. Opt. Soc. Am.* **52**, 131 (1962).

Fig. 10.4 Ripple-tank photos. In one case the waves are simply diffracted by a slit, in the other a series of equally spaced point sources span the aperture and generate a similar pattern. [Photos courtesy PSSC *Physics*, D.C. Heath, Boston, 1960.]

Now, remove a small disk-shaped segment from the center of the screen so that light streams through the aperture. The oscillators which uniformly cover it are removed along with the disk and so the remaining electrons within the screen are no longer affected by them. As a first and certainly approximate approach, *assume that the mutual interaction of the oscillators is essentially negligible*, i.e. the electrons in the screen are completely unaffected by the removal of the electrons in the disk. The field in the region beyond the aperture will then be that which existed prior to the removal of the disk, namely zero, minus the contribution from the disk alone. Except for the sign, it is as if the source and screen were taken away leaving only the oscillators on the disk, rather than vice versa. In other words, the diffraction field, in this approximation, can be pictured as arising exclusively from a set of fictitious noninteracting oscillators distributed uniformly over the region of the aperture. This, of course, is the essence of the Huygens—Fresnel principle.

We can expect, however, that rather than there being no interaction at all between electron-oscillators, there is a short-range effect, since the oscillator fields drop off with distance. In this physically more realistic view, the electrons within the vicinity of the aperture's edge are affected when the disk is removed. For large apertures, the number of oscillators in the disk is very much greater than those along the edge. In such cases, if the point of observation is far away, and in the forward direction, the Huygens—Fresnel principle should, and does, work well (Fig. 10.4). For very small apertures, or at points of observation in the vicinity of the aperture, edge effects become important and we can anticipate difficulties. Indeed, at a point within the aperture itself, the electron-oscillators on the edge are of the greatest significance because of their proximity. Yet these electrons were certainly not unaffected by the removal of the adjacent oscillators of the disk. Thus, the deviation from the Huygens—Fresnel principle should be appreciable.

10.1.2 Fraunhofer and Fresnel Diffraction

Imagine that we have an opaque shield, Σ, containing a single small aperture which is being illuminated by plane waves from a very distant point source, S. The plane of observation σ is a screen parallel with, and very close to, Σ. Under these conditions an image of the aperture is projected onto the screen which is clearly recognizable despite some slight fringing around its periphery. As the plane of observation moves further away from Σ, the image of the aperture, although still easily recognizable, becomes increasingly more structured as the fringes become more prominent. The phenomenon observed is known as *Fresnel* or *near-field* diffraction. Slowly moving the plane of observation out still further results in a continuous change in the fringes. At a very great distance from Σ the projected pattern will have spread out considerably, bearing little or no resemblance to the actual aperture. Thereafter moving σ essentially only

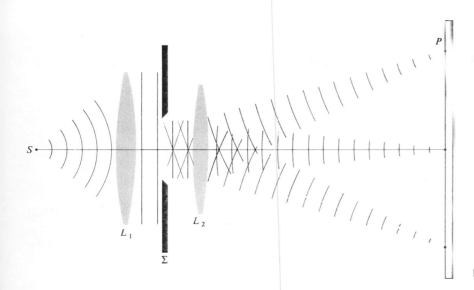

Fig. 10.5 Fraunhofer diffraction.

changes the size of the pattern and not its shape. This is *Fraunhofer* or *far-field* diffraction. If at that point we could sufficiently reduce the wavelength of the incoming radiation, the pattern would revert to the Fresnel case. If λ were decreased even more, so that it approached zero, the fringes would disappear and the image would take on the limiting shape of the aperture as predicted by geometrical optics. Returning to the original set up, if the point source were now moved toward Σ, spherical waves would impinge on the aperture and a Fresnel pattern would exist, even on a distant plane of observation.

In other words, consider a point source S, and a point of observation P, where both are very far from Σ and no lenses are present (Problem 10.1). As long as both the incoming and outgoing waves approach being planar (differing therefrom by a small fraction of a wavelength) over the extent of the diffracting apertures (or obstacles), Fraunhofer diffraction obtains. When S or P or both are too near Σ for the curvature of the wavefronts to be negligible, Fresnel diffraction prevails. The distinction will be made analytically more precise later when we consider the mathematical details.

A practical realization of the Fraunhofer condition, where both S and P are effectively at infinity, is achieved by using an equivalent arrangement to that of Fig. 10.5. The point source S is located at F_1, the principal focus of lens L_1, and the plane of observation is the second focal plane of L_2. In the terminology of geometrical optics the source plane and σ are conjugate planes.

These same ideas can be generalized to any lens system forming an image of an extended source or object (Problem 10.19).* Indeed, that image would be a Fraunhofer diffraction pattern. It is because of these important practical considerations, as well as its inherent simplicity, that we will examine Fraunhofer prior to Fresnel diffraction, even though it is a special case of the latter.

10.1.3 Several Coherent Oscillators

As a simple yet logical bridge between the studies of interference and diffraction, consider the arrangement of Fig. 10.6. The illustration depicts a linear array of N coherent point oscillators (or radiating antennas), which are each identical even to their polarization. For the moment, consider the oscillators to have no intrinsic phase difference, i.e. they each have the same epoch angle. The rays shown are all almost parallel, meeting at some very distant point P. If the spatial extent of the array is comparatively small, the separate wave amplitudes arriving at P will be essentially equal,

---

* A helium–neon laser can be set up to generate magnificent patterns without any auxiliary lenses but this needs plenty of space. If the plane of observation σ is very far from Σ, the diffraction pattern at some point P may be envisioned as arising from the superposition of plane waves coming from every point of the aperture and traveling in the direction of P. By using a large well-corrected lens (L_2) the diffracted waves can be focused on a nearby plane without substantially changing the shape of the pattern. Moreover, with the lens present the incident wave need not be planar so long as the optical path lengths for all rays, from the aperture to P, are essentially equal.

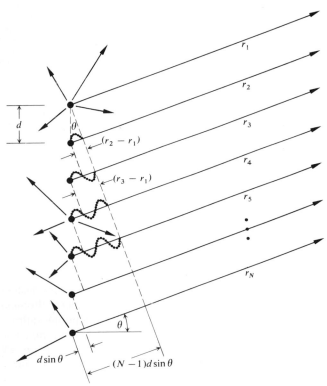

Fig. 10.6 A linear array of in-phase coherent oscillators. Note that at the angle shown $\delta = \pi$ while at $\theta = 0$ δ would be zero.

having traveled nearly equal distances, that is

$$E_0(r_1) = E_0(r_2) = \cdots = E_0(r_N) = E_0(r).$$

The sum of the interfering spherical wavelets yields an electric field at P, given by the real part of

$$E = E_0(r)e^{i(kr_1 - \omega t)} + E_0(r)e^{i(kr_2 - \omega t)} + \cdots + E_0(r)e^{i(kr_N - \omega t)}.$$

(10.1)

It should be clear, from Section 9.1, that we need not be concerned with the vector nature of the electric field for this configuration. Now then

$$E = E_0(r)e^{-i\omega t}e^{ikr_1}[1 + e^{ik(r_2 - r_1)} + e^{ik(r_3 - r_1)} + \cdots + e^{ik(r_N - r_1)}].$$

The phase difference between adjacent sources is obtained from the expression $\delta = k_0\Lambda$ and since $\Lambda = nd\sin\theta$, in a medium of index n, $\delta = kd\sin\theta$. Making use of Fig. 10.6, it follows that $\delta = k(r_2 - r_1)$, $2\delta = k(r_3 - r_1)$ etc. Thus the field at P may be written as

$$E = E_0(r)e^{-i\omega t}e^{ikr_1}[1 + (e^{i\delta}) + (e^{i\delta})^2 + (e^{i\delta})^3 + \cdots (e^{i\delta})^{N-1}].$$

(10.2)

The bracketed geometric series has the value

$$(e^{i\delta N} - 1)/(e^{i\delta} - 1)$$

which can be rearranged into the form

$$\frac{e^{iN\delta/2}[e^{iN\delta/2} - e^{-iN\delta/2}]}{e^{i\delta/2}[e^{i\delta/2} - e^{-i\delta/2}]}$$

or equivalently

$$e^{i(N-1)\delta/2}\left[\frac{\sin N\delta/2}{\sin \delta/2}\right].$$

The field then becomes

$$E = E_0(r)e^{-i\omega t}e^{i[kr_1 + (N-1)\delta/2]}\left(\frac{\sin N\delta/2}{\sin \delta/2}\right).$$

(10.3)

Notice that if we define R to be the distance from the center of the line of oscillators to the point P, that is

$$R = \tfrac{1}{2}(N-1)d\sin\theta + r_1,$$

then Eq. (10.3) takes on the form

$$E = E_0(r)e^{i(kR - \omega t)}\left(\frac{\sin N\delta/2}{\sin \delta/2}\right).$$

(10.4)

Finally, then, the flux-density distribution within the diffraction pattern due to N coherent, identical, distant point sources in a linear array is proportional to $EE^*/2$ for complex E or

$$I = I_0 \frac{\sin^2 (N\delta/2)}{\sin^2 (\delta/2)},$$

(10.5)

where I_0 is the flux density from any single source arriving at P (see Problem 10.2 for a graphical derivation of the irradiance). For $N = 0$, $I = 0$, for $N = 1$, $I = I_0$, and for $N = 2$, $I = 4I_0 \cos^2 (\delta/2)$ in accord with Eq. (9.6). The functional dependence of I on θ is more apparent in the form

$$I = I_0 \frac{\sin^2 [N(kd/2) \sin \theta]}{\sin^2 [(kd/2) \sin \theta]}.$$

(10.6)

The $\sin^2 [N(kd/2) \sin \theta]$ term undergoes rapid fluctuations, while the function modulating it, $\{\sin [(kd/2) \sin \theta]\}^{-2}$, varies relatively slowly. The combined expression gives rise to a series of sharp principal peaks separated by small subsidiary maxima. The principal maxima occur in directions θ_m such that $\delta = 2m\pi$ where $m = 0, \pm 1, \pm 2, \ldots$. Because $\delta = kd\sin\theta$

$$d\sin\theta_m = m\lambda.$$

(10.7)

Since $[\sin^2 N\delta/2]/[\sin^2 \delta/2] = N^2$ for $\delta = 2m\pi$ (from

Fig. 10.7 Interferometric radio telescope at the University of Sydney, Australia ($N = 32$, $\lambda = 21$ cm, $d = 7$ m, 2 m diameter, 700 ft. east–west base line). [Photo courtesy of Prof. W. N. Christiansen.]

L'Hôpital's rule) the principal maxima have values $N^2 I_0$. This is to be expected inasmuch as all of the oscillators are in phase at that orientation. The system will radiate a maximum in a direction perpendicular to the array ($m = 0$, $\theta_0 = 0$ and π). As θ increases, δ increases and I falls off to zero at $N\delta/2 = \pi$, its first minimum. Note that if $d < \lambda$ in Eq. (10.7), only the $m = 0$ or zero-order principal maximum exists. *If we were looking at an idealized line source of electron-oscillators separated by atomic distances, we could expect only that one principal maximum in the light field.*

The antenna array of Fig. 10.7 can then transmit radiation in the narrow beam or lobe corresponding to a principal maximum (the parabolic dishes shown reflect into the forward direction and the radiation pattern is no longer symmetrical around the common axis.) Suppose that we have a system in which we can introduce an intrinsic phase shift of ε between adjacent oscillators. In that case

$$\delta = kd \sin \theta + \varepsilon;$$

the various principal maxima will occur at new angles

$$d \sin \theta_m = m\lambda - \varepsilon/k.$$

Concentrating on the central maximum $m = 0$, its orientation θ_0 can be varied at will by merely adjusting the value of ε.

The principle of reversibility, which states that without absorption, wave motion is reversible, leads to the same field pattern for an antenna used as either a transmitter or receiver. The array, functioning as a radio telescope, can therefore be "pointed" by combining the output from the individual antennas with an appropriate phase shift, ε, introduced

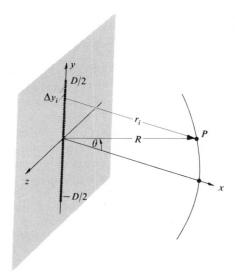

Fig. 10.8 A coherent line source.

between each of them. For a given ε the output of the system corresponds to the signal impinging on the array from a specific direction in space.

Figure 10.7 is a photograph of the first multiple radio interferometer designed by W. N. Christiansen and built in Australia in 1951. It consists of 32 parabolic antennas, each 2 m in diameter, designed to function in phase at the wavelength of the 21 cm hydrogen emission line. The antennas are arranged along an east–west baseline with 7 m separating each one. This particular array utilizes the earth's rotation as the scanning mechanism.

Examine Fig. 10.8 which depicts an idealized line source of electron-oscillators (e.g., the secondary sources of the Huygens–Fresnel principle for a long slit whose width is much less than λ illuminated by plane waves). Each point emits a spherical wavelet which we write as

$$E = \left(\frac{\varepsilon_0}{r}\right) \sin (\omega t - kr)$$

explicitly indicating the inverse r-dependence of the amplitude. The quantity ε_0 is said to be the *source strength*. The present situation is distinct from that of Fig. 10.6 in that now the sources are very weak, their number, N, is tremendously large and the separation between them vanishingly small. A minute, but finite segment of the array Δy_i, will contain $\Delta y_i(N/D)$ sources where D is the entire length of the array.

Imagine then that the array is divided up into M such segments, i.e., i goes from 1 to M. The contribution to the electric field intensity at P from the ith segment is accordingly

$$E_i = \left(\frac{\mathcal{E}_0}{r_i}\right) \sin(\omega t - kr_i)\left(\frac{N\Delta y_i}{D}\right)$$

provided that Δy_i is so small that the oscillators within it have a negligible relative phase difference ($r_i = $ constant) and their fields simply add constructively. We can cause the array to become a continuous (coherent) line source by letting N approach infinity. This description, besides being fairly realistic on a macroscopic scale, also allows the use of the calculus for more complicated geometries. Certainly as N approaches infinity, the source strengths of the individual oscillators must diminish to near zero if the total output is to be finite. We can therefore define a constant $\mathcal{E}_L$ as the *source strength per unit length* of the array, that is

$$\mathcal{E}_L \equiv \frac{1}{D}\lim_{N\to\infty}(\mathcal{E}_0 N). \qquad (10.8)$$

The net field at P from all M segments is

$$E = \sum_{i=1}^{M}\frac{\mathcal{E}_L}{r_i}\sin(\omega t - kr_i)\,\Delta y_i.$$

For a continuous line source Δy_i can become infinitesimal ($M \to \infty$) and the summation is then transformed into a definite integral

$$E = \mathcal{E}_L \int_{-D/2}^{+D/2}\frac{\sin(\omega t - kr)}{r}\,dy, \qquad (10.9)$$

where $r = r(y)$. The approximations used to evaluate Eq. (10.9) must depend on the position of P with respect to the array and will therefore make the distinction between Fraunhofer and Fresnel diffraction. The coherent *optical* line source does not now exist as a physical entity but we will make good use of it as a mathematical device.

10.2 FRAUNHOFER DIFFRACTION

10.2.1 The Single Slit

Return to Fig. 10.8 where now the point of observation is very distant from the coherent line source and $R \gg D$. Under these circumstances $r(y)$ never deviates appreciably from its midpoint value R so that the quantity $(\mathcal{E}_L/R)$ at P is essentially constant for all elements dy. It follows from Eq. (10.9) that the field at P due to the differential segment of the

source dy is

$$dE = \frac{\mathcal{E}_L}{R}\sin(\omega t - kr)\,dy, \qquad (10.10)$$

where $(\mathcal{E}_L/R)\,dy$ is the amplitude of the wave. Notice that the phase is very much more sensitive to variations in $r(y)$ than is the amplitude so that we will have to be more careful about introducing approximations into it. We can expand $r(y)$, in precisely the same manner as was done in Problem (9.4), to get it as an explicit function of y, thus

$$r = R - y\sin\theta + (y^2/2R)\cos^2\theta + \cdots, \qquad (10.11)$$

where θ is measured from the xz-plane. The third term can be ignored so long as its contribution to the phase is insignificant even when $y = \pm D/2$, i.e. $(\pi D^2/4\lambda R)\cos^2\theta$ must be negligible. This will be true for all values of θ when R is adequately large and we again have the Fraunhofer condition. The distance r is then linear in y. Substituting into Eq. (10.10) and integrating leads to

$$E = \frac{\mathcal{E}_L}{R}\int_{-D/2}^{+D/2}\sin[\omega t - k(R - y\sin\theta)]\,dy, \qquad (10.12)$$

and finally

$$E = \frac{\mathcal{E}_L D}{R}\frac{\sin[(kD/2)\sin\theta]}{(kD/2)\sin\theta}\sin(\omega t - kR). \qquad (10.13)$$

To simplify the appearance of things let

$$\beta \equiv (kD/2)\sin\theta \qquad (10.14)$$

so that

$$E = \frac{\mathcal{E}_L D}{R}\left(\frac{\sin\beta}{\beta}\right)\sin(\omega t - kR). \qquad (10.15)$$

The quantity most readily measured is the irradiance (forgetting the constants) $I(\theta) = \langle E^2 \rangle$ or

$$I(\theta) = \frac{1}{2}\left(\frac{\mathcal{E}_L D}{R}\right)^2\left(\frac{\sin\beta}{\beta}\right)^2, \qquad (10.16)$$

where $\langle \sin^2(\omega t - kR)\rangle = \frac{1}{2}$. When $\theta = 0$, $\sin\beta/\beta = 1$ and $I(\theta) = I(0)$ which corresponds to the *principal maximum*. The irradiance resulting from an idealized coherent line source in the Fraunhofer approximation is then

$$I(\theta) = I(0)\left(\frac{\sin\beta}{\beta}\right)^2 \qquad (10.17)$$

or using the *sinc function* (Section 7.2.3, and Table 1 of the Appendix)

$$I(\theta) = I(0)\,\text{sinc}^2\beta.$$

There is symmetry about the y-axis and this expression holds for θ measured in any plane containing that axis.

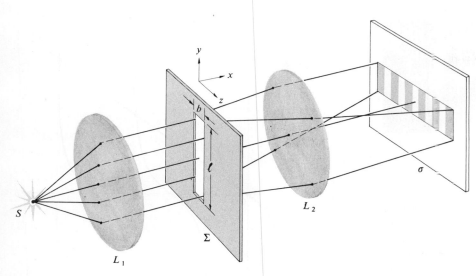

Fig. 10.9 Single-slit Fraunhofer diffraction.

Notice that since $\beta = (\pi D/\lambda) \sin \theta$, when $D \gg \lambda$, the irradiance drops extremely rapidly as θ deviates from zero. This arises from the fact that β becomes very large for large values of length D (a cm or so when using light). The phase of the line source is equivalent, by way of Eq. (10.15), to that of a point source located at the center of the array, a distance R from P. Finally then, a relatively long coherent line source $(D \gg \lambda)$ can be envisioned as a single point emitter radiating predominantly in the $\theta = 0$ direction, i.e. its emission resembles a circular wave in the xz-plane. In contrast, notice

that if $\lambda \gg D$, β is small, $\sin \beta \approx \beta$ and $I(\theta) \approx I(0)$. The irradiance is then constant for *all* θ and the line source resembles a point source emitting spherical waves.

We can now turn our attention to the problem of Fraunhofer diffraction by a slit or elongated narrow rectangular hole (Fig. 10.9). An aperture of this sort might typically have a width of several hundred λ and a length of a few centimeters. The usual procedure to follow in the analysis is to divide the slit into a series of long differential strips (dz by ℓ), parallel to the y-axis, as shown in Fig. 10.10. We

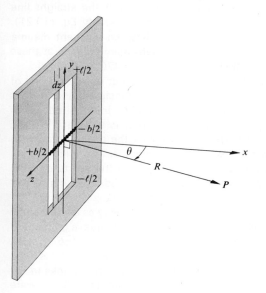

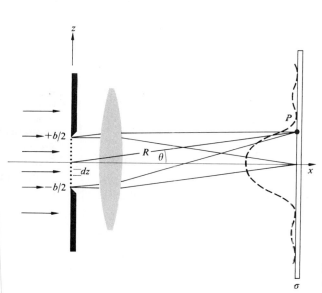

Fig. 10.10 Single-slit geometry.

Fig. 10.11 Diffraction pattern of a single vertical slit under point-source illumination.

immediately recognize however that each such strip is a long coherent line source and can therefore be replaced by a point emitter on the z-axis. In effect, each such emitter radiates a circular wave in the ($y = 0$ or) xz-plane. This is certainly reasonable, since the slit is long and the emerging wavefronts are practically unobstructed in the slit direction. There will therefore be very little diffraction parallel to the edges of the slit. The problem has, as such, been reduced to that of finding the field in the xz-plane due to an infinite number of point sources extending across the width of the slit along the z-axis. We then need only evaluate the integral of the contribution dE from each element dz in the Fraunhofer approximation. But once again, this is equivalent to a coherent line source so that the complete solution for the slit is, as we have seen,

$$I(\theta) = I(0)\left(\frac{\sin \beta}{\beta}\right)^2 \qquad [10.17]$$

provided that

$$\beta = (kb/2)\sin \theta \qquad (10.18)$$

and θ is measured from the xy-plane (see Problem 10.3). We call attention to the fact that here the line source is short, $D = b$, β is not large and although the irradiance falls off rapidly, higher-order subsidiary maxima will be observable (Fig. 10.11). The extrema of $I(\theta)$ occur at values of β which cause $dI/d\beta$ to be zero, that is

$$\frac{dI}{d\beta} = I(0)\frac{2\sin\beta(\beta\cos\beta - \sin\beta)}{\beta^3} = 0. \qquad (10.19)$$

The irradiance has minima, equal to zero, when $\sin\beta = 0$, whereupon

$$\beta = \pm\pi, \pm 2\pi, \pm 3\pi, \ldots. \qquad (10.20)$$

It also follows from Eq. (10.19) that when

$$\beta\cos\beta - \sin\beta = 0$$
$$\tan\beta = \beta. \qquad (10.21)$$

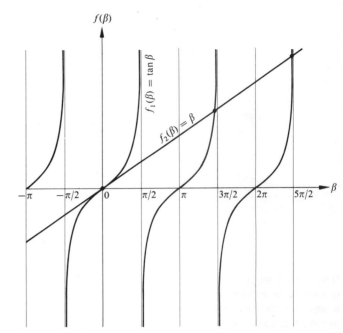

Fig. 10.12 The points of intersection of the two curves are the solutions of Eq. (10.21).

The solutions to this transcendental equation can be gotten graphically, as shown in Fig. 10.12. The points of intersection of the curves $f_1(\beta) = \tan\beta$, with the straight line $f_2(\beta) = \beta$ are common to both and so satisfy Eq. (10.21). Only one such extremum exists between adjacent minima (10.20), so that $I(\theta)$ must have subsidiary maxima at these values of β ($\pm 1.4303\pi, \pm 2.4590\pi, \pm 3.4707\pi, \ldots$).

We should inject a note of caution at this point: one of the frailties of the Huygens–Fresnel principle is that it does not take proper regard of the variations in amplitude, with angle, over the surface of each secondary wavelet. We will come back to this when we consider the *obliquity factor* in Fresnel diffraction where the effect is significant. In Fraunhofer diffraction the distance from the aperture to the plane of observation is so large that we need not be concerned about it provided that θ remains small.

Figure 10.13 is a plot of the flux density as expressed by Eq. (10.17). Envision some point on the curve, e.g. the third subsidiary maximum at $\beta = 3.4707\pi$; since $\beta = (\pi b/\lambda)\sin\theta$, increasing the slit width b requires a decrease in θ if β is to be constant. Under these conditions the pattern shrinks in toward the principal maximum, as it would as well if λ were

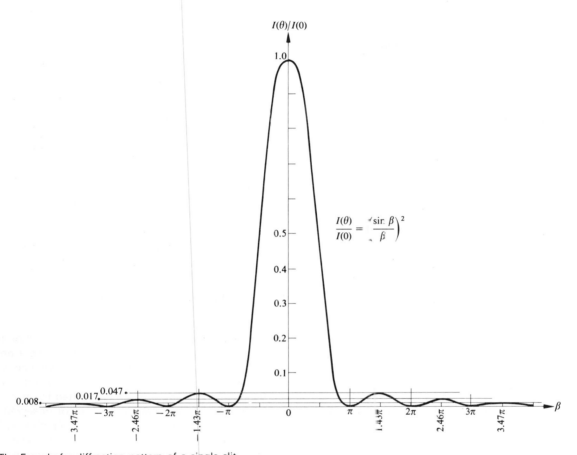

Fig. 10.13 The Fraunhofer diffraction pattern of a single slit.

decreased. If the source emits white light, the higher-order maxima show a succession of colors trailing off into red with increasing θ. Each different colored light-component has its minima and subsidiary maxima at angular positions characteristic of that wavelength (Problem 10.4). Indeed, only in the region about $\theta = 0$ will all of the constituent colors overlap to yield white light.

The point source S in Fig. 10.9 would be imaged at the position of the center of the pattern if the diffracting screen Σ were removed. Under this sort of illumination, the pattern produced with the slit in place is a series of dashes in the yz-plane of the screen σ, much like a spread-out image of S (Fig. 10.11). An incoherent line source (in place of S) positioned parallel to the slit, in the focal plane of the collimator L_1, will broaden the pattern out into a series of bands. Any point on the line source generates an independent

diffraction pattern, each of which is displaced, with respect to the others, along the y-direction. With no diffracting screen present, the image of the line source would be a line parallel to the original slit. With the screen in place the line is spread out, as was the point image of S (Fig. 10.14). Keep in mind that it's the small dimension of the slit which does the spreading out.

The single slit pattern is very easily observed without the use of special equipment. Any number of sources will do, e.g. a distant street light at night, a small incandescent lamp, sunlight streaming through a narrow space in a window shade; almost anything that resembles a point or line source will serve. Probably the best source for our purposes is an ordinary clear *straight-filament* display bulb (the kind where the filament is vertical and about three inches long). You can use your imagination to generate all sorts of single slit

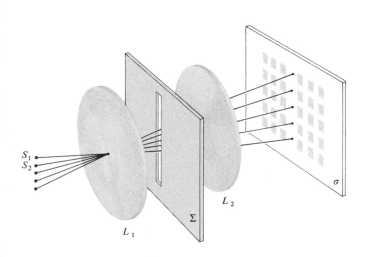

Fig. 10.14 The single-slit pattern with a line source   See first photo of Fig. 10.17.

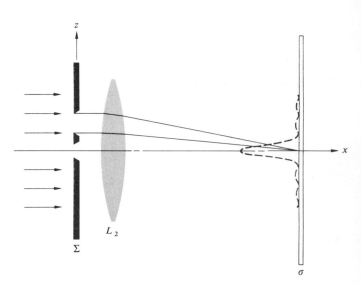

Fig. 10.15 The double-slit set-up.

arrangements, e.g. a comb or fork rotated to decrease the projected space between the tines, or a scratch across a layer of india ink on a microscope slide.   An inexpensive vernier caliper makes a remarkably good variable slit.   Hold the caliper close to your eye with the slit, a few thousandths of an inch wide, parallel to the filament of the lamp.   Focus your eye beyond the slit at infinity so that its lens serves as L_2.

10.2.2 The Double Slit

It might at first seem from Fig. 10.10 that the location of the principal maximum is always to be in line with the center of the diffracting aperture; this however is not generally true.   The diffraction pattern is actually centered about the axis of the lens and has exactly the same shape and location regardless of the slit's position, as long as its orientation is unchanged and the approximations are valid (Fig. 10.15).   All waves traveling parallel to the lens axis converge on the second focal point of L_2, this then is the image of S, and the center of the diffraction pattern.   Suppose now that we have two long slits of width b and center-to-center separation a (Fig. 10.16).   Each aperture, by itself, would generate the same single-slit diffraction pattern on the viewing screen σ.   At any point on σ, the contributions from the two slits overlap and even though each must be essentially equal in amplitude, they may well differ significantly in phase.   Since the same primary wave excites the secondary sources at each slit, the

resulting wavelets will be coherent and interference must occur.   If the primary plane wave is incident on Σ at some angle θ_i (see Problem 10.3) there will be a constant relative phase difference between the secondary sources.   At normal incidence, the wavelets are all emitted in phase.   The interference fringe at a particular point of observation is determined by the differences in the optical path lengths traversed by the overlapping wavelets from the two slits.   As we will see, the consequent flux-density distribution (Fig. 10.17) is a combination of a rapidly varying double-slit interference system, modulated by a single-slit diffraction pattern.

To obtain an expression for the optical disturbance at a point on σ, we have only to slightly reformulate the single-slit analysis.   Each of the two apertures is divided into differential strips (dz by ℓ), which in turn behave as an infinite number of point sources aligned along the z-axis.   The total contribution to the electric field, in the Fraunhofer approximation (10.12), is then

$$E = C \int_{-b/2}^{b/2} F(z)\, dz + C \int_{a-b/2}^{a+b/2} F(z)\, dz, \qquad (10.22)$$

where $F(z) = \sin[\omega t - k(R - z \sin \theta)]$.   The constant-amplitude factor C is the secondary source strength per unit length along the z-axis (assumed to be independent of z

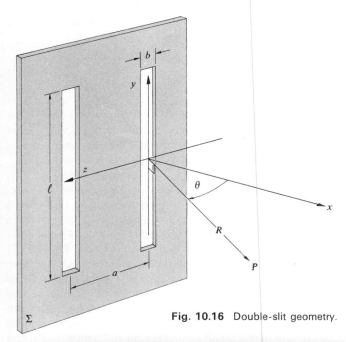

Fig. 10.16 Double-slit geometry.

over each aperture) divided by R which is measured from the origin to P and is taken as constant. We will only be concerned with relative flux densities on σ so that the actual value of C is of little interest to us now. Integration of Eq. (10.22) yields

$$E = bC\left(\frac{\sin\beta}{\beta}\right)[\sin(\omega t - kR) + \sin(\omega t - kR + 2\alpha)] \quad (10.23)$$

with $\alpha \equiv (ka/2)\sin\theta$ and as before $\beta \equiv (kb/2)\sin\theta$. This is just the sum of the two fields at P, one from each slit, as given by Eq. (10.15). The distance from the first slit to P is R, giving a phase contribution of $-kR$. The distance from the second slit to P is $(R - a\sin\theta)$ or $(R - 2\alpha/k)$ yielding a phase term equal to $(-kR + 2\alpha)$ as in the second sine function. The quantity 2β is the phase difference $(k\Lambda)$ between two nearly parallel rays, arriving at a point P on σ, from the edges of one of the slits. The quantity 2α is the phase difference between two waves arriving at P,

Fig. 10.17 Single- and double-slit Fraunhofer patterns. The faint cross hatching arises entirely in the printing process. [Photos courtesy M. Cagnet, M. Francon, and J. C. Thrierr: *Atlas optischer Erscheinungen*, Berlin–Heidelberg–New York: Springer, 1962.]

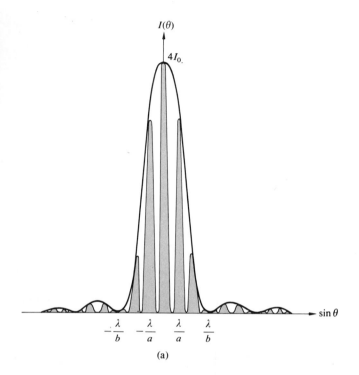

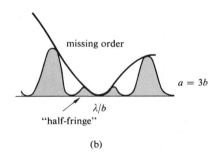

Fig. 10.18 A double slit pattern ($a = 3b$).

one having originated at any point in the first slit, the other coming from the corresponding point in the second slit. Simplifying Eq. (10.23) a bit further, it becomes

$$E = 2bC \left(\frac{\sin \beta}{\beta} \right) \cos \alpha \sin (\omega t - kR + \alpha),$$

which when squared and averaged over a relatively long interval in time is the irradiance

$$I(\theta) = 4I_0 \left(\frac{\sin^2 \beta}{\beta^2} \right) \cos^2 \alpha. \qquad (10.24)$$

In the $\theta = 0$ direction, i.e. when $\beta = \alpha = 0$, I_0 is the flux-density contribution from either slit and $I(0) = 4I_0$ is the total flux density. The factor of four comes from the fact that the amplitude of the electric field is twice what it would be at that point with one slit covered.

If in Eq. (10.24) b gets vanishingly small ($kb \ll 1$), then $(\sin \beta)/\beta \approx 1$, and the equation reduces to the flux-density expression for a pair of long line sources, i.e. Young's experiment, Eq. (9.6). If on the other hand $a = 0$. the two slits coalesce into one, $\alpha = 0$ and Eq. (10.24) becomes $I(0) = 4I_0(\sin^2 \beta)/\beta^2$. This is the equivalent of Eq. (10.17) for single-slit diffraction where the source strength has been doubled. We might then envision the total expression as being generated by a $\cos^2 \alpha$ interference term modulated by a $(\sin^2 \beta)/\beta^2$ diffraction term. If the slits are finite in width but very narrow, the diffraction pattern from either slit will be uniform over a broad central region and bands resembling the idealized Young's fringes will appear within that region. At angular positions (θ-values) where

$$\beta = \pm\pi, \pm2\pi, \pm3\pi, \ldots$$

diffraction effects are such that no light reaches σ, and clearly none is available for interference. At points on σ where

$$\alpha = \pm\pi/2, \pm3\pi/2, \pm5\pi/2, \ldots$$

the various contributions to the electric field will be completely out of phase and cancel, regardless of the actual amount of light made available from the diffraction process.

The irradiance distribution for a double-slit Fraunhofer pattern is illustrated in Fig. 10.18. Notice that it is a combination of Fig. 9.6 and 10.13. The curve is for the particular case when $a = 3b$, i.e. when $\alpha = 3\beta$. You can get a rough idea of what the pattern will look like since if $a = mb$, where m is any number, there will be $2m$ bright fringes (counting "fractional fringes" as well)* within the central diffraction peak (Problem 10.5). It may occur that an interference maximum and a diffraction minimum (zero) correspond to the same θ-value. In that case no light is available at that precise position to partake in the interference process, and the suppressed peak is said to be a *missing order*.

---

* Notice that m need not be an integer. Moreover if m is an integer there will be "half-fringes" as shown in Fig. 10.18(b).

The double-slit pattern is also rather easily observed and the seeing is well worth the effort. A straight filament, tubular bulb is again the best line source. For slits, coat a microscope slide with india ink or if you happen to have some; a colloidal suspension of graphite in alcohol works even better (it's more opaque). Scratch a pair of slits across the dry ink with a razor blade and stand about ten feet from the source. Hold the slits parallel to the filament and close to your eye, which, when focused at infinity, will serve as the needed lens. Interpose red or blue cellophane and observe the change in the width of the fringes. Find out what happens when you cover one and then both of the slits with a microscope slide. Move the slits slowly in the z-direction; then holding them stationary, move your eye in the z-direction; verify that the position of the center of the pattern is indeed determined by the lens and not the aperture.

10.2.3 Diffraction by Many Slits

The procedure for obtaining the irradiance function for a monochromatic wave diffracted by many slits is essentially the same as that used when considering two slits. Here again, the limits of integration must be appropriately altered. Consider the case of N long, parallel, narrow slits each of width b and center-to-center separation a, as illustrated in Fig. 10.19. With the origin of the coordinate system once more at the center of the first slit, the total optical disturbance at a point on the screen σ is given by

$$E = C \int_{-b/2}^{b/2} F(z)\, dz + C \int_{a-b/2}^{a+b/2} F(z)\, dz$$

$$+ C \int_{2a-b/2}^{2a+b/2} F(z)\, dz + \cdots \qquad (10.25)$$

$$+ C \int_{(N-1)a-b/2}^{(N-1)a+b/2} F(z)\, dz$$

where as before, $F(z) = \sin\left[\omega t - k(R - z\sin\theta)\right]$. This applies to the Fraunhofer condition, so that the aperture configuration must be such that all of the slits are close to the origin and the approximation (10.11)

$$r = R - z\sin\theta \qquad (10.26)$$

applies over the entire array. The contribution from the jth slit (where the first one is numbered zero), obtained by evaluating only that one integral in Eq. (10.25), is then

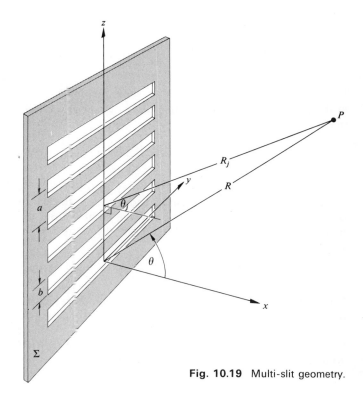

Fig. 10.19 Multi-slit geometry.

$$E_j = \frac{C}{k\sin\theta}\left[\sin\left(\omega t - kR\right)\sin\left(kz\sin\theta\right)\right.$$

$$\left. - \cos\left(\omega t - kR\right)\cos\left(kz\sin\theta\right)\right]_{ja-b/2}^{ja+b/2}$$

provided that we require $\theta_j \approx \theta$. After some manipulation this becomes

$$E_j = bC\left(\frac{\sin\beta}{\beta}\right)\sin\left(\omega t - kR + 2\alpha j\right) \qquad (10.27)$$

recalling that $\beta = (kb/2)\sin\theta$ and $\alpha = (ka/2)\sin\theta$. Notice that this is equivalent to the expression for a line source (10.15) or of course, a single slit, where in accord with Eq. 10.26 and Fig. 10.19, $R_j = R - ja\sin\theta$ so that $-kR + 2\alpha j = -kR_j$. The total optical disturbance as given by Eq. (10.25) is simply the sum of the contributions from each of the slits, that is

$$E = \sum_{j=0}^{N-1} E_j$$

or

$$E = \sum_{j=0}^{N-1} bC\left(\frac{\sin\beta}{\beta}\right)\sin\left(\omega t - kR + 2\alpha j\right). \qquad (10.28)$$

This in turn can be written as the imaginary part of a complex exponential:

$$E = \text{Im}\left[bC\left(\frac{\sin\beta}{\beta}\right)e^{i(\omega t - kR)}\sum_{j=0}^{N-1}(e^{i2\alpha})^j\right]. \quad (10.29)$$

But we have already evaluated this same geometric series in the process of simplifying Eq. (10.2). Equation (10.29) therefore reduces to the form

$$E = bC\left(\frac{\sin\beta}{\beta}\right)\left(\frac{\sin N\alpha}{\sin\alpha}\right)\sin[\omega t - kR + (N-1)\alpha]. \quad (10.30)$$

The distance from the center of the array to the point P is equal to $[R - (N-1)(a/2)\sin\theta]$ and therefore the phase of E at P corresponds to that of a wave emitted from the midpoint of the source. The flux-density distribution function is

$$I(\theta) = I_0\left(\frac{\sin\beta}{\beta}\right)^2\left(\frac{\sin N\alpha}{\sin\alpha}\right)^2. \quad (10.31)$$

Note that I_0 is the flux density in the $\theta = 0$ direction emitted by any one of the slits and that $I(0) = N^2 I_0$. In other words, the waves arriving at P in the forward direction are all in phase and their fields add constructively. Each slit by itself would generate precisely the same flux-density distribution. Superimposed, the various contributions yield a multiple wave interference system modulated by the single-slit diffraction envelope. If the width of each aperture were shrunk to zero, Eq. (10.31) would become the flux-density expression (10.6) for a linear coherent array of oscillators. As in that earlier treatment, (10.17), *principal maxima* occur when $(\sin N\alpha/\sin\alpha) = N$, i.e. when

$$\alpha = 0, \pm\pi, \pm 2\pi, \ldots$$

or equivalently since $\alpha = (ka/2)\sin\theta$

$$a\sin\theta_m = m\lambda \quad (10.32)$$

with $m = 0, \pm 1, \pm 2, \ldots$. This is quite general and gives rise to the same θ-locations for these maxima regardless of the value of $N \geq 2$. Minima, of zero flux-density, exist whenever $(\sin N\alpha/\sin\alpha)^2 = 0$ or when

$$\alpha = \pm\frac{\pi}{N}, \pm\frac{2\pi}{N}, \pm\frac{3\pi}{N}, \ldots, \pm\frac{(N-1)\pi}{N}, \pm\frac{(N+1)\pi}{N}, \ldots . \quad (10.33)$$

Between consecutive principal maxima (i.e. over the range in α of π) there will therefore be $N - 1$ minima. And of course between each pair of minima there will have to be a *subsidiary maximum*. The term $(\sin N\alpha/\sin\alpha)^2$, which we can think of as embodying the interference effects, has a rapidly varying numerator and a slowly varying denominator. The subsidiary maxima are therefore located approximately at points where $\sin N\alpha$ has its greatest value, namely

$$\alpha = \pm\frac{3\pi}{2N}, \pm\frac{5\pi}{2N}, \cdots \quad (10.34)$$

The $N - 2$ *subsidiary maxima* between consecutive principal maxima are clearly visible in Fig. 10.20. We can get some idea of the flux density at these peaks by rewriting Eq. (10.31) as

$$I(\theta) = \frac{I(0)}{N^2}\left(\frac{\sin\beta}{\beta}\right)^2\left(\frac{\sin N\alpha}{\sin\alpha}\right)^2, \quad (10.35)$$

where at the points of interest $|\sin N\alpha| = 1$. For *large* N, α is small and $\sin^2\alpha \approx \alpha^2$. At the first subsidiary peak $\alpha = 3\pi/2N$, in which case

$$I \approx I(0)\left(\frac{\sin\beta}{\beta}\right)^2\left(\frac{2}{3\pi}\right)^2 \quad (10.36)$$

and the flux density has dropped to about $\frac{1}{22}$ of that of the adjacent principal maximum (see Problem 10.6). Since $(\sin\beta)/\beta$ for small β varies slowly it will not differ from one appreciably, close to the zeroth-order principal maximum, so that $I/I(0) \approx \frac{1}{22}$. This flux-density ratio for the next secondary peak is down to $\frac{1}{62}$ and it continues to decrease as α approaches a value halfway between the principal maxima. At that symmetry point, $\alpha \approx \pi/2$, $\sin\alpha \approx 1$ and the flux-density ratio has its lowest value of approximately $1/N^2$. Thereafter $\alpha > \pi/2$ and the flux densities of the subsidiary maxima begin to increase.

Try duplicating Fig. 10.20 using a tubular bulb and homemade slits. You'll probably have difficulty clearly seeing the subsidiary maxima, with the effect that the only perceptible difference between the double and multiple-slit patterns may be an apparent broadening in the dark regions between principal maxima. As in Fig. 10.20, the dark regions will become wider than the bright bands as N increases and the secondary peaks fade out. If we consider each principal maximum to be bounded in width by two adjacent zeroes, then each will extend over a length in θ, $(\sin\theta \approx \theta)$ of approximately $2\lambda/Na$. As N increases, the principal maxima maintain their relative spacing (λ/a) while becoming increasingly more narrow. Figure 10.21 shows the case of six slits where $a = 4b$.

The multiple-slit interference term in Eq. 10.35 has the form $(\sin^2 N\alpha)/N^2\sin^2\alpha$; thus, for large N, $(N^2\sin^2\alpha)^{-1}$

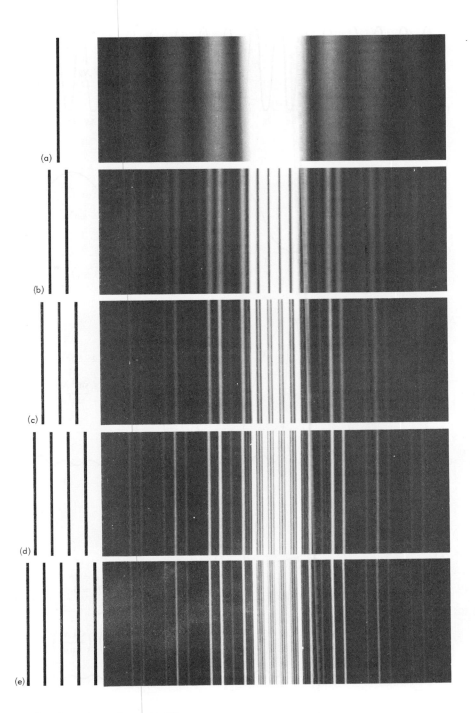

Fig. 10.20 Diffraction patterns for slit systems shown at left.

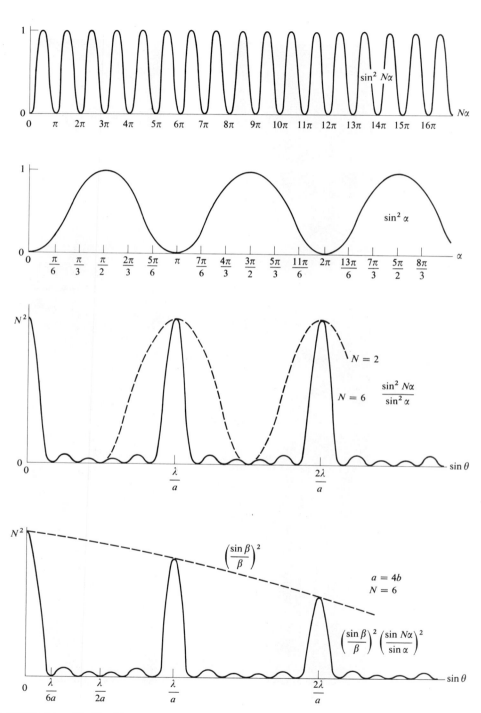

Fig. 10.21 Multiple-slit pattern ($a = 4b$, $N = 6$).

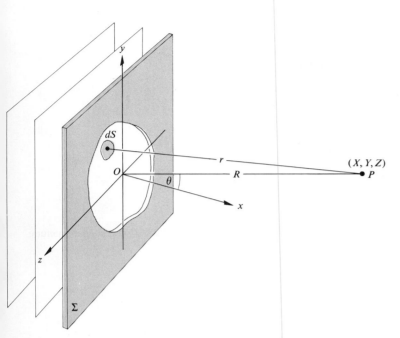

Fig. 10.22 Fraunhofer diffraction from an aperture.

may be envisioned as the curve beneath which $\sin^2 N\alpha$ rapidly varies. Notice that for small α this interference term looks like $\text{sinc}^2 N\alpha$.

10.2.4 The Rectangular Aperture

Consider the configuration depicted in Fig. 10.22. A monochromatic plane wave propagating in the x-direction is incident on the opaque diffracting screen Σ. We wish to find the consequent (far-field) flux-density distribution in space or equivalently at some arbitrary distant point P. According to the Huygens–Fresnel principle, a differential area dS, within the aperture, may be envisioned as being covered with coherent secondary point sources. But dS is much smaller in extent than is λ, so that all of the contributions at P remain in phase and interfere constructively. This is true regardless of θ, i.e. dS emits a spherical wave (Problem 10.7). If $\mathcal{E}_A$ is the source strength per unit area, *assumed constant over the entire aperture*, then the optical disturbance at P due to dS is either the real or imaginary part of

$$dE = \left(\frac{\mathcal{E}_A}{r}\right) e^{i(\omega t - kr)}\, dS. \qquad (10.37)$$

The choice is yours and depends only on whether you like

sine or cosine waves, there being no difference except for a phase shift. The distance from dS to P is

$$r = [X^2 + (Y - y)^2 + (Z - z)^2]^{1/2} \qquad (10.38)$$

and as we have seen the Fraunhofer condition occurs when this distance approaches infinity. As before, it will suffice to replace r by the distance $\overline{OP}$, i.e. R, in the amplitude term as long as the aperture is relatively small. But the approximation for r in the phase needs to be treated a bit more carefully; $k = 2\pi/\lambda$ is a large number. To that end we expand out Eq. (10.38), and by making use of

$$R = [X^2 + Y^2 + Z^2]^{1/2} \qquad (10.39)$$

obtain

$$r = R[1 + (y^2 + z^2)/R^2 - 2(Yy + Zz)/R^2]^{1/2}. \quad (10.40)$$

In the far-field case R is very large in comparison to the dimensions of the aperture and the $(y^2 + z^2)/R^2$ term is certainly negligible. Since P is very far from Σ, θ can still be kept small, even though Y and Z are fairly large and this mitigates any concern about the directionality of the emitters (the obliquity factor). Now

$$r = R[1 - 2(Yy + Zz)/R^2]^{1/2}$$

and dropping all but the first two terms in the binomial expansion we have

$$r = R[1 - (Yy + Zz)/R^2].$$

The total disturbance arriving at P is

$$E = \frac{\mathcal{E}_A e^{i(\omega t - kr)}}{R} \iint\limits_{\text{Aperture}} e^{ik(Yy + Zz)/R}\, dS. \qquad (10.41)$$

Consider the specific configuration shown in Fig. 10.23. Equation (10.41) can now be written as

$$E = \frac{\mathcal{E}_A e^{i(\omega t - kR)}}{R} \int_{-b/2}^{+b/2} e^{ikYy/R}\, dy \int_{-a/2}^{+a/2} e^{ikZz/R}\, dz.$$

where $dS = dy\, dz$. With $\beta' \equiv kbY/2R$ and $\alpha' \equiv kaZ/2R$, we have

$$\int_{-b/2}^{-b/2} e^{ikYy/R}\, dy = b\left(\frac{e^{i\beta'} - e^{-i\beta'}}{2i\beta'}\right) = b\left(\frac{\sin \beta'}{\beta'}\right)$$

and similarly

$$\int_{-a/2}^{+a/2} e^{ikZz/R}\, dz = a\left(\frac{e^{i\alpha'} - e^{-i\alpha'}}{2i\alpha'}\right) = a\left(\frac{\sin \alpha'}{\alpha'}\right),$$

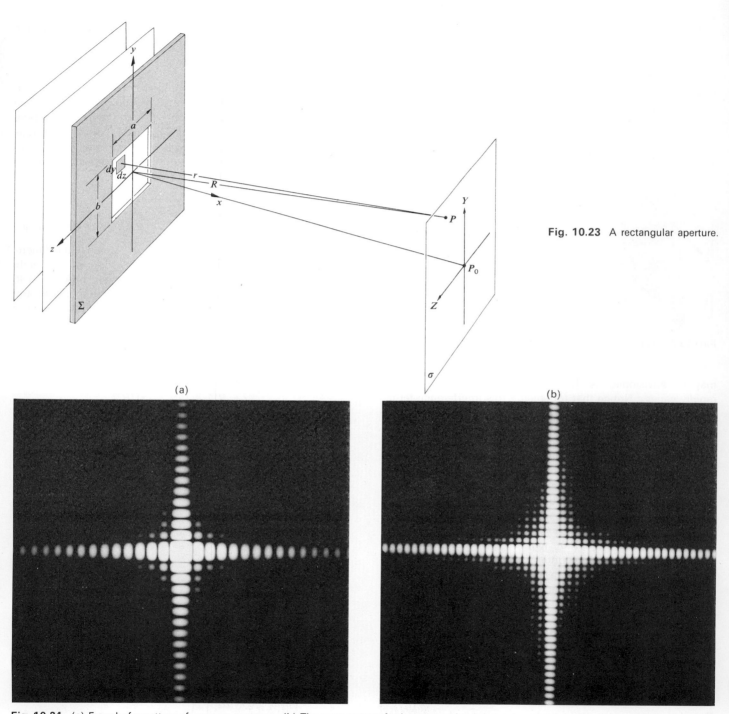

Fig. 10.23 A rectangular aperture.

Fig. 10.24 (a) Fraunhofer pattern of a square aperture. (b) The same pattern further exposed to bring out some of the faint terms. [Photos by E.H.]

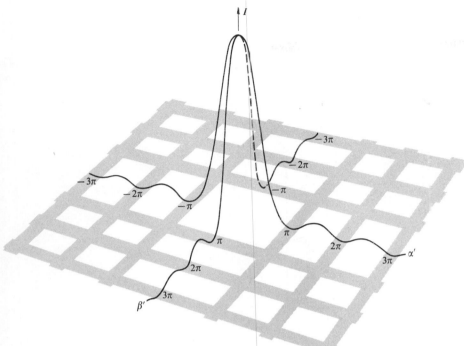

Fig. 10.25 The irradiance distribution for a square aperture.

so that

$$E = \frac{A \mathcal{E}_A e^{i(\omega t - kR)}}{R} \left(\frac{\sin \alpha'}{\alpha'}\right)\left(\frac{\sin \beta'}{\beta'}\right). \tag{10.42}$$

where A is the area of the aperture. Since $I = \langle (\mathrm{Re}\, E)^2 \rangle$

$$I(Y, Z) = I(0) \left(\frac{\sin \alpha'}{\alpha'}\right)^2 \left(\frac{\sin \beta'}{\beta'}\right)^2 \tag{10.43}$$

where $I(0)$ is the irradiance at P_0 i.e. at $Y = 0$, $Z = 0$ (see Fig. 10.24). At values of Y and Z such that $\alpha' = 0$ or $\beta' = 0$, $I(Y, Z)$ assumes the familiar shape of Fig. 10.13. When β' or α' are nonzero integer multiples of π or equivalently when Y and Z are nonzero integer multiples of $\lambda R/b$ and $\lambda R/a$ respectively, $I(Y, Z) = 0$ and we have a rectangular grid of nodal lines as indicated in Fig. 10.25. Notice that the pattern in the Y-, Z-directions varies *inversely* with the y-, z-aperture dimensions. A horizontal, rectangular opening will produce a pattern with a verticle rectangle at its center.

Along the β'-axis, $\alpha' = 0$ and the subsidiary maxima are located approximately halfway between zeroes, i.e. at $\beta'_m = \pm 3\pi/2, \pm 5\pi/2, \pm 7\pi/2, \ldots$. At each subsidiary maximum $\sin \beta'_m = 1$, and of course, along the β'-axis since

$\alpha' = 0$, $(\sin \alpha')/\alpha' = 1$ so that the relative irradiances are approximated simply by

$$\frac{I}{I(0)} = \frac{1}{\beta'^2_m}. \tag{10.44}$$

Similarly along the α'-axis

$$\frac{I}{I(0)} = \frac{1}{\alpha'^2_m}. \tag{10.45}$$

The flux-density ratio* drops off rather rapidly from 1 to $\frac{1}{22}$ to $\frac{1}{62}$ to $\frac{1}{122}$, etc. Even so, the off-axis secondary peaks are

---

* These particular photographs were taken during an undergraduate laboratory session. A 1.5 mW helium–neon laser was used as a plane-wave source. The apparatus was set up in a long darkened room and the pattern was cast directly on 4 × 5 Polaroid (ASA 3000) film. The film was located at about 30 ft from a small aperture so that no focusing lens was needed. The shutter, placed directly in front of the laser, was a student-contrived cardboard guillotine arrangement and therefore no exposure times are available. Any camera shutter (a single-lens reflex with the lens removed and the back open) will serve, but the cardboard one was more fun.

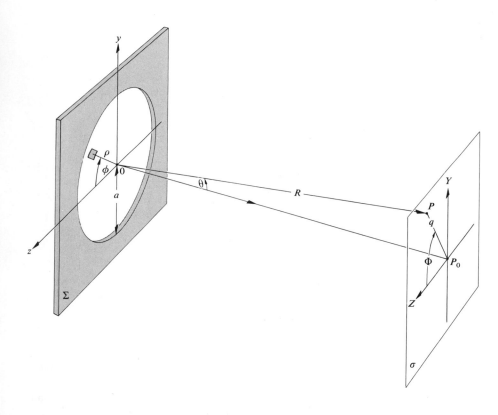

Fig. 10.26   Circular aperture geometry.

still smaller, e.g. the four corner peaks (whose coordinates correspond to appropriate combinations of $\beta' = \pm 3\pi/2$ and $\alpha' = \pm 3\pi/2$) nearest to the central maximum, each have relative irradiances of $(\frac{1}{22})^2$.

10.2.5 The Circular Aperture

Fraunhofer diffraction at a circular aperture is an effect of very great practical significance in the study of optical instrumentation. Envision a typical arrangement; plane waves impinging on a screen Σ containing a circular aperture and the consequent far-field diffraction pattern spread across a distant observing screen σ. By using a focusing lens L_2, σ can be brought in close to the aperture without changing the pattern. Now, if L_2 is positioned within and exactly fills the diffracting opening in Σ, the form of the pattern is essentially unaltered. The light wave reaching Σ is cropped so that only a circular segment propagates through L_2 to form an image in the focal plane. But quite obviously this is the same process that takes place in the eye, a telescope, microscope or camera lens. The image of a distant point source as formed by a perfectly aberration-free converging

lens, is never a point, but rather some sort of diffraction pattern. We are essentially collecting only a fraction of the incident wavefront and cannot therefore hope to form a perfect image. As shown in the last section, the expression for the optical disturbance at P, arising from an arbitrary aperture in the far-field case, is

$$E = \frac{\mathcal{E}_A e^{i(\omega t - kR)}}{R} \iint\limits_{\text{Aperture}} e^{ik(Yy + Zz)/R} \, dS. \qquad [10.41]$$

For a circular opening, symmetry would suggest introducing spherical polar coordinates in both the plane of the aperture and the plane of observation, as shown in Fig. 10.26. Therefore, let

$$z = \rho \cos \phi \qquad y = \rho \sin \phi$$
$$Z = q \cos \Phi \qquad Y = q \sin \Phi$$

and so the differential element of area is now

$$dS = \rho \, d\rho \, d\phi.$$

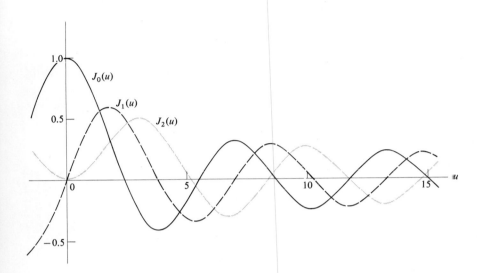

Fig. 10.27 Bessel functions.

Substituting these expressions into Eq. (10.41), it becomes

$$E = \frac{\mathcal{E}_A e^{i(\omega t - kR)}}{R} \int_{\rho=0}^{a} \int_{\phi=0}^{2\pi} e^{i(k\rho q/R)\cos(\phi-\Phi)} \rho \, d\rho \, d\phi \qquad (10.46)$$

Because of the complete axial symmetry, the solution must be independent of Φ. We might just as well solve Eq. (10.46) with $\Phi = 0$ as with any other value, thereby simplifying things slightly.

The portion of the double integral associated with the variable ϕ,

$$\int_{0}^{2\pi} e^{i(k\rho q/R)\cos\phi} \, d\phi,$$

is one that arises quite frequently in the mathematics of physics. It is a unique function in that it cannot be reduced to any of the more common forms, such as the various hyperbolic, exponential or trigonometric functions and indeed, with the exception of these, it is perhaps the most often encountered. The quantity

$$J_0(u) = \frac{1}{2\pi} \int_{0}^{2\pi} e^{iu\cos v} \, dv \qquad (10.47)$$

is known as the *Bessel function* (of the first kind) of order zero. More generally

$$J_m(u) = \frac{i^{-m}}{2\pi} \int_{0}^{2\pi} e^{i(mv + u\cos v)} \, dv \qquad (10.48)$$

represents the Bessel function of order m. Numerical values of $J_0(u)$ and $J_1(u)$ are tabulated for a large range of u in most mathematical handbooks. Just like sine and cosine, the Bessel functions have series expansions and are certainly no more esoteric than these familiar childhood acquaintances. As seen in Fig. 10.27, $J_0(u)$ and $J_1(u)$ are slowly decreasing oscillatory functions that do nothing particularly dramatic.

Equation (10.46) can be rewritten as

$$E = \frac{\mathcal{E}_A e^{i(\omega t - kR)}}{R} 2\pi \int_{0}^{a} J_0(k\rho q/R)\rho \, d\rho. \qquad (10.49)$$

Another general property of Bessel functions, referred to as a recurrence relation, is

$$\frac{d}{du}\left[u^m J_m(u) \right] = u^m J_{m-1}(u).$$

When $m = 1$, this clearly leads to

$$\int_{0}^{u} u' J_0(u') \, du' = u J_1(u) \qquad (10.50)$$

with u' just serving as a dummy variable. If we now return to the integral in Eq. (10.49) and change the variable such that $w = k\rho q/R$, then $d\rho = (R/kq) \, dw$ and

$$\int_{\rho=0}^{\rho=a} J_0(k\rho q/R)\rho \, d\rho = (R/kq)^2 \int_{w=0}^{w=kaq/R} J_0(w)w \, dw.$$

Making use of Eq. (10.50), we get

$$E(t) = \frac{\mathcal{E}_A e^{i(\omega t - kR)}}{R} 2\pi a^2 (R/kaq) J_1(kaq/R). \quad (10.51)$$

The irradiance at point P is $\langle (\text{Re } E)^2 \rangle$ or $\frac{1}{2}EE^*$, that is

$$I = \frac{2\mathcal{E}_A^2 A^2}{R^2} \left[\frac{J_1(kaq/R)}{kaq/R} \right]^2, \quad (10.52)$$

where A is the area of the circular opening. To find the irradiance at the center of the pattern, i.e. at P_0, set $q = 0$. It follows from the above recurrence relation ($m = 1$) that

$$J_0(u) = \frac{d}{du} J_1(u) + \frac{J_1(u)}{u}. \quad (10.53)$$

From Eq. (10.47) we see that $J_0(0) = 1$ and from Eq. (10.48), $J_1(0) = 0$. The ratio of $J_1(u)/u$ as u approaches zero has the same limit (L'Hôpital's rule) as the ratio of the separate derivatives of its numerator and denominator, namely $dJ_1(u)/du$ over one. But that means that the right-hand side of Eq. (10.53) is twice that limiting value, so that $J_1(u)/u = \frac{1}{2}$ at $u = 0$. The irradiance at P_0 is therefore

$$I(0) = \frac{\mathcal{E}_A^2 A^2}{2R^2}, \quad (10.54)$$

which is the same result obtained for the rectangular opening (10.43). If R can be presumed essentially constant over the pattern, we can write

$$I = I(0) \left[\frac{2J_1(kaq/R)}{kaq/R} \right]^2. \quad (10.55)$$

Since $\sin \theta = q/R$, the irradiance can be written as a function of θ:

$$I(\theta) = I(0) \left[\frac{2J_1(ka \sin \theta)}{ka \sin \theta} \right]^2, \quad (10.56)$$

and as such, is plotted in Fig. 10.28. Because of the axial symmetry, the towering central maximum corresponds to a high-irradiance circular spot known as the *Airy disk*. It was Sir George Biddell Airy (1801–92) Astronomer Royal of England, who first derived Eq. (10.56). The central disk is surrounded by a dark ring which corresponds to the first zero of the function $J_1(u)$. From the standard tables $J_1(u) = 0$ when $u = 3.83$ i.e. $kaq/R = 3.83$. The radius q_1 drawn to the center of this first dark ring can be thought of as the extent of the Airy disk. It is given by

$$q_1 = 1.22 \frac{R\lambda}{2a}. \quad (10.57)$$

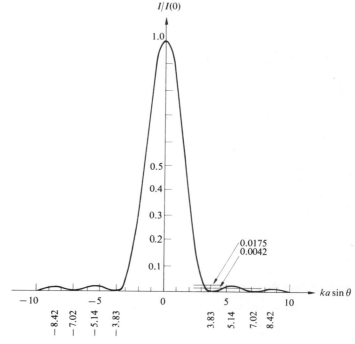

Fig. 10.28 The Airy pattern.

For a lens focused on the screen σ, the focal length $f \approx R$ and so

$$q_1 \approx 1.22 \frac{f\lambda}{D}, \quad (10.58)$$

where D is the aperture diameter i.e. $D = 2a$. (The *diameter* of the Airy disk in the visible spectrum is *very roughly* equal to the $f/\#$ of the lens in millionths of a meter.) As shown in Figs. 10.29 to 10.31, q_1 varies inversely with the hole's diameter. As D approaches λ, the Airy disk can be very large indeed and the circular aperture begins to resemble a point source of spherical waves.

The higher-order zeroes occur at values of kaq/R equal to 7.02, 10.17, etc. The secondary maxima are located where u satisfies the condition

$$\frac{d}{du} \left[\frac{J_1(u)}{u} \right] = 0,$$

which is equivalent to $J_2(u) = 0$. From the tables then, these secondary peaks occur when kaq/R equals 5.14, 8.42, 11.6, etc.; whereupon $I/I(0)$ drops from one to 0.0175, 0.0042 and 0.0016 respectively.

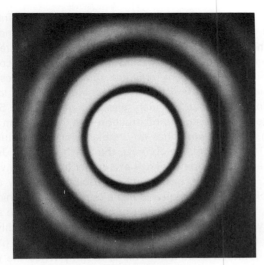

Fig. 10.29 Airy rings (0.5 mm hole diameter). [Photo by E.H.]

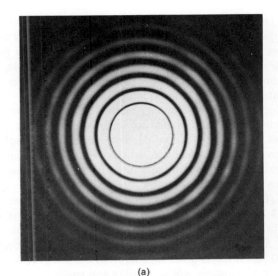

(a)

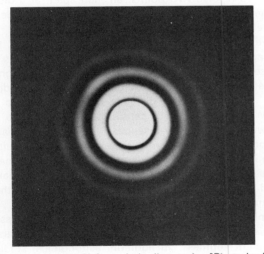

Fig. 10.30 Airy rings (1.0 mm hole diameter). [Photo by E.H.]

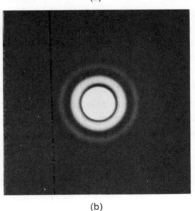

(b)

Fig. 10.31 (a) Airy rings—long exposure (1.5 mm hole diameter). (b) Central Airy disc—short exposure with the same aperture. [Photos by E.H.]

of the pattern, one finds that 84% of the light arrives within the Airy disc and 91% within the bounds of the second dark ring.

10.2.6 Resolution of Imaging Systems

Imagine that we have some sort of lens system which forms an image of an extended object. If the object is self-luminous it is likely that we can regard it as made up of an array of incoherent sources. On the other hand an object seen in reflected light will surely display some phase correlation between its various scattering points. When the point sources

Circular apertures are preferable to rectangular ones as far as lens shapes go, since the circle's irradiance curve is broader around the central peak and drops off more rapidly thereafter. Exactly what fraction of the total light energy incident on σ is confined to within the various maxima is a question of interest, but one somewhat too involved to solve here.* On integrating the irradiance over a particular region

---

* See Born and Wolf, *Principles of Optics*, p. 398 or the very fine elementary text by Towne, *Wave Phenomena*, p. 464.

are in fact incoherent, the lens system will form an image of the object, which consists of a distribution of partially overlapping, yet independent, Airy patterns. In the very finest lenses, where aberrations have been made negligible, the spreading out of each image point due to diffraction represents the ultimate limit on image quality.

Suppose that we simplify matters somewhat and only examine two equal irradiance, incoherent, distant point sources. For example, consider two stars seen through the objective lens of a telescope, where the entrance pupil corresponds to the diffracting aperture. In the previous section we saw that the radius of the Airy disk was given by $q_1 = 1.22 f\lambda/D$. If $\Delta\theta$ is the corresponding angular measure then $\Delta\theta = 1.22\lambda/D$, inasmuch as $q_1/f = \sin\Delta\theta \approx \Delta\theta$. The Airy disk for each star will be spread out over an angular half width $\Delta\theta$ about its geometrical image point, as shown in Fig. 10.32. If the angular separation of the stars is $\Delta\varphi$ and if $\Delta\varphi \gg \Delta\theta$, the images will be distinct and easily resolved. As the stars approach each other their respective images come together, overlap and commingle into a single blend of fringes. Adopting Lord Rayleigh's criterion, the stars are said to be *just resolved* when the center of one Airy disk falls on the first minimum of the Airy pattern of the other star. (We can certainly do a bit better than this, but Rayleigh's criterion, however arbitrary, has the virtue of being particularly uncomplicated.*) The *minimum resolvable angular separation* or *angular limit of resolution* is

$$(\Delta\varphi)_{min} = \Delta\theta = 1.22\lambda/D \qquad (10.59)$$

as depicted in Fig. 10.33. If $\Delta\ell$ is the center-to-center separation of the images, the *limit of resolution* is

$$(\Delta\ell)_{min} = 1.22 f\lambda/D. \qquad (10.60)$$

The *resolving power* for an image-forming system is generally defined as either $1/(\Delta\varphi)_{min}$ or $1/(\Delta\ell)_{min}$.

If the smallest resolvable separation between images is to be reduced, i.e. if the resolving power is to be increased, the wavelength, for instance, might be made smaller. Using ultraviolet rather than visible light in microscopy allows for the perception of finer detail. The electron microscope utilizes equivalent wavelengths of about 10^{-4} to 10^{-5} that of light. This makes it possible to examine objects which

would otherwise be completely obscured by diffraction effects in the visible spectrum. On the other hand, the resolving power of a telescope can be increased by increasing the diameter of the objective lens or mirror. Besides collecting more of the incident radiation, this would also result in a smaller Airy disk and therefore a sharper, brighter image. The Mount Palomar 200 inch telescope has a mirror 5 m in diameter (neglecting the obstruction of a small region at its center). At 550 nm it has an angular limit of resolution of 2.6×10^{-2} s of arc. In contrast, the 250 ft diameter Jodrell Bank radio telescope operates at a rather long 21 cm wavelength. It therefore has a limit of resolution of only about 700 s of arc. The human eye has a pupil diameter which, of course, varies. Taking it under bright conditions to be about 2 mm, with $\lambda = 550$ nm, $(\Delta\varphi)_{min}$ turns out to be roughly one minute of arc. With a focal length of about 20 mm, $(\Delta\ell)_{min}$ on the retina is 6700 nm. This is roughly twice the mean spacing between receptors. The human eye should therefore be able to resolve two points, an inch apart, at a distance of some 100 yards. You will probably not be able to do quite that well; one part in one thousand is more likely.

We should mention, before leaving this section, that a more appropriate criterion for resolving power has been proposed by C. Sparrow. Recall that at the Rayleigh limit there is a central minimum or saddle point between adjacent peaks. Decreasing the distance between the two point sources even further will cause the central dip to grow shallower and ultimately disappear. The angular separation corresponding to that configuration is Sparrow's limit. As shown in Fig. 10.40 the resultant maximum has a broad flat top, i.e. at the origin, which is the center of the peak, the second derivative of the irradiance function is zero; there is no change in slope.

Unlike the Rayleigh rule which rather tacitly assumes incoherence, the Sparrow condition can readily be generalized to coherent sources. In addition, astronomical studies of equal-brightness stars have shown that Sparrow's criterion is by far the more realistic.

10.2.7 The Diffraction Grating

A repetitive array of diffracting elements, either apertures or obstacles, which has the effect of producing periodic alterations in the phase, amplitude, or both, of an emergent wave is said to be a *diffraction grating*. One of the simplest such arrangements is the multiple-slit configuration of Section 10.2.3. It seems to have been invented by the American astronomer David Rittenhouse in about 1785. Some years later Joseph von Fraunhofer quite independently

---

* In Rayleigh's own words: "This rule is convenient on account of its simplicity and it is sufficiently accurate in view of the necessary uncertainty as to what exactly is meant by resolution." See Section 9.8.2 for further discussion.

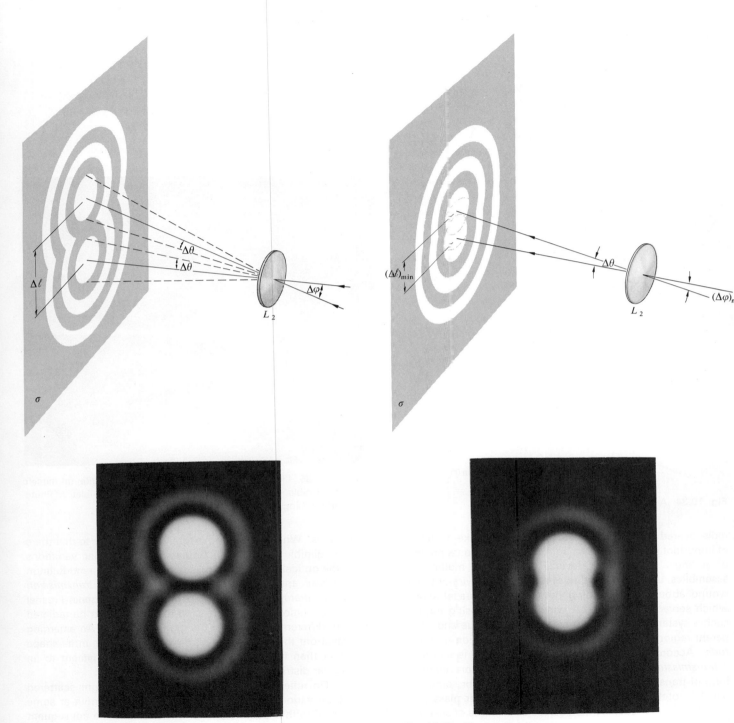

Fig. 10.32 Overlapping images.   **Fig. 10.33** Overlapping images.

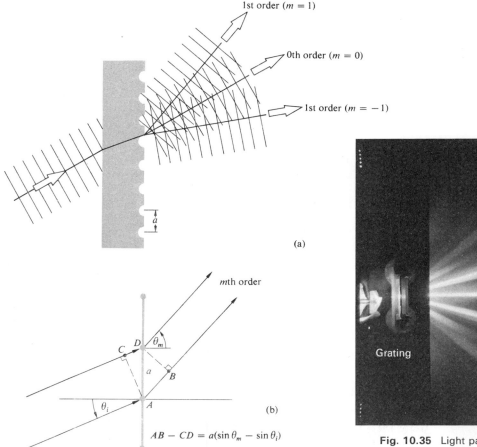

1st order ($m = 1$)

0th order ($m = 0$)

1st order ($m = -1$)

a

(a)

*m*th order

θ_m

C
D

a

B

θ_i

A

(b)

$$AB - CD = a(\sin \theta_m - \sin \theta_i)$$

Fig. 10.34 A transmission grating.

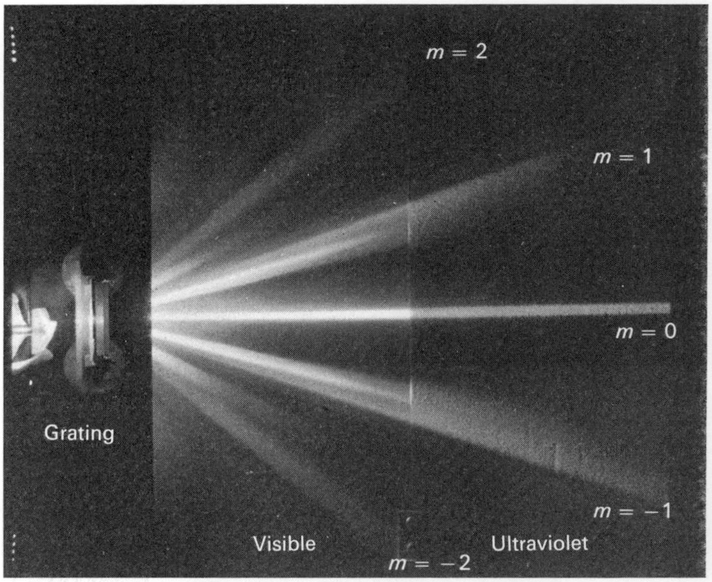

$m = 2$

$m = 1$

$m = 0$

Grating

$m = -1$

Visible   Ultraviolet

$m = -2$

Fig. 10.35 Light passing through a grating. The region on the left is the visible spectrum, that on the right the ultraviolet. [Photo courtesy Klinger Scientific Apparatus Corp.]

rediscovered the principle, and went on to make a number of important contributions to both the theory and technology of gratings. The earliest devices were indeed multiple-slit assemblies, usually consisting of a grid of fine wire or thread wound about and extending between two parallel screws which served as spacers. A wavefront, in passing through such a system, is confronted by alternate opaque and transparent regions, so that it undergoes a modulation in *amplitude*. Accordingly, a multiple-slit configuration is said to be a *transmission amplitude grating*. Another more common form of transmission grating is made by ruling or scratching parallel notches into the surface of a flat, clear glass plate [Fig. 10.34(a)]. Each scratch serves as a source of scattered light and together they form a regular array of parallel line

sources. When the grating is totally transparent, so that there is negligible amplitude modulation, the regular variations in the optical thickness across the grating yield a modulation in *phase* and we have what is known as a *transmission phase grating* (Fig. 10.35). In the Huygens–Fresnel representation you can envision the wavelets to be radiated with different phases over the grating surface. An emerging wavefront therefore contains periodic variations in its shape rather than its amplitude. This in turn is equivalent to an angular distribution of constituent plane waves.

On reflection from this kind of grating, light scattered by the various periodic surface features will arrive at some point P with a definite phase relationship. The consequent interference pattern generated after reflection is quite

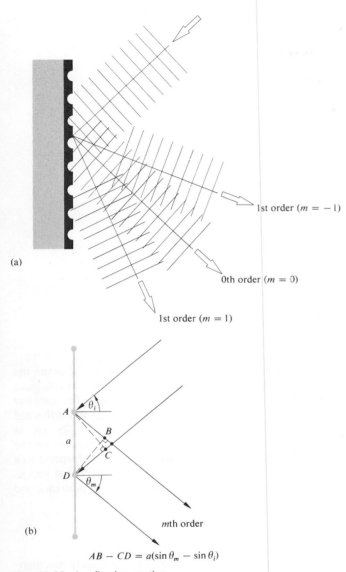

(a)

(b)

$$AB - CD = a(\sin \theta_m - \sin \theta_i)$$

Fig. 10.36 A reflection grating.

similar to that arising from transmission. Gratings designed specifically to function in this fashion are known as *reflection phase gratings* (Fig. 10.36). Contemporary gratings of this sort are generally ruled in thin films of aluminum which have been evaporated onto optically flat glass blanks. The aluminum, being fairly soft, results in less wear on the diamond ruling tool and is also a better reflector in the ultraviolet region.

The manufacture of ruled gratings is extremely difficult and relatively few are made. In actuality most gratings are exceedingly good plastic castings or *replicas* of fine, master ruled gratings.

If you were to look perpendicularly through a transmission grating at a distant parallel line source, your eye would serve as a focusing lens for the diffraction pattern. Recall the analysis of Section 10.2.3 and the expression

$$a \sin \theta_m = m\lambda, \qquad [10.32]$$

which is known as the *grating equation* for normal incidence. The values of m specify the *order* of the various principal maxima. For a source having a broad continuous spectrum, such as a tungsten filament, the $m = 0$, or zeroth-order image corresponds to the undeflected, $\theta_0 = 0$, white light view of the source. The grating equation is λ dependent and so, for any value of $m \neq 0$ the various colored images of the source corresponding to slightly different angles (θ_m), spread out into a continuous spectrum. The regions occupied by the faint subsidiary maxima will show up as bands seemingly devoid of any light. The first-order spectrum $m = \pm 1$ appears on either side of $\theta = 0$ and is followed, along with alternate intervals of darkness, by the higher-order spectra, $m = \pm 2, \pm 3$, etc. Notice that the smaller a becomes in Eq. (10.32), the fewer will be the number of visible orders.

Consider next the somewhat more general situation of oblique incidence as depicted in Figs. 10.34 and 10.36. The grating equation, for both transmission and reflection, becomes

$$a(\sin \theta_m - \sin \theta_i) = m\lambda. \qquad (10.61)$$

This expression applies equally well regardless of the refractive index of the transmission grating itself (Problem 10.8). One of the main disadvantages of the devices examined thus far, and in fact the reason for their present obsolescence, is that they spread the available light energy out over a number of low-irradiance spectral orders. For a grating like that shown in Fig. 10.36 most of the incident light undergoes *specular reflection* as if from a plane mirror. It follows from the grating equation that $\theta_m = \theta_i$ corresponds to the zeroth order, $m = 0$. All of this light is essentially wasted, at least for spectroscopic purposes, since the constituent wavelengths overlap.

In an article in the *Encyclopaedia Britannica* of 1888 Lord Rayleigh suggested that it was at least theoretically possible to shift energy out of the useless zeroth order into one of the higher-order spectra. So motivated, Robert

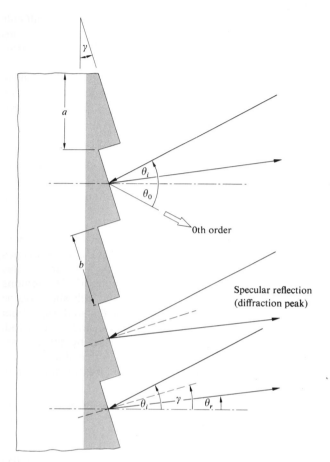

Fig. 10.37 Section of a blazed reflection phase grating.

Williams Wood (1868–1955) succeeded in 1910 in ruling grooves having a controlled shape, as shown in Fig. 10.37. Most modern gratings are of this shaped or *blazed* variety. The angular positions of the nonzero orders, θ_m-values, are determined by a, λ and, of more immediate interest, θ_i. But θ_i and θ_m are measured from the normal to the grating plane and not with respect to the individual groove surfaces. On the other hand, the location of the peak in the single facet diffraction pattern corresponds to *specular reflection* off that face, of each groove. It is governed by the *blaze angle* γ and can be varied independently of θ_m. This is somewhat analogous to the antenna array of Section 10.1.3 where we were able to control the spatial position of the interference pattern (10.6) by adjusting the relative phase shift between sources without actually changing their orientations.

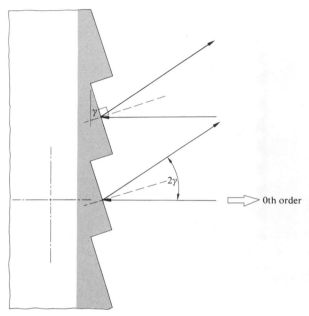

Fig. 10.38 Blazed grating.

Consider the situation depicted in Fig. 10.38 when the incident wave is normal to the plane of a blazed reflection grating i.e. $\theta_i = 0$ and so for $m = 0$, $\theta_0 = 0$. For *specular reflection* $\theta_i - \theta_r = 2\gamma$ (Fig. 10.37), most of the diffracted radiation is now concentrated about $\theta_r = -2\gamma$ (θ_r is negative because the incident and refracted rays are on the same side of the grating normal). This will correspond to a particular nonzero order, on one side of the central image, when $\theta_m = -2\gamma$ i.e. $a \sin(-2\gamma) = m\lambda$ for the desired λ and m.

GRATING SPECTROSCOPY

Quantum mechanics, which evolved in the early nineteen-twenties, had its initial thrust in the area of atomic physics. Predictions were made concerning the detailed structure of the hydrogen atom as manifested by its emitted radiation, and spectroscopy provided the vital proving ground. The need for larger and better gratings became apparent. Grating spectrometers, used over the range from soft x-rays to the far infrared, have enjoyed continued interest. In the hands of the astrophysicist, or rocket borne, they are yielding information concerning the very origins of the universe, information as varied as the temperature of a star, the rotation

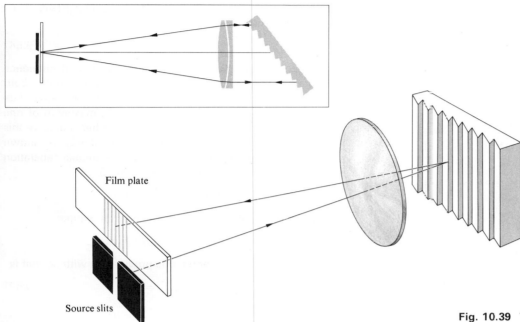

Film plate

Source slits

Fig. 10.39 The Littrow autocollimation mounting.

of a galaxy or the red shift in the spectrum of a quasar. As a further indication of the vitality of the technology, we point out that George R. Harrison and George W. Stroke in recent times have remarkably improved the quality of high-resolution gratings. They used a ruling engine* whose operation was controlled by an interferometrically guided servomechanism.

Let us now examine in some detail a few of the major features of the grating spectrum. Assume an infinitesimally narrow incoherent source. The effective width of an emergent spectral line may be defined as the angular distance between the zeroes on either side of a principal maximum, i.e. $\Delta\alpha = 2\pi/N$, which follows from Eq. (10.33). At oblique incidence we can redefine α as $(ka/2)(\sin\theta - \sin\theta_i)$ and so a *finite* change in α is given by

$$\Delta\alpha = (ka/2)\cos\theta(\Delta\theta) = 2\pi/N, \qquad (10.62)$$

where the angle of incidence is constant, i.e. $\Delta\theta_i = 0$. Thus even when the incident light is monochromatic

$$\Delta\theta = 2\lambda/(Na\cos\theta_m) \qquad (10.63)$$

is the *angular width of a line*, due to *instrumental broadening*. Interestingly enough, the angular linewidth varies inversely with the width of the grating itself, Na. Another quantity of import is the difference in angular position corresponding to a difference in wavelength. The *angular dispersion*, as in the case of a prism, is defined as

$$\mathcal{D} \equiv d\theta/d\lambda. \qquad (10.64)$$

Differentiating the grating equation yields

$$\mathcal{D} = m/a\cos\theta_m. \qquad (10.65)$$

This means that the angular separation between two different frequency lines will increase as the order increases.

Blazed plane gratings with nearly rectangular grooves are most often mounted so that the incident propagation vector is almost normal to either one of the groove faces. This is the condition of *autocollimation*, in which θ_i and θ_m are on the same side of the normal and $\gamma \approx \theta_i \approx -\theta_m$ (see Fig. 10.39) whereupon

$$\mathcal{D}_{\text{auto}} = 2\tan\theta_i/\lambda \qquad (10.66)$$

which is independent of a.

When the wavelength difference between two lines is small enough so that they overlap, the resultant peak becomes

---

* For more details about these marvelous machines see A. R. Ingalls, *Sci. Amer.* **186**, 45 (1952) or the article by E. W. Palmer and J. F. Verrill, *Contemp. Phys.* **9**, 257 (1968).

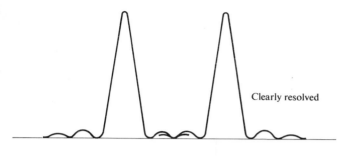

Clearly resolved

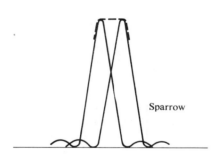

Rayleigh

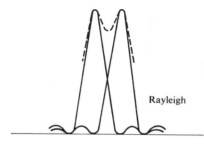

Sparrow

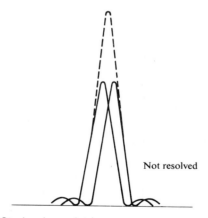

Not resolved

Fig. 10.40 Overlapping point images.

somewhat ambiguous. The chromatic resolving power $\mathscr{R}$ of a spectrometer is defined as

$$\mathscr{R} \equiv \lambda/(\Delta\lambda)_{min} \qquad [9.54]$$

where $(\Delta\lambda)_{min}$ is the least resolvable wavelength difference or *limit of resolution* and λ is the mean wavelength. Lord Rayleigh's criterion for the resolution of two equal flux density fringes requires that the principal maximum of one coincides with the first minimum of the other (compare this to the equivalent statement used in Section 9.8.2). As shown in Fig. 10.40 at the limit of resolution the angular separation is half of the linewidth, or from Eq. (10.63)

$$(\Delta\theta)_{min} = \lambda/Na \cos\theta_m.$$

Applying the expression for the dispersion we get

$$(\Delta\theta)_{min} = (\Delta\lambda)_{min} m/a \cos\theta_m.$$

Combining these two equations provides us with $\mathscr{R}$, that is

$$\lambda/(\Delta\lambda)_{min} = mN \qquad (10.67)$$

or

$$\mathscr{R} = \frac{Na(\sin\theta_m - \sin\theta_i)}{\lambda}. \qquad (10.68)$$

The resolving power is a function of the grating width Na, the angle of incidence and λ. A grating 6 inches wide and containing 15,000 lines per inch will have a total of 9×10^4 lines and a resolving power, in the second order, of 1.8×10^5. In the vicinity of 540 nm the grating could resolve a wavelength difference of 0.003 nm. Notice that the resolving power cannot exceed $2Na/\lambda$, which occurs when $\theta_i = -\theta_m = 90°$. The largest values of $\mathscr{R}$ are obtained when the grating is used in autocollimation, whereupon

$$\mathscr{R}_{auto} = \frac{2Na \sin\theta_i}{\lambda} \qquad (10.69)$$

and again θ_i and θ_m are both on the same side of the normal. For one of Harrison's 260 mm wide blazed gratings at about 75° in a Littrow mount, with $\lambda = 500$ nm, the resolving power just exceeds 10^6.

We now need to consider the problem of overlapping of orders. The grating equation makes it quite clear that a line of 600 nm in the first order will have precisely the same position, θ_m, as a 300 nm line in the second order or a 200 nm line when $m = 3$. If two lines of wavelength λ and $(\lambda + \Delta\lambda)$ in successive orders $(m + 1)$ and m just coincide then

$$a(\sin\theta_m - \sin\theta_i) = (m + 1)\lambda = m(\lambda + \Delta\lambda).$$

That precise wavelength difference is known as the *free spectral range*

$$(\Delta\lambda)_{fsr} = \lambda/m \qquad (10.70)$$

as it was for the Fabry–Perot interferometer. In comparison with that device, whose resolving power was

$$\mathcal{R} = \mathscr{F}m \qquad [9.54]$$

we might take N to be the finesse of a diffraction grating (Problem 10.9).

A high-resolution grating blazed for the first order, so as to have the greatest free spectral range, will require a high groove density (up to about 1200 lines per mm) in order to maintain $\mathcal{R}$. Equation (10.68) shows that $\mathcal{R}$ can be kept constant by ruling fewer lines with increasing spacing such that the grating width Na is constant. But this requires an increase in m and a subsequent decrease in free spectral range, characterized by overlapping orders. If this time N is held constant while a alone is made larger, $\mathcal{R}$ increases as does m so that $(\Delta\lambda)_{fsr}$ again decreases; the angular width of a line is reduced, i.e. the spectral lines become sharper, the coarser the grating becomes, but the dispersion in a given order diminishes, with the effect that the lines in that spectrum approach each other.

Thus far we have considered a particular type of periodic array, namely the *line grating*. A good deal more information is available in the literature* concerning their shapes, mountings, uses etc.

There are a few unlikely household items which can be used as crude gratings, along with a small light source. The grooved surface of a phonograph record works nicely near grazing incidence. And surprisingly enough, under the same conditions an ordinary fine-toothed comb will separate out the constituent wavelengths of white light. This occurs in exactly the same fashion as it would with a more orthodox reflection grating. In a letter to a friend dated May 12, 1673 James Gregory pointed out that sunlight passing through a feather would produce a colored pattern and he asked that his observations be conveyed to Mr. Newton. If you've got one, a feather makes a nice transmission grating.

TWO- AND THREE-DIMENSIONAL GRATINGS

Suppose that the diffracting screen Σ contains a large number,

---

* See F. Kneubühl, "Diffraction Grating Spectroscopy," *Appl. Opt.* **8**, 505 (1969), R. S. Longhurst, *Geometrical and Physical Optics* and the extensive article by G. W. Stroke in the *Encyclopedia of Physics*, Vol. 29, Edited by S. Flügge, p. 426.

N, of identical diffracting objects (apertures or obstacles). These are to be envisioned as distributed over the surface of Σ in a completely random manner. We also require that each and every one be similarly oriented. Imagine the diffracting screen to be illuminated by plane waves which are focused by a perfect lens L_2, after emerging from Σ (see Fig. 10.15). The individual apertures generate identical Fraunhofer diffraction patterns. All of these then overlap on the image plane σ. If there is no regular periodicity in the location of the apertures, we cannot anticipate anything but a random distribution in the relative phases of the waves arriving at an arbitrary point P on σ. We have to be rather careful, however, because there is one exception which occurs when P is on the central axis, i.e. $P = P_0$. All rays, from all apertures, parallel to the central axis will traverse equal optical path lengths before reaching P_0. They will therefore arrive in phase and interfere constructively. Now consider a group of arbitrarily directed parallel rays (not in the direction of the central axis), each one of which is emitted from a different aperture. These will be focused at some point on σ, such that each ray will have an equal probability of arriving with any phase between 0 and 2π. Interference effects amongst these rays should therefore, on the average, negate each other* (Section 7.1.1). We can expect to see the result of N exactly overlapping and *effectively noninteracting* Fraunhofer patterns. The flux density in a particular region on σ will therefore be N times the individual aperture flux density diffracted in that direction. In addition, a bright spot will exist at the very center of the pattern where the flux density is N^2 times that of the individual apertures. If, for example, the screen contains N rectangular holes [Fig. 10.41(a)], the resultant pattern [Fig. 10.41(b)] will resemble Fig. 10.24.† Similarly, an array of circular holes depicted in Fig. 10.41(c) will produce the diffraction rings of Fig. 10.41(d).

As the number of apertures increases, there will be a tendency for the central spot to become so bright as to obscure the rest of the pattern. Note as well that the above considerations apply when all of the apertures are illuminated completely coherently. In actuality, the diffracted flux-

---

* For a statistical treatment, consult J. M. Stone, *Radiation and Optics*, p. 146. Make sure to examine Section 11.3.3(ii) herein, i.e., *The Array Theorem.*

† The radial fiber structure evident in Fig. 10.41(b) arises from the statistical nature of the process. See e.g. Sommerfeld, *Optics*, p. 194. Also take a look at "Diffraction Plates for Classroom Demonstrations" by R. B. Hoover, *Am. J. Phys.* **37**, 871 (1969).

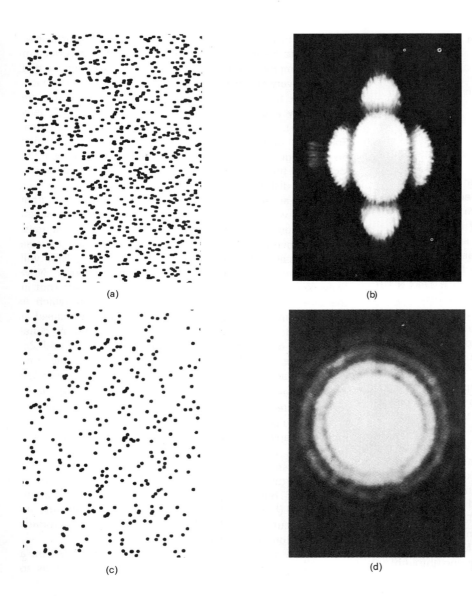

Fig. 10.41   (a) A random array of rectangular apertures.   (b) The resulting white-light Fraunhofer pattern.   (c) A random array of circular apertures. (d) The resulting white-light Fraunhofer pattern. [Photos courtesy The Ealing Corporation and Richard B. Hoover.]

density distribution will be determined by the degree of coherence (see Chapter 12). The pattern will run the gamut from no interference with completely incoherent light, to the case discussed above for completely coherent illumination (Problem 10.20).

The same kind of effects arise from what we might call a two dimensional *phase*-grating. For example, the halo or corona often seen about the sun or moon results from diffraction by random droplets of water vapor, i.e. cloud

particles. If you would like to duplicate the effect, rub a very thin film of talcum powder on a microscope slide and then fog it up with your breath. Look at a white light point source. You should see a pattern of clear, concentric, colored rings (10.56) surrounding a white central disk. If you just see a white blur, you don't have a distribution of roughly equal sized droplets; have another try at the talcum. Strikingly beautiful patterns approximating concentric ring systems can be seen through an ordinary *mesh* nylon stocking. If

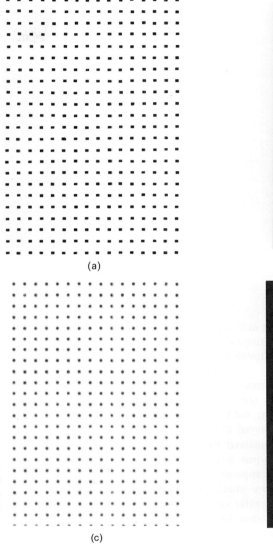

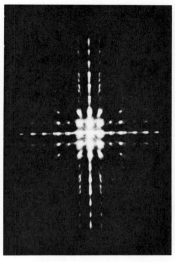

(a)

(b)

(c)

(d)

Fig. 10.42 (a) An ordered array of rectangular apertures. (b) The resulting white-light Fraunhofer pattern. (c) An ordered array of circular apertures.
(d) The resulting white-light Fraunhofer pattern.
[Photos courtesy Richard B. Hoover.]

you are fortunate enough to have mercury-vapor street
lights, you'll have no trouble seeing all of their constituent
visible spectral frequencies. (If not, block out most of a
fluorescent lamp, leaving something resembling a small
source.) Notice the increased symmetry as you increase
the number of layers of nylon. Incidentally, this is precisely
the way Rittenhouse, the inventor of the grating, became
interested in the problem, only he used a silk handkerchief.

Consider the case of a *regular* two-dimensional array
of diffracting elements (Fig. 10.42) under normally incident

plane wave illumination. Each small element behaves as a
coherent source. And because of the regular periodicity
of the lattice of emitters, each emergent wave bears a fixed
phase relation to the others. There will now be certain
directions in which constructive interference prevails.
Obviously, these occur when the distances from each diffracting element to P are such that the waves are nearly in phase
at arrival. The phenomenon can be observed by looking
at a point source through a piece of *square woven*, thin cloth
(like nylon curtain material) or the fine metal mesh of a tea

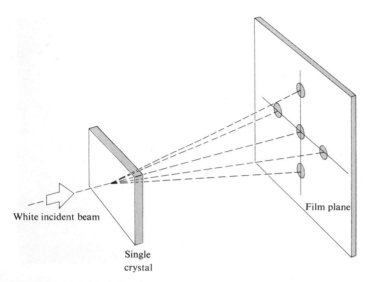

Fig. 10.43 Transmission Laue pattern.

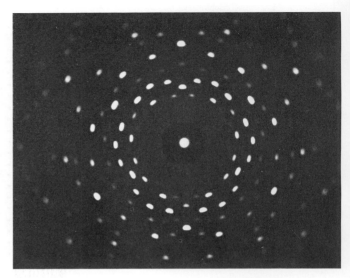

Fig. 10.44 X-ray diffraction pattern for quartz (SiO_2).

strainer (Fig. 10.84). The diffracted image is effectively the superposition of two grating patterns at right angles. Examine the center of the pattern carefully to see its grid-like structure.

As for the possibility of a *three-dimensional grating*, there seems to be no particular conceptual difficulty. A regular spatial array of scattering centers would certainly yield interference maxima in preferred directions. In 1912, Max van Laue (1879–1960) conceived the ingenious idea of using the regularly spaced atoms within a crystal as a three-dimensional grating. It is apparent from the grating equation (10.61) that if λ is very much greater than the grating spacing, only the zeroth order ($m = 0$) is possible. This is equivalent to $\theta_0 = \theta_i$, i.e. specular reflection. Since the spacing between atoms in a crystal is generally several angstroms ($1\text{Å} = 10^{-1}$ nm), light can only be diffracted in the zeroth order.

Von Laue's solution to the problem was to probe the lattice, not with light, but with x-rays whose wavelengths were comparable to the interatomic distances (Fig. 10.43). A narrow beam of white radiation (the broad continuous frequency range emitted by an x-ray tube) was directed onto a thin single crystal. The film plate (Fig. 10.44) revealed a Fraunhofer pattern consisting of an array of precisely located spots. These sites of constructive interference occurred whenever the angle between the beam and a set of atomic

planes within the crystal obeyed Bragg's law:

$$2d \sin \theta = m\lambda. \tag{10.71}$$

Notice that in x-ray work θ is traditionally measured from the plane and not the normal to it. Each set of planes diffracts a particular wavelength into a particular direction. Figure 10.45 rather strikingly shows the analogous behavior in a ripple tank.

Rather than reducing λ to the x-ray range, we could also have scaled everything up by a factor of about a billion and made a lattice of metal balls as a grating for microwaves.

10.3 FRESNEL DIFFRACTION

10.3.1 The Free Propagation of a Spherical Wave

In the Fraunhofer configuration, the diffracting system was relatively small, while the point of observation was very distant. Under these circumstances a few potentially problematic features of the Huygens–Fresnel principle could be completely passed over without concern. But we are now dealing with the near-field region which extends right up to the diffracting element itself and any such approximations would be inappropriate. We therefore return to the Huygens–Fresnel principle in order to reexamine it more closely. At any instant, every point on the primary wavefront is envisioned as a continuous emitter of spherical secondary

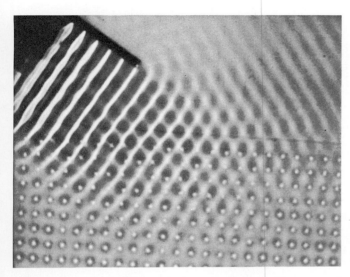

Fig. 10.45 Water waves in a ripple tank reflecting off an array of pegs acting as point scatterers. [Photo courtesy PSSC *Physics*, D. C. Heath, Boston, 1960.]

wavelets. But if each wavelet radiated uniformly in all directions, in addition to generating an ongoing wave, there would also appear a reverse wave traveling back toward the source. No such wave is found experimentally so that we must somehow modify the radiation pattern of the secondary emitters. We now introduce a function $K(\theta)$, known as the *obliquity* or *inclination factor*, in order to describe the directionality of the secondary emissions. Fresnel himself recognized the need to introduce a quantity of this kind, but he did little more than conjecture about its form.[*] It remained for the more analytic Kirchhoff formulation to provide an actual expression for $K(\theta)$ which, as we will see in Section

---

[*] It is interesting to read Fresnel's own words on the matter, keeping in mind that he was talking about light as an elastic vibration of the ether.

> Since the impulse communicated to every part of the primitive wave was directed along the normal, the motion which each tends to impress upon the ether ought to be more intense in this direction than in any other; and the rays which would emanate from it, if acting alone, would be less and less intense as they deviated more and more from this direction.
>
> The investigation of the law according to which their intensity varies about each center of disturbance is doubtless a very difficult matter; . . .

10.4, turns out to be

$$K(\theta) = \tfrac{1}{2}(1 + \cos \theta). \tag{10.72}$$

As shown in Fig. 10.46, θ is the angle made with the normal to the primary wavefront, **k**. This has its maximum value, $K(0) = 1$ in the forward direction and also dispenses with the back wave since $K(\pi) = 0$.

Let us now examine the free propagation of a spherical monochromatic wave emitted from a *point source S*. If the Huygens–Fresnel principle is correct, we should be able to add up the secondary wavelets arriving at a point P and thus obtain the unobstructed primary wave. In the process we will gain some insights, recognize a few shortcomings and develop a very useful technique. Consider the construction shown in Fig. 10.47. The spherical surface corresponds to the primary wavefront at some arbitrary time t' after it has been emitted from S at $t = 0$. The disturbance, having a radius ρ, can be represented by any one of the mathematical expressions describing a harmonic spherical wave, for example,

$$E = \frac{\mathcal{E}_0}{\rho} \cos (\omega t' - k\rho). \tag{10.73}$$

As illustrated, we have divided the wavefront into a number of annular regions. The boundaries of the various regions correspond to the intersections of the wavefront with a series of spheres centered at P of radius $r_0 + \lambda/2$, $r_0 + \lambda$, $r_0 + 3\lambda/2$ etc. These are the *Fresnel* or *half-period zones*. Notice that, for a secondary point source in one zone, there will be a point source in the next adjacent zone which is further from P by an amount $\lambda/2$. Since each zone, although small, is finite in extent we define a ring-shaped differential area element dS as indicated in Fig. 10.48. All of the point sources within dS are coherent and *we assume that each radiates in phase with the primary wave* (10.73). The secondary wavelets travel a distance r to reach P, at a time t, all arriving there with the same phase, $\omega t - k(\rho + r)$. The amplitude of the primary wave at a distance ρ from S is $\mathcal{E}_0/\rho$. We assume, accordingly, that the source strength per unit area $\mathcal{E}_A$ of the secondary emitters on dS is proportional to $\mathcal{E}_0/\rho$ by way of a constant Q, i.e. $\mathcal{E}_A = Q\mathcal{E}_0/\rho$. The contribution to the optical disturbance at P from the secondary sources on dS is, therefore,

$$dE = K \frac{\mathcal{E}_A}{r} \cos [\omega t - k(\rho + r)] \, dS. \tag{10.74}$$

The obliquity factor must vary slowly and may be presumed constant over a single Fresnel zone. To get dS as a function

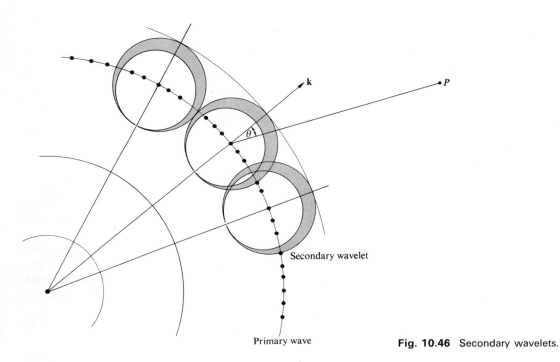

Fig. 10.46 Secondary wavelets.

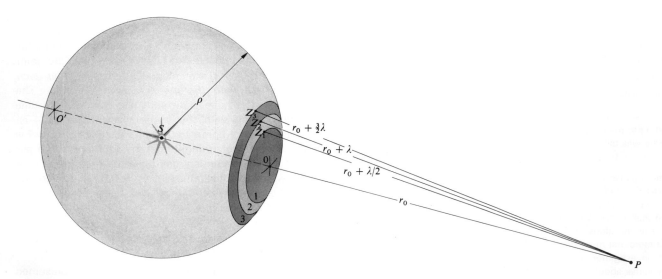

Fig. 10.47 Propagation of a spherical wavefront.

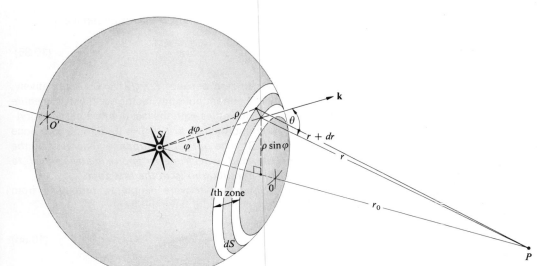

Fig. 10.48 Propagation of a spherical wavefront.

of r begin with

$$dS = \rho \, d\varphi \, 2\pi (\rho \sin \varphi),$$

applying the law of cosines get

$$r^2 = \rho^2 + (\rho + r_0)^2 - 2\rho(\rho + r_0) \cos \varphi.$$

Upon differentiation this yields

$$2r \, dr = 2\rho(\rho + r_0) \sin \varphi \, d\varphi$$

holding ρ and r_0 constant. Making use of the value of $d\varphi$ we find that the area of the element is therefore

$$dS = 2\pi \frac{\rho}{(\rho + r_0)} r \, dr. \qquad (10.75)$$

The disturbance arriving at P from the lth zone is

$$E_l = K_l 2\pi \frac{\mathcal{E}_A \rho}{(\rho + r_0)} \int_{r_{l-1}}^{r_l} \cos \left[\omega t - k(\rho + r) \right] dr$$

Hence

$$E_l = \frac{K_l \mathcal{E}_A \rho \lambda}{(\rho + r_0)} \left[\sin (\omega t - k\rho - kr) \right]_{r=r_{l-1}}^{r=r_l}.$$

Upon the introduction of $r_{l-1} = r_0 + (l-1)\lambda/2$ and $r_l = r_0 + l\lambda/2$ the expression reduces (Problem 10.10) to

$$E_l = (-1)^{l+1} \frac{2K_l \mathcal{E}_A \rho \lambda}{(\rho + r_0)} \sin \left[\omega t - k(\rho + r_0) \right]. \quad (10.76)$$

Observe that the amplitude of E_l alternates between positive and negative values depending on whether l is odd or even. This means that the contributions from adjacent zones are out of phase and tend to cancel. It is here that the obliquity factor makes a crucial difference. As l increases, θ increases and K decreases, so that successive contributions do not in fact completely cancel each other. It is interesting to note that E_l/K_l is independent of any position variables. Although the areas of each zone are almost equal, they do increase slightly as l increases, which means an increased number of emitters. But the mean distance from each zone to P also increases such that E_l/K_l is maintained constant (see Problem 10.11).

The sum of the optical disturbances from all m zones at P is

$$E = E_1 + E_2 + E_3 + \cdots + E_m$$

and since these alternate in sign, we can write

$$E = |E_1| - |E_2| + |E_3| + \cdots \pm |E_m|. \qquad (10.77)$$

If m is odd, the series can be reformulated in two ways, either as

$$E = \frac{|E_1|}{2} + \left(\frac{|E_1|}{2} - |E_2| + \frac{|E_3|}{2} \right) + \left(\frac{|E_3|}{2} - |E_4| + \frac{|E_5|}{2} \right) + \cdots \qquad (10.78)$$

$$+ \left(\frac{|E_{m-2}|}{2} - |E_{m-1}| + \frac{|E_m|}{2} \right) + \frac{|E_m|}{2},$$

or as

$$E = |E_1| - \frac{|E_2|}{2} - \left(\frac{|E_2|}{2} - |E_3| + \frac{|E_4|}{2}\right)$$

$$- \left(\frac{|E_4|}{2} - |E_5| + \frac{|E_6|}{2}\right) + \cdots$$

$$+ \left(\frac{|E_{m-3}|}{2} - |E_{m-2}| + \frac{|E_{m-1}|}{2}\right) - \frac{|E_{m-1}|}{2} + |E_m|. \quad (10.79)$$

There are now two possibilities: either $|E_l|$ is greater than the arithmetic mean of its two neighbors $|E_{l-1}|$ and $|E_{l+1}|$, or it is less than that mean. This is really a question concerning the rate of change of $K(\theta)$. When

$$|E_l| > (|E_{l-1}| + |E_{l+1}|)/2$$

each bracketed term is negative. It follows from Eq. (10.78) that

$$E < \frac{|E_1|}{2} + \frac{|E_m|}{2} \quad (10.80)$$

and from Eq. (10.79) that

$$E > |E_1| - \frac{|E_2|}{2} - \frac{|E_{m-1}|}{2} + |E_m|. \quad (10.81)$$

Since the obliquity factor goes from 1 to 0 over a great many zones, we can neglect any variation between adjacent zones, i.e. $|E_1| \approx |E_2|$ and $|E_{m-1}| \approx |E_m|$. Expression (10.81), to the same degree of approximation, becomes

$$E > \frac{|E_1|}{2} + \frac{|E_m|}{2}. \quad (10.82)$$

We conclude from (10.80) and (10.82) that

$$E \approx \frac{|E_1|}{2} + \frac{|E_m|}{2}. \quad (10.83)$$

This same result is obtained in the case when

$$|E_l| < (|E_{l-1}| + |E_{l+1}|)/2.$$

When the last term $|E_m|$, in the series of Eq. (10.77), corresponds to an even m the same procedure (Problem 10.12) leads to

$$E \approx \frac{|E_1|}{2} - \frac{|E_m|}{2}. \quad (10.84)$$

Fresnel conjectured that the obliquity factor was such that the last contributing zone occurred at $\theta = 90°$, i.e.

$$K(\theta) = 0 \text{ for } \pi/2 \le |\theta| \le \pi.$$

In that case Eqs. (10.83) and (10.84) both reduce to

$$E \approx \frac{|E_1|}{2} \quad (10.85)$$

when $|E_m|$ goes to zero because $K_m(\pi/2) = 0$. Alternatively, using Kirchhoff's correct obliquity factor, we divide the *entire* spherical wave into zones with the last or mth zone surrounding O'. Now θ approaches π, $K_m(\pi) = 0$, $|E_m| = 0$ and once again $E \approx |E_1|/2$. *The optical disturbance generated by the entire unobstructed wavefront is approximately equal to one half of the contribution from the first zone.*

If the primary wave were simply to propagate from S to P in a time t, it would have the form

$$E = \frac{\mathcal{E}_0}{(\rho + r_0)} \cos [\omega t - k(\rho + r_0)]. \quad (10.86)$$

Yet the disturbance synthesized from secondary wavelets, Eqs. (10.76) and (10.85), is

$$E = \frac{K_1 \mathcal{E}_A \rho \lambda}{(\rho + r_0)} \sin [\omega t - k(\rho + r_0)]. \quad (10.87)$$

These two equations must however be exactly equivalent, and we interpret the constants in Eq. (10.87) to make them so. Note that there is some latitude in how we do this. We prefer to have the obliquity factor equal to one in the forward direction, i.e. $K_1 = 1$ (rather than $1/\lambda$), from which it follows that Q must be equal to $1/\lambda$. In that case, $\mathcal{E}_A \rho \lambda = \mathcal{E}_0$ which is fine dimensionally. Keep in mind that $\mathcal{E}_A$ is the secondary wavelet source strength per unit area over the primary wavefront of radius ρ, and $\mathcal{E}_0/\rho$ is the amplitude of that primary wave $E_0(\rho)$. Thus $\mathcal{E}_A = E_0(\rho)/\lambda$. There is one other problem and that is the $\pi/2$ phase difference between Eqs. (10.86) and (10.87). This can be accounted for if we are willing to assume that the secondary sources radiate one quarter of a wavelength out of phase with the primary wave (see Section 3.3.2).

We have found it necessary to modify the initial statement of the Huygens–Fresnel principle but this should not distract us from our rather pragmatic reasons for using it, and these are twofold: (1) The Huygens–Fresnel theory can be shown to be an approximation of the Kirchhoff formulation and as such is no longer merely a contrivance and (2) it yields, in a fairly simple way, many predictions which are in fine agreement with experimental observations. Don't forget that it worked quite well in the Fraunhofer approximation.

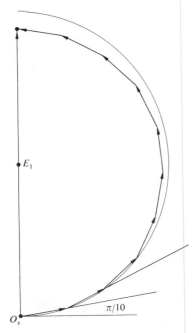

Fig. 10.49 Phasor addition.

10.3.2 The Vibration Curve

We now develop a graphical method for qualitatively analyzing a number of diffraction problems which arise predominantly from circularly symmetric configurations.

Imagine that the first, or polar Fresnel zone in Fig. 10.47 is divided into N subzones by the intersection of spheres, centered on P, of radii

$$r_0 + \lambda/2N, \; r_0 + \lambda/N, \; r_0 + 3\lambda/2N, \ldots, r_0 + \lambda/2.$$

Each subzone contributes to the disturbance at P, the resultant of which is of course just E_1. Since the phase difference across the entire zone, from O to its edge, is π rad (corresponding to $\lambda/2$) each subzone is shifted by π/N rad. Figure 10.49 depicts the vector addition of the subzone phasors, where, for convenience, $N = 10$. The chain of phasors deviates very slightly from the dashed circle, because the obliquity factor shrinks each successive amplitude. When the number of subzones is increased to infinity, i.e. $N \to \infty$, the polygon of vectors blends into a segment of a smooth spiral called a *vibration curve*. For each additional Fresnel zone, the vibration curve swings through *one half-turn* and a phase of π as it spirals inward. As shown in Fig. 10.50,

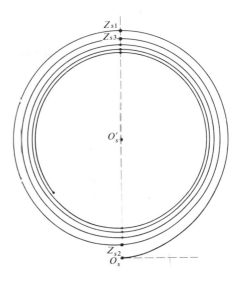

Fig. 10.50 The vibration curve.

the points $O_s, Z_{s1}, Z_{s2}, Z_{s3}, \ldots, O_s'$ on the spiral correspond to points $O, Z_1, Z_2, Z_3, \ldots, O'$ respectively, on the wavefront in Fig. 10.47. Each point $Z_1, Z_2, \ldots, Z_m$ lies on the periphery of a zone and so each point $Z_{s1}, Z_{s2}, \ldots, Z_{sm}$ is separated by a half-turn. We will see later, Eq. (10.91), that the radius of each zone is proportional to the square root of its numerical designation, m. The radius of the hundredth zone will be only ten times that of the first zone. Initially, therefore, the angle θ increases rapidly, and thereafter it gradually slows down as m becomes larger. Accordingly, $K(\theta)$ decreases rapidly only for the first few zones. The result is that as the spiral circulates around with increasing m, it becomes tighter and tighter, deviating from a circle by less for each revolution.

Keep in mind that the spiral is made up of an infinite number of phasors, each shifted by a small phase angle. The relative phase between any two disturbances at P, coming from two points on the wavefront, say O and A, can be depicted as shown in Fig. 10.51. The angle made by the tangents to the vibration curve, at points O_s and A_s, is β and this is the desired phase difference. If the point A is considered to lie on the boundary of a cap-shaped region of the wavefront, the resultant at P from the whole region is $\overline{O_s A_s}$ at an angle δ.

The total disturbance arriving at P from an unimpeded wave is the sum of the contributions from all of the zones between O and O'. The length of the vector from O_s to

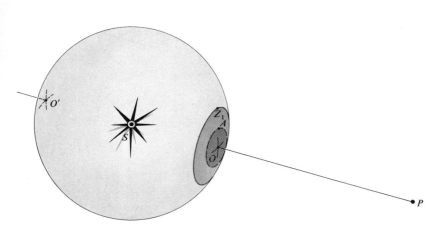

Fig. 10.51   Wavefront and corresponding vibration curve.

O_s' is therefore precisely that amplitude. Note that as expected, the amplitude O_sO_s' is just about one half of the contribution from the first zone, $\overrightarrow{O_sZ_{s1}}$. Observe that $\overrightarrow{O_sO_s'}$ has a phase of 90° with respect to the wave arriving at P from O. A wavelet emitted at O in phase with the primary excitation gets to P still in phase with the primary wave. This means that $\overrightarrow{O_sO_s'}$ is 90° out of phase with the unobstructed primary wave. This, as we have seen, is one of the shortcomings of the Fresnel formulation.

10.3.3 Circular Apertures

i) Spherical Waves

Fresnel's procedure applied to a point source can be used as a semiquantitative method with which to study diffraction at a circular aperture. Envision a monochromatic spherical wave impinging on a screen containing a small hole, as illustrated in Fig. 10.52. We first record the irradiance arriving at a very small sensor placed at point P on the symmetry axis. Our intent is to move the sensor around in space and so get a point-by-point map of the irradiance of the region beyond Σ.

   Let us assume that the sensor at P "sees" an integral number of zones m, filling the aperture. In actuality, the sensor merely records the irradiance at P; the zones having no reality. If m is even, then since $K_m \neq 0$

$$E = (|E_1| - |E_2|) + (|E_3| - |E_4|) + \cdots + (|E_{m-1}| - |E_m|).$$

Because each adjacent contribution is nearly equal

$$E \approx 0$$

and $I \approx 0$. If, on the other hand, m is odd

$$E = |E_1| - (|E_2| - |E_3|)$$
$$- (|E_4| - |E_5|) - \cdots - (|E_{m-1}| - |E_m|)$$

and

$$E \approx |E_1|$$

which is roughly twice the amplitude of the unobstructed wave. This is truly an amazing result. By inserting a screen in the path of the wave, thereby blocking out most of the wavefront, we have increased the irradiance at P by a factor of four. Conservation of energy clearly demands that there be other points where the irradiance has decreased. And, because of the complete symmetry of the set-up, we can expect a circular ring pattern. If m is not an integer, i.e. a fraction of a zone appears in the aperture, the irradiance at P is somewhere between zero and its maximum value. You might see this all a bit more clearly if you imagine that the aperture is expanding smoothly from an initial value of nearly zero. The amplitude at P can be gotten from the vibration curve where A is any point on the edge of the hole. The phasor magnitude O_sA_s is the desired amplitude of the optical field. Return to Fig. 10.51; as the hole increases, A_s moves counterclockwise around the spiral toward Z_{s1} and a maximum. Allowing the second zone in reduces O_sA_s to O_sZ_{s2}, which is nearly zero, and P becomes a dark spot. As the aperture increases, O_sA_s oscillates in length from near zero to a number of successive maxima, which themselves gradually decrease. Finally, when the hole is fairly large,

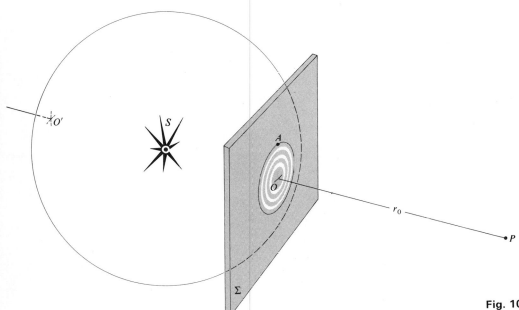

Fig. 10.52 A circular aperture.

the wave is essentially unobstructed, A_s approaches O'_s, and further changes in O_sA_s are imperceptible.

To map the rest of the pattern, we now move the sensor along any line perpendicular to the axis as shown in Fig. 10.53. At P, we assume that two complete zones fill the aperture and $E \approx 0$. At P_1 the second zone has been partially obscured and the third begins to show; E is no longer zero. At P_2 a good fraction of the second zone is hidden, while the third is even more evident. Since the contributions from the first and third zones are in phase, the sensor, placed anywhere on the dotted circle passing through P_2, records a bright spot. As it moves radially outward and portions of successive zones are uncovered, the sensor detects a series of relative maxima and minima. Figure 10.54 shows the diffraction patterns for a number of holes ranging in diameter from 1 mm to 4 mm as they appeared on a screen 1 m away. Starting from the top left and moving right, the first four holes are so small that only a fraction of the first zone is uncovered. The sixth hole uncovers the first and second zones and is therefore black at its center. The ninth hole uncovers the first three zones and is once again bright at its center. Notice that even slightly beyond the geometric shadow at P_3, in Fig. 10.53, the first zone is partially uncovered. Each of the last few contributing segments is only a small fraction of its respective zone and as such is negligible. The sum of all of

the amplitudes of the fractional zones, although small, is therefore still finite. Further into the geometric shadow, however, the entire first zone is obscured, the last terms are again negligible and this time the series does indeed go to zero and darkness.

We can gain a better appreciation of the actual size of the things we are dealing with by computing the number of zones in a given aperture. The area of each zone (from Problem 10.11) is given by

$$A = \frac{\rho}{(\rho + r_0)} \pi r_0 \lambda. \tag{10.88}$$

If the aperture has a radius R, to a good approximation the number of zones within it is simply

$$\frac{\pi R^2}{A} = \frac{(\rho + r_0)R^2}{\rho r_0 \lambda}. \tag{10.89}$$

For example, with a point source 1 m behind the aperture ($\rho \approx 1$ m), a plane of observation 1 m in front of it ($r_0 = 1$ m) and $\lambda = 500$ nm, there are 4 zones when $R = 1$ mm and 400 zones when $R = 1$ cm. When both ρ and r_0 are increased to the point where only a small fraction of a zone appears in the aperture, Fraunhofer diffraction occurs. This is essentially a restatement of the Fraunhofer condition of Section 10.1.2; see Problem 10.1 as well.

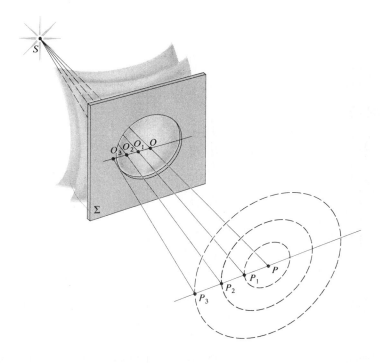

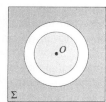

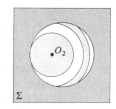

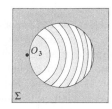

Fig. 10.53 Zones in a circular aperture.

It follows from Eq. (10.89) that the number of zones filling the aperture depends on the distance r_0 from P to O. As P moves in either direction along the central axis, the number of uncovered zones, whether increasing or decreasing, oscillates between odd and even integers. As a result, the irradiance goes through a series of maxima and minima. Clearly, this does not occur in the Fraunhofer configuration where, by definition, more than one zone cannot appear in the aperture.

ii) Plane Waves

Suppose now that the point source has been moved so far from the diffracting screen that the incoming light can be regarded as a plane wave ($\rho \to \infty$). Referring to Fig. 10.55, we now derive an expression for the radius of the mth

zone, R_m. Since $r_m = r_0 + m\lambda/2$ then

$$R_m^2 = (r_0 + m\lambda/2)^2 - r_0^2$$

and so

$$R_m^2 = mr_0\lambda + m^2\lambda^2/4. \qquad (10.90)$$

Under most circumstances the second term in Eq. (10.90) is negligible as long as m is not extremely large, consequently

$$R_m^2 = mr_0\lambda \qquad (10.91)$$

and *the radii are proportional to the square roots of integers.* Using a collimated He–Ne laser ($\lambda = 6328$ Å), the radius of the first zone is 1 mm when viewed from a distance of 1.58 m. Under these particular conditions Eq. (10.91) is applicable as long as $m \ll 10^7$, in which case $R_m = \sqrt{m}$ in millimeters. Figure 10.54 requires a slight modification in

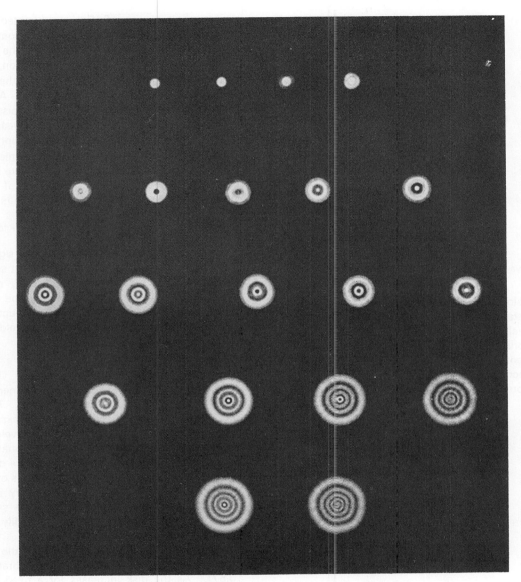

Fig. 10.54 Diffraction patterns for circular apertures of increasing size.

that now the lines $\overline{O_1P_1}$, $\overline{O_2P_2}$, and $\overline{O_3P_3}$ are perpendiculars dropped from the points of observation to Σ.

10.3.4 Circular Obstacles

In 1818 Fresnel entered a competition sponsored by the French Academy. His paper on the theory of diffraction ultimately won first prize and the title *Mémoire Courronné*, but not until it had provided the basis for a rather interesting story. The judging committee consisted of Pierre Laplace, Jean B. Biot, Siméon D. Poisson, Dominique F. Arago and Joseph L. Gay-Lussac, a formidable group indeed. Poisson, who was an ardent protagonist against the wave description of light, deduced a remarkable and seemingly untenable conclusion from Fresnel's theory. He showed that a bright

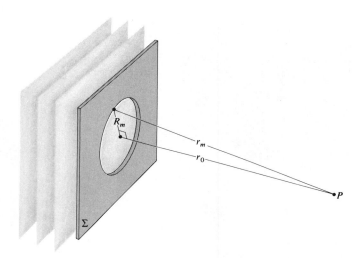

Fig. 10.55 Plane waves incident on a circular hole.

spot would be visible at the center of the shadow of a circular opaque obstacle; a result which he felt proved the absurdity of Fresnel's treatment. We can come to the same conclusion by considering the following somewhat over-simplified argument. Recall that an unobstructed wave yields a disturbance (10.85) given by $E \approx |E_1|/2$. If some sort of obstacle precisely covers the first Fresnel zone, so

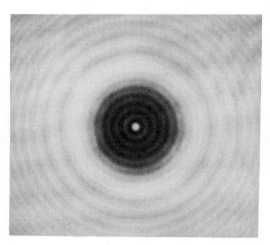

Fig. 10.56 Shadow of a $\frac{1}{8}$ inch diameter ball bearing. The bearing was glued to an ordinary microscope slide and illuminated with a He–Ne laser beam. There are some faint extraneous nonconcentric fringes arising from both the microscope slide and a lens in the beam. [Photo by E.H.]

that its contribution of $|E_1|$ is subtracted out, then $E \approx -|E_1|/2$. It is, therefore, possible that at some point P on the axis, the irradiance will be unaltered by the insertion of that obstruction. This surprising prediction, fashioned by Poisson as the death blow to the wave theory, was almost immediately verified experimentally by Arago; the spot actually existed. Amusingly enough, Poisson's spot, as it is now called, had been observed many years earlier (1723) by Maraldi but this work had long gone unnoticed.

We now examine the problem a bit more closely, since it is quite evident from Fig. 10.56 that there is a good deal of structure in the actual shadow pattern. If the opaque obstacle, be it a disk or sphere, obscures the first m zones, then

$$E = |E_{m+1}| - |E_{m+2}| + \cdots + |E_m|$$

(where, as before, there is no absolute significance to the signs other than that alternate terms must subtract). Unlike the analysis for the circular aperture, E_m now approaches zero because $K_m \to 0$. The series must be evaluated in the same manner as was that of the unobstructed wave (10.78 and 10.79). Repeating that procedure yields

$$E \approx \frac{|E_{m+1}|}{2} \tag{10.92}$$

and the irradiance on the central axis is generally only slightly less than that of the unobstructed wave. *There is a bright spot everywhere along the central axis except immediately behind the circular obstacle.* The wavelets propagating beyond the disk's circumference meet in phase on the central axis. Notice that as P moves close to the disk, θ increases, $K_{m+1} \to 0$ and the irradiance gradually falls off to zero. If the disk is large, the $(m + 1)$th zone is very narrow and any irregularities in the obstacle's surface may seriously obscure that zone. For Poisson's spot to be readily observable, the obstacle must be smooth and circular.

If A is a point on the periphery of the disk or sphere, A_s is the corresponding point on the vibration curve (Fig. 10.57). As the disk increases for a fixed P, A_s spirals in counterclockwise toward O_s' and the amplitude $A_s O_s'$ gradually decreases. The same thing happens as P moves towards a disk of constant size.

Off the axis, the zones covered in Fig. 10.53 for the circular aperture will now be exposed and vice versa. Accordingly, a whole series of concentric bright and dark rings will surround the central spot.

The opaque disk images S at P and would similarly form a crude image of every point in an extended source. R. W.

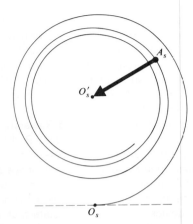

Fig. 10.57 The vibration curve applied to a circular obstruction.

Pohl has shown that a small disk can therefore be used as a crude positive lens.

The diffraction pattern can be seen with little difficulty, but you need a telescope or binoculars. Glue a small ball bearing ($\approx \frac{1}{8}$ or $\frac{1}{4}$ inch diameter) to a microscope slide which then serves as a handle. Place the bearing a few meters beyond the point source and observe it from three or four meters away. Position it directly in front of and completely obscuring the source. You will need the telescope to magnify the image since r_0 is so large. If you can hold the telescope steady, the ring system should be quite clear.

10.3.5 The Fresnel Zone Plate

In our previous considerations we utilized the fact that successive Fresnel zones tended to nullify each other. This suggests that we would observe a tremendous increase in irradiance at P if we remove all of either the even or odd zones. A screen which alters the light, either in amplitude or phase, coming from every other half-period zone is called a *zone plate*.*

Suppose that we construct a zone plate which passes only the first 20 odd zones and obstructs the even zones.

$$E = E_1 + E_3 + E_5 + \cdots + E_{39}$$

and each of these terms is approximately equal. For an unobstructed wavefront, the disturbance at P would be $E_1/2$

---

* Lord Rayleigh seems to have invented the zone plate as witnessed by this entry of April 11, 1871 in his notebook. "The experiment of blocking out the odd Huygens zones so as to increase the light at centre succeeded very well . . ."

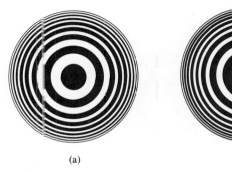

Fig. 10.53 Zone plates.

while with the zone plate in place $E \approx 20E_1$. The irradiance has been increased by a factor of 1600 times. The same result would obviously be true if the even zones were passed instead.

To calculate the radii of the zones shown in Fig. 10.58, refer to Fig. 10.59. The outer edge of the mth zone is marked with the point A_m. By definition, a wave which travels the path $S-A_m-P$ must arrive out of phase by $m\lambda/2$ with a wave which traverses the path $S-O-P$, that is

$$(\rho_m + r_m) - (\rho_0 + r_0) = m\lambda/2. \qquad (10.93)$$

Clearly

$$\rho_m = (R_m^2 + \rho_0^2)^{1/2}$$

and

$$r_m = (R_m^2 + r_0^2)^{1/2}.$$

Expand both these expressions using the binomial series. Since R_m is comparatively small, retaining only the first two terms yields

$$\rho_m = \rho_0 + \frac{R_m^2}{2\rho_0}$$

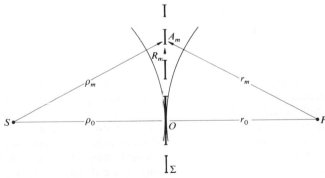

Fig. 10.59 Zone-plate geometry.

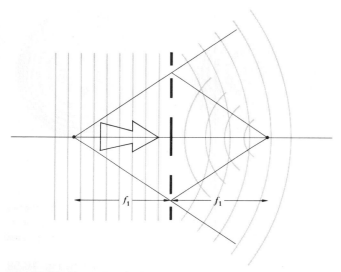

Fig. 10.60 Zone-plate foci.

and

$$r_m = r_0 + \frac{R_m^2}{2r_0}.$$

Finally then, substituting into Eq. (10.93) it becomes

$$\left(\frac{1}{\rho_0} + \frac{1}{r_0}\right) = \frac{m\lambda}{R_m^2}. \tag{10.94}$$

Under plane-wave illumination ($\rho_0 \to \infty$) and Eq. (10.94) reduces to

$$R_m^2 = mr_0\lambda, \tag{10.91}$$

which is an approximation of the exact expression stated by Eq. (10.90). Equation (10.94) has an identical form to that of the thin-lens equation, a fact which is not merely a coincidence since S is actually imaged in converging diffracted light at P. Accordingly, the *primary focal length* is said to be

$$f_1 = \frac{R_m^2}{m\lambda}. \tag{10.95}$$

(Note that the zone plate will show extensive chromatic aberration.) The points S and P are said to be conjugate foci. With a collimated incident beam (Fig. 10.60) the image distance is the primary or *first-order* focal length, which in turn corresponds to a principal maximum in the irradiance distribution. In addition to this real image there is also a virtual image formed of diverging light a distance f_1 in front of Σ. At a distance of f_1 from Σ each ring on the plate is filled by exactly one half-period zone on the wavefront. If we move a sensor along the $S–P$ axis toward Σ, it registers a series of very small irradiance maxima and minima until it arrives at a point $f_1/3$ from Σ. At that *third-order focal point*, there is a pronounced irradiance peak. Additional focal points will exist at $f_1/5$, $f_1/7$, etc., unlike a lens but even more unlike a simple opaque disk.

Following a suggestion by Lord Rayleigh, R. W. Wood constructed a *phase-reversal zone plate*. Rather than blocking out every other zone, he increased the thickness of alternate zones, thereby retarding their phase by π. Since the entire plate is transparent, the amplitude should therefore double and the irradiance increase by a factor of four. In actuality, the device does not work quite that well because the phase is not really constant over each zone. Ideally, the retardation should be made to vary gradually over a zone, jumping back by π at the start of the next zone.*

The usual way to make an optical zone plate is to draw a large-scale version and then photographically reduce it. Plates with hundreds of zones can be made by photographing a Newton's ring pattern, in collimated quasimonochromatic light. Rings of aluminum foil on cardboard work very well for microwaves.

Zone plates can be made of metal with a self-supporting spoked structure so that the transparent regions are devoid of any material. These will function as lenses in the range from ultraviolet to soft x-rays where ordinary glass is opaque.

10.3.6 Fresnel Integrals and the Rectangular Aperture

We now consider a class of problems within the domain of Fresnel diffraction which no longer have the circular symmetry of the previously studied configurations. Consider Fig. 10.61 where dS is an area element situated at some arbitrary point A whose coordinates are (y, z). The location of the origin O is determined by a perpendicular drawn to Σ from the position of the monochromatic point source. The contribution to the optical disturbance at P from the secondary sources on dS has the form given by Eq. (10.74). Making use of what we learned from the freely propagating wave ($\mathcal{E}_A\rho\lambda = \mathcal{E}_0$] we can rewrite that equation as

$$dE_p = \frac{K(\theta)\mathcal{E}_0}{\rho r\lambda} \cos\left[k(\rho + r) - \omega t\right] dS. \tag{10.96}$$

---

* See Ditchburn, *Light*, 2nd edn., p. 232, M. Sussman, "Elementary Diffraction Theory of Zone Plates," *Am. J. Phys.* **28**, 394 (1960), or Ora E. Myers, Jr., "Studies of Transmission Zone Plates," *Am. J. Phys.* **19**, 359 (1951).

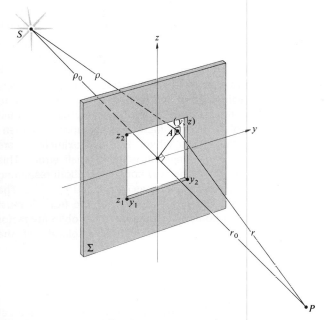

Fig. 10.61 Fresnel diffraction at a rectangular aperture.

The sign of the phase has changed from that of Eq. (10.74) and is written in this way to conform with traditional treatment. *In the case where the dimensions of the aperture are small* in comparison to ρ_0 and r_0, we can set $K(\theta) = 1$ and let $1/\rho r$ equal $1/\rho_0 r_0$ in the amplitude coefficient. Being more careful about approximations introduced into the phase, apply the Pythagorean theorem to triangles SOA and POA to get

$$\rho = (\rho_0^2 + y^2 + z^2)^{1/2}$$

and

$$r = (r_0^2 + y^2 + z^2)^{1/2}.$$

Expand these using the binomial series and form

$$\rho + r \approx \rho_0 + r_0 + (y^2 + z^2)\frac{\rho_0 + r_0}{2\rho_0 r_0}. \quad (10.97)$$

Observe that this is a more sensitive approximation than that used in the Fraunhofer analysis (10.40) where the term quadratic in the aperture variables was neglected. The disturbance at P in the complex representation is

$$E_p = \frac{\mathcal{E}_0 e^{-i\omega t}}{\rho_0 r_0 \lambda} \int_{y_1}^{y_2} \int_{z_1}^{z_2} e^{ik(\rho+r)} \, dy \, dz. \quad (10.98)$$

Following the usual form of derivation, we introduce the

dimensionless variables u and v defined by

$$u \equiv y\left[\frac{2(\rho_0 + r_0)}{\lambda \rho_0 r_0}\right]^{1/2}, \quad v \equiv z\left[\frac{2(\rho_0 + r_0)}{\lambda \rho_0 r_0}\right]^{1/2}. \quad (10.99)$$

Substituting Eq. (10.97) into Eq. (10.98) and utilizing the new variables, we arrive at

$$E_p = \frac{\mathcal{E}_0}{2(\rho_0 + r_0)} e^{i[k(\rho_0 + r_0) - \omega t]} \int_{u_1}^{u_2} e^{i\pi u^2/2} \, du \int_{v_1}^{v_2} e^{i\pi v^2/2} \, dv. \quad (10.100)$$

The term in front of the integral represents the unobstructed disturbance at P divided by 2; let us call it $E_u/2$. The integral itself can be evaluated using two functions, $\mathcal{C}(w)$ and $\mathcal{S}(w)$, where w represents either u or v. These quantities, which are known as the *Fresnel integrals*, are defined by

$$\mathcal{C}(w) \equiv \int_0^w \cos(\pi w'^2/2) \, dw', \quad \mathcal{S}(w) \equiv \int_0^w \sin(\pi w'^2/2) \, dw'. \quad (10.101)$$

Both functions have been extensively studied and their numerical values are well tabulated. Their interest to us at this point derives from the fact that

$$\int_0^w e^{i\pi w'^2/2} \, dw' = \mathcal{C}(w) + i\mathcal{S}(w),$$

and this, in turn, has the form of the integrals in Eq. (10.100). The disturbance at P is then

$$E_p = \frac{E_u}{2}[\mathcal{C}(u) + i\mathcal{S}(u)]_{u_1}^{u_2}[\mathcal{C}(v) + i\mathcal{S}(v)]_{v_1}^{v_2}, \quad (10.102)$$

which can be evaluated using the tabulated values of $\mathcal{C}(u_1)$, $\mathcal{C}(u_2)$, $\mathcal{S}(u_1)$ etc. The mathematics becomes rather involved if we compute the disturbance at all points of the plane of observation leaving the position of the aperture fixed. Instead we will fix the $S-O-P$ line and imagine that we move the aperture through small displacements in the Σ-plane. This has the effect of translating the origin O with respect to the fixed aperture, thereby scanning the pattern over the point P. Each new position of O corresponds to a new set of relative boundary locations y_1, y_2, z_1, and z_2. These in turn mean new values of u_1, u_2, v_1, and v_2 which, when substituted into Eq. (10.102), yield a new E_p. The error encountered in such a procedure is negligible as long as the aperture is displaced by distances which are small compared to ρ_0. This approach is therefore even more appropriate to incident plane waves. In that case if E_0 is the amplitude of the incoming plane wave at Σ, Eq. (10.96) becomes simply

$$dE_p = \frac{E_0 K(\theta)}{r\lambda} \cos (kr - \omega t) \, dS$$

where, as before, $\mathcal{E}_A = E_0/\lambda$. This time with

$$u = y \left(\frac{2}{\lambda r_0}\right)^{1/2}, \qquad v = z \left(\frac{2}{\lambda r_0}\right)^{1/2}, \qquad (10.103)$$

where we have divided the numerator and denominator in Eq. (10.99) by ρ_0 and then let it go to infinity, E_p takes the same form as Eq. (10.102) where E_u is again the unobstructed disturbance. The irradiance at P is $E_p E_p^*/2$, keeping in mind that E_u is complex; hence

$$I_p = \frac{I_0}{4} \{ [\mathcal{C}(u_2) - \mathcal{C}(u_1)]^2 + [\mathcal{S}(u_2) - \mathcal{S}(u_1)]^2 \}$$

$$\times \{ [\mathcal{C}(v_2) - \mathcal{C}(v_1)]^2 + [\mathcal{S}(v_2) - \mathcal{S}(v_1)]^2 \}, \quad (10.104)$$

where I_0 is the unobstructed irradiance at P.

As a simple example, envision a square hole 2 mm on each side under plane-wave illumination at 500 nm. If P is 4 m away and directly opposite point O at the center of the aperture, $u_2 = 1.0$, $u_1 = -1.0$, $v_2 = 1.0$, and $v_1 = -1.0$. The Fresnel integrals are both odd functions, i.e.

$$\mathcal{C}(w) = -\mathcal{C}(-w) \quad \text{and} \quad \mathcal{S}(w) = -\mathcal{S}(-w);$$

consequently

$$I_p = \frac{I_0}{4} \{ [2\mathcal{C}(1)]^2 + [2\mathcal{S}(1)]^2 \}^2$$

and a numerical value is easily obtained. To find the irradiance somewhere else in the pattern, e.g. 0.1 mm to the left of center, move the aperture relative to the OP-line accordingly, whereupon $u_2 = 1.1$, $u_1 = -0.9$, $v_2 = 1.0$, and $v_1 = -1.0$. The resultant I_p will also be equal to that found at 0.1 mm to the right of center. Indeed, because the aperture is square, the same value obtains 0.1 mm directly above and below center as well (Fig. 10.62).

We can approach the limiting case of free propagation by allowing the aperture dimensions to increase indefinitely. Making use of the fact that $\mathcal{C}(\infty) = \mathcal{S}(\infty) = \frac{1}{2}$ and $\mathcal{C}(-\infty) = \mathcal{S}(-\infty) = -\frac{1}{2}$ the irradiance at P, opposite the center of the aperture, is

$$I_p = I_0$$

which is exactly correct. This is rather remarkable considering that when the length $\overline{OA}$ is large, all of the approximations

made in the derivation are no longer applicable. it should be realized however, that a relatively small aperture which satisfies the approximations can still be large enough to effectively show no diffraction in the region opposite its center. For example, with $\rho_0 = r_0 = 1$ m an aperture which subtends an angle of about one or two degrees at P may correspond to values of $|u|$ and $|v|$ of roughly twenty-five to fifty. The quantities $\mathcal{C}$ and $\mathcal{S}$ are then very close to their limiting values of $\frac{1}{2}$. Further increasing the aperture dimensions beyond the point where the approximations are violated can, therefore, only introduce a small error. This implies that we need not be very concerned about restricting the actual aperture size (as long as $r_0 \gg \lambda$ and $\rho_0 \gg \lambda$). The contributions from wavefront regions remote from O must be quite small, a condition attributable to the obliquity factor and the inverse r-dependence of the amplitude of the secondary wavelets.

10.3.7 The Cornu Spiral

Marie Alfred Cornu (1841–1902), professor at the École Polytechnique in Paris, devised an elegant geometrical depiction of the Fresnel integrals, akin to the vibration curve already considered. Figure 10.63, which is known as the *Cornu spiral*, is a plot in the complex plane of the points $B(w) \equiv \mathcal{C}(w) + i\mathcal{S}(w)$ as w takes on all possible values from 0 to $\pm\infty$. This just means that we plot $\mathcal{C}(w)$ on the horizontal or real axis and $\mathcal{S}(w)$ on the vertical or imaginary axis. The appropriate numerical values are taken from the tables. If $d\ell$ be an element of arc length measured along the curve, then

$$d\ell^2 = d\mathcal{C}^2 + d\mathcal{S}^2.$$

From the definitions (10.101)

$$d\ell^2 = (\cos^2 \pi w^2/2 + \sin^2 \pi w^2/2) \, dw^2$$

and

$$d\ell = dw.$$

Values of w correspond to the arc length and are marked off along the spiral in Fig. 10.63. As w approaches $\pm\infty$ the curve spirals into its limiting values at $B^+ = \frac{1}{2} + i\frac{1}{2}$ and $B^- = -\frac{1}{2} - i\frac{1}{2}$. The slope of the spiral is

$$\frac{d\mathcal{S}}{d\mathcal{C}} = \frac{\sin \pi w^2/2}{\cos \pi w^2/2} = \tan \frac{\pi w^2}{2} \qquad (10.105)$$

and so the angle between the tangent to the spiral at any point and the $\mathcal{C}$-axis is $\beta = \pi w^2/2$.

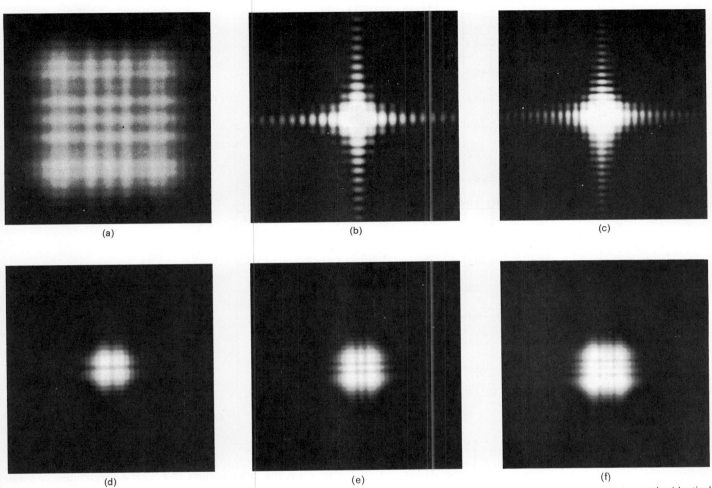

(a)   (b)   (c)

(d)   (e)   (f)

Fig. 10.62 (a) A typical Fresnel pattern for a square aperture. (b)–(f) A series of Fresnel patterns for various square apertures under identical conditions. Note that as the hole increases in size the pattern changes from a spread-out Fraunhofer-like distribution to a far more localized structure. [Photos by E.H.]

The Cornu spiral can be used either as a convenient tool for quantitative determinations or as an aid to gaining a qualitative picture of a diffraction pattern (as was the case with the vibration curve). As an example of its quantitative uses, reconsider the problem of a 2 mm square hole, dealt with in the previous section ($\lambda = 500$ nm, $r_0 = 4$ m and plane-wave illumination). We wish to find the irradiance at P directly opposite the aperture's center where, in this case, $u_1 = -1.0$ and $u_2 = 1.0$. The variable u is measured along the arc, i.e. w is replaced by u on the spiral. Place two points on the spiral at distances from O_s equal to u_1 and u_2; (these

are symmetric with respect to O_s because P is now opposite the aperture's center). Label the two points $B_1(u)$ and $B_2(u)$ respectively, as in Fig. 10.64. The phasor $\mathbf{B}_{12}(u)$ drawn from $B_1(u)$ to $B_2(u)$ is just the complex number $B_2(u) - B_1(u)$,

$$\mathbf{B}_{12}(u) = [\mathscr{C}(u) + i\mathscr{S}(u)]_{u_1}^{u_2},$$

and is the first term in the expression (10.102) for E_p. Similarly for $v_1 = -1.0$ and $v_2 = 1.0$, $B_2(v) - B_1(v)$ is

$$\mathbf{B}_{12}(v) = [\mathscr{C}(v) + i\mathscr{S}(v)]_{v_1}^{v_2},$$

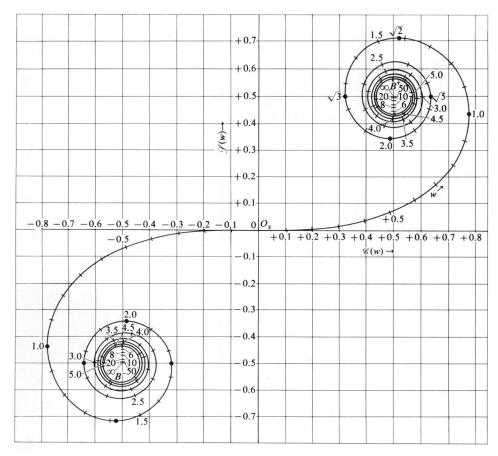

Fig. 10.63 The Cornu spiral.

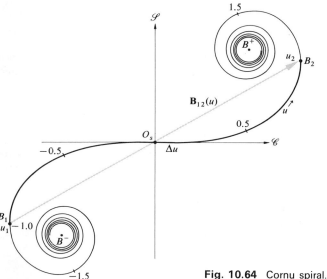

Fig. 10.64 Cornu spiral.

which is the latter portion of E_p. The magnitudes of these two complex numbers are just the lengths of the appropriate $\mathbf{B}_{12}$-phasors which can be read off the curve with a ruler, using either axis as a scale. The irradiance is then simply

$$I_p = \frac{I_0}{4}|\mathbf{B}_{12}(u)|^2|\mathbf{B}_{12}(v)|^2 \qquad (10.106)$$

and the problem is solved. Notice that the arc lengths along the spiral, i.e. $\Delta u = u_2 - u_1$ and $\Delta v = v_2 - v_1$, are proportional to the aperture's overall dimensions in the y- and z-directions, respectively. *The arc lengths are therefore constant regardless of the position of P in the plane of observation.* On the other hand, the phasors $\mathbf{B}_{12}(u)$ and $\mathbf{B}_{12}(v)$ which span the arc lengths are not constant, and they do indeed depend on the location of P.

Maintaining the position of P opposite the center of the diffracting hole, we now suppose the aperture size to be adjustable. As the square hole is gradually opened, Δv and Δu increase accordingly. The endpoints B_1 and B_2 of either of these arc lengths spiral around counterclockwise toward their limiting values of B^- and B^+ respectively. The phasors $\mathbf{B}_{12}(u)$ and $\mathbf{B}_{12}(v)$, which are identical in this instance because of the symmetry, pass through a series of extrema. The central spot in the pattern therefore gradually shifts from relative brightness to darkness and back. All the while, the entire irradiance distribution varies continually from one beautifully intricate display to the next (Fig. 10.62). For any particular aperture size the off-center diffraction pattern can be computed by repositioning P. It is helpful to visualize the arc length as a piece of string, whose measure is equal to either Δv or Δu. Imagine it lying on the spiral, with O_s initially at its midpoint. As P is moved, for example to the left along the y-axis (Fig. 10.61), y_1 and therefore u_1 both become less negative, while y_2 and u_2 increase positively. This has the effect that our Δu-string slides up the spiral. As the distance between the endpoints of the Δu-string changes, $|\mathbf{B}_{12}(u)|$ changes, and the irradiance (10.106) varies accordingly. When P is at the left edge of the geometric shadow $y_1 = u_1 = 0$. As the point of observation moves into the geometric shadow u_1 increases *positively* and the Δu-string is now entirely on the upper half of the Cornu spiral. As u_1 and u_2 continue to increase, the string winds ever more tightly about the B^+-limit. Its ends, B_1 and B_2, become closer to each other with the result that $|\mathbf{B}_{12}(u)|$ becomes quite small and I_p decreases within the geometric shadow region (we will come back to this point in more detail in the next section). The same process applies when we scan in the z-direction; Δv is constant and $\mathbf{B}_{12}(v)$ varies.

If the aperture is completely opened out revealing an unobstructed wave, $u_1 = v_1 = -\infty$ which means that $B_1(u) = B_1(v) = B^-$ and $B_2(u) = B_2(v) = B^+$. The B^-B^+-line makes a 45° angle with the $\mathscr{C}$-axis and has a length equal to $\sqrt{2}$. Consequently, the phasors $\mathbf{B}_{12}(u)$ and $\mathbf{B}_{12}(v)$ each have magnitude $\sqrt{2}$ and phase $\pi/4$, i.e. $\mathbf{B}_{12}(u) = \sqrt{2}\exp(i\pi/4)$ and $\mathbf{B}_{12}(v) = \sqrt{2}\exp(i\pi/4)$. It follows from Eq. (10.103) that

$$E_p = E_u e^{i\pi/2}, \qquad (10.107)$$

and as in Section 10.3.1 we have the unobstructed amplitude except for a $\pi/2$ phase discrepancy.* Finally, using (10.106), $I_p = I_0$.

---

* The phase discrepancy will be resolved by the Kirchhoff theory of Section 10.4.

Table 10.ʳ Fresnel integrals.

w	$\mathscr{C}(w)$	$\mathscr{S}(w)$	w	$\mathscr{C}(w)$	$\mathscr{S}(w)$
0.00	0.0000	0.0000	4.50	0.5261	0.4342
0.10	0.1000	0.0005	4.60	0.5673	0.5162
0.20	0.1999	0.0042	4.70	0.4914	0.5672
0.30	0.2994	0.0141	4.80	0.4338	0.4968
0.40	0.3975	0.0334	4.90	0.5002	0.4350
0.50	0.4923	0.0647	5.00	0.5637	0.4992
0.60	0.5811	0.1105	5.05	0.5450	0.5442
0.70	0.6597	0.1721	5.10	0.4998	0.5624
0.80	0.7230	0.2493	5.15	0.4553	0.5427
0.90	0.7648	0.3398	5.20	0.4389	0.4969
1.00	0.7799	0.4383	5.25	0.4610	0.4536
1.10	0.7638	0.5365	5.30	0.5078	0.4405
1.20	0.7154	0.6234	5.35	0.5490	0.4662
1.30	0.6386	0.6863	5.40	0.5573	0.5140
1.40	0.5431	0.7135	5.45	0.5269	0.5519
1.50	0.4453	0.6975	5.50	0.4784	0.5537
1.60	0.3655	0.6389	5.55	0.4456	0.5181
1.70	0.3238	0.5492	5.60	0.4517	0.4700
1.80	0.3336	0.4508	5.65	0.4926	0.4441
1.90	0.3944	0.3734	5.70	0.5385	0.4595
2.00	0.4882	0.3434	5.75	0.5551	0.5049
2.10	0.5815	0.3743	5.80	0.5298	0.5461
2.20	0.6363	0.4557	5.85	0.4819	0.5513
2.30	0.6266	0.5531	5.90	0.4486	0.5163
2.40	0.5550	0.6197	5.95	0.4566	0.4688
2.50	0.4574	0.6192	6.00	0.4995	0.4470
2.60	0.3890	0.5500	6.05	0.5424	0.4689
2.70	0.3925	0.4529	6.10	0.5495	0.5165
2.80	0.4675	0.3915	6.15	0.5146	0.5496
2.90	0.5624	0.4101	6.20	0.4676	0.5398
3.00	0.6058	0.4963	6.25	0.4493	0.4954
3.10	0.5616	0.5818	6.30	0.4760	0.4555
3.20	0.4664	0.5933	6.35	0.5240	0.4560
3.30	0.4058	0.5192	6.40	0.5496	0.4965
3.40	0.4385	0.4296	6.45	0.5292	0.5398
3.50	0.5326	0.4152	6.50	0.4816	0.5454
3.60	0.5880	0.4923	6.55	0.4520	0.5078
3.70	0.5420	0.5750	6.60	0.4690	0.4631
3.80	0.4481	0.5656	6.65	0.5161	0.4549
3.90	0.4223	0.4752	6.70	0.5467	0.4915
4.00	0.4934	0.4204	6.75	0.5302	0.5362
4.10	0.5738	0.4758	6.80	0.4831	0.5436
4.20	0.5418	0.5633	6.85	0.4539	0.5060
4.30	0.4494	0.5540	6.90	0.4732	0.4624
4.40	0.4383	0.4622	6.95	0.5207	0.4591

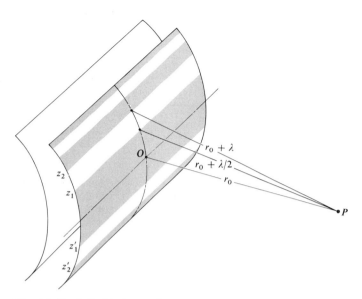

Fig. 10.65 Cylindrical wavefront zones.

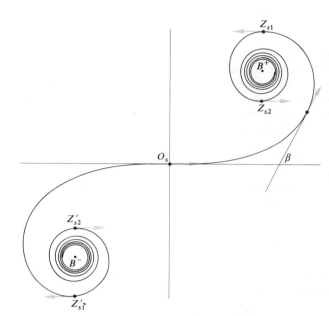

Fig. 10.66 Cornu spiral related to the cylindrical wavefront.

We can construct a more palpable picture of what the Cornu spiral represents by considering Fig. 10.65 which depicts a cylindrical wavefront propagating from a coherent line source. The present procedure is exactly the same as that used in deriving the vibration curve and the reader is referred back to Section 10.3.2 for a more leisurely discussion. Suffice it to say that the wavefront is divided into *half-period strip zones* by its intersection with a family of cylinders having a common axis and radii of $r_0 + \lambda/2$, $r_0 + \lambda$, $r_0 + 3\lambda/2$ etc. *The contributions from these strip zones are proportional to their areas, which decrease rapidly.* This is in contrast to the circular zones whose radii increase, thereby keeping the areas nearly constant. Each strip zone is in turn similarly divided into N subzones which have a relative phase difference of π/N. The vector sum of all of the amplitude contributions from zones above the center line is a spiraling polygon. Letting N go to ∞ and including the contributions generated by the strip zones below the center line causes the polygon to smooth out into a continuous Cornu spiral. This is not surprising since the coherent line source generates an infinite number of overlapping point-source patterns.

Figure 10.66 shows a number of unit tangent vectors at various positions along the spiral. The vector at O_s corresponds to the contribution from the central axis passing through O on the wavefront. The points associated with the boundaries of each strip zone can be located on the spiral

since at those positions the relative phase, β, is either an even or odd multiple of π. For example the point Z_{s1} on the spiral (Fig. 10.66), which is related to z_1 (Fig. 10.65) on the wavefront is, by definition, 180° out of phase with O_s. Therefore Z_{s1} must be located at the top of the spiral where $w = \sqrt{2}$, inasmuch as there $\beta = \pi w^2/2 = \pi$.

It will be helpful as we go along in the treatment to visualize the blocking out of these strip zones when analyzing the effects of obstructions. Quite obviously one could even make an appropriate zone plate which would accomplish this to some advantage and such devices are in use.

10.3.8 Fresnel Diffraction by a Slit

We can treat Fresnel diffraction at a long slit as an extension of the rectangular-aperture problem. We need only elongate the rectangle by allowing y_1 and y_2 to move very far from O, as shown in Fig. 10.67. As the point of observation moves along the y-axis, so long as the vertical boundaries at either end of the slit are still essentially at infinity, $u_1 \approx \infty$, $u_2 \approx -\infty$ and $\mathbf{B}_{12}(u) \approx \sqrt{2}e^{i\pi/4}$. From Eq. (10.106), for either point-source or plane-wave illumination,

$$I_p = \frac{I_0}{2}|\mathbf{B}_{12}(v)|^2, \qquad (10.108)$$

and the pattern is independent of y. The values of z_1 and

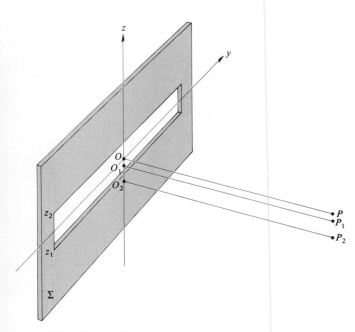

Fig. 10.67 Single-slit geometry.

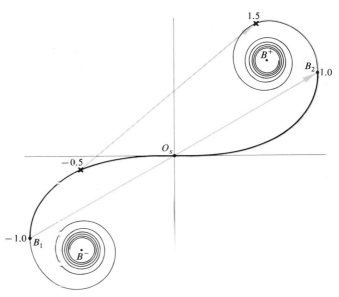

Fig. 10.68 Cornu spiral for the slit.

z_2 which fix the slit width determine the important parameter $\Delta v = v_2 - v_1$, which in turn governs $\mathbf{B}_{12}(v)$. Imagine once again that we have a string of length Δv lying along the spiral. At P, opposite point O, the aperture is symmetric and the string would be centered on O_s (Fig. 10.68). The chord $|\mathbf{B}_{12}(v)|$ need only be measured and substituted into Eq. (10.108) to find I_p. At point P_1, z_1 and therefore v_1 are smaller negative numbers while z_2 and v_2 have increased positively. The arc length Δv (the string) moves up the spiral (Fig. 10.68) and the chord decreases. As the point of observation moves down into the geometric shadow, the string winds about B^+ and the chord goes through a series of relative extrema. If Δv is very small, our imaginary piece of string is small and the chord $|\mathbf{B}_{12}(v)|$ decreases appreciably only when the radius of curvature of the spiral itself is small. This occurs in the vicinity of B^+ or B^- i.e. far out into the geometric shadow. There will, therefore, be light well beyond the edges of the aperture, as long as the aperture is relatively small. Note too that with small Δv there will be a broad central maximum. In fact, if Δv is much less than one, $r_0\lambda$ is much greater than the aperture width and the Fraunhofer condition prevails. This transition of Eq. (10.108) into the form of Eq. (10.17) is more plausible when we realize that for large w the Fresnel integrals have trigonometric representations (see Problem 10.15).

As the slit widens, Δv becomes larger, for a fixed r_0, until a configuration like that in Fig. 10.69 exists for a point opposite the slit's center. If the point of observation is moved vertically either up or down, Δv slides either down or up the spiral. Yet the chord increases in both cases, so that

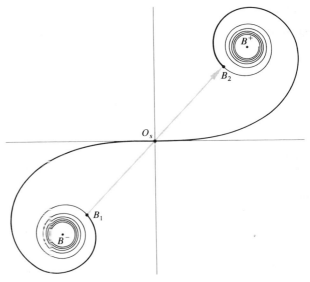

Fig. 10.69 An irradiance minimum in the slit pattern.

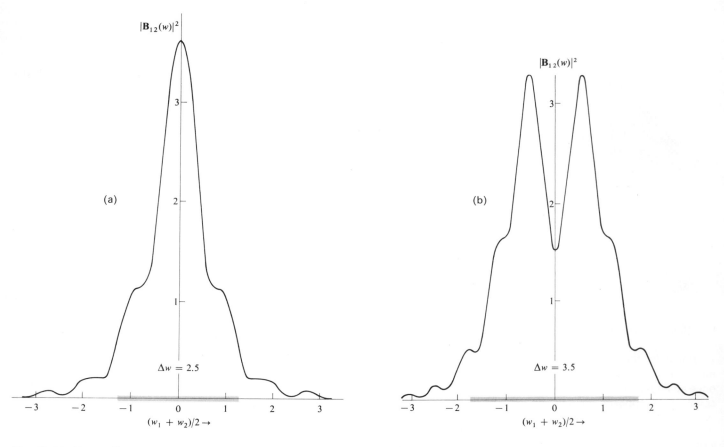

Fig. 10.70 $|\mathbf{B}_{12}(w)|^2$ versus $(w_1 + w_2)/2$ for (a) $\Delta w = 2.5$, and (b) $\Delta w = 3.5$.

the center of the diffraction pattern must be a relative minimum. Fringes now appear within the geometric image of the slit, unlike the Fraunhofer pattern.

Figure 10.70 shows two curves of $|\mathbf{B}_{12}(w)|^2$ plotted against $(w_1 + w_2)/2$ which is the center point of the arc length Δw. (Recall that the symbol w stands for either u or v.) A family of such curves running the range in Δw from about one to ten would cover the region of interest. The curves are computed by first choosing a particular Δw and then reading the appropriate $|\mathbf{B}_{12}(w)|$ values off the Cornu spiral as Δw slides along it. For a long slit

$$I_p = \frac{I_0}{2}|\mathbf{B}_{12}(v)|^2, \qquad [10.108]$$

and since Δz is the slit width which corresponds to Δv, each

curve in Fig. 10.70 *is proportional to the irradiance distribution* for a given slit. For example, Fig. 10.70(a) can be read as $|\mathbf{B}_{12}(v)|^2$ versus $(v_1 + v_2)/2$ for $\Delta v = 2.5$. The abscissa relates to $(z_1 + z_2)/2$, i.e. the displacement of the point of observation from the center of the slit. In Fig. 10.70(b) $\Delta w = 3.5$, which means that a slit having a $\Delta v = 3.5$ clearly has fringes appearing within the geometric image as expected (Problem 10.13). The curves could, of course, be plotted in terms of values of Δz or Δy explicitly, but that would unnecessarily limit them to only one set of configuration parameters ρ_0, r_0 and λ.

As the slit is widened still further, Δv approaches and then surpasses 10. An increasing number of fringes appear within the geometric image and the pattern no longer extends appreciably beyond that image.

The same kind of reasoning applies equally well to the analysis of the rectangular aperture where use can also be made of the curves of Fig. 10.70.

To actually observe Fresnel slit diffraction, form a long narrow space between two fingers held at arm's length. Make a similar parallel slit close to your eye using your other hand. With a *bright* source like the daytime sky or a large lamp illuminating the far slit, observe it through the nearby aperture. After inserting the near slit the far slit will appear to widen and rows of fringes will be clearly evident.

10.3.9 The Semi-infinite Opaque Screen

We now form a semi-infinite planar opaque screen by removing the upper half of Σ in Fig. 10.67. This is done simply enough by letting $z_2 = y_1 = y_2 = \infty$. Remembering the original approximations, we limit the geometry so that the point of observation is close to the screen's edge. Since $v_2 = u_2 = \infty$ and $u_1 = -\infty$, Eq. (10.104) or (10.108) leads to

$$I_p = \frac{I_0}{2}\{[\tfrac{1}{2} - \mathscr{C}(v_1)]^2 + [\tfrac{1}{2} - \mathscr{S}(v_1)]^2\}. \quad (10.109)$$

When the point P is directly opposite the edge $v_1 = 0$, $\mathscr{C}(0) = \mathscr{S}(0) = 0$ and $I_p = I_0/4$. This was to be expected, since half of the wavefront is obstructed, the amplitude of the disturbance is halved and the irradiance drops to one quarter. This occurs at point (3) in Figs. 10.71 and 10.72. Moving into the geometric shadow region to point (2) and then on to (1) and still further, the successive chords clearly decrease monotonically (Problem 10.15). No irradiance oscillations exist within that region; the irradiance merely drops off rapidly. At any point above (3) the screen's edge will be below it, i.e. $z_1 < 0$ and $v_1 < 0$. At about $v_1 = -1.2$ the chord reaches a maximum and the irradiance is a maximum. Thereafter I_p oscillates about I_0, gradually diminishing in magnitude. With sensitive electronic techniques many hundreds of these fringes can be observed.[*]

It is evident that the diffraction pattern of Fig. 10.73 would appear in the vicinity of the edges of a wide *slit* (Δv greater than about 10) as a limiting case. The irradiance distribution suggested by geometrical optics is obtained only when λ goes to zero. Indeed as λ decreases, the fringes move closer to the edge and become increasingly finer in extent.

The straight-edge pattern can be observed using any kind of slit, held up in front of a broad lamp at arm's length,

---

[*] J. D. Barnett and F. S. Harris, Jr., *J. Opt. Soc. Amer.* **52**, 637 (1962).

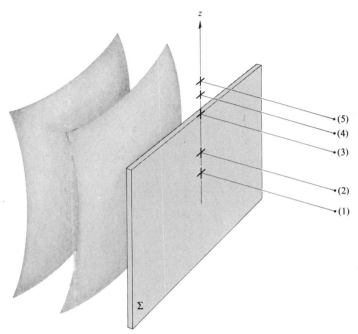

Fig. 10.71 The semi-infinite opaque screen.

as a source. Introduce an opaque obstruction (e.g. a blackened microscope slide or a razor blade) very near your eye. As the edge of the obstruction passes in front of the source slit parallel to it, a series of fringes will appear.

10.3.10 Diffraction by a Narrow Obstacle

Refer back to the discussion describing the single narrow slit; consider the complementary case where the slit is opaque and the screen transparent. Let's envision, for example, a vertical opaque wire. At a point directly opposite the wire's center there will be two separate contributing regions extending from y_1 to $-\infty$ and from y_2 to $+\infty$. On the Cornu spiral these correspond to two arc lengths from u_1 to B^- and from u_2 to B^+. The amplitude of the disturbance at a point F on the plane of observation is the magnitude of the *vector* sum of the two phasors $\overrightarrow{B^- u_1}$ and $\overrightarrow{u_2 B^+}$ illustrated in Fig. 10.74. As with the opaque disk, the symmetry is such that there will always be an illuminated region along the central axis. This can be seen from the spiral since when P is on the central axis, $\overrightarrow{B^- u_1} = \overrightarrow{u_2 B^+}$ and their sum can never be zero. The arc length Δu represents the obscured region of the spiral which increases as the diameter of the wire increases. For thick wires, u_1 approaches B^-, u_2 approaches B^+, the phasors decrease in length and the irradiance on

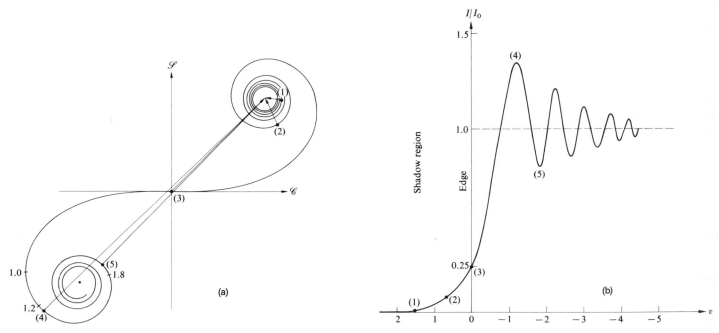

Fig. 10.72 (a) The Cornu spiral for a semi-infinite screen. (b) The corresponding irradiance distribution.

the shadow's axis drops off. This is evident in Fig. 10.75 which shows the patterns actually cast by a thin piece of lead from a mechanical pencil and by a $\frac{1}{8}$ inch diameter rod. Imagine that we have a small irradiance sensor at point P on the plane of observation (or the film plate). As P moves off the central axis to the right, y_1 and u_1 increase negatively while y_2 and u_2, which are positive, decrease. The opaque region Δu, slides down the spiral. When the sensor is at the right edge of the geometrical shadow $y_2 = 0$, $u_2 = 0$, i.e. u_2 is at O_s. Notice that if the wire is thin, i.e. if Δu is small,

the sensor will record a gradual decrease in irradiance as u_2 approaches O_s. On the other hand, if the wire is thick Δu is large and u_1 and u_2 are large. As Δu slides down the spiral, the two phasors revolve through a number of complete rotations going in and out of phase in the process. The resulting additional extrema appearing within the geometrical shadow are evident in Fig. 10.75(b). In fact, the separation between internal fringes varies inversely with the width of the rod just as if the pattern arose from the interference of two waves (Young's experiment) reflected at the rod's edges.

Fig. 10.73 The fringe pattern for a half-screen.

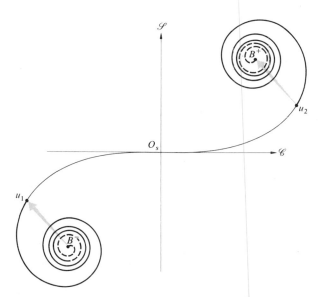

Fig. 10.74 The Cornu spiral as applied to a narrow obstacle.

10.3.11 Babinet's Principle

Two diffracting screens are said to be *complementary* when the transparent regions on one exactly correspond to opaque regions on the other and vice versa. When two such screens are overlapped, the combination is obviously completely opaque. Now then, let E_1 or E_2 be the scalar optical disturbance arriving at P when either complementary screen Σ_1 or Σ_2, respectively, is in place. The total contribution from each aperture is determined by integrating over the area bounded by that aperture. If both *apertures* are present at once, there are no opaque regions at all; the limits of integration go to infinity and we have the unobstructed disturbance E_0, whereupon

$$E_1 + E_2 = E_0 \qquad (10.110)$$

which is the statement of *Babinet's principle*. Take a close look at Figs. 10.69 and 10.74 which depict the Cornu spiral configurations for a transparent slit and a narrow opaque obstacle. If the two arrangements are made complementary, Fig. 10.76 illustrates Babinet's principle quite clearly. The phasor arising from a narrow obstacle $(\overrightarrow{B^-B_1} + \overrightarrow{B_2B^+})$ added to that from a slit $\overrightarrow{B_1B_2}$ yields the unobstructed phasor $\overrightarrow{B^-B^+}$.

The principle implies that when $E_0 = 0$, $E_1 = -E_2$, i.e. these disturbances are precisely equal in magnitude and 180° out of phase. One would, therefore, observe exactly the

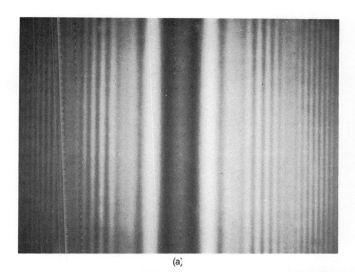

(a)

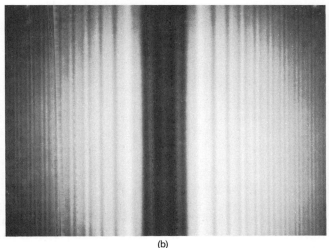

(b)

Fig. 10.75 (a) The shadow pattern cast by the lead from a mechanical pencil. (b) The pattern cast by a $\frac{1}{8}$ inch diameter rod. [Photos by E.H.]

same irradiance distribution with either Σ_1 or Σ_2 in place; an interesting result indeed. It is evident, however, that the principle cannot be exactly true since for an unobstructed wave from a point source there are no zero-amplitude points (i.e. $E_0 \neq 0$ everywhere). Yet if the source is imaged at P_0 by perfect lenses as in Fig. 10.9 (with neither Σ_1 nor Σ_2 present), there will be a large essentially zero-amplitude region beyond the immediate vicinity of P_0 (beyond the Airy disk) in which $E_1 + E_2 = E_0 = 0$. It is therefore only for the case of Fraunhofer diffraction that complementary screens

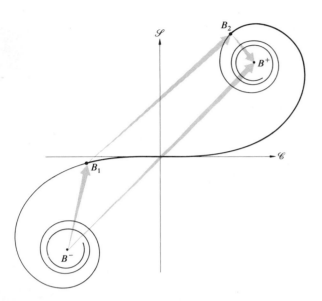

Fig. 10.76 The Cornu spiral illustrating Babinet's principle.

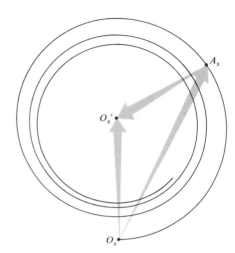

Fig. 10.77 The vibration curve illustrating Babinet's principle.

will generate equivalent irradiance distributions, i.e. $E_1 = -E_2$ (excluding point P_0). Nonetheless Eq. (10.110) is still valid in Fresnel diffraction even though the irradiances obey no simple relationship. This is exemplified by the slit and narrow obstacle of Fig. 10.76. Moreover, for a circular hole and disk refer back to Figs. 10.52 and 10.58 and then examine Fig. 10.77. Equation (10.110) is again clearly applicable even though the diffraction patterns are certainly not equivalent.

The real beauty of Babinet's principle is most evident when applied to Fraunhofer diffraction as shown in Fig. 10.78, where the patterns from complementary screens are almost identical.

10.4 KIRCHHOFF'S SCALAR DIFFRACTION THEORY

We have described a number of diffracting configurations, quite satisfactorily, within the context of the relatively simple Huygens–Fresnel theory. Yet the whole imagery of surfaces covered with fictitious point sources, which was the basis of that analysis, was merely postulated rather than derived from fundamental principles. The Kirchhoff treatment shows that these results are actually derivable from the *scalar* differential wave equation.

The discussion to follow is rather formal and involved. As such, portions of it have been relegated to an appendix

where we can indulge in succinctness and risk sacrificing easy readability for rigor.

In the past, when dealing with a distribution of monochromatic point sources, we computed the resultant optical disturbance at point P, i.e. E_p, by carrying out a superposition of the individual waves. There is, however, another completely different approach which is founded in potential theory. Here one is concerned, not with the sources themselves, but rather with the scalar optical disturbance and its derivatives over an arbitrary closed surface surrounding P. We assume that a Fourier analysis can separate the constituent frequencies so that we need only deal with one such frequency at a time. The monochromatic optical disturbance E is a solution of the differential wave equation

$$\nabla^2 E = \frac{1}{c^2} \frac{\partial^2 E}{\partial t^2}. \tag{10.111}$$

Without specifying the precise spatial nature of the wave it can be written as

$$E = \mathscr{E} e^{-ikct}. \tag{10.112}$$

Here $\mathscr{E}$ represents the complex space part of the disturbance. Substituting into the wave equation it then becomes

$$\nabla^2 \mathscr{E} + k^2 \mathscr{E} = 0. \tag{10.113}$$

This is known as the *Helmholtz equation* and is solved, with the aid of Green's theorem, in Appendix 2. The optical disturbance existing at a point P, expressed in terms of the optical disturbance and its gradient evaluated on an arbitrary

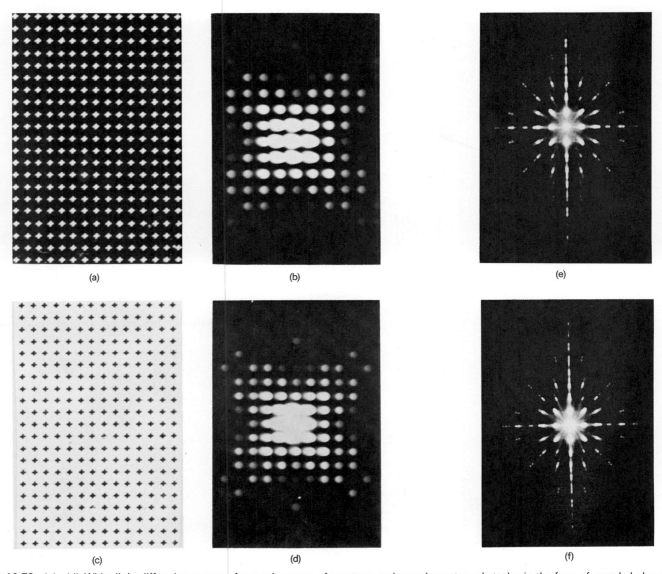

(a)   (b)   (e)

(c)   (d)   (f)

Fig. 10.78 (a)–(d) White light diffraction patterns for regular arrays of apertures and complementary obstacles in the form of rounded plus signs. (e) and (f) Diffraction patterns for a regular array of rectangular apertures and obstacles, respectively. [Photos courtesy The Ealing Corporation and Richard B. Hoover.]

closed surface S, enclosing P, is

$$\mathscr{E}_p = \frac{1}{4\pi}\left[\iint_S \frac{e^{ikr}}{r}\nabla\mathscr{E}\cdot d\mathbf{S} - \iint_S \mathscr{E}\nabla\left(\frac{e^{ikr}}{r}\right)\cdot d\mathbf{S}\right] \quad (10.114)$$

Known as the *Kirchhoff integral theorem*, Eq. (10.114) relates to the geometric configuration illustrated in Fig. 10.79.

We now apply the theorem to the specific instance of an unobstructed spherical wave originating at a point source s, as shown in Fig. 10.80. The disturbance has the form

$$E(\rho, t) = \frac{\mathcal{E}_0}{\rho}e^{i(k\rho - \omega t)}, \quad (10.115)$$

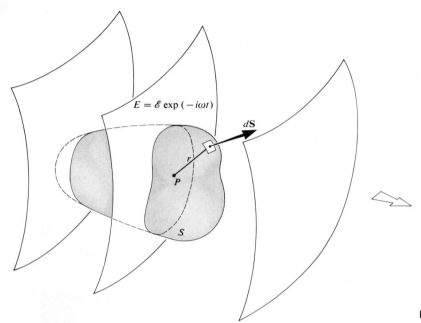

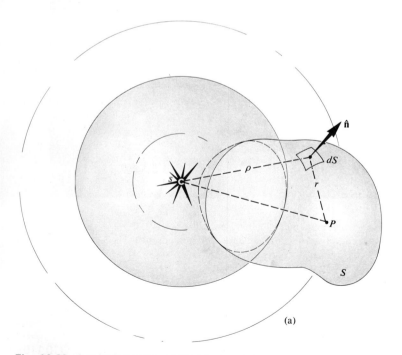

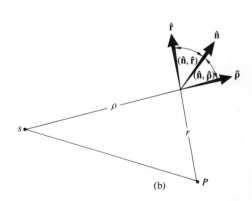

Fig. 10.79 An arbitrary closed surface *S* enclosing point *P*.

Fig. 10.80 A spherical wave emitted from point *s*.

in which case

$$\mathscr{E}(\rho) = \frac{\mathscr{E}_0}{\rho} e^{ik\rho}. \qquad (10.116)$$

If we substitute this into Eq. (10.114) it becomes

$$\mathscr{E}_p = \frac{1}{4\pi} \left[\oiint_S \frac{e^{ikr}}{r} \frac{\partial}{\partial \rho} \left(\frac{\mathscr{E}_0}{\rho} e^{ik\rho} \right) \cos(\hat{\mathbf{n}}, \hat{\boldsymbol{\rho}}) \, dS \right.$$
$$\left. - \oiint_S \frac{\mathscr{E}_0}{\rho} e^{-ik\rho} \frac{\partial}{\partial r} \left(\frac{e^{ikr}}{r} \right) \cos(\hat{\mathbf{n}}, \hat{\mathbf{r}}) \, dS \right],$$

where $d\mathbf{S} = \hat{\mathbf{n}} \, dS$, $\hat{\mathbf{n}}$, $\hat{\mathbf{r}}$ and $\hat{\boldsymbol{\rho}}$ are unit vectors,

$$\boldsymbol{\nabla} \left(\frac{e^{ikr}}{r} \right) = \hat{\mathbf{r}} \frac{\partial}{\partial r} \left(\frac{e^{ikr}}{r} \right)$$

and

$$\boldsymbol{\nabla} \mathscr{E}(\rho) = \hat{\boldsymbol{\rho}} \, \partial \mathscr{E} / \partial \rho.$$

The differentiations under the integral signs are

$$\frac{\partial}{\partial \rho} \left(\frac{e^{ik\rho}}{\rho} \right) = e^{ik\rho} \left(\frac{ik}{\rho} - \frac{1}{\rho^2} \right)$$

and

$$\frac{\partial}{\partial r} \left(\frac{e^{ikr}}{r} \right) = e^{ikr} \left(\frac{ik}{r} - \frac{1}{r^2} \right).$$

When $\rho \gg \lambda$ and $r \gg \lambda$ the $1/\rho^2$ and $1/r^2$ terms can be neglected. This approximation is fine in the optical spectrum but certainly need not be true for microwaves. Proceeding, we write

$$\mathscr{E}_p = -\frac{\mathscr{E}_0 i}{\lambda} \oiint_S \frac{e^{ik(\rho+r)}}{\rho r} \left[\frac{\cos(\hat{\mathbf{n}}, \hat{\mathbf{r}}) - \cos(\hat{\mathbf{n}}, \hat{\boldsymbol{\rho}})}{2} \right] dS, \qquad (10.117)$$

which is known as the *Fresnel–Kirchhoff diffraction formula*.

    Take a long look at Eq. (10.96), which represents the disturbance at P arising from an element dS in the Huygens–Fresnel theory, and compare it with Eq. (10.117). In Eq. (10.117), the angular dependence is contained in the single term $\frac{1}{2}[\cos(\hat{\mathbf{n}}, \hat{\mathbf{r}}) - \cos(\hat{\mathbf{n}}, \hat{\boldsymbol{\rho}})]$ which we shall call the obliquity factor $K(\theta)$, showing it to be equivalent to Eq. (10.72) later on. Notice as well that k can be replaced by $-k$ everywhere, since we certainly could have chosen the phase of Eq. (10.115) to have been $(\omega t - k\rho)$. Now, multiply both sides of Eq. (10.117) by $\exp(-i\omega t)$; the differential element is then

$$dE_p = \frac{K(\theta)\,\mathscr{E}_0}{\rho r \lambda} \cos[k(\rho+r) - \omega t - \pi/2]\,dS. \qquad (10.118)$$

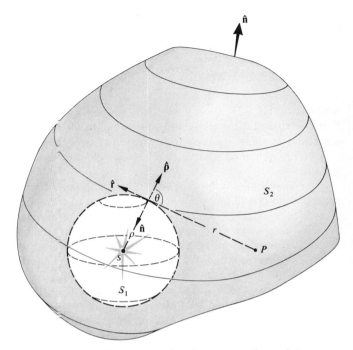

Fig. 10.81 A doubly connected region surrounding point *s*.

This is the contribution to E_p arising from an element of surface area dS a distance r from P. The $\pi/2$ term in the phase comes as a result of the fact that $-i = \exp(-i\pi/2)$. The Kirchhoff formulation therefore leads to the same total result, with the exception that it includes the correct $\pi/2$ phase shift, lacking in the Huygens–Fresnel treatment (10.96).

    We have yet to ensure that the surface S can be made to correspond to the unobstructed portion of the wavefront as it does in the Huygens–Fresnel theory. For the case of a freely propagating spherical wave emanating from the point source *s* we construct the doubly connected region shown in Fig. 10.81. The surface S_2 completely surrounds the small spherical surface S_1. At $\rho = 0$ the disturbance $E(\rho, t)$ has a singularity and is therefore properly excluded from the volume V between S_1 and S_2. The integral must now include both surfaces S_1 and S_2. But we can have S_2 increase outward indefinitely by requiring its radius to go to infinity. In that case, the contribution to the surface integral vanishes. (This is true whatever the form of the incoming disturbance as long as it drops off at least as rapidly as a spherical wave.) The remaining surface S_1 is a sphere centered at the point source. Since, over S_1, n̂ and ρ̂ are antiparallel, it is evident from Fig. 10.80(b) that the angles (n̂, r̂) and (n̂, ρ̂) are θ and 180° respectively. The obliquity factor then becomes

$$K(\theta) = \frac{\cos\theta + 1}{2},$$

which is Eq. (10.72). Clearly, since the surface of integration S_1 is centered at s it does indeed correspond to the spherical wavefront at some instant. *The Huygens–Fresnel principle is therefore directly traceable to the scalar differential wave equation.*

We shan't pursue the Kirchhoff formulation any farther other than to briefly point out how it is applied to diffracting screens. The single closed surface of integration surrounding the point of observation P is generally taken to be the entire screen Σ capped by an infinite hemisphere. There are then three distinct areas with which to be concerned. The contribution to the integral from the region of the infinite hemisphere is zero. Moreover, it is presumed that there is no disturbance immediately behind the opaque screen, so that this second region contributes nothing. The disturbance at P is therefore determined solely by the contributions arising from the aperture and one need only integrate Eq. (10.117) over that area.

The very fine results obtained by using the Huygens–Fresnel principle are now quite justified theoretically, the main limitations being that $\rho \gg \lambda$ and $r \gg \lambda$.

10.5 BOUNDARY DIFFRACTION WAVES

In Section 10.1.1 we took the view that the diffracted wave could be envisioned as arising from a fictitious distribution of secondary emitters spread across the unobstructed portion of the wavefront, i.e. the Huygens–Fresnel principle. There is, however, another completely different and rather appealing possibility. Suppose that an incoming wave sets the electrons on the rear of the diffracting screen Σ into oscillation and these in turn radiate. We anticipate a twofold effect. First, all of the oscillators which are remote from the edge of the aperture radiate back toward the source in such a fashion as to cancel the incoming wave at all points, except within the projection of the aperture itself. In other words, if this were the only contributing mechanism, a perfect geometrical image of the aperture would appear on the plane of observation. There is, however, an additional contribution arising from those oscillators in the vicinity of the aperture's edge. A portion of the energy radiated by these secondary

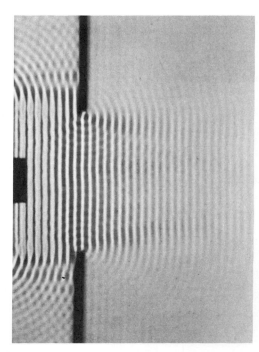

Fig. 10.82 Ripple tank waves passing through a slit. [Photo courtesy PSSC *Physics*, D. C. Heath, Boston 1960.]

sources propagates in the forward direction. The superposition of this scattered wave (known as the *boundary diffraction wave*) and the unobstructed portion of the primary wave (known as the *geometrical wave*) yields the diffraction pattern. A rather cogent reason for contemplating such a scheme appears when one examines the following arrangement. Tear a small hole ($\approx \frac{1}{2}$ cm dia.) of arbitrary shape in a piece of paper and, holding it at arm's length, view an ordinary light bulb some meters distant. Even with your eye in the shadow region, the edges of the aperture will be brightly illuminated. The ripple-tank photograph of Fig. 10.82 also illustrates the process. Notice how each edge of the slit seems to serve as a center for a circular disturbance which then propagates beyond the aperture. There are no electron-oscillators here, which rather implies that these ideas have a certain generality, being applicable to elastic waves as well.

The formulation of diffraction in terms of the interference of a scattered edge wave and a geometrical wave is perhaps more physically appealing than the fictitious emitters of the Huygens–Fresnel principle. It is not, however, a new concept. Indeed it was first propounded by the ubiquitous Thomas Young even before Fresnel's celebrated memoir on diffraction. But in time Fresnel's brilliant successes unfortunately convinced Young to reject his own ideas, and he finally did so in a letter to Fresnel in 1818. Strengthened by Kirchhoff's work, the Fresnel conception of diffraction became generally accepted and has persisted (right up to Section 10.3). The very beginnings of a resurrection of Young's theory came in 1888. At that time Gian Antonio Maggi proved that Kirchhoff's analysis, for a point source at least, was equivalent to two contributing terms. One of these was a geometrical wave but the other, unhappily, was an integral which allowed no clear physical interpretation at the time. In his doctoral thesis (1893) Eugen Maey showed that an edge wave could indeed be extracted from a modified Kirchhoff formulation for a semi-infinite half-plane. Arnold Sommerfeld's rigorous solution of the half-plane problem (see Section 10.1) showed that a cylindrical wave actually does proceed from the screen's edge. It propagates into both the geometrical shadow region and the illuminated region. In the latter, the boundary diffraction wave combines with the geometrical wave, in complete accord with Young's theory. Adalbert (Wojciech) Rubinowicz in 1917, was able to prove that Kirchhoff's formula for a plane or spherical wave can be appropriately decomposed into the two desired waves, thereby revealing the basic correctness of Young's ideas. He also later established that the boundary diffraction wave, to a first approximation, was generated by reflection of the primary wave from the aperture's edge. In 1923, Friedrich Kottler pointed out the equivalence of the solutions of Maggi and Rubinowicz and one now speaks of the Young–Maggi–Rubinowicz theory. Most recently, Kenro Miyamoto and Emil Wolf (1962) have extended the boundary-diffraction theory to the case of arbitrary incident waves.[*] A very useful contemporary approach to the problem has been devised by Joseph B. Keller. He has developed a geometrical theory of diffraction which is closely related to Young's edge wave picture. Along with the usual rays of geometrical optics, he hypothesizes the existence of diffracted rays. Rules governing these diffracted rays, which are analogous to the laws of reflection and refraction, are employed to determine the resultant fields.

---

[*] A fairly complete bibliography can be found in the article by A. Rubinowicz in *Progress in Optics*, Vol. 4, p. 199.

PROBLEMS

10.1 A point source S is a perpendicular distance L away from the center of a circular hole of radius r in an opaque screen. If the distance to the periphery is $(L + \ell)$ show that Fraunhofer diffraction will occur on a very distant screen when

$$\lambda L \gg r^2/2.$$

What is the smallest satisfactory value of L if the hole has a radius of 1 mm, $\ell \leq \lambda/10$ and $\lambda = 500$ nm?

10.2 Using Fig. 10.83 derive the irradiance equation for N coherent oscillators, Eq. (10.5).

10.3* In Section 10.1.3 we talked about introducing an intrinsic phase shift ε between oscillators in a linear array. With this in mind show that Eq. (10.18) becomes

$$\beta = (kb/2)(\sin \theta - \sin \theta_i)$$

when the incident plane wave makes an angle θ_i with the plane of the slit.

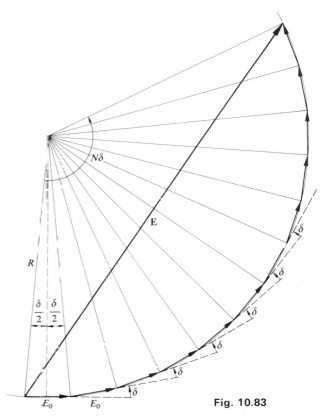

Fig. 10.83

10.4 The angular distance between the center and the first minimum of a single-slit Fraunhofer diffraction pattern is called the *half-angular breadth*; write an expression for it. Find the corresponding *half-linear width* when (a) no focusing lens is present and the slit-viewing screen distance is L, (b) when a lens of focal length f_2 is very close to the aperture. Notice that the half-linear width is also the distance between the successive minima.

10.5 Show that for a double-slit Fraunhofer pattern, if $a = mb$, the number of bright fringes (or parts thereof) within the central diffraction maximum will be equal to $2m$.

10.6 What is the relative irradiance of the subsidiary maxima in a three slit Fraunhofer diffraction pattern? Draw a graph of the irradiance distribution, when $a = 2b$, for two and then three slits.

10.7* Starting with the irradiance expression for a finite slit, shrink the slit down to a minuscule area element and show that it emits equally in all directions.

10.8 Prove that the equation

$$a(\sin \theta_m - \sin \theta_i) = m\lambda \qquad [10.61]$$

when applied to a transmission grating is independent of the refractive index.

10.9 A high-resolution grating of width 260 mm, with 300 lines per mm, at about 75° in autocollimation has a resolving power of just about 10^6 for $\lambda = 500$ nm. Find its free spectral range. How do these values of $\mathcal{R}$ and $(\Delta\lambda)_{fsr}$ compare with those of a Fabry–Perot etalon having a 1 cm air gap and a finesse of 25?

10.10* Perform the necessary mathematical operations needed to arrive at Eq. (10.76).

10.11 Referring to Fig. 10.48, integrate the expression $dS = 2\pi\rho^2 \sin \varphi \, d\varphi$ over the lth zone to get the area of that zone,

$$A_l = \frac{\lambda\pi\rho}{\rho + r_0}\left[r_0 + \frac{(2l - 1)\lambda}{4}\right].$$

Show that the mean distance to the lth zone is

$$r_l = r_0 + \frac{(2l - 1)\lambda}{4},$$

so that the ratio A_l/r_l is constant.

10.12* Derive Eq. (10.84).

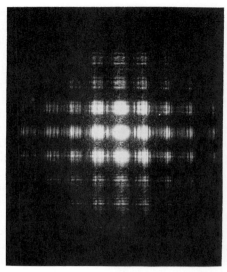

Fig. 10.84

10.13 Use the Cornu spiral to make a rough sketch of $|\mathbf{B}_{12}(w)|^2$ versus $(w_1 + w_2)/2$ for $\Delta w = 5.5$. Compare your results to those of Fig. 10.70.

10.14 The neo-impressionist painter Georges Seurat was a member of the pointillist school. His paintings consisted of an enormous number of closely spaced small dots ($\approx \frac{1}{10}$ inch) of pure pigment. The illusion of color mixing was produced only in the eye of the observer. How far from such a painting should one stand in order to achieve the desired blending of color?

10.15 The Fresnel integrals have the asymptotic forms (corresponding to large values of w) given by

$$\mathscr{C}(w) \approx \tfrac{1}{2} + \left(\frac{1}{\pi w}\right)\sin\left(\frac{\pi w^2}{2}\right),$$

$$\mathscr{S}(w) \approx \tfrac{1}{2} - \left(\frac{1}{\pi w}\right)\cos\left(\frac{\pi w^2}{2}\right).$$

Using this fact, show that the irradiance in the shadow of a semi-infinite opaque screen decreases in proportion to the inverse square of the distance to the edge, as z_1 and therefore v_1 become large.

10.16* Suppose then that we have a laser emitting a diffraction-limited beam ($\lambda_0 = 632.84$ nm) having a 2 mm diameter. How big a light spot would be produced on the surface of the moon a distance of 376×10^3 km away from such a device? Neglect any effects of the earth's atmosphere.

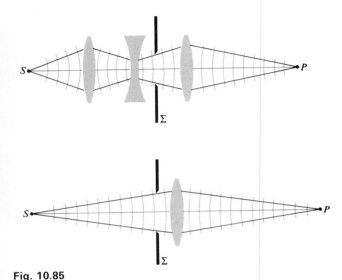

Fig. 10.85

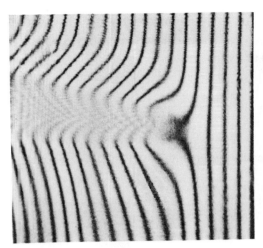

Fig. 10.86

10.17 What would you expect to see on the plane of observation if the half-plane Σ in Fig. 10.71 were semitransparent?

10.18 Imagine that you are looking through a piece of square woven cloth at a point source ($\lambda_0 = 600$ nm) 20 meters away. If you see a square arrangement of bright spots located about the point source (Fig. 10.84), each separated by an apparent nearest-neighbor distance of 12 cm, how close together are the strands of cloth?

10.19 Examine the set-up of Fig. 10.5 in order to determine what is happening in the image space of the lenses, i.e. locate the exit pupil and relate it to the diffraction process. Show that the configurations in Fig. 10.85 are equivalent to that of Fig. 10.5 and will therefore result in Fraunhofer diffraction. Design at least one more such arrangement.

10.20* Imagine an opaque screen containing thirty randomly located circular holes. The light source is such that every aperture is coherently illuminated by its own plane wave. Each wave in turn is completely incoherent with respect to all the others. Describe the resulting far-field diffraction pattern.

10.21 What should be the total number of lines a grating must have in order to just separate the sodium doublet ($\lambda_1 = 5895.9$ Å, $\lambda_2 = 5890.0$ Å) in the third order?

10.22 Plane waves from a collimated He–Ne laser beam ($\lambda_0 = 632.8$ nm) impinge on a 2.5 mm diameter steel rod. Draw a rough *graphical* representation of the diffraction pattern which would be seen on a screen 3.16 m from the rod.

10.23 Referring back to the multiple antenna system of Fig. 10.7, compute the angular separation between successive lobes or principal maxima and the width of the central maximum.

10.24* Figure 10.86 is the pattern generated by placing the tip of a hot soldering iron in one arm of a Michelson interferometer. Qualitatively discuss the optical processes involved in each region of the pattern. Why is the iron's geometric shadow seen in only variations of gray?

10.25* Show that Fraunhofer diffraction patterns have a center of symmetry [i.e. $I(Y, Z) = I(-Y, -Z)$] regardless of the configuration of the aperture as long as there are no phase variations in the field over the region of the hole. Begin with Eq. (10.41). We'll see later (Chapter 11) that this restriction is equivalent to saying that the aperture function is real.

10.26 With the results of Problem 10.25 in mind, discuss the symmetries which would be evident in the Fraunhofer diffraction pattern of an aperture which itself is symmetrical about a line (assuming normally incident quasimonochromatic plane waves).

10.27 From symmetry considerations, create a rough sketch of the Fraunhofer diffraction patterns of an equilateral triangular aperture and an aperture in the form of a plus sign.

10.28 Figure 10.87 shows several aperture configurations. Roughly sketch the Fraunhofer patterns for each. Note that the circular regions should generate Airy-like ring systems centered at the origin.

L ◄ 2 Z S 5

Fig. 10.87

10.29 Make a rough sketch of the irradiance function for a Fresnel diffraction pattern arising from a double slit. What would the Cornu spiral picture look like at point P_0?

10.30* Make a rough sketch of a possible Fresnel diffraction pattern arising from each of the indicated apertures (Fig. 10.88).

Fig. 10.88

10.31* Suppose the slit in Fig. 10.67 is made very wide. What would the Fresnel diffraction pattern look like?

Fourier Optics 11

11.1 INTRODUCTION

In what is to follow we will further extend the discussion of Fourier methods introduced in Chapter 7. It is our intent to provide a strong basic introduction to the subject rather than a complete treatment. Besides the real mathematical power of the method Fourier analysis leads to a marvelous way of treating optical processes in terms of spatial frequencies.* It is always exciting to discover a new bag of analytic toys but it's perhaps even more valuable to unfold yet another way of thinking about a broad range of physical problems— hopefully we shall do both.†

11.2 FOURIER TRANSFORMS

11.2.1 One-Dimensional Transforms

It was seen in Section 7.8 that a one-dimensional function of some space variable $f(x)$ could be expressed as a linear combination of an infinite number of harmonic contributions:

$$f(x) = \frac{1}{\pi} \left[\int_0^\infty A(k) \cos kx \, dk + \int_0^\infty B(k) \sin kx \, dk \right]. \quad [7.56]$$

The weighting factors which determine the significance of the various spatial frequency (k) contributions, that is $A(k)$ and $B(k)$, are the *Fourier cosine and sine transforms of* $f(x)$ given by

$$A(k) = \int_{-\infty}^{+\infty} f(x') \cos kx' \, dx'$$

and

$$B(k) = \int_{-\infty}^{+\infty} f(x') \sin kx' \, dx' \quad [7.57]$$

---

* See Chapter 14 for a nonmathematical discussion.

† As general references for this chapter, see R. C. Jennison, *Fourier Transforms and Convolutions for the Experimentalist*, N. F. Barber, *Experimental Correlograms and Fourier Transforms*, A. Papoulis, *Systems and Transforms with Applications in Optics* and J. W. Goodman, *Introduction to Fourier Optics*.

respectively. Here the quantity x' is a dummy variable over which the integration is carried out so that neither $A(k)$ nor $B(k)$ are explicit functions of x' and the choice of symbol used to denote it is irrelevant. The sine and cosine transforms can be consolidated into a single complex exponential expression as follows: substituting Eq. (7.57) into Eq. (7.56) the latter becomes

$$f(x) = \frac{1}{\pi} \int_0^\infty \cos kx \int_{-\infty}^{+\infty} f(x') \cos kx' \, dx' \, dk$$

$$+ \frac{1}{\pi} \int_0^\infty \sin kx \int_{-\infty}^{+\infty} f(x') \sin kx' \, dx' \, dk.$$

But since $\cos k(x' - x) = \cos kx \cos kx' + \sin kx \sin kx'$ this can be rewritten as

$$f(x) = \frac{1}{\pi} \int_0^\infty \left[\int_{-\infty}^{+\infty} f(x') \cos k(x' - x) \, dx' \right] dk. \quad (11.1)$$

The quantity in the square brackets is an even function of k and therefore changing the limits on the outer integral leads to

$$f(x) = \frac{1}{2\pi} \int_{-\infty}^{+\infty} \left[\int_{-\infty}^{+\infty} f(x') \cos k(x' - x) \, dx' \right] dk. \quad (11.2)$$

Inasmuch as we are looking for an exponential representation, Euler's theorem comes to mind. Consequently observe that

$$\frac{i}{2\pi} \int_{-\infty}^{+\infty} \left[\int_{-\infty}^{+\infty} f(x') \sin k(x' - x) \, dx' \right] dk = 0,$$

because the factor in brackets is an odd function of k. Adding these last two expressions yields the complex form of the Fourier integral

$$f(x) = \frac{1}{2\pi} \int_{-\infty}^{+\infty} \left[\int_{-\infty}^{+\infty} f(x') e^{ikx'} \, dx' \right] e^{-ikx} \, dk. \quad (11.3)$$

Thus we can write

$$f(x) = \frac{1}{2\pi} \int_{-\infty}^{+\infty} F(k) e^{-ikx} \, dk, \quad (11.4)$$

provided that

$$F(k) = \int_{-\infty}^{+\infty} f(x) e^{ikx} \, dx, \quad (11.5)$$

having set $x' = x$ for Eq. (11.5). The function $F(k)$ is said to be the *Fourier transform of* $f(x)$, which is symbolically denoted by

$$F(k) = \mathscr{F}\{f(x)\}. \quad (11.6)$$

Actually there are several equivalent slightly different ways of defining the transform which appear in the literature. For example, the signs in the exponentials could be interchanged or the factor of $1/2\pi$ could be split symmetrically between $f(x)$ and $F(k)$, i.e. each would then have a coefficient of $1/\sqrt{2\pi}$. Note that $A(k)$ is the real part of $F(k)$ while $B(k)$ is its imaginary part, that is

$$F(k) = A(k) + iB(k). \qquad (11.7)$$

Just as $F(k)$ is the transform of $f(x)$, $f(x)$ itself is said to be the *inverse Fourier transform of $F(k)$*, or symbolically

$$f(x) = \mathscr{F}^{-1}\{F(k)\} \qquad (11.8)$$

and $f(x)$ and $F(k)$ are frequently referred to as a Fourier-transform pair. Obviously if f were a function of time rather than space we would merely have to replace x by t and then k, the angular spatial frequency, by ω, the angular temporal frequency, in order to get the appropriate transform pair in the time domain, that is

$$f(t) = \frac{1}{2\pi}\int_{-\infty}^{+\infty} F(\omega)e^{-i\omega t}\,d\omega \qquad (11.9)$$

and

$$F(\omega) = \int_{-\infty}^{+\infty} f(t)e^{i\omega t}\,dt. \qquad (11.10)$$

It should be mentioned that if we write $f(x)$ as a sum of functions its transform (11.5) will apparently be the sum of the transforms of the individual component functions. This can sometimes be quite a convenient way of establishing the transforms of complicated functions which can be constructed from well-known constituents. Figure 11.1 makes this procedure fairly self-evident.

i) Transform of the Gaussian Function

As an example of the method let's examine the Gaussian probability function,

$$f(x) = Ce^{-ax^2}, \qquad (11.11)$$

where $C = \sqrt{a/\pi}$ and a is a constant. If you like, you can imagine this to be the profile of a pulse at $t = 0$. The familiar bell-shaped curve [Fig. 11.2(a)] is quite frequently encountered in optics. It will be germane to a diversity of considerations such as the wave packet representation of individual photons, the cross-sectional irradiance distribution of a laser beam in the TEM_{00} mode and the statistical treatment of thermal light in coherence theory. Its Fourier transform $\mathscr{F}\{f(x)\}$ is gotten by evaluating

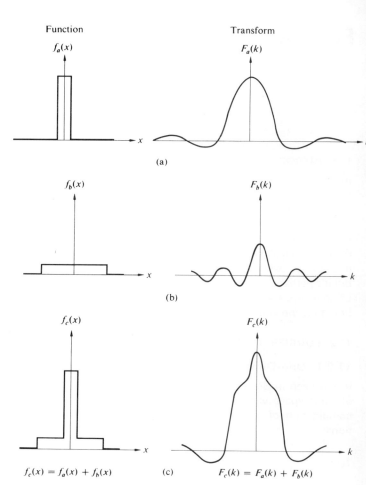

Function $f_a(x)$ — Transform $F_a(k)$

(a)

Function $f_b(x)$ — Transform $F_b(k)$

(b)

Function $f_c(x)$ — Transform $F_c(k)$

$f_c(x) = f_a(x) + f_b(x)$ (c) $F_c(k) = F_a(k) + F_b(k)$

$$F(k) = \mathscr{F}\{f(x)\}$$

Fig. 11.1 A composite function and its Fourier transform.

$$F(k) = \int_{-\infty}^{+\infty} (Ce^{-ax^2})e^{ikx}\,dx.$$

On completing the square, the exponent, $-ax^2 + ikx$, becomes $-(x\sqrt{a} - ik/2\sqrt{a})^2 - k^2/4a$, and letting $x\sqrt{a} - ik/2\sqrt{a} = \beta$ yields

$$F(k) = \frac{C}{\sqrt{a}}e^{-k^2/4a}\int_{-\infty}^{+\infty} e^{-\beta^2}\,d\beta.$$

The definite integral can be found in tables and equals $\sqrt{\pi}$; hence

$$F(k) = e^{-k^2/4a}, \qquad (11.12)$$

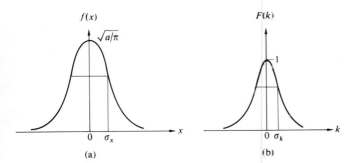

Fig. 11.2 A Gaussian and its Fourier transform.

which *is again a Gaussian function* [Fig. 11.2(b)], this time with k as the variable. The standard deviation is defined as the range of the variable (x or k) over which the function drops by a factor of $e^{-1/2} = 0.607$ of its maximum value. Thus the standard deviations for the two curves are $\sigma_x = 1/\sqrt{2a}$ and $\sigma_k = \sqrt{2a}$. As a increases $f(x)$ becomes narrower while, in contrast, $F(k)$ broadens. In other words, the shorter the pulse length, the broader the spatial frequency bandwidth.

11.2.2 Two-Dimensional Transforms

Thus far the discussion has been limited to one-dimensional functions, but optics generally involves two-dimensional signals: for example, the field across an aperture or the flux-density distribution over an image plane. The Fourier-transform pair can readily be generalized to two dimensions whereupon

$$f(x, y) = \frac{1}{(2\pi)^2} \int\!\!\!\int_{-\infty}^{+\infty} F(k_x, k_y)e^{-i(k_x x + k_y y)}\,dk_x\,dk_y \quad (11.13)$$

and

$$F(k_x, k_y) = \int\!\!\!\int_{-\infty}^{+\infty} f(x, y)e^{i(k_x x + k_y y)}\,dx\,dy. \quad (11.14)$$

The quantities k_x and k_y are the angular spatial frequencies along the two axes. Suppose we were looking at the image of a tiled floor made up alternately of black and white squares aligned with their edges parallel to the x- and y-directions. If the floor were infinite in extent, the mathematical distribution of reflected light could be regarded in terms of a two-dimensional Fourier series. With each tile having a length ℓ, the spatial period along either axis would be 2ℓ and the associated fundamental spatial frequencies would equal π/ℓ. These and their harmonics would certainly be needed to

construct a function describing the scene. If the pattern were finite in extent, the function would no longer be truly periodic and the Fourier integral would have to replace the series. In effect, Eq. (11.13) says that $f(x, y)$ can be constructed out of a linear combination of elementary functions having the form $\exp[-i(k_x x + k_y y)]$, each appropriately weighted in amplitude and phase by a complex factor $F(k_x, k_y)$. The transform simply tells you how much of and with what phase each elementary component must be added to the recipe. In three dimensions, the elementary functions appear as $\exp[-i(k_x x + k_y y + k_z z)]$ or $\exp(-i\mathbf{k} \cdot \mathbf{r})$ which correspond to planar surfaces. Furthermore if f is a wave function, i.e. some sort of three-dimensional wave $f(\mathbf{r}, t)$, these elementary contributions became plane waves that look like $\exp[-i(\mathbf{k} \cdot \mathbf{r} - \omega t)]$. In other words, the disturbance can be synthesized out of a linear combination of plane waves having various propagation numbers and moving in various directions. Similarly in two dimensions the elementary functions are "oriented" in different directions as well. That is to say, for a given set of values of k_x and k_y, the exponent or phase of the elementary functions will be constant along lines

$$k_x x + k_y y = \text{constant} = A$$

or

$$y = -\frac{k_x}{k_y}x + \frac{A}{k_y}. \quad (11.15)$$

The situation is analogous to one where a set of planes normal to and intersecting the xy-plane does so along the lines given by Eq. (11.15) for differing values of A. A vector perpendicular to the set of lines, call it $\mathbf{k}_\alpha$, would have components k_x and k_y. Figure 11.3 shows several of these lines (for a given k_x and k_y) where $A = 0, \pm 2\pi, \pm 4\pi$ etc. The slopes are all equal to $-k_x/k_y$ or $-\lambda_y/\lambda_x$ while the y-intercepts equal $A/k_y = A\lambda_y/2\pi$. The orientation of the constant phase lines is

$$\alpha = \tan^{-1}\frac{k_y}{k_x} = \tan^{-1}\frac{\lambda_x}{\lambda_y}. \quad (11.16)$$

The wavelength, or spatial period λ_α, measured along $\mathbf{k}_\alpha$ is obtained from the similar triangles in the diagram where $\lambda_\alpha/\lambda_y = \lambda_x/\sqrt{\lambda_x^2 + \lambda_y^2}$ and

$$\lambda_\alpha = \frac{1}{\sqrt{\lambda_x^{-2} + \lambda_y^{-2}}}. \quad (11.17)$$

The angular spatial frequency k_α being $2\pi/\lambda_\alpha$ is then

$$k_\alpha = \sqrt{k_x^2 + k_y^2} \quad (11.18)$$

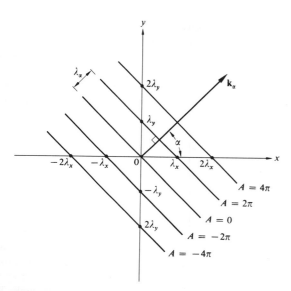

Fig. 11.3 Geometry for Eq. (11.15).

as expected. All of this just means that in order to construct a two-dimensional function, harmonic terms in addition to those of spatial frequency k_x and k_y will generally have to be included as well, and these are oriented in directions other than along the *x*- and *y*-axes.

i)   Transform of the Cylinder Function

The cylinder function

$$f(x, y) = \begin{cases} 1 & \sqrt{x^2 + y^2} \leqslant a \\ 0 & \sqrt{x^2 + y^2} > a \end{cases} \qquad (11.19)$$

[Fig. 11.4(a)] provides an important practical example of the application of Fourier methods to two dimensions. The mathematics will not be particularly simple but the relevance of the calculation to the theory of diffraction by circular apertures and lenses amply justifies the effort. The evident circular symmetry suggests polar coordinates and so let

$$k_x = k_\alpha \cos \alpha$$

$$k_y = k_\alpha \sin \alpha$$

$$x = r \cos \theta$$

$$y = r \sin \theta \qquad (11.20)$$

in which case $dx\, dy = r\, dr\, d\theta$. The transform, $\mathscr{F}\{f(x)\}$, then

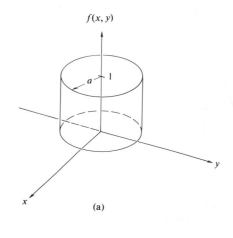

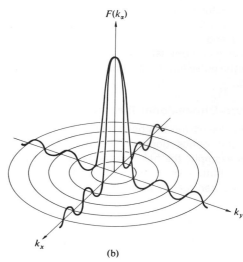

Fig. 11.4 The cylinder or top-hat function and its transform.

reads

$$F(k_\alpha, \alpha) = \int_{r=0}^{a} \left[\int_{\theta=0}^{2\pi} e^{ik_\alpha r \cos(\theta - \alpha)}\, d\theta \right] r\, dr. \qquad (11.21)$$

Inasmuch as $f(x, y)$ is circularly symmetric, its transform must be symmetric as well. This implies that $F(k_\alpha, \alpha)$ is independent of α. The integral can therefore be simplified by letting α equal some constant value which we choose to be zero, whereupon

$$F(k_\alpha) = \int_{0}^{a} \left[\int_{0}^{2\pi} e^{ik_\alpha r \cos \theta}\, d\theta \right] r\, dr. \qquad (11.22)$$

It follows from Eq. (10.47) that

$$F(k_\alpha) = 2\pi \int_0^a J_0(k_\alpha r)\, r\, dr, \qquad (11.23)$$

the $J_0(k_\alpha r)$ being a Bessel function of order zero. Introducing a change of variable, namely $k_\alpha r = w$, we have $dr = k_\alpha^{-1} dw$ and the integral becomes

$$\frac{1}{k_\alpha^2} \int_{w=0}^{k_\alpha a} J_0(w)\, w\, dw. \qquad (11.24)$$

Using Eq. (10.50) the transform takes the form of a first-order Bessel function (see Fig. 10.27), that is

$$F(k_\alpha) = \frac{2\pi}{k_\alpha^2} k_\alpha a J_1(k_\alpha a)$$

or

$$F(k_\alpha) = 2\pi a^2 \left[\frac{J_1(k_\alpha a)}{k_\alpha a} \right]. \qquad (11.25)$$

The similarity between this expression [Fig. 11.4(b)] and the formula for the electric field in the Fraunhofer diffraction pattern of a circular aperture (10.51) is not accidental, as will be seen imminently.

11.2.3 The Dirac Delta Function

There are many physical phenomena which occur over very short durations in time with great intensity, and one is frequently concerned with the consequent response of some system to such stimuli. For example: How will a mechanical device, like a billiard ball, respond to being slammed with a hammer? Or how will a particular circuit behave if the input is a short burst of current? In much the same way we can envision some stimulus which is a sharp pulse in the space, rather than in the time, domain. A bright minute source of light imbedded in a dark background is essentially a highly localized, two-dimensional, spatial pulse—a spike of irradiance. A convenient idealized mathematical representation of this sort of sharply peaked stimulus is the *Dirac delta function* $\delta(x)$. This is a quantity which is zero everywhere except at the origin where it goes to infinity in a manner so as to encompass a unit area, that is

$$\delta(x) = \begin{cases} 0 & x \neq 0 \\ \infty & x = 0 \end{cases} \qquad (11.26)$$

and

$$\int_{-\infty}^{+\infty} \delta(x)\, dx = 1. \qquad (11.27)$$

This is not really a function in the traditional mathematical sense. In fact because it is so singular in nature, it remained the focus of considerable controversy long after it was re-introduced and brought into prominence by P. A. M. Dirac in 1930. Yet physicists, pragmatic as they sometimes are, found it so highly useful that it soon became an established tool despite what seemed a lack of rigorous justification. The precise mathematical theory of the delta function evolved roughly twenty years later, in the early nineteen fifties, principally at the hands of Laurent Schwartz.

Perhaps the most basic operation to which $\delta(x)$ can be applied is the evaluation of the integral

$$\int_{-\infty}^{+\infty} \delta(x) f(x)\, dx.$$

Here the expression $f(x)$ corresponds to any continuous function. Over a tiny interval running from $x = -\gamma$ to $+\gamma$ centered about the origin, $f(x) \approx f(0) \approx$ constant since the function is continuous at $x = 0$. From $x = -\infty$ to $x = -\gamma$ and from $x = +\gamma$ to $x = +\infty$, the integral is zero simply because the δ-function is zero there. Thus the integral equals

$$f(0) \int_{-\gamma}^{+\gamma} \delta(x)\, dx.$$

Because $\delta(x) = 0$ for all x other than 0, the interval can be vanishingly small, i.e., $\gamma \to 0$, and still

$$\int_{-\gamma}^{+\gamma} \delta(x)\, dx = 1$$

from Eq [11.27]. Hence we have the exact result that

$$\int_{-\infty}^{+\infty} \delta(x) f(x)\, dx = f(0). \qquad (11.28)$$

This is often spoken of as the *sifting property* of the δ-function because it manages to extract only the one value of $f(x)$ taken at $x = 0$ from all its other possible values. Similarly with a shift of origin of an amount x_0

$$\delta(x - x_0) = \begin{cases} 0 & x \neq x_0 \\ \infty & x = x_0 \end{cases} \qquad (11.29)$$

and the spike resides at $x = x_0$ rather than $x = 0$. The corresponding sifting property can be appreciated by letting $x - x_0 = x'$, then with $f(x' + x_0) = g(x')$

$$\int_{-\infty}^{+\infty} \delta(x - x_0) f(x)\, dx = \int_{-\infty}^{+\infty} \delta(x') g(x')\, dx' = g(0)$$

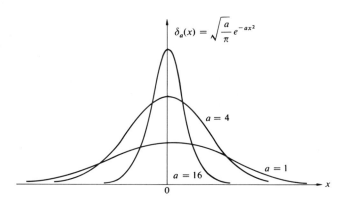

$$\delta_a(x) = \sqrt{\frac{a}{\pi}}\, e^{-ax^2}$$

$a = 4$

$a = 16$

$a = 1$

Fig. 11.5 A sequence of Gaussians.

and

$$\int_{-\infty}^{+\infty} \delta(x - x_0) f(x)\, dx = f(x_0). \qquad (11.30)$$

Formally, rather than worrying about a precise definition of $\delta(x)$ for each value of x, it would be more fruitful to continue along the lines of defining the effect of $\delta(x)$ on some other function $f(x)$. Accordingly, Eq. (11.28) is really the definition of an entire operation which assigns a number $f(0)$ to the function $f(x)$. Incidentally, an operation which performs this service is called a *functional*.

It is possible to construct a number of sequences of pulses each member of which has an ever-decreasing width and a concomitant increasing height such that any one pulse encompasses a unit area. A sequence of square pulses of height a/L and width L/a for which $a = 1, 2, 3, \ldots$ would fit the bill; so would a sequence of Gaussians (11.11),

$$\delta_a(x) = \sqrt{\frac{a}{\pi}}\, e^{-ax^2} \qquad (11.31)$$

as in Fig. 11.5 or a sequence of sinc functions

$$\delta_a(x) = \frac{a}{\pi}\, \text{sinc}\,(ax). \qquad (11.32)$$

Such strongly peaked functions which approach the sifting property, i.e. for which

$$\lim_{a \to \infty} \int_{-\infty}^{+\infty} \delta_a(x) f(x)\, dx = f(0) \qquad (11.33)$$

are known as *delta sequences*. It is often useful, but not actually rigorously correct, to imagine $\delta(x)$ as the convergence

limit of such sequences as $a \to \infty$. The extension of these ideas into two dimensions is provided by the definition

$$\delta(x, y) = \begin{cases} \infty & x = y = 0 \\ 0 & \text{otherwise} \end{cases} \qquad (11.34)$$

and

$$\int\!\!\int_{-\infty}^{+\infty} \delta(x, y)\, dx\, dy = 1, \qquad (11.35)$$

and the sifting property becomes

$$\int\!\!\int_{-\infty}^{+\infty} f(x, y)\delta(x - x_0)\delta(y - y_0)\, dx\, dy = f(x_0, y_0). \qquad (11.36)$$

Another representation of the δ-function follows from Eq. (11.3), the Fourier integral, which can be restated as

$$f(x) = \int_{-\infty}^{+\infty} \left[\frac{1}{2\pi} \int_{-\infty}^{+\infty} e^{-ik(x - x')}\, dk \right] f(x')\, dx',$$

and hence

$$f(x) = \int_{-\infty}^{+\infty} \delta(x - x') f(x')\, dx' \qquad (11.37)$$

provided that

$$\delta(x - x') = \frac{1}{2\pi} \int_{-\infty}^{+\infty} e^{-ik(x - x')}\, dk. \qquad (11.38)$$

Equation (11.37) is identical to Eq. (11.30) since $\delta(x - x') = \delta(x' - x)$. The (divergent) integral of Eq. (11.38) is zero everywhere except at $x = x'$. Evidently

$$\delta(x) = \frac{1}{2\pi} \int_{-\infty}^{+\infty} e^{-ikx}\, dk = \frac{1}{2\pi} \int_{-\infty}^{+\infty} e^{ikx}\, dk \qquad (11.39)$$

which implies, via (11.4), that the delta function can be thought of as the inverse Fourier transform of unity, i.e. $\delta(x) = \mathscr{F}^{-1}\{1\}$. We can imagine a square pulse becoming narrower and taller as its transform, in turn, grows ever broader until finally the pulse is infinitesimal in width and its transform is infinite in extent, i.e., a constant. If the δ-spike is shifted off $x = 0$ to say $x = x_0$, its transform will change phase but not amplitude—that remains equal to one. To see this, evaluate

$$\mathscr{F}\{\delta(x - x_0)\} = \int_{-\infty}^{+\infty} \delta(x - x_0) e^{ikx}\, dx.$$

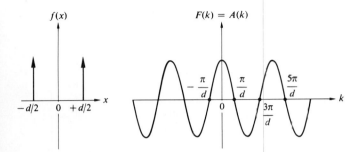

Fig. 11.6 Two delta functions and their sine-function transform.

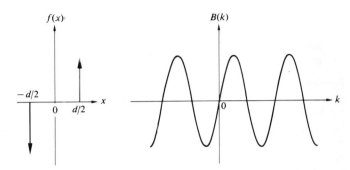

Fig. 11.7 Two delta functions and their cosine-function transform.

From the sifting property (11.30) the expression becomes

$$\mathscr{F}\{\delta(x - x_0)\} = e^{ikx_0}, \qquad (11.40)$$

which means that only the phase is affected, the amplitude being one as it was when $x_0 = 0$. This whole process can be appreciated somewhat more intuitively if we switch to the time domain and think of an infinitesimally narrow pulse (such as a spark) occurring at $t = 0$. This results in the generation of an infinite range of frequency components which are all initially in phase at the instant of creation ($t = 0$). On the other hand, suppose the pulse occurs at a time t_0. Again every frequency is produced but in this situation the harmonic components are all in phase at $t = t_0$. Consequently, if we extrapolate back, the phase of each constituent at $t = 0$ will now have to be different depending on the particular frequency. Besides, we know that all of these components superimpose to yield zero everywhere except at t_0 so that a frequency-dependent phase shift is quite reasonable. This phase shift is evident in Eq. (11.40) for the space domain. Note that it does vary with the spatial frequency k.

We saw earlier (Fig. 11.1) that if the function at hand can be written as a sum of individual functions its transform is simply the sum of the transforms of the component functions. If we have a string of delta functions spread out like the teeth on a comb,

$$f(x) = \sum_j \delta(x - x_j) \qquad (11.41)$$

the transform will simply be a sum of terms such as that of Eq. (11.40):

$$\mathscr{F}\{f(x)\} = \sum_j e^{ikx_j}. \qquad (11.42)$$

In particular if there are two δ-functions, one at $x = d/2$ and the other at $x = -d/2$

$$\mathscr{F}\{f(x)\} = e^{ikd/2} + e^{-ikd/2}$$

which is just

$$\mathscr{F}\{f(x)\} = 2 \cos (kd/2) \qquad (11.43)$$

as in Fig. 11.6. Thus the transform of the sum of these two symmetrical δ-functions is a cosine function and vice versa. The composite is a real even function and $F(k) = \mathscr{F}\{f(x)\}$ will also be real. This should be reminiscent of Young's experiment with infinitesimally narrow slits—we'll come back to it later. If the phase of one of the δ-functions is shifted as in Fig. 11.7, the composite function is asymmetrical and we get

$$\mathscr{F}\{f(x)\} = e^{ikd/2} - e^{-ikd/2} = 2i \sin (kd/2). \qquad (11.44)$$

The real sine transform (11.7) is then

$$B(k) = 2 \sin (kd/2). \qquad (11.45)$$

11.3 OPTICAL APPLICATIONS

11.3.1 Linear Systems

Fourier techniques provide a particularly elegant framework from which to evolve a description of the formation of images. And for the most part, this will be the direction in which we shall be moving although some side excursions are unavoidable in order to develop the needed mathematics.

A key point in the analysis is the concept of a *linear system* which, in turn, is defined in terms of its input–output relations. Suppose then that an input signal $f(y, z)$ passing through some optical system results in an output $g(Y, Z)$. The system is linear if:

1. multiplying $f(y, z)$ by a constant a produces an output $ag(Y, Z)$.

2. when the input is a weighted sum of two functions, $af_1(y, z) + bf_2(y, z)$, the output will similarly have the form $ag_1(Y, Z) + bg_2(Y, Z)$, where $f_1(y, z)$ and $f_2(y, z)$ generate $g_1(Y, Z)$ and $g_2(Y, Z)$ respectively.

Furthermore, a linear system will be *space invariant* if it possesses the property of *stationarity*, i.e. in effect, changing the position of the input merely changes the location of the output without altering its functional form. The idea behind much of this is that the output produced by an optical system can be treated as a linear superposition of the outputs arising from each of the individual points on the object. In fact if we symbolically represent the operation of the linear system as $\mathscr{L}\{\ \}$ the input and output can be written as

$$g(Y, Z) = \mathscr{L}\{f(y, z)\}. \tag{11.46}$$

Using the sifting property of the δ-function (11.36) this becomes

$$g(Y, Z) = \mathscr{L}\left\{\int\int_{-\infty}^{+\infty} f(y', z')\delta(y' - y)\delta(z' - z)\, dy'\, dz'\right\}.$$

The integral expresses $f(y, z)$ as a linear combination of elementary delta functions, each weighted by a number $f(y', z')$. It follows from the linearity conditions that the system operator can equivalently perform on each of the elementary functions; thus

$$g(Y, Z) = \int\int_{-\infty}^{+\infty} f(y', z')\mathscr{L}\{\delta(y' - y)\delta(z' - z)\}\, dy'\, dz'. \tag{11.47}$$

The quantity $\mathscr{L}\{\delta(y' - y)\delta(z' - z)\}$ is the response of the system (11.46) to a delta function located at the point (y', z') in the input space—it's called the *impulse response*. Apparently, if the impulse response of a system is known, the output can be determined directly from the input via Eq. (11.47). If the elementary sources are coherent, the input and output signals will be electric fields; if incoherent, they'll be flux densities.

Consider the self-luminous and, therefore, incoherent source depicted in Fig. 11.8. We can imagine that each point on the object plane, Σ_0, emits light which is processed by the optical system. It emerges to form a spot on the focal or image plane, Σ_i. In addition, we assume that the magnification between object and image planes is one. If $I_0(y, z)$ is the irradiance distribution on the object plane, an element $dy\, dz$ located at (y, z) will emit a radiant flux of $I_0(y, z)\, dy\, dz$.

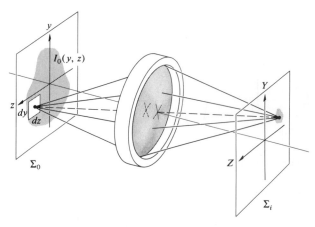

Fig. 11.8 A lens system forming an image.

Because of diffraction (and the possible presence of aberrations), this light is smeared out into some sort of blur spot over a finite area on the image plane rather than focused to a point. The spread of radiant flux is described mathematically by a function $\mathbb{S}(y, z; Y, Z)$ such that the flux density arriving at the image point from $dy\, dz$ is

$$dI_i(Y, Z) = \mathbb{S}(y, z; Y, Z)I_0(y, z)\, dy\, dz. \tag{11.48}$$

This is the patch of light in the image plane at (Y, Z) and $\mathbb{S}(y, z; Y, Z)$ is known as the *point-spread function*. In other words, when the irradiance $I_0(y, z)$ over the source element $dy\, dz$ is 1 W/m², $\mathbb{S}(y, z; Y, Z)\, dy\, dz$ is the profile of the resulting irradiance distribution in the image plane. Because of the incoherence of the source, the flux-density contributions from each of its elements are additive and so

$$I_i(Y, Z) = \int\int_{-\infty}^{+\infty} I_0(y, z)\, \mathbb{S}(y, z; Y, Z)\, dy\, dz. \tag{11.49}$$

In a "perfect," diffraction-limited optical system having no aberrations $\mathbb{S}(y, z; Y, Z)$ would correspond in shape to the diffraction figure of a point source at (y, z). Evidently if we set the input equal to a δ-pulse centered at (y_0, z_0), i.e. $I_0(y, z) = A\delta(y - y_0)\delta(z - z_0)$. Here the constant A of magnitude one carries the needed units (i.e. irradiance times area) and so Eq. (11.49) leads via (11.36) to $I_i(Y, Z) = A\mathbb{S}(y_0, z_0; Y, Z)$. The point-spread function has the identical functional form to that of the image generated by a δ-pulse input. It's the impulse response of the system [compare Eqs. (11.47) and (11.49)] whether optically perfect or not.

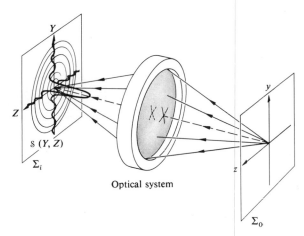

Fig. 11.9 The point-spread function.

In a well-corrected system S, apart from a multiplicative constant, is the Airy irradiance distribution function (10.56) centered on the Gaussian image point.

If the system is space invariant, a point source input can be moved about over the object plane without any effect other than changing the location of its image. Equivalently, one can say that the spread function is the same for any point (y, z). In practice, however, the spread function will vary, but even so the image plane can be divided into small regions over each of which S doesn't change appreciably. Thus if the object, and therefore its image, are small enough, the system can be taken to be space invariant. We can imagine a spread function sitting at every Gaussian image point on Σ_i, each multiplied by a different weighting factor $I_0(y, z)$ but all of the same general shape independent of (y, z). Since the magnification was set at one, the coordinates of any object and conjugate image point have the same magnitude.

What kind of dependence on the image and object space variables will $S(y, z; Y, Z)$ have? The spread function can only depend on (y, z) as far as the location of its center is concerned. Thus the value of $S(y, z; Y, Z)$ anywhere on Σ_i merely depends on the displacement at that location from the particular Gaussian image point $(Y = y, Z = z)$ on which S is centered. In other words

$$S(y, z; Y, Z) = S(Y - y, Z - z). \qquad (11.50)$$

When the object point is on the central axis $(y = 0, z = 0)$, the Gaussian image point is as well and the spread function is then just $S(Y, Z)$ as depicted in Fig. 11.9. Under the circumstances of space invariance and incoherence

$$I_i(Y, Z) = \int\int_{-\infty}^{+\infty} I_0(y, z)\, S(Y - y, Z - z)\, dy\, dz. \qquad (11.51)$$

11.3.2 The Convolution Integral

Equation (11.51) has a characteristic form which is quite common and very important—it's a two-dimensional *convolution integral*. The corresponding one-dimensional expression describing the convolution of two functions $f(x)$ and $h(x)$,

$$g(X) = \int_{-\infty}^{+\infty} f(x)h(X - x)\, dx, \qquad (11.52)$$

is perhaps a bit easier to visualize. The essential features of the process are illustrated in Fig. 11.10. The resulting signal $g(X_1)$, at some point X_1 in the output space, is a linear superposition of all of the individual overlapping contributions that exist at X_1. In other words, each source element dx yields a signal of a particular strength $f(x)\, dx$ which is then smeared out by the system into a region centered about the Gaussian image point $(X = x)$. The output at X_1 is then $dg(X_1) = f(x)h(X_1 - x)\, dx$. The integral sums up all of these contributions from each source element. Of course the elements more remote from a given point on Σ_i contribute less because the spread function generally drops off with displacement. Thus we can imagine $f(x)$ to be a one-dimensional irradiance distribution, such as a series of vertical bands, as in Fig. 11.11(a). If the one-dimensional *line-spread function*, $h(X - x)$, is that of Fig. 11.11(b), the resulting image will simply be a somewhat blurred version of the input [Fig. 11.11(c)].

Let's now examine the convolution a bit more as a mathematical entity. Actually it's a rather subtle beast that performs a process which might certainly not be obvious at first glance; so let's approach it from a slightly different viewpoint. Accordingly, we'll have two ways of thinking about the convolution integral which, we shall show, are equivalent.

Suppose $h(x)$ looks like the asymmetrical function in Fig. 11.12(a). Then $h(-x)$ appears in Fig. 11.12(b) and its shifted form $h(X - x)$ is shown in (c). The convolution of $f(x)$ [depicted in (d)] and $h(x)$ is $g(X)$ as given by Eq. (11.52). This is often written more concisely as just $f(x) \circledast h(x)$. The integral simply says that the area under the product function $f(x)h(X - x)$ for all x is $g(X)$. Evidently the product is nonzero only over the range d wherein $h(X - x)$ is nonzero, i.e. where the two curves overlap [Fig. 11.12(e)].

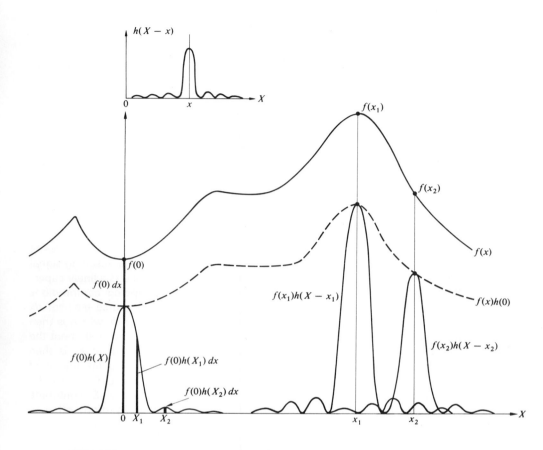

Fig. 11.10 The overlapping of weighted spread functions.

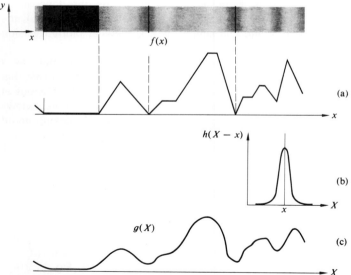

Fig. 11.11 An object, the spread function and the resulting image.

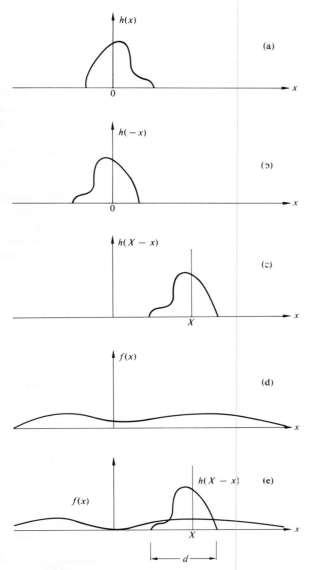

Fig. 11.12 The geometry of the convolution process in the object coordinates.

At a particular point X_1 in the output space the area under the product $f(x)h(X_1 - x)$ is $g(X_1)$. This fairly direct interpretation can be related back to the physically more pleasing view which sees the integral in terms of overlapping point contributions as depicted previously in Fig. 11.10. Remember that there we said that each source element was

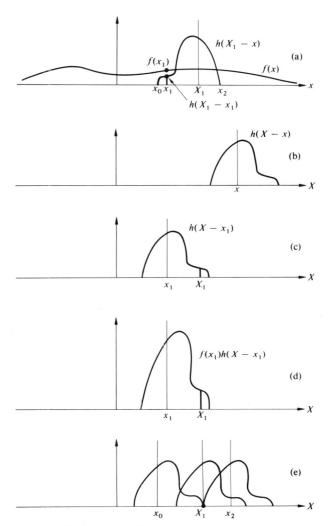

Fig. 11.13 The geometry of the convolution process in the image coordinates.

smeared out in a blur spot on the image plane having the shape of the spread function. Now suppose we take the direct approach and wish to compute the product area of Fig. 11.12(e) at X_1, i.e. $g(X_1)$. A differential element dx centered on any point in the region of overlap [Fig. 11.13(a)], say x_1, will contribute an amount $f(x_1)h(X_1 - x_1)\, dx$ to the area. This same differential element will make an identical contribution when viewed in the overlapping spread-function scheme. To see this, examine Figs. 11.13(b) and (c) which are *now drawn in the output space*. The latter shows the

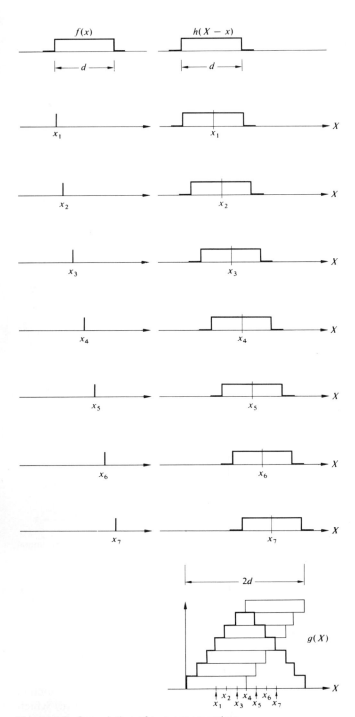

Fig. 11.14 Convolution of two square pulses.

spread function "centered" at $X = x_1$. A source element dx, in this case located on the object at x_1, generates a smeared-out signal proportional to $f(x_1)h(X - x_1)$ as in (d), where $f(x_1)$ is just a number. The piece of this signal that exists at X_1 is $f(x_1)h(X_1 - x_1)\,dx$, which indeed is identical to the contribution made by dx at x_1 in (a). Similarly, each differential element of the product area (at any $x = x'$) in Fig. 11.13(a) has its counterpart in a curve like that of (d) but "centered" on a new point ($X = x'$). Points beyond $x = x_2$ make no contribution because they are not in the overlap region of (a) and equivalently, because they are too far from X_1 for the smear to reach it as shown in (e).

If the functions being convolved are simple enough, $g(X)$ can be determined roughly without any calculations at all. The convolution of two identical square pulses is illustrated, from both of the viewpoints discussed above, in Figs. 11.14 and 11.15. In Fig. 11.14 each impulse com-

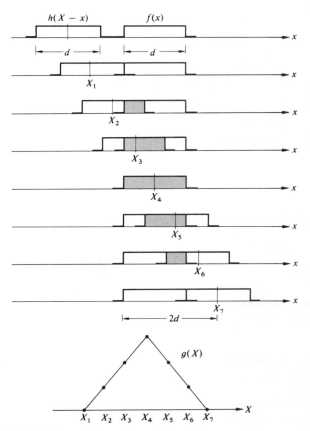

Fig. 11.15 Convolution of two square pulses.

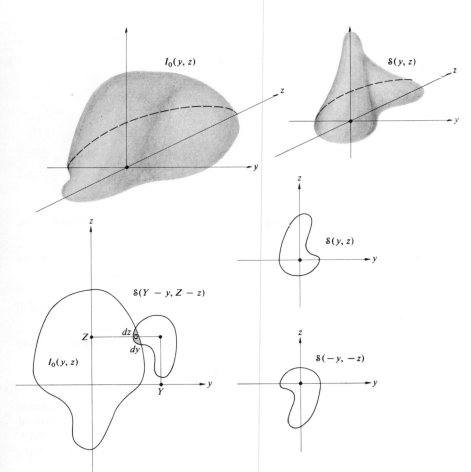

Fig. 11.16 Convolution in two dimensions.

prising $f(x)$ is spread out into a square pulse and summed. In Fig. 11.15 the overlapping area, as h varies, is plotted against X. In both instances the result is a triangular pulse. Incidentally, observe that $(f \circledast h) = (h \circledast f)$ as can be seen by a change of variable ($x' = X - x$) in Eq. (11.52), being careful with the limits (see Problem 11.5).

Figure 11.16 illustrates the convolution of two functions $I_0(y, z)$ and $S(y, z)$ in two dimensions as given by Eq. (11.51). Here the volume under the product curve $I_0(y, z) \ S(Y - y, Z - z)$, i.e. the region of overlap, equals $I_i(Y, Z)$ at (Y, Z); (see Problem 11.6).

i) The Convolution Theorem

Suppose we have two functions $f(x)$ and $h(x)$ with Fourier transforms $\mathscr{F}\{f(x)\} = F(k)$ and $\mathscr{F}\{h(x)\} = H(k)$ respectively.

The *convolution theorem* states that if $g = f \circledast h$

$$\mathscr{F}\{g\} = \mathscr{F}\{f \circledast h\} = \mathscr{F}\{f\} \cdot \mathscr{F}\{h\} \qquad (11.53)$$

or

$$G(k) = F(k)H(k) \qquad (11.54)$$

where $\mathscr{F}\{g\} = G(k)$. The proof is quite straightforward:

$$\mathscr{F}\{f \circledast h\} = \int_{-\infty}^{+\infty} g(X)e^{ikX} \, dX$$

$$= \int_{-\infty}^{+\infty} e^{ikX} \left[\int_{-\infty}^{+\infty} f(x)h(X - x) \, dx \right] dX.$$

Thus

$$G(k) = \int_{-\infty}^{+\infty} \left[\int_{-\infty}^{+\infty} h(X - x)e^{ikX} \, dX \right] f(x) \, dx.$$

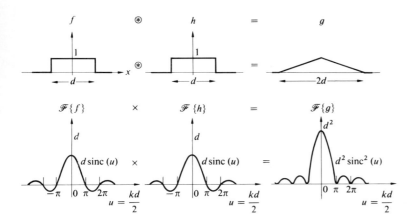

Fig. 11.17 An illustration of the convolution theorem.

If we put $w = X - x$ in the inner integral, then $dX = dw$ and

$$G(k) = \int_{-\infty}^{+\infty} f(x)e^{ikx}\, dx \int_{-\infty}^{+\infty} h(w)e^{ikw}\, dw.$$

Hence

$$G(k) = F(k)H(k),$$

which verifies the theorem. As an example of its application refer to Fig. 11.17. Since the convolution of two identical square pulses ($f \circledast h$) is a triangular pulse (g), the product of their transforms (Fig. 7.16) must be the transform of g, namely

$$\mathscr{F}\{g\} = [d \, \mathrm{sinc}\, (kd/2)]^2. \tag{11.55}$$

As an additional example, convolve a square pulse with the two δ-functions of Fig. 11.7. The transform of the resulting double pulse (Fig. 11.18) is again the product of the individual transforms.

The k-space counterpart of Eq. (11.53), namely the *frequency convolution theorem*, is given by

$$\mathscr{F}\{f \cdot h\} = \frac{1}{2\pi} \mathscr{F}\{f\} \circledast \mathscr{F}\{h\} \tag{11.56}$$

that is, the transform of the product is the convolution of the transforms.

ii) Transform of the Gaussian Wave Packet

As a further example of the usefulness of the convolution theorem, let's evaluate the Fourier transform of a pulse of light in the configuration of the wave packet of Fig. 11.19. Taking a rather general approach notice that since a one-dimensional harmonic wave has the form

$$E(x, t) = E_0 e^{-i(k_0 x - \omega t)}$$

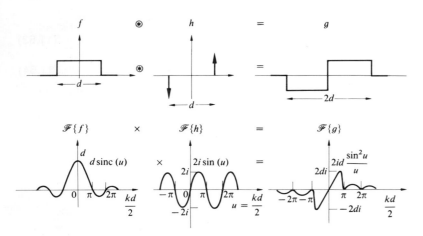

Fig. 11.18 An illustration of the convolution theorem.

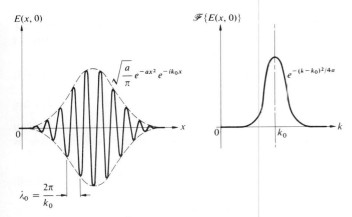

Fig. 11.19 A Gaussian wave packet and its transform.

one need only modulate the amplitude to get a pulse of the desired structure. Assuming the wave's profile to be independent of time, it can be written as

$$E(x, 0) = f(x)e^{-ik_0 x}.$$

Now to determine $\mathscr{F}\{f(x)e^{-ik_0 x}\}$ evaluate

$$\int_{-\infty}^{+\infty} f(x)e^{-ik_0 x}e^{ikx}\, dx. \qquad (11.57)$$

Letting $k' = k - k_0$, we get

$$F(k') = \int_{-\infty}^{+\infty} f(x)e^{ik'x}\, dx = F(k - k_0). \qquad (11.58)$$

In other words, if $F(k) = \mathscr{F}\{f(x)\}$, then $F(k - k_0) = \mathscr{F}\{f(x)e^{-ik_0 x}\}$. For the specific case of a Gaussian envelope (11.11), as in the figure, $f(x) = \sqrt{a/\pi}\, e^{-ax^2}$, that is

$$E(x, 0) = \sqrt{a/\pi}\, e^{-ax^2}e^{-ik_0 x}. \qquad (11.59)$$

From the foregoing discussion and Eq. (11.12) it follows that

$$\mathscr{F}\{E(x, 0)\} = e^{-(k-k_0)^2/4a}. \qquad (11.60)$$

In quite a different way, the transform can be determined from Eq. (11.56). The expression $E(x, 0)$ is now viewed as the product of two functions: $f(x) = \sqrt{a/\pi} \exp(-ax^2)$ and $h(x) = \exp(-ik_0 x)$. One way to evaluate $\mathscr{F}\{h\}$ is to set $f(x) = 1$ in Eq. (11.57). This yields the transform of one with k replaced by $k - k_0$. Since $\mathscr{F}\{1\} = 2\pi\delta(k)$, see Problem 11.4, $\mathscr{F}\{e^{-ik_0 x}\} = 2\pi\delta(k - k_0)$. Thus $\mathscr{F}\{E(x, 0)\}$ is $1/2\pi$ times the convolution of $2\pi\delta(k - k_0)$ with the

Gaussian $e^{-k^2/4a}$ centered on zero. The result* is once again a Gaussian centered on k_0, namely, $e^{-(k-k_0)^2/4a}$.

11.3.3 Fourier Methods in Diffraction Theory

i) Fraunhofer Diffraction

Fourier-transform theory provides a particularly beautiful insight into the mechanism of Fraunhofer diffraction. Let's go back then to Eq. (10.41), rewritten as

$$E(Y, Z) = \frac{\mathcal{E}_A e^{i(\omega t - kR)}}{R} \int\!\!\int_{\text{Aperture}} e^{ik(Yy + Zz)/R}\, dy\, dz. \qquad (11.61)$$

This formula refers to Fig. 10.22 which depicts an arbitrary diffracting aperture in the yz-plane upon which is incident a monochromatic plane wave. The quantity R is the distance from the center of the aperture to the output point where the field is $E(Y, Z)$. The source strength per unit area of the aperture is denoted by $\mathcal{E}_A$. We are talking about electric fields that are of course time varying; ergo, the term $\exp i(\omega t - kR)$ which just references the phase of the net disturbance at the point (Y, Z) to that at the center of the aperture. The $1/R$ corresponds to the drop off of field amplitude with distance from the aperture. The phase term in front of the integral is of little present concern since we are interested in the relative amplitude distribution of the field and it doesn't much matter what the resultant phase is at any particular output point. Thus if we limit ourselves to a small region of output space over which R is essentially constant, everything in front of the integral, with the exception of $\mathcal{E}_A$, can be lumped into a single constant. The $\mathcal{E}_A$ has thus far been assumed invariant over the aperture, but that certainly need not be the case. Indeed, if the aperture

---

* We should actually have used the real part of $\exp(-ik_0 x)$ to start with in this derivation since the transform of the complex exponential is different from the transform of $\cos k_0 x$ and taking the real part afterwards is insufficient. This is the same sort of difficulty one always encounters when forming products of complex exponentials. The final answer (11.60) should, in fact, contain an additional $\exp[-(k + k_0)^2/4a]$ term as well as a multiplicative constant of $\frac{1}{2}$. This second term is usually negligible in comparison, however. Even so, had we used $\exp(+ik_0 x)$ to start with (11.59), only the negligible term would have resulted! Using the complex exponential to represent the sine or cosine in this fashion is *rigorously incorrect*, albeit pragmatically common practice. As a short-cut device, it should be indulged in only with the greatest caution!

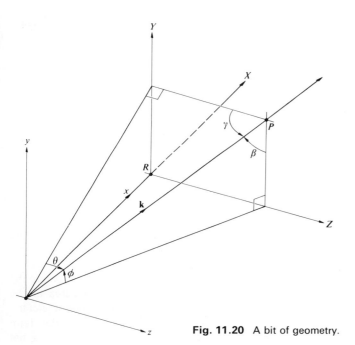

Fig. 11.20 A bit of geometry.

were filled with a bumpy piece of dirty glass, the field emanating from each area element $dy\,dz$ could differ in both amplitude and phase. There would be nonuniform absorption as well as a position-dependent optical path length through the glass that would certainly affect the diffracted field distribution. The variations in $\mathscr{E}_A$ as well as the multiplicative constant can be combined into a single complex quantity

$$\mathscr{A}(y, z) = \mathscr{A}_0(y, z)e^{i\phi(y,z)}, \qquad (11.62)$$

known as the *aperture function*. The amplitude of the field over the aperture is described by $\mathscr{A}_0(y, z)$ while the point-to-point phase variation is represented by $\exp[i\phi(y, z)]$. Accordingly, $\mathscr{A}(y, z)\,dy\,dz$ is proportional to the diffracted field emanating from the differential source element $dy\,dz$. Consolidating this much, Eq. (11.61) can be reformulated more generally as

$$E(Y, Z) = \int\int^{+\infty} \mathscr{A}(y, z)e^{ik(Yy + Zz)/R}\,dy\,dz. \qquad (11.63)$$

The limits on the integral can be extended to $\pm\infty$ because the aperture function is nonzero only over the region of the aperture.

It might be helpful to envision $dE(Y, Z)$ at a given point P as if it were a plane wave propagating in the direction of **k** as in Fig. 11.20 and having an amplitude determined by

$\mathscr{A}(y, z)\,dy\,dz$. To underscore the similarity between Eq. (11.63) and Eq. (11.14) let's define the *spatial frequencies* k_Y and k_Z as

$$k_Y \equiv kY/R = k\sin\phi = k\cos\beta \qquad (11.64)$$

and

$$k_Z \equiv kZ/R = k\sin\theta = k\cos\gamma. \qquad (11.65)$$

For each point on the image plane, there is a corresponding spatial frequency. The diffracted field can now be written as

$$E(k_Y, k_Z) = \int\int_{-\infty}^{+\infty} \mathscr{A}(y, z)e^{i(k_Y y + k_Z z)}\,dy\,dz \qquad (11.66)$$

and we've arrived at the key point: *the field distribution in the Fraunhofer diffraction pattern is the Fourier transform of the field distribution across the aperture (i.e. the aperture function).* Symbolically, this is written as

$$E(k_Y, k_Z) = \mathscr{F}\{\mathscr{A}(y, z)\}. \qquad (11.67)$$

The field distribution in the image plane is the spatial-frequency spectrum of the aperture function. The inverse transform is then

$$\mathscr{A}(y, z) = \frac{1}{(2\pi)^2}\int\int_{-\infty}^{+\infty} E(k_Y, k_Z)e^{-i(k_Y y + k_Z z)}\,dk_Y\,dk_Z, \qquad (11.68)$$

that is

$$\mathscr{A}(y, z) = \mathscr{F}^{-1}\{E(k_Y, k_Z)\}. \qquad (11.69)$$

As we have seen time and again, the more localized the signal, the more spread out is its transform—the same is true in two dimensions. The smaller the diffracting aperture, the larger is the angular spread of the diffracted beam, or equivalently the larger is the spatial frequency bandwidth.

As an illustration of the method, consider the long slit in the y-direction of Fig. 10.10 illuminated by a plane wave. Assuming that there are no phase or amplitude variations across the aperture, $\mathscr{A}(y, z)$ has the form of a square pulse (Fig. 7.16):

$$\mathscr{A}(y, z) = \begin{cases} \mathscr{A}_0 \text{ when } |z| \leqslant b/2 \\ 0 \quad \text{ when } |z| > b/2, \end{cases}$$

where $\mathscr{A}_0$ is no longer a function of y and z. Taking it as a one-dimensional problem

$$E(k_Z) = \mathscr{F}\{\mathscr{A}(z)\} = \mathscr{A}_0\int_{z=-b/2}^{+b/2} e^{ik_Z z}\,dz = \mathscr{A}_0 b \text{ sinc } k_Z b/2.$$

With $k_Z = k\sin\theta$, this is precisely the form derived in

Section 10.2.1. The far-field diffraction pattern of a rectangular aperture (Section 10.2.4) is the two-dimensional counterpart of the slit. With $\mathscr{A}(y, z)$ again equal to $\mathscr{A}_0$ over the aperture (Fig. 10.23).

$$E(k_Y, k_Z) = \mathscr{F}\{\mathscr{A}(y, z)\} = \int_{y=-b/2}^{+b/2} \int_{z=-a/2}^{+a/2} \mathscr{A}_0 e^{i(k_Y y + k_Z z)}\, dy\, dz.$$

Hence,

$$E(k_Y, k_Z) = \mathscr{A}_0 ba\, \text{sinc}\, \frac{bkY}{2R}\, \text{sinc}\, \frac{akZ}{2R}$$

just as in Eq. (10.42) where ba is the area of the hole.

ii) Apodization

The term *apodization* derives from the Greek α, to take away, and $\pi o \delta o \zeta$, meaning foot. It refers to the process of suppressing the secondary maxima (side lobes) or feet of a diffraction pattern. In the case of a circular pupil (Section 10.2.5), the diffraction pattern is a central spot surrounded by concentric rings. The first ring has a flux density of 1.75% that of the central peak—it's small but it can be troublesome. About 16% of the light incident on the image plane is distributed in the ring system. The presence of these side lobes can diminish the resolving power of an optical system to a point where apodization is called for, as is often the case in astronomy and spectroscopy. For example, the star Sirius, which appears as the brightest star in the sky (it's in the constellation *Canis Major*—the big dog), is actually one of a binary system. It's accompanied by a faint white dwarf as they both orbit about their mutual center of mass. Because of the tremendous difference in brightness (10^4 to 1), the image of the faint companion as viewed with a telescope is generally completely obscured by the side lobes of the diffraction pattern of the main star.

Apodization can be accomplished in several ways, e.g. by altering the shape of the aperture or its transmission characteristics.* We already know from Eq. (11.66) that the diffracted field distribution is the transform of $\mathscr{A}(y, z)$. Thus we could effect a change in the side lobes by altering $\mathscr{A}_0(y, z)$ or $\phi(y, z)$. Perhaps the simplest approach is the one where only $\mathscr{A}_0(y, z)$ is manipulated. This can be accomplished physically by covering the aperture with a suitably coated flat glass plate (or so coating the objective lens itself). Suppose that the coating becomes increasingly more opaque as it goes radially out from the center (in the

* For an extensive treatment of the subject, see P. Jacquinot and B. Roizen-Dossier, "Apodization", in Vol III of *Progress in Optics.*

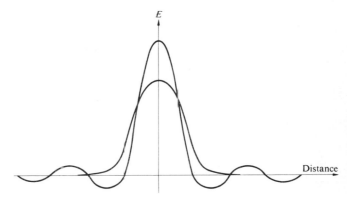

Fig. 11.21 An Airy pattern compared with a Gaussian.

yz-plane towards the edges of a circular pupil. The transmitted field will correspondingly decrease off axis until it is made to become negligible at the periphery of the aperture. In particular, imagine that this drop off in amplitude follows a Gaussian curve. Then $\mathscr{A}_0(y, z)$ is a Gaussian function as is its transform $E(Y, Z)$, and consequently the ring system vanishes. Even though the central peak is broadened, the side lobes are indeed suppressed (Fig. 11.21).

Another rather heuristic but appealing way to look at the process is to realize that the higher spatial frequency contributions go into sharpening up the details of the function being synthesized. As we saw earlier in one dimension (Fig. 7.13), the high frequencies serve to fill in the corners on the square pulse. In the same way, since $\mathscr{A}(y, z) = \mathscr{F}^{-1}\{E(k_Y, k_Z)\}$, sharp edges on the aperture necessitate the presence of appreciable contributions of high spatial frequency in the diffracted field. It follows that making $\mathscr{A}_0(y, z)$ fall off gradually will reduce these high frequencies which in turn is manifest in a suppression of the side lobes.

Apodization is one aspect of the more encompassing technique of *spatial filtering* which is discussed in an extensive yet nonmathematical treatment in Chapter 14.

iii) The Array Theorem

Imagine that we have a screen containing N identical holes as in Fig. 11.22. In each aperture, at the same relative position, we locate a point $O_1, O_2, \ldots, O_N$ at (y_1', z_1'), $(y_2, z_2), \ldots, (y_N, z_N)$ respectively. Each of these, in turn, fixes the origin of a local coordinate system (y', z'). Thus a point (y', z') in the local frame of the jth aperture has coordinates $(y_j + y', z_j + z')$ in the (y, z)-system. Under coherent monochromatic illumination, the resulting Fraunhofer diffraction field $E(Y, Z)$ at some point P on the image

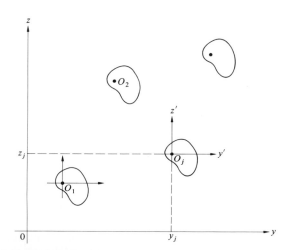

Fig. 11.22 Multiple-aperture configuration.

plane will be a superposition of the individual fields at P arising from each separate aperture, i.e.

$$E(Y, Z) = \sum_{j=1}^{N} \int\limits_{-\infty}^{+\infty}\int \mathscr{A}_I(y',\, z')e^{ik[Y(y_j + y') + Z(z_j + z')]/R}\, dy'\, dz'$$

(11.70)

or

$$E(Y, Z) = \int\limits_{-\infty}^{+\infty}\int \mathscr{A}_I(y',\, z')e^{ik(Yy' + Zz')/R}\, dy'\, dz' \sum_{j=1}^{N} e^{ik(Yy_j + Zz_j)/R},$$

(11.71)

where $\mathscr{A}_I(y',\, z')$ is the individual aperture function applicable to each hole. This can be recast, using Eqs. (11.64) and (11.65) as

$$E(k_Y,\, k_Z) = \int\limits_{-\infty}^{+\infty}\int \mathscr{A}_I(y',\, z')\, e^{i(k_Y y' + k_Z z')}\, dy'\, dz' \sum_{j=1}^{N} e^{i(k_Y y_j)}e^{i(k_Z z_j)}.$$

(11.72)

Notice that the integral is the Fourier transform of the individual aperture function while the sum is the transform (11.42) of an array of delta functions

$$A_\delta = \sum_j \delta(y - y_j)\, \delta(z - z_j).$$

(11.73)

Inasmuch as $E(k_Y,\, k_Z)$ itself is the transform $\mathscr{F}\{\mathscr{A}(y, z)\}$ of the total aperture function for the entire array, we have

$$\mathscr{F}\{\mathscr{A}(y, z)\} = \mathscr{F}\{\mathscr{A}_I(y', z')\} \cdot \mathscr{F}\{A_\delta\}.$$

(11.74)

This equation is a statement of the *array theorem* which says that *the field distribution in the Fraunhofer diffraction pattern of an array of similarly oriented identical apertures equals the Fourier transform of an individual aperture function (i.e. its diffracted field distribution) multiplied by the pattern which would result from a set of point sources arrayed in the same configuration (which is the transform of A_δ).*

This can be seen from a slightly different point of view. The total aperture function may be formed by convolving the individual aperture function with an appropriate array of delta functions, each sitting at one of the coordinate origins (y_1, z_1), (y_2, z_2), etc. Hence

$$\mathscr{A}(y, z) = \mathscr{A}_I(y', z') \circledast A_\delta,$$

(11.75)

whereupon the array theorem follows immediately from the convolution theorem (11.53).

As a simple example, imagine that we have Young's experiment using two slits along the y-direction, of width b and separation a. The individual aperture function for each slit is a step function,

$$\mathscr{A}_I(z') = \begin{cases} \mathscr{A}_{I0} \text{ when } |z'| \leqslant b/2 \\ 0 \quad \text{ when } |z'| > b/2, \end{cases}$$

and so

$$\mathscr{F}\{\mathscr{A}_I(z')\} = \mathscr{A}_{I0}b \text{ sinc } k_z b/2.$$

With the slits located at $z = \pm a/2$,

$$A_\delta = \delta(z - a/2) + \delta(z + a/2)$$

and from Eq. (11.43)

$$\mathscr{F}\{A_\delta\} = 2 \cos k_z a/2.$$

Thus

$$E(k_Z) = 2\mathscr{A}_{I0}b \text{ sinc}\left(\frac{k_z b}{2}\right)\cos\left(\frac{k_z a}{2}\right),$$

which is the same conclusion arrived at earlier in Section 10.2.2. The irradiance pattern is a set of *cosine-squared* interference fringes modulated by a *sinc-squared* diffraction envelope.

11.3.4 Spectra and Correlation

i) Parseval's Formula

Suppose that $f(x)$ is a pulse of finite extent and $F(k)$ is its Fourier transform (11.5). Thinking back to Section 7.8, we recognize the function $F(k)$ as the amplitude of the spatial frequency spectrum of $f(x)$. And $F(k)\, dk$ then connotes the amplitude of the contributions comprising the pulse within the frequency range from k to $k + dk$. Hence it seems that $|F(k)|$ serves as a spectral amplitude density and its square, $|F(k)|^2$, should be proportional to the energy per unit spatial frequency interval. Similarly, in the time domain, if $f(t)$ is a radiated electric field, $|f(t)|^2$ is proportional to the radiant flux or power and the total emitted energy is proportional to $\int_0^\infty |f(t)|^2 \, dt$. With $F(\omega) = \mathscr{F}\{f(t)\}$ it appears that $|F(\omega)|^2$ must be a measure of the radiated energy per unit temporal frequency interval. To be a bit more precise, let's evaluate $\int_{-\infty}^{+\infty} |f(t)|^2 \, dt$ in terms of the appropriate Fourier transforms. Since $|f(t)|^2 = f(t)f^*(t) = f(t) \cdot [\mathscr{F}^{-1}\{F(\omega)\}]^*$,

$$\int_{-\infty}^{+\infty} |f(t)|^2 \, dt = \int_{-\infty}^{+\infty} f(t)\left[\frac{1}{2\pi}\int_{-\infty}^{+\infty} F^*(\omega) e^{+i\omega t}\, d\omega\right] dt.$$

Interchanging the order of integration, we obtain

$$\int_{-\infty}^{+\infty} |f(t)|^2 \, dt = \frac{1}{2\pi}\int_{-\infty}^{+\infty} F^*(\omega)\left[\int_{-\infty}^{+\infty} f(t) e^{i\omega t}\, dt\right] d\omega$$

and so

$$\int_{-\infty}^{+\infty} |f(t)|^2 \, dt = \frac{1}{2\pi}\int_{-\infty}^{+\infty} |F(\omega)|^2 \, d\omega, \qquad (11.76)$$

where $|F(\omega)|^2 = F^*(\omega)F(\omega)$. This is *Parseval's formula*. As expected, the total energy is proportional to the area under the $|F(\omega)|^2$ curve and consequently $|F(\omega)|^2$ is sometimes called the *power spectrum* or *spectral energy distribution*. The corresponding formula for the space domain is

$$\int_{-\infty}^{+\infty} |f(x)|^2 \, dx = \frac{1}{2\pi}\int_{-\infty}^{+\infty} |F(k)|^2 \, dk. \qquad (11.77)$$

ii) The Lorentzian Profile

As an indication of the manner in which these ideas are applied in practice, consider the damped harmonic wave $f(t)$ at $x = 0$ depicted in Fig. 11.23. Here

$$f(t) = \begin{cases} 0 \text{ from } t = -\infty \text{ to } t = 0 \\ f_0 e^{-t/2\tau} \cos \omega_0 t \text{ from } t = 0 \text{ to } t = +\infty. \end{cases}$$

The negative exponential dependence arises quite generally whenever the rate of change of a quantity depends on its instantaneous value. In this case, we might suppose that the power radiated by an atom varies as $e^{-t/\tau}$ and the concomitant emitted electric field as $(e^{-t/\tau})^{1/2}$. In any event, τ is known as the time constant of the oscillation and $\tau^{-1} = \gamma$ is the damping constant. The transform of $f(t)$ is

$$F(\omega) = \int_0^\infty (f_0 e^{-t/2\tau} \cos \omega_0 t) e^{i\omega t}\, dt. \qquad (11.78)$$

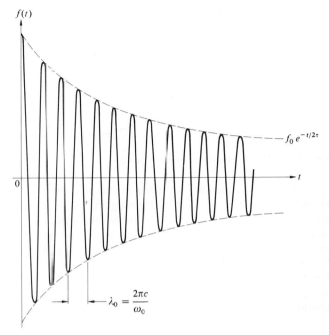

Fig. 11.23 A damped harmonic wave.

The evaluation of this integral is explored in the problems— it's one which is often done incompletely in the literature. One finds on performing the calculation that

$$F(\omega) = \frac{f_0}{2}\left[\frac{1}{2\tau} - i(\omega + \omega_0)\right]^{-1} + \frac{f_0}{2}\left[\frac{1}{2\tau} - i(\omega - \omega_0)\right]^{-1}.$$

At optical frequencies, the first term is negligible and the spectrum $F(\omega)F^*(\omega)$ becomes

$$|F(\omega)|^2 = \frac{f_0^2}{\gamma^2}\frac{\gamma^2/4}{(\omega - \omega_0)^2 + \gamma^2/4}, \qquad (11.79)$$

having a peak value of f_0^2/γ^2 at $\omega = \omega_0$ as shown in Fig. 11.24. At the half-power points $(\omega - \omega_0) = \pm\gamma/2$, $|F(\omega)|^2 = f_0^2/2\gamma^2$ which is half its maximum value. The width of the spectral line between these points is equal to γ.

If $f(t)$ is the radiated field of an atom, τ denotes the *lifetime* of the excited state (of the order of 10^{-8}s) and the curve given by Eq. (11.79) is known as the *resonance* or *Lorentz profile*. The frequency bandwidth arising from the finite duration of the excited state is called the *natural linewidth*.

If the radiating atom suffers a collision, it can lose energy and thereby further shorten the duration of emission.

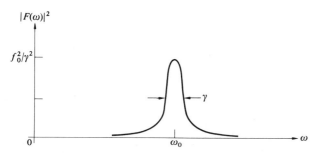

Fig. 11.24 The resonance or Lorentz profile.

The frequency bandwidth increases in the process which is known as *Lorentz broadening*. Here again, the spectrum is found to have a Lorentz profile. Furthermore, because of the random thermal motion of the atoms in a gas, the frequency bandwidth will be increased via the Doppler effect. *Doppler broadening*, as it is called, results in a Gaussian spectrum. The Gaussian drops more slowly in the immediate vicinity of ω_0 and then more quickly away from it than does the Lorentzian profile. These effects can be combined mathematically to yield a single spectrum by convolving the Gaussian and Lorentzian functions. In a low-pressure gaseous discharge, the Gaussian profile is by far the wider and generally predominates.

iii) Auto-and Cross-Correlation

Let's now go back to the derivation of Parseval's formula and follow it through again, this time with a slight modification. We wish to evaluate $\int_{-\infty}^{+\infty} f(t+\tau)f^*(t)\,dt$ using much the same approach as before. Thus, if $F(\omega) = \mathscr{F}\{f(t)\}$

$$\int_{-\infty}^{+\infty} f(t+\tau)f^*(t)\,dt = \int_{-\infty}^{+\infty} f(t+\tau)\left[\frac{1}{2\pi}\int_{-\infty}^{+\infty} F^*(\omega)e^{+i\omega t}\,d\omega\right]dt.$$

$$(11.80)$$

Changing the order of integration, this becomes

$$\frac{1}{2\pi}\int_{-\infty}^{+\infty} F^*(\omega)\left[\int_{-\infty}^{+\infty} f(t+\tau)e^{i\omega t}\,dt\right]d\omega$$

$$= \frac{1}{2\pi}\int_{-\infty}^{+\infty} F^*(\omega)\mathscr{F}\{f(t+\tau)\}\,d\omega.$$

To evaluate the transform within the last integral, notice that

$$f(t+\tau) = \frac{1}{2\pi}\int_{-\infty}^{+\infty} F(\omega)e^{-i\omega(t+\tau)}\,d\omega$$

by a change of variable in Eq. (11.9). Hence,

$$f(t+\tau) = \mathscr{F}^{-1}\{F(\omega)e^{-i\omega\tau}\},$$

so that $\mathscr{F}\{f(t+\tau)\} = F(\omega)e^{-i\omega\tau}$ and Eq. (11.80) becomes

$$\int_{-\infty}^{+\infty} f(t+\tau)f^*(t)\,dt = \frac{1}{2\pi}\int_{-\infty}^{+\infty} F^*(\omega)F(\omega)e^{-i\omega\tau}\,d\omega, \quad (11.81)$$

and both sides are functions of the parameter τ. The left-hand side of this formula is said to be the *autocorrelation* of $f(t)$ denoted by

$$c_{ff}(\tau) \equiv \int_{-\infty}^{+\infty} f(t+\tau)f^*(t)\,dt. \quad (11.82)$$

Taking the transform of both sides, Eq. (11.81) then becomes

$$\mathscr{F}\{c_{ff}(\tau)\} = |F(\omega)|^2. \quad (11.83)$$

This is a form of the *Wiener–Khintchine theorem*. It allows for the determination of the spectrum by way of the autocorrelation of the generating function. The definition of $c_{ff}(\tau)$ applies when the function has finite energy. When it doesn't, things will have to be changed slightly. The integral can also be restated as

$$c_{ff}(\tau) = \int_{-\infty}^{+\infty} f(t)f^*(t-\tau)\,dt \quad (11.84)$$

by a simple change of variable ($t+\tau$ to t). Similarly the *cross-correlation* of the functions $f(t)$ and $h(t)$ is defined as

$$c_{fh}(\tau) = \int_{-\infty}^{+\infty} f^*(t)h(t+\tau)\,dt. \quad (11.85)$$

Changing variables as with Eq. (11.84), and assuming the functions to be real, we can rewrite $c_{fh}(\tau)$ as

$$c_{fh}(\tau) = \int_{-\infty}^{+\infty} f(t)h(t+\tau)\,dt, \quad (11.86)$$

which is obviously similar to the expression for the convolution of $f(t)$ and $h(t)$. Equation (11.86) is often written symbolically as $c_{fh}(\tau) = f(t) \odot h(t)$. Indeed, if either $f(t)$ or $h(t)$ is even, then $f(t) \circledast h(t) = f(t) \odot h(t)$, as we shall see by example presently. Recall that the convolution flips one of the functions over and then sums up the overlap area (Fig. 11.12), i.e. the area under the product curve. In contrast, the correlation sums up the overlap without flipping the function and thus if the function is even, $f(t) = f(-t)$, it isn't changed by being flipped (or folded about the symmetry axis) and the two integrands are identical. For this to obtain either function must be even since $f(t) \circledast h(t) = h(t) \circledast f(t)$. The autocorrelation of a square pulse is therefore equal to the convolution of the pulse with itself which yields a triangular signal as in Fig. 11.15. This same conclusion follows from Eq. (11.83) and Fig. 11.17. The transform of a square pulse is a sinc function so that the power spectrum varies as $\text{sinc}^2 u$. The inverse transform of $|F(\omega)|^2$, i.e. $\mathscr{F}^{-1}\{\text{sinc}^2 u\}$, is $c_{ff}(\tau)$ which, as we have seen, is again a triangular pulse.

It is clearly possible for a function to have infinite energy (11.76) over an integration ranging from $-\infty$ to $+\infty$ and yet still have a finite *average power*

$$\lim_{T \to \infty} \frac{1}{2T} \int_{-T}^{+T} |f(t)|^2 \, dt.$$

Accordingly, we will define a correlation which is divided by the integration interval:

$$C_{fh}(\tau) \equiv \lim_{T \to \infty} \frac{1}{2T} \int_{-T}^{+T} f(t)h(t + \tau) \, dt. \quad (11.87)$$

For example, if $f(t) = A$, i.e. a constant, its autocorrelation

$$C_{ff}(\tau) = \lim_{T \to \infty} \frac{1}{2T} \int_{-T}^{+T} (A)(A) \, dt = A^2$$

and the power spectrum, which is the transform of the autocorrelation, becomes

$$\mathscr{F}\{C_{ff}(\tau)\} = A^2 2\pi \delta(\omega),$$

a single impulse at the origin ($\omega = 0$) which is sometimes referred to as a *dc*-term. Notice that $C_{fh}(\tau)$ can be thought of as the time average of a product of two functions, one of which is shifted by an interval τ. In the next chapter, expressions of the form $\langle f^*(t)h(t + \tau)\rangle$ arise as coherence functions relating electric fields. They are also quite useful in the analysis of noise problems, e.g. that of film grain noise.

We can obviously reconstruct a function from its transform but once the transform is squared, as in Eq. (11.83), we lose information about the signs of the frequency contributions, i.e. their relative phases. In the same way, the autocorrelation of a function contains no phase information and is not unique. To see this more clearly, imagine we have a number of harmonic functions of different amplitude and frequency. If their relative phases are altered, the resultant function changes as does its transform but in all cases the amount of energy available at any frequency must be constant. Thus, whatever the form of the resultant profile, its autocorrelation is unaltered. It is left as a problem to show that when $f(t) = A \sin(\omega t + \varepsilon)$, $C_{ff}(\tau) = (A^2/2) \cos \omega \tau$, which confirms the loss of phase information.

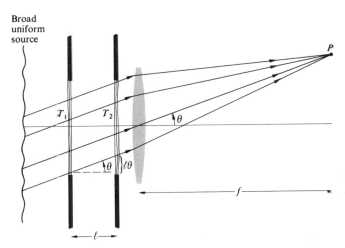

Fig. 11.25 Optical correlation of two functions.

Figure 11.25 shows a means of optically correlating two two-dimensional spatial functions. Each of these signals is represented as a point-by-point variation in the irradiance transmission property of a photographic transparency (T_1 and T_2). For relatively simple signals opaque screens with appropriate apertures could serve instead of transparencies (e.g. for square pulses).[*] The irradiance at any point P on the image is due to a focused bundle of parallel rays which has traversed both transparencies. The coordinates of P, $(\theta f, \varphi f)$, are fixed by the orientation of the ray bundle, i.e. the angles θ and φ. If the transparencies are identical, a ray passing through any point (x, y) on the first film with a transmittance $g(x, y)$ will pass through a corresponding point $(x + X, y + Y)$ on the second film where the transmittance is $g(x + X, y + Y)$. The shifts in coordinate are given by $X = \ell\theta$ and $Y = \ell\phi$, where ℓ is the separation between the transparencies. The irradiance at P is therefore proportional to the autocorrelation of $g(x, y)$, that is

$$c_{ff}(X, Y) = \iint_{-\infty}^{+\infty} g(x, y)g(x + X, y + Z) \, dx \, dy \quad (11.88)$$

and the entire flux density pattern is called a *correlogram*. If the transparencies are different, the image is of course representative of the cross correlation of the functions. Similarly, if one of the transparencies is rotated by 180° with respect to the other, the convolution can be obtained (see Fig. 11.16).

---

[*] See L. S. G. Kovasznay and A. Arman, *Rev. Sci. Instr.* **28**, 793 (1958) and D. McLachlan, Jr., *J. Opt. Soc. Am.* **52**, 454 (1962).

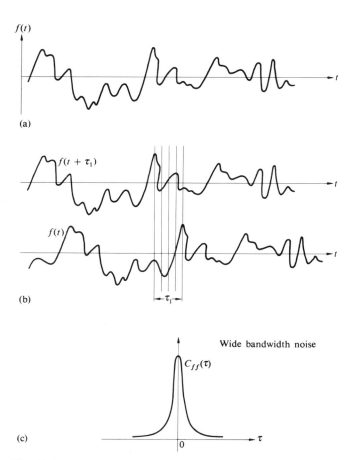

(a)

(b)

(c)

Fig. 11.26 A signal $f(t)$ and its autocorrelation.

Before moving on, let's make sure that we actually do have a good physical feeling for the operation performed by the correlation functions. Accordingly, suppose we have a random noise-like signal (e.g. a fluctuating irradiance at a point in space, a time-varying voltage or electric field) as in Fig. 11.26(a). The autocorrelation of $f(t)$ in effect compares the function with its value at some other time, $f(t + \tau)$. For example, with $\tau = 0$ the integral runs along the signal in time summing up and averaging the product of $f(t)$ and $f(t + \tau)$; in this case it's simply $f^2(t)$. Since at each value of t, $t^2(t)$ is positive, $C_{ff}(0)$ will be a comparatively large number. On the other hand, when the noise is compared with itself shifted by an amount $+\tau_1$, $C_{ff}(\tau_1)$ will be somewhat reduced. There will be points in time where $f(t)f(t + \tau_1)$ is positive and other points where it will be negative so that the

value of the integral drops off [Fig. 11.26(b)]. In other words, by shifting the signal with respect to itself, we have reduced the point-by-point similarity which previously ($\tau = 0$) occurred at any instant. As this shift τ increases, what little correlation existed quickly vanishes as depicted in Fig. 11.26(c). We can assume from the fact that the autocorrelation and the power spectrum form a Fourier transform pair, (11.83), that the broader the frequency bandwidth of the noise the narrower is the autocorrelation. Thus for wide-bandwidth noise even a slight shift markedly reduces any similarity between $f(t)$ and $f(t + \tau)$. Furthermore, if the signal is comprised of a random distribution of rectangular pulses, we can see intuitively that the similarity we spoke of earlier persists for a time commensurate with the width of the pulses. The wider (in time) are the pulses, the more slowly the correlation decreases as τ increases. But this is equivalent to saying that reducing the signal bandwidth broadens $C_{ff}(\tau)$. All of this is in keeping with our previous observation that the autocorrelation tosses out any phase information, which in this case would correspond to the locations in time of the random pulses. Clearly, $C_{ff}(\tau)$ shouldn't be affected by the position of the pulses along t.

In very much the same way the cross-correlation is a measure of the similarity between two different waveforms, $f(t)$ and $h(t)$, as a function of the relative time shift τ. Unlike the autocorrelation there is now nothing special about $\tau = 0$. Once again, for each value of τ we average the product $f(t)h(t + \tau)$ to get $C_{fh}(\tau)$ via Eq. (11.87). For the functions shown in Fig. 11.27, $C_{fh}(\tau)$ would have a positive peak at $\tau = \tau_1$.

11.3.5 Transfer Functions

i) An Introduction to the Concepts

The traditional means of determining the quality of an optical element or system of elements was to evaluate its limit of resolution. The greater the resolution the better the system was presumed to be. In the spirit of this approach one might train an optical system on a resolution target consisting, for instance, of a series of alternating light and dark parallel rectangular bars. We have already seen that an object point is imaged as a smear of light described by the point-spread function $S(Y, Z)$ as in Fig. 11.9. Under incoherent illumination these elementary flux-density patterns overlap and add linearly to create the final image. The one-dimensional counterpart is the *line spread function* $S(Z)$ which corresponds to the flux-density distribution across the image of a geometrical line source having infinitesimal

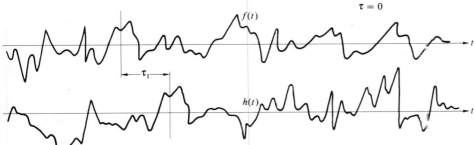

$$\tau = 0$$

$f(t)$

τ_1

$h(t)$

Fig. 11.27 The cross-correlation of $f(t)$ and $h(t)$.

width (Fig. 11.28). Because even an ideally perfect system is limited by diffraction effects, the image of a resolution target will be somewhat blurred (see Fig. 11.11). Thus, as the width of the bars on the target is made narrower, a limit will be reached where the fine-line structure (akin to a *Ronchi ruling*) will no longer be discernible—this then is the resolution limit of the system. We can think of it as a spatial frequency cut-off where each bright and dark bar pair constitutes one cycle on the object (a common measure of which is *line pairs per mm*). An obvious analogy which underscores the shortcomings of this approach would be to evaluate a high-fidelity sound system simply on the basis of its upper-frequency cut-off. Evidently it would be more helpful to have some figure of merit applicable to the entire operating frequency range. The limitations of this scheme became quite apparent with the introduction of detectors such as the plumbicon, image orthicon and vidicon. These tubes have a relatively coarse scanning raster which fixes the resolution limit of the lens-tube system at a fairly low spatial frequency. Accordingly, it would seem reasonable to design the optics preceding such detectors so that it provided the most contrast over this limited frequency range. It would clearly be unnecessary and perhaps, as we shall see, even detrimental to select a mating lens system merely because of its own high limit of resolution.

A highly useful parameter in evaluating the performance of a system is the *contrast* or *modulation*, defined by

$$\text{Modulation} \equiv \frac{I_{max} - I_{min}}{I_{max} + I_{min}}. \qquad (11.89)$$

Figure 11.29 is a plot of image modulation versus spatial frequency for two hypothetical lens systems. Suppose one of these is to be coupled to a detector whose cut-off frequency is indicated in the diagram. Despite the fact that lens 1 has a higher limit of resolution, lens 2 would certainly

provide better performance when coupled to the particular detector.

It should be pointed out that a square bar target provides an input signal which is a series of square pulses and the contrast in the image is actually a superposition of contrast

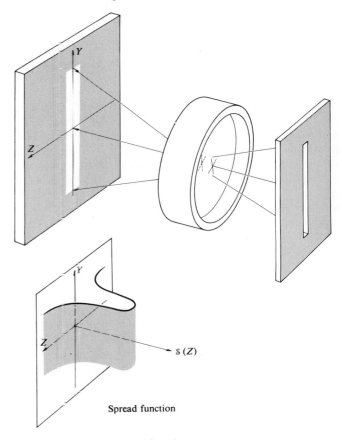

Spread function

Fig. 11.28 The line-spread function.

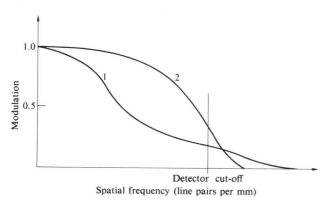

Fig. 11.29 Modulation versus spatial frequency for two lenses.

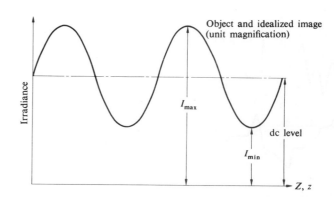

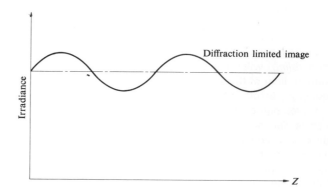

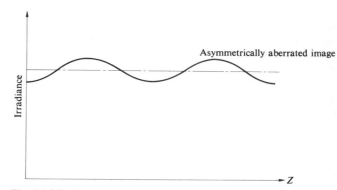

Fig. 11.30 Harmonic input and resulting output.

variations due to the constituent Fourier components. Indeed, one of the keypoints in what is to follow is that *optical elements functioning as linear operators transform a sinusoidal input into an undistorted sinusoidal output*. Despite this, the input and output irradiance distributions as a rule will not be identical. For example, the system's magnification affects the spatial frequency of the output (henceforth the magnification will be taken as one). Diffraction and aberrations reduce the sinusoid's amplitude (contrast). And finally, asymmetric aberrations (e.g. coma) and poor centering of elements produce a shift in the position of the output sinusoid corresponding to the introduction of a phase shift. This latter point can better be appreciated (at least until we do it analytically) using a diagram like that of Fig. 11.30. If the spread function is symmetrical, the image irradiance will be an unshifted sinusoid, whereas an asymmetric spread function will apparently push the output over a bit as in Fig. 11.31. In either case, *regardless of the form of the spread function, the image is harmonic if the object is harmonic*. Consequently, envisioning an object as being composed of Fourier components, the manner in which these individual harmonic components are transformed by the optical system into the corresponding harmonic constituents of the image is the quintessential feature of the process. The function which performs this service is known as the *optical transfer function* or OTF. It is a spatial-frequency dependent complex quantity whose modulus is the *modulation transfer function* (MTF) and whose phase, naturally enough, is the *phase transfer function* (PTF). The former is a measure of the reduction in contrast from object to image over the spectrum. The latter represents the commensurate relative phase shift. Phase shifts in centered optical systems occur

only off axis and often the PTF is of less interest than the MTF. Even so, each application of the transfer function must be studied carefully; there are situations wherein the PTF plays a crucial role. In point of fact, the MTF (i.e. *the ratio of image-to-object modulation for sinusoids of varying spatial frequency*) has become a widely used means of specifying

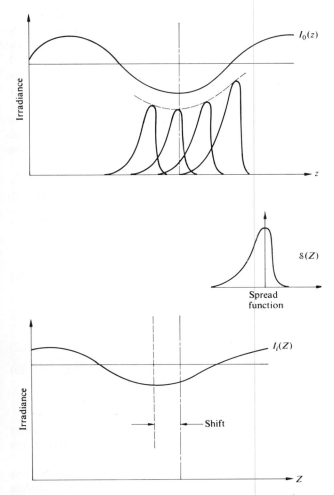

Fig. 11.31 Harmonic input and output with an asymmetric spread function.

the performance of all sorts of elements, conditions and systems from lenses, magnetic tape and films, to telescopes, the atmosphere and the eye, to mention but a few. Moreover, it has the advantage that if the MTF's for the individual independent components in a system are known, the total MTF is often simply their product. This is inapplicable to the cascading of lenses since the aberrations in one lens can compensate for those of another lens in tandem with it and they are therefore not independent. Thus if we photograph an object having a modulation of 0.3 at 30 cycles per mm using a camera whose lens at the appropriate setting has an MTF

of 0.5 at 30 c/mm and a film* like Tri-X with an MTF of 0.4 at 30 c/mm, the image modulation will be 0.3 × 0.5 × 0.4 = 0.06.

ii) A More Formal Discussion

We saw in Eq. (11.51) that the image (under the conditions of space invariance and incoherence) could be expressed as the convolution of the object irradiance and the point-spread function, i.e.

$$I_i(Y, Z) = I_0(y, z) \circledast S(y, z). \qquad (11.90)$$

The corresponding statement in the spatial frequency domain is obtained by a Fourier transform, namely

$$\mathcal{F}\{I_i(Y, Z)\} = \mathcal{F}\{I_0(y, z)\} \cdot \mathcal{F}\{S(y, z)\}. \qquad (11.91)$$

where use was made of the convolution theorem (11.53). This says that the frequency spectrum of the image irradiance distribution equals the product of the frequency spectrum of the object irradiance distribution and the transform of the spread function. Thus, it is multiplication by $\mathcal{F}\{S(y, z)\}$ which produces the alteration in the frequency spectrum of the object converting it into that of the image spectrum. In other words, it is $\mathcal{F}\{S(y, z)\}$ which, in effect, transfers the object spectrum into the image spectrum. This is just the service performed by the OTF and indeed we shall define the *unnormalized* OTF as

$$\mathcal{T}(k_Y, k_Z) \equiv \mathcal{F}\{S(y, z)\}, \qquad (11.92)$$

The modulus of $\mathcal{T}(k_Y, k_Z)$ will affect a change in the amplitudes of the various frequency components of the object spectrum while its phase will, of course, appropriately alter the phase of these components to yield $\mathcal{F}\{I_i(Y, Z)\}$. Bear in mind that in the right-hand side of Eq. (11.90) the only quantity dependent on the actual optical system is $S(y, z)$. And so it's not surprising that the spread function is the spatial counterpart of the optical transfer function.

Let's now verify the statement made earlier that a harmonic input transforms into a somewhat altered harmonic output. To that end suppose

$$I_0(z) = 1 + a \cos(k_z z + \varepsilon) \qquad (11.93)$$

where for simplicity's sake, we'll use a one-dimensional distribution. The 1 is a dc bias which makes sure the

---

* Incidently, the whole idea of treating film as a noise-free linear system is somewhat suspect. For further reading see J. B. De Velis and G. B. Parrent Jr., "Transfer Function for Cascaded Optical Systems," *J. Opt. Soc. Am.* **57,** 1486 (1967).

irradiance doesn't take on any unphysical negative values. Insofar as $f \circledast h = h \circledast f$ it will be more convenient here to use

$$I_i(Z) = \mathbb{S}(z) \circledast I_0(z)$$

and so

$$I_i(Z) = \int_{-\infty}^{+\infty} \{1 + a \cos [k_Z(Z - z) + \varepsilon]\} \mathbb{S}(z) \, dz.$$

Expanding out the cosine, we obtain

$$I_i(Z) = \int_{-\infty}^{+\infty} \mathbb{S}(z) \, dz + a \cos (k_Z Z + \varepsilon) \int_{-\infty}^{+\infty} \cos k_Z z \, \mathbb{S}(z) \, dz$$

$$+ a \sin (k_Z Z + \varepsilon) \int_{-\infty}^{+\infty} \sin k_Z z \, \mathbb{S}(z) \, dz.$$

Referring back to Eq. (7.57) we recognize the second and third integrals as the Fourier cosine and sine transforms of $\mathbb{S}(z)$, respectively, i.e. $\mathscr{F}_c\{\mathbb{S}(z)\}$ and $\mathscr{F}_s\{\mathbb{S}(z)\}$. Hence

$$I_i(Z) = \int_{-\infty}^{+\infty} \mathbb{S}(z) \, dz + \mathscr{F}_c\{\mathbb{S}(z)\} a \cos (k_Z Z + \varepsilon)$$

$$+ \mathscr{F}_s\{\mathbb{S}(z)\} a \sin (k_Z Z + \varepsilon). \qquad (11.94)$$

Recall that the complex transform which we've become so used to working with was defined such that

$$\mathscr{F}\{f(z)\} = \mathscr{F}_c\{f(z)\} + i\mathscr{F}_s\{f(z)\} \qquad (11.95)$$

or

$$F(k_Z) = A(k_Z) + iB(k_Z). \qquad [11.7]$$

And, as in the Argand diagram of Fig. 2.9, it can be expressed as

$$\mathscr{F}\{f(z)\} = |F(k_Z)|e^{i\varphi(k_Z)} = |F(k_Z)|[\cos \varphi + i \sin \varphi],$$

where

$$|F(k_Z)| = [A^2(k_Z) + B^2(k_Z)]^{1/2} \qquad (11.96)$$

and

$$\varphi(k) = \tan^{-1} \frac{B(k_Z)}{A(k_Z)}. \qquad (11.97)$$

In precisely the same way, we apply this to the OTF writing it as

$$\mathscr{F}\{\mathbb{S}(z)\} \equiv \mathscr{T}(k_Z) = \mathscr{M}(k_Z)e^{i\Phi(k_Z)}, \qquad (11.98)$$

where $\mathscr{M}(k_Z)$ and $\Phi(k_Z)$ are the unnormalized MTF and the PTF respectively. It is left as a problem to show that Eq. (11.94) can be recast as

$$I_i(Z) = \int_{-\infty}^{+\infty} \mathbb{S}(z) \, dz + a\mathscr{M}(k_Z) \cos [k_Z Z + \varepsilon - \Phi(k_Z)].$$

$$(11.99)$$

Notice that this is a function of the same form as the input signal (11.93), $I_0(z)$, which is just what we set out to determine. If the line-spread function is symmetrical, i.e. even, $\mathscr{F}_s\{\mathbb{S}(z)\} = 0$, $\mathscr{M}(k_z) = \mathscr{F}_c\{\mathbb{S}(z)\}$ and $\Phi(k_z) = 0$; there is no phase shift as was pointed out in the previous section. For an asymmetric (odd) spread function $\mathscr{F}_s\{\mathbb{S}(z)\}$ is nonzero as is the PTF.

It has now become customary practice to define a set of *normalized transfer functions* by dividing $\mathscr{T}(k_z)$ by its zero spatial frequency value, i.e. $\mathscr{T}(0) = \int_{-\infty}^{+\infty} \mathbb{S}(z) \, dz$. The normalized spread function becomes

$$\mathbb{S}_n(z) = \frac{\mathbb{S}(z)}{\displaystyle\int_{-\infty}^{+\infty} \mathbb{S}(z) \, dz}, \qquad (11.100)$$

while the normalized OTF is

$$T(k_z) \equiv \frac{\mathscr{F}\{\mathbb{S}(z)\}}{\displaystyle\int_{-\infty}^{+\infty} \mathbb{S}(z) \, dz} = \mathscr{F}\{\mathbb{S}_n(z)\}, \qquad (11.101)$$

or in two dimensions

$$T(k_Y, k_Z) = M(k_Y, k_Z)e^{i\Phi(k_Y, k_Z)}, \qquad (11.102)$$

where $M(k_Y, k_Z) \equiv \mathscr{M}(k_Y, k_Z)/\mathscr{T}(0, 0)$. $I_i(Z)$ in Eq. (11.99) would then be proportional to

$$1 + aM(k_Z) \cos [k_Z Z + \varepsilon - \Phi(k_Z)].$$

The image modulation (11.89) becomes $aM(k_Z)$, the object modulation (11.93) is a and the ratio is, as expected, the normalized MTF = $M(k_Z)$.

This discussion is really only an introductory one designed more as a strong foundation than a complete structure. There are many other insights to be explored such as the relationship between the autocorrelation of the pupil function and the OTF and from there the means of computing and measuring transfer functions—but for this the reader is directed to the literature.*

---

* See the series of articles "The Evolution of the Transfer Function" by F. Abbott beginning March 1970 in *Optical Spectra*, the articles "Physical Optics Notebook" by G. B. Parrent, Jr. and B. J. Thompson beginning December 1964 in the *S.P.I.E. Journal* Vol. 3 or "Image Structure and Transfer" by K. Sayanagi, 1967, available from the Institute of Optics, University of Rochester. A number of books are worth consulting for practical emphasis, e.g. *Modern Optics* by E. Brown, *Modern Optical Engineering* by W. Smith and *Applied Optics* by L. Levi. In all of these, be careful of the sign convention in the transforms.

PROBLEMS

11.1 Determine the Fourier transform of the function

$$E(x) = \begin{cases} E_0 \sin k_p x, & |x| \leqslant L \\ 0, & |x| > L. \end{cases}$$

Make a sketch of $\mathscr{F}\{E(x)\}$. Discuss its relationship to Fig. 11.7.

11.2* Determine the Fourier transform of

$$f(x) = \begin{cases} \sin^2 k_p x, & |x| \leqslant L \\ 0, & |x| > L. \end{cases}$$

Make a sketch of it.

11.3 Determine the Fourier transform of

$$F(t) = \begin{cases} \cos^2 \omega_p t, & |t| \leqslant T \\ 0, & |t| > T. \end{cases}$$

Make a sketch of $F(\omega)$, then sketch its limiting form as $T \to \pm\infty$.

11.4* Show that $\mathscr{F}\{1\} = 2\pi\delta(k)$.

11.5 Prove that $f \circledast h = h \circledast f$ directly. Now do it using the convolution theorem.

11.6* Suppose we have two functions, $f(x, y)$ and $h(x, y)$, where both have a value of 1 over a square region in the xy-plane and are zero everywhere else (Fig. 11.32). If $g(X, Y)$ is their convolution, make a plot of $g(X, 0)$.

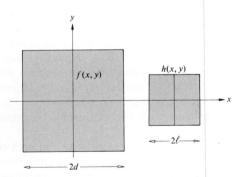

Fig. 11.32

11.7 Referring to the previous problem, justify the fact that the convolution is zero for $|X| \geqslant d + \ell$ when h is viewed as a spread function.

11.8* Use the method illustrated in Fig. 11.14 to convolve the two functions depicted in Fig. 11.33.

Fig. 11.33

11.9* Make a sketch of the resulting function arising from the convolution of the two functions depicted in Fig. 11.34.

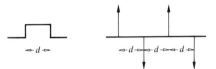

Fig. 11.34

11.10 Beginning with Eq. (11.72) show that the irradiance arising from a randomly located distribution of N identical apertures is given by

$$I \simeq \langle E_1^2 \rangle N,$$

where E_1 is the diffracted field of a single aperture and N is very large.

11.11 Show (for normally incident plane waves) that if an aperture has a center of symmetry, i.e. if the aperture function is even, then the diffracted field in the Fraunhofer case also possesses a center of symmetry.

11.12 Suppose a given aperture produces a Fraunhofer field pattern $E(Y, Z)$. Show that if the aperture's dimensions are altered such that the aperture function goes from $\mathscr{A}(y, z)$ to $\mathscr{A}(\alpha y, \beta z)$, the newly diffracted field will be given by

$$E'(Y, Z) = \frac{1}{\alpha\beta} E\left(\frac{Y}{\alpha}, \frac{Z}{\beta}\right).$$

11.13 Show that when $f(t) = A \sin(\omega t + \varepsilon)$, $C_{ff}(\tau) = (A^2/2) \cos \omega\tau$, which confirms the loss of phase information in the autocorrelation.

11.14 Suppose we have a single slit along the y-direction of width b where the aperture function is constant across it at a value of $\mathscr{A}_0$. What is the diffracted field if we now apodize the slit with a cosine function amplitude mask? In other words, we cause the aperture function to go from $\mathscr{A}_0$ at the center to 0 at $\pm b/2$ via a cosinusoidal drop-off.

Basics of Coherence   12
Theory

Thus far in our discussion of phenomena involving the superposition of waves, we've restricted the treatment to that of either completely coherent or completely incoherent disturbances. This was done primarily as a mathematical convenience since, as is quite often the case, the extremes in a physical situation are the easiest to deal with analytically. In fact, both of these limiting conditions are more conceptual idealizations than actual physical realities. There is a middle ground between these antithetic poles which is of considerable contemporary concern—the domain of *partial coherence*. Even so, the need for extending the theoretical structure is not new, it dates back at least to the mid eighteen-sixties when Emile Verdet demonstrated that a primary source commonly considered as being incoherent, such as the sun, could produce observable fringes when it illuminated the closely spaced pinholes ($\lesssim 0.05$ mm) of Young's experiment (Section 9.3). Theoretical interest in the study of partial coherence lay dormant until it was revived in the nineteen thirties by P. H. van Cittert and then later by Fritz Zernike. And as the technology flourished, advancing from traditional light sources which were essentially optical frequency noise generators to the laser, a new practical impetus was given the subject. Moreover, the recent advent of individual-photon detectors has made it possible to examine related processes associated with the corpuscular aspects of the optical field.

Optical coherence theory is presently an area of active research. Thus, even though much of the excitement in the field is associated with material beyond the level of this book, we shall nonetheless introduce some of the basic ideas.

12.1 INTRODUCTION

Earlier (Section 7.10) we evolved the highly useful picture of quasimonochromatic light as resembling a series of randomly phased finite wave trains (Fig. 7.20). Such a disturbance is nearly sinusoidal although the frequency does vary slowly (in comparison to the rate of oscillation, 10^{15} Hz) about some mean value. Moreover, the amplitude fluctuates as well, but this too is a comparatively slow variation. The average constituent wave train exists roughly for a time Δt which is the *coherence time* given by the inverse of the frequency bandwidth Δv.

It is often convenient, even if a bit artificial, to partition coherence effects into two classifications, *temporal* and *spatial. The former relates directly to the finite bandwidth of the source, the latter to its finite extent in space.*

To be sure, if the light were monochromatic Δv would be zero and Δt infinite, but this is, of course, unattainable. However, over an interval much shorter than Δt an actual wave behaves essentially as if it were monochromatic. In effect the coherence time is loosely *the temporal interval over which we can reasonably predict the phase of the light wave at a given point in space.* This then is what is meant by *temporal coherence,* i.e. if Δt is large, the wave has a high degree of temporal coherence and vice versa.

The same characteristic can be viewed somewhat differently. To that end, imagine that we have two separate points P_1 and P_2 lying on a radius drawn from a quasi-monochromatic point source. If the coherence length, $c \Delta t$, is much larger than the distance (r_{12}) between P_1 and P_2 then a single wave train can extend over the whole separation. The disturbance at P_1 would then be highly correlated with the disturbance occurring at P_2. On the other hand, if this longitudinal separation were very much greater than the coherence length, many wave trains, each with an unrelated phase, would span the gap r_{12}. In that case, the disturbances at the two points in space would be independent at any given time. The degree to which a correlation exists is sometimes spoken of alternatively as the amount of *longitudinal coherence.* Whether we think in terms of coherence time (Δt) or coherence length ($c\Delta t$), the effect still arises from the finite bandwidth of the source.

The idea of *spatial coherence* is most often used to describe effects arising from the finite extent of ordinary light sources. Suppose then that we have a classical broad monochromatic source. Two point radiators on it, separated by a lateral distance which is large compared to λ, will presumably behave quite independently. That is to say, there will be a lack of correlation existing between the phases of the two emitted disturbances. Extended sources of this sort are generally referred to as incoherent but this description is somewhat misleading as we shall see in a moment.

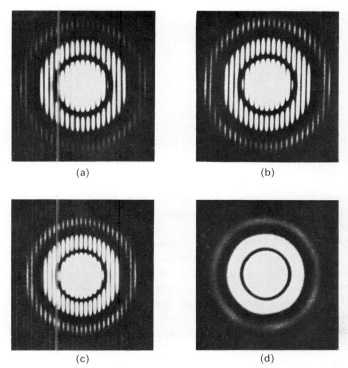

(a)   (b)

(c)   (d)

Fig. 12.1 Double-beam interference from a pair of circular apertures. (a) He—Ne laser light illuminating the holes. (b) Laser light once again but now a 0.5 mm thick glass plate is covering one of the holes. (c) Fringes with collimated mercury-arc illumination but no glass plate. (d) This time the fringes disappear when the plate is inserted using mercury light. [From B. J. Thompson, *J. Soc. Photo. Inst. Engr.* **4**, 7 (1965).]

Usually one is not so much interested in what is happening on the source itself but rather in what is occurring within some distant region of the radiation field. The question to be answered is really: How does the nature of the source and the geometrical configuration of the situation relate to the resulting phase correlation between two laterally spaced points in the light field? This brings to mind Young's experiment where a primary monochromatic source S illuminates two pinholes in an opaque screen. These in turn serve as secondary sources, S_1 and S_2, to generate a fringe pattern on a distant plane of observation, Σ_0 (Fig. 9.5). We already know that if S is an idealized point source, the wavelets issuing from any set of apertures S_1 and S_2 on Σ_a will maintain a constant relative phase; they will be precisely correlated and therefore coherent. A well-defined array of

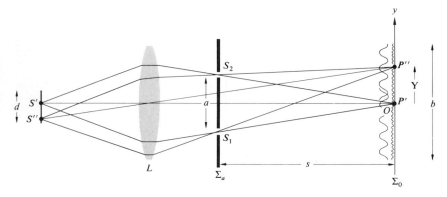

Fig. 12.2 Young's experiment with an extended slit source.

stable fringes results and the field is spatially coherent. At the other extreme, if the pinholes are illuminated by separate thermal sources (even with narrow bandwidths), no correlation exists; no fringes will be observable with existing detectors, and the fields at S_1 and S_2 are said to be incoherent. The generation of interference fringes is then seemingly a very convenient measure of the coherence.

Young's experiment can also be used to demonstrate temporal coherence effects with a finite bandwidth source. Figure 12.1(a) shows the fringe patterns obtained with two small circular apertures illuminated by a He–Ne laser. Prior to taking the photo of Fig. 12.1(b) an optically flat, 0.5 mm thick, piece of glass was positioned over one of the pinholes (say S_1). No change in the form of the pattern (other than a shift in its location) is evident because the coherence length of the laser light far exceeds the optical path-length difference introduced by the glass. On the other hand, when the same experiment is repeated using the light from a collimated mercury arc [Fig. 12.1(c) and (d)] the fringes disappear. Here the coherence length is short enough, and the additional optical path-length difference of the glass long enough for uncorrelated wave trains from the two apertures to arrive at the plane of observation. In other words, of any two coherent wave trains which leave S_1 and S_2, the one from S_1 is now delayed so long in the glass that it falls completely behind the other and arrives at Σ_0 to meet a totally different wave train from S_2.

In both cases of temporal and spatial coherence we are really concerned with one phenomenon, namely the correlation between optical disturbances. That is, we are generally interested in determining the effects arising from relative fluctuations in the fields at two points in space–time. Admittedly the term temporal coherence seems to imply an effect which is exclusively temporal. However, it relates back to the finite extent of the wave train in either space or time and some people even prefer to refer to it as *longitudinal*

spatial rather than temporal coherence. Even so, it does depend intrinsically on the stability of phase in time, and accordingly we will continue to use the term temporal coherence. Spatial coherence or, if you will, *lateral spatial coherence*, is perhaps easier to appreciate because it's so closely related to the concept of the wavefront. Thus if two laterally displaced points reside on the same wavefront at a given time, the fields at those points are said to be spatially coherent (see Section 12.3.1).

12.2 VISIBILITY

The quality of the fringes produced by an interferometric system can be described quantitatively using the *visibility* $\mathscr{V}$ which, as first formulated by Michelson, is given by

$$\mathscr{V}(\mathbf{r}) \equiv \frac{I_{max} - I_{min}}{I_{max} + I_{min}}. \tag{12.1}$$

Here I_{max} and I_{min} are the irradiances corresponding to the maximum and adjacent minimum in the fringe system. If we set up Young's experiment, we could vary the separation of the apertures or the size of the primary monochromatic source, measure $\mathscr{V}$ as it changes in turn and then relate all of this to the idea of coherence. An analytic expression can be derived for the flux-density distribution with the aid of Fig. 12.2.* Here we use a lens L to more effectively localize the fringe pattern, i.e., to make the cones of light diffracted by the finite pinholes more completely overlap on the plane Σ_0. A point source S' located on the central axis would generate the usual pattern given by

$$I = 4I_0 \cos^2\left(\frac{ya\pi}{s\lambda}\right) \tag{12.2}$$

---

* This treatment in part follows that given by Towne in Chapter 11 of *Wave Phenomena*. See Klein, *Optics*, Section 6.3 or Problem 12.3 for different versions.

from Section 9.3. Similarly, a point source above or below S' lying on a line normal to the line $\overline{S_1 S_2}$ would generate the same straight band fringe system slightly displaced in a direction parallel to the fringes. Thus replacing S' by an incoherent line source (normal to the plane of the drawing) effectively just increases the amount of light available. This is something we presumably already knew. In contrast, an off-axis point source, at say S'', will generate a pattern centered about P'', its image point on Σ_0 in the absence of the aperture screen. A spherical wavelet leaving S'' is focused at P''; thus all rays from S'' to P'' traverse equal optical paths and the interference must be constructive, i.e., the central maximum appears at P''. The path difference $\overline{S_1 P''} - \overline{S_2 P''}$ accounts for the displacement $\overline{P'P''}$. Consequently, S'' produces a fringe system identical to that of S' but shifted by an amount $\overline{P'P''}$ with respect to it. Since these source points are incoherent, their irradiances add on Σ_0 rather than their field amplitudes.

The pattern arising from a broad source having a rectangular aperture, normal to the plane of the drawing, can be determined by finding the irradiance due to an incoherent continuous line source parallel to $\overline{S_1 S_2}$. Each differential element of the line source will contribute a fringe system centered about its own image point, a distance Y from the origin on Σ_0. Moreover, its contribution to the flux-density pattern dI is proportional to the differential line element or more conveniently to its image, dY on Σ_0. Thus the contribution to the total irradiance arising from dY is

$$dI = A\, dY \cos^2\left[\frac{a\pi}{s\lambda}(y - Y)\right], \qquad (12.3)$$

where A is an appropriate constant. This, in analogy to Eq. (12.2), is the expression for an entire fringe system of minute irradiance centered at Y. By integrating over the extent b of the image of the line source, we effectively integrate over the source and get the entire pattern:

$$I(y) = A \int_{-b/2}^{b/2} \cos^2\left[\frac{a\pi}{s\lambda}(y - Y)\right] dy. \qquad (12.4)$$

After a good bit of straightforward trigonometric manipulation, this becomes

$$I(y) = \frac{Ab}{2} + \frac{A}{2}\frac{s\lambda}{a\pi}\sin\left(\frac{a\pi}{s\lambda}b\right)\cos\left(2\frac{a\pi}{s\lambda}y\right). \qquad (12.5)$$

The irradiance oscillates about an average value of $\bar{I} = Ab/2$, and so

$$\frac{I(y)}{\bar{I}} = 1 + \left(\frac{\sin a\pi b/s\lambda}{a\pi b/s\lambda}\right)\cos\left(2\frac{a\pi}{s\lambda}y\right) \qquad (12.6)$$

or

$$\frac{I(y)}{\bar{I}} = 1 + \mathrm{sinc}\left(\frac{a\pi b}{s\lambda}\right)\cos\left(2\frac{a\pi}{s\lambda}y\right). \qquad (12.7)$$

It follows that the extreme values of the relative irradiance are given by

$$\frac{I_{max}}{\bar{I}} = 1 + \left|\mathrm{sinc}\left(\frac{a\pi b}{s\lambda}\right)\right| \qquad (12.8)$$

and

$$\frac{I_{min}}{\bar{I}} = 1 - \left|\mathrm{sinc}\left(\frac{a\pi b}{s\lambda}\right)\right|. \qquad (12.9)$$

When b is very small in comparison to the fringe width $(s\lambda/a)$, the sinc function (Table 1) approaches 1 and $I_{max}/\bar{I} = 2$ while $I_{min}/\bar{I} = 0$ (see Fig. 12.3). As b increases, I_{min} begins to differ from zero and the fringes lose contrast until they finally vanish entirely at $b = s\lambda/a$. Between the arguments of π and 2π (i.e. $b = s\lambda/a$ and $b = 2s\lambda/a$), the sinc is negative. As the primary slit source widens beyond $b = s\lambda/a$, the fringes reappear but as if shifted, i.e. previously there was a maximum at $y = 0$, now there will be a minimum.

In actuality, the light diffracted by the apertures is localized (Section 10.2) so that the fringe system does not continue out uniformly indefinitely as y increases. Instead, the pattern of Fig. 12.3(a) will look more like Fig. 12.4.

The visibility of the fringes is simply

$$\mathscr{V} = \left|\mathrm{sinc}\left(\frac{a\pi b}{s\lambda}\right)\right|, \qquad (12.10)$$

which s plotted in Fig. 12.5. Observe that $\mathscr{V}$ is a function of both the source breadth via b and the aperture separation a. Holding either one of these parameters constant and varying the other will cause $\mathscr{V}$ to change in precisely the same way. Note that the visibilities in both Figs. 12.3(a) and 12.4 are equal to one because $I_{min} = 0$. Incidentally, we encountered the *sinc function* before in connection with the diffraction pattern resulting from a rectangular aperture (10.17).

When the primary source is circular, the visibility is a good deal more complicated to calculate. It turns out to be proportional to a first-order Bessel function (Fig. 12.6). This too is quite reminiscent of diffraction, this time at a circular aperture (10.56). These similarities between expressions for $\mathscr{V}$ and the corresponding diffraction patterns

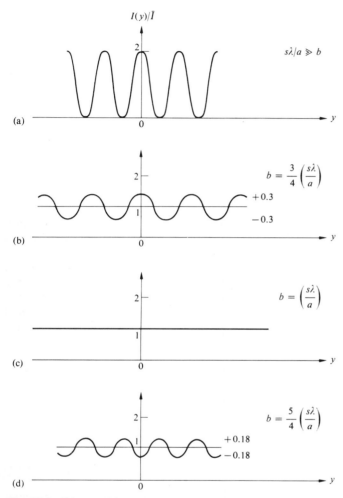

Fig. 12.3 Fringes with varying source slit size.

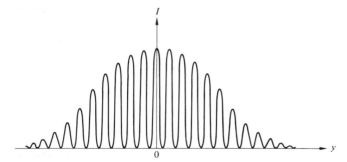

Fig. 12.4 Double beam interference fringes showing the effect of diffraction.

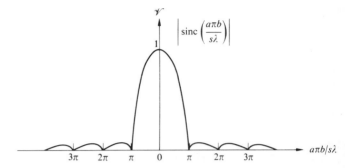

Fig. 12.5 The visibility as given by Eq. (12.10).

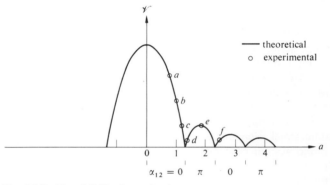

Fig. 12.6 The visibility for a circular source.

for the same shape aperture are not merely fortuitous but rather are a manifestation of something called the van Cittert–Zernike theorem. A discussion of this theorem is the domain of a more advanced treatment—here we merely point out its existence.

Figure 12.7 shows a sequence of fringe systems where the circular incoherent primary source is constant in size but the separation a between S_1 and S_2 is increased. The visibility decreases from Fig. 12.7(a) to (d), then increases for (e) and decreases again at (f). All of the associated $\mathscr{V}$-values are plotted in Fig. 12.6. Note the shift in the peaks, i.e., the change in phase at the center of the pattern for each point on the second lobe of Fig. 12.6 (the Bessel function is negative over that range). In other words, (a), (b), and (c) have a central maximum while (d) and (e) have a central minimum and (f) on the third lobe is back to a maximum. In the same way for a slit source, the domain where sinc $(a\pi b/s\lambda)$ in Eq. (12.7) is positive or negative will yield a maximum or minimum, respectively, in $I(0)/\bar{I}$. These in turn correspond to the odd or even lobes of the visibility

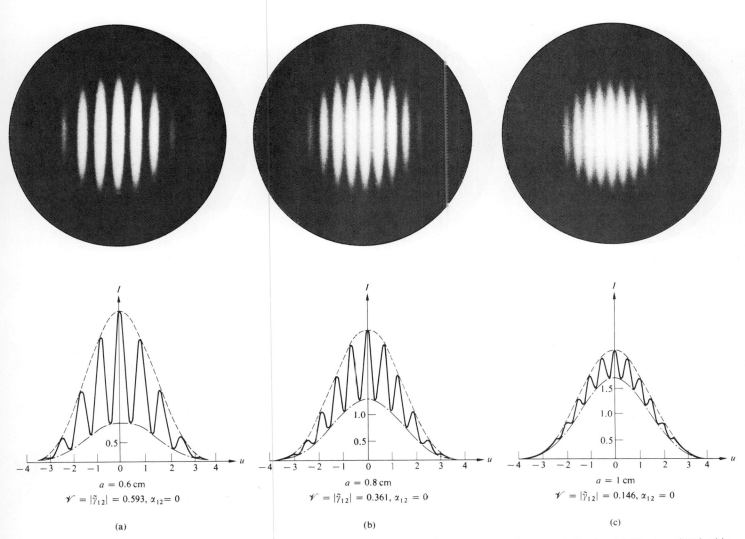

Fig. 12.7 Double-beam interference patterns using partially coherent light. The photos correspond to a variation in visibility associated with changes in *a*, the separation between the apertures. In the theoretical curves $I_{max} \propto 1 + |2J_1(u)/u|$ and $I_{min} \propto 1 - |2J_1(u)/u|$. Several of the symbols will be discussed later. [From B. J. Thompson and E. Wolf, *J. Opt. Soc. Am.* **47**, 895 (1957).]

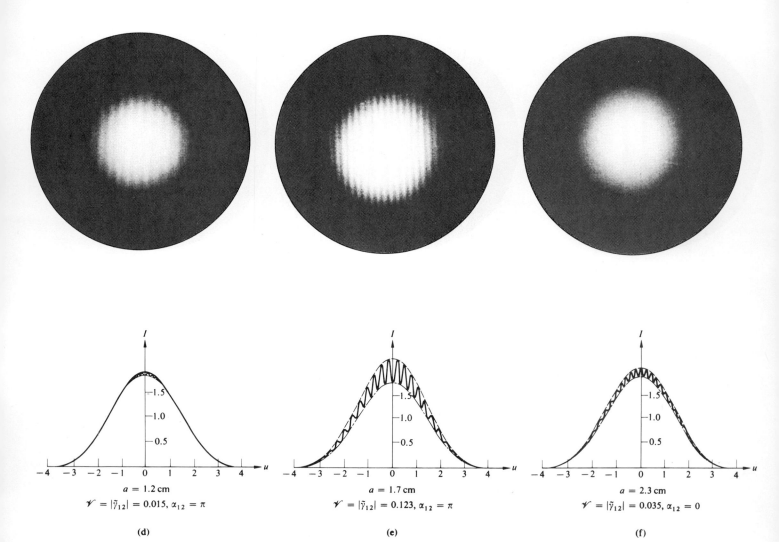

$a = 1.2 \, \text{cm}$

$\mathscr{V} = |\tilde{\gamma}_{12}| = 0.015, \, \alpha_{12} = \pi$

(d)

$a = 1.7 \, \text{cm}$

$\mathscr{V} = |\tilde{\gamma}_{12}| = 0.123, \, \alpha_{12} = \pi$

(e)

$a = 2.3 \, \text{cm}$

$\mathscr{V} = |\tilde{\gamma}_{12}| = 0.035, \, \alpha_{12} = 0$

(f)

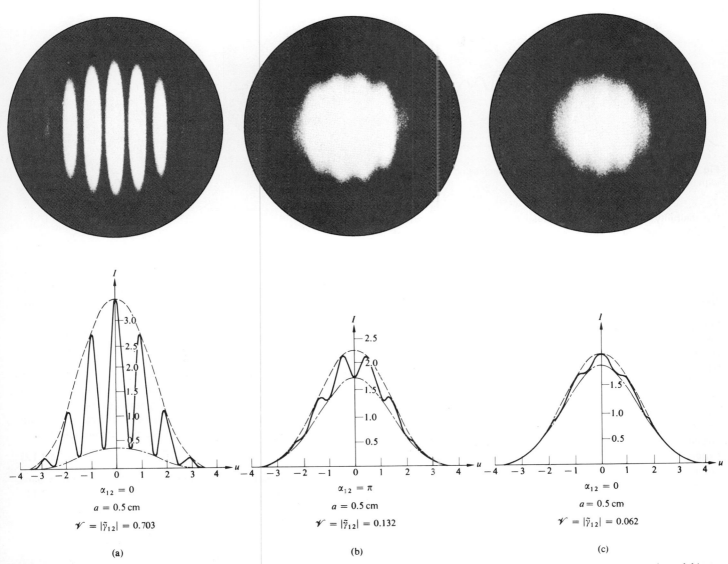

Fig. 12.8 Double beam interference patterns. Here the aperture separation was held constant, thereby yielding a constant number of fringes per unit displacement in each photo. The visibility was altered by varying the size of the primary incoherent source. [From B. J. Thompson, *J. Soc. Photo. Inst. Engr.* **4**, 7 (1965).]

curve of Fig. 12.5. Bear in mind that we could define a complex visibility of magnitude $\mathscr{V}$, having an argument corresponding to the phase shift—we'll come back to this idea later.

Since the width of the fringes is inversely proportional to a, the spatial frequency of the bright and dark bands

increases accordingly from Fig. 12.7(a) to (f). Figure 12.8 results when the separation a is held constant while the primary incoherent source diameter is increased.

We should also mention that the effects of the finite bandwidth will show up in a given fringe pattern as a gradually decreasing value of $\mathscr{V}$ with y as in Fig. 12.9. When

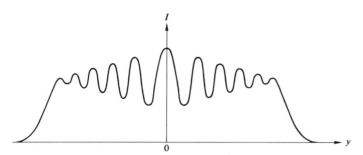

Fig. 12.9 A finite bandwidth results in a decreasing value of $\mathscr{V}$ with increasing y.

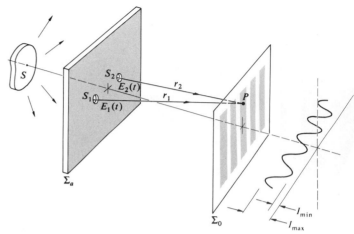

Fig. 12.10 Young's experiment.

the visibility is determined in these cases using the central region of each of a series of patterns, the dependence of $\mathscr{V}$ on aperture separation will again match Fig. 12.6.

12.3 THE MUTUAL COHERENCE FUNCTION AND THE DEGREE OF COHERENCE

Let's now carry the discussion a bit further in a more formal fashion. Again suppose we have a broad, narrow bandwidth source which generates a light field whose complex representation* is $\tilde{E}(\mathbf{r}, t)$. We'll overlook polarization effects and therefore a scalar treatment will do. The disturbances at two points in space S_1 and S_2 are then $\tilde{E}(S_1, t)$ and $\tilde{E}(S_2, t)$ or more succinctly $\tilde{E}_1(t)$ and $\tilde{E}_2(t)$. If these two points are then isolated using an opaque screen with two circular apertures (Fig. 12.10), we're back to Young's experiment. The two apertures serve as sources of secondary wavelets which propagate out to some point P on Σ_0. There the resultant field is

$$\tilde{E}_P(t) = \tilde{K}_1 \tilde{E}_1(t - t_1) + \tilde{K}_2 \tilde{E}_2(t - t_2), \quad (12.11)$$

where $t_1 = r_1/c$ and $t_2 = r_2/c$. This says that the field at the space–time point (P, t) can be determined from the fields which existed at S_1 and S_2 at t_1 and t_2 respectively, these being the instants when the light, which is now overlapping, first emerged from the apertures. The quantities $\tilde{K}_1$ and $\tilde{K}_2$, which are known as propagators, depend on the size of the apertures and their relative locations with respect to P. They mathematically affect the alterations in the field resulting from its having traversed either of the apertures. For example, the secondary wavelets issuing from the pin-

---

* We'll put a wavy line over quantities that are complex just as a continuing reminder.

holes in this set-up are out of phase by $\pi/2$ rad with the primary wave incident on the aperture screen, Σ_a (Section 10.3.1). Clearly someone is going to have to tell $\tilde{E}(\mathbf{r}, t)$ to shift phase beyond Σ_a—that's just what the $\tilde{K}$ factors are for. Moreover, they reflect a reduction in the field which might arise from a number of physical causatives; absorption, diffraction, etc. Here, since there is a $\pi/2$ phase shift in the field, which can be introduced by multiplying by $\exp i\pi/2$, $\tilde{K}_1$ and $\tilde{K}_2$ are purely imaginary numbers.

The resultant irradiance at P measured over some finite interval, which is long compared to the coherence time, is

$$I_P = \langle \tilde{E}_P(t) \tilde{E}_P^*(t) \rangle. \quad (12.12)$$

It should be remembered that Eq. (12.12) is written sans several multiplicative constants. Hence using Eq. (12.11)

$$\begin{aligned}
I_P = &\ \tilde{K}_1 \tilde{K}_1^* \langle \tilde{E}_1(t - t_1) \tilde{E}_1^*(t - t_1) \rangle \\
&+ \tilde{K}_2 \tilde{K}_2^* \langle \tilde{E}_2(t - t_2) \tilde{E}_2^*(t - t_2) \rangle \\
&+ \tilde{K}_1 \tilde{K}_2^* \langle \tilde{E}_1(t - t_1) \tilde{E}_2^*(t - t_2) \rangle \\
&+ \tilde{K}_1^* \tilde{K}_2 \langle \tilde{E}_1^*(t - t_1) \tilde{E}_2(t - t_2) \rangle. \quad (12.13)
\end{aligned}$$

It is now assumed that the wave field is *stationary*, as is almost universally the case in classical optics, i.e., it does not alter its statistical nature with time so that the time average is independent of whatever origin we select. Thus, even though there are fluctuations in the field variables, the time origin can be shifted and the averages in Eq. (12.13) will be unaffected. It shouldn't matter over which particular moment we decide to actually measure I_P. Accordingly, the first two time averages can be rewritten as

$$I_{S_1} = \langle \tilde{E}_1(t)\tilde{E}_1^*(t) \rangle \quad \text{and} \quad I_{S_2} = \langle \tilde{E}_2(t)\tilde{E}_2^*(t) \rangle,$$

where the origin was displaced by amounts t_1 and t_2, respectively. Here the subscripts underscore the fact that these are the irradiances at points S_1 and S_2. Furthermore if we let $\tau = t_2 - t_1$ we can shift the time origin by an amount t_2 in the last two terms of Eq. (12.13) and write them as

$$\tilde{K}_1\tilde{K}_2^*\langle \tilde{E}_1(t + \tau)\tilde{E}_2^*(t) \rangle + \tilde{K}_1^*\tilde{K}_2\langle \tilde{E}_1^*(t + \tau)\tilde{E}_2(t) \rangle.$$

But this is a quantity plus its own complex conjugate and is therefore just twice its real part, i.e., it equals

$$2 \text{ Re } [\tilde{K}_1\tilde{K}_2^*\langle \tilde{E}_1(t + \tau)\tilde{E}_2^*(t) \rangle].$$

The $\tilde{K}$-factors are purely imaginary and so $\tilde{K}_1\tilde{K}_2^* = \tilde{K}_1^*\tilde{K}_2 = |\tilde{K}_1||\tilde{K}_2|$. The time-average portion of this term is a cross-correlation function [Section 11.3.4(iii)] that we denote by

$$\tilde{\Gamma}_{12}(\tau) \equiv \langle \tilde{E}_1(t + \tau)\tilde{E}_2^*(t) \rangle, \quad (12.14)$$

and refer to as the *mutual coherence function* of the light field at S_1 and S_2. If we make use of all of this Eq. (12.13) takes the form

$$I_P = |\tilde{K}_1|^2 I_{S_1} + |\tilde{K}_2|^2 I_{S_2} + 2|\tilde{K}_1||\tilde{K}_2| \text{ Re } \tilde{\Gamma}_{12}(\tau). \quad (12.15)$$

The terms $|\tilde{K}_1|^2 I_{S_1}$ and $|\tilde{K}_2|^2 I_{S_2}$, again overlooking multiplicative constants, are the irradiances at P arising when one or the other of the apertures is open alone, i.e., $\tilde{K}_2 = 0$ or $\tilde{K}_1 = 0$, respectively. Denoting these by $I_P^{(1)}$ and $I_P^{(2)}$ Eq. (12.15) becomes

$$I_P = I_P^{(1)} + I_P^{(2)} + 2|\tilde{K}_1||\tilde{K}_2| \text{ Re } \tilde{\Gamma}_{12}(\tau). \quad (12.16)$$

Note that when S_1 and S_2 are made to coincide, the mutual coherence function becomes

$$\tilde{\Gamma}_{11}(\tau) = \langle \tilde{E}_1(t + \tau)\tilde{E}_1^*(t) \rangle$$

or

$$\tilde{\Gamma}_{22}(\tau) = \langle \tilde{E}_2(t + \tau)\tilde{E}_2^*(t) \rangle.$$

We can imagine that two wave trains emerge from this coalesced source point and somehow pick up a relative phase delay proportional to τ. In the present situation τ becomes zero (since the O.P.D. goes to zero) and these functions reduce to the corresponding irradiances $I_{S_1} = \langle \tilde{E}_1(t)\tilde{E}_1^*(t) \rangle$ and $I_{S_2} = \langle \tilde{E}_2(t)\tilde{E}_2^*(t) \rangle$ on Σ_a. Hence

$$\Gamma_{11}(0) = I_{S_1} \text{ and } \Gamma_{22}(0) = I_{S_2},$$

and these are called *self coherence functions*. Thus

$$I_P^{(1)} = |\tilde{K}_1|^2 \, \Gamma_{11}(0) \quad \text{and} \quad I_P^{(2)} = |\tilde{K}_2|^2 \Gamma_{22}(0).$$

Keeping Eq. (12.16) in mind, observe that

$$|\tilde{K}_1||\tilde{K}_2| = \sqrt{I_P^{(1)}}\sqrt{I_P^{(2)}}/\sqrt{\Gamma_{11}(0)}\sqrt{\Gamma_{22}(0)}.$$

The normalized form of the mutual coherence function is defined as

$$\tilde{\gamma}_{12}(\tau) \equiv \frac{\tilde{\Gamma}_{12}(\tau)}{\sqrt{\Gamma_{11}(0)\,\Gamma_{22}(0)}} = \frac{\langle \tilde{E}_1(t + \tau)\tilde{E}_2^*(t) \rangle}{\sqrt{\langle |\tilde{E}_1|^2\rangle\langle |\tilde{E}_2|^2\rangle}}, \quad (12.17)$$

and it's spoken of as the *complex degree of coherence* for reasons which will be clear imminently. Equation (12.16) can then be recast as

$$I_P = I_P^{(1)} + I_P^{(2)} + 2\sqrt{I_P^{(1)} I_P^{(2)}}\text{Re } \tilde{\gamma}_{12}(\tau), \quad (12.18)$$

which is the *general interference law for partially coherent light*.

For quasimonochromatic light the phase angle difference concomitant with the optical path difference is given by

$$\varphi = \frac{2\pi}{\bar{\lambda}}(r_2 - r_1) = 2\pi\bar{v}\tau, \quad (12.19)$$

where $\bar{\lambda}$ and $\bar{v}$ are the mean wavelength and frequency. Now $\tilde{\gamma}_{12}(\tau)$ is a complex quantity expressible as

$$\tilde{\gamma}_{12}(\tau) = |\tilde{\gamma}_{12}(\tau)|e^{i\Phi_{12}(\tau)}. \quad (12.20)$$

The phase angle of $\tilde{\gamma}_{12}(\tau)$ relates back to Eq. (12.14) and the phase angle between the fields. If we set $\Phi_{12}(\tau) = \alpha_{12}(\tau) - \varphi$, then

$$\text{Re } \tilde{\gamma}_{12}(\tau) = |\tilde{\gamma}_{12}(\tau)| \cos [\alpha_{12}(\tau) - \varphi].$$

Equation (12.18) is then expressible as

$$I_P = I_P^{(1)} + I_P^{(2)} + 2\sqrt{I_P^{(1)} I_P^{(2)}}|\tilde{\gamma}_{12}(\tau)| \cos [\alpha_{12}(\tau) - \varphi]. \quad (12.21)$$

It can be shown from Eq. (12.17) and the Schwarz inequality that $0 \leqslant |\tilde{\gamma}_{12}(\tau)| \leqslant 1$. In fact by comparing Eqs. (12.21) and (9.5), the latter having been derived for the case of complete coherence, it is evident that if $|\tilde{\gamma}_{12}(\tau)| = 1$, I_P would be the same as that generated by two *coherent* waves out of phase at S_1 and S_2 by an amount $\alpha_{12}(\tau)$. If at the other extreme $|\tilde{\gamma}_{12}(\tau)| = 0$, $I_P = I_P^{(1)} + I_P^{(2)}$, there is no interference and the two disturbances are said to be *incoherent*. When $0 < |\tilde{\gamma}_{12}(\tau)| < 1$ we have *partial coherence*; the measure of which is $|\tilde{\gamma}_{12}(\tau)|$ itself, which is known as the *degree of coherence*. In summary then

$$|\tilde{\gamma}_{12}| = 1 \qquad \text{coherent limit}$$

$$|\tilde{\gamma}_{12}| = 0 \qquad \text{incoherent limit}$$

$$0 < |\tilde{\gamma}_{12}| < 1 \qquad \text{partial coherence.}$$

The basic statistical nature of the entire process must be underscored. Clearly $\tilde{\Gamma}_{12}(\tau)$ and, therefore, $\tilde{\gamma}_{12}(\tau)$ are the key quantities in the various expressions for the irradiance distribution; they are the essence of what we previously called the interference term (9.1). It should be pointed out that $\tilde{E}_1(t + \tau)$ and $\tilde{E}_2(t)$ are in fact two disturbances occurring at different points in both space and time. We anticipate, as well, that the amplitudes and phases of these disturbances will somehow fluctuate in time. If these fluctuations at S_1 and S_2 are completely independent then $\tilde{\Gamma}_{12}(\tau) = \langle \tilde{E}_1(t + \tau)\tilde{E}_2^*(t)\rangle$ will go to zero since $\tilde{E}_1$ and $\tilde{E}_2$ can be either positive or negative with equal likelihood and their product averages to zero. In that case no correlation exists and $\tilde{\Gamma}_{12}(\tau) = \tilde{\gamma}_{12}(\tau) = 0$. If the field at S_1 at a time $(t + \tau)$ were perfectly correlated with the field at S_2 at a time t, their relative phase would remain unaltered despite individual fluctuations. The time average of the product of the fields would certainly not be zero just as it would not be zero even if the two were only slightly correlated.

Both $|\tilde{\gamma}_{12}(\tau)|$ and $\alpha_{12}(\tau)$ are slowly varying functions of τ in comparison to $\cos 2\pi\bar{\nu}\tau$ and $\sin 2\pi\bar{\nu}\tau$. In other words, as P is moved across the resultant fringe system, the point-by-point spatial variations in I_P are predominantly due to the changes in φ as $(r_2 - r_1)$ changes.

The maximum and minimum values of I_P occur when the cosine term in Eq. (12.21) is $+1$ and -1 respectively. The visibility at P (Problem 12.4) is then

$$\mathcal{V}_P = \frac{2\sqrt{I_P^{(1)}}\sqrt{I_P^{(2)}}}{I_P^{(1)} + I_P^{(2)}}|\tilde{\gamma}_{12}(\tau)|. \qquad (12.22)$$

Perhaps the most common arrangement occurs when things are adjusted so that $I_P^{(1)} = I_P^{(2)}$, whereupon

$$\mathcal{V}_P = |\tilde{\gamma}_{12}(\tau)|, \qquad (12.23)$$

i.e. the modulus of the complex degree of coherence is identical to the visibility of the fringes (take another look at Fig. 12.7).

It is essential to realize that Eqs. (12.17) and (12.18) clearly suggest the way in which the real parts of $\tilde{\Gamma}_{12}(\tau)$ and $\tilde{\gamma}_{12}(\tau)$ can be determined from direct measurements. When the flux densities of two disturbances are adjusted to be equal, Eq. (12.23) provides an experimental means of obtaining $|\tilde{\gamma}_{12}(\tau)|$ from the resultant fringe pattern. Furthermore, the off-axis shift in the location of the central fringe (from $\varphi = 0$) is a measure of $\alpha_{12}(\tau)$, the apparent relative retardation of the phase of the disturbances at S_1 and S_2. Thus, measurements of the visibility and fringe position yield both the amplitude and phase of the complex degree of coherence.

By the way, it can be shown[*] that $|\tilde{\gamma}_{12}(\tau)|$ will equal 1 for all values of τ and any pair of spatial points if and only if the optical field is strictly monochromatic, and therefore such a situation is unattainable. Moreover, a nonzero radiation field for which $|\tilde{\gamma}_{12}(\tau)| = 0$ for all values of τ and any pair of spatial points cannot exist in free space either.

12.3.1 Temporal and Spatial Coherence

Let's now relate the ideas of temporal and spatial coherence to the above formalism.

If the primary source S in Fig. 12.10 shrinks down to a point source on the central axis having a finite frequency bandwidth, temporal coherence effects will predominate. The optical disturbances at S_1 and S_2 will then be identical. In effect, the mutual coherence (12.14) between the two points will be the self coherence of the field. Hence $\tilde{\Gamma}(S_1, S_2, \tau) = \tilde{\Gamma}_{12}(\tau) = \tilde{\Gamma}_{11}(\tau)$ or $\tilde{\gamma}_{12}(\tau) = \tilde{\gamma}_{11}(\tau)$. The same thing obtains when S_1 and S_2 coalesce and $\tilde{\gamma}_{11}(\tau)$ is sometimes referred to as the *complex degree of temporal coherence* at that point for two instances of time separated by an interval τ. This would be the case in an amplitude-splitting interferometer such as Michelson's where τ equals the path-length difference divided by c. The expression for I_P, i.e. Eq. (12.18), would then contain $\tilde{\gamma}_{11}(\tau)$ rather than $\tilde{\gamma}_{12}(\tau)$.

Suppose a light wave is divided into two identical disturbances of the form

$$\tilde{E}(t) = E_0 e^{i\phi(t)} \qquad (12.24)$$

by an amplitude-splitting interferometer which later recombines them to generate a fringe pattern. Then

$$\tilde{\gamma}_{11}(\tau) = \frac{\langle \tilde{E}(t + \tau)\tilde{E}^*(t)\rangle}{|\tilde{E}|^2} \qquad (12.25)$$

or

$$\tilde{\gamma}_{11}(\tau) = \langle e^{i\phi(t + \tau)} e^{-i\phi(t)}\rangle.$$

---

[*] The proofs are given in Beran and Parrent, *Theory of Partial Coherence*, Section 4.2.

Hence

$$\tilde{\gamma}_{11}(\tau) = \lim_{T \to \infty} \frac{1}{T} \int_0^T e^{i[\phi(t+\tau) - \phi(t)]} \, dt \qquad (12.26)$$

and

$$\tilde{\gamma}_{11}(\tau) = \lim_{T \to \infty} \frac{1}{T} \int_0^T (\cos \Delta\phi + i \sin \Delta\phi) \, dt,$$

where $\Delta\phi(t + \tau) - \phi(t)$. For a strictly monochromatic plane wave of infinite coherence length $\phi(t) = \mathbf{k} \cdot \mathbf{r} - \omega t$, $\Delta\phi = -\omega\tau$ and

$$\tilde{\gamma}_{11}(\tau) = \cos \omega\tau - i \sin \omega\tau = e^{-i\omega\tau}.$$

Hence $|\tilde{\gamma}_{11}| = 1$; the argument of $\tilde{\gamma}_{11}$ is just $-2\pi\nu\tau$ and we have complete coherence. In contradistinction, for a quasimonochromatic wave where τ is greater than the coherence length, $\Delta\phi$ will be random, varying between 0 and 2π such that the integral averages to zero, $|\tilde{\gamma}_{11}(\tau)| = 0$ corresponding to complete incoherence. A path difference (O.P.D.) of 60 cm, produced when the two arms of a Michelson interferometer differ in length by 30 cm, corresponds to a time delay between the recombining beams of $\tau \approx 2$ ns. This is roughly the coherence time of a good isotope discharge lamp and the visibility of the pattern under this sort of illumination will be quite poor. If white light is used instead, $\Delta\nu$ is large, Δt very small and the coherence length is less than one wavelength. In order for $\tau < \Delta t$, i.e. in order that the visibility be good, the O.P.D. will have to be a small fraction of a wavelength. The other extreme

is laser light where Δt can be so long that a value of $c\tau$ which will cause an appreciable decrease in visibility would require an impractically large interferometer.

We see that $\tilde{\Gamma}_{11}(\tau)$, being a measure of temporal coherence, must be intimately related to the coherence time and therefore the bandwidth of the source. Indeed, the Fourier transform of the self-coherence function, $\tilde{\Gamma}_{11}(\tau)$, is the power spectrum which describes the spectral energy distribution of the light (Section 11.3.4).

If we go back to Young's experiment (Fig. 12.10) with a very narrow-bandwidth extended source, spatial coherence effects will predominate. The optical disturbances at S_1 and S_2 will differ and the fringe pattern will depend on $\tilde{\Gamma}(S_1, S_2, \tau) = \tilde{\Gamma}_{12}(\tau)$. By examining the region about the central fringe where $(r_2 - r_1) = 0$, $\tau = 0$ and $\tilde{\Gamma}_{12}(0)$ and $\tilde{\gamma}_{12}(0)$ can be determined. This latter quantity is the *complex degree of spatial coherence* of the two points at the same instant in time. $\tilde{\Gamma}_{12}(0)$ plays a central role in the description of the Michelson stellar interferometer to be discussed forthwith.

12.4 COHERENCE AND STELLAR INTERFEROMETRY

12.4.1 The Michelson Stellar Interferometer

In 1890 A. A. Michelson, following an earlier suggestion by Fizeau, proposed an interferometric device (Fig. 12.11) which is of interest here both because it was the precursor

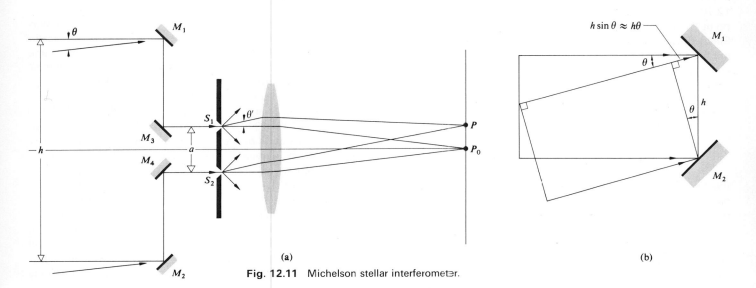

(a)   (b)

Fig. 12.11 Michelson stellar interferometer.

to some important modern techniques and because it lends itself to an interpretation in terms of coherence theory. The function of the stellar interferometer, as it is called, is to measure the small angular dimensions of remote astronomical bodies.

Two widely spaced movable mirrors M_1 and M_2 collect rays, presumed to be parallel, from a very distant star. The light is then channeled via mirrors M_3 and M_4 through apertures S_1 and S_2 of a mask and thence into the objective of a telescope. The optical paths $M_1M_3S_1$ and $M_2M_4S_2$ are made equal so that the relative phase-angle difference between a disturbance at M_1 and M_2 is the same as that between S_1 and S_2. The two apertures generate the usual Young's experiment fringe system in the focal plane of the objective. Actually, the mask and openings are not really necessary; the mirrors alone could serve as apertures. Suppose we now point the device so that its central axis is directed toward one of the stars in a closely spaced double-star configuration. Because of the tremendous distances involved, the rays reaching the interferometer from either star are well collimated. Furthermore we assume, at least for the moment, that the light has a narrow linewidth centered about a mean wavelength of $\bar{\lambda}_0$. The disturbances arising at S_1 and S_2 from the axial star are in phase and a pattern of bright and dark bands form centered on P_0. Similarly, rays from the other star arrive at some angle θ, but this time the disturbances at M_1 and M_2 (and therefore at S_1 and S_2) are out of phase by approximately $\bar{k}_0 h\theta$ or, if you will, retarded by a time $h\theta/c$, as indicated in Fig. 12.10(b). The resulting fringe system is centered about a point P shifted by an angle θ' from P_0 such that $h\theta/c = a\theta'/c$. Since these stars behave as though they were incoherent point sources, the individual irradiance distributions simply overlap. The separation between the fringes set up by either star is equal and dependent solely on a. Yet the visibility varies with h. Thus if h is increased from nearly zero until $\bar{k}_0 h\theta = \pi$, i.e. until

$$h = \frac{\bar{\lambda}_0}{2\theta}, \qquad (12.27)$$

the two fringe systems take on an increasing relative displacement until finally the maxima from one star overlap the minima from the other, at which point, if their irradiances are equal, $\mathscr{V} = 0$. Hence, when the fringes vanish, one need only measure h to determine the angular separation between the stars, θ. Notice that the appropriate value of h varies inversely with θ.

Note that even though the source points, the two stars, are assumed to be completely uncorrelated, the resulting optical fields at any two points (M_1 and M_2) are not necessarily incoherent. For that matter, as h becomes very small, the light from each point source arrives with essentially zero relative phase at M_1 and M_2; $\mathscr{V}$ approaches 1 and the fields at those locations are highly coherent.

In much the same way as with a double star system, the angular diameter (θ) of certain single stars can be measured. Once again the fringe visibility corresponds to the degree of coherence of the optical field at M_1 and M_2. If the star is presumed to be a circular distribution of incoherent point sources such that it has a uniform brilliance, its visibility is equivalent to that already plotted in Fig. 12.6. Earlier, we alluded to the fact that $\mathscr{V}$ for this sort of source was given by a first-order Bessel function and in fact it is expressible as

$$\mathscr{V} = |\tilde{\gamma}_{12}(0)| = 2 \left| \frac{J_1(\pi h\theta/\bar{\lambda}_0)}{\pi h\theta/\bar{\lambda}_0} \right|. \qquad (12.28)$$

Recall that $J_1(u)/u = \frac{1}{2}$ at $u = 0$ and the maximum value of $\mathscr{V}$ is 1. The first zero of $\mathscr{V}$ occurs when $\pi h\theta/\bar{\lambda}_0 = 3.83$ as in Fig. 10.28. Equivalently, the fringes disappear when

$$h = 1.22 \frac{\bar{\lambda}_0}{\theta} \qquad (12.29)$$

and, as before, one simply measures h to find θ.

In Michelson's arrangement, the two outrigged mirrors were movable on a long girder which was mounted on the 100 inch reflector of the Mt. Wilson Observatory. Betelgeuse (α Orionis) was the first star whose angular diameter was measured using the device. It's the orange-looking star in the upper left of the constellation Orion. In fact its name is a contraction for the Arabic phrase meaning *the armpit of the central one*, i.e. Orion. The fringes formed by the interferometer, one cold December night in 1920, were made to vanish at $h = 121$ inches and with $\bar{\lambda}_0 = 570$ nm, $\theta = 1.22(570 \times 10^{-9})/121(2.54 \times 10^{-2}) = 22.6 \times 10^{-8}$ rad or 0.047 seconds of arc. Using its known distance, determined from parallax measurements, the star's diameter turned out to be about 240 million miles or roughly 280 times that of the sun. Actually, Betelgeuse is an irregular variable star whose maximum diameter is so tremendous that it's larger than the orbit of Mars about the sun. The main limitation on the use of the stellar interferometer is due to the inconveniently long mirror separations required for all but the largest stars. And this is true as well in radio-astronomy where an analogous set-up has been widely used to measure the extent of celestial sources of radio-frequency emissions.

Incidentally, we assume, as is often done, that "good" coherence means a visibility of 0.88 or better. For a disk source this occurs when $\pi h\theta/\bar{\lambda}_0$ in Eq. (12.28) equals *one*, that is when

$$h = 0.32\frac{\bar{\lambda}_0}{\theta}. \qquad (12.30)$$

For a narrow-bandwidth source of diameter D a distance R away, there is an *area of coherence* equal to $\pi(h/2)^2$ over which $|\tilde{\gamma}_{12}| \geqslant 0.88$. Since $D/R = \theta$

$$h = 0.32\frac{R\bar{\lambda}_0}{D}. \qquad (12.31)$$

These expressions are very handy for estimating the required physical parameters in an interference or diffraction experiment. For example, if we put a red filter over a 1 mm diameter disk-shaped flashlight source and stand back 20 m from it, then

$$h = 0.32(20)\,(600 \times 10^{-9})/10^{-3} = 3.8\ \text{mm},$$

where the mean wavelength is taken as 600 nm. This means that a set of apertures, spaced at about h or less should produce nice fringes. Evidently the area of coherence increases with R and this is why you can always find a distant bright street light to use as a convenient source.

12.4.2 Correlation Interferometry

Let's return for a moment to the representation of a disturbance emanating from a thermal source as discussed in Section 7.10. Here the word *thermal* connotes a light field arising predominantly from the superposition of spontaneously emitted waves issuing from a great many independent atomic sources.[*] A quasimonochromatic optical field can be represented by

$$E(t) = E_0(t) \cos\left[\varepsilon(t) - 2\pi\bar{\nu}t\right]. \qquad [7.65]$$

The amplitude is a relatively slowly varying function of time, as is the phase. For that matter, the wave might undergo tens of thousands of oscillations before either the amplitude (i.e. the envelope of the field vibrations) or the phase would change appreciably. Thus, just as the coherence time is a measure of the fluctuation interval of the phase, it is also a measure of the interval over which $E_0(t)$ is fairly predictable.

---

[*] Thermal light is sometimes spoken of as *Gaussian light* because the amplitude of the field follows a Gaussian probability distribution.

Large fluctuations in ε are generally accompanied by correspondingly large fluctuations of E_0. Presumably a knowledge of these amplitude fluctuations of the field could be related to the phase fluctuations and therefore to the correlation (i.e. coherence) functions. Accordingly, at two points in space–time where the phases of the field are correlated, we could expect the amplitudes to be related as well.

When a fringe pattern exists for the Michelson stellar interferometer, it is because the fields at M_1 and M_2, the apertures, are somehow correlated; that is $\tilde{\Gamma}_{12}(0) = \langle \tilde{E}_1(t)\,\tilde{E}_2^*(t)\rangle \neq 0$. If we could measure the field amplitudes at these points, their fluctuations would likewise show an interrelationship. Since this isn't practicable because of the high frequencies involved, we might instead measure and compare the fluctuations in irradiance at the locations of M_1 and M_2 and from this, in some as yet unknown way, infer $|\tilde{\gamma}_{12}(0)|$. In other words, if there are values of τ for which $\tilde{\gamma}_{12}(\tau)$ is nonzero, the field at the two points is partially coherent and a correlation between the irradiance fluctuations at these locations is implied. This is the essential idea behind a series of remarkable experiments conducted in the years 1952 to 1956 by R. Hanbury-Brown in collaboration with R. Q. Twiss and others. The culmination of their work was the so-called *correlation interferometer*.

Thus far we have evolved only an intuitive justification for the phenomenon rather than a firm theoretical treatment. Such an analysis, however, is beyond the scope of this discussion and we shall have to content ourselves with merely outlining its salient features.[*] Just as in Eq. (12.14), we are interested in determining the cross-correlation function, this time, of the irradiances at two points in a partially coherent field, $\langle I_1(t+\tau)I_2(t)\rangle$. The contributing wave trains, which are again represented by complex fields, are presumed to have been randomly emitted in accord with Gaussian statistics with the final result that

$$\langle I_1(t+\tau)I_2(t)\rangle = \langle I_1\rangle\langle I_2\rangle + |\tilde{\Gamma}_{12}(\tau)|^2 \qquad (12.32)$$

or

$$\langle I_1(t+\tau)I_2(t)\rangle = \langle I_1\rangle\langle I_2\rangle[1 + |\tilde{\gamma}_{12}(\tau)|^2]. \qquad (12.33)$$

The instantaneous irradiance fluctuations $\Delta I_1(t)$ and $\Delta I_2(t)$ are given by the variations of the instantaneous irradiances $I_1(t)$ and $I_2(t)$ about their mean values $\langle I_1(t)\rangle$ and $\langle I_2(t)\rangle$,

---

[*] For a complete discussion, see e.g. L. Mandel, "Fluctuations of Light Beams," *Progress in Optics* Vol II, p. 193, or Francon, *Optical Interferometry*, p. 182.

$I(t)$

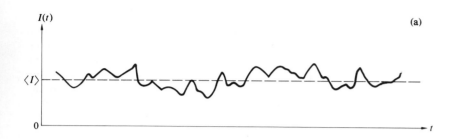

(a)

$\langle I \rangle$

0

t

$\Delta I = I - \langle I \rangle$

(b)

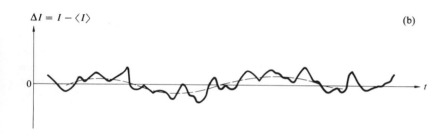

0

t

Fig. 12.12 Irradiance variations.

as in Fig. 12.12. Consequently if we use

$$\Delta I_1(t) = I_1(t) - \langle I_1 \rangle, \qquad \Delta I_2(t) = I_2(t) - \langle I_2 \rangle$$

and the fact that

$$\langle \Delta I_1(t) \rangle = 0 \qquad \text{and} \qquad \langle \Delta I_2(t) \rangle = 0,$$

Eqs. (12.32) and (12.33) become

$$\langle \Delta I_1(t + \tau)\Delta I_2(t) \rangle = |\tilde{\Gamma}_{12}(\tau)|^2 \qquad (12.34)$$

or

$$\langle \Delta I_1(t + \tau)\Delta I_2(t) \rangle = \langle I_1 \rangle \langle I_2 \rangle |\tilde{\gamma}_{12}(\tau)|^2 \qquad (12.35)$$

(Problem 12.8). These are the desired cross-correlations of the irradiance fluctuations. They exist as long as the field is partially coherent at the two points in question. Incidentally, these expressions correspond to linearly polarized light. When the wave is unpolarized, a multiplicative factor of $\frac{1}{2}$ must be introduced on the right-hand side.

The validity of the principle of correlation interferometry was first established in the radio-frequency region of the spectrum where signal detection was a fairly straightforward matter. Soon after, in 1956, Hanbury-Brown and Twiss proposed the optical stellar interferometer illustrated in Fig. 12.13. But the only suitable detectors which could be used at optical frequencies were photoelectric devices whose very operation is keyed to the quantized nature of the light field.

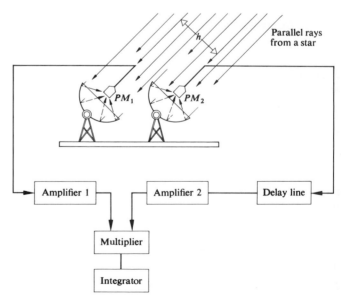

Fig. 12.13 Stellar correlation interferometer.

Thus

> ... it was by no means certain that the correlation would be fully preserved in the process of photo-electric emission. For these reasons a laboratory experiment was carried out as described below.*

That experiment is shown in Fig. (12.14). Filtered light from an Hg arc was passed through a rectangular aperture and different portions of the emerging wavefront were sampled by two photomultipliers, PM_1 and PM_2. The degree of coherence was altered by moving PM_1, i.e. by varying h. The signals from the two photomultipliers were presumably proportional to the incident irradiances $I_1(t)$ and $I_2(t)$. These were then filtered and amplified such that the steady, or dc, component of each of the signals (being proportional to $\langle I_1 \rangle$ and $\langle I_2 \rangle$) was removed, leaving only the fluctuations, i.e. $\Delta I_1(t) = I_1(t) - \langle I_1 \rangle$ and $\Delta I_2(t) = I_2(t) - \langle I_2 \rangle$. The two signals were then multiplied together in the correlator and the time average of the product which was proportional to $\langle \Delta I_1(t) \Delta I_2(t) \rangle$ was finally recorded. The values of $|\tilde{\gamma}_{12}(0)|^2$ for various separations, h, as deduced experimentally via Eq. (12.35), were in fine agreement with those calculated from theory. For the given geometry, the

---

* Taken from R. Hanbury-Brown and R. Q. Twiss, "Correlation Between Photons in Two Coherent Beams of Light," *Nature* **127**, 27 (1956).

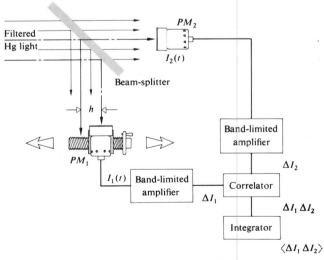

Fig. 12.14 Hanbury-Brown and Twiss experiment.

correlation most definitely existed and moreover, it was preserved through photoelectric detection.

The irradiance fluctuations have a frequency bandwidth of roughly the bandwidth $(\Delta \nu)$ of the incident light, i.e. $(\Delta t)^{-1}$ and that's about 100 MHz or more. This is a lot better than trying to follow the field alternations at 10^{15} Hz. Even so fast circuitry with roughly a 100 MHz pass bandwidth is called for. In actuality the detectors have a finite resolving time T so that the signal currents $\mathscr{I}_1$ and $\mathscr{I}_2$ are actually proportional to averages of $I_1(t)$ and $I_2(t)$ over T and not their instantaneous values. In effect, the measured fluctuations are smoothed out as illustrated by the dashed curve of Fig. 12.12(b). For $T > \Delta t$, which is normally the case, this just leads to a reduction, by a factor of $\Delta t/T$, in the correlation actually observed:

$$\langle \Delta \mathscr{I}_1(t) \Delta \mathscr{I}_2(t) \rangle = \langle \mathscr{I}_1 \rangle \langle \mathscr{I}_2 \rangle \frac{\Delta t}{T} |\tilde{\gamma}_{12}(0)|^2. \qquad (12.36)$$

For example, in the preceding laboratory arrangement, the filtered mercury light had a coherence time of about 1 ns while the electronics had a reciprocal pass bandwidth or effective integration time of ≈ 40 ns. Note that Eq. (12.36) isn't any different conceptually from Eq. (12.35) — it's just been made a bit more realistic.

Shortly after their successful laboratory experiment, Hanbury-Brown and Twiss constructed the stellar interferometer of Fig. 12.13. Searchlight mirrors were used to collect starlight and focus it onto two photomultipliers. One arm contained a delay line so that the mirrors could physically be located at the same height and yet have any differences in the arrival times of the light compensated for. The measurement of $\langle \Delta \mathscr{I}_1(t) \Delta \mathscr{I}_2(t) \rangle$ at various separations of the detectors allowed the square of the modulus of the degree of coherence, $|\tilde{\gamma}_{12}(0)|^2$, to be deduced and this in turn yielded the angular diameter of the source, just as it did with the Michelson stellar interferometer. This time, however, the separation h could practically be very large because one no longer had to worry about messing up the phase of the waves, as was the case in the Michelson device. There, a slight shift in a mirror of a fraction of a wavelength was fatal. Here, in contrast, the phase was discarded so that the mirrors didn't even have to be of high optical quality. The star Sirius was the first to be examined and it was found to have an angular diameter of 0.0069 seconds of arc. More recently, a correlation interferometer with a baseline of 618 feet has been constructed at Narrabri, Australia. For certain stars, angular diameters of as little as

0.0005 seconds of arc can be measured with this instrument—that's a long way from the angular diameter of Betelgeuse (0.047 seconds of arc).*

The electronics involved in irradiance correlation could be greatly simplified if the incident light were very nearly monochromatic and of considerably higher flux density. Laser light isn't thermal and doesn't display the same statistical fluctuations, but it can nonetheless be used to generate *pseudothermal*† light. A pseudothermal source is composed of an ordinary bright source (a laser is most convenient) and a moving medium of *nonuniform* optical thickness, such as a rotating ground glass disk. If the scattered beam emerging from a stationary piece of ground glass is examined with a *sufficiently slow detector*, the inherent irradiance fluctuations will be smoothed out completely. By setting the ground glass in motion, irradiance fluctuations appear with a simulated coherence time commensurate with the disk's speed. In effect, one has an extremely brilliant thermal source of variable Δt (from say 1 s to 10^{-5} s) which can be used to examine a whole range of coherence effects. For example, Fig. 12.15 shows the correlation function, which is proportional to $[2J_1(u)/u]^2$, for a pseudothermal circular aperture source determined from irradiance fluctuations. The experiment set-up resembled that of Fig. 12.14 although the electronics was considerably simpler.‡

PROBLEMS

12.1 Suppose we set up a fringe pattern using a Michelson interferometer with a mercury vapor lamp as the source. Switch on the lamp in your mind's eye and discuss what will happen to the fringes as the mercury vapor pressure builds to its steady state value.

* For a discussion of the photon aspects of irradiance correlation see Garbuny, *Optical Physics*, Section 6.2.5.2 or Klein, *Optics*, Section 6.4.

† See W. Martienssen and E. Spiller, "Coherence and Fluctuations in Light Beams," *Amer. J. Phys.* **32**, 919 (1964) and A. B. Haner and N. R. Isenor, "Intensity Correlations from Pseudothermal Light Sources," *Amer. J. Phys.* **38**, 748 (1970). Both of these articles are well worth studying.

‡ A good overall reference for this chapter is the review article by L. Mandel and E. Wolf, "Coherence Properties of Optical Fields," *Revs. Modern Phys.* **37**, 231 (1965). This is rather heavy reading. Take a look at K. I. Kellermann, "Intercontinental Radio Astronomy," *Sci. Amer.* **226**, 72 (February 1972).

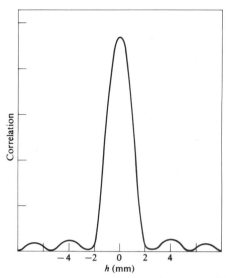

Fig. 12.15 A correlation function for a pseudothermal source. [From A. B. Haner and N. R. Isenor, *Amer. J. Phys.*, **38**, 748 (1970).]

12.2* Imagine that we've got the arrangement depicted in Fig. 12.2. If the separation between fringes (max. to max.) is 1 mm and if the projected width of the source slit on the screen is 0.5 mm, compute the visibility.

12.3 Referring to the slit source and pinhole screen arrangement of Fig. 12.16, show that

$$I(y) \propto w + \frac{\sin(\pi a/\lambda d)\, w}{\pi a/\lambda d}\cos(2\pi a y/\lambda s).$$

12.4 Carry out the details leading to the expression for the visibility given by Eq. (12.22).

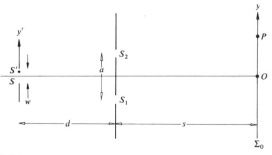

Fig. 12.16

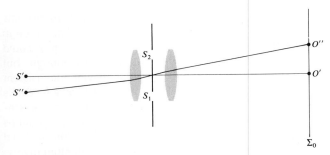

Fig. 12.17

12.5 Under what circumstances will the irradiance on Σ_0 in Fig. 12.17 be equal to $4I_0$, where I_0 is the irradiance due to either incoherent point source alone?

12.6* Suppose we set up Young's experiment with a small circular hole of diameter 0.1 mm in front of a sodium lamp ($\bar{\lambda}_0 = 589.3$ nm) as the source. If the distance from the source to the slits is 1 m, how far apart will the slits be when the fringe pattern disappears?

12.7 Taking the angular diameter of the sun viewed from the earth to be about $1/2°$, determine the diameter of the corresponding area of coherence neglecting any variations in brightness across the surface.

12.8 Show that Eqs. (12.34) and (12.35) follow from Eqs. (12.32) and (12.33).

Some Aspects of the Quantum Nature of Light   13

Man's understanding of the physical world has changed in a most profound manner since the beginning of this century. We have come to appreciate fundamental similarities between all of the various forms of radiant energy and matter. Optics, which was traditionally the study of light, is broadening its domain to encompass the entire electromagnetic spectrum. Moreover, the advent of quantum mechanics has brought with it yet another extension into what might be called *matter optics* (e.g. electron and neutron diffraction).

Our main purpose in this chapter is to conceptually weave some of the ideas of quantum mechanics into the fabric of optics.

13.1 QUANTUM FIELDS

The nineteenth-century physicist envisioned the electromagnetic field as a disturbance of the all-pervading ether medium. If two charges interacted, it was because the ether in which they were imbedded was distorted by their presence and the resulting *strain* was transmitted from one to the other. Maxwell's field equations described this measurable disturbance of the medium without explicitly discussing the ether itself. Light was then simply a wave train consisting of oscillatory mechanical stresses within the ether. Since there were electromagnetic waves, there had to be a transmitting medium—it was as clear as that. Yet curiously enough, even after the Michelson—Morley experiment (Section 9.10.3) and Einstein's special theory of relativity had put aside the ether hypothesis, Maxwell's equations remained. Even though the entire imagery had to be changed, the validity of those equations persisted. There seemed little conceptual alternative; the field itself had to be a physical entity, independent of any medium and capable of traversing otherwise empty space. An electromagnetic wave was seen as a disturbance propagated in the electromagnetic field.

In the early part of this century it began to become evident (for reasons we shall see imminently) that although Maxwell's equations seemed to be the truth, they could not be the whole truth. The field was real enough, but experiments were starting to reveal behavior inconsistent with the representation of the field exclusively as a fluid-like continuum. The electromagnetic field displayed particle-like properties in that it was emitted and absorbed in lumps or photons and not at all continuously. Even into the mid nineteen-twenties, in the formative years of quantum theory, fields and particles were envisioned as separate entities. But it soon became evident, with the melding of quantum theory and relativity, that each particle, material or otherwise, could be envisioned as a quantized manifestation of a distinct field (e.g. the photon is a quantum of the electromagnetic field). As with the photon, material particles can be created and destroyed. Their corresponding fields can transport all observable physical characteristics such as energy, charge, mass, etc. while advancing through space as waves. Within the context of quantum field theory, as this description is called, particles are viewed essentially as localized packets of field energy. Another far-reaching distinction between this and the classical picture is in the consideration of interactions. Quantum field theory maintains that all interactions arise from the creation and annihilation of particles. To wit, forces, in the classical sense, are envisioned as due to the exchange of quanta or lumps of the field in question. Charged particles can interact by absorbing and emitting, in a mutual exchange, quanta of the electromagnetic field, i.e. photons. Presumably the gravitational interaction is similarly the result of an exchange of quanta of the gravitational field—gravitons.

This then is something of a cursory view of the direction taken by contemporary quantum field theory.* In the next few sections we will consider some of the experiments which led to the development of the quantum mechanical photon picture.

13.2 BLACKBODY RADIATION—PLANCK'S QUANTUM HYPOTHESIS

At the turn of the nineteenth century, the electromagnetic theory of light, fashioned by Maxwell and meticulously verified by Hertz, was firmly established as one of the cornerstones of science. But periods of contentment in

---

* As with all theories it is in a continuous state of evolution and certain aspects are sure to undergo change. Nonetheless, in the rhetoric of our times, *this is where it's at.*

442

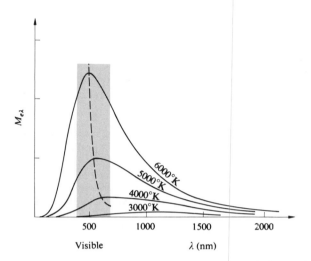

Fig. 13.1 Blackbody radiation curves. The hyperbola passing through peak points corresponds to Wien's law.

physics are usually short lived and Max Planck in 1900 unleashed a conceptual whirlwind which ultimately led to a radical change in the picture of the physical universe. Planck, who had been a student of Helmholtz and Kirchhoff, was working on a theoretical analysis of a seemingly obscure phenomenon known as *blackbody radiation*. We know that if an object of some sort is in thermal equilibrium with its environment, it must emit as much radiant energy as it absorbs. It follows that a good absorber is a good emitter. *A perfect absorber, one which absorbs all radiant energy incident upon it, regardless of wavelength, is said to be a blackbody.* Generally, one approximates a blackbody in the laboratory by a hollow insulated enclosure (an oven) which contains a small hole in one wall. Radiant energy entering the hole has little chance of being reflected out again so that the enclosure acts as a near perfect absorber. On the other hand, if the oven is heated it can serve as a source emitting energy through the hole. In accord with common experience, we can anticipate that the spectral distribution of the emitted radiant energy will be dependent on the oven's absolute temperature T. As the temperature increases, the hole will initially radiate predominantly infrared and then gradually it will take on a faint reddish glow that gets brighter and brighter shifting to yellow, white and finally blue-white. Experimental investigations (notably by O. Lummer and E. Pringsheim, 1899) resulted in spectral curves similar to those of Fig. 13.1. The quantity $M_{e\lambda}$ which is plotted as the

ordinate is known as the *spectral flux density* or *spectral exitance.* It corresponds to the emitted power per unit area per unit wavelength interval leaving the hole. Were we to make such measurements, at least in principle, we could determine the exitance (in W/m^2) from the blackbody at a given wavelength λ using some sort of power meter. But in actuality, any such meter would accept a range of wavelengths $\Delta\lambda$ centered about λ, and so we introduce the notion of *spectral* exitance. The curves of $M_{e\lambda}$ versus λ can be plotted so that the area beneath them is measured in W/m^2. Notice now the peaks in the curves shift toward the shorter wavelengths as T increases.

In 1879 Josef Stefan (1835–93) observed that the total radiant flux density (or exitance, M_e) of a blackbody was proportional to the fourth power of its absolute temperature. A few years later, Ludwig Boltzmann (1844–1906) derived that relationship in a combined application of Maxwell's theory and thermodynamic arguments. The *Stefan–Boltzmann law*, as it is now called, is

$$M_e = \sigma T^4, \tag{13.1}$$

where the Stefan–Boltzmann constant σ is equal to $(5.6697 \pm 0.0029) \times 10^{-8}$ W m^{-2} °K^{-4}. The last notable success in applying classical theory to the problem of blackbody radiation came in 1893 at the hands of the German physicist and Nobel laureate Wilhelm Carl Werner Otto Fritz Franz Wien (1864–1928), known to his friends as Willy. He was able to show that the wavelength, λ_{max}, at which $M_{e\lambda}$ (the flux density *per unit wavelength interval* emerging from the blackbody) is a maximum, varies as

$$\lambda_{max} T = 2.8978 \times 10^{-3} \text{ m °K}. \tag{13.2}$$

As T increases λ_{max} decreases and the peaks are displaced as we have already pointed out in connection with Fig. 13.1. Accordingly, the expression (13.2) is known as *Wien's displacement law.*

It was at this point in time that classical theory began to falter. All attempts to fit the entire radiation curve (Fig. 13.1) with some theoretical expression based on electromagnetism led only to the most limited successes. Wien produced a formula which agreed with the observed data fairly well in the short wavelength region but deviated from it substantially at large λ. Lord Rayleigh [John William Strutt (1824–1919)] and later Sir James Jeans (1877–1946) developed a description in terms of the standing wave modes of the field within the enclosure. But the resulting *Rayleigh–Jeans formula* only matched the experimental curves in the very long wavelength region. The failure of classical theory was totally inexplicable; a turning point in the history of physics had arrived.

Planck's approach to the problem was a rather systematic and practical one. He first matched the observed data with an empirical expression. Then he set about finding a physical justification for that expression within the framework of thermodynamics. In effect his model pictured the atoms in the walls of the oven to be in thermal equilibrium with the enclosed radiation field. He presumed that the atoms behaved as electrical oscillators absorbing and emitting radiant energy. He further assumed that all oscillator frequencies, v, were possible and thus all frequencies should be present in the emitted spectrum. But now he introduced a completely unprecedented *ad hoc* assumption whose only justification was a pragmatic one—it worked. Planck asserted that *an atomic resonator could absorb or emit only discrete amounts of energy which were proportional to its oscillatory frequency. Moreover, each such energy value had to be an integral multiple of what he called an "energy element" hv.* Thus all possible oscillator energies $\mathscr{E}_m$ are given by

$$\mathscr{E}_m = mhv, \tag{13.3}$$

where m is a positive integer and h is a constant to be determined by fitting the actual data. After applying statistical arguments, which are of little concern here, Planck derived the following formula for the spectral exitance:*

$$M_{e\lambda} = \frac{2\pi h c^2}{\lambda^5}\left[\frac{1}{e^{hc/\lambda kT} - 1}\right], \tag{13.4}$$

where k, in this instance, is Boltzmann's constant. *Planck's radiation law*, as given by Eq. (13.4), is in extremely good agreement with experimental results when h is chosen appropriately. The presently accepted value of Planck's constant is

$$h = (6.6256 \pm 0.0005) \times 10^{-34} \text{ J s.}$$

Thus Planck had indeed managed to fit the blackbody radiation curve, but even more significantly he had uncovered a still greater find—the quantum of energy.

The hypothesis that energy was emitted and absorbed in quanta of hv (which initially seemed only a computational contrivance) has proved to be a fundamental statement of the nature of things.* Moreover, the quantity h rather than simply being a particular curve-fitting parameter has shown itself to be a universal constant of the greatest importance. Nonetheless, we should point out that the true significance of Planck's work went unappreciated for several years and even he was rather cautious, as witnessed by this commentary on the derivation.†

> It is true that we shall not thereby prove that this hypothesis represents the only possible or even the most adequate expression of the elementary dynamical law of the vibration of oscillators. On the contrary I think it very probable that it may be greatly improved as regards form and contents . . . and as long as no contradiction in itself or with experiment is discovered in it, and as long as no more adequate hypothesis can be advanced to replace it, it may justly claim a certain importance.

13.3 THE PHOTOELECTRIC EFFECT—EINSTEIN'S PHOTON CONCEPT

It's rather ironical that Heinrich Hertz, who helped to establish the classical wave picture of radiant energy, was an unwitting contributor to its ultimate reformulation. This came by way of his discovery of the *photoelectric effect*, whose description first appeared in 1887 in a paper entitled "On an Effect of Ultraviolet Light upon the Electric Discharge." While engaged in his now famous experiments on electromagnetic waves (Section 3.5.4) he noticed that the spark induced in his receiving circuit was stronger when the terminals of the gap were illuminated by the light coming from the primary spark. He was able to establish that the effect was most pronounced when ultraviolet light impinged on the negative terminal of the gap, but he did not pursue the work any further. Later in 1889, Wilhelm Hallwachs (1859–1922) showed that negative particles were released from similarly illuminated metal surfaces such as zinc, sodium, potassium, etc. Thereafter Philipp Eduard Anton von Lenard (1862–1947), who was a colleague of Hertz, measured the charge-to-mass ratio of these particles, thus confirming that the spark enhancement observed by Hertz was the result of the emission of electrons (now referred to as *photoelectrons*). Using devices which were similar in principle to that depicted in Fig. 13.2 a number of researchers began to accrue data on the *photoelectric effect*,

---

* Don't confuse this with spectral energy density which equals $4M_{e\lambda}/c$.

---

* Planck's original derivation leads to erroneous photon energies of mhv but it was later correctly reformulated by Bose and Einstein.

† M. Planck and M. Masius, *The Theory of Heat Radiation*.

The early experiments of J. Elster and H. Geitel in 1889 had revealed that photoelectrons were frequently forcibly ejected from the illuminated metal surfaces under study. Electrons apparently emerged with small but finite speeds ranging from zero to some maximum value, v_{max}. By making the collecting plate negative with respect to the illuminated plate a retarding force could be exerted on the electrons. The retarding voltage which would stop even the most energetic electrons from reaching the collector, thereby bringing the photocurrent to zero, is known as the *stopping potential* V_0. Thus

$$\tfrac{1}{2}m_0 v_{max} = q_e V_0, \tag{13.5}$$

where m_0 is the rest mass of the electron. Figure 13.3(a) depicts the manner in which the *photocurrent* i_p varies as the retarding voltage V is altered. There is nothing about Fig. 13.3(a) which is at variance with the classical picture. The distribution in energy of the emerging electrons, which manifests itself in the gradual drop off of the curve, can satisfactorily be attributed to differences in the energy binding the various electrons to the metal. Electrons do not spontaneously escape from metal surfaces so that such binding is quite reasonable.

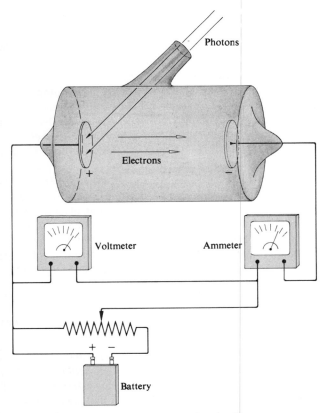

Fig. 13.2 Set-up to observe the photoelectric effect.

i.e. *the process whereby electrons are liberated from materials under the action of radiant energy.* It soon became apparent that the photoelectric effect was another instance in which classical electromagnetic theory was paradoxically impotent. This protracted dilemma was finally resolved by Einstein in a brilliant paper appearing in the *Annalen der Physik* of 1905.* It was there that he boldly extended Planck's quantum hypothesis and in so doing gave impetus to the sweeping reinterpretation of classical physics which was to take place later in the twenties. Let's now set the scene (c. 1905) so that we can appreciate how insightful Einstein's work actually was in light of the limited existent data.

---

* 1905 was a good year for Einstein. It was then at the age of about 26 that he published his theories of special relativity, Brownian motion and the photoelectric effect. Nonetheless, he once confided in a friend that the latter was the result of five years of thinking about Planck's hypothesis.

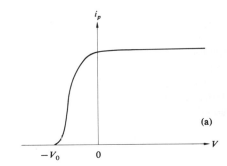

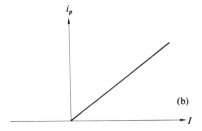

Fig. 13.3 (a) Photocurrent versus voltage.
(b) Photocurrent versus irradiance.

In 1893 it was observed that i_p was directly proportional to the incident irradiance, I, as indicated in Fig. 13.3(b). This too represented no departure from the classical scheme. Increasing I increases the total energy absorbed by the surface and should thus yield a proportionately larger number of emitted photoelectrons.

In contrast, it had early been established that there was no discernible time delay between the instant the plate was illuminated and the initiation of photoemission. This behavior is completely incomprehensible within the context of the classical description. For example, if $I = 10^{-10}$ W/m² (at $\lambda_0 = 500$ nm) theory predicts (Problem 13.10) that it might take about 10 hours before electrons could accumulate the amount of energy they had been observed to possess. To the contrary, Elster and Geitel, working with an even smaller irradiance, found no measurable time lag whatever.

In 1902 Lenard discovered that for a given metal the stopping potential, and therefore the maximum kinetic energy, was independent of the radiant flux density arriving at the plate as shown schematically in Fig. 13.4. He determined that even though the incident irradiance was varied seventy fold it did not alter V_0 by even one percent. This result led to yet another conundrum. It was well known that the maximum kinetic energy of the photoelectrons depended on the source being used. Yet Lenard's work showed that this energy was independent of I. One could only conclude that the maximum kinetic energy varied in some way with the frequency of the source and not with the total incident energy—a perplexing result indeed. Furthermore, recall that Hertz, in his original experiment, pointed out that ultraviolet radiation rather than visible light was the effective stimulus.

This implied that as the frequency of the radiant energy increased a threshold value was reached after which photoelectrons were emitted. But this too was classically inexplicable; whether or not emission takes place should depend on I and not v.

The quintessence of Planck's original hypothesis was that the energy of the radiation field could only *change* by discrete quanta, i.e. integer multiples of hv. This was a consequence of the fact that he had quantized the energy of the electric oscillators. Going far beyond this, Einstein proposed that *the radiation field itself was quantized and thus energy could be absorbed from it only in quanta of hv (called photons)*. The mechanism of the photoelectric effect now becomes quite clear. Envision an electron, within the interior of the material, which has absorbed a photon hv. In rising to the surface it will lose some of that energy and in escaping from the surface it will lose even more. Let the total energy spent in leaving the material be Φ. The difference between hv and Φ appears in the form of kinetic energy:

$$hv = \frac{mv^2}{2} + \Phi. \tag{13.6}$$

When the electron happens to be at the surface, Φ has its minimum value Φ_0. Known as the *work function*, Φ_0 simply corresponds to the energy needed by an electron to break free of the surface (see Table 13.1). In that special case

$$hv = \frac{mv_{max}^2}{2} + \Phi_0, \tag{13.7}$$

this being a statement of *Einstein's photoelectric equation*. The lowest, or *threshold frequency* (v_0), capable of promoting emission would just barely eject the electrons. To wit, $v_{max} \approx 0$ and

$$v_0 = \Phi_0/h. \tag{13.8}$$

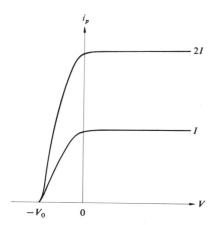

Fig. 13.4 The stopping potential is independent of the irradiance.

Table 13.1 Photoelectric threshold frequencies and work functions for a few metals.

Metal		v_0(THz)	Φ_0(eV)
Cesium	Cs	460	1.9
Beryllium	Be	940	3.9
Titanium	Ti	990	~ 4.1
Mercury	Hg	1100	4.5
Nickel	Ni	1210	5.0
Platinum	Pt	1530	6.3

In the photon picture, an electron literally absorbs a blast of energy as opposed to a gradual trickle. Accordingly, there will be no appreciable time delay in the emission. The interrelationship between irradiance and photocurrent is also quite understandable. An increase in I corresponds to more photons of the same energy and thus an increase in i_p but not in V_0.

The quantum theory rather neatly accounts for the existence of a threshold frequency, the dependence of $(mv_{max}^2/2)$ on v, the lack of a time lag, the independence of V_0 on I and the relationship of I to i_p. Even so since quantitative data were scanty and the photon so radical an idea it remained unaccepted by many.

The photoelectric equation went even further than accounting for all of the known observations; it also represented one of the great prognostications of all times. After it had been published, a great flurry of experimental work brought with it all sorts of confirmation. The proportionality between I and i_p was extended over a range of 5×10^7 in irradiance. Ernest O. Lawrence and J. W. Beams (1928) used a Kerr cell to create pulses of light and therewith found that if a time lag existed in the emission of electrons it had to be less than* 3×10^{-9} s. In 1916 the American physicist Robert Andrews Millikan (1868–1953) published an extensive and remarkably accurate study of the relationship of Einstein's equation and the photoelectric effect. His own words on the subject are quite enlightening:

> I spent ten years of my life testing the 1905 equation of Einstein's and contrary to all my expectations, I was compelled in 1915 to assert its unambiguous experimental verification in spite of its unreasonableness since it seemed to violate everything that we knew about the interference of light.

A representation of Millikan's results is shown in Fig. 13.5. Note that since $v_0 = \Phi_0/h$ we can write

$$\frac{mv_{max}^2}{2} = h(v - v_0), \qquad (13.9)$$

which means that a plot of maximum kinetic energy $(q_e V_0)$ versus v for any given material should be a straight line having a slope h and an intercept of $-\Phi_0$. These predictions were completely confirmed by Millikan.† The amazing fact

---

* E. O. Lawrence and J. W. Beams, "The Element of Time in the Photoelectric Effect," *Phys. Rev.* **32**, 478 (1928).

† In 1923, two years after Einstein received the Nobel prize for his work on the photoelectric effect, Millikan was awarded the same honor, in part for his experimental efforts on that subject.

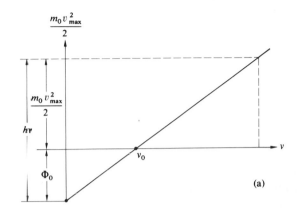

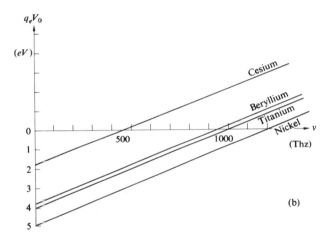

Fig. 13.5 Some of Millikan's results.

that the slope actually turned out to be equal to h is a tribute to the insight of Planck and the genius of Einstein. Different metals have characteristic values of Φ_0 and v_0 but in all cases the slope of the line remained constant at h as predicted.

The quantization of the electromagnetic field had been established; all of physics, and particularly optics, would never quite be the same again.*

---

* Notwithstanding the great influence the photoelectric effect had on the photon historically, it is nonetheless possible to explain that effect without resorting to a quantization of the electromagnetic field. Indeed one can treat the field classically imparting the quantum nature to the matter alone. See the article by W. E. Lamb, Jr. and M. O. Scully in *Polarization, Matter and Radiation, Jubilee Volume in Honor of Alfred Kastler.*

13.4 PARTICLES AND WAVES

According to Maxwell's electromagnetic theory (see Chapter 3) the energy $\mathscr{E}$ and momentum p of an electromagnetic wave are related by the expression

$$\mathscr{E} = cp. \qquad (13.10)$$

Alternatively, the energy and momentum of a particle of rest mass m_0 are related by way of the formula

$$\mathscr{E} = c(m_0^2 c^2 + p^2)^{1/2}, \qquad (13.11)$$

whose origins are in the special theory of relativity. Inasmuch as the photon is a creature of both these disciplines, we can expect either equation to be equally applicable; indeed they must be identical. It follows that *the rest mass of a photon is equal to zero*. The photon's total energy, as with any particle, is given by the relativistic expression $\mathscr{E} = mc^2$ where

$$m = \frac{m_0}{\sqrt{1 - v^2/c^2}}. \qquad (13.12)$$

Thus, since it has a finite relativistic mass m and since $m_0 = 0$, it follows that *a photon can only exist at a speed c*; the energy $\mathscr{E}$ is purely kinetic.

The fact that the photon possesses inertial mass leads to some rather interesting results, e.g. the *gravitational red shift* (Problem 13.13) and the deflection of starlight by the sun (Problem 13.16). The red shift was actually observed under laboratory conditions in 1960 by R. V. Pound and G. A. Rebka Jr. at Harvard University. In brief, if a particle of mass m moves upward a height d in the earth's gravitational field it will do work in overcoming the field and thus decrease in energy by an amount mgd. Therefore, if the photon's initial energy is $h\nu_i$, its final energy after traveling a vertical distance d will be given by

$$h\nu_f = h\nu_i - mgd \qquad (13.13)$$

and so $\nu_f < \nu_i$, ergo the name red shift. Pound and Rebka, using gamma-ray photons, were able to confirm that quanta of the electromagnetic field behave as if they had a mass $m = \mathscr{E}/c^2$.

From Eq. (13.10) the momentum of a photon can be written as

$$p = \frac{\mathscr{E}}{c} = \frac{h\nu}{c} \qquad (13.14)$$

or

$$p = h/\lambda. \qquad (13.15)$$

If we had a perfectly monochromatic beam of light of wavelength λ each constituent photon would possess a momentum of h/λ or equivalently

$$\mathbf{p} = \hbar \mathbf{k}. \qquad [3.60]$$

We can arrive at this same end by way of a somewhat different route. Momentum quite generally is the product of mass and speed, thus

$$p = mc = \frac{\mathscr{E}}{c},$$

and we're back to Eq. (13.14). The momentum relationship, $p = h/\lambda$, for photons was confirmed in 1923 by Arthur Holly Compton (1892–1962). In a classic experiment he irradiated electrons with x-ray quanta and studied the frequency of the scattered photons. By applying the laws of conservation of momentum and energy relativistically as if the collisions were between particles, Compton was able to account for an otherwise inexplicable decrease in the frequency of the scattered radiant energy.

A few years later in France, Louis Victor, Prince de Broglie (b. 1891) in his doctoral thesis drew a marvelous analogy between photons and matter particles. He proposed that every particle, and not just the photon, should have an associated wave nature. Thus since $p = h/\lambda$ the *wavelength of a particle having a momentum mv would then be*

$$\lambda = h/mv. \qquad (13.16)$$

Because $h = 6.6 \times 10^{-34}$ is small and because of the relative enormity of the momenta of macroscopic entities, such bodies have miniscule wavelengths. For example a one gram pebble moving at 1 cm/s has a wavelength of 6.6×10^{-29} m, roughly 10^{22} times shorter than that of red light. In contrast let's compute the voltage needed to impart a wavelength of 1 Å to an electron; this is of the order of the spacing between atoms. Starting from rest, the electron has a kinetic energy of $mv^2/2$ after traversing a potential difference of V, that is

$$q_e V = \frac{mv^2}{2}.$$

Using Eq. (13.14) we can write

$$V = \frac{h^2}{2mq_e\lambda^2} = \frac{(6.6 \times 10^{-34}\,\text{J s})^2}{2(9.1 \times 10^{-31}\,\text{kg})(1.6 \times 10^{-19}\,\text{C})(10^{-10}\,\text{m})^2}$$

or

$$V = 150\,\text{V}.$$

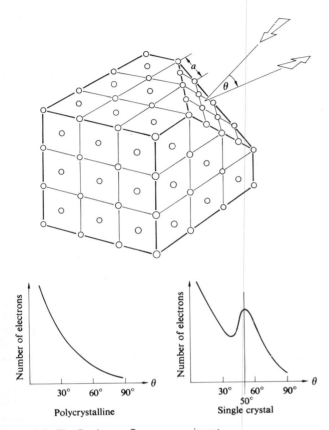

Fig. 13.6 The Davisson–Germer experiment.

In this instance the lattice spacing a is 2.15 Å and so $\lambda = 2.15 \sin 50°$ or 1.65 Å in fine agreement with the value of 1.67 Å computed from the de Broglie equation (13.16). Amazingly enough, a beam of electrons had thus been diffracted in a manner completely analogous to a light wave bouncing off a reflection grating. The first observation of electron diffraction which was made by Davisson and Germer was quite accidental; they were neither looking for it nor did they at first realize it had happened. In contrast, Thomson had set out deliberately to verify diffraction. Taking a somewhat different approach, he passed a beam of high-speed electrons through a thin polycrystalline foil (100 nm thick) and observed a diffraction pattern made up of concentric rings (Fig. 13.7). In 1928 E. Rupp diffracted a beam of slow electrons (70 eV) at grazing incidence off a metal optical grating (1300 lines per cm) and observed first-second- and third-order images. A few years later in 1930 I. Estermann and Otto Stern demonstrated the occurrence of diffraction effects using beams of both helium atoms and molecular hydrogen.

In recent times it has become possible to produce a remarkable range of interference and diffraction patterns using electrons, as witness the photographs of Fig. 13.8.

Out of the long list of material particles which have been observed to display wave properties, neutrons are amongst the most useful. Because they carry no charge, slow or *thermal neutrons* can have long wavelengths and yet be immune to the electrical forces which strongly disturb low-momentum electrons. The diffraction of thermal neutrons (generally originating from nuclear reactors) is now a routinely used procedure in the study of atomic structure (Fig. 13.9).

Not long ago (1969) a beam of neutral potassium atoms was used to observe diffraction arising from a macroscopic slit (23×10^{-6} m wide). The resulting pattern was in accord with de Broglie's hypothesis and the scalar Fresnel diffraction theory.*

We are limited by our language to a list of words much as our worldly experiences limit the concepts those words bring to mind. Our senses have read the environment and in so doing provided the basis for our understanding of it. In what seemed a logical extension, we have tried, a bit naïvely, to use macroscopic imagery to describe submicroscopic entities. But electrons do not behave as miniscule billiard balls any more than light can be pictured in

An electron so accelerated has an energy of 150 eV (1 eV = 1.602×10^{-19} J) and a wavelength of 1 Å which is just about that of a typical x-ray photon.

Experimental verification of de Broglie's hypothesis came in the years 1927–8 as a result of the efforts of Clinton Joseph Davisson (1881–1958) and Lester Germer (b. 1896) in the United States and Sir George Paget Thomson (b. 1892) in Great Britain. Davisson and Germer used a nickel crystal (face-centered cubic structure) as a three-dimensional diffraction grating for electrons. When a 54 eV beam was incident, perpendicular to the cut face of the crystal as shown in Fig. 13.6, a strong reflection appeared at 50° to the normal. Making use of the grating equation,

$$a \sin \theta_m = m\lambda, \qquad [10.32]$$

we find that the first-order ($m = 1$) maximum corresponds to

$$a \sin \theta_1 = \lambda.$$

---

* J. Leavitt and F. Bills, "Single-slit Diffraction Pattern of a Thermal Atomic Potassium Beam," *Am. J. Phys.* **37**, 905 (1969).

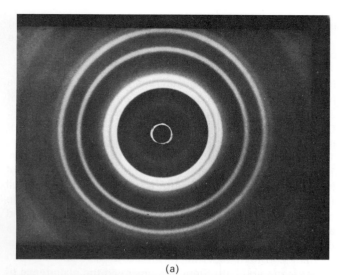

(a)

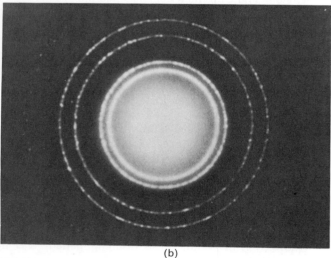

(b)

Fig. 13.7 (a) Diffraction pattern arising from x-rays passing through a thin polycrystalline aluminum foil. (b) Diffraction pattern arising from electrons passing through the same aluminum foil. [From the PSSC film *Matter Waves*.]

terms of scaled-down rolling ocean waves. *Particles and waves are macroscopic concepts which gradually lose their relevance as we approach the submicroscopic domain.*

13.5 PROBABILITY AND WAVE OPTICS

The fundamental wave nature of optical phenomena was established well over a hundred years ago. Its mainstay

was the work of Young, Fresnel and many others who studied the processes of interference, diffraction and polarization. During the intervening century, our conception of light has metamorphosed from that of a rudimentary mechanical ether wave to the contemporary photon description. Yet the concept that light is somehow inherently oscillatory has persisted throughout this transition period. And so we might again press the point and ask *what is it that oscillates when we envisage light as a stream of photons*; or for that matter, what aspect of an electron vibrates? The answer to this will obviously give us some clue as to how quanta display interference effects.

The Danish physicist Niels Henrik David Bohr (1885–1962) provided an essential link between classical and quantum physics in what has become known as the *correspondence principle.* Briefly stated, *any new theory must agree with the results of the classical theory it supersedes in the domain where the latter is known to be effective.*[*] Thus while only quantum theory can explain blackbody radiation, the photoelectric effect, Compton scattering, electron diffraction and a myriad of other observations, it must also account for what might be called classical behavior. The entire range of familiar effects such as Snell's law, the reflection law, the Doppler formula[†] etc. which are usually treated in terms of electromagnetic theory must also be understandable within the context of the photon description. The quantum theory is not just an esoteric addendum; it must encompass all confirmed observations which have gone before it, no matter how mundane.

Imagine, if you will, a monochromatic light source illuminating an optical element of some kind followed by an observation screen. Presumably in many cases one could calculate, using classical wave optics, the flux-density distribution appearing on the screen. Suppose then that we have such a case, for example, a plane wave incident on a double slit arrangement. The irradiance $I(\theta)$ represents the average energy density per unit time at the plane of observation; in this instance, the familiar fringe pattern of Young's experiment. Thus the average number of photons impinging on a small area element dA, in a time interval dt, will be $(I\,dA\,dt)/h\nu$ where I, of course, varies from one point

---

[*] Although it might seem little more than obvious here, the correspondence principle becomes a powerful tool when interpreted as a mathematical limiting process. For example, classical physics is the correspondence limit of quantum physics as h is envisioned to approach zero, thereby making quantized phenomena continuous.

[†] See Section 4.6, also A. Sommerfeld, *Optics* p. 82.

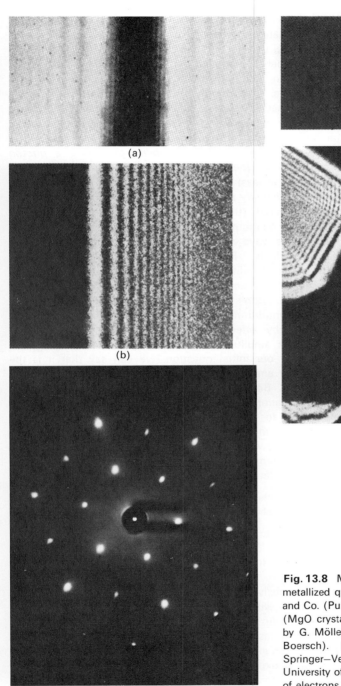

(a)

(b)

(e)

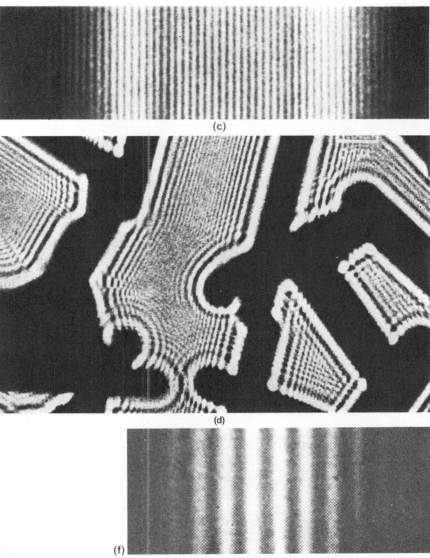

(c)

(d)

(f)

Fig. 13.8 Matter-wave diffraction. (a) Fresnel electron diffraction pattern of a 2 μ diameter metallized quartz filament. [Photo from O. E. Klemperer, *Electron Physics*, Butterworths and Co. (Publishers) Ltd., London (1972).] (b) Fresnel electron diffraction at a half plane (MgO crystal). (c) Interference fringes observed with an electron biprism arrangement by G. Möllenstedt. (d) Fresnel diffraction of electrons by zinc oxide crystals (After H. Boersch). [The last three photos are from *Handbuch der Physik*, edited by S. Flügge, Springer–Verlag, Heidelberg.] (e) Electron diffraction by a UO₂ crystal. [Photo courtesy University of California's Los Alamos Scientific Laboratory.] (f) Double beam interference of electrons. [Photo by C. Jönsson, reprinted from J. Orear, *Fundamental Physics*, John Wiley, New York (1967).] The faint cross hatching in this photo arises purely in the printing process, it's a Moiré effect from rescreening.

(a)

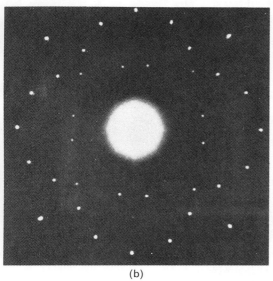

(b)

Fig. 13.9 Diffraction patterns generated by (a) neutrons, (b) x-ray photons incident on a single crystal of NaCl. A polycrystalline specimen would produce a great many randomly oriented dot patterns of this sort which would blend into the ring systems of Fig. 13.7. [Photo (a) by E. O. Wollan which along with (b) is from Lapp and Andrews, *Nuclear Radiation Physics* 3rd ed., Prentice-Hall, Inc., Englewood Cliffs, N.J. (1963).]

to the next over the surface of the screen. Keep in mind that we can only detect the emission or absorption of a photon, i.e. its interaction with matter. There is no way to predict where a particular photon will arrive on the plane of observation, although some regions are more likely sites than others. Accordingly, if a total of N photons strike the screen in each interval dt we can say that each photon has a probability equal to $(I\,dA\,dt)/h\nu N$ of arriving at the given area element dA. *The irradiance as computed classically is therefore related to the probability of finding a photon somewhere on the screen.* It is convenient at this point to introduce, at least conceptually, a complex quantity known as the *probability amplitude*, i.e. a quantity whose absolute value squared (the so-called *wave-intensity*) yields the probability distribution. It is this probability amplitude propagating as a wave which describes the whole range of interference effects. For example, in Young's experiment the photon's probability amplitude for reaching its final state is the sum of two amplitudes, each of these being associated with the photon's passage through one of the slits. The various contributing amplitudes in a given situation overlap and thereby effectively interfere, yielding the resultant probability amplitude and from that the irradiance. In answer to our initial question, we can say that it is the probability amplitude associated with the photon which is oscillating. Bear in mind that the same kind of discomforting reinterpretation of familiar ideas which we are encountering now had to be made when Maxwell's electromagnetic theory first emerged on the scene.

Let's now briefly examine the implications of a rather famous statement made by the renowned British physicist and Nobel laureate Paul Adrien Maurice Dirac (b. 1902):

> . . . each photon interferes only with itself. Interference between different photons never occurs.*

This is in accord with the conclusion that each photon possesses a distinct wave nature. Evidently the wave properties of light are not attributable to the beam acting as a whole. In Young's experiment each photon somehow simultaneously interacts with both slits; close either one and the fringes will disappear. Presumably since each photon interferes with itself, the same fringe pattern would gradually occur, one flash at a time, even if we shone a single photon a day at the slits. This remarkable conclusion was actually confirmed experimentally by Geoffrey I. Taylor, a student at the University of Cambridge in 1909. Using a lightproof

---

* P. A. M. Dirac, *Quantum Mechanics* 4th ed., p. 9.

box, a gas flame illuminating an entrance slit and a number of attenuating smoked glass screens, he set about photographing the diffraction pattern in the shadow of a needle. By drastically reducing the incoming flux density, he was able to obtain exposure times of up to about 3 months. In such cases the energy density in the box was so low that there was usually only one photon at a time in the region beyond the entrance slit. Nonetheless the customary array of diffraction fringes appeared and moreover

> In no case was there any diminution in the sharpness of the pattern . . .[†]

Much of the foregoing discussion can be applied to material particles as well. In fact, the same dynamical equations determine the interrelationship of ν, λ and v with p and $\mathscr{E}$ for all particles, material or otherwise. Consequently from Eq. (13.11) we find that

$$p = (\mathscr{E}^2 - m_0^2 c^4)^{1/2}/c \qquad (13.17)$$

while $\lambda = h/p$ leads to

$$\lambda = hc/(\mathscr{E}^2 - m_0^2 c^4)^{1/2}. \qquad (13.18)$$

Since $p = mv$, $v = pc^2/(mc^2) = pc^2/\mathscr{E}$ and

$$v = c[1 - (m_0^2 c^4/\mathscr{E}^2)]^{1/2}. \qquad (13.19)$$

Evidently one of the main distinguishing characteristics of the photon is just its zero rest mass. In that case, the above equations simply become $p = \mathscr{E}/c$, $\lambda = hc/\mathscr{E} = c/\nu$ and $v = c$.

In a way analogous to that of the photon, the probability amplitude or de Broglie wave for a *matter field* is represented by the function $\psi(x, y, z, t)$ (also referred to as the *wave function*). The probability of finding a particle of finite rest mass is then proportional to the wave-intensity $|\psi|^2$. One determines the wave function for a particular circumstance involving material particles from the *Schrödinger equation*. Once again it is the probability amplitude of the particle which is oscillatory, propagates through space as a wave, and partakes in interference.

13.6 FERMAT, FEYNMAN AND PHOTONS

In classically treating interference and diffraction problems with coherent waves one generally sums all of the electric-field contributions at a given point—these quite frequently being written in complex form. The square of the absolute value of this sum is proportional to the irradiance and is consequently proportional to the probability of finding a photon at the point in question. We will now qualitatively generalize these remarks along the lines of Richard Feynman's elegant variational formulation of quantum mechanics.[*] Suppose then that a particle (photon, electron etc.) is emitted from a source point s and is later detected at point p. The probability of arrival, P, is equal to the square of the absolute value of a complex quantity Φ which, as before, is said to be the probability amplitude, i.e. $P = |\Phi|^2$. Unlike the classical treatment where the field was expressed in complex form as a convenience, Φ must be complex in the quantum-mechanical formulation. Consequently it has an amplitude and a phase, the latter being a function of both the spatial position of p and of time. The event can occur by several alternative routes 1, 2, 3, . . . and it was postulated by Feynman that in such cases *each path contributes to the total probability amplitude*. In other words

$$\Phi = \Phi_1 + \Phi_2 + \Phi_3 + \cdots \qquad (13.20)$$

and so

$$P = |\Phi_1 + \Phi_2 + \Phi_3 + \cdots|^2. \qquad (13.21)$$

It was further postulated that *the magnitudes of these individual probability amplitudes are all equal*, that is

$$|\Phi_1| = |\Phi_2| = |\Phi_3| = \cdots, \qquad (13.22)$$

whereas their phases are not equal and indeed depend on the particular paths. Note that a value of $P = 1$ means that the particle will arrive at p with complete certainty while $P = 0$ means that it will most definitely not reach p. Generally then, P will range in value between 0 and 1. Equation (13.21) evidently introduces the phenomenon of interference into the scheme, whether it be for photons or electrons. In contrast, if we were dealing with classical particles, such as a stream of B-B pellets, P would equal $|\Phi_1|^2 + |\Phi_2|^2 + |\Phi_3|^2 + \cdots$ and there would be no interference, i.e. P would be independent of the individual phases. As with incoherent light one then adds irradiances rather than amplitudes.

Let's now turn to the idealized Young's experiment of Fig. 13.10 consisting of two extremely small pinholes. In that case

$$P = |\Phi_1 + \Phi_2|^2 \qquad (13.23)$$

where there are effectively two paths, one through each

[†] G. I. Taylor, "Interference Fringes with Feeble Light," *Proc. Camb. Phil. Soc.* **15**, 114 (1909).

[*] R. P. Feynman "Space–Time Approach to Non-Relativistic Quantum Mechanics," *Rev. Mod. Phys.* **20**, 367 (1948).

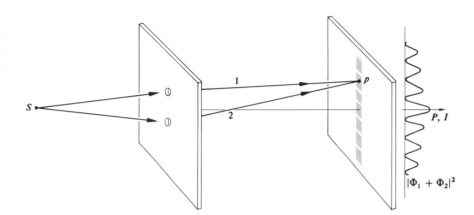

Fig. 13.10 Double-beam experiment.

aperture. If the phases of the probability amplitudes at p differ by an odd multiple of π they will interfere destructively, that is,

$$P = (|\Phi_1| - |\Phi_2|)^2 = 0. \tag{13.24}$$

On the other hand if they are in phase, constructive interference results at p whereupon

$$P = (|\Phi_1| + |\Phi_2|)^2 = 4|\Phi_1|^2, \tag{13.25}$$

which is equivalent to

$$I = 4I_0 \cos^2 \frac{\delta}{2} \tag{9.6}$$

for $\delta = 0, \pi, 2\pi, \ldots$. The phases of the probability amplitudes at p depend on the path lengths traversed along each route and so P can clearly have any value between these extremes as well. In the same way if we were shooting B-B pellets through two small holes, the probability of their arriving at p would be the sum $|\Phi_1|^2 + |\Phi_2|^2$. Here $|\Phi_1|^2$ and $|\Phi_2|^2$ are simply the individual probabilities of arrival with either hole 1 or hole 2 open respectively as indicated in Figs. 13.11 and 13.12. The resulting distribution of B-B pellets is just the superposition of the two separate patterns for each aperture; there are no fringes and no interference.

If the screen had N such apertures, rather than just two, the probability of a photon reaching p would be

$$P = |\sum_{i=1}^{N} \Phi_i|^2. \tag{13.26}$$

For a large aperture, as for example a lens or mirror, the summation becomes an integral over the area of the aperture. Incidentally, Feynman has shown that, for material particles,

the total value of the probability amplitude for all paths is the wave function satisfying Schrödinger's equation.[*]

We now go back to the picture of a single ray of light leaving a source and reflecting off a mirror, ultimately to arrive at a sensor. The probability of a photon encountering the sensor is determined by Φ which in turn is composed of contributions from each of the possible paths. All of this talk about paths should bring to mind Fermat's principle (Section 4.2.4) which maintains that the actual path taken by a ray is stationary. Everything fits together rather nicely when we realize that the relative differences in path length and phase of the corresponding probability amplitudes at the sensor are small only for paths near the stationary one $(\theta_i = \theta_r)$. These probability amplitudes interfere constructively, thereby providing the predominant contribution to P. This is then the quantum-mechanical basis for Fermat's principle. Probability amplitudes associated with paths remote from the stationary one will have large phase-angle differences resulting in relatively little cumulative effect on P. This discussion is reminiscent of the Cornu spiral (Section 10.3.7) which in quite an analogous fashion can be thought of as the diagrammatic sum of a great number of phasors each of different amplitude but the same phase angle. Suppose that we wish to determine I or equivalently P at a point on the central axis of e.g. a long slit. In that case contributions from remote areas of the aperture correspond to the tightly wound regions of the spiral and therefore contribute

---

[*] To see how these ideas are related to Hamilton's principle function, the principle of least action and the WKB approximation, refer e.g. to D. B. Beard and G. B. Beard, *Quantum Mechanics with Applications*, p. 44 and S. Borowitz, *Fundamentals of Quantum Mechanics*, p. 165.

Fig. 13.11 Lower hole covered in double-beam set up.

Fig. 13.12 Upper hole covered in double-beam set up.

little to the complex number (phasor) B_{12}. Recall [Eqs. (10.106) or (10.108)] that I is proportional to $|B_{12}|^2$ just as it is proportional to $|\Phi|^2$. Equation (13.20) can similarly be envisioned pictorially in terms of the addition of a number of equal-amplitude phasors, in which case P is proportional to the square of the magnitude of the resultant. Phasors corresponding to probability amplitudes for paths in the vicinity of a stationary one differ in phase by very little and therefore add almost along a straight line, thus making a major contribution. Where the relative phases of successive phasors is large, the curve spirals around with little effect on $|\Phi|$. The analogy can even be extended if we now visualize the Cornu spiral as if it were composed of a great number of equal-amplitude phasors whose phase angles are ever increasing as they get farther from the center of the spiral [from Eq. (10.105) $\beta = \pi w^2/2$]. In any event the phasor representation of the contributing probability amplitudes is a handy device to keep in mind.

13.7 ABSORPTION, EMISSION AND SCATTERING

Let's now take a brief look at the quantum-mechanical aspects of a few important interactions occurring between light and matter. Suppose that a photon of frequency v_i collides with and is absorbed by an atom. Energy is transmitted to a bound electron resulting in the excitation of the atom. The absorption probability is greatest when the frequency of the incident photon is equal to an excitation energy of the atom (see Section 8.5.1). In dense gases, liquids and solids, absorption occurs over a range or band of frequencies and the energy is generally dissipated by way of intermolecular collisions. In contrast, the excited atoms of a low-pressure gas can reradiate a photon of the same frequency (v_i) in a random direction, a process first observed by R. W. Wood in 1904 and known as *resonance radiation*. Accordingly, there is preponderant scattering at frequencies coincident with the excitation energies of the atoms. The effect is easily demonstrated using Wood's technique which incorporates an evacuated glass bulb containing a bit of pure metallic sodium. Gradually heating the bulb increases the sodium vapor pressure within it. If a region of the vapor is then illuminated with a strong beam of light from a sodium arc, that portion will glow with the characteristic yellow resonance radiation of Na.

Scattering can also occur, but with less likelihood, at frequencies other than those corresponding to the atom's stable energy levels. In such cases a photon will be reradiated without any appreciable time delay and most often with the same energy as that of the absorbed quantum. The process is known as *elastic* or *coherent scattering* because there is a phase relationship between the incident and scattered fields. This is the *Rayleigh scattering* we talked about in Section 8.5.1.

It is also possible that an excited atom will not return to its initial state after the emission of a photon. This kind of behavior had been observed and studied extensively by George Stokes prior to the advent of quantum theory. Since the atom drops down to an interim state, it emits a photon of lower energy than the incident primary photon in what is usually referred to as a *Stokes transition*. If the process takes place rapidly (roughly 10^{-7} s) it is called *fluorescence*, whereas if there is an appreciable delay (in some cases seconds, minutes or even many hours), it is known as *phosphorescence*. Using ultraviolet quanta to generate a fluorescent emission of visible light has become an accepted occurrence in our everyday lives. Any number of common-place materials (e.g. detergents, organic dyes, tooth enamel, etc.) will emit characteristic visible photons so that they appear to glow under ultraviolet illumination; ergo the widespread use of the phenomenon for commercial display purposes.

13.7.1 The Spontaneous Raman Effect

If quasimonochromatic light is scattered from a substance it will thereafter consist mainly of light of the same frequency. Yet it is possible to observe very weak additional components having higher and lower frequencies (side bands). More-over, the difference between the side bands and the incident frequency v_i is found to be characteristic of the material and therefore suggests an application to spectroscopy. The *spontaneous Raman effect*, as it is now called, was predicted in 1923 by Adolf Smekal and observed experimentally in 1928 by Sir Chandrasekhara Vankata Raman (1888–1970), then professor of physics at the University of Calcutta. The effect was difficult to put to actual use because one needed strong sources (usually Hg discharges were used) and large samples. Often the ultraviolet from the source would further complicate matters by decomposing the specimen. And so it is not surprising that little sustained interest was aroused by the promising practical aspects of the Raman effect. The situation was changed dramatically when the laser became a reality. *Raman spectroscopy* is now a unique and powerful analytical tool.

To appreciate how the phenomenon operates, let's review the germane features of molecular spectra. A molecule can absorb radiant energy in the far-infrared and

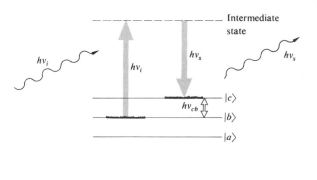

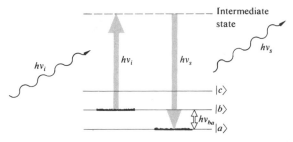

Fig. 13.13 Spontaneous Raman scattering.

microwave regions converting it to rotational kinetic energy. Furthermore it can absorb infrared photons (i.e. ones within a wavelength range from roughly 10 mm^{-2} down to about 700 nm) transforming that energy into vibrational motion of the molecule. Finally a molecule can absorb energy in the visible and ultraviolet regions through the mechanism of electron transitions much like those of an atom. Suppose then that we have a molecule in some vibrational state which, using quantum-mechanical notation, we call $|b\rangle$ as indicated diagrammatically in Fig. 13.13(a). This need not necessarily be an excited state. An incident photon of energy hv_i is absorbed raising the system to some intermediate or virtual state whereupon it immediately makes a Stokes transition emitting a (scattered) photon of energy $hv_s < hv_i$. In conserving energy the difference $hv_i - hv_s = hv_{cb}$ goes into exciting the molecule to a higher vibrational energy level $|c\rangle$. It is possible that electronic or rotational excitation results as well. Alternatively, if the initial state is an excited one (just heat the sample) the molecule, after absorbing and emitting a photon, may drop back to an even lower state [Fig. 13.13(b)] thereby making an *anti-Stokes* transition. In this instance $hv_s > hv_i$ which means that some vibrational energy of the molecule ($hv_{ba} = hv_s - hv_i$) has

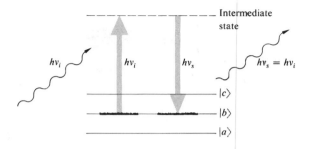

Fig. 13.14 Rayleigh scattering.

been converted into radiant energy. In either case the resulting differences between v_s and v_i correspond to specific energy-level differences for the substance under study and as such yield insights into its molecular structure. Figure 13.14, for comparison's sake, depicts Rayleigh scattering where $v_s = v_i$.

The laser is an ideal source for spontaneous Raman scattering. It is bright, quasimonochromatic and available in a wide range of frequencies. Figure 13.15 illustrates a typical laser-Raman system. Complete research instruments of this sort are commercially available including the laser (usually helium–neon, argon or krypton), focusing lens

systems and photon-counting electronics. The double scanning monochromator provides the needed discrimination between v_i and v_s since unshifted laser light (v_i) is scattered along with the Raman spectra (v_s). The increased sensitivity provided by the laser has made it possible to observe Raman scattering associated with molecular rotation and even electron motion.

13.7.2 The Stimulated Raman Effect

In 1962 Eric J. Woodbury and Won K. Ng rather fortuitously discovered a remarkable related effect known as *stimulated Raman scattering*. They had been working with a million-watt pulsed ruby laser incorporating a nitrobenzene Kerr cell shutter (see Section 8.11.3). They found that about 10 percent of the incident energy at 694.3 nm was shifted in wavelength and appeared as a *coherent* scattered beam at 766.0 nm. It was subsequently determined that the corresponding frequency shift of about 40 THz was characteristic of one of the vibrational modes of the nitrobenzene molecule as were other new frequencies also present in the scattered beam. Stimulated Raman scattering can occur in solids, liquids or dense gases under the influence of focused high-energy laser pulses (Fig. 13.16). The effect is schematically depicted in Fig. 13.17. Here two photon beams are simultaneously incident on a molecule; one corresponding to the

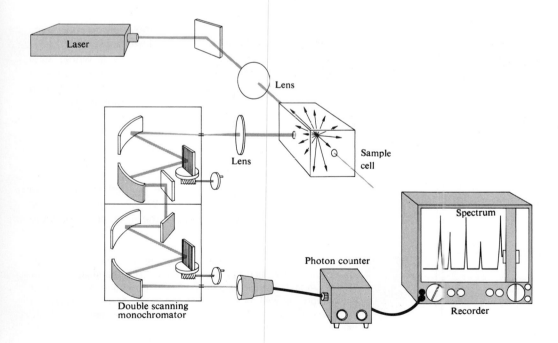

Fig. 13.15 A laser–Raman system.

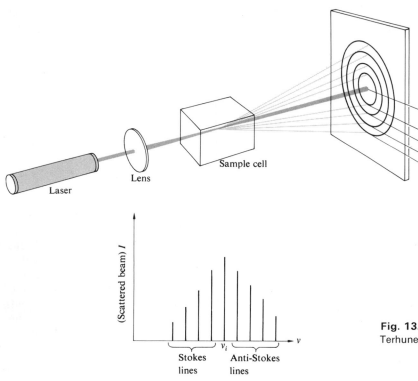

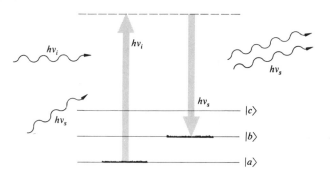

Fig. 13.6 Stimulated Raman scattering. [See R. W. Minck, R. W. Terhune and C. C. Wang, *Proc. IEEE* **54**, 1357 (1966).]

laser frequency v_i, the other having the scattered frequency v_s. In the original set-up the scattered beam was reflected back and forth through the specimen, but the effect can occur without a resonator. The laser beam loses a photon hv_i while the scattered beam gains a photon hv_s and is subsequently *amplified*. The remaining energy $(hv_i - hv_s = hv_{ba})$ is transmitted to the sample. The chain reaction in which a large portion of the incident beam is converted into stimula-

ted Raman light can only occur above a certain high-threshold flux density of the exciting laser beam.

Stimulated Raman scattering provides a whole new range of high-flux density coherent sources extending from the infrared to the ultraviolet. It should be mentioned that in principle each spontaneous scattering mechanism (e.g. Rayleigh and Brillouin scattering) has its stimulated counterpart.*

PROBLEMS

13.1 Suppose that we measure the emitted exitance from a small hole in a furnace to be 22.8 W/cm² using an optical pyrometer of some sort. Compute the internal temperature of the furnace.

---

* For further reading on these subjects you might try the review-tutorial paper by Nicolaas Bloembergen, "The Stimulated Raman Effect," *Am. J. Phys.* **35**, 989 (1967). It contains a fairly good bibliography as well as a historical appendix. Many of the papers in *Lasers and Light* also deal with this material and are highly recommended reading.

Fig. 13.17 Energy-level diagram of stimulated Raman scattering.

13.2* When the sun's spectrum is photographed, using rockets to range above the earth's atmosphere, it is found to have a peak in its spectral exitance at roughly 465 nm. Compute the sun's surface temperature assuming it to be a blackbody. This approximation yields a value which is about 400° K too high.

13.3 Beginning with Eq. (13.4) show that the exitance per unit frequency interval for a blackbody is given by

$$M_{ev} = \frac{2\pi h v^3}{c^2} \left[\frac{1}{e^{hv/kT} - 1} \right]. \tag{13.27}$$

13.4 Compute the wavelength of a 0.15 kg baseball moving at 25 m/s. Compare this with the wavelength of a hydrogen atom ($m_0 = 1.673 \times 10^{-27}$ kg) having a speed of 10^3 m/s.

13.5* Determine the energy of a 500 nm (green) photon in both joules and electron volts. Make the same calculation for a 1 MHz radio wave.

13.6 Write an expression for the wavelength of a photon in angstroms ($1\text{Å} = 10^{-10}$ m) in terms of its energy in eV.

13.7 Figure 13.18 shows the *spectral irradiance* impinging on a horizontal surface, for a clear day, at sea level, with the sun at the zenith. What is the most energetic photon we can expect to encounter (in eV and J)?

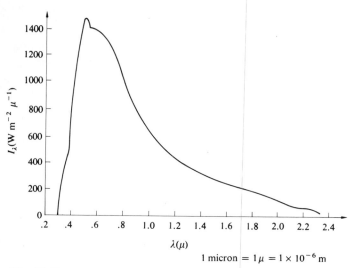

$\lambda(\mu)$

1 micron $= 1\mu = 1 \times 10^{-6}$ m

Fig. 13.18

13.8* Suppose we have a 100 W yellow light bulb (550 nm) 100 m away from a 3 cm diameter shuttered aperture. Assuming the bulb to have a 2.5% conversion to radiant power, how many photons will pass through the aperture if the shutter is opened for $\frac{1}{1000}$ s?

13.9 The *solar constant* is the radiant flux density at a spherical surface centered on the sun having a radius equal to that of the earth's mean orbital radius; it has a value of 0.133–0.14 W/cm². If we assume an average wavelength of about 700 nm, how many photons at most will arrive on each square meter per second of a solar cell panel just above the atmosphere?

13.10 With respect to the photoelectric effect, imagine that we have an incident beam with an irradiance of 10^{-10} W/m² at a wavelength of 500 nm. What is the energy per quantum? Supposing the target atoms to have radii of 10^{-10} m, how long would it take for any one of them to accumulate the energy of a single photon using the classical wave picture? In 1916 Rayleigh showed classically that an atomic oscillator absorbs radiant energy with an effective area of the order of λ^2 at resonance. How does this help?

13.11 The work function for outgassed polycrystalline sodium is 2.28 eV. What is the minimum frequency a photon must have in order to liberate an electron? What will be the maximum kinetic energy of an electron ejected by a 400 nm photon?

13.12* Suppose that we have a beam of light of a given flux density incident on a photoelectric tube. Draw a plot of i_p versus V showing what we might expect to happen to the stopping potential as the frequency is increased from v_1 to v_2 to v_3.

13.13 To examine the *gravitational red shift* consider a photon of frequency v which is emitted from a star having a mass M and a radius R. Show that at the star's surface the energy of the photon is given by

$$\mathscr{E} = h v \left(1 - \frac{GM}{c^2 R} \right).$$

When it arrives at the earth, having essentially escaped the gravitational pull of the star, the photon will have a lower frequency. Show that the frequency shift is then

$$\Delta v = \frac{GM}{c^2 R} v.$$

The effect is quite noticeable for the class of stars known as *white dwarfs*. (This problem should have been analyzed using general relativity but the answer would have been the same.)

13.14 Compute the fractional gravitational red shift, i.e. $\Delta\nu/\nu$, for the sun ($M = 1.991 \times 10^{30}$ kg and $R = 6.960 \times 10^8$ m). How much of a change would occur in the frequency and wavelength of a photon of $\lambda_0 = 650$ nm emitted from the sun? (See previous problem.)

13.15 Show that a photon moving upward a distance d in the earth's gravitational field (Section 13.4) will undergo a frequency decrease equal to

$$\Delta\nu = -gd\nu/c^2.$$

Compute the value of $\Delta\nu/\nu$ if $d = 20$ m. Pound and Rebka actually measured that shift in a vertical tower at Harvard University using the extreme sensitivity of the Mössbauer effect.

13.16 The following problem concerns itself with the bending of a beam of light as it passes a massive body such as the sun. It should actually be solved using general rather than special relativity because of the presence of gravity. As a result our simple approach yields half the correct answer. Be that as it may, let us plunge on. Show that the force component acting on the photon transverse to its initial direction of motion (Fig. 13.19) is given by

$$F_\perp = \frac{GMm}{R^2}\cos^3\theta.$$

Since $c\,dt = ds = d(R\tan\theta)$, show that the total transverse component of momentum received by the photon is

$$p_\perp = \frac{2GMm}{cR}.$$

Inasmuch as $p_{\parallel} = mc$, compute ϕ for the sun ($R = 6.960 \times 10^8$ m and $M = 1.991 \times 10^{30}$ kg).

13.17* Imagine that we accelerate a beam of electrons through a potential difference of 100 V and then cause it to pass through a slit 0.1 mm wide. Determine the angular width of the central diffraction maximum ($m_0 = 9.108 \times 10^{-31}$ kg). How do things change if we decrease the beam's energy?

13.18 A thermal neutron is one which is in thermal equilibrium with matter at a given temperature. Compute the wavelength of such a neutron at 25°C ($\approx$ room temperature). Recall from kinetic theory that the average kinetic energy would be equal to $\frac{3}{2}kT$. (Boltzmann's constant $k = 1.380 \times 10^{-23}$ J/°K and $m_0 = 1.675 \times 10^{-27}$ kg).

13.19 In Young's experiment can we imagine that an incident photon splits and passes through both slits? Discuss your conclusion.

13.20* Suppose we have a laser beam of radius a and wavelength λ. Using the uncertainty principle ($\Delta x\,\Delta p_x \sim h$) make an approximate calculation of the radius q of the smallest spot the beam will make on a screen a distance R away.

13.21 What is the *photon flux* Π of a 1000 W continuous CO_2 laser emitting at 10,600 nm in the IR?

13.22 Derive the dispersion relation, i.e. $\omega = \omega(k)$, for the de Broglie wave of a particle of mass m_0 nonrelativistically in a region where it has constant potential energy U.

13.23* Derive an expression for the dispersion relation of a free ($U = 0$) relativistically moving particle of rest mass m_0.

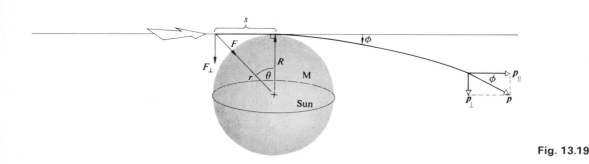

Fig. 13.19

13.24 Assuming that the de Broglie wave for a particle in a region where its potential energy is constant is given by

$$\psi(x, t) = C_1 e^{-i(\omega t + kx)} + C_2 e^{-i(\omega t - kx)},$$

use the results of Problem (13.22) to show that

$$i\hbar\frac{\partial \psi}{\partial t} = -\frac{\hbar^2}{2m}\frac{\partial^2 \psi}{\partial x^2} + U\psi.$$

This is a form of the famous Schrödinger equation of quantum mechanics.

Sundry Topics from 14
Contemporary Optics

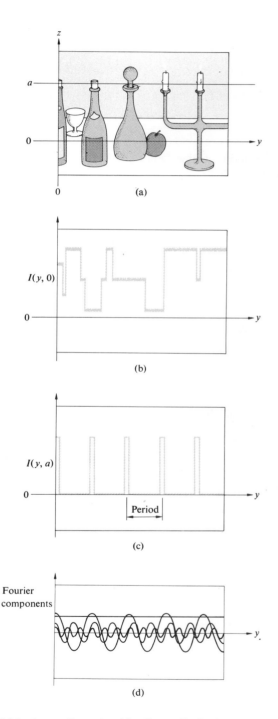

14.1 IMAGERY—THE SPATIAL DISTRIBUTION OF OPTICAL INFORMATION

The manipulation of all sorts of data via optical techniques has already become a technological *fait accompli*. The literature of the nineteen-sixties reflects, in a diversity of areas, this far reaching interest in the methodology of optical data processing. Practical applications have been made in the fields of TV and photographic image enhancement, radar and sonar signal processing (phased and synthetic array antenna analysis) as well as in pattern recognition (e.g. aerial photointerpretation and fingerprint studies) to list only a very few.

Our concern here is to develop the nomenclature and some of the ideas necessary for an appreciation of this contemporary thrust in optics—it is a vital part of the "wave of the future."

14.1.1 Spatial Frequencies

In electrical processes one is most frequently concerned with signal variations in time, e.g. the moment-by-moment alteration in voltage which might appear across a pair of terminals at some fixed location in space. By comparison, in optics we are most often concerned with information spread across a region of space at a fixed location in time. For example, we can think of the scene depicted in Fig. 14.1 (a) as a two-dimensional flux-density distribution. It might be an illuminated transparency, a TV picture or an image projected on a screen; in any event there is presumably some function $I(y, z)$ which assigns a value of I to each point in the picture. To simplify matters a bit, suppose we scan across the screen on a horizontal line ($z = 0$) and plot point-by-point variations in irradiance with distance, as in Fig. 14.1 (b). The function $I(y, 0)$ can be synthesized out of harmonic functions using the techniques of Fourier analysis treated in Chapters 7 and 11. In this instance, the function

Fig. 14.1 A two-dimensional irradiance distribution.

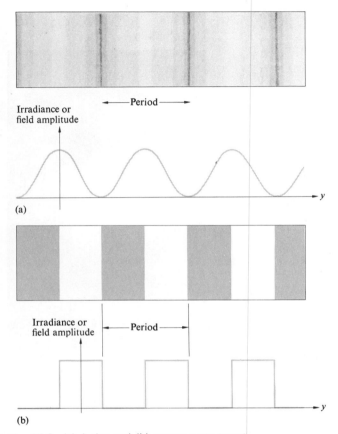

Fig. 14.2 (a) A sine and (b) square wave target.

is rather complicated and it would take many terms to adequately represent it. Yet if the functional form of $I(y, 0)$ is known, the procedure is straightforward enough. Scanning across another line, e.g. $z = a$, we get $I(y, a)$ which is drawn in Fig. 14.1(c) and which just happens to turn out to be a series of equally spaced square pulses. This function is one which was considered at length in Section 7.7 and a few of its constituent Fourier components are roughly sketched in Fig. 14.1(d). If the peaks in (c) are separated, center to center, by e.g. 1 cm intervals, the spatial period equals 1 cm per cycle and its reciprocal, which is the spatial frequency, equals 1 cycle per cm.

Quite generally we can transform the information associated with any scan line into a series of sinusoidal functions of appropriate amplitude and spatial frequency. In the case of either of the simple sine- or square-wave targets of Fig. 14.2 each such horizontal scan line is identical

and the patterns are effectively one dimensional. The spatial-frequency spectrum of Fourier components needed to synthesize the square wave is shown in Fig. 7.15. On the other hand, $I(y, z)$ for the wine bottle candelabra scene is two dimensional and we have to think in terms of two-dimensional Fourier transforms (Section 11.2.2). We might mention as well that, at least in principle, we could have recorded the amplitude of the electric field at each point of the scene and then performed a similar decomposition of that signal into its Fourier components.

Recall (Section 11.3.3) that the far-field or Fraunhofer diffraction pattern is, in fact, identical to the Fourier transform of the aperture function $\mathscr{A}(y, z)$. The aperture function is proportional to $\mathcal{E}_A(y, z)$, the source strength per unit area (10.37) over the input or object plane. In other words, if the field distribution on the object plane is given by $\mathscr{A}(y, z)$, its two-dimensional Fourier transform will appear as the field distribution $E(Y, Z)$ on a very distant screen. As in Fig. 10.10 we can introduce a lens (L_t) after the object in order to shorten the distance to the image plane. That objective lens is commonly referred to as the *transform lens* since we can imagine it as if it were an optical computer capable of generating instant Fourier transforms. Now, suppose we illuminate a somewhat idealized transmission grating with a spatially coherent, quasimonochromatic wave such as the plane wave emanating from a laser or a collimated, filtered Hg arc source (Fig. 14.3). In either case, the amplitude of the field is assumed to be fairly constant over the incident wavefront. The aperture function is then a periodic step function (Fig. 14.4), i.e. as we move from point to point on the object plane the amplitude of the field is either zero or a constant. If a is the grating spacing it is also the spatial period of the step function and its reciprocal is the fundamental spatial frequency of the grating. The central spot ($m = 0$) in the diffraction pattern is the dc term corresponding to a zero spatial frequency—it's the bias level that arises from the fact that the input $\mathscr{A}(y)$ is everywhere positive. This bias level can be shifted by constructing the step-function pattern on a uniform gray background. As the spots in the image (or in this case the transform) plane get farther from the central axis their associated spatial frequencies (m/a) increase in accord with the grating equation $\sin \theta_m = \lambda(m/a)$. A coarser grating would have a larger value of a so that a given order (m) would be concomitant with a lower frequency, (m/a), and the spots would all be closer to the central or optical axis.

Had we used as an object a transparency resembling the sine target [Fig. 14.2(a)] such that the aperture function

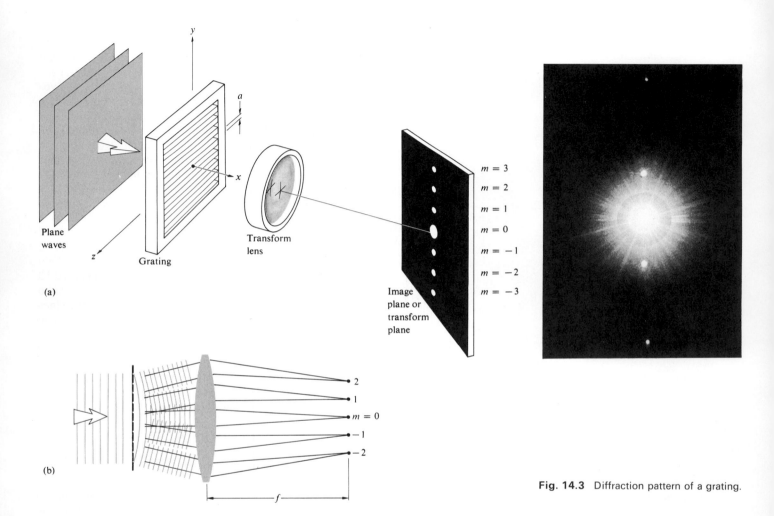

(a)

(b)

Fig. 14.3 Diffraction pattern of a grating.

varied sinusoidally, there would ideally have only been three spots on the transform plane, these being the zero-frequency central peak and the first order or fundamental ($m = \pm 1$) on either side of the center.   Extending things into two dimensions, a crossed grating (or mesh) yields the diffraction pattern shown in Fig. 14.5. Note that in addition to the obvious periodicity horizontally and vertically across the mesh, it is also repetitive, e.g. along diagonals.   A more involved object such as a transparency of the surface of the moon would generate an extremely complex diffraction pattern. Because of the simple periodic nature of the grating we could think of its Fourier-series components but now we will certainly have to think in terms of Fourier transforms.

In any case *each spot of light in the diffraction pattern denotes the presence of a specific spatial frequency which is proportional to its distance from the optical axis (zero-frequency location)*. Frequency components of positive and negative sign appear diametrically opposite each other about the central axis. If we could measure the electric field at each point in the transform plane, we would indeed observe the transform of the aperture function, but this is not practicable. Instead, what will be detected is the flux-density distribution where at each point the irradiance is proportional to the time average of the electric field squared or equivalently to the square of the amplitude of the particular spatial frequency contribution at that point.

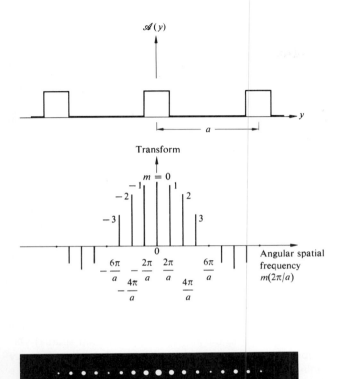

$\mathscr{A}(y)$

Transform

$m = 0$

Diffraction pattern

Fig. 14.4 Square wave and its transform.

Fig. 14.5 Diffraction pattern of a crossed grating.

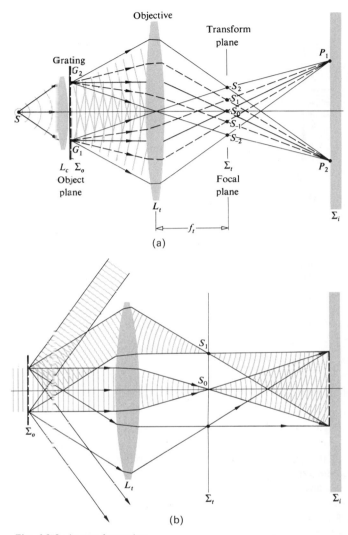

(a)

(b)

Fig. 14 6 Image formation.

14.1.2   Abbe's Theory of Image Formation

Consider the system depicted in Fig. 14.6(a) which is just
an elaborated version of Fig. 14.3(b). Plane monochromatic
wavefronts emanating from the collimating lens (L_c) are
diffracted by a grating. The result is a distorted wavefront
which we resolve into a new set of plane waves, each corresponding to a given order $m = 0, \pm 1, \pm 2, \ldots$ or spatial
frequency and each traveling in a specific direction [Fig.
14.6(b)]. The objective lens (L_t) serves as a *transform lens*
forming the Fraunhofer diffraction pattern of the grating on

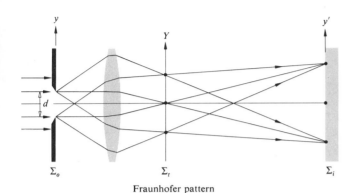

Fraunhofer pattern

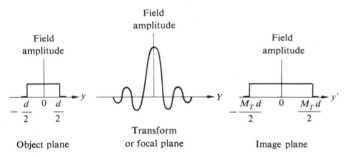

Object plane

Transform
or focal plane

Image plane

Fig. 14.7 The image of a slit.

the transform plane Σ_t (which is also the back focal plane of L_t). The waves, of course, propagate beyond Σ_t and arrive at the image plane Σ_i. There they overlap and interfere to form an inverted image of the grating. Accordingly, points G_1 and G_2 are imaged at P_1 and P_2 respectively. The objective lens forms two distinct patterns of interest. One is the Fourier transform on the focal plane conjugate to the plane of the source, and the other is the image of the object, formed on the plane conjugate to the object plane. Figure 14.7 shows the same set-up for a long, narrow, horizontal slit coherently illuminated.

We can envision the points S_0, S_1, S_2 etc. in Fig. 14.6(a) as if they were point emitters of Huygens' wavelets and the resulting diffraction pattern on Σ_i is then the grating's image. In other words, *the image arises from a double diffraction process*. Alternatively, we can imagine that the incoming wave is diffracted by the object and the resulting diffracted wave is then diffracted once again by the objective lens. If that lens were not there, the diffraction pattern of the object would appear on Σ_i in place of the image.

These ideas were first propounded by Professor Ernst Abbe (1840–1905) in 1873.* His interest at the time concerned the theory of microscopy, whose relationship to the above discussion is clear if we consider L_t as a microscope objective. Moreover if the grating is replaced by a piece of some thin translucent material (i.e. the specimen being examined) which is illuminated by light from a small source and condenser, the system certainly resembles a microscope.

Carl Zeiss (1816–88), who in the mid-eighteen hundreds was running a small microscope factory in Jena, realized the shortcomings of the trial-and-error development techniques of that era. In 1866 he enlisted the services of Ernst Abbe, then lecturer at the University of Jena, to establish a more scientific approach to microscope design. Abbe soon found by experimentation that a larger aperture resulted in higher resolution even though the apparent cone of incident light filled only a small portion of the objective. Somehow the surrounding "dark space" contributed to the image. Consequently he took the approach that the then well-known diffraction process which occurs at the edge of a lens (leading to the Airy pattern for a point source) was not operative in the same sense as it was for an incoherently illuminated telescope objective. Specimens, whose size was of the order of λ, were apparently scattering light into the "dark space" of the microscope objective. Observe that if, as in Fig. 14.6(b), the aperture of the objective is not large enough to collect all of the diffracted light, the image does not correspond exactly to that object. Rather it relates to a fictitious object whose complete diffraction pattern matches the one collected by L_t. We know from the previous section that these lost portions of the outer region of the Fraunhofer pattern are associated with the higher spatial frequencies. And, as we shall see presently, their removal will result in a loss in image sharpness and resolution.

Practically speaking, unless the grating considered earlier has an infinite width it cannot actually be strictly periodic. This means that it has a continuous Fourier spectrum dominated by the usual discrete Fourier-series terms, the others being much smaller in amplitude. Complicated,

---

* An alternative and yet ultimately equivalent approach was put forth in 1896 by Lord Rayleigh. He envisaged each point on the object as a coherent source whose emitted wave was diffracted by the lens into an Airy pattern. Each of these in turn was centered on the ideal image point (on Σ_i) of the corresponding point source. Thus Σ_i was covered with a distribution of somewhat overlapping and interfering Airy patterns.

irregular objects clearly display the continuous nature of their Fourier transforms. In any event, it should be emphasized that *unless the objective lens has an infinite aperture it functions as a low-pass filter rejecting spatial frequencies above a given value and passing all those below* (the former being those which extend beyond the physical boundary of the lens). Consequently, all practical lens systems will be limited in their ability to reproduce the high spatial frequency content of an actual object under coherent illumination.* It might be mentioned as well that there is a basic non-linearity associated with optical imaging systems operating at high spatial frequencies.†

14.1.3 Spatial Filtering

Suppose we actually set up the system shown in Fig. 14.6(a) using a laser as a plane-wave source. If the points S_0, S_1, S_2, etc. are to be the sources of a Fraunhofer pattern, the image screen must presumably be located at $x = \infty$ (although 30 or 40 ft will often do). At the risk of being repetitious, recall that the reason for using L_t originally was to bring the diffraction pattern of the object in from infinity and so we now introduce an *imaging lens* L_i for just that purpose (Figs. 14.8 and 14.9). The transform lens causes the light from the object to converge in the form of a diffraction pattern on the plane Σ_t, i.e. it produces on Σ_t a two-dimensional Fourier transform of the object. To wit, the spatial-frequency spectrum of the object is spread across the transform plane. Thereafter L_i (the *inverse transform lens*) projects the diffraction pattern of the light distributed over Σ_t onto the image plane. In other words, it diffracts the diffracted beam which effectively means that it generates an inverse transform. Thus an inverse transform of the data on Σ_t appears as the final image. Quite frequently in practice L_t and L_i are identical ($f_t = f_i$) well-corrected multielement lenses [for quality work these might have resolutions of about 150 line pairs/mm—one line pair being a period in Fig. 14.2(b)]. For less demanding applications two projector objectives of large aperture (about 100 mm) having convenient focal lengths of roughly 30 or 40 cm serve quite nicely. One of these lenses is then merely turned around so that both their back focal planes coincide with Σ_t. Incidentally, the

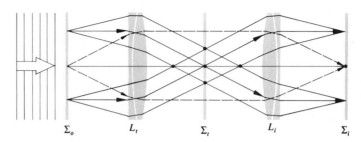

Fig. 14.8 Object, transform and image planes.

input or object plane need not be located a focal length away from L_t; the transform still appears on Σ_t. Moving Σ_0 affects only the phase of the amplitude distribution, and that is generally of little interest. The device shown in Figs. 14.8 and 14.9 is often referred to as a *coherent optical computer*. It allows us to insert obstructions, i.e. masks or filters, into the transform plane and in so doing partially or completely block out certain spatial frequencies stopping them from reaching the image plane. *This process of altering the frequency spectrum of the image is known as spatial filtering.* And herein lie some of the most beautiful, exciting and promising aspects of contemporary optics.

From our earlier discussion of Fraunhofer diffraction we know that a long narrow slit at Σ_0, regardless of its orientation and location, generates a transform at Σ_t consisting of a series of dashes of light lying along a straight line perpendicular to the slit (Fig. 10.11) and *passing through the origin*. Consequently, if the straight-line object is described by $y = mz + b$ the diffraction pattern lies along the line $Y = -Z/m$ or equivalently from Eqs. (11.64) and (11.65) $k_Y = -k_Z/m$. With this and the Airy pattern in mind we should be able to anticipate some of the gross structure of the transforms of various objects. Be aware as well that these transforms are centered about the zero-frequency optical axis of the system. For example, a transparent plus sign whose horizontal line is thicker than its vertical one has a two-dimensional transform again shaped more or less like a plus sign. The thick horizontal line generates a series of short vertical dashes while the thin vertical element produces a line of long horizontal dashes. Remember that object elements with small dimensions diffract through relatively large angles. Along with Abbe, one could think of this entire subject in these terms rather than using the concepts of spatial frequency filtering and transforms which represent the more modern influence of communication theory.

---

* Refer to H. Volkmann, "Ernst Abbe and His Work," *Appl. Opt.* **5** 1720 (1966) for a more detailed account of Abbe's many accomplishments in optics.

† R. J. Becherer and G. B. Parrent, Jr., "Nonlinearity in Optical Imaging Systems," *J. Opt. Soc. Am.* **57**, 1479 (1967).

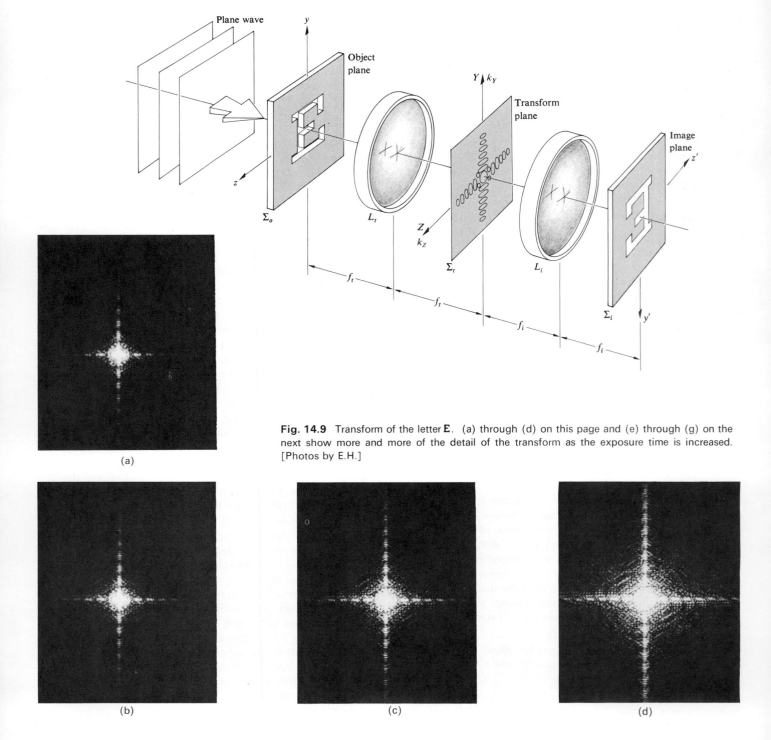

Fig. 14.9 Transform of the letter **E**. (a) through (d) on this page and (e) through (g) on the next show more and more of the detail of the transform as the exposure time is increased. [Photos by E.H.]

(a)

(b)   (c)   (d)

| (e) | (f) | (g) |

Fig. 14.9. Continued.

The vertical portions of the symbol E in Fig. 14.9 generate the broad frequency spectrum appearing as the horizontal pattern. Note that all parallel line sources on a given object correspond to a single linear array on the transform plane. This, in turn, passes through the origin on Σ_t (the intercept is zero) just as in the case of the grating. A transparent figure **5** will generate a pattern consisting of both a horizontal and vertical distribution of spots extending over a relatively large frequency range. There will also be a comparatively low-frequency, concentric ring-like structure.

The transforms of disks and rings and the like will obviously be circularly symmetric. Similarly a horizontal elliptical aperture will generate vertically oriented concentric elliptical bands. Most often far-field patterns possess a center of symmetry (see Problems 10.25 and 11.11).

We are now in a position to better appreciate the process of spatial filtering and to that end we will consider an experiment very similar to one published in 1906 by A. B. Porter. Figure 14.10(a) shows a fine wire mesh whose periodic pattern is disrupted by a few particles of dust. With

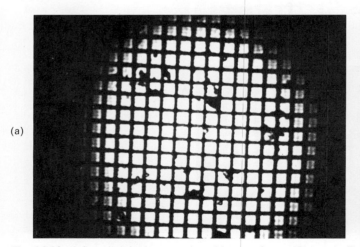

(a)

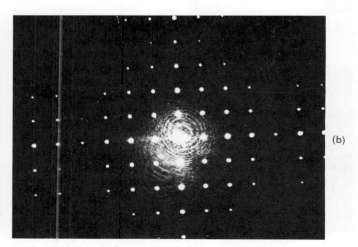

(b)

Fig. 14.10 A fine slightly dusty mesh and its transform. [Photos in Figs. 14.10–14.12 from D. Dutton, M. P. Givens and R. E. Hopkins *Spectra-Physics Laser Technical Bulletin Number* 3.]

Altered image   Filtered transform   Altered image   Filtered transform

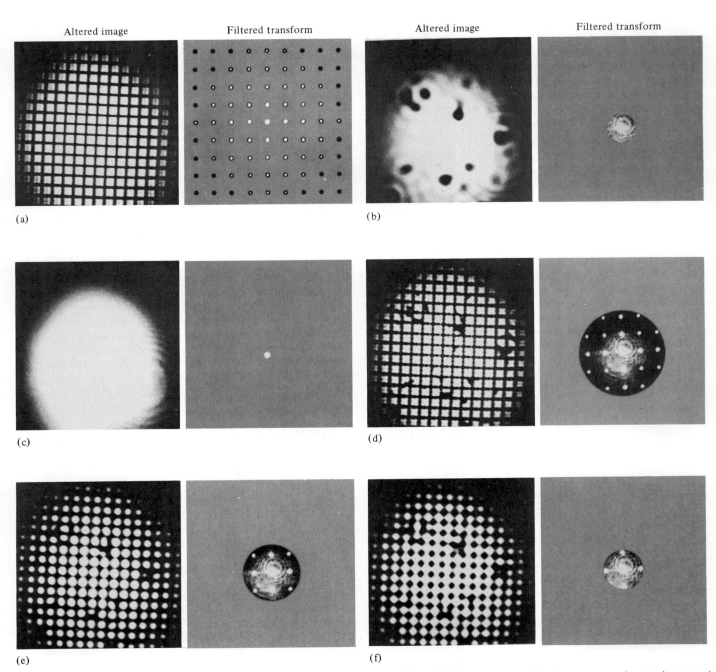

(a)

(b)

(c)

(d)

(e)

(f)

Fig. 14.11 Images resulting when various portions of the diffraction pattern of Fig. 14.10 (b) are obscured by the accompanying masks or spatial filters.

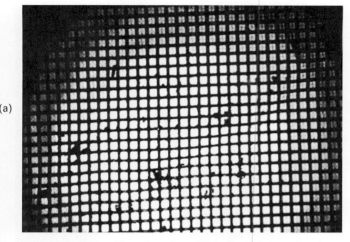

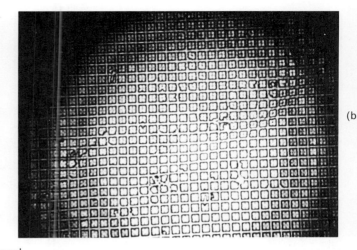

(a)   (b)

Fig. 14.12 (b) Is a filtered version of (a) where the zeroth order was removed.

the mesh at Σ_0, Fig. 14.10(b) shows the transform as it would appear on Σ_t. Now the fun starts—since the transform information relating to the dust is located in an irregular cloud-like distribution about the center point, we can easily eliminate it by inserting an opaque mask at Σ_t. If the mask has holes at each of the principal maxima, thus passing on only those frequencies, the image appears dustless, Fig. 14.11(a). At the other extreme, if we just pass the cloud-like pattern near center, very little of the periodic structure appears, leaving an image consisting of essentially just the dust particles 14.11(b). Passing only the zero-order central spot generates a uniformly illuminated (dc) field just as if the mesh were no longer in position. Observe that as more and more of the higher frequencies are eliminated, the detail of the image deteriorates markedly [Fig. 14.11(d), (e) and (f)]. This can be understood quite simply by remembering how a function, with what we might call "sharp edges", was synthesized out of harmonic components. The square wave of Fig. 7.13 serves to illustrate the point. It is evident that the addition of higher harmonics serves predominantly to square up the corners and flatten out the peaks and troughs of the profile. In this way, the high spatial frequencies contribute to the sharp edge detail between light and dark regions of the image. The removal of the high-frequency terms causes a rounding out of the step function and a consequent loss of resolution in the two-dimensional case.

What would happen if we took out the dc component [Fig. 14.11(c)] by passing everything but the central spot? A point on the original image which appears black in the photo denotes a near-zero irradiance and perforce a near-zero field amplitude. Presumably all of the various optical field components completely cancel each other at that point —ergo, no light. Yet with the removal of the dc term the point in question must certainly then have a nonzero field amplitude. When squared ($I \propto E_0^2/2$) this will generate a nonzero irradiance. It follows that regions which were originally black in the photo will now appear whitish while regions which were white will become grayish as in Fig. 14.12.

Let's now examine some of the possible applications of this technique. Figure 14.13(a) shows a composite photo of the moon consisting of film strips pieced together to form a single mosaic. The video data were telemetered to earth by Lunar Orbiter 1. Clearly the grating-like regular discontinuities between adjacent strips in the object photo generate the broad bandwidth vertical frequency distribution evident in Fig. 14.13(c). When these frequency components are blocked, the enhanced image shows no sign of having been a mosaic. In very much the same way, one can suppress extraneous data in bubble chamber photos of subatomic particle tracks.* These photos are made difficult to analyze because of the presence of the unscattered beam tracks (Fig. 14.14) which, since they are all parallel, are easily removed by spatial filtering.

---

* D. G. Falconer, "Optical Processing of Bubble Chamber Photographs," *Appl. Opt.* **5**, 1365 (1966), includes some additional uses for the coherent optical computer.

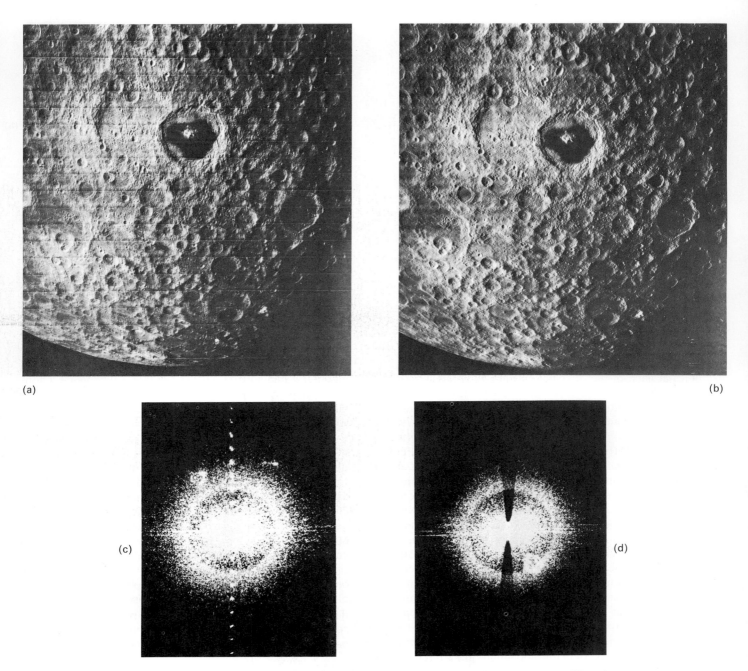

Fig. 14.13 Spatial filtering. (a) A Lunar Orbiter composite photo of the moon. (b) Filtered version of the photo sans horizontal lines. (c) A typical unfiltered transform (power spectrum) of a moonscape. (d) Diffraction pattern with the vertical dot pattern filtered out. [Photos courtesy D. A. Ansley, W. A. Blikken, The Conductron Corporation and N.A.S.A.]

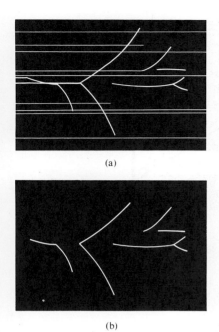

(a)

(b)

Fig. 14.14 Unfiltered and filtered bubble-chamber tracks.

Fig. 14.15 A self-portrait of K. E. Bethke consisting of only black and white regions as in a halftone. When the high frequencies are filtered out, shades of gray appear and the sharp boundaries vanish. [From R. A. Phillips, *Amer. J. Phys.* **37**, 536 (1969).]

Now consider the familiar half-tone or facsimile process by which a printer can create the illusion of various tones of gray while using only black ink and white paper (take a close look at a newspaper photo). If a transparency* of such a facsimile is inserted at Σ_0 in Fig. 14.8 its frequency spectrum will appear on Σ_t. Once again the relatively high-frequency components arising from the half-tone mesh can easily be eliminated. This yields an image in shades of gray (Fig. 14.15) showing none of the discontinuous nature of the original. One could construct a precise filter to obstruct only the square mesh frequencies by actually using a negative transparency of the transform of the basic checkerboard array. Alternatively, it usually suffices to use a low-pass circular aperture filter and in so doing inadvertently discard some of the high-frequency detail of the original scene; at least as long as the mesh frequency is comparatively high. The same procedure can be used to remove the graininess of highly enlarged photographs, which is of value e.g. in aerial photo reconnaissance. In contrast, we could sharpen up the details in a slightly blurred photo by emphas-

izing its high-frequency components. This could be done with a filter which preferentially absorbed the low-frequency portion of the spectrum. A great deal of effort beginning in the nineteen-fifties has gone into the study of photographic image enhancement and the ensuing successes have been notable indeed. Prominent amongst the contributors is A. Maréchal of the Institut d'Optique, Université de Paris who has combined absorbing and phase-shifting filters to reconstitute the detail in badly blurred photographs. These filters are transparent coatings deposited on optical flats so as to retard the phase of various portions of the spectrum (Section 14.1.4).

As this work in optical data processing continues into the coming decades, we will surely see the replacement of the photographic stages, in increasingly many applications, by real-time electro-optical devices (e.g. arrays of ultrasonic light modulators forming a multichannel input are already in use).* The coherent optical computer will reach

---

* Polaroid 55 P/N film is satisfactory for medium resolution work while Kodak 649 plates are good where higher resolution is required of the transparency.

---

* We have only touched on the subject of optical data processing; a more extensive discussion of these matters is given e.g. by Goodman in *Introduction to Fourier Optics*, Chapter 7. That text also includes a good reference list for further reading in the journal literature. Also see P. F. Mueller, "Linear Multiple Image Storage," *Appl. Opt.* **8**, 267 (1969). Here, as in much of modern optics, the frontiers are fast moving and obsolescence is a hard rider.

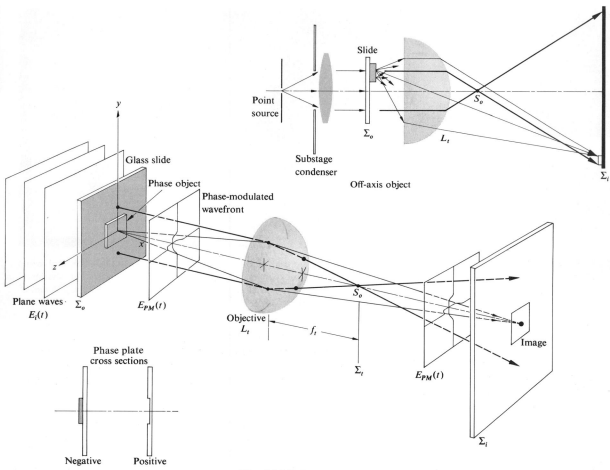

Fig. 14.16 Phase-contrast set-up.

14.1.4 Phase Contrast

It was mentioned rather briefly in the last section that the reconstructed image could be altered by introducing a phase-shifting filter. Probably the best-known example of this technique dates back to 1934 and the work of the Dutch physicist Fritz Zernike who invented the method of *phase contrast* and applied it in the *phase-contrast microscope*.

An object can be "seen" because it stands out from its surroundings—it has a color, tone, or a lack of color which provides contrast with the background. This kind of structure is known as an *amplitude object* because it is observable by dint of variations which it causes in the amplitude of the light wave. The wave which is either reflected or transmitted by such an object becomes *amplitude modulated* in the process. In contradistinction, it is often desirable to "see" *phase objects*, i.e. ones which are transparent, thereby providing practically no contrast with their environs and altering only the phase of the detected wave. The optical thickness of such objects generally varies from point to point as either the refractive index or the actual thickness or both vary. Obviously since the eye cannot detect phase variations such objects are invisible. This is the problem which led biologists to develop techniques for staining

a certain maturity, becoming an even more powerful tool when the input, filtering and output functions are performed electro-optically. A continuous stream of real-time data could flow into and out of such a device.

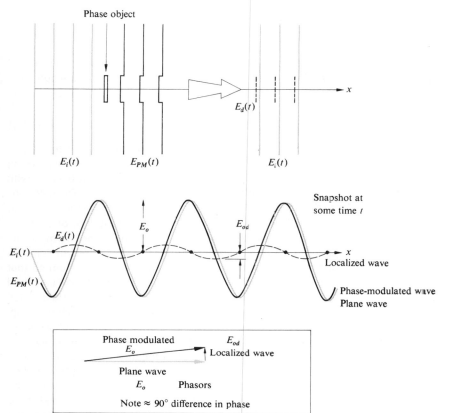

Fig. 14.17 Wavefronts in the phase-contrast process.

transparent microscope specimens and in so doing convert phase objects into amplitude objects. But this approach is unsatisfactory in many respects as, for example, when the stain kills the specimen whose life processes are under study, as is all too often the case.

Recall that diffraction occurs when a portion of the surface of constant phase is obstructed in some way, that is when a region of the wavefront is altered (either in amplitude or phase, i.e. shape). Suppose then that a plane wave passes through a transparent particle which retards the phase of a region of the front. The emerging wave is no longer perfectly planar but contains a small indentation corresponding to the area retarded by the specimen; the wave is *phase modulated*.

Taking a rather simplistic view of things, we can imagine the phase-modulated wave $E_{PM}(\mathbf{r}, t)$ (Fig. 14.16) to consist of the original incident plane wave $E_i(x, t)$ plus a localized disturbance $E_d(\mathbf{r}, t)$. (The symbol $\mathbf{r}$ means that E_{PM} and E_d depend on x, y, and z, i.e. they vary over the yz-

plane whereas E_i is uniform and does not.) Indeed, if the phase retardation is very small, the localized disturbance is a wave of very small amplitude, E_{0d}, lagging by just about $\lambda_0/4$ as in Fig. 14.17. There the difference between $E_{PM}(\mathbf{r}, t)$ and $E_i(x, t)$ is shown to be $E_d(\mathbf{r}, t)$. The disturbance $E_i(x, t)$ is called the direct or *zeroth-order wave* while $E_d(\mathbf{r}, t)$ is the *diffracted wave*. The former produces a uniformly illuminated field at Σ_i which is unaffected by the object while the latter carries all of the information about the optical structure of the particle. After broadly diverging from the object, these higher-order spatial frequency terms (see Section 14.1.2) are caused to converge on the image plane. The direct and diffracted waves recombine out of phase by $\pi/2$ to again form the phase-modulated wave. Since the amplitude of the reconstructed wave $E_{PM}(\mathbf{r}, t)$ is everywhere the same on Σ_i, even though the phase varies from point to point, the flux density is uniform and no image is perceptible. Likewise, the zeroth-order spectrum of a phase grating will be $\pi/2$ out of phase with the higher-order spectra.

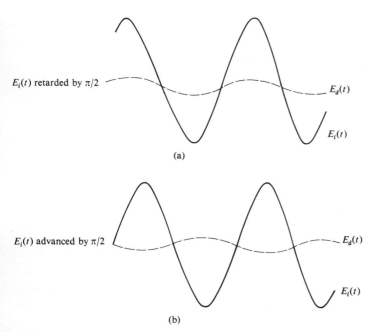

$E_i(t)$ retarded by $\pi/2$

$E_d(t)$

$E_i(t)$

(a)

$E_i(t)$ advanced by $\pi/2$

$E_d(t)$

$E_i(t)$

(b)

Fig. 14.18 Effect of phase shifts.

If we could somehow shift the relative phase between the diffracted and direct beams by an additional $\pi/2$ prior to their recombination, they would still be coherent and could then interfere either constructively or destructively (Fig. 14.18). In either case, the reconstructed wavefront over the region of the image would then be amplitude modulated—the image would be visible.

We can see this in a very simple analytical way where

$$E_i(x, t)|_{x=0} = E_0 \sin \omega t \qquad (14.1)$$

is the incoming monochromatic light wave at Σ_o without the specimen in place. The particle will induce a position-dependent phase variation $\phi(y, z)$ such that the wave just leaving it is

$$E_{PM}(\mathbf{r}, t)|_{x=0} = E_0 \sin [\omega t + \phi(y, z)]. \qquad (14.2)$$

This is a constant-amplitude wave which is essentially the same on the conjugate image plane. That is, there are some losses but if the lens is large and aberration free and we neglect the orientation and size of the image Eq. (14.2) will suffice to represent the PM wave on either Σ_o or Σ_i. Reformulating that disturbance as

$$E_{PM}(y, z, t) = E_0 \sin \omega t \cos \phi + E_0 \cos \omega t \sin \phi \qquad (14.3)$$

and limiting ourselves to *very small values* of ϕ it becomes

$$E_{PM}(y, z, t) = E_0 \sin \omega t + E_0 \phi(y, z) \cos \omega t.$$

The first term is independent of the object while the second term obviously isn't. Thus, as above, if we change their relative phase by $\pi/2$, i.e. either change the cosine to sine or vice versa, we get

$$E_{AM}(y, z, t) = E_0[1 + \phi(y, z)] \sin \omega t, \qquad (14.4)$$

which is an amplitude-modulated wave. Observe that $\phi(y, z)$ can be expressed in terms of a Fourier expansion, thereby introducing the spatial frequencies associated with the object. Incidentally this discussion is precisely analogous to the one proposed in 1936 by E. H. Armstrong for converting AM radio waves to FM [$\phi(t)$ could be thought of as a frequency modulation wherein the zeroth-order term is the carrier]. An electrical bandpass filter was used to separate the carrier from the remaining information spectrum so that the $\pi/2$ phase shift could be accomplished. Zernike's method of doing essentially the same thing is as follows. He inserted a spatial filter in the transform plane Σ_t of the objective (Fig. 14.16) which was capable of inducing the $\pi/2$ phase shift. Observe that the direct light actually forms a small image of the source on the optical axis at the location of Σ_t. The filter could then be a small circular indentation of depth d etched in a transparent glass plate of index n_g. Ideally, only the direct beam would pass through the indentation and in so doing it would take on a *phase advance* with respect to the diffracted wave of $(n_g - 1)d$ which is made to equal $\lambda_0/4$. A filter of this sort is known as a *phase plate* and since its effect corresponds to Fig. 14.18(b), i.e. destructive interference, phase objects which are thicker or have higher indices appear dark against a bright background. If, instead, the phase plate had a small raised disk at its center, the opposite would be true. The former case is called *positive-phase contrast*; the latter, *negative-phase contrast*.

In actual practice a brighter image is obtained by using a broad, rather than a point, source along with a substage condenser. The emerging plane waves illuminate an annular diaphragm (Fig. 14.19) which, since it is the source plane, is conjugate to the transform plane of the objective. The zeroth-order waves, shown in the figure, pass through the object according to the tenets of geometrical optics. They then traverse the thin annular region of the phase plate located at Σ_t. That region of the plate is quite small and so the cone of diffracted rays, for the most part, misses it. By making the annular region absorbing as well (a thin metal film will do), the very large uniform zeroth-order term (Fig. 14.20) is

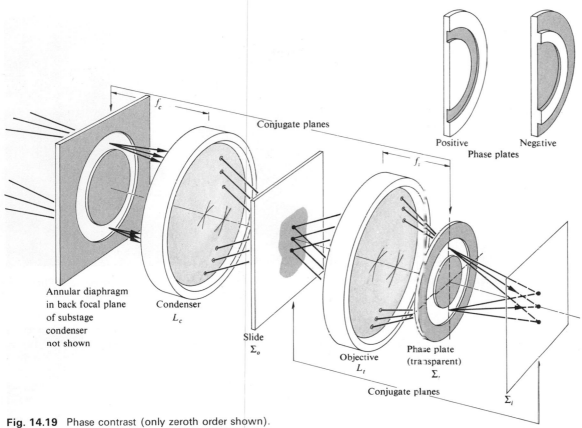

Fig. 14.19   Phase contrast (only zeroth order shown).

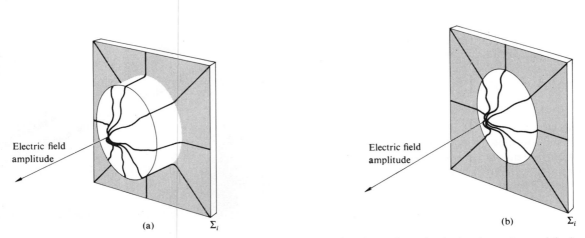

Fig. 14.20   Field amplitude over a circular region on the image plane. In one case there is no absorption in the phase plate and the irradiance would be a small ripple on a great plateau. With the zeroth order attenuated the contrast increases.

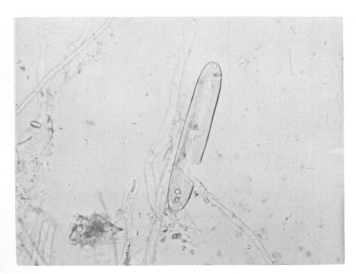
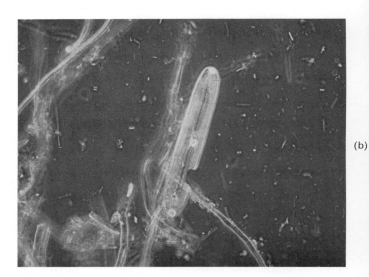

Fig. 14.21   (a) A conventional photomicrograph of diatoms, fibres and bacteria.   (b) A phase photomicrograph of the same scene.   [Photos by T. J. Lowery and R. Hawley.]

reduced with respect to the higher orders and the contrast improves. Or, if you like, E_0 is reduced to a value comparable with that of the diffracted wave E_{0d}. Generally a microscope will come with an assortment of these phase plates having different absorptions.

In the parlance of modern optics (the still-blushing bride of communications theory), phase contrast is simply the process whereby we introduce a $\pi/2$ phase shift in the zeroth-order spectrum of the Fourier transform of a phase object (and perhaps attenuate its amplitude as well) through the use of an appropriate spatial filter.

The phase-contrast microscope, which earned Zernike the Nobel prize in 1953, has found extensive application (Fig. 14.21); perhaps the most fascinating of which is the study of the life functions of otherwise invisible organisms.

14.1.5   The Dark-Ground and Schlieren Methods

Suppose we go back to Fig. 14.16 where we were examining a phase object and this time rather than retard and attenuate the central zeroth order, we remove it completely with an opaque disk at S_o. Without the object in place the image plane would then be completely dark—ergo the name *dark ground*. With the object in position only the localized diffracted wave will appear at Σ_i to form the image. (This can also be accomplished in microscopy by illuminating the object obliquely so that no direct light enters the objective lens.) Observe that by eliminating the dc contribution, the

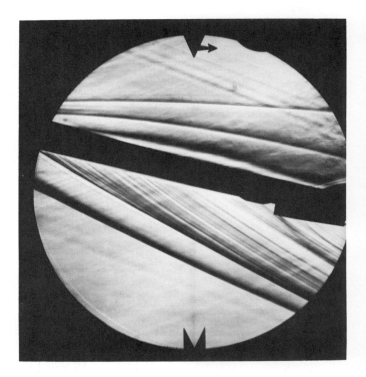

Fig. 14.22   A schlieren photo taken in a wind tunnel.   [Courtesy the Boeing Co.]

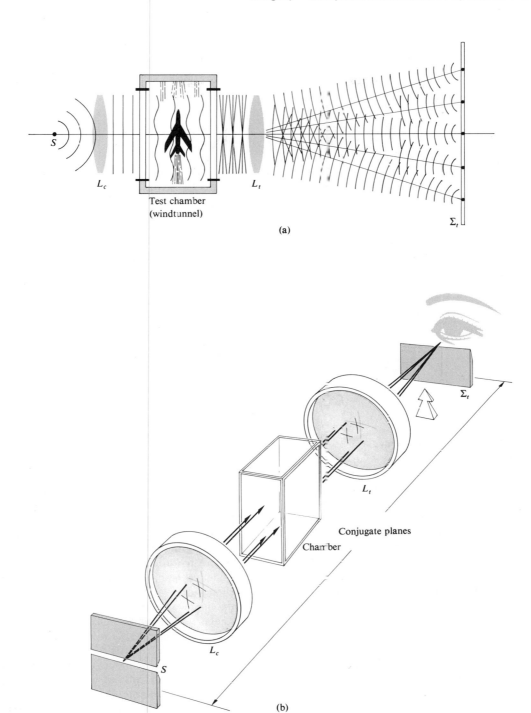

Fig. 14.23 A schlieren set-up.

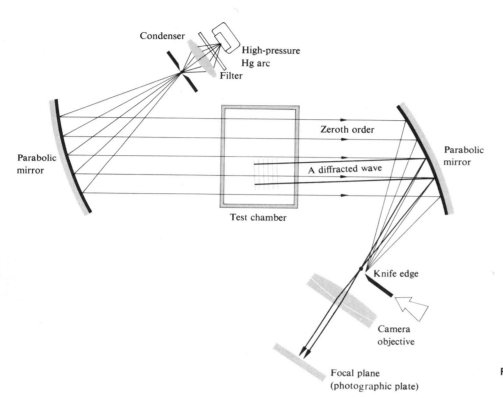

Fig. 14.24 A schlieren set-up using mirrors.

amplitude distribution (as in Fig. 14.20) will be lowered and portions which were near zero prior to filtering will become negative.   Inasmuch as irradiance is proportional to the amplitude squared, this will result in somewhat of a contrast reversal from that which would have been seen in phase contrast (see Section 4.1.3).   In general this technique has not been as satisfactory as the phase-contrast method which generates a flux-density distribution across the image that is directly proportional to the phase variations induced across the object.

In 1864 A. Toepler introduced a procedure for examining defects in lenses, which has come to be known as the *schlieren method.**   We will discuss it here because of the widespread current usage of the method in a broad range of fluid dynamics studies and furthermore because it is another beautiful example of the application of spatial filtering. Schlieren systems are particularly useful in ballistics, aero-

dynamics and ultrasonic wave analysis (Fig. 14.22), indeed wherever it is desirable to examine pressure variations as revealed by refractive-index mapping.

Suppose that we set up any one of the possible arrangements for viewing Fraunhofer diffraction, e.g. Figs. 10.5 or 10.85.   But now, instead of using an aperture of some sort as the diffracting amplitude object, we insert a phase object, e.g. a gas-filled chamber (Fig. 14.23).   Again a Fraunhofer pattern is formed in Σ_t and if that plane is followed by the objective lens of a camera, an image of the chamber is formed on the film plane.   We could then photograph any amplitude objects within the test area but, of course, phase objects would still be invisible.   Imagine that we now introduce a knife edge at Σ_t, raising it from below until it obstructs (sometimes only partially) the zeroth-order light and therefore all the higher orders on the bottom side as well.   Just as in the dark-ground method phase objects are then perceptible.   Inhomogeneities in the test chamber windows and flaws in the lenses are also noticeable.   For this reason and because of the large field of view usually required, mirror systems (Fig. 14.24) have now become commonplace.

---

* The word *Schlieren* in German means streaks or striae.   It's frequently capitalized because all nouns are in German and not because there was a Mr. Schlieren.

Quasimonochromatic illumination is generally made use of when resulting data are to be analyzed electronically, e.g. with a photodetector. Sources with a broad spectrum on the other hand, allow us to exploit the considerable color sensitivity of photographic emulsions and a number of color schlieren systems have been devised.

14.2   LASERS AND LASER LIGHT

During the early ninteen-fifties a remarkable device known as the *maser* came into being through the efforts of a number of scientists. Principal amongst these men were Charles Hard Townes of the U.S.A. and Alexandr Mikhailovich Prokhorov and Nikolai Gennadievich Basov of the U.S.S.R., all of whom shared the 1964 Nobel Prize in Physics for their work. The maser, which is an acronym for Microwave Amplification by Stimulated Emission of Radiation, is, as the name implies, an extremely low-noise, microwave amplifier.[*] It functioned in what was then a rather unconventional way, making direct use of the quantum-mechanical interaction of matter and radiant energy. Almost immediately after its inception speculation arose as to whether or not the same technique could be extended into the optical region of the spectrum. In 1958 Townes and Arthur L. Schawlow prophetically set forth the general physical conditions which would have to be met in order to achieve Light Amplification by Stimulated Emission of Radiation. And then in July of 1960 Theodore H. Maiman announced the first successful operation of an optical maser or *laser*—certainly one of the great milestones in the history of optics, and indeed in the history of science, had been achieved.

14.2.1   The Laser

Speaking first in generalities, suppose we have a collection of atoms, as for example, in a solid, gas or liquid. Recall that each atom (taken as a system composed of a nucleus and electron cloud) possesses a certain amount of internal energy, and each tends to maintain its lowest energy configuration. This is said to be the *ground state* for that particular kind of atom. Furthermore, each atom can exist in specific, well-defined configurations corresponding to higher energies than the ground state. Any of these are termed *excited states*.

In a conventional light source, such as a tungsten lamp, energy is *pumped* into the reacting atoms, in this case located

---

[*] See James P. Gorden, "The Maser," *Sci. Amer.* **199,** 42 (December 1958).

within the filament. These are consequently "raised" into excited states. Each can then drop back *spontaneously* (i.e. without external inducement) to the ground state emitting the absorbed energy in the form of a randomly directed photon. Atoms in this kind of source radiate essentially independently. The photons in the emitted stream bear no particular phase relationship with each other and the light is incoherent. It varies in phase from point to point and moment to moment.

Now imagine that light impinges on an atomic system of some sort. If an incident photon is energetic enough, it may be absorbed by an atom, raising the latter to an excited state. It was pointed out by Einstein in 1917 that an excited atom can revert to a lower state (which need not necessarily be the ground state) through photon emission via two distinctive mechanisms. In one instance the atom emits energy spontaneously, while in the other it is triggered into emission by the presence of electromagnetic radiation of the proper frequency. The latter process is known as *stimulated emission* and it is a key to the operation of the laser. In either situation the emerging photon will carry off the energy difference ($h v_{if}$) between the initial higher state $|i\rangle$ and the final lower state $|f\rangle$, that is

$$\mathscr{E}_i - \mathscr{E}_f = h v_{if} \qquad (14.5)$$

where $\mathscr{E}_i$ and $\mathscr{E}_f$ are the energies of the two states.

If an incident electromagnetic wave is to trigger an excited atom into stimulated emission, it must have the frequency v_{if}. A remarkable feature of this process is that *the emitted photon is in phase with, has the polarization of, and propagates in the same direction as, the stimulating wave.* Thus the photon is said to be in the same *radiation mode* as the incident wave and tends to add to it, increasing its flux density. However, since most of the atoms are ordinarily in the ground state, absorption is usually far more likely than stimulated emission. But this raises an intriguing point: What would happen if a substantial percentage of the atoms could somehow be excited into an upper state leaving the lower state all but empty? For obvious reasons this is known as *population inversion.* An incident photon of the proper frequency could then trigger an avalanche of stimulated photons—*all in phase.* The initial wave would continue to build so long as there were no dominant competitive processes (such as scattering) and provided the population inversion could be maintained. In effect, energy (electrical, chemical, optical etc.) would be pumped in to sustain the inversion and a beam of light would be extracted after sweeping across the active medium.

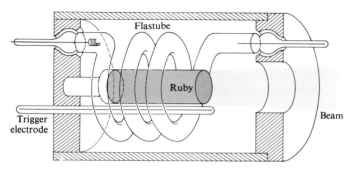

Fig. 14.25 A typical ruby-laser configuration.

i) The First (Pulsed Ruby) Laser

To see how all of this is accomplished in practice, let's take a look at Maiman's original device (Fig. 14.25). The first operative laser had as its active medium a small, cylindrical, synthetic, pale pink ruby, i.e., an Al_2O_3 crystal containing about 0.05 percent (by weight) of Cr_2O_3. Ruby, which is still one of the most common of the crystalline laser media, had been used earlier in maser applications and was suggested for use in the laser by Schawlow. The rod's end faces were polished flat, parallel and normal to the axis. Then both were silvered (one only partially) to form a resonant cavity. It was surrounded by a helical gaseous discharge flashtube which provided broadband *optical pumping*. Ruby appears red because the chromium atoms have absorption bands in the blue and green regions of the spectrum Fig. 14.26(a). Firing the flashtube generates an intense burst of light lasting for a few milliseconds. Much of this energy is lost in heat but many of the Cr^{3+} ions are excited into the absorption bands. A simplified energy-level diagram appears in Fig. 14.26(b). The excited ions rapidly relax; giving up energy to the crystal lattice and making nonradiative transitions, they preferentially drop "down" to an especially long-lived interim state. They remain in this so-called *metastable state* for up to several milliseconds before randomly, and in most cases spontaneously, dropping down to the ground state. This is accompanied by the emission of the characteristic red fluorescent radiation of ruby. The emission occurs in a relatively broad spectral range centered about 694.3 nm; it emerges in all directions and is incoherent. However, when the pumping rate is increased somewhat, a population inversion occurs and the first few spontaneously emitted photons stimulate a chain reaction. One quantum triggers the rapid, in-phase emission of another, dumping energy from the metastable atoms into the evolving light

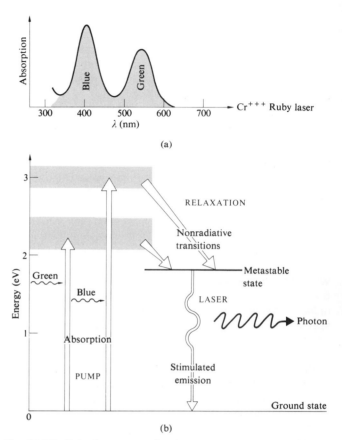

Fig. 14.26 Ruby-laser energy levels.

wave. The wave continues to grow as it sweeps back and forth across the active medium (provided enough energy is available to overcome losses at the mirrored ends). Since one of those reflecting surfaces was partially silvered, an intense pulse of red laser light (lasting about 0.5 ms and having a linewidth of about 0.01 nm) emerges from that end of the ruby rod. Notice how neatly everything works out. The broad absorption bands make the initial excitation rather easy, while the long lifetime of the metastable state facilitates the population inversion. The atomic system in effect consists of (1) the absorption bands, (2) the metastable state, and (3) the ground state. Accordingly it is spoken of as a *three-level* laser.

ii) Optical Resonant Cavities

The resonant cavity, which in this case is of course a Fabry–Perot etalon, plays a most significant role in the operation of the laser. In the early stages of the laser process, spontaneous photons are emitted in every direction as are the concommitant stimulated photons. But all of these, with the singular exception of those propagating very nearly along the cavity axis, quickly pass out of the sides of the ruby. In contrast, the axial beam continues to build as it bounces back and forth across the active medium. This accounts for the amazing degree of collimation of the issuing laser beam which is then effectively a coherent plane wave. Though the medium acts to amplify the wave, the *optical feedback* provided by the cavity converts the system into an oscillator and hence into a light generator—the acronym is thus somewhat of a misnomer.

In addition, the disturbance propagating within the cavity takes on a standing-wave configuration determined by the separation (d) of the mirrors. The cavity resonates (i.e., standing waves exist within it) when there is an integer number (m) of half wavelengths spanning the region between the mirrors. Thus

$$m = \frac{d}{\lambda/2}$$

and

$$v_m = \frac{mv}{2d}. \tag{14.6}$$

There are therefore an infinite number of possible oscillatory *longitudinal cavity modes*, each with a distinctive frequency v_m. Consecutive modes are separated by a constant difference,

$$\Delta v = \frac{v}{2d}, \tag{14.7}$$

which is the free spectral range of the etalon [Eq. (9.57)] and, incidentally, the inverse of the round trip time. For a gas laser 1 m long $\Delta v = 150$ MHz. The resonant modes of the cavity are considerably narrower in frequency than the bandwidth of the normal spontaneous atomic transition. These modes, whether the device is constructed so that there is one or more, will be the ones which are sustained in the cavity and hence the emerging beam is restricted to a region close to those frequencies (Fig. 14.27). In other words, the radiative transition makes available a relatively broad range of frequencies out of which the cavity will select and amplify only certain narrow bands and, if desired, even only one such band. This is the origin of the laser's extreme quasimonochromaticity.

In addition to the longitudinal or axial modes of oscillation, which correspond to standing waves set up along the cavity or z-axis, transverse modes can be sustained as well.

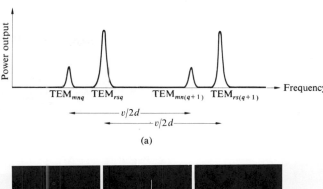

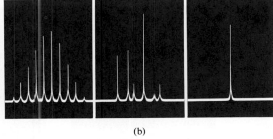

Fig. 14.27 Laser modes: (a) illustrates the nomenclature, (b) depicts three operation configurations for a c-w gas laser, first showing several longitudinal modes under a roughly Gaussian envelope, then several longitudinal and transverse modes and finally a single longitudinal mode.

Since the fields are very nearly normal to z these are known as TEM_{mn} modes (transverse electric and magnetic). The m and n subscripts are the integer number of transverse nodal lines in the x- and y-directions across the emerging beam. That is to say the beam is segmented in its cross section into one or more regions. Each such array is associated with a given TEM mode as shown in Fig. 14.28. The lowest order or TEM_{00} transverse mode is perhaps the most widely used, and this for several compelling reasons: the flux density is ideally Gaussian over the beam's cross section (Fig. 14.29); there are no phase shifts in the electric field across the beam as there are in other modes, and so it is completely spatially coherent (Fig. 14.30); the beam's angular divergence is the smallest and it can be focused down to the smallest sized spot. Note that the amplitude in this mode is actually not constant over the wavefront and it is consequently an inhomogeneous wave.

A complete specification of each mode has the form TEM_{mnq}, where q is the longitudinal mode number. For each transverse mode (m, n) there can be many longitudinal modes (i.e., values of q). Often, however, it's unnecessary to work

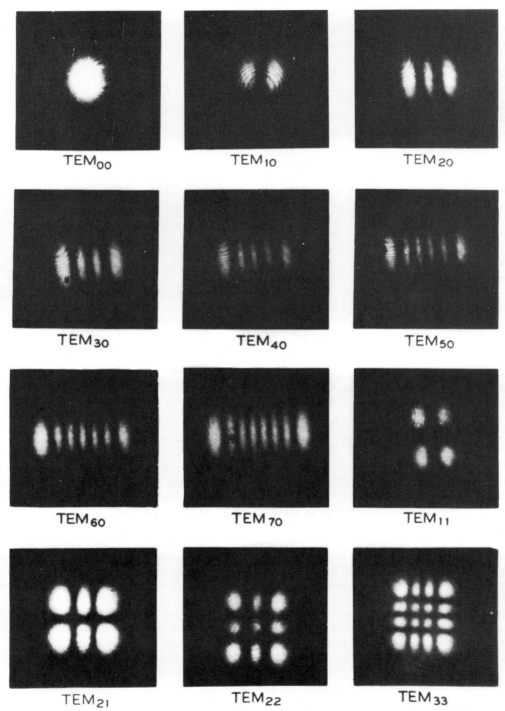

Fig. 14.28 Mode patterns (without the faint interference fringes this is what the beam looks like in cross section). [Photos courtesy Bell Telephone Laboratories.]

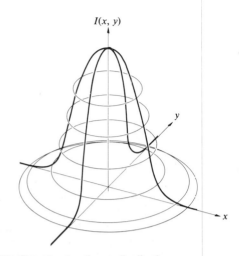

Fig. 14.29 Gaussian irradiance distribution.

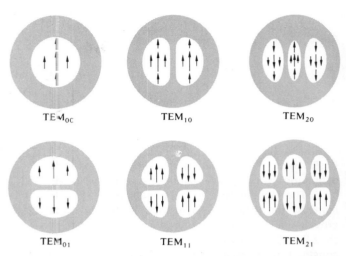

Fig. 14.30 Mode configurations (rectangular symmetry). Circularly symmetric modes are also observable but any slight asymmetry (such as Brewster windows) destroys them.

with a particular longitudinal mode and the q subscript is usually simply dropped.*

There are several additional cavity arrangements which are of considerably more practical significance than is the original plane-parallel set-up, which at the moment is mainly of historical interest. For example, if the planar mirrors are replaced by identical concave spherical mirrors separated by a distance very nearly equal to their radius of curvature, we have the *confocal* resonator. Thus the focal points are almost coincident on the axis midway between mirrors— ergo the name confocal. If one of the spherical mirrors is made planar, the cavity is termed a *hemispherical* resonator. Both these configurations are considerably easier to align than is the plane-parallel form. The actual selection of a resonator configuration is governed by the specific requirements of the system—there is no universally best arrangement.

The decay of energy in a cavity is expressed in terms of the Q or *quality factor* of the resonator. The origin of the expression dates back to the early days of radio engineering when it was used to describe the performance of an oscillating (tuning) circuit. A high Q, low-loss circuit meant a narrow bandpass and a sharply tuned radio. If an optical cavity is somehow disrupted, as for example by the displacement or removal of one of the mirrors, the laser action

generally ceases. When this is done deliberately in order to delay the onset of oscillation in the laser cavity, it's known as *Q-spoiling* or *Q-switching*. The power output of a laser is self limited in the sense that the population inversion is continuously depleted via stimulated emission by the radiation field within the cavity. However, if oscillation is prevented, the number of atoms pumped into the (long lived) metastable state can be considerably increased, thereby creating a very extensive population inversion. When the cavity is switched on at the proper moment, a tremendously powerful *giant pulse* (perhaps up to several hundred megawatts) will emerge as the atoms drop down to the lower state almost in unison. A great many *Q-switching* arrangements utilizing various control schemes, as for example, bleachable absorbers which become transparent under illumination, rotating prisms and mirrors, mechanical choppers, ultrasonic cells, or electro-optic shutters such as Kerr or Pockels cells, are in use today.

iii) The Helium–Neon Laser

Maiman's announcement of the first operative laser came at a New York news conference on July 7, 1960.* By February of 1961 Ali Javan and his associates W. R. Bennett, Jr., and

---

* Take a look at R. A. Phillips and R. D. Gehrz, "Laser Mode Structure Experiments for Undergraduate Laboratories," *Am. J. Phys.* **38**, 429 (1970).

* His initial paper, which would have made his findings known in a more traditional fashion, was rejected for publication by *Physical Review Letters*—this to their everlasting chagrin.

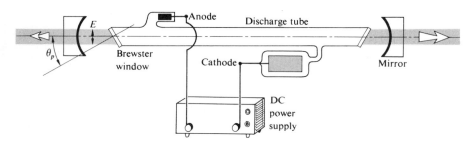

Fig. 14.31   A laser configuration.

D. R. Herriott had reported the successful operation of a continuous wave (c-w) helium–neon, gas laser at 1152.3 nm. The modern He–Ne laser (depicted in Fig. 14.31) is currently the most popular device of its kind; most often providing a few milliwatts of continuous power in the visible (632.8 nm). It's appeal arises primarily because it's easy to construct, relatively inexpensive, fairly reliable and in most cases can be operated by a flick of a single switch. Pumping is usually accomplished by electrical discharge (via either dc, ac or electrodeless rf excitation). Free electrons and ions are accelerated by an applied field and, as a result of collisions, cause further ionization and excitation of the gaseous medium (typically a mixture of about 0.8 torr of He and about 0.1 torr of Ne). Many helium atoms, after dropping down from several upper levels, accumulate in the long-lived 2^1S- and 2^3S-states. These are metastable states (Fig. 14.32) from which there are no allowed radiative transitions. The excited He atoms inelastically collide with and transfer energy to ground state Ne atoms, raising them in turn to the $3s_2$- and $2s_2$-states. These are the upper laser levels and there then exists a population inversion with respect to the lower $3p_4$- and $2p_4$-states. Spontaneous photons initiate stimulated emission and the chain reaction begins. The dominant laser transitions correspond to 1152.3 nm and 3391.2 nm in the infrared and, of course, the ever popular 632.8 nm in the visible (bright red). The p-states drain off into the 1s-state, thus themselves remaining uncrowded and thereby continuously sustaining the inversion. The 1s-level is metastable, so that 1s-atoms return to the ground state after losing energy to the walls of the enclosure. This is why the plasma tube's diameter inversely affects the gain and is, accordingly, a significant design parameter. In contrast to the ruby, where the laser transition is down to the ground state, stimulated emission in the He–Ne laser occurs between two upper levels. The significance of this, for example, is that since the $2p_4$-state is ordinarily only sparsely occupied, a population inversion is very easily obtained and this without having to half empty the ground state.

Return to Fig. 14.31 which pictures the relevant features of a typical He–Ne laser. The mirrors here are coated with a multilayered dielectric film having a reflectance of over 99%. The laser output is made linearly polarized by the inclusion of Brewster end windows (i.e. plates tilted at the polarization angle) terminating the discharge tube. If these end faces were instead normal to the axis, reflection losses (4% at each interface) would become unbearable. By tilting them at the polarization angle, the windows presumably have 100% transmission for light whose electric field component is parallel to the plane of incidence (the plane of the drawing). This polarization state rapidly becomes dominant since the

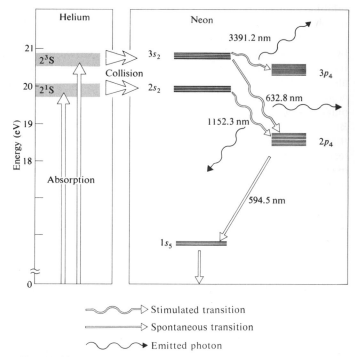

Fig. 14.32   He–Ne laser energy levels.

normal component is partially reflected off axis at each transit of the windows. Linearly polarized light in the plane of incidence soon becomes the preponderant stimulating mechanism in the cavity, to the ultimate exclusion of the orthogonal polarization.* The beam issuing from a He—Ne laser is therefore most frequently linearly polarized.

iv) A Survey of Laser Developments

Laser technology is so dynamic a field that what was a laboratory breakthrough two or three years ago may be a commonplace off-the-shelf item today. The whirlwind will certainly not pause. to allow descriptive terms like "the smallest," "the largest," "the most powerful," etc. to be applicable for very long. With this in mind, we briefly survey the existing scene without trying to anticipate the wonders which will surely come after this type is set. Laser beams have already been bounced off the moon, they have spot welded detached retinas, generated fusion neutrons, stimulated seed growth, served as communications links, guided milling machines, missiles, ships and grating engines, carried color TV pictures, drilled holes in diamonds, levitated tiny objects,† and intrigued countless amongst the curious. And this is just the first decade—only the beginning.

Along with ruby there are a great many other *solid-state lasers* whose outputs range in wavelength from roughly 500 to 2500 nm. For example, the trivalent rare earths Nd^{3+}, Ho^{3+}, Gd^{3+}, Tm^{3+}, Er^{3+}, Pr^{3+} and Eu^{3+} undergo laser action in a host of hosts such as $CaWO_4$, Y_2O_3, $SrMoO_4$, LaF_3, yttrium aluminum garnet (YAG for short) and glass, to name only a few. Of these, neodymium-doped glass and neodymium-doped YAG are of particular importance. Both constitute high-powered laser media operating at approximately 1060 nm. Nd:YAG lasers generating in excess of a kilowatt of continuous power have been constructed. Tremendous power outputs in pulsed systems have been obtained by operating several lasers in tandem. The first laser in the train serves as a Q-switched oscillator that fires into the next stage which functions as an amplifier; and

---

* Half of the output power of the laser is *not* lost in reflections at the Brewster windows when the transverse $\mathscr{P}$-state light is scattered. Energy simply isn't continuously channeled into that polarization component by the cavity. If it's reflected out of the plasma tube, it's not present to stimulate further emission.

† See M. Lubin and A. Fraas, "Fusion by Laser," *Sci. Amer.* **224**, 21 (June 1971) and A. Ashkin, "The Pressure of Laser Light," *Sci. Amer.* **226**, 63 (February 1972).

there may be one or more such amplifiers in the system. By reducing the feedback of the cavity, a laser will no longer be self-oscillatory but it will amplify an incident wave which has triggered stimulated emission. Thus the amplifier is, in effect, an active medium which is pumped, but for which the end faces are only partially reflecting or even nonreflecting. Ruby systems of this kind delivering a few GW (gigawatts, i.e. 10^9 W) in the form of pulses lasting several nanoseconds are available commercially and a glass system generating 25 TW (terawatts, i.e. 10^{12} W) in 50 J pulses has been reported. Systems of this sort are being used in an effort to produce controlled thermonuclear fusion reactions.

A large group of *gas lasers* operate across the spectrum from the far IR to the UV. Primary amongst these are helium—neon, argon, and krypton, as well as several molecular gas systems such as carbon dioxide and molecular nitrogen (N_2). Argon "lases" mainly in the green, blue-green and violet (predominantly at 488.0 and 514.5 nm) in either pulsed or continuous operation. Although its output is usually several watts c-w, it has gone as high as 150 W c-w. The argon ion laser is similar in some respects to the He—Ne laser although it evidently differs in its usually greater power, shorter wavelength, broader linewidth and higher price. All of the noble gases (He, Ne, A, Kr, Xe) have been made to lase individually as have the gaseous ions of many other elements, but the former grouping has been studied most extensively.

The CO_2 molecule, which lases between vibrational modes, emits in the infrared at 10.6μ (1 micron = 10^{-6} m) with typical c-w power levels of from watts to several kilowatts Its efficiency can be an unusually high 15% when aided by additions of N_2 and He. While it once took a discharge tube nearly two hundred meters long to generate 10 kW c-w, considerably smaller "table models" are now available commercially. For the moment at least, the record output belongs to an experimental gas-dynamic laser utilizing thermal pumping on a mixture of CO_2, N_2 and H_2O to generate 60 kW c-w at 10.6μ in multimode operation—shades of the infamous death ray!

The pulsed nitrogen laser operates at 337.1 nm in the ultraviolet as does the c-w helium—cadmium laser. A number of metal vapors (e.g. Zn, Hg, Sn, Pb) have displayed laser transitions in the visible, but problems such as maintaining uniformity of the vapor in the discharge region have handicapped their practical exploitation. The He—Cd laser, which is a recent arrival on the scene, emits at 325.0 nm and 441.6 nm. These are transitions of the cadmium ion arising after excitation resulting from collisions with metastable helium atoms.

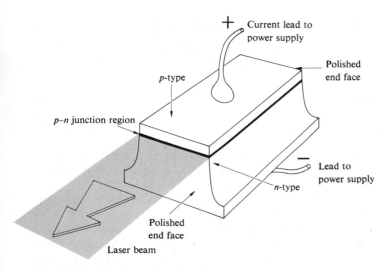

Fig. 14.33 A GaAs *p–n* junction laser.

The gallium arsenide *semiconductor laser* was invented in 1962. Here transitions occur between the conduction and valence bands and stimulated emission results in the immediate vicinity of the *p–n* junction (Fig. 14.33). Semiconductor lasers, made of a number of materials, span the spectrum from IR to UV. The GaAs laser, which is particularly important, is extremely small, being simply a semiconductor chip about the size of a pinhead (sans power supply).[*]

The first *liquid laser* was operated in January of 1963.[†] All of the early devices of this sort were exclusively *chelates* (i.e., metallo-organic compounds formed of a metal ion with organic radicals). That original liquid laser contained an alcohol solution of europium benzoylacetonate emitting at 613.1 nm. The discovery of laser action in nonchelate organic liquids was made in 1966. It came with the fortuitous lasing (at 755.5 nm) of a chloroaluminum phthalocyanine solution during a search for stimulated Raman emission in that substance.[‡] A great many fluorescent dye solutions of such families as e.g. the fluoresceins, coumarins and rhodamines have since been made to lase at frequencies from the IR

---

[*] For a discussion of heterostructure diode lasers refer to M. B. Panish and I. Hayashi, "A New Class of Diode Lasers," *Sci. Amer.* **225,** 32 (July 1971).

[†] See Adam Heller, "Laser Action in Liquids," *Phys. Today* (November 1967), p. 35 for a more detailed account.

[‡] P. Sorokin, "Organic Lasers," *Sci. Amer.* **220,** 30 (February 1969).

into the UV. These have usually been pulsed although c-w operation has been obtained. There are so many organic dyes that it would seem possible to build such a laser at any frequency in the visible. Moreover, these devices are distinctive in that they inherently can be tuned continuously over a range of wavelengths (of perhaps 70 nm or so, although a pulsed system tunable over 170 nm exists). Indeed, there are other arrangements which will vary the frequency of a primary laser beam, i.e. the beam enters with one color and emerges with another (Section 14.4), but in the case of the dye laser, the primary beam itself is tuned internally. This is accomplished, for example, by changing the concentration or the length of the dye cell or by adjusting a diffraction grating reflector at the end of the cavity. Several multicolor dye laser systems, which can easily be switched from one dye to another and thereby operate over a very broad frequency range, are available commercially.

A *chemical laser* is one which is pumped utilizing energy released via a chemical reaction. The first of this kind was operated in 1964 but it was not until 1969 that a continuous-wave chemical laser was developed. One of the most promising of these is the deuterium fluoride–carbon dioxide (DF–CO_2) laser. It is self-sustaining, in that it requires no external power source. In brief, the reaction $F_2 + D_2 \rightarrow 2DF$, which occurs on the mixing of these two fairly common gases, generates enough energy to pump a CO_2 laser.

There are solid-state, gaseous, liquid, vapor (e.g. H_2O) and semiconductor lasers; a great, hulking giant, driven by a rocket engine which blasts out 60 kW of c-w IR and, at the other extreme, a tiny lasing GaAs chip that easily fits on a fingertip. Lasers run the gamut in size, power and wavelength. They've been pumped optically, electrically, thermally and chemically.

14.2.2   The Light Fantastic

Laser light, using the latter term rather loosely to cover IR as well as UV, differs somewhat in nature from one type of laser to another. There are, however, several extremely remarkable features which are displayed, to varying degrees, by all laser emissions.

Quite apparent is the fact that most laser beams are exceedingly directional, or if you will, highly collimated. One need only blow some smoke into the otherwise invisible, visible-laser beam to see (via scattering) a fantastic thread of light stretched across the room. A He–Ne beam in the TEM_{00} mode generally has a divergence of only about one minute of arc or less, which is determined solely by diffraction at the exit aperture. Recall that the emission approximates a

Gaussian irradiance distribution, i.e., the flux density drops off from a maximum at the central axis of the beam and has no side lobes. The typical He–Ne beam is quite narrow, usually issuing at no more than a few millimeters in diameter. Since the beam consists of nearly plane waves, it is of course *spatially coherent*. And in fact, the directionality may be thought of as a manifestation of that coherence.

Laser light is quasimonochromatic, generally having an exceedingly narrow frequency bandwidth (see Section 7.10). In other words, it is *temporally coherent*.

Another attribute is the high flux or *radiant power* that can be delivered in that narrow frequency band. As we've seen, the laser is distinctive in that it emits all its energy in the form of a single, narrow, diffraction-limited beam. In contrast, a 100 W incandescent light bulb may pour out considerably more radiant energy in toto than a low-power c-w laser, but the light is incoherent, spread over a large solid angle, and has a broad bandwidth as well. A good lens* can totally intercept a laser beam and focus essentially all of its energy into a minute diffraction-limited spot (whose diameter varies directly with λ and the focal length, and inversely with the beam diameter). Spot diameters of just a few thousandths of an inch can readily be attained with convenient short focal length lenses. And a spot diameter of a few hundred-millionths of an inch is possible in principle. Thus flux densities can be generated in a focused laser beam of over 10^{17} W/cm² in contrast to, say, an oxyacetylene flame having roughly 10^3 W/cm². To get a better feel for these power levels, note that a focused CO_2 laser beam of a few kilowatts c-w can burn a hole through a quarter inch stainless steel plate in about ten seconds. By comparison, a pinhole and filter positioned in front of an ordinary source will certainly produce spatially and temporally coherent light, but only at a minute fraction of the total power output.

Laser light is, of course, available as a continuous wave or in the form of pulses. As a matter of fact, extremely short-duration *sub-picosecond pulses* are of considerable current interest† (the shortest yet reported being 0.3×10^{-12} s). A picosecond pulse corresponds to a wavetrain only 0.3 mm in extent—that's just about several hundred wavelengths long. As such they could serve as probes on an atomic

scale or to study chemical processes which occur in an interval of from 10^{-9} s to 10^{-13} s, e.g. molecular lifetimes. There are many other potential applications of picosecond pulses: for example, they might be used for high-density communications, data manipulation, or high-resolution optical radar.

A rather striking and easily observable manifestation of the spatial coherence of laser light is its granular appearance on reflection from a diffuse surface. Using a He–Ne laser (632.8 nm), expand the beam a bit by passing it through a simple lens and project it onto a wall or a piece of paper. The illuminated disk appears speckled with bright and dark regions that sparkle and shimmer in a dazzling psychedelic dance. Squint and the grains grow in size; step toward the screen and they shrink; take off your eyeglasses and the pattern stays in perfect focus. In fact, if you are nearsighted the diffraction fringes caused by dust on the lens blur out and disappear but the speckles do not. Hold a pencil at varying distances from your eye so that the disk appears just above it. At each position, focus on the pencil; wherever you focus, the granular display is crystal clear. Indeed, look at the pattern through a telescope and as you adjust the scope from one extreme to the other, the ubiquitous granules remain perfectly distinct even though the wall is completely blurred.

The spatially coherent light scattered from a diffuse surface fills the surrounding region with a *stationary* interference pattern (just as in the case of the wavefront splitting arrangements of Section 9.3). At the surface the granules are exceedingly small and they increase in size with distance. At any location in space the resultant field is the superposition of many contributing scattered wavelets. These must have a constant relative phase determined by the optical path length from each scatterer to the point in question, if the interference pattern is to be sustained. Figure 14.34 illustrates this point rather nicely. It shows a cement block illuminated in one case by laser light and in the other by collimated light from an Hg arc lamp, both of about the same spatial coherence. Yet while the laser's coherence length is much greater than the height of the surface features, the coherence length of the Hg light is not. In the former case, the speckles in the photo are large and they obscure the surface structure; in the latter, despite its spatial coherence, the speckle pattern is not observable in the photo and the surface features predominate. Because of the rough texture the optical path-length difference between two wavelets arriving at a point in space, scattered from different surface bumps, are generally greater than the coherence length of the mercury light. This means that the relative phases of the

---

* Spherical aberration is usually the main problem since laser beams are, as a rule, both quasimonochromatic and incident along the axis of the lens.

† Take a look at "Ultrafast laser pulses" by A. De Maria, W. Glenn and M. Mack, *Phys. Today* (July 1971), p. 19.

 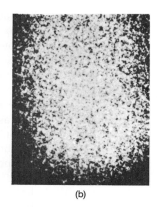

(a) (b)

Fig. 14.34 Speckle patterns (a) A cement block illuminated by a mercury arc and (b) a He—Ne laser. [From B. J. Thompson, *J. Soc. Phot. Inst. Engr.* **4**, 7 (1965).]

overlapping wave trains change rapidly and randomly in time, washing out the large-scale interference pattern.

A real system of fringes is formed of the scattered waves which converge in front of the screen. The fringes can be viewed by intersecting the interference pattern with a sheet of paper at a convenient location. After forming the real image in space, the rays proceed to diverge and any region of the image can therefore be viewed directly with the eye appropriately focused. In contrast, rays which initially diverge appear to the eye as if they had originated behind the scattering screen and thus form a virtual image.

It seems that as a result of chromatic aberration, normal and farsighted eyes tend to focus red light behind the screen. Contrarily, a nearsighted person observes the real field in front of the screen (regardless of wavelength). Thus if the viewer moves her head to the right, the pattern will move to the right in the first instance (where the focus is beyond the screen) and to the left in the second (focus in front). The pattern will follow the motion of your head if you're viewing it very close to the surface. The same apparent parallax motion can be seen by looking through a window; outside objects will seem to move with your head, inside ones opposite to it. The brilliant, narrow-bandwidth, spatially coherent laser beam is ideally suited for observing the granular effect although other means are certainly possible.* In unfiltered sunlight the grains are minute on

---

* For further reading on this effect, see L. I. Goldfischer, *J. Opt. Soc. Am.* **55**, 247 (1965), D. C. Sinclair, *J. Opt. Soc. Am.* **55**, 575 (1965), J. D. Rigden and E. I. Gordon, *Proc. IRE* **50**, 2367 (1962), B. M. Oliver, *Proc. IEEE* **51**, 220 (1963).

the surface and multicolored. The effect is easy to observe on a smooth, flat-black material (e.g. poster-painted paper) but you can see it on a fingernail or a worn coin as well.

Although it provides a marvelous demonstration, both esthetically and pedagogically, the granular effect can be a real practical nuisance in coherently illuminated systems. For example, in holographic imagery the speckle pattern corresponds to troublesome background noise.

14.3 HOLOGRAPHY

The technology of photography has been with us for a long time and we've all grown accustomed to seeing the three-dimensional world compressed into the flatness of a scrap-book page. The depthless TV pitchman who smiles out of a myriad of phosphorescent flashes, although inescapably there, seems no more palpable than a postcard image of the Eiffel Tower. Both share the severe limitation of being simply irradiance mappings. In other words, when the image of a scene is ordinarily reproduced, by whatever traditional means, what we ultimately see is not an accurate reproduction of the light field which once inundated the object, but rather a point-by-point record of just the square of the field's amplitude. The light reflecting off a photograph carries with it information about the irradiance but nothing about the phase of the wave which once emanated from the object. Indeed, if both the amplitude and phase of the original wave could be reconstructed somehow, the resulting light field (presuming the frequencies are the same) would be indistinguishable from the original. This means that you would then see (and could photograph) the reformed image in perfect three-dimensionality, exactly as if the object were there before you, actually generating the wave.

14.3.1 Methods

Dennis Gabor had been thinking along these lines for a number of years prior to 1947 when he began conducting his now famous experiments in holography at the Research Laboratory of the British Thomson—Houston Company. His original set-up, depicted in Fig. 14.35, was a two-step lens-less imaging process in which he first photographically recorded an interference pattern, generated by the interaction of scattered quasimonochromatic light from an object and a coherent reference wave. The resulting pattern was something he called a *hologram* after the Greek word *holos* meaning whole. The second step in the procedure was the *reconstruction* of the optical field or image and this was

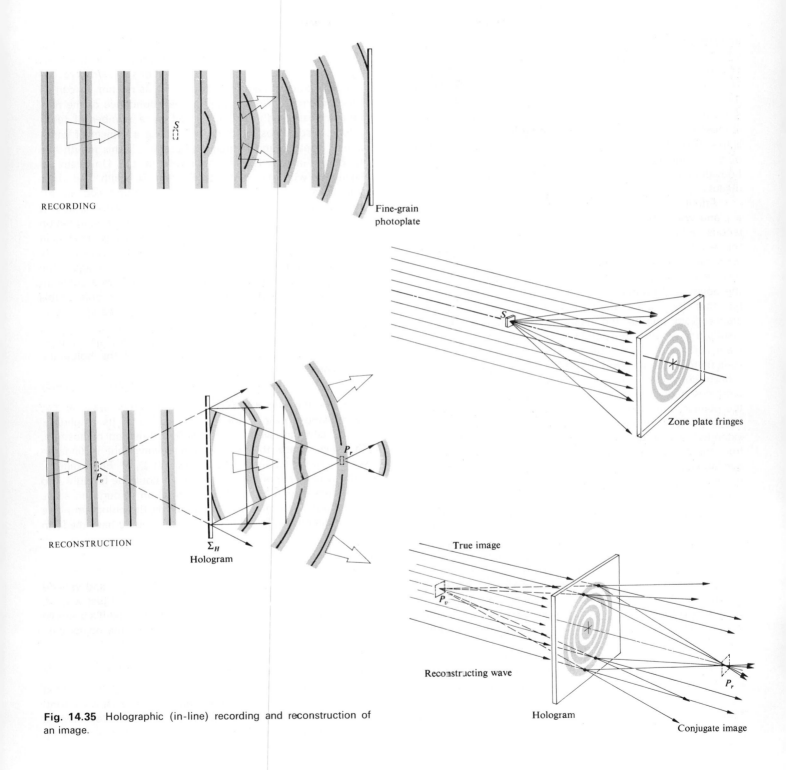

Fig. 14.35 Holographic (in-line) recording and reconstruction of an image.

done via the diffraction of a coherent beam by a transparency, which was the developed hologram. In a way quite reminiscent of Zernike's phase-contrast technique (Section 14.1.4), the hologram was formed when the unscattered *background* or *reference wave* interfered with the diffracted wave from the small semitransparent object, *S*—which was, in those early days, often a piece of microfilm. The key point is that the interference pattern or hologram contains, by way of the fringe configuration, information corresponding to both the amplitude and phase of the wave scattered by the object.

Admittedly, it's not at all obvious that by now shining a plane wave through the processed hologram one could reconstruct an image of the original object. Suffice it to say for the moment that if the object were very small the scattered wave would be nearly spherical and the interference pattern a series of concentric rings (centered about an axis through the object and normal to the plane wave). Except for the fact that the circular fringes would vary gradually in irradiance from one to the next, the resulting flux-density distribution would correspond to a conventional Fresnel zone plate (Section 10.3.5). Recall that a zone plate functions something like a lens in that it diffracts collimated light into a beam converging to a real focal point, P_r. In addition, it produces a diverging wave which appears to come from the point P_v and constitutes a virtual image. Thus we can imagine, albeit rather simplistically, that each point on an extended object generates its own zone plate displaced from the others and that the ensemble of all such partially overlapping zone plates forms the hologram.* During the reconstruction step, each constituent zone plate forms both a real and virtual image of a single object point and in this way, point by point, the hologram regenerates the original light field. When the reconstructing beam has the same wavelength as the initial recording beam (which need not necessarily be the case, and quite often isn't), the virtual image is undistorted and appears at the location formerly occupied by the object. Thus it is the virtual image field which actually corresponds to the original object field. As such, the virtual image is sometimes spoken of as the *true image* while the other is the real or, perhaps more fittingly, the *conjugate image*.

Gabor's research, which won him the 1971 Nobel Prize for physics, had as its motivation an improvement in electron microscopy. His work initially generated some in-

---

* See M. P. Givens, ''Introduction to Holography,'' *Am. J. Phys.* **35,** 1056 (1967).

terest, but all in all it remained in a state of quasi-unnoticed oblivion for about fifteen years. In the early nineteen-sixties, there was a resurgence of interest in Gabor's *wavefront reconstruction* process and, in particular, in its relation to certain radar problems. Soon, aided by an abundance of the new coherent laser light and extended by a number of technological advances, holography became a subject of widespread research and tremendous promise. This rebirth had its origin in the Radar Laboratory of the University of Michigan with the work of Emmett N. Leith and Juris Upatnieks. Among other things, they introduced an improved arrangement for generating holograms which is illustrated in Fig. 14.36. Unlike Gabor's *in-line* configuration where the conjugate image was inconveniently located in front of the true image, the two were now satisfactorily separated off axis as shown in the diagram. Once again, the hologram is an interference pattern arising from a coherent reference wave and a wave scattered from the object (this type is sometimes referred to as a *side band Fresnel hologram*).

The process can be treated analytically as follows; suppose that the *xy*-plane is the plane of the hologram, Σ_H. Then

$$E_B(x, y) = E_{0B} \cos [2\pi ft + \phi(x, y)] \qquad (14.8)$$

describes the planar background or reference wave at Σ_H, overlooking considerations of polarization. Its amplitude, E_{0B}, is constant while the phase is a function of position. This just means that the reference wavefront is tilted in some known manner with respect to Σ_H. For example, if the wave were oriented such that it could be brought into coincidence with Σ_H by a single rotation through an angle of θ about *y*, the phase at any point on the hologram plane would depend on its value of *x*. Thus ϕ would have the form

$$\phi = \frac{2\pi}{\lambda} x \sin \theta = kx \sin \theta$$

being, in that particular case, independent of *y* and varying linearly with *x*. For the sake of simplicity we'll just write it, quite generally, as $\phi(x, y)$ and keep in mind that it's a simple known function. The wave scattered from the object can, in turn, be expressed as

$$E_O(x, y) = E_{0O}(x, y) \cos [2\pi ft + \phi_O(x, y)], \qquad (14.9)$$

where both the amplitude and phase are now complicated functions of position corresponding to an irregular wavefront. From the communications-theoretic point of view, this is an amplitude- and phase-modulated carrier wave bearing

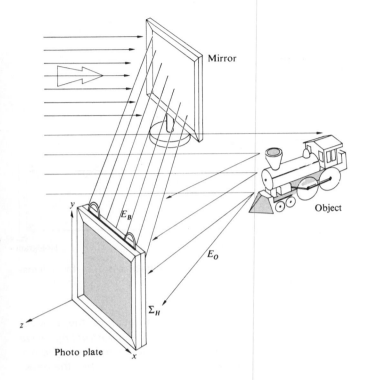

all of the available information about the object. Note that this information is coded in spatial rather than temporal variations of the wave. The two disturbances E_B and E_O superimpose and interfere to form an irradiance distribution which is recorded by the photographic emulsion. The resulting irradiance, except for a multiplicative constant, is $I(x, y) = \langle (E_B + E_O)^2 \rangle$ which, from Section 9.1, is given by

$$I(x, y) = \frac{E_{0B}^2}{2} + \frac{E_{0O}^2}{2} + E_{0B}E_{0O}\cos(\phi - \phi_O). \quad (14.10)$$

Observe that the phase of the object wave determines the location on Σ_H of the irradiance maxima and minima. Moreover, the contrast or fringe visibility

$$\mathscr{V} \equiv (I_{max} - I_{min})/(I_{max} + I_{min}) \quad [12.1]$$

across the hologram plane, which is

$$\mathscr{V} = 2E_{0B}E_{0O}/(E_{0B}^2 + E_{0O}^2), \quad (14.11)$$

contains the appropriate information about the object wave's amplitude.

Once more, in the parlance of communications theory, we might observe that the film plate serves as both the storage device and detector or mixer. It produces, over its surface, a distribution of opaque regions corresponding to a modulated spatial waveform. Accordingly, the third or difference frequency term in Eq. (14.10) is both amplitude

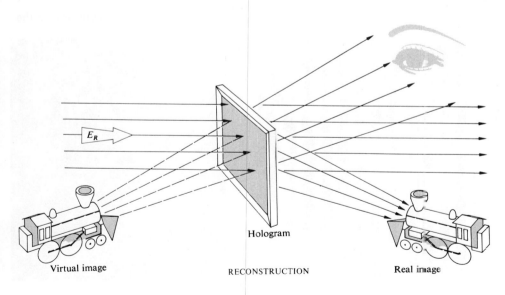

Fig. 14.36 Holographic (side-band) recording and reconstruction of an image.

and phase modulated by way of the position dependence of $E_{oo}(x, y)$ and $\phi_o(x, y)$.

We've shown the configuration utilizing diffusely reflected light from an opaque object but it could equally well be rearranged a bit, as in Fig. 14.37, to get sideband Fresnel holograms from transparent objects. Figure 14.38(b) is an enlarged view of a portion of the fringe pattern which constitutes the hologram for a simple, essentially two-dimensional, semitransparent object. Were the two interfering waves perfectly planar [as in Fig. 14.38(a)], the evident variations in fringe position and irradiance, which represent the information, would be absent, yielding the traditional Young's pattern (Section 9.3). The sinusoidal transmission-grating configuration [Fig. 14.38(a)] may be thought of as the carrier waveform which is then modulated by the signal. Furthermore. we can imagine that the coherent superposition of countless zone-plate patterns, one arising from each point on a large object, have metamorphosed into the modulated fringes of Fig. 14.38(b). When the amount of modulation is further greatly increased, as it would be for a large three-dimensional diffusely reflecting object, the fringes lose the kind of symmetry still discernible in Fig. 14.38(b) and become considerably more complicated. Incidentally, holograms are often covered with extraneous swirls and concentric ring systems that arise from diffraction by dust and the like on the optical elements.

The amplitude transmittance of the processed hologram can be made proportional to $I(x, y)$. In that case, the *final*

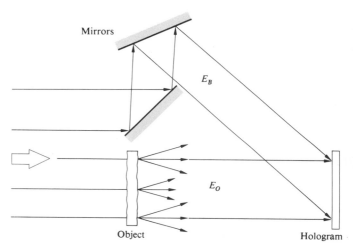

Fig. 14.37 A side-band Fresnel holographic set-up for a transparent object.

emerging wave, $E_F(x, y)$, is proportional to the product $I(x, y)E_R(x, y)$, where $E_R(x, y)$ is the *reconstructing wave* incident on the hologram. Thus if the reconstructing wave, of frequency ν, is incident obliquely on Σ_H as was the background wave, we can write

$$E_R(x, y) = E_{0R} \cos\left[2\pi\nu t + \phi(x, y)\right]. \qquad (14.12)$$

The final wave (except for a multiplicative constant) is the

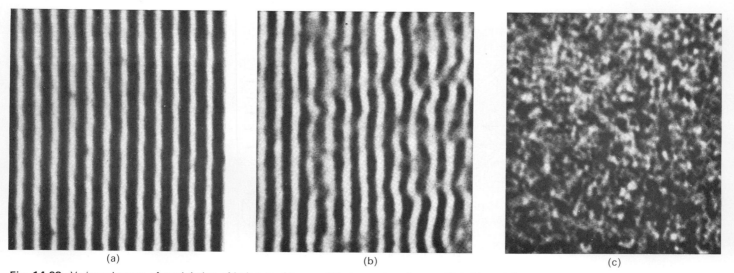

(a)   (b)   (c)

Fig. 14.38 Various degrees of modulation of hologram fringes. [Photo courtesy Emmett N. Leith and *Scientific American*.]

product of Eqs. (14.10) and (14.12):

$$E_F(x, y) = \tfrac{1}{2} E_{0R}(E_{0B}^2 + E_{0O}^2) \cos[2\pi\nu t + \phi(x, y)]$$
$$+ \tfrac{1}{2} E_{0R} E_{0B} E_{0O} \cos(2\pi\nu t + 2\phi - \phi_O)$$
$$+ \tfrac{1}{2} E_{0R} E_{0B} E_{0O} \cos(2\pi\nu t + \phi_O). \qquad (14.13)$$

Three terms describe the light issuing from the hologram; the first can be rewritten as

$$\tfrac{1}{2}(E_{0B}^2 + E_{0O}^2) E_R(x, y),$$

and is an amplitude-modulated version of the reconstructing wave. In effect, each portion of the hologram functions as a diffraction grating and this is the *zeroth-order*, undeflected, direct beam. Since it contains no information about the phase of the object wave, ϕ_O, it is of little concern here.

The next two or *side band waves* are the sum and difference terms, respectively. These are the two *first-order waves* diffracted by the grating-like hologram. The first of these, i.e., the sum term, represents a wave which, except for a multiplicative constant, has the same amplitude as the object wave $E_{0O}(x, y)$. Moreover, its phase contains a $2\phi(x, y)$ contribution which, as you recall, arose from tilting the background and reconstructing wavefronts with respect to Σ_H. It's this phase factor which provides the angular separation between the real and virtual images. Furthermore, rather than containing the phase of the object wave, the sum term contains its negative. Thus it's a wave carrying all of the appropriate information about the object but in a way which is not quite right. Indeed, this is the real image formed in converging light in the space beyond the hologram, i.e. between it and the viewer. The negative phase is manifest in an inside-out image something like the pseudoscopic effect occurring when the elements of a photographic stereo pair are interchanged. Bumps appear as indentations and object points which were in front of and nearer to Σ_H are now imaged nearer to but beyond Σ_H. Thus a point on the original subject closest to the observer appears farthest away in the real image. The scene is turned in on itself along one axis in a way that perhaps must be seen to be appreciated. For example, imagine you are looking down the holographic conjugate image of a bowling alley. The "back" row of pins, even though partially obscured by the "front" rows, are nonetheless imaged closer to the viewer than is the one-pin. Despite this, bear in mind that it's not as if you were looking at the array from behind. No light from the very backs of the pins was ever recorded—you're seeing an inside-out front view. As a consequence the conjugate

image is usually of limited utility although it can be made to have a normal configuration by forming a second hologram with the real image as the object.

The difference term in Eq. (14.13), except for a multiplicative constant, has precisely the form of the object wave $E_O(x, y)$. If you were to peer into (not at) the illuminated hologram, as if it were a window looking out onto the scene beyond, you would "see" the object exactly as if it were truly sitting there. You could move your head a bit and look around an item in the foreground in order to see the view it had previously been obstructing. In other words, in addition to complete three-dimensionality, parallax effects are apparent as they are in no other reproducing technique (Fig. 14.39) Imagine that you are viewing the holographic image of a magnifying glass focused on a page of print. As you move your eye with respect to the hologram plane, the words being magnified by the lens (which is itself just an image) actually change just as they would in "real" life with a "real" lens and "real" print. In the case of an extended scene having considerable depth, your eyes would have to refocus as you viewed different regions of it at various distances. In precisely the same way, a camera lens would have to be readjusted if you were photographing those regions of the virtual image.

There are several other extremely interesting features which holograms display. For example, if you were standing close to a window you could obscure all of it with, say, a piece of cardboard except for a tiny area through which you could then peer and still see the objects beyond. The same is true of a hologram since each small fragment of it contains information about the entire object, at least as seen from that vantage point, and can reproduce, albeit with diminishing resolution, the entire image.

The zone-plate interpretation has been applicable to the various holographic schemes we've considered thus far and this regardless of whether the diffracted wave was of the *near-* or *far*-field variety (i.e. whether we had Fresnel or Fraunhofer holograms, respectively). Indeed it applies generally where the interferogram results from the super-positioning of the scattered spherical wavelets from each object point and a coherent plane or even spherical reference wave (provided the latter's curvature is different from that of the wavelets). An inherent failing, which these schemes therefore have in common, arises from the fact that the zone plate radii, R_m, vary as $m^{1/2}$ via Eq. (10.91). Thus the zone fringes are more densely packed farther from the center of each zone lens (i.e. at larger values of m). This is tantamount to an increasing spatial frequency of bright and dark rings

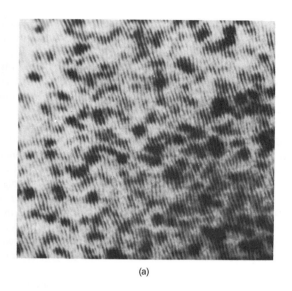

(a)

(b)

(c)

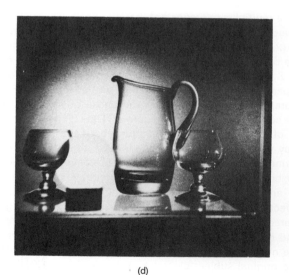

(d)

Fig. 14.39 (b), (c), (d). Three different views photographed from the same holographic image generated by the hologram in (a). [Photos from Smith, *Principles of Holography*.]

which must be recorded by the photographic plate. But since film, no matter how fine grained, is limited in its spatial frequency response, there will be a cut-off beyond which it cannot record data. All of this represents a built-in limitation on resolution. In contrast, if the mean frequency of the fringes could be made constant, the limitations imposed by the photographic medium would be considerably reduced and the resolution correspondingly increased. So long as it could record the average spatial fringe frequency even coarse emulsions, such as Polaroid P/N, could be used without extensive loss of resolution. Figure 14.40 shows an arrangement which accomplishes just this by having the

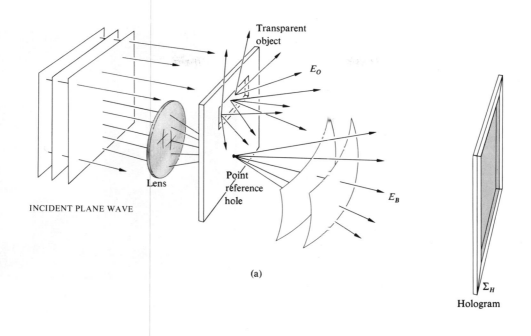

(a)

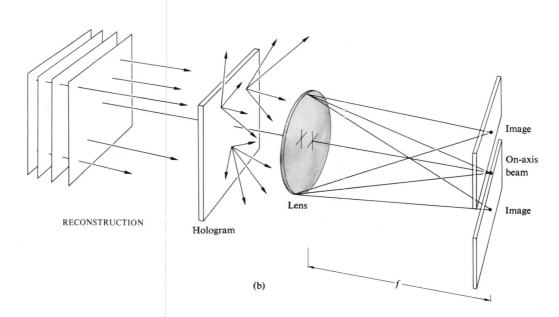

(b)

Fig. 14.40 Lensless Fourier transform holography (a transparent object).

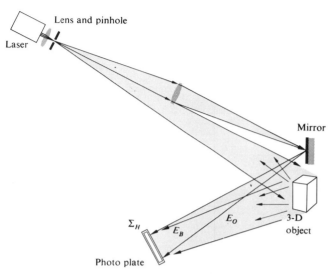

Fig. 14.41 Lensless Fourier transform holography (an opaque object).

Fig. 14.42 A reconstruction of a Fourier transform hologram. [From G. W. Stroke, D. Brumm and A. Funkhauser, *J. Opt. Soc. Am.* **55**, 1327 (1965).]

diffracted object wavelets interfere with a spherical reference wave of about the same curvature. The resulting interferogram is known as a *Fourier transform* hologram (in this specific instance, it's of the high-resolution *lensless* variety). This scheme is designed to have the reference wave cancel the quadratic (zone-lens type) dependence of the phase with position on Σ_H. But that will occur precisely only for a planar two-dimensional object. In the case of a three-dimensional object (Fig. 14.41) this only happens over one plane and the resulting hologram is therefore a composite of both types, i.e. a zone lens and Fourier transform. Unlike the other arrangements, both images generated by a Fourier-transform hologram are virtual, in the same plane and oriented as if reflected through the origin (Fig. 14.42).

The grating-like nature of all previous holograms is evident here as well. In fact, if you look through a Fourier-transform hologram at a small white-light source (a flashlight in a dark room works beautifully), you see the two mirror images but they are extremely vague and surrounded by bands of spectral colors. The similarity with white light which has passed through a grating is unmistakable.*

---

* See DeVelis and Reynolds, *Theory and Applications of Holography*, Stroke, *An Introduction to Coherent Optics and Holography*, Goodman, *Introduction to Fourier Optics*, Smith, *Principles of Holography* or perhaps *The Engineering Uses of Holography*, edited by E. R. Robertson and J. M. Harvey.

14.3.2 Recent Developments

For years holography was an invention in search of application, that notwithstanding certain obvious possibilities like the all too inevitable 3-D billboard. Fortunately, several significant technological developments have in recent times begun what will surely be an ongoing extension of the scope and utility of holography. The early efforts in the field were typified by countless images of toy cars and trains, chesspieces and statuettes—small objects resting on giant blocks of granite. They had to be small because of limited laser power and coherence length; while the ever-present massive granite platform served to isolate the slightest vibrations which might blur the fringes and thereby degrade or obliterate the stored data. A loud sound or gust of air could result in deterioration of the reconstructed image by causing the photo plate, object, or mirrors to shift several millionths of an inch during the exposure, which itself might last of the order of a minute or so. That was the still-life era of holography. But now, using new, more sensitive films and the short duration (~ 40 ns) high-power light flashes from a single-mode pulsed ruby laser, even portraiture and stop-action holography have become a reality* (Fig. 14.43).

---

* L. D. Siebert, *Appl. Phys. Letters* **11**, 326 (1967) and R. G. Zech and L. D. Siebert, *Appl. Phys. Letters* **13**, 417 (1968).

Fig. 14.43 A reconstruction of a holographic portrait.   [Photo courtesy L. D. Siebert.]

i) Volume Holograms

Yuri Nikolayevitch Denisyuk of the Soviet Union, in 1962, introduced a scheme for generating holograms which was conceptually similar to the early (1891) color photographic process of Gabriel Lippmann. In brief, the object wave is reflected from the subject and propagates backward, overlapping the incoming coherent background wave. In so doing, the two waves set up a three-dimensional pattern of standing waves.   The concomitant spatial distribution of fringes is, in part, recorded by the photoemulsion throughout its entire thickness to form what has become known as a *volume hologram*. Several variations have since been introduced, but the basic ideas are the same; rather than generating a two-dimensional grating-like structure, the volume hologram is a three-dimensional grating. In other words, it's a three-dimensional, modulated, periodic array of phase or amplitude objects which represent the data. It can be recorded in several media, for example, in thick photoemulsions wherein the amplitude objects are grains of deposited silver; in photochromic glass; with halogen crystals like KBr which respond to irradiation via color-center variations; or with a ferroelectric crystal such as lithium niobate which undergoes local alterations in its index of refraction, thus forming what might be called a phase volume hologram. In any event, one

is left with a volume array of data, however stored in the medium which, in the reconstruction process, behaves very much like a crystal being irradiated by x-rays. It scatters the incident (reconstructing) wave according to Bragg's law (Section 10.2.7). This isn't very surprising since both the scattering centers and λ have simply been scaled up proportionately. One important feature of volume holograms is the interdependence [via Bragg's law, $2d \sin \theta = m\lambda$ (10.71)] of the wavelength and the scattering angle, i.e. only a given color light will be diffracted at a particular angle by the hologram. Another significant property is that by successively altering the incident angle (or the wavelength), a single-volume medium can store a great many coexisting holograms at one time. This latter property makes such systems extremely appealing as densely packed memory devices. A single lithium niobate crystal is capable of easily storing thousands of holograms and any one of them could be replayed by addressing the crystal with a laser beam at the appropriate angle. Imagine a 3-D holographic motion picture; a library; or everyone's vital statistics—beauty marks, credit cards, taxes, bad habits, income, life history, etc. all recorded on a handful of small transparent crystals.

Multicolored reconstructions have been formed using (black and white) volume holographic plates. Two, three or more different colored and mutually incoherent overlapping laser beams are used to generate separate, cohabitating, component holograms of the object and this can be done one at a time or all at once. When these are illuminated simultaneously by the various constituent beams, a multicolored image results.

Another important and highly promising scheme devised by G. W. Stroke and A. E. Labeyrie is known as *white-light reflection holography*. Here, the reconstructing wave is an ordinary white-light beam from say, a flashlight or projector, having a wavefront similar to the original quasimonochromatic background wave. When illuminated on the same side as the viewer, only the specific wavelength which enters the volume hologram at the proper Bragg angle is reflected off to form a reconstructed 3-D virtual image. Thus if the scene were recorded in red laser light, only red light is presumably reflected as an image. It is of pedagogical interest to point out, however, that the emulsion may shrink during the fixing process and if it is not swollen back to its original form chemically (with say triethylnolamine) the spacing of the Bragg planes, d, decreases. That means that at a given angle θ, the reflected wavelength will decrease proportionately. Hence, a scene recorded in He–Ne red might play back in orange or even green when reconstructed by a beam of white light.

If several overlapping holograms corresponding to different wavelengths are stored, a multicolored image will result. The advantages of using an ordinary source of white light to reconstruct full-color 3-D images are obvious and far reaching.

ii) Holographic Interferometry

One of the most innovative and practical of recent holographic advances is in the area of interferometry. Three distinctive approaches have proved to be quite useful in a wealth of nondestructive testing situations where, for example, one might wish to study microinch distortions in an object resulting from strain, vibration, heat etc. In the *double exposure* technique, one simply makes a hologram of the undisturbed object and then, before processing, exposes the hologram for a second time to the light coming from the now distorted object. The ultimate result is two overlapping reconstructed waves which proceed to form a fringe pattern indicative of the displacements suffered by the object, i.e. the changes in optical path length (Fig. 14.44). Variations in index such as those arising in wind tunnels and the like will generate the same sort of pattern.

In the *real-time* method, the subject is left in its original

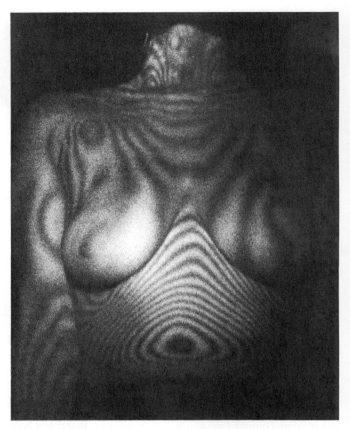

Fig. 14.44 Double exposure holographic interferogram. [From S. M. Zivi and G. H. Humberstone, "Chest Motion Visualized by Holographic Interferometry," *Medical Research Eng.* p.5 (June 1970).]

position throughout; a processed hologram is formed and the resulting virtual image is made to precisely overlap the object (Fig. 14.45). Any distortions which arise during subsequent testing show up, on looking through the hologram, as a system of fringes which can be studied as they evolve in real time. The method applies to both opaque and transparent objects. Motion pictures can be taken to form a continuous record of the response.

The third method is the *time-average* approach and is particularly applicable to rapid, small amplitude, oscillatory systems. Here the film plate is exposed for a relatively long duration, during which time the vibrating object has executed a number of oscillations. The resulting hologram

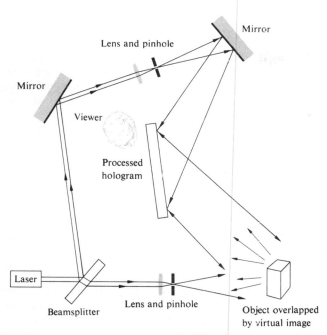

Fig. 14.45 Real-time holographic interferometry.

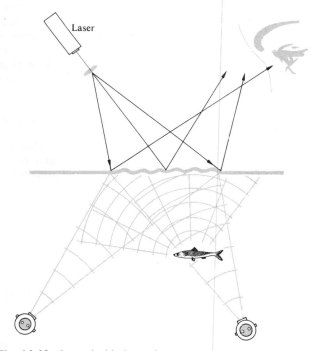

Fig. 14.46 Acoustical holography.

can be thought of as a superposition of a multiplicity of images with the effect that a standing-wave pattern emerges. Bright areas reveal undeflected or stationary nodal regions while contour lines trace out areas of constant vibrational amplitude.

iii) More to Come

Two recent developments seem particularly worthy of note, if only briefly; one is *acoustical* holography, the other, *computer-generated* holography.

In acoustical holography, an ultra high-frequency sound wave (ultrasound) is used to create the hologram initially and a laser beam then serves to form a recognizable reconstructed image. In one application, the stationary ripple pattern on the surface of a water body produced by submerged coherent transducers corresponds to a hologram of the object beneath (Fig. 14.46). Photographing it creates a hologram that can be illuminated optically to form a visual image. Alternatively, the ripples can be irradiated from above with a laser beam to produce an instantaneous reconstruction in reflected light.

The advantages of acoustical techniques reside in the fact that sound waves can propagate considerable distances in dense liquids and solids where light cannot. Thus acoustical holograms can record such diverse things as underwater submarines and internal body organs.* In the case of Fig. 14.46 one would see something that resembled an x-ray motion picture of the fish.

It is possible to synthesize, point by point, a hologram of a fictitious object. In other words, in the most direct approach holograms can be produced by calculating, via digital computer, the irradiance distribution which would arise were some object appropriately illuminated in a hypothetical recording session. A computer-controlled plotter drawing or cathode ray tube read out of the interferogram is then photographed, thence to serve as the actual hologram. The result upon illumination is a three-dimensional reconstructed image of an object which never had any real existence in the first place.†

---

* See A. F. Metherell, "Acoustical Holography," *Sci. Amer.* **221**, 36 (October 1969). Refer to A. L. Dalisa et al., "Photoanodic Engraving of Holograms on Silicon," *Appl. Phys. Letters* **17**, 208 (1970) for another interesting use of surface relief patterns.

† See e.g. L. B. Lesem, P. M. Hirsch and J. A. Jordan, Jr., "The Promise of the Kinoform," *Optical Spectra* (December 1970), p. 18.

14.4 NONLINEAR OPTICS

Generally, the domain of *nonlinear optics* is understood to encompass those phenomena for which electric and magnetic field intensities of higher powers than the first play a dominant role. The Kerr effect (Section 8.11.3) which is a quadratic variation of refractive index with applied voltage, and thereby electric field, is typical of several long-known nonlinear effects.

The usual classical treatment of the propagation of light—superposition, reflection, refraction, etc. presumes a linear relationship between the electromagnetic light field and the responding atomic system comprising the medium. But just as an oscillatory mechanical device (e.g. a weighted spring) can be overdriven into nonlinear response through the application of large enough forces, so too we might anticipate that an extremely intense beam of light could generate appreciable nonlinear optical effects. The electric fields associated with light beams from ordinary or, if you will, traditional sources are far too small for such behavior to be easily observable. It was for this reason, coupled with an initial lack of technical prowess, that the subject had to patiently await the advent of the laser in order that sufficient brute force could be brought to bear in the optical region of the spectrum. As an example of the kinds of fields readily obtainable with the current technology, consider that a good lens can focus a laser beam down to a spot having a diameter of about 10^{-3} inch or so, which corresponds to an area of roughly 10^{-9} m^2. A 200 megawatt pulse, from say a Q-switched ruby laser, would then produce a flux density of 20×10^{16} W/m^2. It follows (Problem 14.6) from Section 3.4.1 that the corresponding electric-field amplitude is given by

$$E_0 = 27.4 \left(\frac{I}{n}\right)^{1/2}. \tag{14.14}$$

In this particular case, for $n \approx 1$, the field amplitude is about 1.2×10^8 V/m. This is more than enough to cause the breakdown of air (roughly 3×10^6 V/m) and just several orders of magnitude less than the typical fields holding a crystal together, the latter being roughly about the same as the cohesive field on the electron in a hydrogen atom (5×10^{11} V/m). The availability of these and even greater (10^{12} V/m) fields have made possible a wide range of important new nonlinear phenomena and devices. We shall limit this discussion to the consideration of several nonlinear phenomena associated with passive media (i.e. media which act essentially as catalysts without making their own

characteristic frequencies evident). Specifically, we'll consider optical rectification, optical harmonic generation, frequency mixing and self-focusing of light. In contrast, stimulated Raman, Rayleigh and Brillouin scattering (Section 13.8) exemplify nonlinear optical phenomena arising in active media which do impose their characteristic frequencies on the light wave.*

As you may recall (Section 3.3.1), the electromagnetic field of a light wave propagating through a medium exerts forces on the loosely bound, outer or valence electrons. Ordinarily these forces are quite small and in a linear isotropic medium the resulting electric polarization is parallel with and directly proportional to the applied field. In effect, the polarization follows the field; if the latter is harmonic, the former will be harmonic as well. Consequently, one can write

$$P = \epsilon_0 \chi E, \tag{14.15}$$

where χ is a dimensionless constant known as the electric susceptibility, and a plot of P versus E is a straight line. Quite obviously in the extreme case of very high fields, we can expect that P will become saturated, i.e. it simply cannot increase linearly indefinitely with E (just as in the familiar case of ferromagnetic materials where the magnetic moment becomes saturated at fairly low values of H). Thus we can anticipate a gradual increase of the ever-present, but usually insignificant, nonlinearity as E increases. Since the directions of **P** and **E** coincide in the simplest case of an isotropic medium, we can express the polarization more effectively as a series expansion:

$$P = \epsilon_0(\chi E + \chi_2 E^2 + \chi_3 E^3 + \cdots). \tag{14.16}$$

The usual linear susceptibility, χ, is much greater than the coefficients of the nonlinear terms χ_2, χ_3, etc. and hence the latter contribute noticeably only at high-amplitude fields. Now suppose that a light wave of the form

$$E = E_0 \sin \omega t$$

is incident on the medium. The resulting electric polarization

$$P = \epsilon_0 \chi E_0 \sin \omega t + \epsilon_0 \chi_2 E_0^2 \sin^2 \omega t + \epsilon_0 \chi_3 E_0^3 \sin^3 \omega t + \cdots \tag{14.17}$$

---

* For a more extensive treatment than is possible here, see N. Bloembergen, *Nonlinear Optics* or G. C. Baldwin, *An Introduction to Nonlinear Optics*.

can be rewritten as

$$P = \epsilon_0 \chi E_0 \sin \omega t + \frac{\epsilon_0 \chi_2}{2} E_0^2 (1 - \cos 2\omega t)$$

$$+ \frac{\epsilon_0 \chi_3}{4} E_0^3 (3 \sin \omega t - \sin 3\omega t) + \cdots \quad (14.18)$$

As the harmonic light wave sweeps through the medium, it creates what might be thought of as a polarization wave, i.e. an undulating redistribution of charge within the material in response to the field. If only the linear term were effective, the electric polarization wave would correspond to an oscillatory current following along with the incident light. The light thereafter reradiated in such a process would be the usual refracted wave generally propagating with a reduced speed v (Section 3.3.2) and having the same frequency as the incident light. In contrast, the presence of higher-order terms in Eq. (14.17) implies that the polarization wave certainly does have the same harmonic profile as the incident field. In fact, Eq. (14.18) can be likened to a Fourier series representation of the distorted profile of $P(t)$.

14.4.1 Optical Rectification

The second term in Eq. (14.18) has two components of great interest. First there is a *dc* or *constant bias polarization* varying as E_0^2. Consequently, if an intense plane-polarized beam traverses an appropriate (piezoelectric) crystal, the presence of the quadratic nonlinearity will, in part, be manifest by a constant electric polarization of the medium. A voltage difference, proportional to the beam's flux density, will accordingly appear across the crystal. This effect, in analogy to its radiofrequency counterpart, is known as *optical rectification*.

14.4.2 Second Harmonic Generation

The $\cos 2\omega t$ term (14.18) corresponds to a variation in electric polarization at twice the fundamental frequency, i.e. at twice that of the incident wave. The reradiated light which arises from the driven oscillators also has a component at this same frequency, 2ω, and the process is spoken of as *second-harmonic generation* or SHG for short. In terms of the photon representation we can envision two identical photons of energy $\hbar\omega$ coalescing within the medium to form a single photon of energy $\hbar 2\omega$. Peter A. Franken and several coworkers at the University of Michigan in 1961 were the first to experimentally observe SHG. They focused a 3 kW pulse of red (694.3 nm) ruby laser light onto a quartz crystal. Just about one part in 10^8 of this incident wave was converted to the 347.15 nm ultraviolet second harmonic.

Notice that, for a given material, if $P(E)$ is an odd function, i.e. if reversing the direction of the **E**-field simply reverses the direction of **P**, the even powers of E in Eq. 14.16 must vanish. But this is just what happens in an isotropic medium such as glass or water—there are no special directions in a liquid. Moreover, in crystals like calcite which are so structured as to have what's known as a *center of symmetry* or an *inversion center*, a reversal of all of the coordinate axes must leave the interrelationships between physical quantities unaltered. Thus no even harmonics can be produced by materials of this sort. Third-harmonic generation (THG), however, can exist and has been observed, for example, in calcite. The requirement for SHG that a crystal not have inversion symmetry is also necessary for it to be piezoelectric. Under pressure a piezoelectric crystal [such as quartz, potassium dihydrogen phosphate (KDP) or ammonium dihydrogen phosphate (ADP)] undergoes an asymmetric distortion of its charge distribution thus producing a voltage. Of the 32 crystal classes, 20 are of this kind and may therefore be useful in SHG. The simple scalar expression (14.16) is actually not an adequate description of a typical dielectric crystal. Things are a good deal more complicated because the field components in several different directions in a crystal can affect the electric polarization in any one direction. A complete treatment requires that **P** and **E** be related, not by a single scalar, but by a group of quantities arranged in the particular form of a tensor, namely the susceptibility tensor.[*]

A major difficulty in generating copious amounts of second-harmonic light arises from the frequency dependence of the refractive index, i.e. dispersion. At some initial point where the incident or ω-wave generates the second harmonic or 2ω-wave, the two are coherent. As the ω-wave propagates through the crystal, it continues to generate additional contributions of second-harmonic light which all combine totally constructively only if they maintain a proper phase relationship. Yet the ω-wave travels at a phase velocity v_ω which is ordinarily different from the phase velocity, $v_{2\omega}$, of the 2ω-wave. Thus the newly emitted second harmonic periodically falls out of phase with some of the previously generated 2ω-waves. When the irradiance

---

[*] Incidentally, there is nothing extraordinary about this kind of behavior—it comes up all the time. There are inertia tensors, demagnetization coefficient tensors, stress tensors, etc.

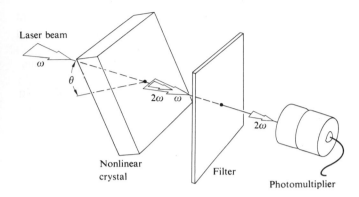

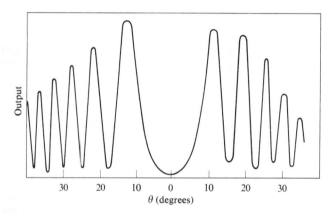

Fig. 14.47 Second harmonic generation as a function of θ for a 0.78 mm thick quartz plate. Peaks occur when the effective thickness is an even multiple of ℓ_c. [From P. D. Maker, R. W. Terhune, M. Nisenoff and C. M. Savage, *Phys. Rev. Letters* **8**, 21 (1962).]

of the second harmonic, $I_{2\omega}$, emerging from a plate of thickness ℓ is computed* it turns out to be

$$I_{2\omega} \propto \frac{\sin^2 [2\pi (n_\omega - n_{2\omega}) \ell / \lambda_0]}{(n_\omega - n_{2\omega})^2} \qquad (14.19)$$

(see Fig. 14.47). This yields the result that $I_{2\omega}$ has its maximum value when $\ell = \ell_c$, where

$$\ell_c = \frac{1}{4} \frac{\lambda_0}{|n_\omega - n_{2\omega}|}. \qquad (14.20)$$

---

* See e.g. B. Lengyel, *Introduction to Laser Physics*, Chapter VII. This is a fine elementary treatment.

This is commonly known as the *coherence length* (although a different name would perhaps be better) and it's usually of the order of only about $20\lambda_0$. Despite this, efficient SHG can be accomplished by a procedure known as *index matching*, which negates the undesirable effects of dispersion; in short, one arranges things so that $n_\omega = n_{2\omega}$. A commonly used SHG material is KDP. It is piezoelectric, transparent and also negatively uniaxially birefringent. Furthermore, it has the interesting property that if the fundamental light is a linearly polarized *ordinary wave*, the resulting second harmonic will be an *extraordinary wave*. As can be seen from Fig. 14.48, if light propagates within a KDP crystal at the specific angle θ_0 with respect to the optic axis, the index, $n_{o\omega}$, of the ordinary fundamental wave will precisely equal the index of the extraordinary second harmonic $n_{e2\omega}$. The second-harmonic wavelets will then interfere constructively thereby increasing the conversion efficiency by several orders of magnitude to roughly 20%. Second-harmonic generators, which are simply appropriately cut and oriented crystals, are available commercially, but do keep in mind that θ_0 is a function of λ and each such device performs at one frequency. In recent times, a continuous 1 W second-harmonic beam at 532.3 nm was obtained by placing a barium sodium niobate crystal within the cavity of a 1 W 1.06μ laser. The fact that the ω-wave sweeps back and forth through the crystal increases the net conversion efficiency.

14.4.3 Frequency Mixing

Another situation of considerable practical interest involves the *mixing* of two or more primary beams of different frequencies within a nonlinear dielectric. The process can most easily be appreciated by substituting a wave of the form

$$E = E_{01} \sin \omega_1 t + E_{02} \sin \omega_2 t \qquad (14.21)$$

into the simplest expression for P given by Eq. (14.16). The second-order contribution is then

$$\epsilon_0 \chi_2 (E_{01}^2 \sin^2 \omega_1 t + E_{02}^2 \sin^2 \omega_2 t + 2 E_{01} E_{02} \sin \omega_1 t \sin \omega_2 t).$$

The first two terms can be expressed as functions of $2\omega_1$ and $2\omega_2$ respectively, while the last quantity gives rise to sum and difference terms, $\omega_1 + \omega_2$ and $\omega_1 - \omega_2$.

As for the quantum picture, the photon of frequency $\omega_1 + \omega_2$ simply corresponds to a coalescing of the two original photons into a new photon, just as it did in the case of SHG, where both quanta had the same frequency. The energy and momentum of the annihilated photons is carried off by the created sum photon. The generation of an

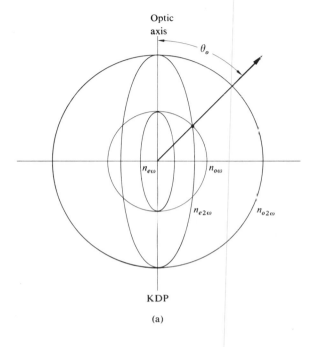

Optic
axis

θ_o

$n_{e\omega}$ $n_{o\omega}$

$n_{e2\omega}$ $n_{o2\omega}$

KDP

(a)

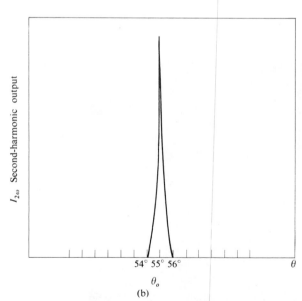

$I_{2\omega}$ Second-harmonic output

54° 55° 56° θ

θ_o

(b)

Fig. 14.48 Refractive index surface for KDP. (b) $I_{2\omega}$ versus crystal orientation in KDP. [From Maker *et al.*]

$\omega_1 - \omega_2$ difference photon is a little more involved. Conservation of energy and momentum requires that on interacting with an ω_2-photon only the higher-frequency ω_1-photon vanishes, thereby creating two new quanta, one an ω_2-photon and the other a difference-photon.

As an application of this phenomenon, suppose we beat, within a nonlinear crystal, a strong wave of frequency ω_p called the *pump light* with a weak *signal wave* of lower frequency ω_s which is to be amplified. Pump light is thereby converted into both signal light and a difference wave called *idler light*, of frequency $\omega_i = \omega_p - \omega_s$. Now if the idler light is then made to beat with the pump light, the latter is converted into additional amounts of idler and signal light. In this way both the signal and idler waves are amplified. This is actually an extension into the optical-frequency region of the well-known concept of *parametric amplification*, whose use in the microwave spectrum dates back to the late nineteen-forties. The first *optical-parametric oscillator* which was operated in 1965 is depicted in Fig. 14.49. The flat parallel end faces of a nonlinear crystal (lithium niobate) were coated to form an optical Fabry–Perot cavity. The signal and idler frequencies (both about 1000 nm) corresponded to two of the resonant frequencies of the cavity. When the flux density of the pumping light was high enough, energy was transferred from it into the signal and idler oscillatory modes with the consequent build up of those modes and emission of coherent radiant energy at those frequencies. This transfer of energy from one wave to another within a lossless medium typifies parametric processes. By changing the refractive index of the crystal (via temperature, electric field, etc.) the oscillator becomes tunable. Various oscillator configurations have since evolved, using other nonlinear materials as well, such as barium sodium niobate. The optical parametric oscillator is a laser-like, broadly tunable, source of coherent radiant energy in the IR to the UV.

14.4 4   Self-Focusing of Light

When a dielectric is subjected to an electric field which varies in space, i.e. when there is a gradient of the field parallel to **P**, an internal force will result. This has the effect of altering the density, changing the permittivity and thereby varying the refractive index, and this in both linear and nonlinear isotropic media. Suppose then that we shine an intense laser beam with a transverse Gaussian flux-density distribution onto a specimen. The induced refractive-index variations will cause the medium in the region of the beam to function much as if it were a positive lens. Accordingly,

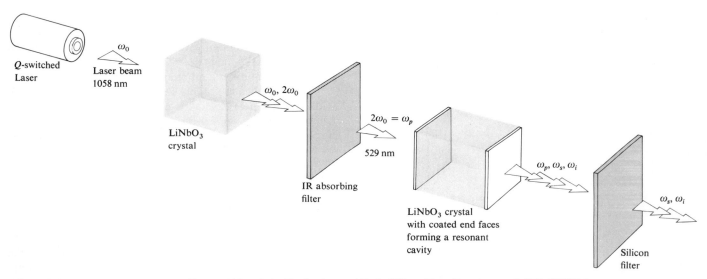

Fig. 14.49 An optical parametric oscillator. [After J. A. Giordmaine and R. C. Miller, *Phys. Rev. Letters* **4**, 973 (1965).]

the beam contracts, the flux density increases even more and the contraction continues in a process known as *self-focusing*. The effect can be sustained until the beam reaches a limiting filament diameter (of about 5×10^{-6} m) being totally internally reflected as if it were in a fiber optical element imbedded within the medium.*

PROBLEMS

14.1 What would the pattern look like for a laser beam diffracted by the three crossed gratings of Fig. 14.50?

14.2 Make a rough sketch of the Fraunhofer diffraction pattern that would arise if a transparency of Fig. 14.51(a) served as the object. How would you filter it to get Fig. 14.51(b)?

14.3 Repeat the previous problem using Fig. 14.52 instead.

14.4* Repeat the previous problem using Fig. 14.53 this time.

14.5 Returning to Fig. 14.10, what kind of spatial filter would produce each of the patterns shown in Fig. 14.54?

---

* See J. A. Giordmaine, "Nonlinear Optics," *Physics Today* (January 1969), p. 39.

Fig. 14.50

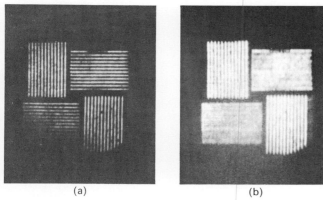

Fig. 14.51 [Photos courtesy R. A. Phillips.]

(a) (b)

Fig. 14.52 [Photos courtesy R. A. Phillips.]

(a)

LOOK AT YOUR FUTURE

PPG representatives will interview at
University of Minnesota on

(b)

LOOK AT YOUR FUTURE

PPG representatives will interview at

Fig. 14.53 [Photos courtesy R. A. Phillips.]

(a)

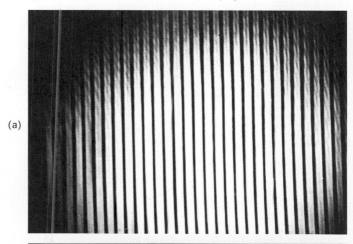

(b)

Fig. 14.54 [Photos courtesy D. Dutton, M. P. Givens and R. E. Hopkins.]

14.6 Show that the maximum electric field intensity, E_{max}, which exists for a given irradiance I is

$$E_{max} = 27.4 \left(\frac{I}{n}\right)^{1/2} \text{ in units of V/m}$$

where n is the refractive index of the medium.

14.7* The arrangement shown in Fig. 14.55 is used to convert a collimated laser beam into a spherical wave. The pinhole cleans up the beam, i.e. it eliminates diffraction effects due to dust and the like on the lens. How does it manage it?

14.8 What would happen to the speckle pattern if a laser beam were projected onto a suspension such as milk rather than a smooth wall?

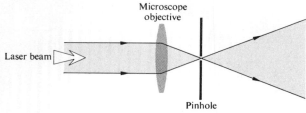

Fig. 14.55

Appendix 1
Electromagnetic Theory

MAXWELL'S EQUATIONS IN DIFFERENTIAL FORM

The set of integral expressions which have come to be known as Maxwell's equations are

$$\oint_C \mathbf{E} \cdot d\mathbf{l} = -\iint_A \frac{\partial \mathbf{B}}{\partial t} \cdot d\mathbf{S} \qquad [3.5]$$

$$\oint_C \mathbf{B} \cdot d\mathbf{l} = \mu \iint_A \left(\mathbf{J} + \epsilon \frac{\partial \mathbf{E}}{\partial t} \right) \cdot d\mathbf{S} \qquad [3.13]$$

$$\oiint_A \mathbf{E} \cdot d\mathbf{S} = \frac{1}{\epsilon} \iiint_V \rho \, dV \qquad [3.7]$$

and

$$\oiint_A \mathbf{B} \cdot d\mathbf{S} = 0, \qquad [3.9]$$

where the units, as usual, are MKS.

Maxwell's equations can be written in a differential form which are more useful for deriving the wave aspects of the electromagnetic field. This transition can readily be accomplished by making use of two theorems from vector calculus, namely Gauss's divergence theorem,

$$\oiint_A \mathbf{F} \cdot d\mathbf{S} = \iiint_V \mathbf{\nabla} \cdot \mathbf{F} \, dV \qquad (A1.1)$$

and Stokes' theorem

$$\oint_C \mathbf{F} \cdot d\mathbf{l} = \iint_A \mathbf{\nabla} \times \mathbf{F} \cdot d\mathbf{S}. \qquad (A1.2)$$

Here the quantity $\mathbf{F}$ is not one fixed vector, but a function which depends on the position variables. It is a rule which associates a single vector, e.g. in Cartesian coordinates, $\mathbf{F}(x, y, z)$ with each point (x, y, z) in space. Vector-valued functions of this kind such as $\mathbf{E}$ and $\mathbf{B}$ are known as vector fields.

Applying Stokes' theorem to the electric field intensity we have

$$\oint \mathbf{E} \cdot d\mathbf{l} = \iint \mathbf{\nabla} \times \mathbf{E} \cdot d\mathbf{S}. \qquad (A1.3)$$

Upon comparing this with Eq. (3.5) it follows that

$$\iint \mathbf{\nabla} \times \mathbf{E} \cdot d\mathbf{S} = -\iint \frac{\partial \mathbf{B}}{\partial t} \cdot d\mathbf{S}. \qquad (A1.4)$$

This result must be true for all surfaces bounded by the path C. This can only be the case if the integrands are themselves equal, .e. if

$$\mathbf{\nabla} \times \mathbf{E} = -\frac{\partial \mathbf{B}}{\partial t}. \qquad (A1.5)$$

A similar application of Stokes' theorem to $\mathbf{B}$, using Eq. (3.13) results in

$$\mathbf{\nabla} \times \mathbf{B} = \mu \left(\mathbf{J} + \epsilon \frac{\partial \mathbf{E}}{\partial t} \right). \qquad (A1.6)$$

Gauss's divergence theorem applied to the electric field intensity yields

$$\oiint \mathbf{E} \cdot d\mathbf{S} = \iiint \mathbf{\nabla} \cdot \mathbf{E} \, dV. \qquad (A1.7)$$

If we make use of Eq. (3.7) this becomes

$$\iiint_V \mathbf{\nabla} \cdot \mathbf{E} \, dV = \frac{1}{\epsilon} \iiint_V \rho \, dV, \qquad (A1.8)$$

and since this is to be true for any volume (i.e. for an arbitrary closed domain) the two integrands must be equal. Consequently, at any point (x, y, z, t) in space—time

$$\mathbf{\nabla} \cdot \mathbf{E} = \rho / \epsilon. \qquad (A1.9)$$

In the same fashion Gauss's divergence theorem applied to the $\mathbf{B}$-field and combined with Eq. 3.9 yields

$$\mathbf{\nabla} \cdot \mathbf{B} = 0. \qquad (A1.10)$$

Equations (A1.5), (A1.6), (A1.9) and (A1.10) are Maxwell's equations in differential form. Refer back to Eqs. (3.18) through (3.21) for the simple case of Cartesian coordinates and *free space* ($\rho = J = 0, \epsilon = \epsilon_0, \mu = \mu_0$).

ELECTROMAGNETIC WAVES

To derive the electromagnetic wave equation in its most general form, we must again consider the presence of some medium. We saw in Section 3.3.1 that there is a need to introduce the *polarization* vector **P** which is a measure of the overall behavior of the medium in that it is the resultant electric dipole moment per unit volume. Since the field within the material has been altered, we are led to define a new field quantity, the *displacement* **D**:

$$\mathbf{D} = \epsilon_0 \mathbf{E} + \mathbf{P}. \qquad (A1.11)$$

Clearly then,

$$\mathbf{E} = \frac{\mathbf{D}}{\epsilon_0} - \frac{\mathbf{P}}{\epsilon_0}.$$

The internal electric field **E** is the difference between the field $\mathbf{D}/\epsilon_0$, which would exist in the absence of polarization, and the field $\mathbf{P}/\epsilon_0$ arising from polarization.

For a homogeneous, linear, isotropic dielectric, **P** and **E** are in the same direction and are mutually proportional. It follows that **D** is therefore also proportional to **E**:

$$\mathbf{D} = \epsilon \mathbf{E}. \qquad (A1.12)$$

Like **E**, **D** extends throughout space and is in no way limited to the region occupied by the dielectric as is **P**. The lines of **D** begin and end on free, movable charges. Those of **E** begin and end on either free charges or bound polarization charges. If no free charge is present, as might be the case in the vicinity of a polarized dielectric or in free space, the lines of **D** close on themselves.

Since in general the response of optical media to **B**-fields is only slightly different from that of a vacuum, we need not describe the process in detail. Suffice it to say that the material will become polarized. We can define a *magnetic polarization* or *magnetization* vector **M** as the magnetic dipole moment per unit volume. In order to deal with the influence of the magnetically polarized medium, we introduce an auxiliary vector **H**, traditionally known as the *magnetic field intensity*

$$\mathbf{H} = \mu_0^{-1} \mathbf{B} - \mathbf{M}. \qquad (A1.13)$$

For a homogeneous linear (nonferromagnetic), isotropic medium **B** and **H** are parallel and proportional:

$$\mathbf{H} = \mu^{-1} \mathbf{B}. \qquad (A1.14)$$

Along with Eqs. (A1.12) and (A1.14) there is one more *constitutive equation*,

$$\mathbf{J} = \sigma \mathbf{E}. \qquad (A1.15)$$

Known as *Ohm's law*, it is a statement of an experimentally determined rule which holds for conductors at constant temperatures. The electric-field intensity and therefore the force acting on each electron in a conductor determines the flow of charge. The constant of proportionality relating **E** and **J** is the conductivity of the particular medium, σ.

Consider the rather general environment of a linear (nonferroelectric and nonferromagnetic), homogeneous, isotropic medium, which is physically at rest. By making use of the constitutive relations, Maxwell's equations can be rewritten as

$$\nabla \cdot \mathbf{E} = \rho / \epsilon \qquad [A1.9]$$

$$\nabla \cdot \mathbf{B} = 0 \qquad [A1.10]$$

$$\nabla \times \mathbf{E} = -\frac{\partial \mathbf{B}}{\partial t} \qquad [A1.5]$$

and

$$\nabla \times \mathbf{B} = \mu \sigma \mathbf{E} + \mu \epsilon \frac{\partial \mathbf{E}}{\partial t}. \qquad (A1.16)$$

If these expressions are somehow to yield a wave equation (2.61), we had best form some second derivatives with respect to the space variables. Taking the curl of Eq. (A1.16) we obtain

$$\nabla \times (\nabla \times \mathbf{B}) = \mu \sigma (\nabla \times \mathbf{E}) + \mu \epsilon \frac{\partial}{\partial t}(\nabla \times \mathbf{E}), \qquad (A1.17)$$

where, since **E** is assumed to be a well-behaved function, the space and time derivatives can be interchanged. Equation (A1.5) can be substituted to obtain the needed second derivative with respect to time:

$$\nabla \times (\nabla \times \mathbf{B}) = -\mu \sigma \frac{\partial \mathbf{B}}{\partial t} - \mu \epsilon \frac{\partial^2 \mathbf{B}}{\partial t^2} \qquad (A1.18)$$

The vector triple product can be simplified by making use of the operator identity

$$\nabla \times (\nabla \times \ ) = \nabla(\nabla \cdot \ ) - \nabla^2 \qquad (A1.19)$$

so that

$$\nabla \times (\nabla \times \mathbf{B}) = \nabla(\nabla \cdot \mathbf{B}) - \nabla^2 \mathbf{B},$$

where in Cartesian coordinates

$$(\nabla \cdot \nabla)\mathbf{B} = \nabla^2 \mathbf{B} \equiv \frac{\partial^2 \mathbf{B}}{\partial x^2} + \frac{\partial^2 \mathbf{B}}{\partial y^2} + \frac{\partial^2 \mathbf{B}}{\partial z^2}.$$

Since the divergence of **B** is zero, Eq. (A1.18) becomes

$$\nabla^2 \mathbf{B} - \mu \epsilon \frac{\partial^2 \mathbf{B}}{\partial t^2} - \mu \sigma \frac{\partial \mathbf{B}}{\partial t} = 0. \qquad (A1.20)$$

A similar equation is satisfied by the electric-field intensity. Following essentially the same procedure as above, take the curl of Eq. (A1.5)

$$\nabla \times (\nabla \times \mathbf{E}) = -\frac{\partial}{\partial t}(\nabla \times \mathbf{B}).$$

Eliminating **B** this becomes

$$\nabla \times (\nabla \times \mathbf{E}) = -\mu\sigma\frac{\partial \mathbf{E}}{\partial t} - \mu\epsilon\frac{\partial^2 \mathbf{E}}{\partial t^2},$$

and then by making use of Eq. (A1.19) we arrive at

$$\nabla^2 \mathbf{E} - \mu\epsilon\frac{\partial^2 \mathbf{E}}{\partial t^2} - \mu\sigma\frac{\partial \mathbf{E}}{\partial t} = \nabla(\rho/\epsilon),$$

having utilized the fact that

$$\nabla(\nabla \cdot \mathbf{E}) = \nabla(\rho/\epsilon).$$

For an uncharged medium ($\rho = 0$) and

$$\nabla^2 \mathbf{E} - \mu\epsilon\frac{\partial^2 \mathbf{E}}{\partial t^2} - \mu\sigma\frac{\partial \mathbf{E}}{\partial t} = 0. \qquad (A1.21)$$

Equation (A1.20) and (A1.21) are known as the *equations of telegraphy.*[*]

In nonconducting media, $\sigma = 0$ and these equations become

$$\nabla^2 \mathbf{B} - \mu\epsilon\frac{\partial^2 \mathbf{B}}{\partial t^2} = 0 \qquad (A1.22)$$

$$\nabla^2 \mathbf{E} - \mu\epsilon\frac{\partial^2 \mathbf{E}}{\partial t^2} = 0 \qquad (A1.23)$$

and similarly

$$\nabla^2 \mathbf{H} - \mu\epsilon\frac{\partial^2 \mathbf{H}}{\partial t^2} = 0 \qquad (A1.24)$$

and

$$\nabla^2 \mathbf{D} - \mu\epsilon\frac{\partial^2 \mathbf{D}}{\partial t^2} = 0. \qquad (A1.25)$$

In the special nonconducting medium of a vacuum (free space) where

$$\rho = 0, \qquad \sigma = 0, \qquad K_e = 1, \qquad K_m = 1,$$

these equations become simply

$$\nabla^2 \mathbf{E} = \mu_0\epsilon_0\frac{\partial^2 \mathbf{E}}{\partial t^2} \qquad (A1.26)$$

and

$$\nabla^2 \mathbf{B} = \mu_0\epsilon_0\frac{\partial^2 \mathbf{B}}{\partial t^2}. \qquad (A1.27)$$

Both of these expressions describe coupled space- and time-dependent fields and both have the form of the differential wave equation (see Section 3.2 for further discussion).

---

[*] For a pair of parallel wires which might serve as a telegraph line, the finite wire resistance results in a power loss and Joule heating. An electromagnetic wave advancing along the line has less and less energy available to it. The first-order time derivatives in Eqs. (A1.20) and (A1.21) arise from the conduction current and lead to the dissipation or damping.

Appendix 2
The Kirchhoff Diffraction Theory

To solve the Helmholtz equation (10.113) suppose that we have two scalar functions U_1 and U_2 for which Green's Theorem is

$$\iiint_V (U_1 \nabla^2 U_2 - U_2 \nabla^2 U_1)\, dV = \oiint_S (U_1 \nabla U_2 - U_2 \nabla U_1) \cdot d\mathbf{S}.$$

$$(A2.1)$$

It is clear that if U_1 and U_2 are solutions of the Helmholtz equation, i.e. if

$$\nabla^2 U_1 + k^2 U_1 = 0$$

and

$$\nabla^2 U_2 + k^2 U_2 = 0,$$

then

$$\oiint_S (U_1 \nabla U_2 - U_2 \nabla U_1) \cdot d\mathbf{S} = 0. \qquad (A2.2)$$

Let $U_1 = \mathscr{E}$, the space portion of an unspecified scalar optical disturbance (10.112). And let

$$U_2 = \frac{e^{ikr}}{r},$$

where r is measured from a point P. Both of these choices clearly satisfy the Helmholtz equation. There is a singularity at point P, where $r = 0$, so that we surround it by a small sphere in order to exclude P from the region enclosed by S, see Fig. A2.1. Equation (A2.2) now becomes

$$\oiint_S \left[\mathscr{E} \, \nabla \left(\frac{e^{ikr}}{r} \right) - \frac{e^{ikr}}{r} \nabla \mathscr{E} \right] \cdot d\mathbf{S}$$

$$+ \oiint_{S'} \left[\mathscr{E} \, \nabla \left(\frac{e^{ikr}}{r} \right) - \frac{e^{ikr}}{r} \nabla \mathscr{E} \right] \cdot d\mathbf{S} = 0.$$

$$(A2.3)$$

Now expand out the portion of the integral corresponding to S'. On the small sphere, the unit normal $\hat{\mathbf{n}}$ points toward the origin at P, and

$$\nabla \left(\frac{e^{ikr}}{r} \right) = \left(\frac{1}{r^2} - \frac{ik}{r} \right) e^{ikr} \hat{\mathbf{n}},$$

since the gradient is directed radially outward. In terms of the solid angle ($dS = r^2 \, d\Omega$) measured at P, the integral over S' becomes

$$\oiint_{S'} \left(\mathscr{E} - ik\mathscr{E}r + r\frac{\partial \mathscr{E}}{\partial r} \right) e^{ikr} \, d\Omega, \qquad (A2.4)$$

where $\nabla \mathscr{E} \cdot d\mathbf{S} = -(\partial \mathscr{E}/\partial r)r^2 \, d\Omega$. As the sphere surrounding P shrinks, $r \to 0$ on S' and $\exp(ikr) \to 1$. Because of the continuity of $\mathscr{E}$ its value at any point on S' approaches its value at P, i.e. $\mathscr{E}_p$. The last two terms in Eq. (A2.4) go to zero and the integral becomes $4\pi\mathscr{E}_p$. Finally then, Eq. (A2.3) becomes

$$\mathscr{E}_p = \frac{1}{4\pi} \left[\oiint_S \frac{e^{ikr}}{r} \nabla \mathscr{E} \cdot d\mathbf{S} - \oiint_S \mathscr{E} \, \nabla \left(\frac{e^{ikr}}{r} \right) \cdot d\mathbf{S} \right], \qquad [10.114]$$

which is known as the *Kirchhoff integral theorem*.

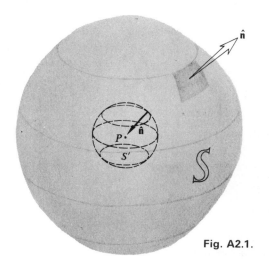

Fig. A2.1.

Table 1

Table 1 *

(Sin u)/u

u	0.00	0.01	0.02	0.03	0.04	0.05	0.06	0.07	0.08	0.09
0.0	1.000000	0.999983	0.999933	0.999850	0.999733	0.999583	0.999400	0.999184	0.998934	0.998651
0.1	0.998334	0.997985	0.997602	0.997186	0.996737	0.996254	0.995739	0.995190	0.994609	0.993994
0.2	0.993347	0.992666	0.991953	0.991207	0.990428	0.989616	0.988771	0.987894	0.986984	0.986042
0.3	0.985067	0.984060	0.983020	0.981949	0.980844	0.979708	0.978540	0.977339	0.976106	0.974842
0.4	0.973546	0.972218	0.970858	0.969467	0.968044	0.966590	0.965105	0.963588	0.962040	0.960461
0.5	0.958851	0.957210	0.955539	0.953836	0.952104	0.950340	0.948547	0.946723	0.944869	0.942985
0.6	0.941071	0.939127	0.937153	0.935150	0.933118	0.931056	0.928965	0.926845	0.924696	0.922518
0.7	0.920311	0.918076	0.915812	0.913520	0.911200	0.908852	0.906476	0.904072	0.901640	0.899181
0.8	0.896695	0.894182	0.891641	0.889074	0.886480	0.883859	0.881212	0.878539	0.875840	0.873114
0.9	0.870363	0.867587	0.864784	0.861957	0.859104	0.856227	0.853325	0.850398	0.847446	0.844471
1.0	0.841471	0.838447	0.835400	0.832329	0.829235	0.826117	0.822977	0.819814	0.816628	0.813419
1.1	0.810189	0.806936	0.803661	0.800365	0.797047	0.793708	0.790348	0.786966	0.783564	0.780142
1.2	0.776699	0.773236	0.769754	0.766251	0.762729	0.759188	0.755627	0.752048	0.748450	0.744833
1.3	0.741199	0.737546	0.733875	0.730187	0.726481	0.722758	0.719018	0.715261	0.711488	0.707698
1.4	0.703893	0.700071	0.696234	0.692381	0.688513	0.684630	0.680732	0.676819	0.672892	0.668952
1.5	0.664997	0.661028	0.657046	0.653051	0.649043	0.645022	0.640988	0.636942	0.632885	0.628815
1.6	0.624734	0.620641	0.616537	0.612422	0.608297	0.604161	0.600014	0.595858	0.591692	0.587517
1.7	0.583332	0.579138	0.574936	0.570725	0.566505	0.562278	0.558042	0.553799	0.549549	0.545291
1.8	0.541026	0.536755	0.532478	0.528194	0.523904	0.519608	0.515307	0.511001	0.506689	0.502373
1.9	0.498053	0.493728	0.489399	0.485066	0.480729	0.476390	0.472047	0.467701	0.463353	0.459002
2.0	0.454649	0.450294	0.445937	0.441579	0.437220	0.432860	0.428499	0.424137	0.419775	0.415414
2.1	0.411052	0.406691	0.402330	0.397971	0.393612	0.389255	0.384900	0.380546	0.376194	0.371845
2.2	0.367498	0.363154	0.358813	0.354475	0.350141	0.345810	0.341483	0.337161	0.332842	0.328529
2.3	0.324220	0.319916	0.315617	0.311324	0.307036	0.302755	0.298479	0.294210	0.289947	0.285692
2.4	0.281443	0.277202	0.272967	0.268741	0.264523	0.260312	0.256110	0.251916	0.247732	0.243556
2.5	0.239389	0.235231	0.231084	0.226946	0.222817	0.218700	0.214592	0.210495	0.206409	0.202334
2.6	0.198270	0.194217	0.190176	0.186147	0.182130	0.178125	0.174132	0.170152	0.166185	0.162230
2.7	0.158289	0.154361	0.150446	0.146546	0.142659	0.138786	0.134927	0.131083	0.127253	0.123439
2.8	0.119639	0.115854	0.112084	0.108330	0.104592	0.100869	0.097163	0.093473	0.089798	0.086141
2.9	0.082500	0.078876	0.075268	0.071678	0.068105	0.064550	0.061012	0.057492	0.053990	0.050506
3.0	0.047040	0.043592	0.040163	0.036753	0.033361	0.029988	0.026635	0.023300	0.019985	0.016689
3.1	0.013413	0.010157	0.006920	0.003704	0.000507	−0.002669	−0.005825	−0.008960	−0.012075	−0.015169
3.2	−0.018242	−0.021294	−0.024325	−0.027335	−0.030324	−0.033291	−0.036236	−0.039160	−0.042063	−0.044943
3.3	−0.047802	−0.050638	−0.053453	−0.056245	−0.059014	−0.061762	−0.064487	−0.067189	−0.069868	−0.072525
3.4	−0.075159	−0.077770	−0.080358	−0.082923	−0.085465	−0.087983	−0.090478	−0.092950	−0.095398	−0.097823
3.5	−0.100224	−0.102601	−0.104955	−0.107285	−0.109591	−0.111873	−0.114131	−0.116365	−0.118575	−0.120761
3.6	−0.122922	−0.125060	−0.127173	−0.129262	−0.131326	−0.133366	−0.135382	−0.137373	−0.139339	−0.141282
3.7	−0.143199	−0.145092	−0.146960	−0.148803	−0.150622	−0.152416	−0.154186	−0.155930	−0.157650	−0.159345
3.8	−0.161015	−0.162661	−0.164281	−0.165877	−0.167448	−0.168994	−0.170515	−0.172011	−0.173482	−0.174929
3.9	−0.176350	−0.177747	−0.179119	−0.180466	−0.181788	−0.183086	−0.184358	−0.185606	−0.186829	−0.188027

* Adapted from L. Levi, *Applied Optics*.

Table 1 515

Table 1 (*continued*)

(Sin u)/u

u	0.00	0.01	0.02	0.03	0.04	0.05	0.06	0.07	0.08	0.09
4.0	−0.189201	−0.190349	−0.191473	−0.192573	−0.193647	−0.194698	−0.195723	−0.196724	−0.197700	−0.198652
4.1	−0.199580	−0.200483	−0.201361	−0.202216	−0.203046	−0.203851	−0.204633	−0.205390	−0.206124	−0.206833
4.2	−0.207518	−0.208179	−0.208817	−0.209430	−0.210020	−0.210586	−0.211128	−0.211647	−0.212142	−0.212614
4.3	−0.213062	−0.213487	−0.213888	−0.214267	−0.214622	−0.214955	−0.215264	−0.215550	−0.215814	−0.216055
4.4	−0.216273	−0.216469	−0.216642	−0.216793	−0.216921	−0.217028	0.217112	−0.217174	−0.217214	−0.217232
	−0.217229	−0.217204	−0.217157	−0.217089	−0.217000	−0.216889	−0.216757	−0.216604	−0.216430	−0.216235
4.6	−0.216020	−0.215784	−0.215527	−0.215250	−0.214953	−0.214635	−0.214298	−0.213940	−0.213563	−0.213166
4.7	−0.212750	−0.212314	−0.211858	−0.211384	−0.210890	−0.210377	−0.209846	−0.209296	−0.208727	−0.208140
4.8	−0.207534	−0.206911	−0.206269	−0.205609	−0.204932	−0.204236	−0.203524	−0.202794	−0.202046	−0.201282
4.9	−0.200501	−0.199702	−0.198887	−0.198056	−0.197208	−0.196344	−0.195464	−0.194568	−0.193656	−0.192728
5.0	−0.191785	−0.190826	−0.189853	−0.188864	−0.187860	−0.186841	−0.185808	−0.184760	−0.183699	−0.182622
5.1	−0.181532	−0.180428	−0.179311	−0.178179	−0.177035	−0.175877	−0.174706	−0.173522	−0.172326	−0.171117
5.2	−0.169895	−0.168661	−0.167415	−0.166158	−0.164888	−0.163607	−0.162314	−0.161010	−0.159695	−0.158369
5.3	−0.157032	−0.155684	−0.154326	−0.152958	−0.151579	−0.150191	−0.148792	−0.147384	−0.145967	−0.144540
5.4	−0.143105	−0.141660	−0.140206	−0.138744	−0.137273	−0.135794	−0.134307	−0.132812	−0.131309	−0.12979
5.5	−0.128280	−0.126755	−0.125222	−0.123683	−0.122137	−0.120584	−0.119024	−0.117459	−0.115887	−0.114310
5.6	−0.112726	−0.111137	−0.109543	−0.107943	−0.106338	−0.104728	−0.103114	−0.101495	−0.099871	−0.098243
5.7	−0.096611	−0.094976	−0.093336	−0.091693	−0.090046	−0.088396	−0.086743	−0.085087	−0.083429	−0.081768
5.8	−0.080104	−0.078438	−0.076770	−0.075100	−0.073428	−0.071755	−0.070080	−0.068404	−0.066726	−0.065048
5.9	−0.063369	−0.061689	−0.060009	−0.058329	−0.056648	−0.054967	−0.053287	−0.051606	−0.049927	−0.048248
6.0	−0.046569	−0.044892	−0.043216	−0.041540	−0.039867	−0.038195	−0.036524	−0.034856	−0.033189	−0.031525
6.1	−0.029863	−0.028203	−0.026546	−0.024892	−0.023240	−0.021592	−0.019947	−0.018305	−0.016667	−0.015032
6.2	−0.013402	−0.011775	−0.010152	−0.008533	−0.006919	−0.005309	−0.003703	−0.002103	−0.000507	0.001083
6.3	0.002669	0.004249	0.005824	0.007393	0.008956	0.010514	0.012066	0.013612	0.015151	0.016684
6.4	0.018211	0.019731	0.021244	0.022751	0.024250	0.025743	0.027228	0.028706	0.030177	0.031640
6.5	0.033095	0.034543	0.035983	0.037414	0.038838	0.040253	0.041661	0.043059	0.044449	0.045831
6.6	0.047203	0.048567	0.049922	0.051268	0.052604	0.053931	0.055249	0.056558	0.057857	0.059146
6.7	0.060425	0.061695	0.062955	0.064204	0.065444	0.066673	0.067892	0.069101	0.070299	0.071487
6.8	0.072664	0.073830	0.074986	0.076130	0.077264	0.078386	0.079498	0.080598	0.081688	0.082765
6.9	0.083832	0.084887	0.085930	0.086962	0.087982	0.088991	0.089987	0.090972	0.091945	0.092906
7.0	0.093855	0.094792	0.095717	0.096629	0.097530	0.098418	0.099293	0.100157	0.101008	0.101846
7.1	0.102672	0.103485	0.104286	0.105074	0.105849	0.106611	0.107361	0.108098	0.108822	0.109533
7.2	0.110232	0.110917	0.111589	0.112249	0.112895	0.113528	0.114149	0.114756	0.115350	0.115931
7.3	0.116498	0.117053	0.117594	0.118122	0.118637	0.119138	0.119627	0.120102	0.120563	0.121012
7.4	0.121447	0.121869	0.122277	0.122673	0.123055	0.123423	0.123779	0.124121	0.124449	0.124765
7.5	0.125067	0.125355	0.125631	0.125893	0.126142	0.126378	0.126600	0.126809	0.127005	0.127188
7.6	0.127358	0.127514	0.127658	0.127788	0.127905	0.128009	0.128100	0.128178	0.128243	0.128295
7.7	0.128334	0.128360	0.128373	0.128373	0.128361	0.128335	0.128297	0.128247	0.128183	0.128107
7.8	0.128018	0.127917	0.127803	0.127677	0.127539	0.127388	0.127224	0.127049	0.126861	0.126661
7.9	0.126448	0.126224	0.125988	0.125739	0.125479	0.125207	0.124923	0.124627	0.124320	0.124000

Table 1

Table 1 (*continued*)

(Sin u)/u

u	0.00	0.01	0.02	0.03	0.04	0.05	0.06	0.07	0.08	0.09
8.0	0.123670	0.123328	0.122974	0.122609	0.122232	0.121845	0.121446	0.121036	0.120615	0.120183
8.1	0.119739	0.119286	0.118821	0.118345	0.117859	0.117363	0.116855	0.116338	0.115810	0.115272
8.2	0.114723	0.114165	0.113596	0.113018	0.112429	0.111831	0.111223	0.110605	0.109978	0.109341
8.3	0.108695	0.108040	0.107376	0.106702	0.106019	0.105327	0.104627	0.103918	0.103200	0.102473
8.4	0.101738	0.100994	0.100243	0.099483	0.098714	0.097938	0.097154	0.096362	0.095562	0.094755
8.5	0.093940	0.093117	0.092287	0.091450	0.090606	0.089755	0.088896	0.088031	0.087159	0.086280
8.6	0.085395	0.084503	0.083605	0.082701	0.081790	0.080874	0.079951	0.079023	0.078089	0.077149
8.7	0.076203	0.075253	0.074296	0.073335	0.072369	0.071397	0.070421	0.069439	0.068453	0.067463
8.8	0.066468	0.065468	0.064465	0.063457	0.062445	0.061429	0.060410	0.059386	0.058359	0.057328
8.9	0.056294	0.055257	0.054217	0.053173	0.052127	0.051077	0.050025	0.048970	0.047913	0.046853
9.0	0.045791	0.044727	0.043660	0.042592	0.041521	0.040449	0.039375	0.038300	0.037223	0.036145
9.1	0.035066	0.033985	0.032904	0.031821	0.030738	0.029654	0.028569	0.027484	0.026399	0.025313
9.2	0.024227	0.023141	0.022055	0.020970	0.019884	0.018799	0.017714	0.016630	0.015547	0.014464
9.3	0.013382	0.012301	0.011222	0.010143	0.009066	0.007990	0.006916	0.005843	0.004772	0.003703
9.4	0.002636	0.001570	0.000507	−0.000554	−0.001612	−0.002669	−0.003722	−0.004774	−0.005822	−0.006868
9.5	−0.007911	−0.008950	−0.009987	−0.011021	−0.012051	−0.013078	−0.014101	−0.015121	−0.016138	−0.017150
9.6	−0.018159	−0.019164	−0.020165	−0.021161	−0.022154	−0.023142	−0.024126	−0.025106	−0.026081	−0.027051
9.7	−0.028017	−0.028977	−0.029933	−0.030884	−0.031830	−0.032771	−0.033707	−0.034637	−0.035562	−0.036482
9.8	−0.037396	−0.038304	−0.039207	−0.040104	−0.040995	−0.041881	−0.042760	−0.043633	−0.044500	−0.045361
9.9	−0.046216	−0.047064	−0.047906	−0.048741	−0.049570	−0.050392	−0.051208	−0.052017	−0.052819	−0.053614
10.0	−0.054402	−0.055183	−0.055957	−0.056724	−0.057484	−0.058237	−0.058982	−0.059720	−0.060450	−0.061173
10.1	−0.061888	−0.062596	−0.063296	−0.063988	−0.064673	−0.065350	−0.066019	−0.066680	−0.067333	−0.067978
10.2	−0.068615	−0.069244	−0.069865	−0.070477	−0.071082	−0.071678	−0.072266	−0.072845	−0.073416	−0.073979
10.3	−0.074533	−0.075078	−0.075615	−0.076143	−0.076663	−0.077174	−0.077677	−0.078170	−0.078655	−0.079131
10.4	−0.079599	−0.080057	−0.080507	−0.080947	−0.081379	−0.081802	−0.082216	−0.082620	−0.083016	−0.083403
10.5	−0.083781	−0.084149	−0.084509	−0.084859	−0.085200	−0.085532	−0.085855	−0.086169	−0.086473	−0.086768
10.6	−0.087054	−0.087331	−0.087599	−0.087857	−0.088106	−0.088346	−0.088576	−0.088797	−0.089009	−0.089212
10.7	−0.089405	−0.089589	−0.089764	−0.089929	−0.090085	−0.090232	−0.090370	−0.090498	−0.090617	−0.090727
10.8	−0.090827	−0.090919	−0.091001	−0.091073	−0.091137	−0.091191	−0.091236	−0.091272	−0.091299	−0.091316
10.9	−0.091324	−0.091324	−0.091314	−0.091295	−0.091267	−0.091229	−0.091183	−0.091128	−0.091064	−0.090990
11.0	−0.090908	−0.090817	−0.090717	−0.090608	−0.090490	−0.090364	−0.090228	−0.090084	−0.089931	−0.089770
11.1	−0.089599	−0.089420	−0.089233	−0.089037	−0.088832	−0.088619	−0.088397	−0.088167	−0.087929	−0.087682
11.2	−0.087427	−0.087163	−0.086891	−0.086612	−0.086324	−0.086027	−0.085723	−0.085411	−0.085091	−0.084763
11.3	−0.084426	−0.084083	−0.083731	−0.083371	−0.083004	−0.082630	−0.082247	−0.081857	−0.081460	−0.081055
11.4	−0.080643	−0.080223	−0.079796	−0.079362	−0.078921	−0.078473	−0.078017	−0.077555	−0.077086	−0.076609
11.5	−0.076126	−0.075636	−0.075140	−0.074637	−0.074127	−0.073611	−0.073088	−0.072559	−0.072023	−0.071481
11.6	−0.070934	−0.070379	−0.069819	−0.069253	−0.068681	−0.068103	−0.067519	−0.066929	−0.066334	−0.065733
11.7	−0.065127	−0.064515	−0.063898	−0.063275	−0.062647	−0.062014	−0.061376	−0.060733	−0.060084	−0.059431
11.8	−0.058773	−0.058111	−0.057443	−0.056771	−0.056095	−0.055414	−0.054728	−0.054039	−0.053345	−0.052646
11.9	−0.051944	−0.051238	−0.050528	−0.049814	−0.049096	−0.048375	−0.047650	−0.046921	−0.046189	−0.045453

Table 1 517

Table 1 (*continued*)

(Sin u)/u

u	0.00	0.01	0.02	0.03	0.04	0.05	0.06	0.07	0.08	0.09
12.0	−0.044714	−0.043972	−0.043227	−0.042479	−0.041727	−0.040973	−0.040216	−0.039456	−0.038694	−0.037929
12.1	−0.037161	−0.036391	−0.035618	−0.034844	−0.034067	−0.033288	−0.032506	−0.031723	−0.030938	−0.030152
12.2	−0.029363	−0.028573	−0.027781	−0.026988	−0.026193	−0.025398	−0.024600	−0.023802	−0.023003	−0.022202
12.3	−0.021401	−0.020599	−0.019796	−0.018992	−0.018188	−0.017384	−0.016578	−0.015773	−0.014967	−0.014161
12.4	−0.013355	−0.012549	−0.011743	−0.010937	−0.010131	−0.009326	−0.008521	−0.007716	−0.006912	−0.006109
12.5	−0.005306	−0.004504	−0.003702	−0.002902	−0.002103	−0.001304	−0.000507	0.000289	0.001083	0.001877
12.6	0.002668	0.003459	0.004248	0.005035	0.005820	0.006603	0.007385	0.008164	0.008942	0.009717
12.7	0.010491	0.011262	0.012030	0.012797	0.013560	0.014321	0.015080	0.015836	0.016589	0.017339
12.8	0.018087	0.018831	0.019572	0.020311	0.021046	0.021778	0.022506	0.023231	0.023953	0.024671
12.9	0.025386	0.026097	0.026804	0.027507	0.028207	0.028903	0.029594	0.030282	0.030966	0.031645
13.0	0.032321	0.032992	0.033658	0.034321	0.034978	0.035632	0.036281	0.036925	0.037564	0.038199
13.1	0.038829	0.039454	0.040075	0.040690	0.041300	0.041905	0.042506	0.043101	0.043690	0.044275
13.2	0.044854	0.045428	0.045996	0.046559	0.047117	0.047669	0.048215	0.048756	0.049291	0.049820
13.3	0.050344	0.050861	0.051373	0.051879	0.052379	0.052873	0.053361	0.053843	0.054319	0.054788
13.4	0.055252	0.055709	0.056160	0.056605	0.057043	0.057476	0.057901	0.058321	0.058733	0.059140
13.5	0.059540	0.059933	0.060320	0.060700	0.061073	0.061440	0.061800	0.062154	0.062500	0.062840
13.6	0.063174	0.063500	0.063820	0.064132	0.064438	0.064737	0.065029	0.065314	0.065593	0.065864
13.7	0.066128	0.066385	0.066636	0.066879	0.067115	0.067344	0.067566	0.067781	0.067989	0.068190
13.8	0.068384	0.068570	0.068750	0.068922	0.069087	0.069245	0.069396	0.069540	0.069677	0.069806
13.9	0.069929	0.070044	0.070152	0.070253	0.070346	0.070433	0.070512	0.070584	0.070649	0.070707
14.0	0.070758	0.070801	0.070838	0.070867	0.070889	0.070904	0.070912	0.070913	0.070907	0.070893
14.1	0.070873	0.070846	0.070811	0.070770	0.070721	0.070666	0.070603	0.070534	0.070457	0.070374
14.2	0.070284	0.070186	0.070082	0.069971	0.069854	0.069729	0.069598	0.069460	0.069315	0.069163
14.3	0.069005	0.068840	0.068668	0.068490	0.068305	0.068114	0.067916	0.067712	0.067501	0.067283
14.4	0.067060	0.066829	0.066593	0.066350	0.066101	0.065845	0.065584	0.065316	0.065042	0.064762
14.5	0.064476	0.064183	0.063885	0.063581	0.063271	0.062954	0.062633	0.062305	0.061971	0.061632
14.6	0.061287	0.060936	0.060580	0.060218	0.059851	0.059478	0.059100	0.058717	0.058328	0.057933
14.7	0.057534	0.057129	0.056719	0.056304	0.055884	0.055459	0.055029	0.054594	0.054154	0.053710
14.8	0.053260	0.052806	0.052347	0.051884	0.051416	0.050944	0.050467	0.049985	0.049500	0.049010
14.9	0.048516	0.048017	0.047515	0.047008	0.046497	0.045983	0.045464	0.044942	0.044416	0.043886
15.0	0.043353	0.042815	0.042275	0.041730	0.041183	0.040632	0.040077	0.039520	0.038959	0.038395
15.1	0.037828	0.037257	0.036684	0.036108	0.035529	0.034948	0.034363	0.033776	0.033187	0.032595
15.2	0.032000	0.031403	0.030803	0.030202	0.029598	0.028992	0.028383	0.027773	0.027161	0.026547
15.3	0.025931	0.025313	0.024693	0.024072	0.023450	0.022825	0.022199	0.021572	0.020944	0.020314
15.4	0.019683	0.019051	0.018418	0.017783	0.017148	0.016512	0.015875	0.015237	0.014599	0.013960
15.5	0.013320	0.012680	0.012040	0.011399	0.010758	0.010116	0.009475	0.008833	0.008191	0.007549
15.6	0.006907	0.006266	0.005624	0.004983	0.004342	0.003702	0.003062	0.002422	0.001783	0.001145
15.7	0.000507	−0.000130	−0.000766	−0.001401	−0.002035	−0.002668	−0.003300	−0.003931	−0.004561	−0.005190
15.8	−0.005817	−0.006443	−0.007067	−0.007690	−0.008311	−0.008931	−0.009549	−0.010166	−0.010780	−0.011393
15.9	−0.012004	−0.012613	−0.013219	−0.013824	−0.014427	−0.015027	−0.015625	−0.016221	−0.016814	−0.017405

Table 1 (*continued*)

(Sin u)/u

u	0.00	0.01	0.02	0.03	0.04	0.05	0.06	0.07	0.08	0.09
16.0	−0.017994	−0.018580	−0.019163	−0.019744	−0.020322	−0.020898	−0.021470	−0.022040	−0.022607	−0.023170
16.1	−0.023731	−0.024289	−0.024843	−0.025395	−0.025943	−0.026488	−0.027030	−0.027568	−0.028103	−0.028634
16.2	−0.029162	−0.029686	−0.030207	−0.030724	−0.031237	−0.031747	−0.032252	−0.032754	−0.033252	−0.033746
16.3	−0.034236	−0.034722	−0.035204	−0.035682	−0.036156	−0.036626	−0.037091	−0.037552	−0.038009	−0.038461
16.4	−0.038909	−0.039352	−0.039792	−0.040226	−0.040656	−0.041081	−0.041502	−0.041918	−0.042330	−0.042737
16.5	−0.043139	−0.043536	−0.043928	−0.044315	−0.044698	−0.045076	−0.045448	−0.045816	−0.046179	−0.046536
16.6	−0.046889	−0.047236	−0.047578	−0.047915	−0.048247	−0.048574	−0.048895	−0.049212	−0.049522	−0.049828
16.7	−0.050128	−0.050423	−0.050713	−0.050997	−0.051275	−0.051548	−0.051816	−0.052078	−0.052335	−0.052586
16.8	−0.052831	−0.053071	−0.053306	−0.053535	−0.053758	−0.053975	−0.054187	−0.054393	−0.054594	−0.054789
16.9	−0.054978	−0.055161	−0.055339	−0.055511	−0.055677	−0.055837	−0.055992	−0.056141	−0.056284	−0.056421
17.0	−0.056553	−0.056678	−0.056798	−0.056912	−0.057021	−0.057123	−0.057220	−0.057310	−0.057395	−0.057474
17.1	0.057548	−0.057615	−0.057677	−0.057732	−0.057782	−0.057826	−0.057865	−0.057897	−0.057924	−0.057944
17.2	−0.057959	−0.057968	−0.057972	−0.057969	−0.057961	−0.057947	−0.057927	−0.057902	−0.057870	−0.057833
17.3	−0.057790	−0.057742	−0.057688	−0.057628	−0.057562	−0.057491	−0.057414	−0.057331	−0.057243	−0.057149
17.4	−0.057049	−0.056944	−0.056834	−0.056717	−0.056596	−0.056468	−0.056336	−0.056197	−0.056054	−0.055905
17.5	−0.055750	−0.055590	−0.055425	−0.055254	−0.055078	−0.054897	−0.054710	−0.054518	−0.054321	−0.054119
17.6	−0.053912	−0.053699	−0.053481	−0.053258	−0.053031	−0.052798	−0.052560	−0.052317	−0.052069	−0.051816
17.7	−0.051558	−0.051296	−0.051028	−0.050756	−0.050479	−0.050198	−0.049911	−0.049620	−0.049324	−0.049024
17.8	−0.048719	−0.048410	−0.048096	−0.047778	−0.047455	−0.047128	−0.046796	−0.046461	−0.046121	−0.045776
17.9	−0.045428	−0.045075	−0.044718	−0.044358	−0.043993	−0.043624	−0.043251	−0.042875	−0.042494	−0.042110
18.0	−0.041722	−0.041330	−0.040934	−0.040535	−0.040132	−0.039726	−0.039316	−0.038902	−0.038485	−0.038065
18.1	−0.037642	−0.037215	−0.036785	−0.036351	−0.035915	−0.035475	−0.035033	−0.034587	−0.034139	−0.033687
18.2	−0.033233	−0.032775	−0.032315	−0.031853	−0.031387	−0.030919	−0.030449	−0.029976	−0.029500	−0.029022
18.3	−0.028541	−0.028059	−0.027574	−0.027086	−0.026597	−0.026105	−0.025612	−0.025116	−0.024619	−0.024119
18.4	−0.023618	−0.023114	−0.022610	−0.022103	−0.021594	−0.021085	−0.020573	−0.020060	−0.019546	−0.019030
18.5	−0.018512	−0.017994	−0.017474	−0.016953	−0.016431	−0.015908	−0.015384	−0.014859	−0.014333	−0.013806
18.6	−0.013278	−0.012750	−0.012220	−0.011691	−0.011160	−0.010629	−0.010098	0.009566	−0.009033	−0.008501
18.7	−0.007968	−0.007435	−0.006901	−0.006368	−0.005834	−0.005301	−0.004767	−0.004234	−0.003701	−0.003168
18.8	−0.002635	−0.002102	−0.001570	−0.001038	−0.000507	0.000024	0.000554	0.001083	0.001612	0.002140
18.9	0.002668	0.003194	0.003720	0.004245	0.004769	0.005292	0.005813	0.006334	0.006853	0.007371
19.0	0.007888	0.008404	0.008918	0.009431	0.009942	0.010452	0.010960	0.011466	0.011971	0.012474
19.1	0.012976	0.013475	0.013973	0.014468	0.014962	0.015454	0.015944	0.016431	0.016917	0.017400
19.2	0.017881	0.018360	0.018836	0.019310	0.019782	0.020251	0.020717	0.021181	0.021643	0.022102
19.3	0.022558	0.023011	0.023462	0.023910	0.024355	0.024797	0.025236	0.025672	0.026105	0.026535
19.4	0.026962	0.027386	0.027807	0.028224	0.028638	0.029049	0.029457	0.029861	0.030262	0.030659
19.5	0.031053	0.031444	0.031831	0.032214	0.032594	0.032970	0.033342	0.033711	0.034076	0.034437
19.6	0.034794	0.035148	0.035497	0.035843	0.036185	0.036522	0.036856	0.037186	0.037512	0.037833
19.7	0.038151	0.038464	0.038774	0.039079	0.039379	0.039676	0.039968	0.040256	0.040540	0.040820
19.8	0.041095	0.041365	0.041632	0.041893	0.042151	0.042404	0.042652	0.042896	0.043135	0.043370
19.9	0.043600	0.043826	0.044047	0.044263	0.044475	0.044682	0.044885	0.045082	0.045275	0.045464

Solutions to
Selected Problems

CHAPTER 2

2.1 $(0.003)(2.54 \times 10^{-2})/580 \times 10^{-9}$ = Number of waves = 131.2.
$c = v\lambda$, $\lambda = c/v = 3 \times 10^8/10^{10}$, $\lambda = 3$ cm.

Waves extend 393.6 cm.

2.3 $\psi = A \sin 2\pi(\kappa x - vt)$, $\psi_1 = 4 \sin 2\pi(0.2x - 3t)$

a) $v = 3$ b) $\lambda = 1/0.2$ c) $\tau = 1/3$
d) $A = 4$ e) $v = 15$ f) positive x

   $\psi = A \sin (kx + \omega t)$, $\psi_2 = (1/2.5) \sin (7x + 3.5t)$

a) $v = 3.5/2\pi$ b) $\lambda = 2\pi/7$ c) $\tau = 2\pi/3.5$
d) $A = 1/2.5$ e) $v = \frac{1}{2}$ f) negative x

2.5 $v_y = -\omega A \cos (kx - \omega t + \epsilon)$, $a_y = -\omega^2 y$. Simple harmonic motion since $a_y \propto y$.

2.6

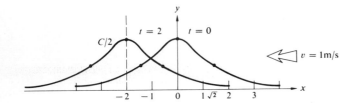

$y(x, t) = C/[2 + (x + vt)^2]$.

2.8 $\psi = A \exp i(k_x x + k_y y + k_z z)$
$k_x = k\alpha$ $k_y = k\beta$ $k_z = k\gamma$
$|\mathbf{k}| = [(k\alpha)^2 + (k\beta)^2 + (k\gamma)^2]^{1/2} = k[\alpha^2 + \beta^2 + \gamma^2]^{1/2}$.

2.9 $30°$ corresponds to $\frac{1}{12}\lambda$ or $(1/12)3 \times 10^8/6 \times 10^{14} = 41.6$ nm.

2.10 $\psi = A \sin 2\pi \left(\frac{x}{\lambda} \pm \frac{t}{\tau} \right)$

$\psi = 60 \sin 2\pi \left(\frac{x}{400 \times 10^{-9}} - \frac{t}{1.33 \times 10^{-15}} \right)$

$\lambda = 400$ nm $v = 400 \times 10^{-9}/1.33 \times 10^{-15} = 3 \times 10^8$ m/s

$v = (1/1.33) \times 10^{+15}$Hz, $\tau = 1.33 \times 10^{-15}$s.

2.12 $\lambda = h/mv = 6.6 \times 10^{-34}/6(1) = 1.1 \times 10^{-34}$m.

CHAPTER 3

3.1 $E_y = 2 \cos [2\pi \times 10^{14}(t - x/c) + \pi/2]$

   $E_y = A \cos [2\pi v(t - x/v) + \pi/2]$ from Eq. (2.26)

a) $v = 10^{14}$Hz, $v = c$ and $\lambda = c/v = 3 \times 10^8/10^{14} = 3 \times 10^{-6}$m, moves in positive x-direction, $A = 2$ V/m, $\varepsilon = \pi/2$ linearly polarized in y-direction.

b) $B_x = 0$, $B_y = 0$, $B_z = \frac{2}{c} \cos [2\pi \times 10^{14}(t - x/c) + \pi/2]$.

3.2 $E_z = 0$, $E_y = E_x = E_0 \sin (kz - \omega t)$ or   cosine; $B_z = 0$, $B_y = -B_x = E_y/c$, or if you like

   $\mathbf{E} = E_0(\hat{\imath} + \hat{\jmath}) \sin (kz - \omega t)$. $\mathbf{B} = \frac{E_0}{c}(\hat{\jmath} - \hat{\imath}) \sin (kz - \omega t)$.

3.4 Thermal agitation of the molecular dipoles causes a marked reduction in K_e but has little effect on n. At optical frequencies n is predominantly due to electronic polarization, rotations of the molecular dipoles having ceased to be effective at much lower frequencies.

3.5 From Eq. (3.36), for a single resonant frequency we get

$$n = \left[1 + \frac{Nq_e^2}{\epsilon_0 m_e} \left(\frac{1}{\omega_0^2 - \omega^2} \right) \right]^{1/2},$$

since for low-density materials $n \approx 1$ the second term is $\ll 1$ and we need only retain the first two terms of the binomial expansion of n. Thus $\sqrt{1 + x} \approx 1 + x/2$ and

$$n \approx 1 + \frac{1}{2} \frac{Nq_e^2}{\epsilon_0 m_e} \left(\frac{1}{\omega_0^2 - \omega^2} \right).$$

3.6 The normal order of the spectrum for a glass prism is R, O, Y, G, B, V with red (R) deviated the least and violet (V) deviated the most. For a fuchsin prism, there is an absorption band in the green and so the indices for yellow and blue on either side (n_Y and n_B) of it are extremes as in Fig. 3.14; that is, n_Y is the maximum, n_B the minimum and $n_Y > n_O > n_R > n_V > n_B$. Thus the spectrum in order of increasing deviation is B, V, black band, R, O, Y.

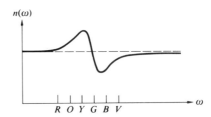

3.7 The phase angle is retarded by an amount $(n \Delta y 2\pi/\lambda) - \Delta y 2\pi/\lambda$ or $(n - 1) \Delta y \omega/c$. Thus

$$E_p = E_0 \exp i\omega[t - (n - 1)\Delta y/c - y/c]$$

or

$$E_p = E_0 \exp [-i\omega(n - 1) \Delta y/c] \exp i\omega(t - y/c)$$

if $n \approx 1$ or $\Delta y \ll 1$. Since $e^x \approx 1 + x$ for small x,

$$\exp [-i\omega(n - 1)\Delta y/c] \approx 1 - i\omega(n - 1)\Delta y/c$$

and since $\exp (-i\pi/2) = -i$

$$E_p = E_u + \frac{\omega(n - 1)\Delta y}{c} E_u e^{-i\pi/2}.$$

3.8 $\langle \cos^2 (\mathbf{k} \cdot \mathbf{r} - \omega t) \rangle = \frac{1}{T} \int_t^{t+T} \cos^2 (\mathbf{k} \cdot \mathbf{r} - \omega t') \, dt'$.

Let $\mathbf{k} \cdot \mathbf{r} - \omega t' = x$; then

$$\langle \cos^2 (\mathbf{k} \cdot \mathbf{r} - \omega t) \rangle = \frac{1}{-\omega t} \int \cos^2 x \, dx = \frac{1}{-\omega T} \int \frac{1 + \cos 2x}{2} \, dx$$

$$= -\frac{1}{\omega T} \left[\frac{x}{2} + \frac{\sin 2x}{4} \right]_{\mathbf{k} \cdot \mathbf{r} - \omega t}^{\mathbf{k} \cdot \mathbf{r} - \omega (t + T)}$$

$$= \frac{1}{2} - \frac{1}{4\omega T} \{ \sin [2\mathbf{k} \cdot \mathbf{r} - 2\omega (t + T)]$$

$$- \sin 2(\mathbf{k} \cdot \mathbf{r} - \omega t) \}.$$

When $T = \tau$, $\omega = 2\pi/\tau$ and the second term vanishes yielding $\langle \cos^2 (\mathbf{k} \cdot \mathbf{r} - \omega t) \rangle = \frac{1}{2}$, and when $T \gg \tau$, $1/4\omega T$ becomes $\ll 1$ and the second term again vanishes. The other expression follows in the same way.

3.10 $u = \dfrac{(\text{power}) \, (t)}{(\text{volume})} = \dfrac{(10^{-3} \text{W}) \, (t)}{(\pi r^2) \, (ct)} = \dfrac{10^{-3} \text{W}}{\pi (10^{-3})^2 (3 \times 10^8)}$

$$u = \frac{10^{-5}}{3\pi} \text{J/m}^3 = 1.06 \times 10^{-6} \text{ J/m}^3.$$

3.12 $h = 6.63 \times 10^{-34}$, $E = h\nu$

$$\frac{I}{h\nu} = \frac{19.88 \times 10^{-2}}{(6.63 \times 10^{-34}) \, (100 \times 10^6)} = 3 \times 10^{24} \text{ photons/m}^2 \text{ s}.$$

All photons in volume V cross unit area in one second

$$V = (ct) \, (1 \text{ m}^2) = 3 \times 10^8 \text{ m}^3$$

$$3 \times 10^{24} = V(\text{density})$$

$$\text{density} = 10^{16} \text{ photons/m}^3.$$

3.14 $P_e = iV = (0.25) \, (3.0) = 0.75$ W. This is the electrical power dissipated. The power available as light is

$$P_l = (0.01)P_e = 75 \times 10^{-4} \text{ W}.$$

a) Photon flux $= P_l/h\nu = 75 \times 10^{-4} \lambda/hc$

$$= 75 \times 10^{-4} (550 \times 10^{-9})/(6.63 \times 10^{-34}) 3 \times 10^8$$

$$= 2.08 \times 10^{16} \text{ photons/s}.$$

b) There are 2.08×10^{16} in volume $3 \times 10^8 (1\text{s}) \, (10^{-3} \text{ m}^2)$;

$$\therefore \frac{2.08 \times 10^{16}}{3 \times 10^5} = \text{photons/m}^3 = 0.69 \times 10^{11}.$$

c) $1 = 75 \times 10^{-4} \text{ W}/10 \times 10^{-4} \text{ m}^2 = 7.5 \text{ W/m}^2.$

3.16 Imagine two concentric cylinders of radius r_1 and r_2 surrounding the wave. The energy flowing per second through the first cylinder must pass through the second cylinder, that is,

$$\langle S_1 \rangle 2\pi r_1 = \langle S_2 \rangle 2\pi r_2 \text{ and so } \langle S \rangle 2\pi r = \text{constant}$$

and $\langle S \rangle$ varies inversely with r. Therefore, since $\langle S \rangle \propto E_0^2$, E_0 varies as $\sqrt{1/r}$.

3.18 $\left\langle \dfrac{dp}{dt} \right\rangle = \dfrac{1}{c} \left\langle \dfrac{dW}{dt} \right\rangle$, $A = $ area. $\mathscr{P} = \dfrac{1}{A} \left\langle \dfrac{dp}{dt} \right\rangle = \dfrac{1}{Ac} \left\langle \dfrac{dW}{dt} \right\rangle = \dfrac{I}{c}.$

3.20 $\mathscr{E} = 300 \text{ W}(100 \text{ s}) = 3 \times 10^4 \text{ J},$

$$p = \mathscr{E}/c = 3 \times 10^4/3 \times 10^8 = 10^{-4} \text{ kg} \cdot \text{m/s}.$$

3.22 $F = \dfrac{dp}{dt} = ma = \dfrac{1}{c} \dfrac{dW}{dt} = \dfrac{10W}{3 \times 10^8} = 3.3 \times 10^{-8} \text{ n}$

$$a = 3.3 \times 10^{-8}/100 \text{ kg} = 3.3 \times 10^{-10} \text{ m/s}^2$$

$$v = at = \frac{1}{3} \times 10^{-9}(t) = 10 \text{ m/s}$$

$$t = 3 \times 10^{10} \text{ s}, \quad 1 \text{ year} = 3.2 \times 10^7 \text{ s}.$$

CHAPTER 4

4.1
$$n_i \sin \theta_i = n_t \sin \theta_t$$
$$\sin 30° = 1.52 \sin \theta_t$$
$$\theta_t = \sin^{-1} (1/3.04)$$
$$\theta_t = 19° \, 13'.$$

4.3

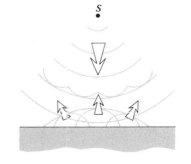

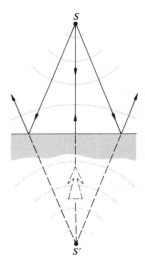

4.5 $n_{ti} = \dfrac{r_t}{r_i} = \dfrac{c/v_t}{c/v_i} = \dfrac{v_i}{v_t} = \dfrac{v\lambda_i}{v\lambda_t} = \dfrac{\lambda_i}{\lambda_t}$

therefore $\lambda_t = \lambda_i 3/4 = $ **9 cm**

$$\sin \theta_i = n_{ti} \sin \theta_t$$
$$\sin^{-1} [\tfrac{3}{4} (0.707)] = \theta_t = 32°.$$

4.7

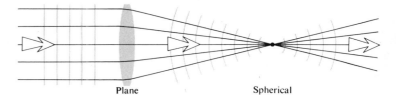

Plane   Spherical

4.8

θ_i (degrees)	θ_t (degrees)
0	0
10	6.7
20	13.3
30	19.6
40	25.2
50	30.7
60	35.1
70	38.6
80	40.6
90	41.8

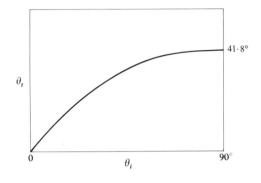

4.9 Let τ be the time for the wave to move along a ray from b_1 to b_2, from a_1 to a_2, and from a_1 to a_3. Thus $a_1 a_2 = b_1 b_2 = v_i \tau$ and $a_1 a_3 = v_t \tau$.

$$\sin \theta_i = \overline{b_1 b_2}/\overline{a_1 b_2} = v_i/\overline{a_1 b_2}$$

$$\sin \theta_t = \overline{a_1 a_3}/\overline{a_1 b_2} = v_t/\overline{a_1 b_2}$$

$$\sin \theta_r = \overline{a_1 a_2}/\overline{a_1 b_2} = v_i/\overline{a_1 b_2}$$

$$\frac{\sin \theta_i}{\sin \theta_t} = \frac{v_i}{v_t} = \frac{n_t}{n_i} = n_{ti} \quad \text{and} \quad \theta_i = \theta_r.$$

4.10

$$n_i \sin \theta_i = n_t \sin \theta_t$$
$$n_i (\hat{\mathbf{k}}_i \times \hat{\mathbf{u}}_n) = n_t (\hat{\mathbf{k}}_t \times \hat{\mathbf{u}}_n),$$

where $\hat{\mathbf{k}}_i$, $\hat{\mathbf{k}}_t$ are unit propagation vectors. Thus

$$n_t (\hat{\mathbf{k}}_t \times \hat{\mathbf{u}}_n) - n_i (\hat{\mathbf{k}}_i \times \hat{\mathbf{u}}_n) = 0$$
$$(n_t \hat{\mathbf{k}}_t - n_i \hat{\mathbf{k}}_i) \times \hat{\mathbf{u}}_n = 0.$$

Let $n_t \hat{\mathbf{k}}_t - n_i \hat{\mathbf{k}}_i = \mathbf{\Gamma} = \Gamma \hat{\mathbf{u}}_n$.

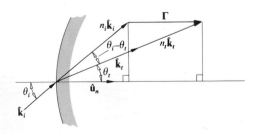

Γ is often referred to as the *astigmatic constant*; Γ = the difference between the projections of $n_t \hat{\mathbf{k}}_t$ and $n_i \hat{\mathbf{k}}_i$ on $\hat{\mathbf{u}}_n$, i.e. take dot product $\mathbf{\Gamma} \cdot \hat{\mathbf{u}}_n$:

$$\Gamma = n_t \cos \theta_t - n_i \cos \theta_i.$$

4.11 Since $\theta_i = \theta_r$, $\hat{\mathbf{k}}_{ix} = \hat{\mathbf{k}}_{rx}$ and $\hat{\mathbf{k}}_{iy} = -\hat{\mathbf{k}}_{ry}$ and since $(\hat{\mathbf{k}}_i \cdot \hat{\mathbf{u}}_n)\hat{\mathbf{u}}_n = \hat{\mathbf{k}}_{iy}$, $\hat{\mathbf{k}}_i - \hat{\mathbf{k}}_r = 2(\hat{\mathbf{k}}_i \cdot \hat{\mathbf{u}}_n)\hat{\mathbf{u}}_n$.

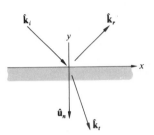

4.12 Since $\overline{SB'} > \overline{SB}$ and $\overline{B'P} > \overline{BP}$, the shortest path corresponds to B' coincident with B in the plane of incidence.

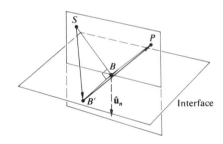

Interface

4.14

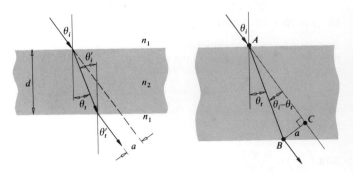

$$n_1 \sin \theta_i = n_2 \sin \theta_t$$
$$\theta_t = \theta_t'$$
$$n_2 \sin \theta_t' = n_1 \sin \theta_i'$$
$$n_1 \sin \theta_i = n_1 \sin \theta_t' \text{ and } \theta_i = \theta_t'.$$
$$\cos \theta_t = d/AB$$
$$\sin(\theta_i - \theta_t) = a/AB$$
$$\sin(\theta_i - \theta_t) = \frac{a}{d} \cos \theta_t$$
$$\frac{d \sin(\theta_i - \theta_t)}{\cos \theta_t} = a.$$

4.16 Rather than propagating from point S to point P in a straight line the ray traverses a path which crosses the plate at a sharper angle. Although in so doing the path lengths in air are slightly increased, the decrease in time spent within the plate more than compensates. This being the case, we might expect the displacement a to increase with n_{21}. As n_{21} gets larger for a given θ_i, θ_t decreases, $(\theta_i - \theta_t)$ increases and from the results of Problem 4.14 a clearly increases.

4.17 From Eq. (4.40)

$$r_{\parallel} = \frac{1.52 \cos 30° - \cos 19°13'}{\cos 19°13' + 1.52 \cos 30°}$$

where from Problem 4.1 $\theta_t = 19°\,13'$. Similarly

$$t_{\parallel} = \frac{2 \cos 30}{\cos 19°13' + 1.52 \cos 30°}$$

$$r_{\parallel} = \frac{1.32 - 0.944}{0.944 + 1.32} = 0.165$$

$$t_{\parallel} = \frac{1.732}{0.944 + 1.32} = 0.765.$$

4.18
$$\oint_C \mathbf{E} \cdot d\mathbf{l} = -\iint_A \frac{\partial \mathbf{B}}{\partial t} \cdot d\mathbf{S}. \qquad [3.5]$$

This reduces in the limit to

$$E_{2x}(\overline{BC}) - E_{1x}(\overline{AD}) = 0,$$

since area $\to 0$ and $\partial \mathbf{B}/\partial t$ is finite. Thus $E_{2x} = E_{1x}$.

4.19 Starting with Eq. (4.34) divide top and bottom by n_i and replace n_{ti} with $\sin \theta_i / \sin \theta_t$ to get

$$r_{\perp} = \frac{\sin \theta_t \cos \theta_i - \sin \theta_i \cos \theta_t}{\sin \theta_t \cos \theta_i + \sin \theta_i \cos \theta_t},$$

which is equivalent to Eq. (4.42). Equation (4.44) follows in exactly the same way. To find $r_{\parallel}$ start the same way with Eq. (4.40) and get

$$r_{\parallel} = \frac{\sin \theta_i \cos \theta_i - \cos \theta_t \sin \theta_t}{\cos \theta_t \sin \theta_t + \sin \theta_i \cos \theta_i}.$$

There are several routes that can be taken now: one is to rewrite $r_{\parallel}$ as

$$r_{\parallel} = \frac{[\sin \theta_i \cos \theta_t - \sin \theta_t \cos \theta_i](\cos \theta_i \cos \theta_t - \sin \theta_i \sin \theta_t)}{[\sin \theta_i \cos \theta_t + \sin \theta_t \cos \theta_i](\cos \theta_i \cos \theta_t + \sin \theta_i \sin \theta_t)}$$

and so

$$r_{\parallel} = \frac{\sin(\theta_i - \theta_t)\cos(\theta_i + \theta_t)}{\sin(\theta_i + \theta_t)\cos(\theta_i - \theta_t)} = \frac{\tan(\theta_i - \theta_t)}{\tan(\theta_i + \theta_t)}.$$

We can find $t_{\parallel}$, which has the same denominator, in a similar way.

4.20 $[E_{0i}]_{\perp} + [E_{0i}]_{\perp} = [E_{0t}]_{\perp}$; tangential field in incident medium equals that in transmitting medium

$$[E_{0t}/E_{0i}]_{\perp} - [E_{0r}/E_{0i}]_{\perp} = 1, \qquad t_{\perp} - r_{\perp} = 1.$$

Alternatively, from Eqs. (4.42) and (4.44),

$$\frac{+\sin(\theta_i - \theta_t) + 2 \sin \theta_t \cos \theta_i}{\sin(\theta_i + \theta_t)} \overset{?}{=} 1$$

$$\frac{\sin \theta_i \cos \theta_t - \cos \theta_i \sin \theta_t + 2 \sin \theta_t \cos \theta_i}{\sin \theta_i \cos \theta_t + \cos \theta_i \sin \theta_t} = 1.$$

4.23 $1.00029 \sin 88.7° = n \sin 90°$

$$(1.00029)(0.99974) = n, \quad n = 1.00003.$$

4.24
$$\theta_i + \theta_t = 90° \text{ when } \theta_i = \theta_p$$
$$n_i \sin \theta_p = n_t \sin \theta_t = n_t \cos \theta_p$$
$$\tan \theta_p = n_t/n_i = 1.52, \quad \theta_p = 56°\,40' \qquad [8.25]$$

4.25 $\tan \theta_p = n_t/n_i = n_2/n_1$, $\tan \theta_p' = n_1/n_2$, $\tan \theta_p = 1/\tan \theta_p'$.

$$\frac{\sin \theta_p}{\cos \theta_p} = \frac{\cos \theta_p'}{\sin \theta_p'} \quad \therefore \sin \theta_p \sin \theta_p' - \cos \theta_p \cos \theta_p' = 0$$

$$\cos(\theta_p + \theta_p') = 0, \quad \theta_p + \theta_p' = 90°.$$

4.26 From Eq. (4.94)

$$\tan \gamma_r = r_{\perp}[E_{0i}]_{\perp}/r_{\parallel}[E_{0i}]_{\parallel}$$

$$\tan \gamma_r = \frac{r_{\perp}}{r_{\parallel}} \tan \gamma_i$$

and from Eqs. (4.42) and (4.43)

$$\tan \gamma_r = -\frac{\cos(\theta_i - \theta_t)}{\cos(\theta_i + \theta_t)} \tan \gamma_i$$

4.28

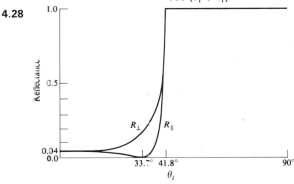

4.29 $T_\perp = \left(\dfrac{n_t \cos \theta_t}{n_i \cos \theta_i}\right) t_\perp^2$. From Eq. (4.44) and Snell's law,

$$T_\perp = \left(\frac{\sin \theta_i \cos \theta_t}{\sin \theta_t \cos \theta_i}\right)\left(\frac{4 \sin^2 \theta_t \cos^2 \theta_i}{\sin^2 (\theta_i + \theta_t)}\right) = \frac{\sin 2\theta_i \sin 2\theta_t}{\sin^2 (\theta_i + \theta_t)}.$$

Similarly for $T_\parallel$.

4.31 If Φ_i is the incident radiant flux or power and T is the transmittance across the first air–glass boundary, the transmitted flux is then $T\Phi_i$. From Eq. (4.68) at normal incidence the transmittance from glass to air is also T. Thus a flux $T\Phi_i T$ emerges from the first slide and $\Phi_i T^{2N}$ from the last one. Since $T = 1 - R$, $T_t = (1 - R)^{2N}$ from Eq. (4.67).

$$R = (0.5/2.5)^2 = 4\%, \quad T = 96\%$$

$$T_t = (0.96)^6 \approx 78.3\%.$$

4.32 $T = \dfrac{I(y)}{I_0} = e^{-\alpha y}$, $T_1 = e^{-\alpha}$, $T = (T_1)^y$. $T_t = (1 - R)^{2N}(T_1)^d$.

4.33 At $\theta_i = 0$, $R = R_\parallel = R_\perp = \left(\dfrac{n_t - n_i}{n_t + n_i}\right)^2$.   [4.67]

As $n_{ti} \to 1$, $n_t \to n_i$ and clearly $R \to 0$.
At $\theta_i = 0$,

$$T = T_\parallel = T_\perp \frac{4 n_t n_i}{(n_t + n_i)^2}$$

and since $n_t \to n_i$, $\lim\limits_{n_{ti} \to 1} T = 4 n_i^2 / (2 n_i)^2 = 1$.

From Problem 4.29, i.e. Eqs. (4.100) and (4.101) and the fact that as $n_t \to n_i$ Snell's law says that $\theta_t \to \theta_i$, we have

$$\lim_{n_{ti} \to 1} T_\parallel = \frac{\sin^2 2\theta_i}{\sin^2 2\theta_i} = 1, \quad \lim_{n_{ti} \to 1} T_\perp = 1.$$

From Eq. (4.43) and the fact that $R_\parallel = r_\parallel^2$ and $\theta_t \to \theta_i$, $\lim\limits_{n_{ti} \to 1} R_\parallel = 0$.

Similarly from Eq. (4.42) $\lim\limits_{n_{ti} \to 1} R_\perp = 0$.

4.35 For $\theta_i > \theta_c$, Eq. (4.70) can be written

$$r_\perp = \frac{\cos \theta_i - i(\sin^2 \theta_i - n_{ti}^2)^{1/2}}{\cos \theta_i + i(\sin^2 \theta_i - n_{ti}^2)^{1/2}},$$

$$r_\perp r_\perp^* = \frac{\cos^2 \theta_i + \sin^2 \theta_i - n_{ti}^2}{\cos^2 \theta_i + \sin^2 \theta_i - n_{ti}^2} = 1.$$

Similarly $r_\parallel r_\parallel^* = 1$.

4.36

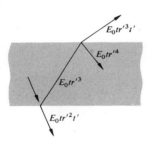

4.38 From Eq. (4.45)

$$t_\parallel'(\theta_p') t_\parallel(\theta_p) = \left[\frac{2 \sin \theta_p \cos \theta_p'}{\sin (\theta_p + \theta_p') \cos (\theta_p' - \theta_p)}\right]$$

$$\times \left[\frac{2 \sin \theta_p' \cos \theta_p}{\sin (\theta_p + \theta_p') \cos (\theta_p - \theta_p')}\right]$$

$$= \frac{\sin 2\theta_p' \sin 2\theta_p}{\cos^2 (\theta_p - \theta_p')} \quad \text{since } \theta_p + \theta_p' = 90°$$

$$= \frac{\sin^2 2\theta_p}{\cos^2 (\theta_p - \theta_p')} \quad \text{since } \sin 2\theta_p' = \sin 2\theta_p$$

$$= \frac{\sin^2 2\theta_p}{\cos^2 (2\theta_p - 90°)} = 1.$$

$$t_\parallel = \frac{2 \sin \theta_2 \cos \theta_1}{\sin (\theta_1 + \theta_2) \cos (\theta_1 - \theta_2)}$$

$$t_\parallel' = \frac{2 \sin \theta_1 \cos \theta_2}{\sin (\theta_1 + \theta_2) \cos (\theta_2 - \theta_1)}$$

$$t_\parallel t_\parallel' = \frac{\sin 2\theta_1 \sin 2\theta_2}{\sin^2 (\theta_1 + \theta_2) \cos^2 (\theta_1 - \theta_2)} = T_\parallel \text{ from Eq. (4.100).}$$

Similarly $t_\perp t_\perp' = T_\perp$

$$r_\parallel^2 = \left[\frac{\tan (\theta_1 - \theta_2)}{\tan (\theta_1 + \theta_2)}\right]^2 = \left[\frac{-\tan (\theta_2 - \theta_1)}{\tan (\theta_1 + \theta_2)}\right]^2.$$

$$r_\parallel'^2 = \left[\frac{\tan (\theta_2 - \theta_1)}{\tan (\theta_1 + \theta_2)}\right]^2 = r_\parallel^2 = R_\parallel.$$

4.39

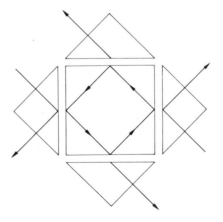

Can be used as a mixer to get various proportions of the two incident waves in the emitted beams. This could be done by adjusting gaps. [For some further remarks, see H. A. Daw and J. R. Izatt, *J. Opt. Soc. Am.* **55**, 201 (1965).]

4.40 From Fig. 4.37 the obvious choice is silver. Note that in the vicinity of 300 nm, $n_I \approx n_R \approx 0.6$, in which case Eq. (4.83) yields $R \approx 0.18$. Just above 300 nm n_I increases rapidly while n_R decreases quite strongly with the result that $R \approx 1$ across the visible and then some.

4.41 Light traverses the base of the prism as an evanescent wave which propagates along the adjustable coupling gap. Energy moves into the dielectric film when the evanescent wave meets certain requirements. The film acts like a waveguide which will support characteristic vibration configurations or modes. Each mode has associated with it a given speed and polarization. The evanescent wave will couple into the film when it matches a mode configuration.

CHAPTER 5

5.1 From (5.2), $\ell_o + \ell_i 3/2 = $ constant, $5 + (6)3/2 = 14$. Therefore $2\ell_o + 3\ell_i = 28$ when $\ell_o = 6$, $\ell_i = 5.3$, $\ell_o = 7$, $\ell_i = 4.66$ Note that the arcs centered on S and P have to intercept for physically meaningful values of ℓ_o and ℓ_i.

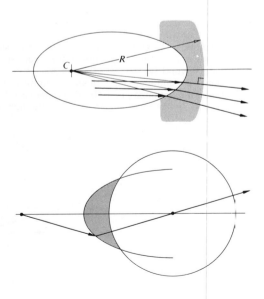

5.2 From Fig. 5.4(b) a plane wave impinging on a concave elliptical surface becomes spherical. If the second spherical surface has that same curvature, the wave will have all rays normal to it and emerge unaltered.

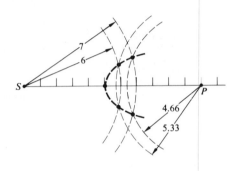

5.4 First surface:

$$\frac{n_1}{s_o} + \frac{n_2}{s_i} = \frac{n_2 - n_1}{R}.$$

$$\frac{1}{1.2} + \frac{1.5}{s_i} = \frac{0.5}{0.1}.$$

$s_i = 0.36$ m (real image 0.36 m to the right of first vertex). Second surface $s_o = 0.20 - 0.36 = -0.16$ m (virtual object distance).

$$\frac{1.5}{-0.16} + \frac{1}{s_i} = \frac{-0.5}{-0.1}, \quad s_i = 0.069.$$

Final image is real ($s_i > 0$), inverted ($M_T < 0$) and 6.9 cm to the right of the second vertex.

5.5 $s_o + s_i = s_o s_i / f$ to minimize $s_o + s_i$,

$$\frac{d}{ds_o}(s_o + s_i) = 0 = 1 + \frac{ds_i}{ds_o} \text{ or } \frac{d}{ds_o}\left(\frac{s_o s_i}{f}\right) = \frac{s_i}{f} + \frac{s_o}{f}\frac{ds_i}{ds_o} = 0.$$

Thus

$$\frac{ds_i}{ds_o} = -1 \quad \text{and} \quad \frac{ds_i}{ds_o} = -\frac{s_i}{s_o}, \quad \therefore s_i = s_o.$$

The separation would be maximum if either were ∞, but both could not be. Hence, $s_i = s_o$ is the condition for a minima. From Gaussian equation, $s_o = s_i = 2f$.

5.6 From (5.8), $1/8 + 1.5/s_i = 0.5/-20$. At first surface, $s_i = -10$ cm. Virtual image 10 cm to left of first vertex. At second surface, object is *real* 15 cm from second vertex.

$$1.5/15 + 1/s_i = -0.5/10, \quad s_i = -20/3 = -6.66 \text{ cm}.$$

Virtual, to left of second vertex.

5.8 $1/5 + 1/s_i = 1/10$, $s_i = -10$ cm virtual, $M_T = -s_i/s_o = 10/5 = 2$ erect. Image is 4 cm high. Or $-5(x_i) = 100$, $x_i = -20$, $M_T = -x_i/f = 20/10 = 2$.

5.9

$$1/s_o + 1/s_i = 1/f$$

s_o	0	f	∞	$2f$	$3f$	$-f$	$-2f$	$f/2$
s_i	0	∞	f	$2f$	$f3/2$	$f/2$	$f2/3$	$-f$

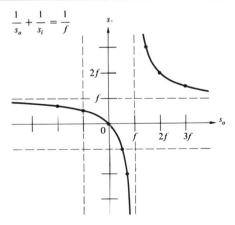

$$\frac{1}{s_o} + \frac{1}{s_i} = \frac{1}{f}$$

5.10 $s_i < 0$ because image is virtual. $1/100 + 1/-50 = 1/f$, $f = -100$ cm. Image is 50 cm to the right as well. $M_T = -s_i/s_o = 50/100 = 0.5$. Ant's image is half sized and erect ($M_T > 0$).

5.12

$$\frac{1}{f} = (n_l - 1)\left(\frac{1}{R_1} - \frac{1}{R_2}\right),$$

$$\frac{1}{f} = 0.5\left(\frac{1}{\infty} - \frac{1}{10}\right) = -\frac{0.5}{10},$$

$f = -20$ cm, $\mathscr{D} = 1/f = -1/0.2 = -5$ diopters.

5.13

$$\frac{1}{f} = (n_{lm} - 1)\left(\frac{1}{R_1} - \frac{1}{R_2}\right),$$

$$\frac{1}{f_w} = \frac{(n_{lm} - 1)}{(n_l - 1)}\frac{1}{f_a} = \frac{1.5/1.33 - 1}{1.5 - 1}\frac{1}{f_a} = \frac{0.125}{0.5}\frac{1}{f_a},$$

$f_w = 4f_a$.

5.15 $1/f = 1/f_1 + 1/f_2$, $1/50 = 1/f_1 - 1/50$, $f_1 = 25$ cm.
If R_{11} and R_{12}, and R_{21} and R_{22} are the radii of the first and second lenses,

$$1/f = (n_l - 1)(1/R_{11} - 1/R_{12}), \quad 1/25 = 0.5(2/R_{11}),$$

$$R_{11} = -R_{12} = -R_{21} = 25 \text{ cm},$$

$$1/f = (n_l - 1)(1/R_{21} - 1/R_{22}),$$

$$-1/50 = 0.55(1/-25 - 1/R_{22}),$$

$$R_{22} = -275 \text{ cm}.$$

5.16 $M_{T_1} = -s_{i1}/s_{o1} = -f_1/(s_{o1} - f_1)$

$M_{T_2} = -s_{i2}/s_{o2} = -s_{i2}/(d - s_{i1})$

$M_T = f_1 s_{i2}/(s_{o1} - f_1)(d - s_{i1})$.

From (5.30), on substituting for s_{i1}, we have

$$M_T = \frac{f_1 s_{i2}}{(s_{o1} - f_1)d - s_{o1}f_1}.$$

5.17 First lens $1/s_{i1} = 1/30 - 1/30 = 0$, $s_{i1} = \infty$.
Second lens $1/s_{i2} = 1/-20 - 1/\infty$,
$s_{i2} = -20$ cm, virtual 10 cm to the left of first lens.

$$M_T = (-\infty/30)(+20/\infty) = \tfrac{2}{3}.$$

From (5.34)

$$M_T = \frac{30(-20)}{10(30 - 30) - 30(30)} = \frac{2}{3}.$$

5.19

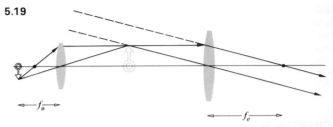

5.21 The angle subtended by L_1 at S is $\tan^{-1} 3/12 = 14°$. To find the image of the diaphragm in L_1 we use Eq. (5.23): $x_o x_i = f^2$, $(-6)(x_i) = 81$, $x_i = -13.5$ cm, so that the image is 4.5 cm behind L_1. The magnification is $-x_i/f = 13.5/9 = 1.5$ and thus the image (of the edge) of the hole is $(0.5)(1.5) = 0.75$ cm in radius. Hence the angle subtended at S is $\tan^{-1} 0.75/16.5 = 2.6°$. The image of L_2 in L_1 is gotten from $(-4)(x_i) = 81$, $x_i = -20.2$ cm, i.e., the image is 11.2 cm to the right of L_1. $M_T = 20.2/9 = 2.2$; hence, the edge of L_2 is imaged 4.4 cm above the axis. Thus its subtended angle at S is $\tan^{-1} 4.4/(12 + 11.2)$ or $9.8°$. Accordingly, the diaphragm is the A.S. and the entrance pupil (its image in L_1) has a diameter of 1.5 cm at 4.5 cm behind L_1. The image of the diaphragm in L_2 is the exit pupil. Consequently, $\tfrac{1}{2} + 1/s_i = \tfrac{1}{3}$ and $s_i = -6$, i.e. 6 cm in front of L_2. $M_T = \tfrac{6}{2} = 3$ so that the exit pupil diameter is 3 cm.

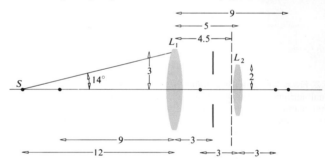

5.22 Either the margin of L_1 or L_2 will be the A.S.; thus, since no lenses are to the left of L_1, either its periphery or P_1 corresponds to the entrance pupil. Beyond (to the left of) point A, L_1 subtends the smallest angle and is the entrance pupil; nearer in (to the right of A) P_1 marks the edge of the entrance pupil. In the former case P_2 is the exit pupil; in the latter (since there are no lenses to the right of L_2), the exit pupil is the edge of L_2 itself.

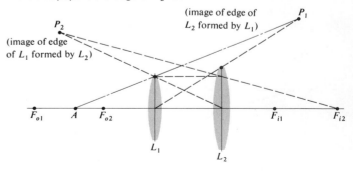

5.23 The A.S. is either the edge of L_1 or L_2. Thus the entrance pupil is either marked by P_1 or P_2. Beyond F_{o1}, P_1 subtends the smaller angle; thus Σ_1 locates the A.S. The image of the A.S. in the lenses to its right, L_2, locates P_3 as the exit pupil.

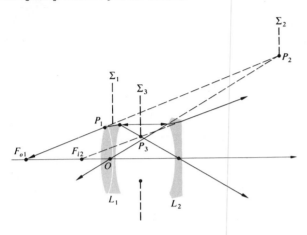

5.24

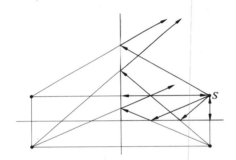

5.26 $1/s_o + 1/s_i = -2/R$. Let $R \to \infty$: $1/s_o + 1/s_i = 0$, $s_o = -s_i$, and $M_T = +1$. Image is virtual, same size and erect.

5.27 From (5.49), $1/100 + 1/s_i = -2/80$, $s_i = -28.5$ cm. Virtual $(s_i < 0)$, erect $(M_T > 0)$ and minified. (Check with Table 5.5).

5.29 Image on screen must be real $\therefore$ s_i is $+$

$$\frac{1}{25} + \frac{1}{100} = -\frac{2}{R}, \quad \frac{5}{100} = -\frac{2}{R}, \quad R = -40 \text{ cm}.$$

5.30 The image is erect and minified. That implies (Table 5.5) a convex spherical mirror.

5.31 No—although she might be looking at you.

5.32 The mirror is parallel to the plane of the painting and so the girl's image should be directly behind her and not off to the right.

5.34 $f = -R/2 = 30$ cm, $1/20 - 1/s_i = 1/30$, $1/s_i = 1/30 - 1/20$.

$s_i = -60$ cm, $M_T = -s_i/s_o = 60/20 = 3$.

Image is virtual $(s_i < 0)$, erect $(M_T > 0)$ located 60 cm behind mirror and 9 inches tall.

5.36

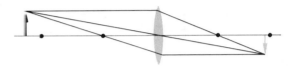

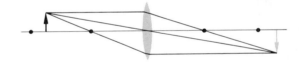

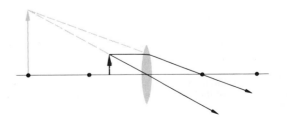

5.37 Draw the chief ray from the tip to L_1 such that when extended it passes through the center of the entrance pupil. From there it goes through the center of the A.S. and then it bends at L_2 so as to extend through the center of the exit pupil. A marginal ray from S extends to the edge of the entrance pupil, is bent at L_1 so it just misses edge of A.S. and then bends at L_2 so as to pass by edge of exit pupil.

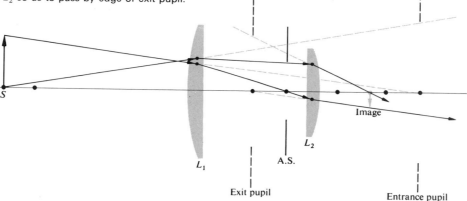

5.38

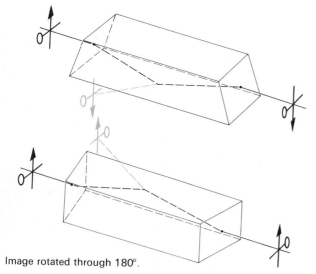

Image rotated through 180°.

5.39 From Eq. (5.61)

$$\text{N.A.} = (2.624 - 2.310)^{1/2} = 0.550,$$

$$\theta_{max} = \sin^{-1} 0.550 = 33° \; 22'.$$

Maximum acceptance angle is $2\theta_{max} = 66° \; 44'$. A ray at 45° would quickly leak out of the fiber, i.e., very little energy fails to escape even at the first reflection.

5.40 $M_T = -f/x_o = -1/x_o \mathscr{D}$. For the eye $\mathscr{D} \approx 58.6$ diopters.

$$x_o \doteq 230{,}000 \times 1.61 = 371 \times 10^3 \text{ km}$$

$$M_T = -1/3.71 \times 10^6 (58.6) = 4.6 \times 10^{-11}$$

$$y_i = 2160 \times 1.61 \times 10^3 \times 4.6 \times 10^{-11} = 0.16 \text{ mm}.$$

5.42 $1/20 + 1/s_{io} = 1/4, \quad s_{io} = 5$ m.

$$1/0.3 + 1/s_{ie} = 1/0.6, \quad s_{ie} = -0.6 \text{ m}.$$

$$M_{To} = -5/10 = -0.5$$

$$M_{Te} = -(-0.6)/0.5 = +1.2$$

$$M_{To} M_{Te} = -0.6.$$

5.45 Ray 1 in the figure misses the eye-lens and there is, therefore, a decrease in the energy arriving at the corresponding image point. This is vignetting.

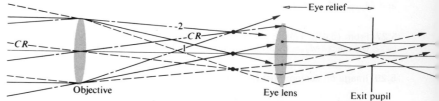

5.46 Rays which would have missed the eye lens in the previous problem are made to pass through it by the field lens. Note how the field lens bends the chief rays a bit so that they cross the optical axis slightly closer to the eye lens thereby moving the exit pupil and shortening the eye relief. (For more on the subject, see *Modern Optical Engineering* by Smith.)

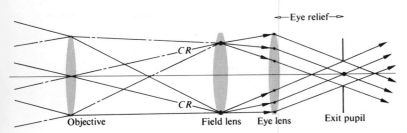

Objective   Field lens   Eye lens   Exit pupil

CHAPTER 6

6.2 From Eq. (6.8)

$$1/f = 1/f' + 1/f' - d/f'f' = 2/f' - 2/3f', \quad f = 3f'/4.$$

From Eq. (6.9),

$$\overline{H_{11}H_1} = (3f'/4)(2f'/3)/f' = f'/2.$$

From Eq. (6.10),

$$\overline{H_{22}H_2} = -(3f'/4)(2f'/3)/f' = -f'/2.$$

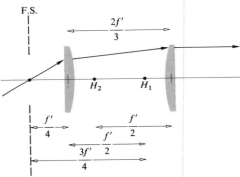

6.3 From Eq. (6.2) $1/f = 0$ when $-(1/R_1 - 1/R_2) = (n_l - 1)d/n_l R_2 R_2$. Thus $d = n_l(R_1 - R_2)/(n_l - 1)$.

6.4 $1/f = 0.5[1/6 - 1/10 + 0.5(3)/1.5(6)10]$

$$= 0.5[10/60 - 6/60 + 1/60]$$

$$f = +24,$$

$$h_1 = -24(0.5)(3)/10(1.5) = -2.4,$$

$$h_2 = -24(0.5)(3)/6(1.5) = -4.$$

6.6 $h_1 = n_{i1}(1 - a_{11})/-a_{12} = (\mathcal{D}_2 d_{21}/n_{t1})f$

$$= -(n_{t1} - 1)d_{21}f/R_2 n_{t1}, \text{ from Eq. (5.64) where } n_{t1} = n_l;$$

$$h_2 = n_{t2}(a_{22} - 1)/-a_{12} = -(\mathcal{D}_1 d_{21}/n_{t1})f \text{ from Eq. (5.63)}$$

$$= -(n_{t1} - 1)d_{21}f/R_1 n_{t1}.$$

6.7 $\mathcal{A} = \mathcal{R}_2 \mathcal{T}_{21} \mathcal{R}_1$, but for the planar surface

$$\mathcal{R}_2 = \begin{bmatrix} 1 & -\mathcal{D}_2 \\ 0 & 1 \end{bmatrix}$$

and $\mathcal{D}_2 = (n_{t1} - 1)/-R_2$ but $R_2 = \infty$

$$\mathcal{R}_2 = \begin{bmatrix} 1 & 0 \\ 0 & 1 \end{bmatrix}$$

which is the unit matrix, hence $\mathcal{A} = \mathcal{T}_{21}\mathcal{R}_1$.

6.8 $\mathcal{D}_1 = (1.5 - 1)/0.5 = 1$ and $\mathcal{D}_2 = (1.5 - 1)/-(-0.25) = 2$

$$\mathcal{A} = \begin{bmatrix} 1 - 2(0.3)/1.5 & -1 + 2(1)(0.3)/1.5 - 2 \\ 0.3/1.5 & -1(0.3)/1.5 + 1 \end{bmatrix} = \begin{bmatrix} 0.6 & -2.6 \\ 0.2 & 0.8 \end{bmatrix}$$

$$|\mathcal{A}| = 0.6(0.8) - (0.2)(-2.6) = 0.48 + 0.52 = 1.$$

6.10 See E. Slayter, *Optical Methods in Biology*. $\overline{PC}/\overline{CA} = (n_1/n_2)R/R = n_1/n_2$ while $\overline{CA}/\overline{P'C} = n_1/n_2$. Therefore triangles ACP and ACP' are similar, using the sine law

$$\frac{\sin \angle PAC}{\overline{PC}} = \frac{\sin \angle APC}{\overline{CA}} \quad \text{or} \quad n_2 \sin \angle PAC = n_1 \sin \angle APC,$$

but $\theta_i = \angle PAC$, thus $\theta_t = \angle APC = \angle P'AC$ and the refracted ray appears to come from P'.

6.11 From Eq. (5.6) let $\cos \varphi = 1 - \varphi^2/2$; then
$$\ell_o = [R^2 + (s_0 + R)^2 - 2R(s_0 + R) + R(s_0 + R)\varphi^2]^{1/2},$$

$$\ell_o^{-1} = [s_o^2 + R(s_o + R)\varphi^2]^{-1/2},$$

$$\ell_i^{-1} = [s_i^2 - R(s_i - R)\varphi^2]^{-1/2},$$

where the first two terms of the binomial series are used,

$$\ell_o^{-1} \approx s_o^{-1} - (s_o + R)h^2/2s_o^3 R \quad \text{where } \varphi \approx h/R,$$

$$\ell_i^{-1} \approx s_i^{-1} + (s_i - R)h^2/2s_i^3 R.$$

Substituting into Eq. (5.5) leads to Eq. (6.40).

6.12

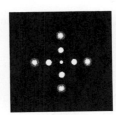

CHAPTER 7

7.1 $E_0^2 = 36 + 64 + 2 \cdot 6 \cdot 8 \cos \pi/2 = 100, \quad E_0 = 10$

$$\tan \alpha = \tfrac{8}{6}, \quad \alpha = 53.1° = 0.93 \text{ rad.}$$

$$E = 10 \sin(120\pi t + 0.93).$$

7.3 $\dfrac{1 \text{ m}}{500 \text{ nm}} = 0.2 \times 10^7 = 2{,}000{,}000$ waves.

In the glass

$$\frac{0.05}{\lambda_0/n} = \frac{0.05(1.5)}{500 \text{ nm}} = 1.5 \times 10^5;$$

in air

$$\frac{0.95}{\lambda_0} = 0.19 \times 10^7;$$

total 2,050,000 waves.
O.P.D. $= [(1.5)(0.05) + (1)(0.95)] - (1)(1)$
O.P.D. $= 1.025 - 1.000 = 0.025$ m

$$\frac{\Lambda}{\lambda_0} = \frac{0.025}{500 \text{ nm}} = 5 \times 10^4 \text{ waves.}$$

7.6 $E = E_1 + E_2 = E_{01}\{\sin[\omega t - k(x + \Delta x) + \sin(\omega t - kx)]\}$.
Since $\sin\beta + \sin\gamma = 2\sin\frac{1}{2}(\beta + \gamma)\cos\frac{1}{2}(\beta - \gamma)$,

$$E = 2E_{01}\cos\frac{k\,\Delta x}{2}\sin\left[\omega t - k\left(x + \frac{\Delta x}{2}\right)\right].$$

7.7 $E = E_0 \text{ Re } [e^{i(kx + \omega t)} - e^{i(kx - \omega t)}]$

$\qquad = E_0 \text{ Re } [e^{ikx}(e^{i\omega t} - e^{-i\omega t})]$

$\qquad = E_0 \text{ Re } [e^{ikx}2i\sin\omega t]$

$\qquad = E_0 \text{ Re } [2i\cos kx \sin\omega t - 2\sin kx \sin\omega t]$

and
$\qquad E = -2E_0 \sin kx \sin\omega t.$

Standing wave with node at $x = 0$.

7.8
$$\frac{\partial E}{\partial x} = -\frac{\partial B}{\partial t}.$$

Integrate to get

$$B(x, t) = -\int\frac{\partial E}{\partial x}dt = -2E_0 k\cos kx \int\cos\omega t\, dt$$

$$= \frac{2E_0 k}{\omega}\cos kx \sin\omega t.$$

But $E_0 k/\omega = E_0/c = B_0$; thus $B(x, t) = 2B_0 \cos kx \sin\omega t$.

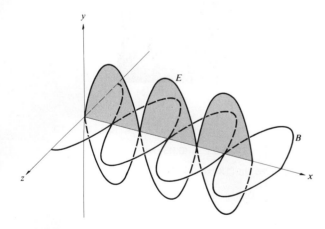

7.10 $E = E_0 \cos\omega_c t + E_0\alpha\cos\omega_m t \cos\omega_c t$

$$= E_0\cos\omega_c t + \frac{E_0\alpha}{2}\left[\cos(\omega_c - \omega_m) + \cos(\omega_c + \omega_m)\right].$$

Audible range $\nu_m = 20$ Hz to 20×10^3 Hz. Maximum modulation frequency $\nu_m(\text{max}) = 20 \times 10^3$ Hz.

$$\nu_c - \nu_m(\text{max}) \leq \nu \leq \nu_c + \nu_m(\text{max})$$

$$\Delta\nu = 2\nu_m(\text{max}) = 40 \times 10^3 \text{ Hz.}$$

7.11 $v = \omega/k = ak, \qquad v_g = d\omega/dk = 2ak = 2v.$

7.12

$$v = \sqrt{\frac{g\lambda}{2\pi}} = \sqrt{g/k}$$

$$v_g = v + k\frac{dv}{dk} \qquad\qquad [7.38]$$

$$\frac{dv}{dk} = -\frac{1}{2k}\sqrt{\frac{g}{k}} = -\frac{v}{2k}$$

$$v_g = v/2.$$

7.14

$$v_g = v + k\frac{dv}{dk} \quad\text{and}\quad \frac{dv}{dk} = \frac{dv}{d\omega}\frac{d\omega}{dk} = v_g\frac{dv}{d\omega}.$$

Since $v = c/n$

$$\frac{dv}{d\omega} = \frac{dv}{dn}\frac{dn}{d\omega} = -\frac{c}{n^2}\frac{dn}{d\omega}$$

$$v_g = v - \frac{v_g ck}{n^2}\frac{dn}{d\omega}$$

$$v_g = \frac{v}{1 + (ck/n^2)(dn/d\omega)} = \frac{c}{n + \omega(dn/d\omega)}.$$

7.15

$$\omega \gg \omega_i, \qquad n^2 = 1 - \frac{Nq_e^2}{\omega^2\epsilon_0 m_e}\sum f_i = 1 - \frac{Nq_e^2}{\omega^2\epsilon_0 m_e}.$$

Using the binomial expansion, we have

$$(1 - x)^{1/2} \approx 1 - \frac{1}{2}x \qquad\text{for}\qquad x \ll 1.$$

$$n = 1 - Nq_e^2/\omega^2\epsilon_0 m_e 2 \qquad dn/d\omega = Nq_e^2/\epsilon_0 m_e\omega^3$$

$$v_g = \frac{c}{n + \omega(dn/d\omega)}$$

$$= \frac{c}{1 - Nq_e^2/\omega^2\epsilon_0 m_e 2 + Nq_e^2/\epsilon_0 m_e\omega^2}$$

$$= \frac{c}{1 + Nq_e^2/\epsilon_0 m_e\omega^2 2}$$

and $v_g < c$,

$$v = c/n = \frac{c}{1 - Nq_e^2/\epsilon_0 m_e\omega^2 2}.$$

Binomial expansion

$$(1 - x)^{-1} \approx 1 + x, \qquad x \ll 1$$

$$v = c[1 + Nq_e^2/\epsilon_0 m_e \omega^2 2]$$

$$vv_g = c^2.$$

7.16

$$\int_0^\lambda \sin akx \sin bkx \, dx$$

$$= \frac{1}{2k}\left[\int_0^\lambda \cos\left[(a - b)kx\right]k \, dx - \int_0^\lambda \cos\left[(a + b)kx\right] k \, dx\right]$$

$$= \frac{1}{2k}\left.\frac{\sin(a - b)kx}{a - b}\right|_0^\lambda - \frac{1}{2k}\left.\frac{\sin(a + b)kx}{a + b}\right|_0^\lambda = 0 \text{ if } a \neq b.$$

Whereas if $a = b$,

$$\int_0^\lambda \sin^2 akx \, dx = \frac{1}{2k}\int_0^\lambda (1 + \cos 2akx) \, k \, dx = \frac{\lambda}{2}.$$

The other integrals are similar.

7.17 Even function, therefore $B_m = 0$.

$$A_0 = \frac{2}{\lambda}\int_{-\lambda/a}^{\lambda/a} dx = \frac{2}{\lambda}\left(\frac{\lambda}{a} + \frac{\lambda}{a}\right) = \frac{4}{a},$$

$$A_m = \frac{2}{\lambda}\int_{-\lambda/a}^{\lambda/a} (1) \cos mkx \, dx = \left.\frac{2}{mk\lambda}\sin mkx\right]_{-\lambda/a}^{\lambda/a},$$

$$A_m = \frac{2}{m\pi}\sin\frac{m2\pi}{a}.$$

7.18

$$f'(x) = \frac{1}{\pi}\int_0^a E_0 L \frac{\sin kL/2}{kL/2}\cos kx \, dk$$

$$= \frac{E_0 L}{\pi 2}\int_0^a \frac{\sin(kL/2 + kx)}{kL/2} dk + \frac{E_0 L}{\pi 2}\int_0^a \frac{\sin(kL/2 - kx)}{kL/2} dk.$$

Let $kL/2 = w$, $\frac{L}{2}dk = dw$, $kx = wx'$.

$$f'(x) = \frac{E_0}{\pi}\int_0^a \frac{\sin(w + wx')}{w} dw + \frac{E_0}{\pi}\int_0^a \frac{\sin(w - wx')}{w} dw.$$

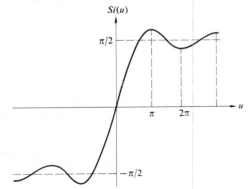

Let $w + wx' = t$, $dw/w = dt/t$.
$0 \leq w \leq a$ and $0 \leq t \leq (x' + 1)a$
let $w - wx' = -t$ in other integral
$0 \leq w \leq a$ and $0 \leq t \leq (x' - 1)a$.

$$f(x) = \frac{1}{\pi}\int_0^{(x' + 1)a} \frac{\sin t}{t} dt - \frac{1}{\pi}\int_0^{(x' - 1)} \frac{\sin t}{t} dt$$

$$f(x) = \frac{1}{\pi}\text{Si}\left[a(x' + 1)\right] - \frac{1}{\pi}\text{Si}\left[a(x' - 1)\right], \quad x' = 2x/L.$$

7.19 By analogy with Eq. (7.61)

$$A(\omega) = \frac{\Delta t}{2}E_C \, \text{sinc}\,(\omega_p - \omega)\frac{\Delta t}{2}.$$

From Table 1 sinc $(\pi/2) = 63.7\%$. Not quite 50% actually,

$$\text{sinc}\left(\frac{\pi}{1.65}\right) = 49.8\%.$$

$$\left|(\omega_p - \omega)\frac{\Delta t}{2}\right| < \frac{\pi}{2} \quad \text{or} \quad -\frac{\pi}{\Delta t} < (\omega_p - \omega) < \frac{\pi}{\Delta t};$$

thus appreciable values of $A(\omega)$ lie in a range $\Delta\omega \sim 2\pi/\Delta t$ and $\Delta v \Delta t \sim 1$. Irradiance is proportional to $A^2(\omega)$ and $[\text{sinc}\,(\pi/2)]^2 = 40.6\%$.

7.20 $\Delta x = c \Delta t$, $\Delta x \sim c/\Delta v$. But $\Delta\omega/\Delta k_0 = \bar{\omega}/\bar{k}_0 = c$; thus $|\Delta v/\Delta\lambda_0| = \bar{v}/\bar{\lambda}_0$,

$$\Delta x \sim \frac{c\bar{\lambda}_0}{\Delta\lambda_0 \bar{v}}, \quad \Delta x \sim \bar{\lambda}_0^2/\Delta\lambda_0.$$

Or try using the uncertainty principle:

$$\Delta x \sim \frac{h}{\Delta p} \quad \text{where} \quad p = h/\lambda \text{ and } \Delta\lambda_0 \ll \bar{\lambda}_0.$$

7.21 $\Delta x = c \Delta t = 3 \times 10^8 \text{ m/s} \, 10^{-8} \text{ s} = 3 \text{ m}.$

$$\Delta\lambda_0 \sim \bar{\lambda}_0^2/\Delta x = (500 \times 10^{-9} \text{ m})^2/3 \text{ m},$$

$$\Delta\lambda_0 \sim 8.3 \times 10^{-14} \text{ m} = 8.3 \times 10^{-5} \text{ nm},$$

$$\Delta\lambda_0/\bar{\lambda}_0 = \Delta v/\bar{v} = 8.3 \times 10^{-5}/500 = 1.6 \times 10^{-7}$$

$$\sim 1 \text{ part in } 10^7.$$

7.22 $\Delta v = 54 \times 10^3 \text{ Hz}$,

$$\Delta v/\bar{v} = \frac{(54 \times 10^3)(10,600 \times 10^{-9} \text{ m})}{(3 \times 10^8 \text{ m/s})} = 1.91 \times 10^{-9}.$$

$$\Delta x = c\Delta t \sim c/\Delta v,$$

$$\Delta x \sim \frac{(3 \times 10^8 \text{ m/s})}{(54 \times 10^3 \text{ Hz})} = 5.55 \times 10^3 \text{ m}.$$

7.24 $\Delta x = c \Delta t = 3 \times 10^8 \times 10^{-10} = 3 \times 10^{-2} \text{ m}$,

$$\Delta v \sim 1/\Delta t = 10^{10} \text{ Hz},$$

$$\Delta\lambda_0 \sim \bar{\lambda}_0^2/\Delta x \text{ (see Problem 7.20)}$$

$$= (632.8 \text{ nm})^2/3 \times 10^{-2} \text{ m} = 0.013 \text{ nm}.$$

$\Delta v = 10^{15}$ Hz, $\Delta x = c \times 10^{-15} = 300$ nm,

$\Delta \lambda_0 \sim \bar{\lambda}_0^2/\Delta x = 1334.78$ nm.

CHAPTER 8

8.1 a) $\mathbf{E} = \hat{i}E_0 \cos(kx - \omega t) + \hat{j}E_0 \cos(kx - \omega t + \pi)$. Equal amplitudes, E_y lags E_x by π. Therefore $\mathscr{P}$-state at 135° or $-45°$.

b) $\mathbf{E} = \hat{i}E_0 \cos(kz - \omega t - \pi/2) + \hat{j}E_0 \cos(kz - \omega t + \pi/2)$. Equal amplitudes, E_y lags E_x by π. Therefore same as (a).

c) E_x leads E_y by $\pi/4$. They have equal amplitudes. Therefore it is an ellipse tilted at $+45°$ and is left-handed.

d) E_y leads E_x by $\pi/2$. They have equal amplitudes. Therefore it is an $\mathscr{R}$-state.

8.2 $\mathbf{E}_x = \hat{i} \cos \omega t$, $\quad \mathbf{E}_y = \hat{j} \sin \omega t$.

Left-handed circular standing wave.

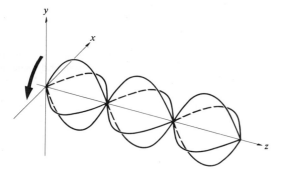

8.3 $\mathbf{E}_{\mathscr{R}} = \hat{i}E_0 \cos(kz - \omega t) + \hat{j}E_0 \sin(kz - \omega t)$

$\mathbf{E}_{\mathscr{L}} = \hat{i}E_0' \cos(kz - \omega t) - \hat{j}E_0' \sin(kz - \omega t)$

$\mathbf{E} = \mathbf{E}_{\mathscr{R}} + \mathbf{E}_{\mathscr{L}} = \hat{i}(E_0 + E_0') \cos(kz - \omega t)$
$+ \hat{j}(E_0 - E_0') \sin(kz - \omega t)$.

Let $E_0 + E_0' = E_{0x}''$ and $E_0 - E_0' = E_{0y}''$; then $\mathbf{E} = \hat{i}E_{0x}'' \cos(kz - \omega t) + \hat{j}E_{0y}'' \sin(kz - \omega t)$. From Eqs. (8.11) and (8.12) it is clear that we have an ellipse where $\varepsilon = -\pi/2$ and $\alpha = 0$.

8.4 In natural light each filter passes 32% of the incident beam. Half of the incoming flux density is in the form of a $\mathscr{P}$-state parallel to the extinction axis and effectively none of this emerges. Thus, 64% of the light parallel to the transmission axis is transmitted. In the present problem 32% I_i enters the second filter and 64% (32% I_i) = 21% I_i leaves it.

8.7 From the figure (upper right) it follows that

$I = \frac{1}{2}E_{01}^2 \sin^2 \theta \cos^2 \theta = \frac{E_{01}^2}{8}(1 - \cos 2\theta)(1 + \cos 2\theta)$

$= \frac{E_{01}^2}{8}(1 - \cos^2 2\theta) = \frac{E_{01}^2}{8}[1 - (\frac{1}{2} \cos 4\theta + \frac{1}{2})]$

$= \frac{E_{01}^2}{16}(1 - \cos 4\theta) = \frac{I_1}{8}(1 - \cos 4\theta)$, $\theta = \omega t$.

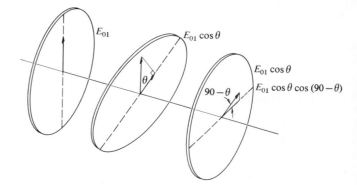

8.8 No. The crystal performs as if it were two oppositely oriented specimens in series. Two similarly oriented crystals in series would behave like one thick specimen and thus separate the o- and e-rays even more.

8.10 Light scattered from the paper passes through the polaroids and becomes linearly polarized. Light from the upper left filter has its $\mathbf{E}$-field parallel to the principal section (which is diagonal across the second and fourth quadrants) and is therefore an e-ray. Notice how the letters P and T are shifted downward in an *extraordinary* fashion. The lower right filter passes an o-ray so that the letter C is undeviated. Note that the ordinary image is closer to the blunt corner.

8.11 (a) and (c) are two aspects of the previous problem. (b) shows double refraction because the polaroid's axis is at roughly 45° to the principal section of the crystal. Thus both an o- and an e-ray will exist.

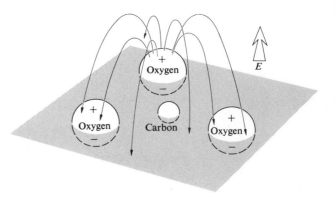

8.13 When $\mathbf{E}$ is perpendicular to the CO_3 plane the polarization will be less than when it is parallel. In the former case, the field of each polarized oxygen atom tends to reduce the polarization of its neighbors. In other words, the induced field, as shown in the figure, is down while $\mathbf{E}$ is up. When $\mathbf{E}$ is in the carbonate plane two dipoles reinforce the third and vice versa. A reduced polarizability leads to a lower dielectric constant, a lower refractive index and a higher speed. Thus $v_{\parallel} > v_{\perp}$.

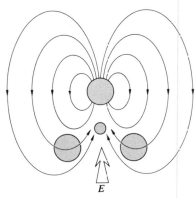

8.14 $n_o = 1.6584$, $n_e = 1.4864$.
Snell's law:

$$\sin \theta_i = n_o \sin \theta_{to} = 0.766$$

$$\sin \theta_i = n_e \sin \theta_{te} = 0.766$$

$$\sin \theta_{to} \approx 0.463, \quad \theta_{to} \approx 27° \, 35';$$

$$\sin \theta_{te} \approx 0.516, \quad \theta_{te} \approx 31° \, 4';$$

$$\Delta \theta \approx 3° \, 29'.$$

8.16 Calcite $n_o > n_e$. Two spectra will be visible when (b) or (c) are used in a spectrometer. The indices are computed in the usual way using

$$n = \frac{\sin \frac{1}{2}(\alpha + \delta_m)}{\sin \frac{1}{2}\alpha},$$

where δ_m is the angle of minimum deviation of either beam.

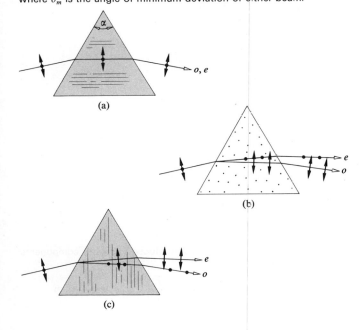

8.17 E_x leads E_y by $\pi/2$. They were initially in phase and $E_x > E_y$. Therefore the wave is left-handed, elliptical and horizontal.

8.18

$$\sin \theta_c = \frac{n_{\text{balsam}}}{n_0} = \frac{1.55}{1.658} = 0.935,$$

$$\theta_c \sim 69°.$$

8.20

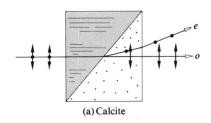

(a) Calcite

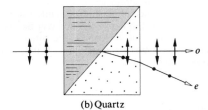

(b) Quartz

c) Undesired energy in the form of one of the $\mathscr{P}$-states can be disposed of without local heating problems.

d) The Rochon transmits an undeviated beam (the o-ray) which is therefore achromatic as well.

8.23

$$\Delta \varphi = \frac{2\pi}{\lambda_0} d \, \Delta n$$

but
$\Delta \varphi = (1/4)(2\pi)$ because of the fringe shift.
Therefore $\Delta \varphi = \pi/2$ and

$$\frac{\pi}{2} = \frac{2\pi d \, (0.005)}{589.3 \times 10^{-9}}$$

$$d = \frac{589.3 \times 10^{-9}}{2(10^{-2})} = 2.94 \times 10^{-5} \, \text{m}.$$

8.24 The $\mathscr{R}$-state incident on the glass screen drives the electrons in circular orbits and they reradiate reflected circular light whose E-field rotates in the same direction as that of the incoming beam. But the propagation direction has been reversed on reflection so that although the incident light is in an $\mathscr{R}$-state, the reflected light is left-handed. It will therefore be completely absorbed by the right-circular polarizer. This is illustrated in the following figure (next page).

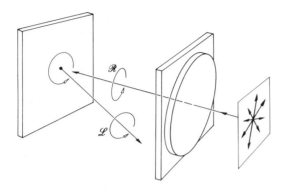

8.25 Yes. If the amplitudes of the $\mathscr{P}$-states differ. The transmitted beam, in a pile-of-plates polarizer, especially for a small pile.

8.27 Place the photoelastic material between circular polarizers with both retarders facing it (as in Fig. 8.52). Under circular illumination no orientation of the stress axes is preferred over any other and they will thus all be indistinguishable. Only the birefringence will have an effect and so the isochromatics will be visible. If the two polarizers are different, i.e. one an $\mathscr{R}$, the other an $\mathscr{L}$, regions where Δn leads to $\Delta\varphi = \pi$ will appear bright. If they are the same, such regions appear dark.

8.29 $V_{\lambda/2} = \lambda_0/2n_0^3 r_{63}$ [8.44]

$$= 550 \times 10^{-9}/2(1.58)^3 5.5 \times 10^{-12}$$

$$= 10^5/2(3.94) = 12.7 \text{ kV.}$$

$$\mathbf{E}_1 \cdot \mathbf{E}_2^* = 0, \quad \mathbf{E}_2 = \begin{bmatrix} e_{21} \\ e_{22} \end{bmatrix}$$

$$\mathbf{E}_1 \cdot \mathbf{E}_2^* = (1)(e_{21})^* + (-2i)(e_{22})^* = 0$$

$$\mathbf{E}_2 = \begin{bmatrix} 2 \\ i \end{bmatrix}$$

$\mathbf{E}_1$ is $\mathbf{E}_2$ is

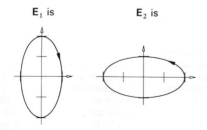

8.31

$$\begin{bmatrix} 1 & 0 & 0 & 0 \\ 0 & 1 & 0 & 0 \\ 0 & 0 & 0 & -1 \\ 0 & 0 & 1 & 0 \end{bmatrix} \begin{bmatrix} 1 & 0 & 0 & 0 \\ 0 & 1 & 0 & 0 \\ 0 & 0 & 0 & -1 \\ 0 & 0 & 1 & 0 \end{bmatrix} = \begin{bmatrix} 1 & 0 & 0 & 0 \\ 0 & 1 & 0 & 0 \\ 0 & 0 & -1 & 0 \\ 0 & 0 & 0 & -1 \end{bmatrix}$$

$$\begin{bmatrix} 1 & 0 & 0 & 0 \\ 0 & 1 & 0 & 0 \\ 0 & 0 & -1 & 0 \\ 0 & 0 & 0 & -1 \end{bmatrix} \begin{bmatrix} 1 \\ 0 \\ 0 \\ 1 \end{bmatrix} = \begin{bmatrix} 1 \\ 0 \\ 0 \\ -1 \end{bmatrix}$$

$$\begin{bmatrix} 1 & 0 & 0 & 0 \\ 0 & 1 & 0 & 0 \\ 0 & 0 & -1 & 0 \\ 0 & 0 & 0 & -1 \end{bmatrix} \begin{bmatrix} 1 \\ 0 \\ 0 \\ -1 \end{bmatrix} = \begin{bmatrix} 1 \\ 0 \\ 0 \\ 1 \end{bmatrix}$$

$$\begin{bmatrix} 1 & 0 & 0 & 0 \\ 0 & 1 & 0 & 0 \\ 0 & 0 & 0 & 1 \\ 0 & 0 & -1 & 0 \end{bmatrix} \begin{bmatrix} 1 & 0 & 0 & 0 \\ 0 & 1 & 0 & 0 \\ 0 & 0 & 0 & 1 \\ 0 & 0 & -1 & 0 \end{bmatrix} = \begin{bmatrix} 1 & 0 & 0 & 0 \\ 0 & 1 & 0 & 0 \\ 0 & 0 & -1 & 0 \\ 0 & 0 & 0 & -1 \end{bmatrix}.$$

8.32

$$\begin{bmatrix} 1 & 0 & 0 & 0 \\ 0 & 1 & 0 & 0 \\ 0 & 0 & 0 & -1 \\ 0 & 0 & 1 & 0 \end{bmatrix} \frac{1}{2}\begin{bmatrix} 1 & 0 & 1 & 0 \\ 0 & 0 & 0 & 0 \\ 1 & 0 & 1 & 0 \\ 0 & 0 & 0 & 0 \end{bmatrix} = \frac{1}{2}\begin{bmatrix} 1 & 0 & 1 & 0 \\ 0 & 0 & 0 & 0 \\ 0 & 0 & 0 & 0 \\ 1 & 0 & 1 & 0 \end{bmatrix}$$

$$\frac{1}{2}\begin{bmatrix} 1 & 0 & 1 & 0 \\ 0 & 0 & 0 & 0 \\ 0 & 0 & 0 & 0 \\ 1 & 0 & 1 & 0 \end{bmatrix} \begin{bmatrix} 1 \\ 0 \\ 0 \\ 0 \end{bmatrix} = \frac{1}{2}\begin{bmatrix} 1 \\ 0 \\ 0 \\ 1 \end{bmatrix}$$

$$\frac{1}{2}\begin{bmatrix} 1 & 0 & 1 & 0 \\ 0 & 0 & 0 & 0 \\ 0 & 0 & 0 & 0 \\ 1 & 0 & 1 & 0 \end{bmatrix} \begin{bmatrix} 1 \\ 0 \\ 1 \\ 0 \end{bmatrix} = \frac{1}{2}\begin{bmatrix} 1 \\ 0 \\ 0 \\ 1 \end{bmatrix}$$

$$\frac{1}{2}\begin{bmatrix} 1 & 0 & 0 & 1 \\ 0 & 0 & 0 & 0 \\ 0 & 0 & 0 & 0 \\ 1 & 0 & 0 & 1 \end{bmatrix} \begin{bmatrix} 1 \\ 0 \\ 0 \\ 1 \end{bmatrix} = \frac{1}{2}\begin{bmatrix} 1 \\ 0 \\ 0 \\ 1 \end{bmatrix}$$

$$\frac{1}{2}\begin{bmatrix} 1 & 0 & 1 & 0 \\ 0 & 0 & 0 & 0 \\ 0 & 0 & 0 & 0 \\ 1 & 0 & 1 & 0 \end{bmatrix} \begin{bmatrix} 1 \\ 0 \\ 0 \\ -1 \end{bmatrix} = \frac{1}{2}\begin{bmatrix} 1 \\ 0 \\ 0 \\ 1 \end{bmatrix}$$

$$\frac{1}{2}\begin{bmatrix} 1 & 0 & 0 & 1 \\ 0 & 0 & 0 & 0 \\ 0 & 0 & 0 & 0 \\ 1 & 0 & 0 & 1 \end{bmatrix} \begin{bmatrix} 1 \\ 0 \\ 0 \\ -1 \end{bmatrix} = \begin{bmatrix} 0 \\ 0 \\ 0 \\ 0 \end{bmatrix}$$

8.34

$$\begin{bmatrix} te^{i\varphi} & 0 \\ 0 & te^{i\varphi} \end{bmatrix},$$

where a phase increment of φ is introduced into both components as a result of traversing the plate.

$$\begin{bmatrix} 1 & 0 \\ 0 & 1 \end{bmatrix} \qquad \begin{bmatrix} 0 & 0 \\ 0 & 0 \end{bmatrix}.$$

8.35

$$\begin{bmatrix} t^2 & 0 & 0 & 0 \\ 0 & t^2 & 0 & 0 \\ 0 & 0 & t^2 & 0 \\ 0 & 0 & 0 & t^2 \end{bmatrix} \begin{bmatrix} 1 & 0 & 0 & 0 \\ 0 & 0 & 0 & 0 \\ 0 & 0 & 0 & 0 \\ 0 & 0 & 0 & 0 \end{bmatrix}.$$

8.36 $V = \dfrac{I_p}{I_p + I_u} = \dfrac{(S_1^2 + S_2^2 + S_3^2)^{1/2}}{S_0}$ [8.56] and [8.29]

$$I_p = (S_1^2 + S_2^2 + S_3^2)^{1/2}$$

$$I - I_p = I_u.$$

a) $S_0 - (S_1^2 + S_2^2 + S_3^2)^{1/2} = I_u$

$$\begin{bmatrix} 4 \\ 0 \\ 0 \\ 0 \end{bmatrix} + \begin{bmatrix} 1 \\ 0 \\ 0 \\ 1 \end{bmatrix} = \begin{bmatrix} 5 \\ 0 \\ 0 \\ 1 \end{bmatrix}$$

$$5 - (0 + 0 + 1)^{1/2} = I_u.$$

CHAPTER 9

9.1 $E_1 \cdot E_2 = \frac{1}{2}(E_1 e^{-i\omega t} + E_1^* e^{i\omega t}) \cdot \frac{1}{2}(E_2 e^{-i\omega t} + E_2^* e^{i\omega t})$,
where Re $(z) = \frac{1}{2}(z + z^*)$.

$$E_1 \cdot E_2 = \tfrac{1}{4}[E_1 \cdot E_2 e^{-2i\omega t} + E_1^* \cdot E_2^* e^{2i\omega t} + E_1 \cdot E_2^* + E_1^* \cdot E_2].$$

The last two terms are time independent while

$$\langle E_1 \cdot E_2 e^{-2i\omega t} \rangle \to 0 \text{ and } \langle E_1^* \cdot E_2^* e^{2i\omega t} \rangle \to 0$$

because of the $1/T\omega$ coefficient. Thus

$$I_{12} = 2\langle E_1 \cdot E_2 \rangle = \tfrac{1}{2}(E_1 \cdot E_2^* + E_1^* \cdot E_2).$$

9.2 The largest value of $(r_1 - r_2)$ is equal to a. Thus if $\varepsilon_1 = \varepsilon_2$, $\delta = k(r_1 - r_2)$ varies from 0 to ka. If $a \gg \lambda$, cos δ and therefore I_{12} will have a great many maxima and minima and therefore average to zero over a large region of space. In contrast, if $a \ll \lambda$, δ varies only slightly from 0 to $ka \ll 2\pi$. Hence I_{12} does not average to zero and from Eq. (9.6) I deviates little from $4I_0$. The two sources effectively behave as a single source of double the original strength.

9.3 A bulb at S would produce fringes. We can imagine it as made up of a very large number of incoherent point sources. Each of these would generate an independent pattern, all of which would then overlap. Bulbs at S_1 and S_2 would be incoherent and could not generate detectable fringes.

9.4 $r_2^2 = a^2 + r_1^2 - 2ar_1 \cos(90 - \theta)$. The contribution to $\cos \delta/2$ from the third term in the Maclaurin expansion will be negligible if

$$\frac{k}{2}\left(\frac{a^2}{2r_1}\cos^2\theta\right) \ll \pi/2;$$

therefore $r_1 \gg a^2/\lambda$.

9.8 $\Delta y = s\lambda/2d\alpha(n - n')$.

9.9 $\Delta y = (s/a)\lambda$, $a = 10^{-2}$ cm, $a/2 = 5 \times 10^{-3}$ cm.

9.11 Eq. (9.25) $m = 2n_f d/\lambda_0 = 10{,}000$. A minimum, therefore central dark region.

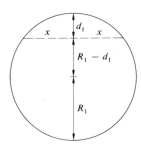

9.12 $x^2 = d_1[(R_1 - d_1) + R_1]$,

$$x^2 = 2R_1 d_1 - d_1^2.$$

Similarly

$$x^2 = 2R_2 d_2 - d_2^2.$$

$$c = d_1 - d_2 = \frac{x^2}{2}\left[\frac{1}{R_1} - \frac{1}{R_2}\right], \quad d = m\frac{\lambda_f}{2}.$$

As $R_2 \to \infty$, x_m approaches Eq. (9.27).

9.13 One plane and one spherical mirror.

9.15 $\Delta z = \lambda_f/2\alpha$, $\alpha = \lambda_0/2n_f \Delta x$,

$$\alpha = 5 \times 10^{-5} \text{ rad} = 10.2 \text{ seconds}.$$

9.18 $E_r E_r^* = E_t E_t^* = E_0^2 (tt')^2/(1 - r^2 e^{-i\delta})(1 - r^2 e^{+i\delta})$

$$I = I_i(tt')^2/(1 - r^2 e^{-i\delta} - r^2 e^{i\delta} + r^4).$$

9.20 At near normal incidence $(\theta_i \approx 0)$ Fig. 4.23(e) indicates that the relative phase shift between an internally and externally reflected beam is π rad. That means a total relative phase difference of

$$\frac{2\pi}{\lambda_f}[2(\lambda_f/4)] + \pi$$

or 2π. The waves are in phase and interfere constructively.

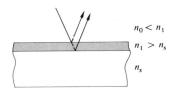

$n_0 < n_1$
$n_1 > n_s$
n_s

9.21 $\delta = k(r_1 - r_2) + \pi$ (Lloyd's mirror)
$\delta = k\{a/2 \sin\alpha - [\sin(90 - 2\alpha)]a/2 \sin\alpha\} + \pi$
$\delta = ka(1 - \cos 2\alpha)/2 \sin\alpha + \pi$, maximum occurs for
$\delta = 2\pi$ when $\sin\alpha(\lambda/a) = (1 - \cos 2\alpha) = 2\sin^2\alpha$.
First maximum $\alpha = \sin^{-1}(\lambda/2a)$.

9.22 $n_a = 1$ $n_s = n_g$ $n_1 = \sqrt{n_g}$

$$\sqrt{1.54} = 1.24$$

$$d = \tfrac{1}{4}\lambda_f = \tfrac{1}{4}\frac{\lambda_0}{n_1} = \frac{540}{1.24} \text{ nm}.$$

No relative phase shift between two waves.

9.24 $I = I_{max} \cos^2 \delta/2$

$I = I_{max}/2$ when $\delta = \pi/2$ $\therefore \gamma = \pi$.

Separation between maxima is 2π

$\mathscr{F} = 2\pi/\gamma = 2$.

9.25 a) $R = 0.80$ $\therefore F = 4R/(1 - R)^2 = 80$

b) $\gamma = 4 \sin^{-1} 1/\sqrt{F} = 0.448$

c) $\mathscr{F} = 2\pi/0.448$ d) $C = 1 + F$

9.26 The fringes are generally a series of fine jagged bands which are fixed with respect to the glass.

9.29

$$\frac{2}{1 + F(\Delta\delta/4)^2} = 0.81 \left[1 + \frac{1}{1 + F(\Delta\delta/2)^2} \right]$$

$$F^2(\Delta\delta)^4 - 15.5F(\Delta\delta)^2 - 30 = 0.$$

9.30

$$E = \tfrac{1}{2}mv^2 \quad v = 0.42 \times 10^6 \text{ m/s}$$

$$\lambda = h/mv = 1.73 \times 10^{-9}$$

$$\Delta y = s\lambda/a = 3.46 \text{ mm}.$$

CHAPTER 10

10.1 $(L + \ell)^2 = L^2 + r^2$ $\therefore L = (r^2 - \ell^2)/2\ell \approx r^2/2\ell$, $\ell L = r^2/2$, so for $\lambda \gg \ell$, $\lambda L \gg r^2/2$ $\therefore L = (1 \times 10^{-3})^2 10/\lambda^2 = 10$ m.

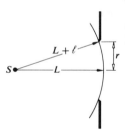

10.2 $E_0/2 = R \sin(\delta/2)$

$E = 2R \sin(N\delta/2)$ chord length

$E = [E_0 \sin(N\delta/2)]/\sin(\delta/2)$

$I = E^2$.

10.4

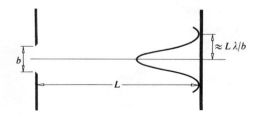

$\beta = \pm\pi$

$\sin\theta = \pm\lambda/b$

$\theta \approx \pm\lambda/b$

$L\theta \approx \pm L\lambda/b$

$L\theta \approx \pm f_2\lambda/b$.

10.5 $\alpha = \dfrac{ka}{2} \sin\theta, \quad \beta = \dfrac{kb}{2} \sin\theta$

$a = mb$, $\alpha = m\beta$, $\alpha = m2\pi$

N = number of fringes $= \alpha/\pi = m2\pi/\pi = 2m$.

10.6 $\alpha = 3\pi/2N = \pi/2$ [10.34]

$$I(\theta) = \frac{I(0)}{N^2} \left(\frac{\sin\beta}{\beta} \right)^2 \quad\quad \text{from Eq. (10.35)}$$

$$I/I(\theta) \approx \tfrac{1}{9}.$$

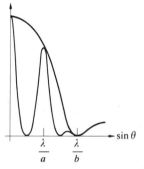
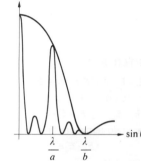

10.8 $\sin\theta_i = n \sin\theta_n$

Optical path length difference $= m\lambda$

$a \sin\theta_m - na \sin\theta_n = m\lambda$.

$a(\sin\theta_m - \sin\theta_i) = m\lambda$.

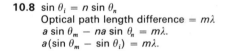
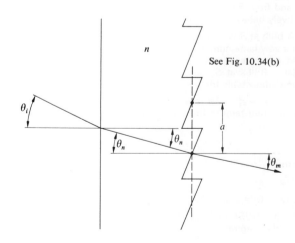

See Fig. 10.34(b)

10.9 $\mathscr{R} = mN = 10^6, N = 78 \times 10^3$
$$\therefore m = 10^6/78 \times 10^3$$
$$\Delta\lambda_{fsr} = \lambda/m = 500 \text{ nm}/(10^6/78 \times 10^3) = 39 \text{ nm}.$$

$$\mathscr{R} = \mathscr{F}m = \mathscr{F}\frac{2n_f d}{\lambda} = 10^6 \qquad \text{[9.54]}$$

$$\Delta\lambda_{fsr} = \lambda^2/2n_f d = 0.0125 \text{ nm}. \qquad \text{[9.56]}$$

10.11 $A = 2\pi\rho^2 \displaystyle\int_0^\varphi \sin\varphi \, d\varphi = 2\pi\rho^2(1 - \cos\varphi)$

$$\cos\varphi = [\rho^2 + (\rho + r_0)^2 - r^2]/2\rho(\rho + r_0)$$

$$r_l = r_0 + l\lambda/2.$$

Area of first l zones

$$A = 2\pi\rho^2 - \pi\rho(2\rho^2 + 2\rho r_0 - l\lambda r_0 - l^2\lambda^2/4)/(\rho + r_0)$$

$$A_l = A - A_{l-1} = \frac{\lambda\pi\rho}{\rho + r_0}\left[r_0 + \frac{(2l-1)\lambda}{4}\right].$$

10.13

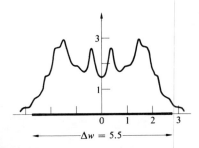

$\Delta w = 5.5$

10.14 1 part in 1000
3 yd ≈ 100 inches.

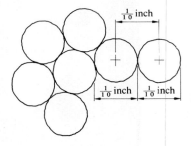

10.15 $I = \dfrac{I_0}{2}\{[\tfrac{1}{2} - \mathscr{C}(v_1)]^2 + [\tfrac{1}{2} - \mathscr{S}(v_1)]^2\}$

$$I = \frac{I_0}{2}\left(\frac{1}{\pi v_1}\right)^2\left[\sin^2\left(\frac{\pi v_1^2}{2}\right) + \cos^2\left(\frac{\pi v_1^2}{2}\right)\right] = \frac{I_0}{2}\left(\frac{1}{\pi v_1}\right)^2.$$

10.17 Fringes in both the clear and shadow region [(see M. P. Givens and W. L. Goffe *Am. J. Phys.* **34**, 248 (1966)].

10.18 $y = L\lambda/d$
$$d = 12 \times 10^{-6}/12 \times 10^{-2} = 10^{-4} \text{ m}.$$

10.19 Converging spherical wave in image space is diffracted by the exit pupil.

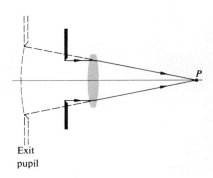

Exit pupil

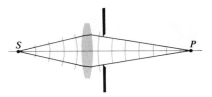

10.21 $\mathscr{R} = \lambda/\Delta\lambda = 5892.9/5.9 = 999$
$$N = \mathscr{R}/m = 333.$$

10.22 $u = y[2/\lambda r_0]^{1/2}$
$$\Delta u = \Delta y \times 10^3 = 2.5.$$

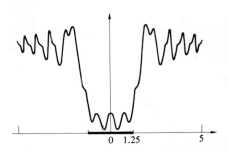

10.23 $d \sin\theta_m = m\lambda,$ $\qquad\qquad \theta = N\delta/2 = \pi$
$7 \sin\theta = (1)(0.21)$ $\qquad \delta = 2\pi/N = kd\sin\theta$
$\sin\theta = 0.03$ $\qquad\qquad \sin\theta = 0.0009$
$\theta = 1.7°$ $\qquad\qquad\quad \theta = 3 \text{ min}.$

10.26 If the aperture is symmetrical about a line the pattern will be symmetrical about a line parallel to it. Moreover, the pattern will be symmetrical about yet another line perpendicular to the aperture's symmetry axis. This follows from the fact that Fraunhofer patterns have a center of symmetry.

10.27

[Photos by E.H.]

10.28

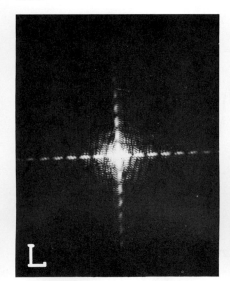

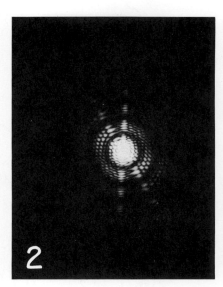

[Photos by E.H.]

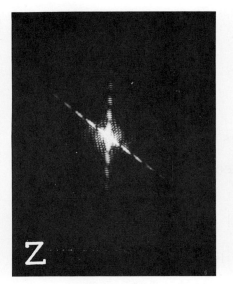

[Photos by E.H.]

10.29

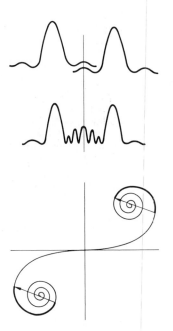

CHAPTER 11

11.1 $\Xi_0 \sin k_p x = E_0(e^{ik_p x} - e^{-ik_p x})/2i$

$$F(k) = \frac{E_0}{2i}\left[\int_{-L}^{+L} e^{i(k+k_p)x}\,dx - \int_{-L}^{+L} e^{i(k-k_p)x}\,dx\right]$$

$$F(k) = -\frac{iE_0 \sin(k+k_p)L}{(k+k_p)} + \frac{iE_0 \sin(k-k_p)L}{(k-k_p)}$$

$$F(k) = iE_0 L\,[\operatorname{sinc}(k-k_p)L - \operatorname{sinc}(k+k_p)L].$$

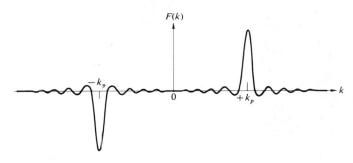

11.3 $\cos^2 \omega_p t = \frac{1}{2} + \frac{1}{2}\cos 2\omega_p t = \frac{1}{2} + \dfrac{e^{2i\omega_p t} + e^{-2i\omega_p t}}{4}.$

$$F(\omega) = \frac{1}{2}\int_{-T}^{+T} e^{i\omega t}\,dt + \frac{1}{4}\int e^{i(\omega+2\omega_p)t}\,dt + \frac{1}{4}\int e^{i(\omega-2\omega_p)t}\,dt$$

$$F(\omega) = \frac{1}{\omega} \sin \omega T + \frac{1}{2(\omega + 2\omega_p)} \sin (\omega + 2\omega_p) T$$

$$+ \frac{1}{2(\omega - 2\omega_p)} \sin (\omega - 2\omega_p) T$$

$$F(\omega) = T \operatorname{sinc} \omega T + \frac{T}{2} \operatorname{sinc} (\omega + 2\omega_p) T + \frac{T}{2} \operatorname{sinc} (\omega - 2\omega_p) T.$$

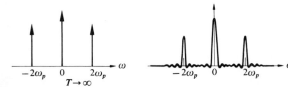

11.5 $\int_{x=-\infty}^{x=+\infty} f(x) h(X-x)\, dx = -\int_{x'=+\infty}^{x'=-\infty} f(X-x') h(x')\, dx'$

$$= \int_{-\infty}^{+\infty} h(x') f(X-x')\, dx'$$

where $x' = X - x$, $dx = -dx'$.

$$f \circledast h = h \circledast f$$

or

$$\mathcal{F}\{f \circledast h\} = \mathcal{F}\{f\} \cdot \mathcal{F}\{h\} = \mathcal{F}\{h\} \cdot \mathcal{F}\{f\} = \mathcal{F}\{h \circledast f\}.$$

11.7 A point on the edge of $f(x, y)$ e.g. at $(x = d, y = 0)$ is spread out into a square 2ℓ on a side centered on $X = d$. Thus it extends no farther than $X = d + \ell$ and so the convolution must be zero at $X = d + \ell$ and beyond.

11.10 $I = \langle E_I^2 \rangle |\sum_j e^{i(k_Y y_j + k_Z z_j)}|^2$.

$$|\sum_j \cos (k_Y y_j + k_Z z_j) + i \sin (k_Y y_j + k_Z z_j)|^2$$

$$= \sum_j^N \cos^2 (k_Y y_j + k_Z z_j) + \sum_j^N \sin^2 (k_Y y_j + k_Z z_j)$$

$$+ 2\sum\sum_{j \neq l} \cos (k_Y y_j + k_Z z_j) \cos (k_Y y_l + k_Z z_l)$$

$$+ 2\sum\sum_{j \neq l} \sin (k_Y y_j + k_Z z_j) \sin (k_Y y_l + k_Z z_l).$$

If there are many randomly positioned apertures, the two double sum terms will tend to zero since the sines and cosines take on values from -1 to $+1$. The squared terms, however, sum up to N times one. Thus

$$I \approx \langle E_I^2 \rangle N.$$

11.11 $\mathscr{A}(y, z) = \mathscr{A}(-y, -z)$.

$$E(Y, Z, t) \propto \int\!\!\int \mathscr{A}(y, z) e^{i(k_Y y + k_Z z)}\, dy\, dz.$$

Change Y to $-Y$, Z to $-Z$, y to $-y$, z to $-z$, then k_Y goes to $-k_Y$ and k_Z to $-k_Z$.

$$E(-Y, -Z) \propto \int\!\!\int \mathscr{A}(-y, -z) e^{i(k_Y y + k_Z z)}\, dy\, dz$$

$$\therefore E(-Y, -Z) = E(Y, Z).$$

11.12 From Eq. (11.63)

$$E(Y, Z) = \int\!\!\int \mathscr{A}(y, z) e^{ik(Yy + Zz)/R}\, dy\, dz$$

$$E'(Y, Z) = \int\!\!\int \mathscr{A}(\alpha y, \beta z) e^{ik(Yy + Zz)/R}\, dy\, dz;$$

now let $y' = \alpha y$ and $z' = \beta z$:

$$E'(Y, Z) = \frac{1}{\alpha\beta} \int\!\!\int \mathscr{A}(y', z') e^{ik[(Y/\alpha)y' + (Z/\beta)z']}\, dy'\, dz'$$

or

$$E'(Y, Z) = \frac{1}{\alpha\beta} E(Y/\alpha, Z/\beta).$$

11.13

$$C_{ff} = \lim_{T \to \infty} \frac{1}{2T} \int_{-T}^{+T} A \sin (\omega t + \varepsilon) A \sin (\omega t - \omega\tau + \varepsilon)\, dt$$

$$= \lim_{T \to \infty} \frac{A^2}{2T} \int [\tfrac{1}{2} \cos (\omega\tau) - \tfrac{1}{2} \cos (2\omega t - \omega\tau + 2\varepsilon)]\, dt,$$

since $\cos \alpha - \cos \beta = -2 \sin \tfrac{1}{2}(\alpha + \beta) \sin \tfrac{1}{2}(\alpha - \beta)$. Thus

$$C_{ff} = \frac{A^2}{2} \cos (\omega\tau).$$

11.14 $E(k_z) = \int_{-b/2}^{+b/2} \mathscr{A}_0 \cos (\pi z/b) e^{ik_z z}\, dz$

$$= \mathscr{A}_0 \int \cos \frac{\pi z}{b} \cos k_z z\, dz + i\mathscr{A}_0 \int \cos \frac{\pi z}{b} \sin k_z z\, dz$$

$$E(k_z) = \mathscr{A}_0 \cos \frac{bk_z}{2} \left[\frac{1}{\left(\frac{\pi}{b} - k_z\right)} + \frac{1}{\left(\frac{\pi}{b} + k_z\right)} \right].$$

CHAPTER 12

12.1 At low pressures, the intensity emitted from the lamp is low, the bandwidth is narrow and the coherence length large. The fringes will initially display a high contrast although they'll be fairly faint. As the pressure builds, the coherence length will decrease, the contrast will drop off and the fringes might even vanish entirely.

12.3 The irradiance at Σ_0 arising from a point source is $4I_0 \cos^2 (\delta/2) = 2I_0(1 + \cos \delta)$.

For a differential source element of width dy' at point S', y' from the axis, the O.P.D. to P at y via the two slits is

$$\Lambda = (\overline{S'S_1} + \overline{S_1P}) - (\overline{S'S_2} + \overline{S_2P})$$

$$= (\overline{S'S_1} - \overline{S'S_2}) + (\overline{S_1P} - \overline{S_2P})$$

$$= ay'/d + ay/s \text{ from Section 9.3.}$$

The contribution to the irradiance from dy' is then

$$dI \propto (1 + \cos k_0 \Lambda)\, dy'$$

$$I \propto \int_{-w/2}^{+w/2} (1 + \cos k_0 \Lambda)\, dy'$$

$$I \propto w + \frac{d}{k_0 a}\left[\sin\left(\frac{ay}{s} + \frac{aw}{2d}\right) - \sin\left(\frac{ay}{s} - \frac{aw}{2d}\right)\right]$$

$$I \propto w + \frac{d}{k_0 a}[\sin(k_0 ay/s)\cos(k_0 aw/2d)$$

$$+ \cos(k_0 ay/s)\sin(k_0 aw/2d)$$

$$- \sin(k_0 ay/2)\cos(k_0 aw/2d)$$

$$+ \cos(k_0 ay/s)\sin(k_0 aw/2d)]$$

$$I \propto w + \frac{d2}{k_0 a}\sin(k_0 aw/2d)\cos(k_0 ay/s).$$

12.4 $\mathscr{V} = \dfrac{I_{max} - I_{min}}{I_{max} + I_{min}}$

$$I_{max} = I_P^{(1)} + I_P^{(2)} + 2\sqrt{I_P^{(1)} I_P^{(2)}}\,|\tilde{\gamma}_{12}|$$

$$I_{min} = I_P^{(1)} + I_P^{(2)} - 2\sqrt{I_P^{(1)} I_P^{(2)}}\,|\tilde{\gamma}_{12}|$$

$$\mathscr{V}_P = \frac{4\sqrt{I_P^{(1)} I_P^{(2)}}\,|\tilde{\gamma}_{12}|}{2(I_P^{(1)} + I_P^{(2)})}.$$

12.5 When $S''S_1 O' - S'S_1 O' = \lambda/2, 3\lambda/2, 5\lambda/2$, etc., the irradiance due to S' is given by

$$I' = 4I_0 \cos^2(\delta'/2) = 2I_0(1 + \cos\delta'),$$

while the irradiance due to S'' is

$$I'' = 4I_0 \cos^2(\delta''/2) = 4I_0 \cos^2(\delta' + \pi)/2 = 2I_0(1 - \cos\delta').$$

Hence $I' + I'' = 4I_0$.

12.7 $\theta = \frac{1}{2}^\circ = 0.0087$ rad
$h = 0.32\bar{\lambda}_0/\theta$ using $\bar{\lambda}_0 = 550$ nm
$h = 0.32\,(550\text{ nm})/0.0087$
$h = 2 \times 10^{-2}$ mm.

12.8 $I_1(t) = \Delta I_1(t) + \langle I_1 \rangle$;
hence
$$\langle I_1(t + \tau) I_2(t)\rangle = \langle[\langle I_1\rangle + \Delta I_1(t+\tau)][\langle I_2\rangle + \Delta I_2(t)]\rangle$$

since $\langle I_1\rangle$ is independent of time.

$$\langle I_1(t + \tau) I_2(t)\rangle = \langle I_1\rangle\langle I_2\rangle + \langle\Delta I_1(t+\tau)\,\Delta I_2(t)\rangle$$

if we recall that $\langle\Delta I_1(t)\rangle = 0$. Eq. (12.34) follows by comparison with Eq. (12.32).

CHAPTER 13

13.1 $M_e = \sigma T^4$ [13.1]

$$(22.8\text{ W cm}^2)(10^4\text{ cm}^2/\text{m}^2) = (5.7 \times 10^{-8}\text{ W m}^{-2}\,^\circ\text{K}^{-4})T^4$$

$$T = \left[\frac{22.8 \times 10^4}{5.7 \times 10^{-8}}\right]^{1/4} = 1.414 \times 10^3$$

$$T = 1414\,^\circ\text{K}.$$

13.3 $v = c/\lambda,\ dv = -c\,d\lambda/\lambda^2.$

Since $M_{e\lambda}$ and M_{ev} are to be positive and since an increase in λ yields a decrease in v, we write

$$M_{e\lambda}\,d\lambda = -M_{ev}\,dv$$

and

$$M_{ev} = -M_{e\lambda}\,d\lambda/dv = M_{e\lambda}\lambda^2/c.$$

13.4

$$\lambda = \frac{h}{mv} = \frac{6.63 \times 10^{-34}\text{ J s}}{(0.15\text{ kg})(25\text{ m/s})}.$$

Baseball $\lambda = \dfrac{6.63 \times 10^{-34}}{3.75} = 1.76 \times 10^{-34}\text{ m}$

Hydrogen $\lambda = \dfrac{6.63 \times 10^{-34}}{(1.67 \times 10^{-27})(10^3)} = 3.96 \times 10^{-10}\text{ m}.$

13.6

$$\lambda = \frac{c}{v} = \frac{hc}{hv} = \frac{(6.63 \times 10^{-34})(3 \times 10^8)}{(1.6 \times 10^{-19})hv[\text{in eV}]}$$

$$\lambda = \frac{12.39 \times 10^{-7}\text{ m}}{hv[\text{in eV}]} = \frac{12,390\,\text{Å}}{hv[\text{in eV}]}.$$

The usual mnemonic is

$$\lambda = \frac{12,345\,\text{Å}}{hv[\text{in eV}]}.$$

13.7 $\lambda(\text{min}) = 300$ nm
$hv = hc/\lambda$

$$= \frac{(6.63 \times 10^{-34}\text{ J s})(3 \times 10^8\text{ m/s})}{300 \times 10^{-9}\text{ m}}$$

$$\mathscr{E} = 6.63 \times 10^{-19}\text{ J} = 4.14\text{ eV}.$$

13.9 $Nhv = (1.4 \times 10^3\text{ W/m}^2)(1\text{ m}^2)(1\text{ s})$

$$N = \frac{1.4 \times 10^3(700 \times 10^{-9})}{(6.63 \times 10^{-34})(3 \times 10^8)} = \frac{980 \times 10^{20}}{19.89}$$

$$N = 49.4 \times 10^{20}.$$

13.10 $hv = \dfrac{hc}{\lambda} = \dfrac{(6.63 \times 10^{-34})(3 \times 10^8)}{500 \times 10^{-9}} = 3.98 \times 10^{-19}\text{ J}$

$$hv = 2.5\text{ eV}.$$

Energy per second $= \pi r^2 I = (3.14)(10^{-20})(10^{-10})$
$$= 3.14 \times 10^{-30}\text{ J/s}$$

$$(T)(3.14 \times 10^{-30}\text{ J/s}) = 3.98 \times 10^{-19}\text{ J}$$

$T = 1.27 \times 10^{11}\text{ s}$ (1 yr $= 3.154 \times 10^7$ s), $T \sim 4000$ years
$\lambda^3 = 25 \times 10^{-14}\text{ m}^2.$ $\lambda^2 I = 25 \times 10^{-24}\text{ J/s}$

$$\bar{t} = \frac{3.98 \times 10^{-19}}{2.5 \times 10^{-23}} = 1.59 \times 10^4\text{ s} \quad (3.6 \times 10^3\text{ s/hr})$$

$\bar{t} = 4.4$ hr (still impossible).

It would take twice as long if $hv = 5$ eV which means (Problem 13.6)

$$\lambda = \frac{12345\text{Å}}{5} = 247 \text{ nm (ultraviolet)}$$

13.11 $v_0 = \Phi_0/h$ [13.8]

$$v_0 = \frac{2.28(1.6 \times 10^{-19})}{6.63 \times 10^{-34}} = 5.5 \times 10^{14} \text{ Hz} = 550 \text{ THz}$$

$$v = c/\lambda = 3 \times 10^8/400 \times 10^{-9} = 750 \times 10^{12} \text{ Hz}.$$

$$\frac{mv^2_{\max}}{2} = h(v - v_0) = h200 \times 10^{12}$$ [13.9]

$$= 13.26 \times 10^{-20} \text{ J}.$$

13.13 The photon's gravitational potential energy $U = -GMm/R$ where m is photon mass but $m = hv/c^2$; thus

$$U = -GMhv/Rc^2.$$

Ergo $\mathscr{E} = hv - GMhv/Rc^2$

$$\mathscr{E} = hv\left(1 - \frac{GM}{c^2R}\right).$$

At the earth $\mathscr{E} = hv_e$ and

$$v_e = v - \frac{GM}{c^2R}v.$$

Since $\Delta v = v - v_e$

$$\Delta v = \frac{GM}{c^2R}v.$$

13.14 $\dfrac{\Delta v}{v} = \dfrac{(6.67 \times 10^{-11} \text{ nm}^2/\text{kg}^2)(1.99 \times 10^{30} \text{ kg})}{(3 \times 10^8 \text{ m/s})^2(6.96 \times 10^8 \text{ m})}$

$$\frac{\Delta v}{v} = 2.12 \times 10^{-6}$$

$$\Delta v = \frac{2.12 \times 10^{-6}(3 \times 10^8)}{650 \times 10^{-9}} = 9.8 \times 10^8 \text{ Hz}$$

or

$$\frac{\Delta\lambda}{\lambda} = \frac{\Delta v}{v} \qquad \Delta\lambda = \Delta v\,\lambda/v$$

$$\Delta\lambda = 2.12 \times 10^{-6}(650 \times 10^{-9})$$

$$\Delta\lambda = 13.8 \times 10^{-13} = 0.0014 \text{ nm}.$$

13.15 $hv_f = hv_i - mgd$ [13.13]

$$\Delta v = -mgd/h = -\frac{hv}{c^2}\frac{gd}{h}$$

$$\Delta v = -gdv/c^2$$

$$\frac{\Delta v}{v} = -\frac{(9.8 \text{ m/s}^2)(20 \text{ m})}{(3 \times 10^8 \text{ m/s})^2} = 2.18 \times 10^{-15}.$$

13.16 $F = GMm/r^2 = GMm/R^2 \sec^2\theta$

$F_\perp = F\cos\theta = GMm\cos\theta/R^2 \sec^2\theta$

$dt = R\sec^2\theta\,d\theta/c.$

$$p_\perp = \int F_\perp\,dt = \frac{GMm}{cR}\int_{-\pi/2}^{\pi/2}\cos\theta\,d\theta = 2GMm/cR.$$

$$\tan\varphi = p_\perp/p_{\parallel} = 2GM/c^2R \approx \varphi$$

$$\varphi = \frac{2(6.67 \times 10^{-11} \text{ nm}^2/\text{kg}^2)(1.99 \times 10^{30} \text{ kg})}{(3 \times 10^8 \text{ m/s})^2(6.96 \times 10^8 \text{ m})}$$

$$\varphi = 24.5 \times 10^{-5} \text{ degrees} = 0.88 \text{ seconds of arc}.$$

13.18 $\frac{3}{2}kT = 6.17 \times 10^{-21} \text{ J} = 3.85 \times 10^{-2} \text{ eV}$

$$p = [2m_0(3kT/2)]^{1/2} = 4.55 \times 10^{-24}$$

$$\lambda = h/p = 1.85 \text{ Å}.$$

13.19 No—splitting a photon would result in two lower-frequency pieces which we could presumably separate and detect.

13.21 $\Pi = \dfrac{1000\text{ W}}{hv} = \dfrac{1000(10600 \times 10^{-9})}{6.63 \times 10^{-34}(3 \times 10^8)}$

$$= 5.06 \times 10^{22} \text{ photons/s}.$$

13.22

$$\mathscr{E} = \frac{p^2}{2m_0} + U, \quad hv = \frac{h^2}{\lambda^2 2m_0} + U, \quad \hbar\omega = \hbar^2k^2/2m_0 + U.$$

13.24 $\psi = C_1 e^{-i(\omega t + kx)} + C_2 e^{-i(\omega t - kx)}$

$$\frac{\partial\psi}{\partial t} = -i\omega\psi$$

$$\frac{\partial\psi}{\partial x} = -ikC_1 e^{-i(\omega t + kx)} + ikC_2 e^{-i(\omega t - kx)}$$

$$\frac{\partial^2\psi}{\partial x^2} = -k^2C_1 e^{-i(\omega t + kx)} - k^2C_2 e^{-i(\omega t - kx)} = -k^2\psi.$$

Using the dispersion relation of Problem 13.22, we obtain

$$\hbar\omega\psi = \hbar^2k^2\psi/2m_0 + U\psi$$

$$i\hbar\frac{\partial\psi}{\partial t} = \frac{-\hbar^2}{2m_0}\frac{\partial^2\psi}{\partial x^2} + U\psi.$$

CHAPTER 14

14.1

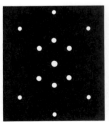

Diffraction pattern

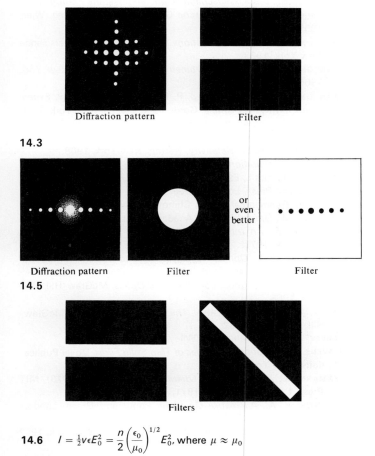

Diffraction pattern   Filter

14.3

Diffraction pattern   Filter   or even better   Filter

14.5

Filters

14.6 $I = \frac{1}{2}v\epsilon E_0^2 = \frac{n}{2}\left(\frac{\epsilon_0}{\mu_0}\right)^{1/2} E_0^2$, where $\mu \approx \mu_0$

$E_0^2 = 2(\mu_0/\epsilon_0)^{1/2}I/n$ $(\mu_0/\epsilon_0)^{1/2} = 376.730\,\Omega$

$E_0 = 27.4(I/n)^{1/2}$.

14.8 The inherent motion of the medium would cause the speckle pattern to vanish.

Bibliography

ANDREWS, C. L., *Optics of the Electromagnetic Spectrum*, Prentice-Hall, Englewood Cliffs, New Jersey, 1960.

BAKER, B. B. and E. J. COPSON, *The Mathematical Theory of Huygens' Principle*, Oxford University Press, London, 1969.

BALDWIN, G. C., *An Introduction to Nonlinear Optics*, Plenum Press, New York, 1969.

BARBER, N. F., *Experimental Correlograms and Fourier Transforms*, Pergamon, Oxford, 1961.

BARTON, A. W., *A Textbook On Light*, Longmans, Green, London, 1939.

BEARD, D. B. and G. B. BEARD, *Quantum Mechanics With Applications*, Allyn and Bacon, Boston, 1970.

BERAN, M. J. and G. B. PARRENT, JR., *Theory of Partial Coherence*, Prentice-Hall, Englewood Cliffs, New Jersey, 1964.

BLOEMBERGEN, N., *Nonlinear Optics*, W. A. Benjamin, New York, 1965.

BLOOM, A. L., *Gas Lasers*, Wiley, New York, 1968.

BORN, M. and E. WOLF, *Principles of Optics*, Pergamon, Oxford, 1970.

BOROWITZ, S., *Fundamentals of Quantum Mechanics*, Benjamin, New York, 1967.

BROUWER, W., *Matrix Methods in Optical Instrument Design*, Benjamin, New York, 1964.

BROWN, E. B., *Modern Optics*, Reinhold, New York, 1965.

CAJORI, F., *A History of Physics*, Macmillan, New York, 1899.

CHANG, W. S. C., *Principles of Quantum Electronics, Lasers: Theory and Applications*, Addison-Wesley, Reading, Mass., 1969.

CONRADY, A. E., *Applied Optics and Optical Design*, Dover Publications, New York, 1929.

COULSON, C. A., *Waves*, Oliver and Boyd, Edinburgh, 1949.

CRAWFORD, F. S., JR., *Waves*, McGraw-Hill, New York, 1965.

DAVIS, H. F., *Introduction to Vector Analysis*, Allyn and Bacon, Boston, 1961.

DAVIS, S. P., *Diffraction Grating Spectrographs*, Holt, Rinehart and Winston, New York, 1970.

DEVELIS, J. B. and G. O. REYNOLDS, *Theory and Applications of Holography*, Addison-Wesley, Reading, Mass., 1967.

DIRAC, P. A. M., *Quantum Mechanics*, Oxford University Press, London, 1958.

DRUDE, P., *The Theory of Optics*, Longmans, Green, London, 1939.

DITCHBURN, R. W., *Light*, Wiley, New York, 1963.

FLÜGGE, J., ed., *Die wissenschafliche und angewandte Photographie; Band 1, Das photographische Objektiv*, Springer-Verlag, Wien, 1955.

FRANÇON, M., *Modern Applications of Physical Optics*, Interscience, New York, 1963.

FRANÇON, M., *Optical Interferometry*, Academic Press, New York, 1966.

FRANÇON, M., N. KRAUZMAN, J. P. MATHIEU, and M. MAY, *Experiments in Physical Optics*, Gordon and Breach, New York, 1970.

FRANK, N. H., *Introduction to Electricity and Optics*, McGraw-Hill, New York, 1950.

FRENCH, A. P., *Special Relativity*, Norton, New York, 1968.

FRENCH, A. P., *Vibrations and Waves*, Norton, New York, 1971.

FROOME, K. D. and L. ESSEN, *The Velocity of Light and Radio Waves*, Academic Press, London, 1969.

FRY, G. A., *Geometrical Optics*, Chilton, Philadelphia, Pa., 1969.

GARBUNY, M., *Optical Physics*, Academic Press, New York, 1965.

GHATAK, A. K., *An Introduction to Modern Optics*, McGraw-Hill, New York, 1971.

GOLDWASSER, E. L., *Optics, Waves, Atoms, and Nuclei: An Introduction*, Benjamin, New York, 1965.

GOODMAN, J. W., *Introduction to Fourier Optics*, McGraw-Hill, New York, 1968.

HARDY, A. C. and F. H. PERRIN, *The Principles of Optics*, McGraw-Hill, New York, 1932.

HARVEY, A. F., *Coherent Light*, Wiley, London, 1970.

HEAVENS, O. S., *Optical Properties of Thin Solid Films*, Dover Publications, New York, 1955.

HERMANN, A., *The Genesis of Quantum Theory (1899–1913)*, MIT Press, Cambridge, Mass., 1971.

HOUSTON, R. A., *A Treatise On Light*, Longmans, Green, London, 1938.

HUYGENS, C., *Treatise on Light*, Dover Publications, New York, 1962 (1690).

JACKSON, J. D., *Classical Electrodynamics*, Wiley, New York, 1962.

JENKINS, F. A. and H. E. WHITE, *Fundamentals of Optics*, McGraw-Hill, New York, 1957.

JENNISON, R. C., *Fourier Transforms and Convolutions for the Experimentalist*, Pergamon Press, Oxford, 1961.

JOHNSON, B. K., *Optics and Optical Instruments*, Dover Publications, New York, 1947.

KLEIN, M. V., *Optics*, Wiley, New York, 1970.

LENGYEL, B. A., *Introduction to Laser Physics*, Wiley, New York, 1966.

LENGYEL, B. A., *Lasers, Generation of Light by Stimulated Emission*, Wiley, New York, 1962.

LEVI, L., *Applied Optics*, Wiley, New York, 1968.

LIPSON, S. G. and H. LIPSON, *Optical Physics*, Cambridge University Press, London, 1969.

LONGHURST, R. S., *Geometrical and Physical Optics*, Wiley, New York, 1967.

MACH, E., *The Principles of Physical Optics, An Historical and Philosophical Treatment*, Dover Publications, New York, 1926.

MAGIE, W. F., *A Source Book in Physics*, McGraw-Hill, New York, 1935.

MARTIN, L. C. and W. T. WELFORD, *Technical Optics*, Sir Isaac Pitman & Sons, Ltd., London, 1966.

MEYER, C. F., *The Diffraction of Light, X-rays and Material Particles*, University of Chicago Press, Chicago, 1934.

MEYER-ARENDT, J. R., *Introduction to Classical and Modern Optics*, Prentice-Hall, Englewood Cliffs, New Jersey, 1972.

Military Standardization Handbook—Optical Design. MIL-HDBK-141, 5 October 1962.

MINNAERT, M., *The Nature of Light and Colour in the Open Air*, Dover Publications, New York, 1954.

MORGAN, J., *Introduction to Geometrical and Physical Optics*, McGraw-Hill, 1953.

NEWTON, I., *Optiks*, Dover Publications, New York, 1952 (1704).

NOAKES, G. R., *A Text-Book of Light*, Macmillan, London, 1944.

NUSSBAUM, A., *Geometric Optics: An Introduction*, Addison-Wesley, Reading, Mass., 1968.

O'NEILL, E. L., *Introduction to Statistical Optics*, Addison-Wesley, Reading, Mass., 1963.

PALMER, C. H., *Optics, Experiments and Demonstrations*, John Hopkins Press, Baltimore, Md., 1962.

PAPOULIS, A., *The Fourier Integral and Its Applications*, McGraw-Hill, New York, 1962.

PAPOULIS, A., *Systems and Transforms with Applications in Optics*, McGraw-Hill, New York, 1968.

PEARSON, J. M., *A Theory of Waves*, Allyn and Bacon, Boston, 1966.

PLANCK, M. and M. MASIUS, *The Theory of Heat Radiation*, Blakiston, Philadelphia, 1914.

ROBERTSON, E. R. and J. M. HARVEY, eds., *The Engineering Uses of Holography*, Cambridge University Press, London, 1970.

ROBERTSON, J. K., *Introduction to Optics Geometrical and Physical*, Van Nostrand, Princeton, N.J., 1957.

RONCHI, V., *The Nature of Light*, Harvard University Press, Cambridge, Mass., 1971.

ROSSI, B., *Optics*, Addison-Wesley, Reading, Mass., 1957.

RUECHARDT, E., *Light Visible and Invisible*, University of Michigan Press, Ann Arbor, Mich., 1958.

SANDERS, J. H., *The Velocity of Light*, Pergamon, Oxford, 1965.

SCHAWLOW, A. L., intr., *Lasers and Light; Readings from Scientific American*, Freeman, San Francisco, 1969.

SCHRÖDINGER, E. C., *Science Theory and Man*, Dover Publications, New York, 1957.

SEARS, F. W., *Optics*, Addison-Wesley, Reading, Mass., 1949.

SHAMOS, M. H., ed., *Great Experiments in Physics*, Holt, New York, 1959.

SHURCLIFF, W. A., *Polarized Light: Production and Use*, Harvard University Press, Cambridge, Mass., 1962.

SHURCLIFF, W. A. and S. S. BALLARD, *Polarized Light*, Van Nostrand, Princeton, New Jersey, 1964.

SINCLAIR, D. C. and W. E. BELL, *Gas Laser Technology*, Holt, Rinehart and Winston, New York, 1969.

SLAYTER, E. M., *Optical Methods in Biology*, Wiley, New York, 1970.

SMITH, H. M., *Principles of Holography*, Wiley, New York, 1969.

SMITH, W. J., *Modern Optical Engineering*, McGraw-Hill, New York, 1966.

Société Française de Physique, ed., *Polarization, Matter and Radiation. Jubilee Volume in Honor of Alfred Kastler*, Presses Universitaires de France, Paris, 1969.

SOMMERFELD, A., *Optics*, Academic Press, New York, 1964.

SOUTHALL. J. P. C., *Introduction to Physiological Optics*, Dover Publications, New York, 1937.

SOUTHALL, J. P. C., *Mirrors, Prisms and Lenses*, Macmillan, New York, 1933.

STONE, J. M., *Radiation and Optics*, McGraw-Hill, New York, 1963.

STROKE, G. W., *An Introduction to Coherent Optics and Holography*, Academic Press, New York, 1969.

STRONG, J., *Concepts of Classical Optics*, Freeman, San Francisco, 1958.

SYMON, K. R., *Mechanics*, Addison-Wesley, Reading, Mass., 1960.

TOLANSKY, S., *An Introduction to Interferometry*, Longmans, Green, London, 1955.

TOLANSKY, S., *Curiosities of Light Rays and Light Waves*, American Elsevier, New York, 1965.

TOLANSKY, S., *Multiple-Beam Interferometry of Surfaces and Films*, Oxford University Press, London, 1948.

TOLANSKY, S., *Revolution in Optics*, Penguin Books, Baltimore, 1968.

TOWNE, D. H., *Wave Phenomena*, Addison-Wesley, Reading, Mass., 1967.

VALASEK, J., *Optics, Theoretical and Experimental*, Wiley, New York, 1949.

VAN HEEL, A. C. S., ed., *Advanced Optical Techniques*, American Elsevier, New York, 1967.

VAN HEEL, A. C. S. and C. H. F. VELZEL, *What is Light?*, McGraw-Hill, New York, 1968.

VASICEK, A., *Optics of Thin Films*, North-Holland, Amsterdam, 1960.

WAGNER, A. F., *Experimental Optics*, Wiley, New York, 1929.

WEBB, R. H., *Elementary Wave Optics*, Academic Press, New York, 1969.

WILLIAMS, W. E., *Applications of Interferometry*, Methuen, London, 1941.

WOLF, E., ed., *Progress in Optics*, North-Holland, Amsterdam.

WOOD, R. W., *Physical Optics*, Dover Publications, New York, 1934.

YARIV, A., *Quantum Electronics*, Wiley, New York, 1967.

YOUNG, H. D., *Fundamentals of Optics and Modern Physics*, McGraw-Hill, New York, 1968.

Indexes

Index of Tables

Index